Nitrogen in Crop Production

Nitrogen in Crop Production

Proceedings of a symposium held 25–27 May 1982 at Sheffield, Alabama. The symposium was sponsored by the National Fertilizer Development Center of the Tennessee Valley Authority, American Society of Agronomy, Crop Science Society of America, and Soil Science Society of America.

ROLAND D. HAUCK, ***editor-in-chief***

Editorial Committee

J. D. Beaton	**R. G. Hoeft**
C. A. I. Goring	**G. W. Randall**
R. D. Hauck	**D. A. Russel**

Managing Editor: RICHARD C. DINAUER

Assistant Editor: SUSAN ERNST

Editor-in-Chief ASA Publications: DWAYNE R. BUXTON

Published by

American Society of Agronomy

Crop Science Society of America

Soil Science Society of America

Madison, Wisconsin USA

1984

Meeting logo by Pam Taylor
Cover design by Wray Dillard
Section title page artwork by Don Sherwin

American Society of Agronomy, Inc.
Crop Science Society of America, Inc.
Soil Science Society of America, Inc.
677 South Segoe Road, Madison, Wisconsin 53711 USA

Library of Congress Cataloging in Publication Data

Nitrogen in crop production.

Includes bibliographies and index.
1. Nitrogen fertilizers—Congresses. 2. Plants, Effect of nitrogen on—Congresses. 3. Crops—Nutrition—Congresses. I. Hauck, Roland Daniel, 1926- . II. National Fertilizer Development Center (U.S.). III. American Society of Agronomy.
S651.N58 1984 631.8'4 84-28325
ISBN 0-89118-081-8

Printed in the United States of America

CONTENTS

DEDICATION

VICTOR JAMES KILMER
1913–1981

This book, *Nitrogen in Crop Production,* is dedicated to Victor James Kilmer (1913–1981) in recognition of his many contributions and outstanding leadership in the fields of soil science and agronomy.

A native of Wisconsin, Kilmer obtained his B.S. and M.S. degrees in soil science at the University of Wisconsin-Madison. During World War II he spent four years in the armed forces, mostly in the Pacific. His entire professional career was devoted to service in U.S. Governmental agencies, starting in the Soil Conservation Service as a soil scientist at the Upper Mississippi Valley Experiment Station at LaCrosse, Wisconsin in 1941. After the war he worked as a soil surveyor, also in Wisconsin. In 1947, he transferred to the National Soil Survey Laboratory at Beltsville, Maryland, at that time a part of the Bureau of Plant Industry, Soils, and Agricultural Engineering.

In 1955, Kilmer shifted his professional interests from soil classification to soil and water management by joining the Soil and Water Research Division of the Agricultural Research Service. In 1961, he joined the Office of Agricultural and Chemical Development of the Tennessee Valley Authority at Muscle Shoals, Alabama, serving first as assistant to the manager and then, until his retirement in 1979, as Chief of the national important Soils and Fertilizer Research Branch.

Kilmer's greatest accomplishment was probably in fostering meaningful and needed research in soils and fertilizers in his own organization and also in institutions cooperating with TVA. His own research of greatest impact dealt with losses of plant nutrients from soils through erosion, runoff, and leaching. Kilmer was a key individual in the 20-year joint effort of the American Society of Agronomy, Crop Science Society of America, and Soil Science Society of America and TVA's National Fertilizer Development Center in conducting symposia dealing with plant nutrition, fertilizers, and crop production. This joint effort resulted in an important series of highly authoritative and widely used books, of which *Nitrogen in Crop Production* is the latest.

Kilmer was honored as Fellow, both of the American Society of Agronomy and the Soil Science Society of America. He served as President of the Soil Science Society of America during 1976–1977. Kilmer was highly regarded, both as a professional soil scientist and as a person.

March 1984

LOUIS B. NELSON (retired)
Director, Division of Agricultural Development
National Fertilizer Development Center
Tennessee Valley Authority
Muscle Shoals, Alabama

FOREWARD

The realization of this book stems from a national symposium, Nitrogen in Crop Production, held in Sheffield, Alabama, 25–27 May 1982, jointly sponsored by the Tennessee Valley Authority National Fertilizer Development Center, the American Society of Agronomy, the Crop Science Society of America, and the Soil Science Society of America. The symposium was organized to provide a comprehensive summary of current knowledge about nitrogen as it relates to crop production. Each of the presentations had a practical orientation that focused on alternative means to improve plant use of nitrogen in different cropping systems. Those presentations as well as several additional closely related topics comprise the chapters of this volume. Our understanding of the role of nitrogen and the continual improvement of management alternatives for its optimal use in crop production as documented here represents an integrated effort of nearly 100 scientists and technologists. Because of their unique effort, we expect that this book will be of interest to farm managers, agronomists, crop and soil scientists, crop ecologists, and cooperative extension specialists.

We want to express our appreciation to the editorial committee, headed by Dr. R. D. Hauck; the authors and reviewers; and the ASA Headquarters staff for the time and effort they have spent in making this publication possible.

November 1984

K. J. FREY, *president*
American Society of Agronomy

W. F. KEIM, *president*
Crop Science Society of America

D. R. NIELSEN, *president*
Soil Science Society of America

PREFACE

Nitrogen in Crop Production provides an authoritative review of the principles and practices of nitrogen use in agricultural cropping systems. It was planned as a companion work to *Nitrogen in Agricultural Soils* (Agronomy no. 22, 1982), with focus on the interrelationships of nitrogen and other crop production factors. Topics discussed include (i) how plants use nitrogen, (ii) sources and supply of plant-available nitrogen, and (iii) the management of crops, fertilizers and fertilizer amendments, manures and other waste products, plant residues, and soils for maximum, economic crop production. Other topics discussed are the relationship between nitrogen use and plant diseases, insect invasion, water stress, and weed infestation. Separate chapters are devoted to crop quality and the quality of the environment, as affected by nitrogen use. The last section of the book describes recommended nitrogen management practices for regions of the United States differing in climate, soils, and cropping systems.

Authors of chapters were asked to prepare overviews rather than comprehensive reviews of their assigned topics and to cite mainly key articles and reviews. Where data on a particular topic were lacking and definite conclusions could not justifiably be drawn, a statement of informed opinion was solicited. The restrictions placed on the authors were necessary because of the many topics that the editorial committee thought desirable to discuss. Even so, relevant topics that might have been included were omitted. We apologize for these omissions and for the omission of references to important work that might have been included, had volume size not been a practical consideration.

The editorial committee expresses appreciation to the authors and the organizations they represent. We acknowledge Richard C. Dinauer, Susan Ernst, David M. Kral, Rodney A. Briggs, Matthias Stelly, and other members of the headquarters staff for assistance in editing and preparing the manuscripts for publication. We pay special tribute to Victor J. Kilmer, whose administrative support during the planning of this book was most helpful; to George Stanford, coauthor of Chapter 17, and to A. G. Norman, author of the Perspective, whose untimely deaths occurred while the book was in progress.

March 1984

The Editorial Committee

R. D. HAUCK, editor-in-chief
Tennessee Valley Authority, Muscle Shoals, Alabama

J. D. BEATON
Potash and Phosphate Institute, Cochrane, Alberta

C. A. I. GORING
The Dow Chemical Company, Midland, Michigan

R. G. HOEFT
University of Illinois, Urbana, Illinois

GYLES W. RANDALL
University of Minnesota, Waseca, Minnesota

D. A. RUSSEL (retired)
Tennessee Valley Authority

CONTRIBUTORS

Frank P. Achorn Senior Scientist, National Fertilizer Development Center, Tennessee Valley Authority, Muscle Shoals, Alabama

Fred Adams Professor of Soil Chemistry, Department of Agronomy and Soils, Auburn University, Auburn, Alabama

Gary W. Akin Agronomist, National Fertilizer Development Center, Tennessee Valley Authority, Muscle Shoals, Alabama

Samuel R. Aldrich Professor of Soil Fertility in Extension, Emeritus, University of Illinois, Urbana, Illinois

Seward E. Allen Agronomist, National Fertilizer Development Center, Tennessee Valley Authority, Muscle Shoals, Alabama

V. Allan Bandel Professor of Soils, Agronomy Department, University of Maryland, College Park, Maryland

Stanley A. Barber Professor of Agronomy, Department of Agronomy, Purdue University, West Lafayette, Indiana

Frederick E. Below Research Assistant/Research Associate, Department of Agronomy, University of Illinois, Urbana, Illinois

B. R. Bock Research Soil Chemist, National Fertilizer Development Center, Tennessee Valley Authority, Muscle Shoals, Alabama

L. V. Boone Agronomist, Department of Agronomy, University of Illinois, Urbana, Illinois

Fred C. Boswell Professor and Soil Scientist, Agronomy Department, University of Georgia, Experiment, Georgia

D. R. Bouldin Professor of Soil Science, Department of Agronomy, Cornell University, Ithaca, New York

F. E. Broadbent Professor of Soil Microbiology, Department of Land, Air, and Water Resources, University of California, Davis, California

Michael F. Broder Agricultural Engineer, National Fertilizer Development Center, Tennessee Valley Authority, Muscle Shoals, Alabama

Michael A. Cole Associate Professor of Agronomy, Department of Agronomy, University of Illinois, Urbana, Illinois

E. L. Deckard Professor of Agronomy, Agronomy Department, North Dakota State University, Fargo, North Dakota

Theodore M. DeJong Assistant Professor, Pomology Department, University of California, Davis, California

R. B. Diamond Coordinator, Fertilizer Evaluations, International Fertilizer Development Center, Muscle Shoals, Alabama

J. W. Doran Soil Scientist, Agricultural Research Service, U.S. Department of Agriculture, University of Nebraska, Lincoln, Nebraska

A. I. Dow Extension Soil Scientist (retired), Washington State University, Prosser, Washington. Currently Soil Management Consultant

David W. Emerich Professor, Department of Biochemistry, University of Missouri, Columbia, Missouri

Harold J. Evans	Professor, Laboratory for Nitrogen Fixation Research, Oregon State University, Corvallis, Oregon
Richard H. Fox	Associate Professor of Soil Science, Department of Agronomy, The Pennsylvania State University, University Park, Pennsylvania
Bryant R. Gardner	Soil Scientist, Yuma Valley Agricultural Center, University of Arizona, Yuma, Arizona
J. W. Gilliam	Professor, Soil Science Department, North Carolina State University, Raleigh, North Carolina
Paul M. Giordano	Research Soil Chemist, National Fertilizer Development Center, Tennessee Valley Authority, Muscle Shoals, Alabama
Sham S. Goyal	Postgraduate Researcher, Plant Growth Laboratory, University of California, Davis, California
Robert C. Gray	Agronomist (retired), National Fertilizer Development Center, Tennessee Valley Authority, Muscle Shoals, Alabama
D. D. Gull	Associate Professor, Vegetable Crops Department, University of Florida, Gainesville, Florida
R. H. Hageman	Professor, Department of Agronomy, University of Illinois, Urbana, Illinois
James E. Harper	Plant Physiologist, Agricultural Research Service, U.S. Department of Agriculture, Department of Agronomy, University of Illinois, Urbana, Illinois
Roland D. Hauck	Research Soil Scientist, National Fertilizer Development Center, Tennessee Valley Authority, Muscle Shoals, Alabama
Walter A. Hill	Professor, Department of Agricultural Sciences, Tuskegee Institute, Tuskegee Institute, Alabama
R. G. Hoeft	Professor of Soil Fertility, Department of Agronomy, University of Illinois, Urbana, Illinois
Ray C. Huffaker	Professor of Agronomy, Plant Growth Laboratory, University of California, Davis, California
T. L. Jackson	Professor of Soil Science, Department of Soil Science, Oregon State University, Corvallis, Oregon
S. D. Klausner	Soil Scientist, Department of Agronomy, Cornell University, Ithaca, New York
Ellery L. Knake	Professor of Weed Science in Agronomy, Department of Agronomy, University of Illinois, Urbana, Illinois
Thor Kommedahl	Professor, Department of Plant Pathology, University of Minnesota, St. Paul, Minnesota
L. T. Kurtz	Professor of Soil Fertility, Agronomy Department, University of Illinois, Urbana, Illinois
R. J. Lambert	Professor of Plant Genetics, Department of Agronomy, University of Illinois, Urbana, Illinois
J. O. Legg	Soil Scientist (retired), Agricultural Research Service, U.S. Department of Agriculture. Currently Adjunct Professor, Agronomy Department, University of Arkansas, Fayetteville, Arkansas
S. J. Locascio	Horticulturist, Vegetable Crops Department, University of Florida, Gainesville, Florida

John B. Martin Agronomist, National Fertilizer Development Center, Tennessee Valley Authority, Muscle Shoals, Alabama

Adolfo Martinez Agricultural Economist, International Fertilizer Development Center, Muscle Shoals, Alabama

D. N. Maynard Professor and Chairman, Vegetable Crops Department, University of Florida, Gainesville, Florida

Marshal D. McGlamery Professor of Weed Science, Agronomy Department, University of Illinois, Urbana, Illinois

J. J. Meisinger Soil Scientist, Agricultural Research Service, U.S. Department of Agriculture, Beltsville Agricultural Research Center, Beltsville, Maryland

M. J. Messmer Plant Breeder, Garst Seed Company, Ames, Iowa. Formerly Graduate Student, University of Illinois, Urbana, Illinois

L. W. Murdock Extension Professor, University of Kentucky, Princeton, Kentucky

Darrell W. Nelson Professor of Agronomy, Department of Agronomy, Purdue University, West Lafayette, Indiana

R. A. Olson Professor, Agronomy Department, University of Nebraska, Lincoln, Nebraska

T. R. Peck Professor of Soil Chemistry, Extension, Agronomy Department, University of Illinois, Urbana, Illinois

G. A. Peterson Professor of Agronomy, Department of Agronomy, University of Nebraska, Lincoln, Nebraska

Donald A. Phillips Professor, Department of Agronomy and Range Science, University of California, Davis, California

J. F. Power Soil Scientist, Agricultural Research Service, U.S. Department of Agriculture, University of Nebraska, Lincoln, Nebraska

P. F. Pratt Professor of Soil Science, Department of Soil and Environmental Sciences, University of California, Riverside, California

Gyles W. Randall Soil Scientist and Professor, Southern Experiment Station, University of Minnesota, Waseca, Minnesota

Roy S. Rauschkolb Formerly Extension Soils Specialist, University of California, Davis, California. Currently Director of Cooperative Extension, College of Agriculture, University of Arizona, Tucson, Arizona

W. S. Reid Professor of Soil Science, Department of Agronomy, Cornell University, Ithaca, New York

F. M. Rhoads Professor of Soil Science, Agricultural Research and Education Center, University of Florida, Quincy, Florida

Robert L. Roth Agricultural Engineer, Yuma Valley Agricultural Center, University of Arizona, Yuma, Arizona

Darrell A. Russel Agriculturist (retired), National Fertilizer Development Center, Tennessee Valley Authority, Muscle Shoals, Alabama

L. E. Schrader Professor of Plant Physiology, Department of Agronomy, University of Wisconsin, Madison, Wisconsin

J. Mark Scriber Associate Professor, Department of Entomology, University of Wisconsin, Madison, Wisconsin

Joseph H. Sherrard Research Associate, Department of Agronomy, University of Illinois, Urbana, Illinois

Lee E. Sommers Professor of Agronomy, Agronomy Department, Purdue University, West Lafayette, Indiana

George Stanford Soil Scientist, Agricultural Research Service, U.S. Department of Agriculture, Beltsville, Maryland. Deceased 28 January 1981

Paul J. Stangel Deputy Managing Director, International Fertilizer Development Center, Muscle Shoals, Alabama

C. Y. Tsai Professor of Genetics, Department of Botany and Plant Pathology, Purdue University, West Lafayette, Indiana

Billy B. Tucker Regents Professor-Agronomy, Agronomy Department, Oklahoma State University, Stillwater, Oklahoma

T. C. Tucker Professor, Soils, Water, and Engineering Department, University of Arizona, Tucson, Arizona

Fred T. Turner Professor of Soils and Plant Nutrition, Agriculture Research and Extension Center, Texas A&M University, Beaumont, Texas

Peter van Berkum Biochemist, Nitrogen Fixation and Soybean Genetics Laboratory, Agricultural Research Service, U.S. Department of Agriculture, Beltsville Agricultural Research Center, Beltsville, Maryland

Regis D. Voss Professor of Agronomy, Agronomy Department, Iowa State University, Ames, Iowa

L. F. Welch Professor of Soil Fertility, Department of Agronomy, University of Illinois, Urbana, Illinois

B. R. Wells Professor, Agronomy Department, University of Arkansas, Fayetteville, Arkansas

K. L. Wells Extension Professor-Soils, Department of Agronomy, University of Kentucky, Lexington, Kentucky

D. G. Westfall Professor, Department of Agronomy, Colorado State University, Fort Collins, Colorado

W. J. Wiltbank Professor (Horticulturist), Department of Fruit Crops, University of Florida, Gainesville, Florida

Factors for converting non-SI units to acceptable SI units

Non-SI Units		Acceptable Units
Multiply	By	To obtain
acre	4.05×10^3	square meter, m^2
acre	0.405	hectare, ha (10^4 m^2)
acre	4.05×10^{-3}	square kilometer, km^2 (10^6 m^2)
Angstrom unit	0.1	nanometer, nm (10^{-9} m)
atmosphere	0.101	megapascal, MPa (10^6 Pa)
bar	0.1	megapascal, MPa (10^6 Pa)
British thermal unit	1.05×10^3	joule, J
calorie	4.19	joule, J
calorie per square centimeter minute (irradiance)	698	watt per square meter, W m^{-2}
calorie per square centimeter (langley)	4.19×10^4	joules per square meter, J m^{-2}
cubic feet	0.028	cubic meter, m^3
cubic feet	28.3	liter, L (10^{-3} m^3)
cubic inch	1.64×10^{-5}	cubic meter, m^3
curie	3.7×10^{10}	becquerel, Bq
degrees (angle)	1.75×10^{-2}	radian, rad
dyne	10^{-5}	newton, N
erg	10^{-7}	joule, J
foot	0.305	meter, m
foot-pound	1.36	joule, J
gallon	3.78	liter, L (10^{-3} m^3)
gallon per acre	9.35	liter per hectare, L ha^{-1}
gauss	10^{-4}	tesla, T
gram per cubic centimeter	1.00	megagram per cubic meter, Mg m^{-3}
gram per square decimeter hour (transpiration)	27.8	milligram per square meter second, mg m^{-2} s^{-1} (10^{-3} g m^{-2} s^{-1})
inch	25.4	millimeter, mm (10^{-3} m)
micromole (H_2O) per square centimeter second (transpiration)	180	milligram (H_2O) per square meter second, mg m^{-2} s^{-1} (10^{-3} g m^{-2} s^{-1})
micron	1.00	micrometer, μm (10^{-6} m)
mile	1.61	kilometer, km (10^3 m)
mile per hour	0.477	meter per second, m s^{-1}
milligram per square decimeter hour (apparent photosynthesis)	0.0278	milligram per square meter second, mg m^{-2} s^{-1} (10^{-3} g m^{-2} s^{-1})
milligram per square centimeter second (transpiration)	10 000	milligram per square meter second, mg m^{-2} s^{-1} (10^{-3} g m^{-2} s^{-1})
millimho per centimeter	0.1	siemen per meter, S m^{-1}
ounce	28.4	gram, g (10^{-3} kg)
ounce (fluid)	2.96×10^{-2}	liter, L (10^{-3} m^3)
pint (liquid)	0.473	liter, L (10^{-3} m^3)
pound	454	gram, g (10^{-3} kg)
pound per acre	1.12	kilogram per hectare, kg ha^{-1}
pound per acre	1.12×10^{-3}	megagram per hectare, Mg ha^{-1}
pound per bushel	12.87	kilogram per cubic meter, kg m^{-3}
pound per cubic foot	16.02	kilogram per cubic meter, kg m^{-3}
pound per cubic inch	2.77×10^4	kilogram per cubic meter, kg m^{-3}
pound per square foot	47.9	pascal, Pa
pound per square inch	6.90×10^3	pascal, Pa
quart (liquid)	0.946	liter, L (10^{-3} m^3)
quintal (metric)	10^2	kilogram, kg
rad	1.00	0.01 Gy
roentgen	1.00	2.58×10^{-4} C (coulomb) kg^{-1}
square centimeter per gram	0.1	square meter per kilogram, m^2 kg^{-1}
square feet	9.29×10^{-2}	square meter, m^2
square inch	645	square millimeter, mm^2 (10^{-6} m^2)
square mile	2.59	square kilometer, km^2
square millimeter per gram	10^{-3}	square meter per kilogram, m^2 kg^{-1}
temperature (°F − 32)	0.556	temperature, °C
temperature (°C + 273)	1	temperature, K
tonne (metric)	10^3	kilogram, kg
ton (2000 lb)	907	kilogram, kg
ton (2000 lb) per acre	2.24	megagram per hectare, Mg ha^{-1}

Importance of Nitrogen to Crop Production

section I

1

Adolfo Martinez and R. B. Diamond
International Fertilizer Development Center
Muscle Shoals, Alabama

Nitrogen Use in World Crop Production

Estimates of N use on a national and regional basis are readily available from several national and international sources. These estimates are usually highly aggregated and do not provide information related to specific N use by different groups of crops or even principal N sources for many countries.

Knowledge of N use by crops is important to government planners, researchers, agriculturalists, and commercial entrepreneurs engaged in the fertilizer and agricultural sectors. Such data would provide guidelines for determining future crop production goals, future fertilizer sales, and policies and strategies that are needed to influence production in the various crop sectors.

This study was undertaken to determine what information is available on use of N by crops in the literature, to obtain additional estimates, and to compile the data within one source. No attempt is made to analyze why the extent of use, levels of use, or sources of N are as indicated. It is hoped that this compilation of data will provide the impetus for people in other countries to develop estimates of use of nutrients by crops and that these data may be useful for various types of analyses.

I. DATA USED

Data used in preparation of this chapter were derived from a more detailed report entitled *Fertilizer Use in Crop Production* that was prepared by IFDC. Sources of data include the following:[1]

[1] The untiring efforts of the library staffs of the Tennessee Valley Authority (TVA) and the International Fertilizer Development Center (IFDC) are gratefully acknowledged for assistance with the literature search. Also see the *Reference* section, this chapter.

1) A questionnaire developed by IFDC and mailed to approximately 300 parties around the world.
2) Fertilizer-related country reports prepared by IFDC during the 1974–1981 period.
3) A publication entitled *Provisional Statistics on Fertilizer Use by Major Crops* by the Food and Agriculture Organization of the United Nations (FAO), Rome, Italy, June 1980.
4) Reports from the United Nations Industrial Development Organization (UNIDO), the Industrial Development Centre for Arab States (IDCAS), ISMA, the U.S. Department of Agriculture (USDA), several publications from the Economic and Social Commission for Asia and the Pacific (ESCAP), and several FAO fertilizer yearbooks.
5) Data on N use in silviculture collected by Dr. George Bengtson, associate dean, School of Forestry, Oregon State University.

In most countries, available data on N use by crops or groups of crops are the result of *estimates* and/or rough calculations made by government officials and/or other persons familiar with the fertilizer sector. Therefore, reliability of data varies among countries. Very few countries collect these kinds of data by use of a systematic survey; among those who do are Sweden, the United Kingdom, and the United States (for six major crops).

II. PRESENTATION OF DATA

Available estimates of N use by crops for various countries are presented and discussed. Table 1 contains a summary of the estimated percentage of N use distribution among crop groups and the principal N sources used for 67 countries. Table 2 shows the estimated percentage of selected harvested crop areas that received N fertilizer in 37 countries. Table 3 presents the estimated rates of N applied to fertilizer areas for selected crops in 37 countries.

Unfortunately, the year for which data were available varies among countries. Also, data for certain crops and/or crop groups were not available for many of the countries included. Regardless of these limitations, data as presented here provide general information and guidelines with respect to N use in crop production throughout the world. A brief description of data, on a regional basis, is presented in the following paragraphs. Available data on N use in silviculture are discussed at the end of this section.

A. North Africa

Countries in North Africa from which data were obtained include: Egypt (1975),[2] Morocco (1979), and Tunisia (1975). Table 1 shows that 61% of the total N used in Egypt was utilized in the production of cereal crops. For Morocco, the only available data on N use were for cereal crops, which

[2] Numbers in parentheses indicate the year for which the data were reported.

Table 1—Estimated percentage distribution of N use among crop groups and principal N sources in selected countries.

Country	Year	Cereals	Roots & tubers	Vege-tables	Fiber crops†	Fruits	Stimu-lants‡	Oil crops	Indus-trial crops§	Pas-tures	Not identi-fied	Principal N products¶
		% of total N										
North Africa												
Egypt	75	61	2	7	17	7	--	1	5	--	--	Urea, CAN, AN
Morocco	79	30	--	--	--	--	--	--	--	--	70	Urea, AS, AN
Tunisia	75	27	--	4	--	21	--	--	--	--	48	AN
Sahel West Africa												
Chad	78	2	--	--	97	--	--	--	1	--	--	Urea, NPK
Gambia	76	91	--	--	9	--	--	--	--	--	--	Urea, NPK
Mali	76	56	--	1	43	--	--	--	--	--	--	Urea, NPK
Niger	79	92	--	1	6	--	--	1	--	--	--	Urea, NPK
Senegal	75	56	--	--	13	--	--	31	--	--	--	Urea, NPK
Upper Volta	75	38	--	--	37	--	--	--	25	--	--	Urea, NPK
West Africa Coast												
Benin	76	2	--	--	96	--	--	2	--	--	--	Urea, AS, NPK
Cameroon	77	13	--	--	--	--	5	33	--	--	49	Urea, CAN, AS, NPK
Ivory Coast	73	5	--	--	20	54	5	3	13	--	--	AS, NPK
Nigeria	79	71	8	2	9	--	3	--	6	--	1	CAN, AS, Urea, NPK
East & South Africa												
Burundi	76	--	--	--	12	--	88	--	--	--	--	Urea, AS, NPK
Central African Republic	76	--	--	--	17	--	25	--	--	--	58	Urea, AS, NPK
Ethiopia	76	63	--	--	13	--	--	13	9	--	2	Urea, AS, DAP
Lesotho	79	95	--	--	--	--	--	--	--	--	5	CAN, NPK
Madagascar	77	50	--	--	10	--	10	--	30	--	--	Urea, AS
Malawi	76	17	--	2	3	--	25	--	22	--	31	Urea, AS, NPK
Mauritius	77	1	1	--	--	--	3	--	95	--	--	CAN, NPK
Somalia	76	12	--	--	5	60	--	--	23	--	--	Urea, MAP, DAP
Sudan	75	27	--	2	46	--	--	--	6	--	19	Urea, AS
Tanzania	75	29	--	--	--	--	29	--	6	--	37	Urea, AS, NPK

(continued on next page)

Table 1—Continued.

Country	Year	Cereals	Roots & tubers	Vege- tables	Fiber crops†	Fruits	Stimu- lants‡	Oil crops	Indus- trial crops§	Pas- tures	Not identi- fied	Principal N products¶
		% of total N										
North America												
Canada	79	65	--	--	--	--	--	13	--	--	22	Urea, AN, UAN, NPK
Mexico	77	65	--	--	5	3	1	8	12	--	6	Urea, AS, AN, NPK
United States#	80	61	--	--	3	1	--	1	--	--	34	AA, NPK, UAN, Urea
Central America & Caribbean												
Cuba	75	11	1	--	--	5	2	--	55	--	26	Urea, AS
Dominican Republic	79	20	--	--	--	--	--	--	68	--	12	Urea, AS, NPK
Panama	77	57	--	--	--	--	--	--	--	--	43	
Andean Region												
Colombia	78	30	12	--	7	--	29	--	--	--	22	Urea, AN, AS, NPK
Ecuador	80	37	4	--	--	--	--	8	20	--	31	Urea, AS, NPK
Peru	80	28	7	--	20	--	1	--	17	--	27	Urea, AS, AN
Venezuela	80	65	--	--	4	--	4	4	14	--	9	Urea, AS, NPK
Eastern South America												
Argentina	79	36	--	--	--	--	--	--	25	2	37	Urea, AS, AN, NPK
Brazil	78	25	3	7	3	3	26	6	15	--	12	Urea, MAP, DAP, NPK
Chile	79	50	--	--	--	--	--	--	--	--	50	SN, UREA
Paraguay	78	26	--	--	--	--	--	41	26	--	7	Urea, NPK
Central & South Asia												
Afghanistan	75	75	--	--	18	--	--	--	--	--	7	Urea
Burma	79	84	--	--	5	--	--	3	2	--	6	Urea
Bangladesh	80	92	1	--	2	--	--	1	1	--	3	Urea
India	79	58	2	--	6	--	--	--	7	--	27	Urea, AS, DAP, NPK
Pakistan	78	69	--	1	15	1	--	--	11	--	3	Urea, CAN, NP
Sri Lanka	80	50	--	--	--	--	23	10	--	--	17	Urea, AS
Southeast Asia												
Indonesia	80	85	--	4	--	--	1	--	4	--	6	Urea, AS
Malaysia	80	27	--	3	--	--	--	--	69	--	1	Urea, AS, NPK

Philippines	80	55	--	4	--	4	--	--	34	--	3	Urea, AS, NPK
Thailand	80	55	--	13	--	10	--	--	21	--	1	AS, AC, Urea, NPK
Middle East Region												
Cyprus	77	13	17	6	--	22	--	--	--	--	42	MAP, NPK
Israel	76	15	--	5	18	44	--	--	--	--	18	Urea, AS, AQA
Turkey	79	58	6	9	--	9	--	2	14	--	2	Urea, AS, CAN, NPK
Yemen Dem.	76	50	3	--	20	10	--	3	--	--	14	Urea
Northeast Asia												
Japan	79	35	3	16	5	16	--	--	8	--	17	Urea, NPK
South Korea	75	63	3	16	--	3	--	3	2	--	10	Urea, NPK
Europe												
Austria	77	48	5	--	--	4	--	--	5	33	5	Urea, CAN, AN, AS
Denmark	80	52	--	--	--	--	--	--	12	35	1	CAN, NPK
Finland	79	50	3	--	--	--	--	2	--	38	7	Urea, CAN, NPK
France	79	51	6	2	--	4	--	2	--	--	35	Urea, AN, CAN, UAN, NPK
Germany, F.R.	79	48	2	--	--	--	--	1	5	--	44	CAN, AS, ANS, UAN, NPK
Hungary	80	62	4	3	--	6	--	4	--	5	16	CAN, AN
Netherlands	79	9	11	--	--	--	--	--	--	75	5	CAN, NPK
Portugal	78	44	10	--	--	11	--	--	--	--	35	Urea, CAN, AS, NPK
Spain	79	45	7	--	3	28	--	6	4	--	7	CAN, AS, ANS, NPK
Sweden	77	47	2	--	--	--	--	6	3	15	27	Urea, CAN, NPK
Switzerland	80	22	--	1	--	1	--	--	--	41	35	Urea, CAN, NPK
United Kingdom	80	32	3	--	--	--	--	2	2	60	1	AN, CAN, NPK
Oceania												
Australia	79	43	2	6	4	6	--	2	24	--	13	Urea, AN, AS, NPK
Fiji	80	1	--	--	--	--	--	--	98	--	1	AS

† Includes cotton, jute (*Corchorus* sp.)
‡ Includes coffee, tea, and cocoa.
§ Includes sugarcane, sugarbeets, rubber, and tobacco.
¶ AA = Anhydrous ammonia
AC = Ammonium chloride
AN = Ammonium nitrate
AQA = Aqua ammonia
CAN = Calcium ammonium nitrate
DAP = Diammonium phosphate
AS = Ammonium sulfate
MAP = Monoammonium phosphate
NP = Fertilizers containing N and P_2O_5
NPK = Compound fertilizer containing three primary nutrients, N, P_2O_5, and K_2O
SN = Sodium nitrate
UAN = Urea ammonium nitrate
# Only the single crops of cotton, citrus, soybean, and sugarcane are included in the crop groups of fiber, fruit, oil, and industrial crops, respectively.

used approximately 30% of the total. In Tunisia, cereal crops received 27% of the total N used. The main fertilizer product used in the three reporting countries was ammonium nitrate (AN). Urea, calcium ammonium nitrate (CAN), and ammonium sulfate (AS) were also used.

Data on the percentage of crop areas fertilized with N and its application rates were available only from Egypt. Table 2 shows that 93% of the fiber crop area and 36% of the barley (*Hordeum vulgare* L.) area received fertilizer. Table 3 shows the application rates for different crops, which ranged from 219 kg N/ha for barley to 54 kg N/ha for oil crops. The generally high rates of N application in Egypt result from the large portion of cropped area that is under irrigation.

B. Sahel of West Africa

Table 1 shows data obtained from the following Sahelian countries of western Africa: Chad (1978), Gambia (1976), Mali (1976), Niger (1979), Senegal (1975), and Upper Volta (1975). This table shows that cereal crops received more N than any other crop group in all of these countries except Chad. Cereal crops receive between 38 and 92% of the total N. Fiber crops —cotton (*Gossypium hirsutum* L.)—received an important part of the total N used in Chad (97%), Mali (43%), and Upper Volta (37%). Groundnuts (*Arachis hypogaea* L.) (oil crop) and sugarcane (*Saccharum officinarum* L.) (industrial crop) received 31 and 25% of N in Senegal and Upper Volta, respectively. Urea was the main source of N in the area; it is followed by NPK products.

Data on the percentage of crop areas fertilized with N and average application rates were available from Mali, Senegal, and Upper Volta. Table 2 shows that Mali and Senegal fertilized > 90% of the area growing fiber crops, while Upper Volta fertilized 35% of the area planted to fiber crops and 100% of the area in sugarcane. Also, Mali fertilized 80% of the maize (*Zea mays* L.) area, and Senegal fertilized 70% of the rice (*Oryza sativa* L.) area.

In Upper Volta, the average application rate of N for sugarcane was 164 kg/ha; in Mali, rice received 118 kg/ha N. In Senegal, 37 kg of N/ha was applied to rice and to fiber crops within the fertilized areas.

C. West Africa Coast

Countries on the western Africa coast from which data were obtained include: Benin (1976), Cameroon (1977), Ivory Coast (1973), and Nigeria (1979). In Benin, fiber crops received 96% of the N fertilizer used. It was applied in the form of urea and NPK products. In Cameroon, 33% of the total N used was for oil crops, and it was applied as urea, AS, CAN, and NPK products. In the Ivory Coast, fruit crops received 54% of the N used; it was applied in the form of AS and NPK products. In Nigeria, cereal crops received 71% of the N fertilizer used; it was applied as CAN, AS, urea, and NPK products.

The percentage of crop areas fertilized and the average application rate data were available only from Cameroon and Nigeria. In Nigeria, the greatest percentage of crop areas fertilized were: 58% of wheat (*Triticum aestivum* L.), 50% of sugarcane, and 49% of cotton. In Cameroon, 95% of the area in sugarcane and 26% of the area in rice were fertilized. Average N application rates in Cameroon were 125 kg/ha for coffee (*Coffea arabica* L.), 84 kg/ha for sugarcane, and 70 kg/ha for maize. In Nigeria, wheat received 80 kg/ha; maize, rice, and sugarcane received about 50 kg/ha; and tuber crops received 21 kg/ha.

D. Central, East, and South Africa

Data were obtained from the following countries located in southern and eastern Africa: Burundi (1976), Central African Republic (1976), Ethiopia (1976), Lesotho (1979), Madagascar (1977), Malawi (1976), Mauritius (1977), Somalia (1976), Sudan (1975), and Tanzania (1975). Stimulant crops received 88% of the total N in Burundi, 25% in Central African Republic, 29% in Tanzania, and 25% in Malawi. In Ethiopia, Lesotho, and Madagascar, cereal crops received 50% or more, while in Sudan and Tanzania, those crops received 27 to 29% of the total N fertilizer used. In Mauritius, sugarcane received 95% of the total N applied, while in Somalia and Sudan, fruit crops and fiber crops received 60 and 46% of the total N fertilizer used, respectively.

Urea was the main N fertilizer product used in the area, followed by AS, NPK products, and CAN. Limited quantities of ammonium phosphates were used.

Data on percentages of crop areas fertilized with N and average application rates were available for Burundi, Ethiopia, Mauritius, Somalia, Sudan, and Tanzania. Table 2 shows that in Burundi, Sudan, and Tanzania, 100% of the area growing tea [*Camellia sinensis* (L.) Ktze.] was fertilized. Also, 100% of the areas with sugar crops in Ethiopia and Somalia and with fiber crops in Ethiopia and Sudan received fertilizer. In Sudan, 100% of the rice area was fertilized. Only in Mauritius was 50% of the maize fertilized.

Average application rates in the fertilized areas were 248 kg N/ha in Mauritius for maize, and 233 kg N/ha in Tanzania for coffee. Sugar crops in the area received N in amounts ranging from 104 kg/ha in Mauritius to 221 kg/ha in Sudan.

E. North America

Data from North America include the following countries: Canada (1979), Mexico (1977), and the United States (1980). Table 1 shows that cereal crops received most of the N fertilizer—65% in Canada and Mexico and 61% in the United States. In Canada and Mexico, 13 and 8% of N, respectively, was used for oil crops; in the United States, $>$ 12% was used for industrial crops (12% of total N was used for sugarcane).

Table 2—Estimated proportion of harvested crop areas receiving N fertilizer in various countries.

Region/Country	Year	Proportion of harvested areas fertilized with N									
		Wheat	Maize	Rice	Barley	Tuber crops	Coffee	Tea	Sugar crops	Fiber crops	Oil crops
		% of harvested crop area									
Africa											
Egypt	75	--	--	--	36	--	--	--	--	93	--
Mali	76	--	80	12	--	--	--	--	--	90	90
Senegal	75	--	--	70	--	--	--	--	--	97	28
Upper Volta	75	--	--	12	--	--	--	--	100	35	--
Cameroon	77	--	--	26	--	--	1	--	95	--	--
Nigeria	79	58	20	34	--	25	--	--	50	49	--
Burundi	76	--	--	--	--	--	20	100	--	--	--
Ethiopia	76	5	3	--	2	--	--	--	100	100	9
Mauritius	77	--	50	--	--	70	--	--	--	--	--
Somalia	76	--	20	75	--	--	--	--	100	75	100
Sudan	75	--	--	100	--	--	--	100	--	100	--
Tanzania	75	--	12	--	--	--	8	100	--	--	--
North America											
Canada	79	60	--	--	--	--	--	--	--	--	75
Mexico	77	99	41	65	52	78	21	--	--	45	47
United States	80	67	96	100	--	--	--	--	47†	71†	23†
Latin America											
Dominican Republic	79	--	15	80	--	11	--	--	55	98	19
Colombia	78	10	15	99	--	99	26	--	--	56	--
Paraguay	78	2	30	6	--	--	--	--	--	13	33

(continued on next page)

Table 2—Continued.

Region/Country	Year	Proportion of harvested areas fertilized with N									
		Wheat	Maize	Rice	Barley	Tuber crops	Coffee	Tea	Sugar crops	Fiber crops	Oil crops
		% of harvested crop area									
Asia											
Bangladesh	80	80	--	53	--	50	--	--	41	32	28
India	79	90	50	75	--	95	--	--	80	50	--
Pakistan	78	78	73	86	--	100	--	--	95	86	10
Israel	76	65	2	--	--	--	--	--	100	100	100
Turkey	79	67	79	96	--	80	--	--	--	--	51
Japan	79	99	99	99	--	99	--	--	99	99	--
Europe											
Austria	77	100	100	--	100	100	--	--	100	--	--
Denmark	80	100	--	--	--	--	--	--	--	100	100
Finland	79	100	--	--	100	--	--	--	--	--	100
France	79	99	99	99	--	99	--	--	--	--	99
Germany, F.R.	79	100	100	--	--	10	--	--	100	--	100
Hungary	80	100	100	100	--	100	--	--	100	--	100
Netherlands	79	100	100	--	--	100	--	--	--	100	--
Portugal	78	99	85	100	--	98	--	--	--	--	--
Spain	79	80	90	100	--	90	--	--	88	9	46
Sweden	77	100	--	--	100	96	--	--	100	--	100
Switzerland	79	99	99	--	--	--	--	--	--	--	--
United Kingdom	80	99	--	--	98	98	--	--	93	--	99
Oceania											
Australia	79	70	70	85	--	95	--	--	100	100	25

† Only the single crops of sugarcane, cotton, and soybeans are included in the crop groups of sugar, fiber, and oil crops, respectively.

Table 3—Estimated rates of N applied to fertilized areas of selected crops in various countries.

Region/Country	Year	Average N applied to fertilized areas									
		Wheat	Maize	Rice	Barley	Tuber crops	Coffee	Tea	Sugar crops	Fiber crops	Oil crops
		kg/ha									
Africa											
Egypt	75	122	151	75	219	181	--	--	207	132	54
Mali	76	11	--	118	--	--	--	--	--	28	--
Senegal	75	--	--	37	--	17	--	--	--	37	10
Upper Volta	75	--	30	--	--	--	--	--	164	14	--
Cameroon	77	--	70	--	--	--	125	--	84	--	--
Nigeria	79	80	49	48	--	21	--	--	56	24	--
Burundi	76	--	--	--	--	--	22	85	--	--	--
Ethiopia	76	18	40	--	20	--	--	--	127	5	13
Mauritius	77	--	248	--	--	170	--	47	104	--	--
Somalia	76	--	10	32	--	--	--	--	143	36	10
Sudan	75	--	--	69	--	--	--	--	221	109	--
Tanzania	75	--	33	--	--	--	233	91	--	--	--
North America											
Canada	79	57	--	--	--	--	--	--	--	--	30
Mexico	77	133	104	90	46	59	101	--	130	121	66
United States	80	66	146	133	--	--	--	--	212†	80†	18†
Latin America											
Dominican Republic	79	--	45	96	--	101	--	--	83	61	24
Colombia	78	45	51	192	--	143	163	--	--	50	--
Paraguay	78	30	30	51	--	--	--	--	--	15	10

(continued on next page)

Table 3—Continued.

Region/Country	Year	Average N applied to fertilized areas									
		Wheat	Maize	Rice	Barley	Tuber crops	Coffee	Tea	Sugar crops	Fiber crops	Oil crops
		kg/ha									
Asia											
Bangladesh	80	37	--	21	--	35	--	--	16	9	12
India	79	30	20	25	--	80	--	--	80	40	--
Pakistan	78	67	67	60	--	150	--	--	100	64	25
Israel	76	120	200	--	--	--	--	--	120	150	50
Turkey	79	53	70	132	--	121	--	--	--	--	3
Japan	79	90	51	93	--	99	--	--	221	280	--
Europe											
Austria	77	90	120	--	80	120	--	--	140	--	--
Denmark	80	150	--	--	--	--	--	--	121	--	120
Finland	79	103	--	--	87	98	--	--	--	--	98
France	79	119	141	144	--	152	--	--	--	--	132
Germany, F.R.	79	120	180	--	--	90	--	--	160	--	143
Hungary	80	118	136	102	--	115	--	--	90	--	74
Netherlands	79	113	108	--	--	178	--	--	130	--	--
Portugal	78	75	100	100	--	120	--	--	--	--	--
Spain	79	50	150	180	--	180	--	--	150	30	45
Sweden	77	99	--	--	67	93	--	--	141	--	135
Switzerland	79	83	119	--	--	--	--	--	--	--	--
United Kingdom	80	146	--	--	88	189	--	--	147	--	257
Oceania											
Australia	79	7	51	82	--	90	--	--	195	116	61

† Only the single crops of sugarcane, cotton, and soybeans are included in the crop groups of sugar, fiber, and oil crops, respectively.

In the United States, straight materials supplied 78% of the N. Anhydrous ammonia (AA) provided about 40% of the N, urea ammonium nitrate (UAN) solutions 18%, urea 9%, and AN 8%. Fluid fertilizers (AA, UAN, NP's, and NPK's) furnished about 65% of all N used in the United States. Urea, AN, and granular NPK appeared to be the more important sources of N in Canada and Mexico.

Table 2 shows the percentages of crop areas receiving fertilizer. In Canada, 60% of the wheat area and 75% of the oil crops area received N fertilizer. Data for Mexico show that the percentage of areas fertilized ranged from 21% for coffee to 99% for wheat. In the United States, 100% of the rice area and 96% of the maize area received N fertilizer. Because portions of the wheat and cotton were grown in drought-prone areas, only about two-thirds of each crop received N fertilizers. Almost one-fourth of the soybean [*Glycine max* (L.) Merr.] area in the United States received N fertilizer.

In Canada, 57 kg N/ha was applied to wheat and 30 kg N/ha to oil crops. In Mexico, application rates of N varied from about 130 kg/ha for wheat and sugarcane to 46 kg/ha for barley. In the United States, sugarcane received 212 kg N/ha; rice and maize received $>$ 100 kg N/ha; and oil crops received only 18 kg N/ha.

F. Central America and Caribbean

Countries in Central America and the Caribbean from which data were obtained include: Cuba (1975), the Dominican Republic (1979), and Panama (1977). Table 1 shows that in Cuba and the Dominican Republic, sugarcane received 55 and 68%, respectively, of the total N used. In Panama, 57% of the total N was applied to cereal crops. Products used to supply N in the Dominican Republic and Cuba include: urea, AS, and NPK's.

Tables 2 and 3 show the percentage of crop areas fertilized with N and the average application rates for the Dominican Republic. It should be noted that 98% of the fiber crop, 80% of the rice crop, and 55% of the sugarcane crop area received fertilizer N. Data on average N application rates indicate that the rates varied from 24 kg/ha for oil crops to about 100 kg/ha for rice and tuber crops.

G. Andean Region

Countries in the Andean region from which data were obtained include: Colombia (1978), Ecuador (1980), Peru (1980), and Venezuela (1980). Table 1 shows that in all four countries of this region cereal crops received more N than any other crop group. About 65% of the N was used for cereal production in Venezuela and from 28 to 37% in the other three countries. Major quantities of N were used on industrial crops in Ecuador, Peru, and Venezuela; on stimulants, roots, and tuber crops in Colombia; and on fiber crops in Peru.

Urea and AS were the principal sources of N in all of these countries.

In addition, NPK products supplied significant amounts of N in Colombia, Ecuador, and Venezuela, while AN was utilized in Colombia and Peru.

Data on the percentage of crop areas fertilized with N and its application rates were available from Colombia only. Tables 2 and 3 show that 99% of the rice, roots, and tuber [mainly potatoes (*Solanum tuberosum* L.)] areas received fertilizer N. Rates of application were 192 and 143 kg N/ha, respectively. Also, 56% of the fiber crops area and 26% of the coffee area were fertilized with N. Application rates were equal to 50 and 163 kg N/ha, respectively.

H. Other South America Countries

Data for four other South American countries appear on Table 1. This table shows that in Argentina (1979) and Chile (1979) cereal crops receive more N than any other crop group. Approximately 36% of the total N was used for cereal production in Argentina and 50% in Chile. In Brazil (1978), stimulant crops—namely coffee—and cereals each received about 25% of the total N. In Paraguay (1978), oil crops—mainly soybeans—received 41% of the total N, while cereal and industrial crops each received 26% of the total N.

Urea and NPK products were widely used in Argentina, Brazil, and Paraguay, while straight materials supplied most of the N in Chile. Considerable amounts of locally produced sodium nitrate were also used in Chile.

Data on percentages of crop areas fertilized with N and average application rates were available from Paraguay only. Table 2 shows that 33% of the oil crops area, and 30% of the maize area received fertilizer. Table 3 shows that the average application rates for these crops were 10 and 30 kg N/ha, respectively.

I. Central and South Asia

Data were collected from the following countries in central and southern Asia: Afghanistan (1975), Burma (1979), Bangladesh (1980), India (1979), Pakistan (1978), and Sri Lanka (1980). In all of these countries, cereals were the crops receiving most of the N fertilizer used. The percentage of the total N used that was applied to cereal crops ranged from 50% in Sri Lanka to 92% in Bangladesh. Urea was the main product used. Also, significant quantities of AS were used in India and Sri Lanka, and NPK's and diammonium phosphate (DAP) supplied significant quantities of N in India.

Table 2 shows the estimated percentage of crop areas receiving fertilizer in Bangladesh, India, and Pakistan. It reveals that high percentages of the tuber, sugar, and cereal crop areas received fertilizer. Table 3 shows the average rates of N applied within fertilized areas for different crops. In India, tubers and sugarcane received 80 kg N/ha and cereals received 20 to 30 kg N/ha. In Pakistan, cereal crops were apparently fertilized with N at about twice the rate in India, and the reported rates of N in fertilized areas were 150 and 200 kg/ha for tubers and sugar crops, respectively. This table shows that in Bangladesh, wheat received 37 and rice 35 kg/ha. Other crops received lower amounts.

J. Southeast Asia

Data were obtained from the following southeastern Asian countries: Indonesia (1980), Malaysia (1980), the Philippines (1980), and Thailand (1980). Cereal crops received between 49 and 85% of the total N used in Indonesia, the Philippines, and Thailand, but only 27% of that used in Malaysia. The industrial crops received 69% of the total N in Malaysia and 34 and 27% in the Philippines and Thailand, respectively.

Urea was the main fertilizer product used in the area, followed by AS and NPK's. Thailand used significant quantities of ammonium chloride (AC). Almost all N used in Indonesia was supplied as urea.

K. Middle East Region

Middle East countries from which data were obtained include: Cyprus (1977), Israel (1976), Turkey (1979), and the Yemen Democratic Republic (1976). In Cyprus and Israel, fruit crops received 22 and 44% of the total N used, respectively. In Turkey and the Yemen Democratic Republic, cereal crops received 58 and 50% of the total N used, respectively. Urea, AS, and NPK products were widely used in countries of this area. In Israel, aqua ammonia (AQA) was also used.

Table 2 shows the estimated percentage of crop areas fertilized in Israel and Turkey. It shows that in Israel, 100% of the crop area of sugar crops, fiber crops, and oil crops received fertilizer, while in Turkey, 96% of the rice area and 80% of the tuber crop area were fertilized. Table 3 shows that the average N application rates in Israel were: maize—200 kg/ha; fiber crops—150 kg/ha; and sugar and wheat—120 kg/ha. In Turkey, rice received 132 kg N/ha, and tuber crops received 121 kg N/ha.

L. Northeast Asia

Data were collected from Japan (1979) and South Korea (1975) of northeastern Asia. Table 1 shows that in both countries, cereal crops received more N than any other crop group. Japan used 35% and Korea 65% of total N consumption for cereal crops. About 16% of total N was used for vegetables in each country, and in Japan, a similar quantity was used for fruits. The N was supplied to crops mainly as urea and NPK products.

Table 2 shows that in Japan, 99% of the crop areas reported received fertilizer. Table 3 shows the average rates applied to different crops. They ranged from 280 kg N/ha for fiber crops to 51 kg N/ha for maize. Sugar crops received about 221 kg and wheat, rice, and tubers received 90 to 99 kg N/ha.

M. Europe

Data collected from Europe include the following countries: Austria (1977), Denmark (1980), Finland (1979), France (1979), the Federal Republic of Germany (1979), Hungary (1980), the Netherlands (1979), Portugal (1978), Spain (1979), Sweden (1977), Switzerland (1980), and the United Kingdom (1980). Table 1 shows that except for the Netherlands, Switzerland, and the United Kingdom, where the greatest portion of N was used for pastures, all countries used more N on their cereal crops than any other crop group. The portion of N received by cereal crops ranged from 9% in the Netherlands to 62% in Hungary. In the Netherlands, Switzerland, and the United Kingdom, pastures received 75, 41, and 60% of the total N used, respectively. Cereals and pastures together received from 81 to 92% of all N in Austria, Denmark, Finland, and the United Kingdom, and possibly France and the Federal Republic of Germany (pastures were not identified in the data sources). Hungary, Sweden, and Switzerland used almost two-thirds of the total N for cereals and pastures. The most common N sources in Europe were NPK's and CAN, although many countries also used significant quantities of urea, ammonium nitrate sulfate (ANS), AN, and UAN.

Table 2 shows the estimated percentage of crop areas receiving fertilizer; with very few exceptions, ≥ 90% of the identified crop areas received N fertilizer. Table 3 presents the estimated application rates for different crops. With few exceptions among countries, most crops except barley received ≥ 100 kg N/ha.

N. Oceania

Table 1 presents data obtained from Australia (1979) and Fiji (1980). It shows that Australia used 43% of the total N for cereal crops and 24% for industrial crops, while Fiji used 98% of the total N for industrial crops, namely sugarcane. Principal fertilizer products utilized included urea, AN, AS, and NPK's in Australia and AS in Fiji.

The estimated percentages of crop areas fertilized in Australia were 70 to 85% for cereal crops and 100% of the crop areas in sugar and fiber crops. About 95% of the crop area in tubers was fertilized with N. The rates of N applied in fertilized areas were 195 kg/ha for sugar crops, 116 kg/ha for fiber crops, 90 kg/ha for tuber crops, and 61 kg/ha for oil crops. It is estimated that only 7 kg N/ha was used for wheat.

O. Silviculture Fertilization

A separate survey was conducted for N fertilization in silviculture (G. W. Bengtson, School of Forestry, Oregon State Univ., Corvallis, Oreg.). Of 41 countries to which survey questionnaires were sent, 14 responders submitted data, 9 responses indicated no significant use or no data available,

Table 4—Nitrogen fertilizer use in silviculture, worldwide, 1979–1980.†

Country	Area fertilized and quantity of N applied according to N source‡										Total area fertilized	Total N applied	Average N rate
	Urea		AN		AS		DAP		Mixtures				
	ha	t	ha	t	ha	t	ha	t	ha	t	ha	t	kg/ha
Australia	0	0	0	0	0	0	0	0	8 197	402	8 197	402	49
Austria	0	0	11 000	560	0	0	0	0	0	0	11 000	550	50
Denmark	0	0	0	0	0	0	0	0	8 700	713	8 700	713	82
United Kingdom	2 000	700	0	0	0	0	0	0	0	0	2 000	700	350
Germany, F.R.§	0	0	3 050	378	0	0	0	0	2 800	160	5 850	538	92
Finland	15 000	2 271	27 700	4 155	0	0	0	0	1 000	600	43 700	7 026	161
France	0	0	2 750	575	0	0	2 500	75	0	0	5 250	650	124
Ireland	25	5	25	5	0	0	0	0	0	0	50	10	200
Japan	0	0	0	0	0	0	0	0	48 706	3 800	48 706	3 800	78
New Zealand	5 000	2 000	1 000	350	100	100	800	400	0	0	6 900	2 850	413
Sweden¶	32	5	131 644	18 973	0	0	0	0	0	0	131 676	18 978	144
United States#	93 365	18 200	1 633	267	2 000	200	20 021	1 589	0	0	117 019	20 256	173
Total††	115 422	23 181	178 802	25 263	2 100	300	23 321	2 064	69 403	5 675	389 048	56 473	145

† Results of a survey conducted by G. W. Bengtson, School of Forestry, Oregon State Univ.

‡ Period of July 1979–June 1980. Reports of "no significant use of N fertilizers in forestry" or "no data available" were received from Belgium, Brazil, Chile, Costa Rica, India, Mexico, the Netherlands, Yugoslavia, and Venezuela. No response to questionnaire was received from Belgium, Canada, Colombia, Czechoslovakia, German Democratic Republic, Greece, Israel, Italy, Kenya, Malawi, Nigeria, Norway, Philippines, Poland, South Africa, Taiwan, Turkey, or Zambia.

§ Includes Bavia, Baden-Wurtemberg and Niedersachsen only.

¶ Data for Sweden are for calendar year 1979 (not 1979–1980).

For the United States, 15 593 ha received both urea and DAP. This area is counted once under urea. N applied as DAP on these 15 593 ha is included in DAP entry.

†† The total figures exclude information from South Korea, which reported 300 000 ha fertilized with 1920 t of N from mixtures. This results in about 6 kg N/ha, which possibly indicates either spot applications within stated areas or an error in the figures reported. Also, information was received from the USSR, which reported that 228 000 ha of nurseries, plantations, and forests is fertilized with AN and urea. The quantity used was not reported.

and 18 countries failed to respond. The data are summarized in Table 4. Sweden and the United States used 69% of the total N reported for silviculture, and these two countries accounted for 64% of the fertilized area reported in the table—the fertilized areas reported by both South Korea and the USSR (but not included in this table) far overshadow all others. The next greatest areas fertilized were in Finland and Japan. New Zealand and the United Kingdom reported the highest rates of N fertilization: 413 and 350 kg N/ha, respectively. Only in Sweden and New Zealand were significant portions of total N used for silviculture: 7.4 and 12.7%, respectively. In other countries, the proportion of total N used for silviculture was usually $<$ 0.5%; AN accounted for 45% and urea for 41% of the N used, respectively.

III. SUMMARY

There are few estimates of N use by crops, and those that exist are incomplete in terms of countries and crops or crop groups within countries. Nitrogen use estimates by crops were compiled from literature sources and new estimates from questionnaires. It is recognized that the reliability of data varied among countries, and in most cases the estimates are only rough approximations.

In Africa, selected information on N use by crops was available for 23 countries. In $>$ 50% of those countries, cereal crops receive a greater portion of N than any other crop group for which N use was identified. Fiber crops received significant portions and sometimes the greatest portion of N in Sahel and western coastal Africa. In eastern and southern Africa, large portions of N were used for stimulant and industrial crops. Generally, small portions of cropped areas of cereals were fertilized, while $>$ 90% of areas of stimulants, sugar, and fiber crops were fertilized with N. Estimated rates of N application were also greater for the commercial crops. Urea and NPK's were the principal fertilizers used in most countries, but CAN or AN and AS were frequently used in northern, eastern, and southern Africa.

In North America, data were available from Canada, Mexico, and the United States to account for 78, 94, and 66% of total N use, respectively. From 61 to 65% of the total N was used for cereal crop production in each country. Oil crops in Canada and oil and industrial crops in Mexico each received about 10% of the total N, while in the United States, cotton was the only crop, other than cereals, identified as receiving as much as 3% of the total N. Portions of crop areas receiving N vary greatly among countries and crops. Generally, $\geq$ 50% of the cropped areas for which estimates were obtained received N fertilizer. Exceptions were coffee in Mexico (20%) and oil crops in the United States (23%). Generally, sugar crops received the largest and oil crops received the lowest rate of N per hectare within the fertilized areas. Urea, AN, and NPK's were the principal N sources in Canada and Mexico, but in the United States, AA for direct application supplied 40%, NPK's 22%, and UAN solution 18% of all N. Since some

NPK's also were applied in fluid form, it is estimated that 65% of all N was applied as fluid fertilizers in the United States. Fluids such as AA, UAN, and NPK's were also used in Canada and Mexico.

In Latin America, estimates of N use by crops were available from only 11 countries. Cereals received more N than any other group of crops in seven of these countries, while stimulant and industrial crops received more N in the other four countries. In the three countries for which information was available on both percentage of cropped areas fertilized and application rates, < 30% of the wheat and maize was fertilized with N, and the average rates were only 30 to 50 kg N/ha. Urea, AS, and NPK's were the principal N sources used in Latin America.

In Asia, including the Middle East, data were available for 16 countries. Cereal crops received the largest portion of N in 13 of those countries. Of the three remaining countries, most N was used for industrial crops in Malaysia and for fruits in Cyprus and Israel. Significant portions of the total N were also used for industrial crops in Pakistan, the Philippines, Thailand, and Turkey, while fruits and vegetables received significant portions of N in Japan, South Korea, and Turkey. Major portions of N were used for fiber crops in Afghanistan, Israel, Pakistan, and the Yemen Democratic Republic. Data were available on areas fertilized and rates of application for only five countries. It was estimated that 99% of all crop groups reported were fertilized with N in Japan and 100% of the sugar, fiber, and oil crops in Israel. In Japan, the rates of application ranged from 51 kg N/ha for maize to well over 200 kg/ha for sugar and fiber crops. It was estimated that from 53 to 90% of the wheat and rice were fertilized with N in Bangladesh, India, and Pakistan at rates of 21 to 37 kg N/ha in Bangladesh, 25 to 30 kg N/ha in India, and 60 to 67 kg N/ha in Pakistan. Approximately two-thirds of the wheat received N in Turkey at an average rate of 53 kg N/ha. Urea and AS were the principal sources of N in Asia, but NPK's were important in several countries.

In Europe, cereal crops were reported to receive the largest portion of N among crop groups in 9 of the 12 countries for which data were available. In most countries, the crop group that received either the second greatest or the greatest proportion of N was pastures. It appeared that in northern Europe, > 80% of all N was used for cereals and pastures. For most crop groups that were identified, > 90% of the areas were fertilized. The only notable exceptions were for fiber and oil crops in Spain. Rates of N application for cereals were usually > 100 kg N/ha, except for wheat in Portugal and Spain. Tuber, sugar, and oil crops were generally fertilized at rates considerably more than 120 kg N/ha. The principal N sources in Europe appeared to be CAN and NPK's, although AN, ANS, AS, and urea were widely used. Fluid fertilizers were used in some countries, notably France, Germany, and Spain.

In Oceania, general crop data were available from only Australia and Fiji. About 43% of total N was used on cereals in Australia and 24% on industrial crops. Almost all N was used for industrial crops in Fiji. About 70% of maize and wheat was fertilized and almost all tuber, sugar, and fiber crops were fertilized with N in Australia. The reported rate of application of N for wheat appeared to be extremely low, but high rates of N were used for sugar and fiber crops. Principal sources of N were urea, AS, AN, and NPK's.

In silviculture, the N used was a significant part of the total N consumption only in New Zealand (12.7%) and Sweden (7.4%).

REFERENCES

Economic and Social Commission for Asia and the Pacific. 1980 and 1981. *News in Brief,* several issues. Econ. & Social Comm. for Asia & the Pacific, Bangkok, Thailand.

Falusi, A. O., and L. B. Williams. 1981. Nigeria fertilizer sector: present situation and future prospects. Tech. Bull. no. T-18. IFDC, Muscle Shoals, Ala.

Food and Agriculture Organization of the United Nations. 1980. Provisional statistics on fertilizer use by major crops. FAO, Rome, Italy.

International Fertilizer Development Center. 1974–1981. Various reports. IFDC, Muscle Shoals, Ala.

ISMA. 1980. Fertilizer consumption report 1978/79. ISMA, Ltd. (now IFA, Ltd.), Paris, France.

Martinez, A., and R. B. Diamond. 1982. Fertilizer use statistics in crop production. IFDC-T-24. IFDC, Muscle Shoals, Ala.

U.S. Department of Agriculture. 1980. 1981 Fertilizer situation. Economics and Statistics Service, USDA, Washington, D.C.

Van Dierendonck, F. J. E., K. P. Krishnaswami, and M. C. Verghese. 1978. Identification of facilities needed to expand fertilizer production and supplies in 23 least developed countries. United Nations Ind. Dev. Organization, Vienna, Austria.

2

Paul J. Stangel
International Fertilizer Development Center
Muscle Shoals, Alabama

World Nitrogen Situation— Trends, Outlook, and Requirements

Nitrogen is the most common and widely used fertilizer nutrient. Produced primarily as NH_3, it can be applied as such or further processed into a variety of liquid or solid N fertilizers. The most common solid N fertilizers are urea, ammonium nitrate (AN), and ammonium sulfate (AS). Ammonium sulfate is frequently produced as a by-product of the chemical, fiber, and steel industries. Some ammonium chloride (AC) is also produced, mainly as a by-product of the soda-ash industry. Urea is the most popular solid N fertilizer, and urea and anhydrous ammonia (AA) are the dominant N products traded on the international market.

The objectives of this paper are to highlight the current trends in N production and consumption, identify those factors that are likely to influence future development of the N sector, and predict the N situation of the next 15 to 20 yr.

I. CURRENT COMPOSITION OF THE N INDUSTRY

The current composition of the N industry is strongly influenced by location, ownership, type and cost of feedstock, plant capacity, and government policies pertaining to the fertilizer sector. The following sections describe the current situation and the immediate prospects in terms of these factors.

A. N Capacity

Nitrogen fertilizers were produced in 72 countries in 1982, primarily as NH_3 and its derivatives. The total installed NH_3 capacity in the world was

Nitrogen in Crop Production.

Table 1—World NH_3 capacity installed by region, 1970–1985.†

Region	1970		1980		1985	
	t × 10³ N	Percent of total	t × 10³ N	Percent of total	t × 10³ N	Percent of total
North America	13 855	27.5	17 601	18.8	19 024	17.2
Latin America	1 336	2.6	4 398	4.7	6 840	6.2
Western Europe	11 978	23.7	14 791	15.8	15 299	13.8
Eastern Europe	5 865	11.6	10 611	11.3	12 094	10.9
USSR	7 073	14.0	20 205	21.6	25 385	23.1
Africa	528	1.1	2 162	2.3	3 136	2.8
Asia	9 341	18.5	23 345	25.0	28 314	25.6
Oceania	514	1.0	446	0.5	446	0.4
Total	50 490	100.0	93 559	100.0	110 538	100.0

† TVA Capacity Files, 1982.

98.6 million tonnes of N as of May 1982. It was 93.6 million tonnes of N in 1980 and only 50.5 million tonnes in 1970 (Table 1).

The world's NH_3 capacity historically has been centered in the developed countries. As recently as 1970, North America and western Europe were the leading producers of NH_3 and accounted for 51.2% of world capacity. Since then these regions have been replaced by Asia, the Union of the Soviet Socialist Republics (USSR), and eastern Europe as the world centers for NH_3 capacity. These latter areas collectively accounted for 58% of world capacity in 1982.

World capacity is expected to reach 110.5 million tonnes by 1985 with the USSR's share rising to 25.4 million tonnes or 23.1% of total world capacity. Asia, currently in the midst of a major expansion, will have 28.3 million tonnes of N capacity by 1985, or 25.6% of total world capacity.

The leading NH_3-producing nations are the USSR, the United States, the People's Republic of China (PRC), and India, which in 1982 collectively accounted for about 50% of world capacity (Table 2). Major expansions in NH_3 capacity now occurring in several countries will be on stream by 1985. The more notable of these are: USSR (5.1 million tonnes of N); PRC (1.1

Table 2—World's leading N producers—1980 (TVA Capacity Files, 1982).

Country	NH_3 capacity	World capacity†
	t × 10³ N	%
USSR	19 779	20.8
United States	15 446	16.2
PRC	8 049	8.5
India	5 054	5.3
Romania	3 238	3.4
Federal Republic of Germany	2 940	3.1
Japan	2 926	3.1
France	2 846	3.0
Netherlands	2 207	2.3
Mexico	2 183	2.3
Canada	2 083	2.2

† Based on 95 179 000 t N of installed capacity.

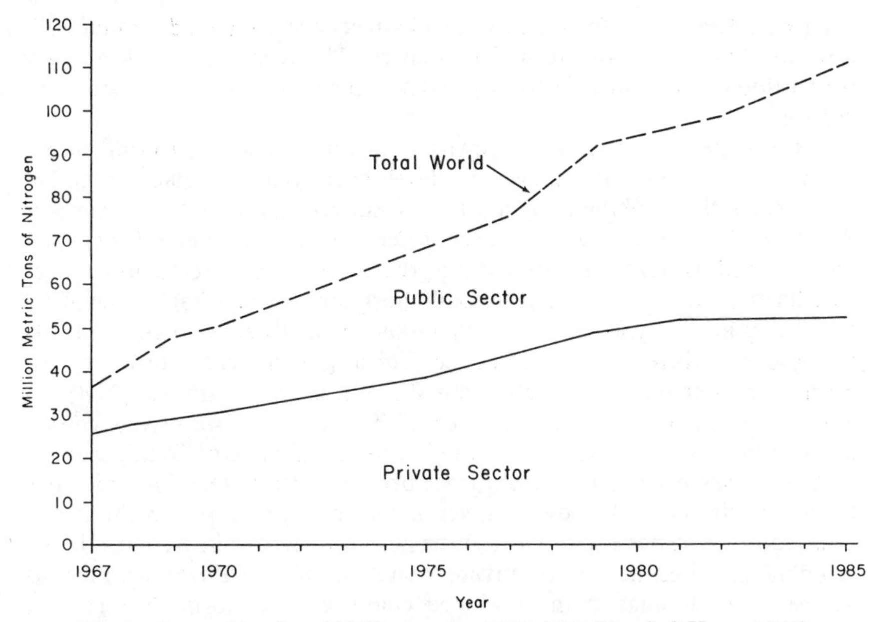

Fig. 1—Ownership of world NH_3 capacity, 1967–1985 (Harris & Harre, 1979).

million tonnes of N each); India and Indonesia (1.0 million tonnes of N each); and Trinidad, Brazil, and Algeria (0.5 million tonnes of N each). Although some expansion may also occur in North America (Canada), western Europe (the Netherlands), and eastern Europe, it is likely to be balanced with the dismantling of obsolete facilities or idling of some inefficient units now in operation, particularly in the United States and Japan. Therefore, current expansions in these regions are not likely to add significantly to overall world NH_3 capacity.

By 1985, slightly more than 50% of world NH_3 capacity will have been contracted and installed since 1972 and, because of the combined effect of rapid escalation in capital costs and construction, units built since then will have much higher production costs than units built before 1972. Once these units become the dominant factor in world production and N supplies are again more nearly in balance, one can expect a rapid and significant increase in the international price of N fertilizer.

B. Ownership

There has been a dramatic shift from private to public ownership of the world N industry (Fig. 1). In 1967, 70% of the N industry was privately held, and only 30% of total capacity was publicly owned (state owned). However, by 1975 the proportion that was publicly owned had climbed to 45%; it is estimated to be 54% in 1982 and forecast to reach 60% by 1986 (Harris & Harre, 1979). The major reason for this shift has been the large expansions that have occurred in the centrally planned and developing countries. The reason for state ownership in centrally planned economies is obvious: The state owns all heavy industry. The reason for such a heavy proportion of state-owned capacity in developing countries, however, is less apparent.

The large increase in public ownership seems to be due to four major factors. First, national planners in developing countries place great importance on the availability of fertilizer to support national food programs (Mudahar & Hignett, 1981). Some consider it the key to economic development as well as political stability and, therefore, want direct control over such an important commodity. Second, most donors providing financial aid to developing countries do so through loans made directly to governments and not to fertilizer companies per se. Third, government controls placed on the fertilizer sector generally limit the return on investment (ROI). In many cases this ROI is below what could be expected from private capital invested in other sectors. Fourth, the international price of fertilizer in recent years has been too low to attract private capital. This low price has automatically resulted in direct government involvement and is the major cause for state ownership of most of the new fertilizer facilities. Thus, in developing countries the flow of private capital into new fertilizer facilities has almost been eliminated; in developed countries that normally serve the international market, private investment in new facilities has been severely restricted. An ever-increasing percentage of N moving in international trade is from state-owned N facilities.

The shift from private to public ownership has major implications for those engaged in international trade. State-owned facilities frequently serve as instruments in carrying out government policies. Most state-owned units do not operate under the normal commercial procedures for buying and selling. They are subsidized; they operate according to production quotas established by the government; and they seldom have the authority to develop long-term commitments in international markets. As a result, publicly owned units tend to enter the international market sporadically, seldom attempt to cultivate customer loyalty, usually sell surplus production at prices expedient for quick sales (sometimes below cost when markets are depressed), and occasionally sell at unusually high prices when supplies are limited. Increasingly, N from state-owned companies is bartered for other commodities on the international market (Lastigzon, 1981) as dictated by other elements within the government. Therefore, many of the decisions are beyond the control of the companies directly involved. As a result of this shift in ownership, many privately owned companies have found it increasingly difficult to compete in international markets.

C. NH_3 Feedstock

Ammonia is produced from a variety of feedstock sources including natural gas, naphtha, refinery gas, liquid petroleum (LPG), liquid natural gas (LNG), heavy fuel oil, coal, and electrolytic H. Natural gas is the dominant source of feedstock; it served 71% of the world's NH_3 capacity in

Table 3—World NH_3 capacity by source of feedstock, 1973–1985 (TVA Capacity Files, 1982).

Feedstock	1973		1975		1980		1985	
	Capacity	Percent of total†	Capacity	Percent of total†	Capacity	Percent of total†	Capacity	Percent of total†
	t × 10³		t × 10³		t × 10³		t × 10³	
Natural gas	35 502	64	46 804	70	67 125	72	81 952	74
Naphtha	6 560	11	7 277	11	8 380	9	11 395	10
Refinery gas (LPG, LNG)	8 034	13	5 806	9	7 949	8	6 479	6
Coal‡	5 055	8	4 922	7	5 897	6	6 082	5
Fuel oil	1 530	3	1 629	2	3 409	4	4 563	4
Electrolytic H	803	1	708	1	799	1	799	1
Total	57 484	100	67 146	100	93 559	100	111 270	100

† Calculated.
‡ Assumes 4 million tonnes of NH_3 capacity in PRC based on coal.

Table 4—Percentage by region of total NH_3 capacity based on naphtha—1973–1985 (Stangel, 1980).

Region/Country	Naphtha-based NH_3 capacity†			
	1973	1975	1980	1985
	% of total			
North America	--	--	--	--
Central and South America	2	4	4	4
Western Europe	33	30	19	12
Eastern Europe	--	--	--	--
USSR	--	--	--	3
Africa	--	--	1	1
Asia	64	65	60	66
Communist Asia	1	1	16	14
Oceania	--	--	--	--
Total	100	100	100	100

† Total naphtha-based NH_3 capacity (t × 10³): 1973—6 560; 1975—7 277; 1908—10 000; 1985—11 395.

1980 (Table 3). Naphtha is a significant but distant second choice as NH_3 feedstock, and 11% of the world's capacity was based on this feedstock in 1980. Sixty percent of the world's 10 million tonnes of naphtha capacity was located in Asia—mainly in Japan, India, North Korea, and South Korea (Tables 4 and 5). Another 19% (1.9 million tonnes in 1980) was located in western Europe—mainly in France, Spain, Portugal, and Italy. For reasons that will be explained in the next section, use of naphtha as a feedstock for NH_3 is likely to decline in the immediate years.

The United Nations Industrial Development Organization (UNIDO, 1978) predicts that natural gas will continue to be the dominant feedstock for NH_3 production for the remainder of this century. They estimate it will account for 74% of the total installed capacity by 1985 and for 64% by the year 2000. They further estimate that by the year 2000 naphtha will account

Table 5—Feedstock source for NH_3 plants in selected Asian countries (TVA Capacity Files, 1982).

Country	Installed capacity†	Natural gas	Naphtha	Fuel oil	Coal	Electro	Other
		Feedstock					
	t × 10^3	% of total capacity					
Japan	2 926‡	17	61	3	2	--	17
South Korea	1 255‡	--	100	--	--	--	--
North Korea	755	--	43	--	25	--	32
PRC	10 099	27	16	--	50	--	7
Taiwan	547	85	--	8	--	--	7
India	7 152	36	28	28	8	--	--
Indonesia	1 806§	97	--	3	--	--	--
Philippines	101	--	27	--	--	--	73
Malaysia	315§	87	--	--	--	--	13
Turkey	1 485	--	92	--	--	--	8
Sri Lanka	147	--	100	--	--	--	--
Israel	66	--	100	--	--	--	--
Syria	313	--	100	--	--	--	--
Australia	446	44	--	--	--	--	56

† As of 1982.
‡ Current exporter of N.
§ Will have potential to export by 1985.

for only 5%, while heavy fuel oil will rise to 15% and coal will climb to 17% as feedstocks for NH_3 plants. The choice of feedstock strongly affects the capital costs of NH_3 plants (IFDC, 1979). Using natural gas units as a base of 1.0, naphtha units require 1.1, oil-based units 1.5, and coal-based units nearly 2.0 times as much capital investment.

D. Feedstock Prices

The rapid escalation in the price of crude oil from 1973 to 1981 had a decided impact on the price of natural gas and naphtha—the feedstocks for about 85% of the world NH_3 capacity. The international price of crude oil increased twelvefold between 1972 and 1982 (Table 6). The price of naphtha, a downstream product from refined oil and an intermediate widely used in the manufacture of motor fuel and a number of petrochemicals, is highly sensitive to changes in the price of crude oil (Anon., 1982a, b). For example, when the Organization of Petroleum Exporting Countries (OPEC) doubled the price of crude oil between 1979 and the end of 1980, prices of naphtha tripled, reaching a high of $390/t on the European spot market during the third quarter of 1980. Spot prices of naphtha have declined since then, and they reached a 3-yr low of $255/t in the European spot market as of 16 Mar. 1982 (Anon., 1982c); this price is still relatively high, however, and is the energy equivalent of $5.81/M scf (1000 standard cubic feet; 28.317 m^3) of natural gas. As a result of the present price structure, naphtha-based NH_3 units are not competitive with comparable units based on natural gas from fields where the opportunity costs for gas are below the energy equivalent for oil.

Table 6—Prices of oil and natural gas in the United States, 1970–1982 (Beck, 1982).

Year	Oil prices†		Natural gas prices‡
	Domestic	Imports	
1970	3.18	2.16	0.171
1971	3.39	2.31	0.182
1972	3.39	2.47	0.186
1973	3.89	3.30	0.216
1974	6.74	10.81	0.304
1975	7.56	11.19	0.445
1976	8.14	12.15	0.580
1977	8.57	13.81	0.790
1978	8.96	14.12	0.905
1979	12.51	18.72	1.178
1980	21.19	30.60	1.496
1981	31.91	33.07	1.767
1982	29.20§	30.50§	2.350§

† U.S. $/barrel.
‡ U.S. $/M scf (average of regulated and unregulated gas).
§ Estimated.

In the past, natural gas prices were not linked to the international price of crude oil as closely as naphtha prices were. There are several reasons for this situation:

1) Until recently, only limited quantities of natural gas moved in international trade. The quantities now moving in international trade come primarily from rather large gas fields that are only a fraction of the total available for NH_3 production. A field only one-sixth the size required for an LNG facility can support 1000 t/d NH_3 complex. There are many fields smaller than this located in very remote areas that can and do supply feedstock to NH_3 units. Under such circumstances the opportunity cost for such gas is at or near zero.

2) Common practice required the signing of long-term contracts (10–20 yr) to supply a specific amount of natural gas to a given production unit at guaranteed prices for a specific time with no provision for periodic price adjustments.

3) Some governments have a tendency to regulate the domestic price of natural gas, frequently at gathering costs and at only a fraction of the international price of oil.

4) Low-cost naphtha was widely used in almost all segments of the petrochemical industry.

The rapid rise in energy costs over the past decade has changed these practices. Increasingly, the natural gas prices to NH_3 plants are linked to the international price of oil. Additionally, opportunity costs for naphtha have risen so rapidly that it no longer serves as a feedstock for NH_3 plants except in special cases.

As a result of these policies, the N sector in many countries finds itself in a very difficult economic situation. There are a large number of NH_3 plants still using natural gas as a feedstock at prices contracted for in the 1960's and, therefore, at prices (energy equivalency) well below the international price of crude oil. For example, some NH_3 units serving the international market are operating on feedstock priced at between \$0.50 and \$1.50 (United States), or between \$0.25 and \$0.90 (Middle East and southeast Asia) and, in some cases even below \$0.15/M scf. New gas (only a fraction of the total amount used) used in the United States for the manufacture of NH_3 is reported to be selling for nearly \$4.00/M scf. This is equivalent to \$167/t (\$22.70/barrel) of oil—roughly 70% of the 1982 international price. Natural gas would have to be priced at \$5.64/M scf to equal 1982 crude oil prices of \$32.00/barrel. Although all natural gas in the United States is scheduled to be deregulated by 1985, much of it is still under long-term contract or is now regulated by the U.S. Government. The average wellhead price in 1982 was only \$2.35/M scf; this represents a doubling in price since 1979 and is nearly 15 times the average price in 1970 (Table 6).

E. Impact of Energy Prices on Structure of Industry

High energy prices combined with the trend toward state ownership of most of the world's NH_3 capacity is having a major impact on the structure of the N industry. Nowhere has this been more evident than in the naphtha-based capacity of Asia (Japan) and western Europe and the natural gas-based capacity of North America (the United States).

1. NAPHTHA-BASED CAPACITY

Naphtha-based NH_3 capacity aimed at export markets, particularly in Japan and to a lesser extent in western Europe, was seriously affected by the rapid increase in energy prices. This has led to a restructuring of the industries in these areas.

a. Japan—In 1973 Japan had 4.6 million tonnes of installed NH_3 capacity; most of this capacity was based on imported feedstock (Stangel, 1976). High costs of energy have placed great economic pressure on these units in recent years. As of April 1982, 1.7 million tonnes of NH_3 capacity had permanently closed. Upon close inspection of remaining capacity (Tables 5 and 7), it is the author's opinion that if world energy prices remain at or climb above the 1982 level, Japan will continue its policy of closing uneconomical units. Therefore, an additional 1 million tonnes of capacity will probably close by 1985, leaving Japan with only 1.9 million tonnes. This will serve a domestic demand for fertilizer of 0.7 million tonnes, an industrial demand of 0.4 million tonnes, and an export capacity (including 0.2 million tonnes of by-product N) of about 0.5 million tonnes of N. On the other hand, if energy prices stabilize at the 1982 levels or decline slightly over the short term, and if N prices improve, Japan may decide to retain its present export position—at least for another 3 to 5 yr. The overall operating rates of these units are likely to be determined by the level of fertilizer trade that Japan wishes to maintain with the PRC and the amount of economic aid in the form of fertilizers that it may be willing to supply to other countries. However, such moves seem doubtful and temporary at best. Recent

Table 7—Japan's NH_3 capacity operational as of 14 Apr. 1980 (Stangel, 1980).

Company	Plant location	Capacity, t × 10^3 N
Fertilizer use		
Mitsui-Toatsu Chemical†	Chiba	40.8
Mitsui-Toatsu Chemical	Osaka	414.1
Mitsubishi Gas Co.	Negata	216.5
Mitsubishi Chemical	Kurosaki	270.6
Mitsubishi Chemical†	Mizushima	74.3
Showa Denko†	Kowasaki	161.5
Nissan Chemical†	Toyama	134.1
Ube Industrial†	Sakai	192.4
Sumitomo	Niihama	225.8
Ube Ammonia	Ube	253.1
Kashima Ammonia‡	Kashima	257.1
Nippon Kasei‡	Iwaki	270.6
Subtotal		2510.9
Industrial use		
Asahi Chemical§	Mizushima	216.5
Shin Nettetsu†,¶	Tobata	43.5
Asabi Glass†,#	Chiba	37.4
Asabi Glass†,#	Kita-Kyushu	47.3
Tokuyama Soda†,††	Tokuyama	53.8
Toyo Soda†,††	Nan-Yo	46.7
Subtotal		445.2
Grand total		2956.1

† Likely to close by 1985 (1 089 000 million tonnes of N).
‡ One of these units closed 30 June 1982.
§ Chemical fiber. ¶ Steel. # Glass. †† Soda ash.

reports (Anon., 1982d) indicate that the government is prepared to allow imports of naphtha and also subsidize the domestic-oriented units. This may be a first step toward the long-awaited total withdrawal by Japan from the N export market.

b. Other Locations—Closures of some naphtha-based plants will also occur elsewhere, possibly in North and South Korea, the Philippines (Table 5), and western Europe (Table 4). The N industry in South Korea appears to be particularly vulnerable at this time since units are almost entirely naphtha based. Total closures are not likely to exceed 500 000 t of N capacity in these countries since most units are aimed primarily at serving domestic markets (except in South Korea) and some success has already been achieved in converting units to using cheaper feedstocks. For example, France, Portugal, and Spain have switched the majority of their naphtha-based units to natural gas, propane, butane, or refinery gas. It is assumed that India, South Korea, and the Philippines will attempt to make similar adjustments. Nevertheless, some permanent closings of naphtha-based plants will occur. These closures of plants and shifts in feedstock are likely to remove 1.5 million tonnes of N from the world N supply and could greatly affect the supply/demand balance by late 1984.

2. NATURAL GAS-BASED CAPACITY

While a large portion of the world's NH_3 capacity is based on natural gas, the increase in energy prices has had an impact primarily on the industry of the United States and a much more modest effect on industries of other countries.

The U.S. N industry has not yet fully adjusted to the realities of the current cost of energy (Curtis, 1982; Anon., 1982e). In contrast with Japan, which by 1980 had closed the majority of its small plants and a significant number of uneconomic, modern-size units (7660 t/d NH_3), the U.S. industry has only recently begun to face the realities of the present situation. As recently as 1977, 23% (about 5 million tonnes) of a total of 22 million tonnes of NH_3 capacity consisted of production units with a capacity of 600 t/d or less (Table 8). Since 1977, 24% (nearly 5.2 million tonnes) of the total capacity has either been temporarily idled (2.3 million tonnes) or permanently closed (2.9 million tonnes). Of this 5.2 million tonnes, a surprising 74% are NH_3 units of 660 t/d or larger.

The situation as of May 1982 could be easily explained on the basis of differential pricing of natural gas to the various NH_3 complexes (Curtis, 1982). Most of the small production units now operating in the United States were commissioned in the early-to-mid-1960's; at that time the cost of natural gas was low (usually < \$0.25/M scf), and contracts were commonly made at a fixed price for 15 to 20 yr with no provision for periodic renegotiations. On the other hand, the larger NH_3 plants (1000–1250 t/d NH_3), for the most part, were commissioned in the late 1960's and early 1970's, and many gas contracts were more closely tied to the rising cost of energy and included a provision for periodic (usually every 5 yr) readjustment to current energy prices. As a result, it is uneconomic to operate many of these large plants under the present price structure; therefore, they are temporarily idled while the seemingly less-efficient, small units continue to operate because they use feedstock that is priced at a fraction of the cost of that used by the larger units.

It seems highly probable that these smaller NH_3 units will soon have to be closed and dismantled as the long-term gas contracts begin to expire. This means that approximately 3.7 million tonnes of NH_3 capacity is likely

Table 8—Classification of U.S. NH_3 capacity† by size and operating status.‡

Status	NH_3 Capacity		
	< 600 t/d	> 600 t/d	Total
		$t \times 10^3$ NH_3	
Idle	264	2 035	2 299
Closed	1 103	1 812	2 915
Operating	3 672	13 114	16 786
Total	5 039	16 961	22 000

† TVA Capacity Files, 1982.
‡ Using 1977 capacity as the base point.

to close over the next 3 to 4 yr; thus, total NH_3 capacity in the United States will be reduced to approximately 16 million tonnes of N. It is difficult to predict how many of the large plants are likely to be closed permanently and dismantled, but some capacity will fall victim to the irregularities of the present pricing structure. Over the long term (5 yr), most forecasters feel that western Europe, not the United States, will eventually have to make the major readjustment (Anon., 1982e) in plant capacity. Natural gas costs to some production units in western Europe are about $4.50/M scf or about 79% on an energy-equivalency basis of the 1982 international price of crude oil. On the other hand, feedstock costs in the United States range from $1.00/M scf on the Gulf Coast to $4.00/M scf for some wells serving only intrastate trade. On the average, the cost of natural gas to U.S. producers was $2.63/M scf in the year ended 30 June 1982, according to The Fertilizer Institute (TFI), or about 58% of the cost paid by some European NH_3 producers. Based on feedstock costs alone, the production cost of NH_3 ranges from $170/t for western Europe to $95/t for the United States and $35/t in the Arab Gulf and the USSR. This differential in feedstock costs goes a long way to explain the large amount of idle capacity in the United States and western Europe as well as the large surge of imports into the United States from the USSR, the Middle East, Mexico, and the Caribbean.

With the scheduled deregulation of gas prices in the United States by 1985 and the development of new gas reserves, natural gas supplies are likely to be plentiful for the remainder of this century. As a result, gas prices are likely to rise only moderately in the United States over the next 5 yr and intrastate gas prices may actually decline slightly from their 1982 levels. This fact makes it very likely that U.S. producers will have a cost advantage over western Europe. Therefore, it would appear that the NH_3 producers of western Europe and Japan will be the major victims of the recent escalation of energy costs. Both of these may withdraw from the export market. Japan and possibly western Europe may actually become major importers of N possibly as early as 1985.

II. CONSUMPTION, PRODUCTION, AND TRADE OF N

The production and consumption of N fertilizer have increased steadily over the past 10 yr.[1]

A. Consumption

Nitrogen consumption was 31.8 million tonnes in 1970 and climbed to 60.3 million tonnes in 1980 (Fig. 2). Thus, the average growth rate was 7%/yr. This growth rate was slowed somewhat in 1980 because N prices rose relative to food grain prices. Actual declines in consumption in 1980, al-

[1] The fertilizer year as used for consumption and production extends from 1 July of the year indicated to 30 June of the following year. For example, the 1970 year extends from 1 July 1970 to 30 June 1971.

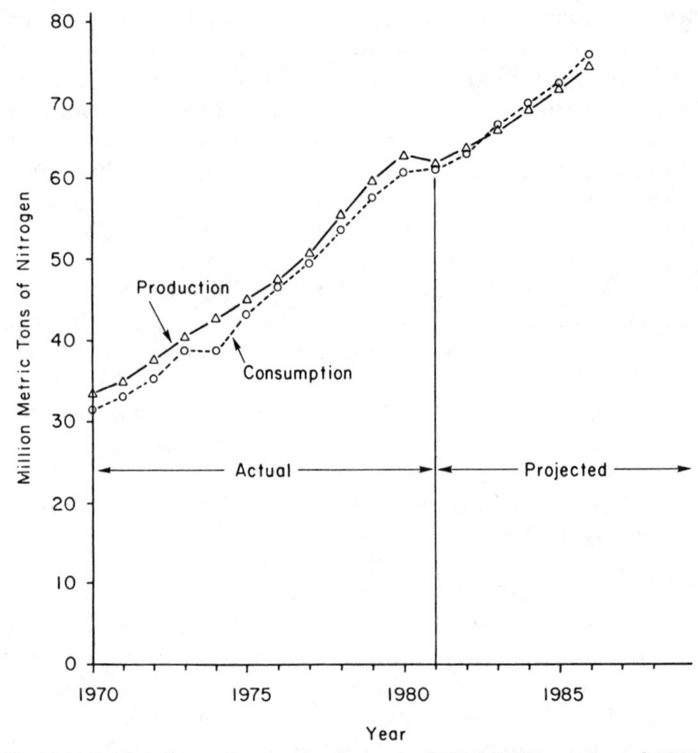

Fig. 2—World N production and consumption actual 1970–1980, projected 1981–1986 (FAO, 1975, 1981a; FAO/FIAC, 1982).

though slight, were experienced in western Europe and other developed economies (Japan and Oceania). On the other hand, the decline was more than offset by sizeable gains over 1979 in consumption in Asia, Africa, Latin America, and North America. Preliminary data for 1981 indicate that world consumption of N fertilizer was about the same as in 1980, while production declined by 0.6 million tonnes. Recent forecasts by the FAO/UNIDO/World Bank Working Group suggest that N consumption in the various regions should change markedly by 1986 (Table 9). They report that major shifts in N consumption have occurred by region over the past 10 yr. For example, in 1970 the developed market economies were the world's leading N consumers and represented a commanding 50.7% of total N use; by 1980 they accounted for only 39.6% of the world total. They were surpassed by the centrally planned countries which in 1980 consumed nearly 41% of the world total. Dramatic increases in N consumption were recorded in the centrally planned countries of Asia where the portion of the world total doubled from 10% in 1970 to nearly 20% by 1980. The centrally planned economies are likely to hold this commanding position through 1986. Although fertilizer consumption is forecast to increase in the developed market economies as well, this growth will be at a reduced rate compared with that of either the developing or centrally planned economies.

Table 9—Regional changes in percentage of total N consumption (1970, 1980, and projected 1986).

Region	1970†	1979‡	1986§
	% of total N		
Developed market economies			
North America	24.6	19.5	18.6
Western Europe	21.6	17.6	15.3
Oceania	0.6	0.4	0.5
Other	3.9	2.1	1.9
Subtotal	50.7	39.6	36.3
Developing market economies			
Africa	0.8	1.0	1.1
Latin America	4.2	4.7	5.4
Near East	2.5	3.2	3.3
Far East	9.0	10.8	12.2
Other	0.1	0.1	0.1
Subtotal	16.6	19.8	22.1
Centrally planned economies			
Asia	9.9	19.6	20.9
Eastern Europe and USSR	22.9	21.0	20.7
Subtotal	32.8	40.6	41.6
Grand Total (%)	100.0	100.0	100.0
Total N (t)	28 677 255.0	57 200 000.0	74 210.0

† FAO, 1975. ‡ FAO, 1981a. § FAO, 1981b.

Therefore, the position of the developed market economies relative to world N consumption is expected to decline even further.

1. USE PER HECTARE

There are major differences in N use per hectare among economic classifications of regions or countries (FAO, 1981a; FAO, 1982). For example, average N use (on arable land) for the world in 1980 was 42 kg/ha; for the centrally planned and developed market economies, average N use was 58 and 66 kg/ha, respectively (Fig. 3). For the developing market economies, however, average N use was only 18 kg/ha (Fig. 3). Vast differences existed within regions of these economic groupings. For example, western European countries had an average N use level of 104 kg N/ha, led by the Netherlands (564 kg N/ha) and the Federal Republic of Germany (199 kg N/ha) (Fig. 4). This level was closely matched by the 117 kg N/ha for the centrally planned countries of Asia where the PRC (106 kg N/ha) accounted for the major portion of this use. Fertilizer use in most regions of the developing market economies was low, ranging from only 5 kg N/ha for Africa to 25 kg N/ha for the countries of the Far East. In spite of the average low levels of N use, some developing countries are using N at a remarkably high level of intensity. The Republic of Korea (199 kg N/ha) and Egypt (176 kg N/ha) serve as prime examples (Fig. 4).

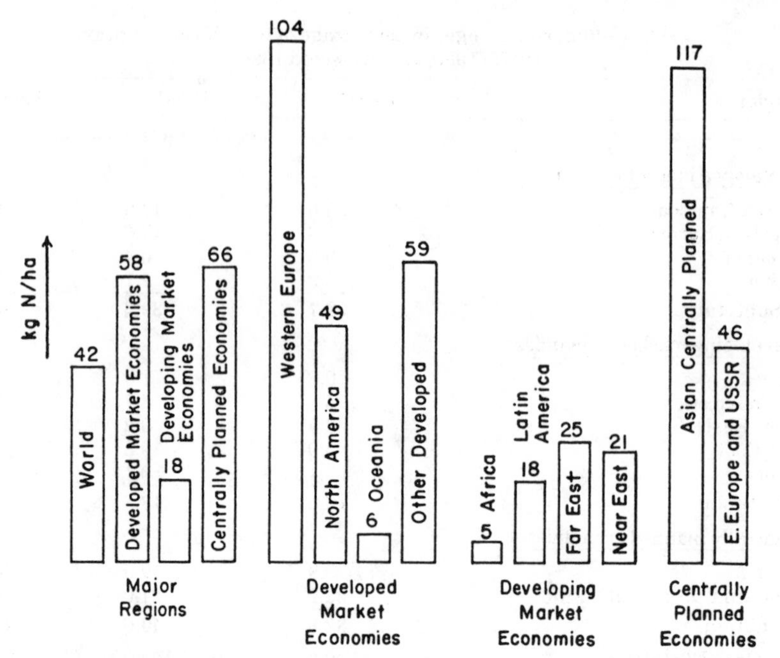

Fig. 3—Average N use for the major regions of the world in 1980 (kg N/ha arable land and permanent crops) (FAO, 1982a).

2. KEY N PRODUCTS

Urea has replaced AN as the major N product produced and consumed in the world today (Fig. 5). Urea accounted for slightly more than 35% of the N produced and consumed in 1980. However, urea is not dominant in all regions of the world (Lastigzon, 1981). Its major dominance is in Asia where urea accounts for 63% of the N consumed. Urea is not the prime fertilizer in either the centrally planned or developed market economies. Only in eastern Europe does urea enjoy major billing, accounting for 25% of the total N consumed. In other regions, urea comprises 10% or less of the total N consumed (Fig. 6). In the United States, for example, N solutions and AA are the dominant forms in which N is consumed (IFDC, 1979). In western Europe, Oceania, and Africa, a major share is consumed as calcium ammonium nitrate (CAN) or in compound fertilizers.

B. Production

Global production of N nearly doubled during the past decade (FAO, 1975, 1981a; FAO/FIAC, 1982). Data for 1980 indicate N production reached 62.7 million tonnes, up 5.2% from the 1979 total of 59.7 million tonnes, and above the 33.2 million tonnes recorded for 1970 (Fig. 2). The developed market economies, mainly North America and western Europe, accounted for the major share (62.2%) of world production in 1970 (Table 10). The centrally planned economies (29.7%) and the developing market economies (8.1%) produced the remaining 37.8%.

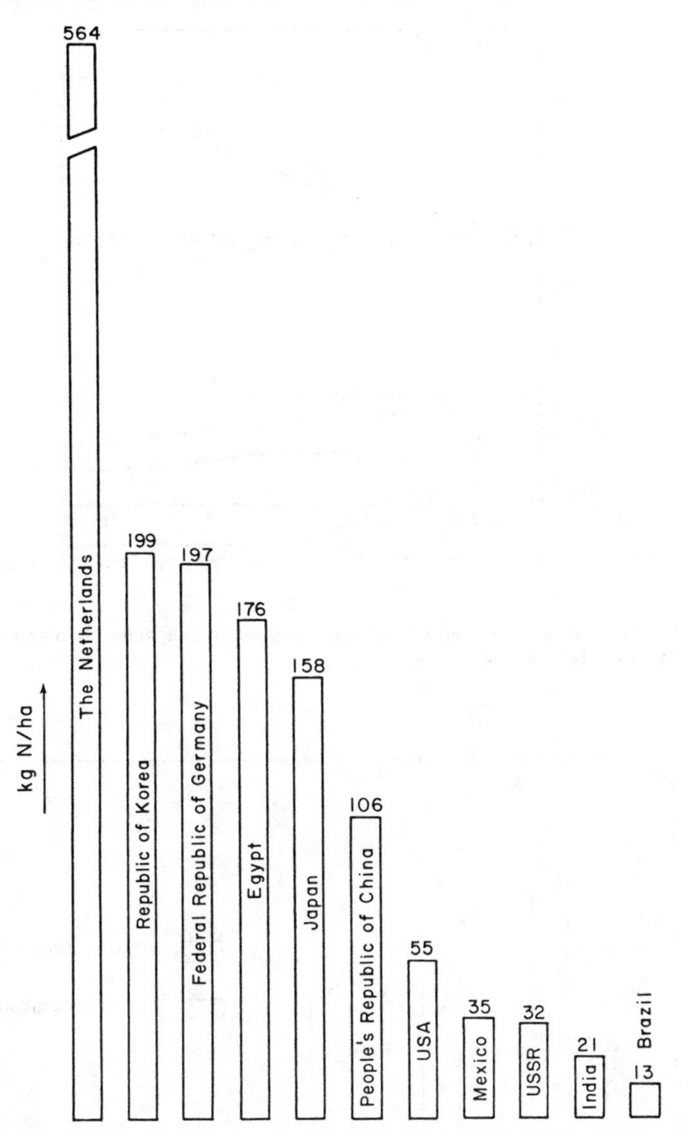

Fig. 4—Average N use for some countries in 1979 (kg N/ha arable land and permanent crops) (FAO, 1981a).

Since 1970, N production in both the centrally planned and developing market economies has increased more rapidly than for the world as a whole. Most of this increase in the share of total production, coming at the expense of the developed economies, has been due to production from new plants coming on stream in the Near East and Far East regions of the developing economies as well as the PRC and the USSR of the centrally planned economies. The growth of N production in Asia has been particularly impressive. Starting from only 8.1% of total world N production in 1970, the Asian

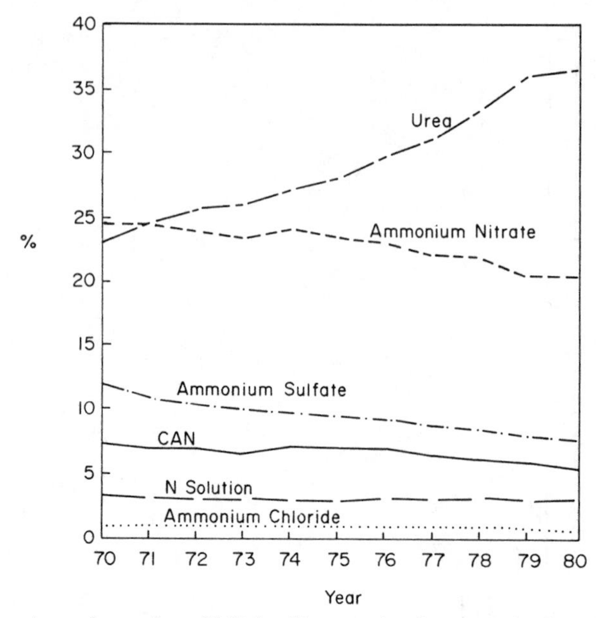

Fig. 5—Percentage share of world N fertilizer production capacity by major product, 1970–1980 (Lastigzon, 1981).

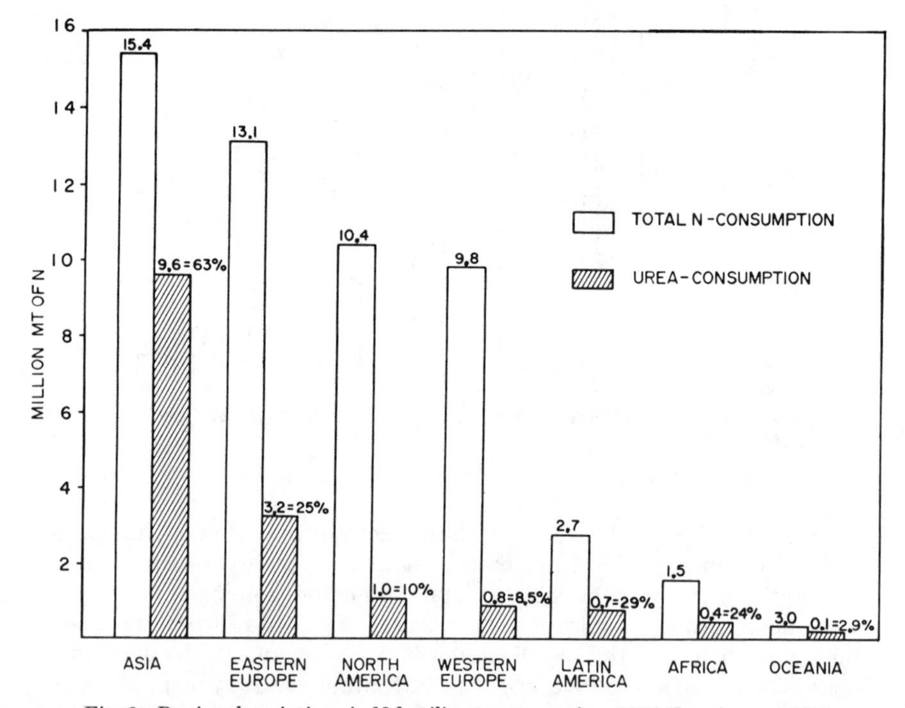

Fig. 6—Regional variations in N fertilizer consumption, 1979 (Lastigzon, 1981).

Table 10—Regional changes in percentage of total N production (1970, 1980, and projected 1986).

Economic classes and regions	1970†	1979‡	1986§
		% of total	
Developed market economies			
North America	27.4	21.5	17.1
Western Europe	26.5	19.7	13.8
Oceania	0.5	0.3	0.3
Other	7.8	3.4	1.7
Subtotal	62.2	44.9	32.9
Developing market economies			
Africa	0.1	0.2	0.5
Latin America	2.5	2.5	6.3
Near East	0.9	3.1	4.2
Far East	4.6	7.8	10.7
Subtotal	8.1	13.6	21.7
Centrally planned economies			
Asia	4.6	16.2	17.3
Eastern Europe/USSR	25.1	25.3	28.1
Subtotal	29.7	41.5	45.4
Grand total (%)	100.0	100.0	100.0
Total N (t)	30 146 644.0	59 764 473.0	73 540 000.0

† FAO, 1975. ‡ FAO, 1981a. § FAO, 1981b.

region of the centrally planned economies increased its share of the world total to 17% in 1981 and is projected to reach 19% by 1986. It is estimated that centrally planned economies will account for 45.4% and developing market economies 21.7% of the total world N production by 1986. The developed market economies are expected to account for the remaining 32.9% of total N production.

In summary, one of the major changes in N production since 1970 appears to have been a greater diversification of N supply and particularly large increases in the regions having indigenous sources of natural gas, large populations, and major demands for food. There is every indication that this diversification in supply will continue through the 1980's.

C. Indicator Countries

Six countries (the United States, PRC, USSR, India, the Netherlands, and Japan) accounted for 59% of world N production, 58.1% of its consumption, 43.6% of the imports, and 42% of the N exports in 1980 (Table 11). Three of these—the United States, USSR, and PRC—account for 50% of the world's production and consumption and 30% of the exports and imports of N. Japan and the Netherlands, while relatively minor producers and consumers of N, supply a considerable amount of exports (12.9% of world total) and therefore, play an important role in international N trade. Obviously, events that influence N production or use in any of these six

Table 11—Key indicator countries—their share of world's production, consumption, import, and export of N fertilizers, 1980.†

Country	Production	Consumption	Imports	Exports
	% of world total			
United States	18.8	17.7	16.8	21.2
PRC	15.9	19.5	13.7	0.0
USSR	16.3	13.5	0.8	7.9
India	3.5	5.8	11.5	0.0
Netherlands	2.6	0.8	0.8	9.1
Japan	1.9	0.8	0.0	3.8
Total	59.0	58.1	43.6	42.0

† Calculated from TVA and FAO unpublished data.

countries will have a significant impact on N availability throughout the world. Therefore, some discussion of the specific situations in each of these countries would seem appropriate.

1. THE UNITED STATES

The United States is the world's largest producer, exporter, and importer of N fertilizer (Table 11). It is second only to the PRC in N consumption. As a result, events that affect N production or consumption in the United States have a strong impact on the N industry worldwide. High costs of N production; low sale prices for fertilizer, N, and feed grains; and stagnant consumption of N have had the combined effect of idling a considerable portion (5.1 million tonnes) of N capacity in the United States. These factors, coupled with the deregulation of natural gas prices scheduled to be completed by 1985, have made it unattractive to invest in new N facilities. While the United States historically has been a net exporter of N (Fig. 7), the current situation in the United States has changed so that it will become a major net importer from 1982 onward. Preliminary figures for the first quarter of 1982 indicate that the United States had a net import balance of 400 000 t of N (TVA, 1982, private communication). Most of these imports come from the USSR, the Netherlands, Mexico, and Trinidad. Barring a major shift upward in the price of feed grains and fertilizer, the continued idling of 3 to 4 million tonnes of NH_3 capacity is likely to keep the United States a major net importer of N for at least the next 3 to 4 yr. This could reach 2.5 to 3.5 million tonnes/yr by 1984 if N consumption in the United States, which is now stagnant, resumes its normal growth rate of 4 to 5%/yr.

2. PEOPLE'S REPUBLIC OF CHINA

The PRC has surprised the world with the rapid advancements it has made in both N production and consumption. It is the world's leading consumer and third only to the United States and the USSR in N production (Table 11). This is the result of an ambitious program to build fertilizer factories and increase N use.

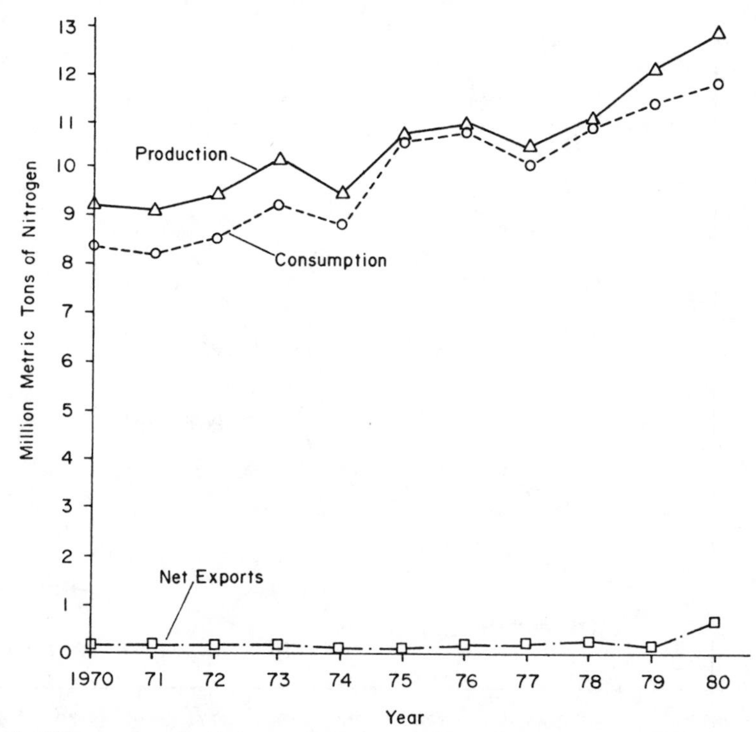

Fig. 7—Nitrogen production, consumption, and net import/export, United States, 1970–1980 (FAO, 1975, 1981b).

Boosted largely by the purchase, construction, and successful operation of 13 1000-t/d NH_3 plants and companion urea units, the PRC's production of N has risen dramatically in recent years (Feng, 1982). Production which in 1970 was estimated to be 1.5 million tonnes of N climbed to nearly 10.0 million by 1980–1981 (Fig. 8). The PRC reportedly has an additional 1 million tonnes of N capacity under construction, which is scheduled to come on stream before 1985.

Equally dramatic increases have also occurred in N consumption. Nitrogen consumption was only 3.2 million tonnes in 1970 but was 11.8 million tonnes in 1980. The PRC imported nearly 1.8 million tonnes in 1980 (mainly from Japan, the Middle East, western Europe, and the United States) and is likely to remain in an import position in the foreseeable future (Table 12).

3. UNION OF SOVIET SOCIALIST REPUBLICS

The USSR is the world's leader in installed NH_3 capacity, second in N production, and a major exporter in N fertilizer (Tables 2 and 11). Nitrogen production, which in 1970 was 5.4 million tonnes, climbed to 10.2 million tonnes in 1980. Nitrogen consumption also increased but not quite so dramatically as production. Nitrogen consumption was 4.6 million tonnes

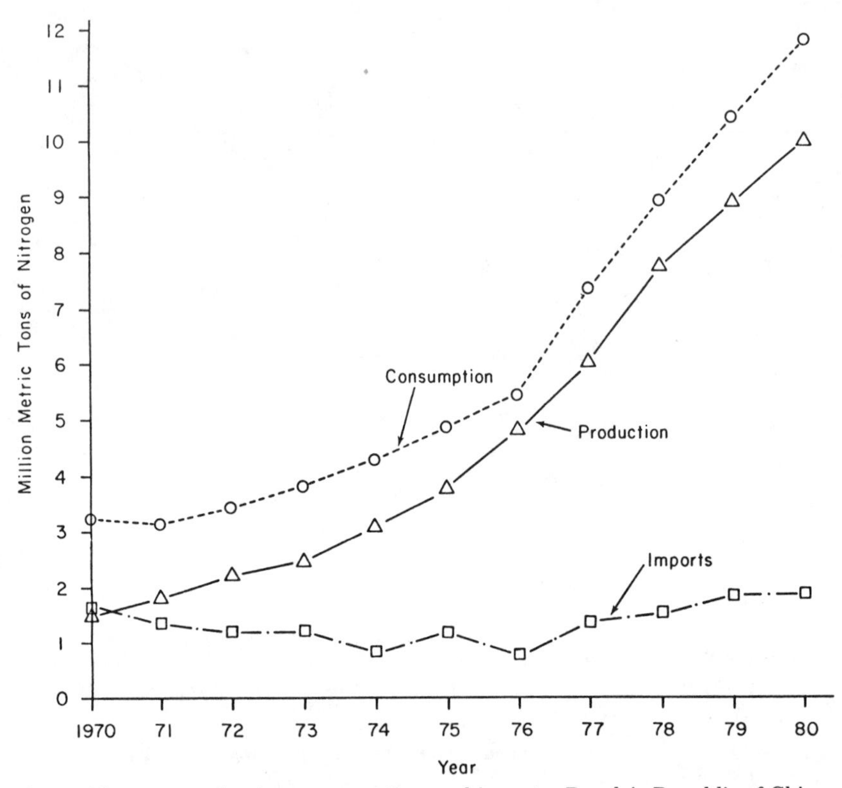

Fig. 8—Nitrogen production, consumption, and imports, People's Republic of China—1970–1980 (FAO, 1975, 1981b).

in 1970; it increased approximately 80% by the end of the decade and totaled nearly 8.3 million tonnes by 1980 (Fig. 9). As a result of this growing surplus in production, the USSR has emerged as a significant exporter of N fertilizer; exports have totaled slightly more than 1 million tonnes of N since 1980. This total export promises to increase sharply in the immediate years as additional capacity (3.1 million tonnes) comes on stream by 1984. Many of these exports are destined for the United States and western Europe (mainly as NH_3) and India (as urea). The main reason for the rapid expansion of its N industry is to help achieve self-sufficiency in feed grain pro-

Table 12—Imports of urea by People's Republic of China—1978–1979 (Stangel, 1980).

Source	Amount
	t of product
Japan	1 035 000
Western Europe	600 000
United States (Alaska)	120 000
USSR	50 000
Middle East	325 000
Total	2 130 000

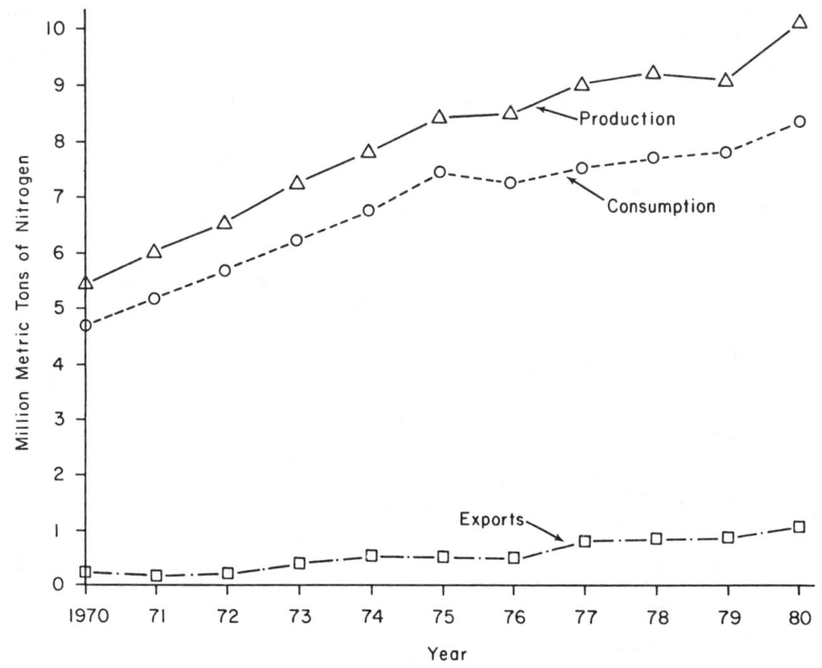

Fig. 9—Nitrogen production, consumption, and exports, USSR—1970–1980 (FAO, 1975, 1981b).

duction [some of the exported N is exchanged for phosphoric acid from the United States and wheat (*Triticum aestivum* L.) and maize (*Zea mays* L.) from India and Thailand, respectively]. Nevertheless, domestic consumption of N has lagged behind government targets. This lag, the USSR's need to earn foreign exchange to finance imports of feed grain, and its ample supply of natural gas lead one to conclude that the USSR will be a major exporter of N fertilizer for the foreseeable future.

4. INDIA

India has experienced a remarkable growth in production and consumption over the past decade. Dramatic increases have been particularly evident in fertilizer consumption. For example, N use, which in 1970 totaled 1.3 million tonnes, nearly tripled by 1980 and reached 3.5 million tonnes (Fig. 10). Preliminary estimates indicate consumption reached 3.7 million tonnes of N for 1981. This represents an annual growth rate of $> 15\%$. Analysts attribute this increase not only to favorable weather conditions but also to improved government policies that enhance availability of N fertilizer and make it profitable not only to use fertilizer but also to produce and distribute it as well. Current forecasts (Patel & Pandey, 1982) call for N consumption to reach 7.4 million tonnes by 1984 and 8.4 million tonnes by 1987. The present world recession coupled with the increased scarcity of foreign exchange in India may make it difficult for the fertilizer sector to reach these targets.

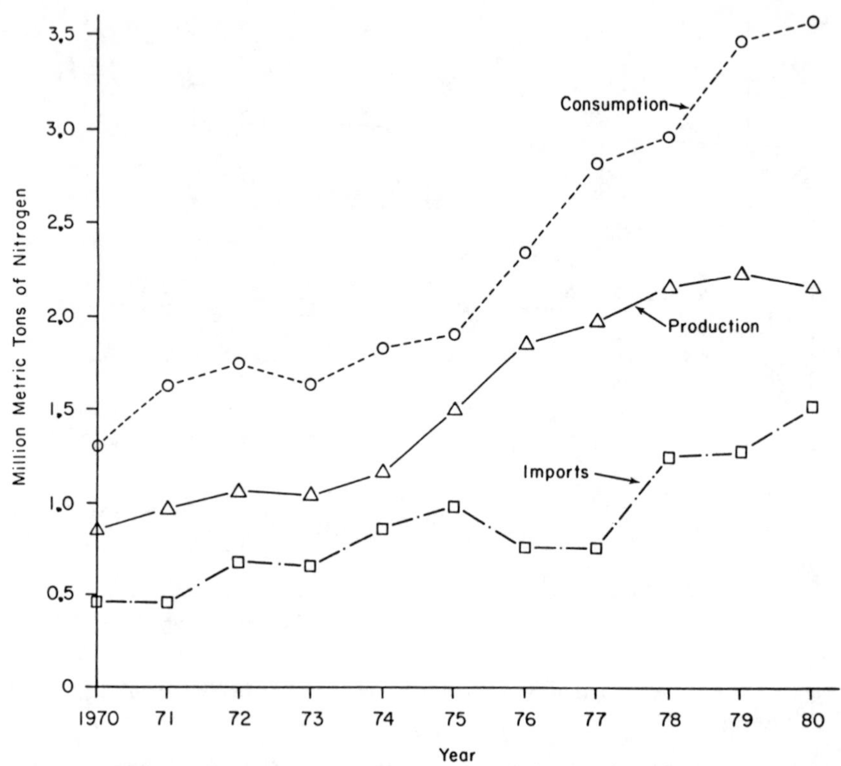

Fig. 10—Nitrogen production, consumption, and imports, India—1970-1980 (FAO, 1975, 1981b).

Production of N has also increased sharply over the past decade, from 0.8 million tonnes in 1970 to 2.2 million tonnes by 1980. This increase, however, was insufficient to meet demand, and the deficit has required a steady increase in imports, which in 1980 reached 1.5 million tonnes (Fig. 10). These imports come mainly from eastern Europe, the USSR, and western Europe with some also coming from Japan and the United States (Table 13). Prospects are high that India will remain a major importer of N fertilizer for the foreseeable future.

Table 13—Imports of urea by India—1979 (Stangel, 1980).

Source	Amount
	t of product
Eastern Europe	590 000
Japan	26 000
USSR	315 000
United States	160 000
Western Europe	525 000
Indonesia	200 000
Middle East (Iraq)	100 000
Total	1 916 000

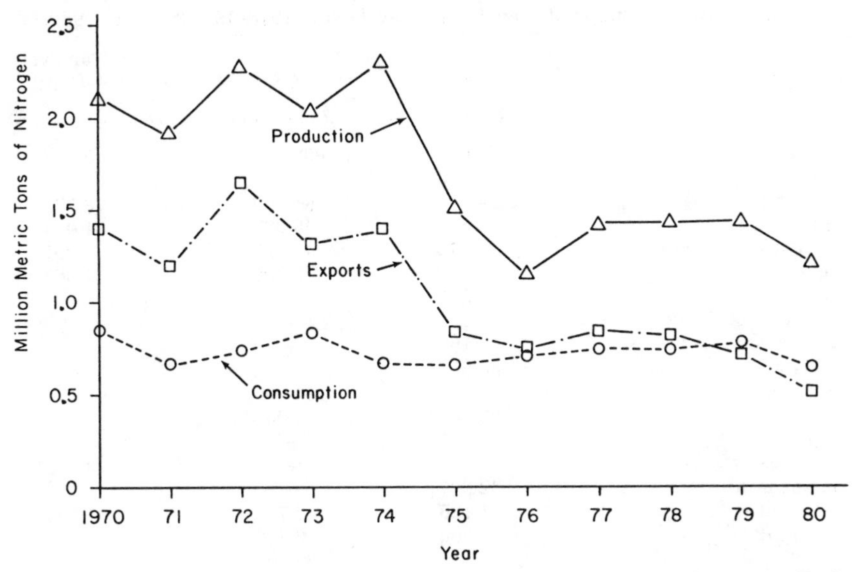

Fig. 11—Nitrogen production, consumption, and exports, Japan—1970–1980 (FAO, 1975, 1981b).

5. JAPAN

Until recently, Japan was the leading net exporter of N. From 1970 to 1974, exports to various countries around the world ranged from 1.0 to 1.6 million tonnes/yr of N. For reasons already explained, many N plants were closed. Fertilizer N exports have subsequently dropped sharply since 1974 and now average between 0.6 and 0.7 million tonnes/yr of N (Fig. 11). The majority of the exports are destined for the PRC, which in 1978–1979 imported nearly 90% of the urea and 50% of the AS that Japan shipped abroad (Table 14). Nitrogen consumption has remained static over the past 8 to 10 yr, ranging between 0.6 and 0.8 million tonnes/yr. There are indications that, because of continued high costs of production, Japan may announce another series of plant closures. If this occurs, the PRC may lose a major and highly dependable supplier of fertilizer N. It does not seem likely that Japan will export anything but token quantities of urea beyond 1985. The main exports are likely to be by-product AS and AC. It is entirely possible that Japan could become a net importer of N by 1985; it will probably begin with NH_3 for industrial purposes and follow with urea for the agricultural sector.

6. THE NETHERLANDS

In recent years, the Netherlands has emerged as a major exporter of N fertilizers; exports rose steadily from 0.6 million tonnes in 1970 to nearly 1.2 million tonnes in 1980. It has been a major supplier of N to western Europe, the United States, and more recently to India and the PRC. Recent increases in prices of natural gas, now reported to be $4.50/M scf, have raised the

Table 14—Actual exports of N fertilizers from Japan—1978–1979 (Stangel, 1980).

Destination	Urea	AS	High analysis N–P–K
	t of product		
PRC	970 600	368 401	5 523
Vietnam	30 000†	--	--
Thailand	--	61 100	76 573‡
Philippines	16 500†	45 000	21 045
Malaysia/Singapore	15 772	44 828	901
Indonesia	--	5 000	4 067
Brunei	350	--	--
Burma	22 281†	--	--
Hong Kong	150	--	--
Pakistan	36 675†	--	--
Sri Lanka	--	104 200‡	--
India	26 305†	45 038†	--
Nepal	14 951†	--	--
Afghanistan	--	--	3 845†
Bangladesh	32 927†	--	--
New Guinea	1 135	2 700	1 895
New Zealand	7 200	39 320	--
Fiji	--	28 500	--
Tonga	100	--	--
United States (Hawaii)	40	--	--
Bolivia	801†	--	1 516†
Ecuador	--	1 320	--
Paraguay	300†	--	1 000†
Sudan	5 000†	--	--
Others	--	1 320	166
Total	1 181 087	746 727	116 531

† Financed through concessional Japanese aid.

cost of production of N. This increased cost, however, does not seem to have dampened their prospects of serving the export market since at least one new NH_3 plant (0.3 million tonnes/yr of N) is expected to come on stream before 1985. It is entirely possible that the Netherlands' production units may fill part of the void left in international trade when Japan leaves the export market.

III. FUTURE OUTLOOK

The fertilizer sector worldwide is now in a major state of change. This change is being fueled by at least four major forces: (i) the growing awareness by most major developing countries that an economically sound agriculture is a key link to the overall economic development of a country; (ii) the policy of many developing countries to achieve self-sufficiency or near self-sufficiency in the production of food as a basis for developing a sound agriculture; (iii) recognition that fertilizer is an important input in agriculture and wherever possible the fertilizer supply should be based on domestic production using indigenous resources; and (iv) the rapid rise in the cost of energy in general and fertilizer raw materials in particular.

Nitrogen is on the leading edge of this change. It is the most widely used fertilizer nutrient today, and the demand for it is likely to grow in both relative and absolute importance in the foreseeable future. The location and quantities of N use will be determined primarily by the importance that developing countries place on increasing the production of food and fiber as well as by their success, in real terms, in improving the economic standards of their population.

Nitrogen is also the most capital- and energy-intensive commercial fertilizer to produce. The location and quantities of N production will be dictated by a combination of factors including the price of fertilizer, the cost of feedstock, whether facilities are publicly or privately held, proximity of production units to major N markets, and progress in improving efficiency in N production.

A. Forecasts of N Demand

The need for N fertilizer will rise in accordance with the demand for food and fiber. Several studies have estimated fertilizer demand (Reidinger, 1976; Unico, 1977; Harris & Harre, 1979; FAO, 1981b; Verghese, 1978; UNIDO, 1978, 1980). Only the reports of UNIDO, FAO, and Verghese have contained projections to 1990 or to the year 2000. Unfortunately, the vast majority of these studies were done prior to the 1979 increase in energy prices.

Estimates of N use for 1985 range from 66 to 78 million tonnes. The FAO estimate of 73.1 million tonnes appears to be the most attainable. However, this estimate is about 13 million tonnes above the 1981 level of consumption, and it assumes an end to the present economic recession.

Nitrogen consumption is likely to remain near its present level of 60 million tonnes of N as long as the world recession continues. A prolonged recession could result in an actual consumption level for 1985 well below the current FAO estimates; FAO estimates N use to be about 90 million tonnes by 1990. A UNIDO estimate for the same year is 92.3 million tonnes.

Only UNIDO has developed forecasts for the year 2000. The initial UNIDO projection estimated N demand to be 164 million tonnes. This estimate was later revised downward to 111 million tonnes of N. Verghese, using essentially UNIDO data, estimated N demand to be about 134 million tonnes by the year 2000.

1. LOCATION OF FUTURE DEMAND

The major growth markets in N demand over the next 10 yr are likely to be in North America, the centrally planned economies, and the developing countries, particularly in Asia. The most notable of these will be India, the PRC, Indonesia, Brazil, and possibly the United States and the USSR.

Because of the demand for food and fiber caused by burgeoning populations and the desire to minimize the drain on foreign exchange, India, Indonesia, the PRC, and Brazil will provide the major growth in N consumption through the remainder of this century. Actual growth will depend on a number of factors. The most important of these factors are the continued expansion of irrigation facilities, the development of fertilizer-responsive plant varieties that can be profitably grown under rainfed conditions, and the continued easy access to fertilizer and profitability of its use by farmers. The future growth of N demand in the United States and the USSR is less clear.

The growth in N use in the United States over the past 10 yr has been closely tied to the ability of this country to increase exports of agricultural products. A strong dollar, record harvests of food and fiber in many parts of the world, and heavy subsidies of agricultural exports by other food surplus countries are making it doubtful whether the United States can maintain its present level, much less continue to increase agricultural exports. If agricultural exports stagnate, N demand in the United States may also stagnate. It is more likely that N consumption will increase, but at rates much lower than those in the 1970's.

Nitrogen demand in the USSR may also stagnate, but for different reasons. In contrast to the United States, the USSR lacks a favorable rainfall pattern to sustain good agricultural productivity. Unlike India, the PRC, or Indonesia, the USSR has a very small percentage of its agricultural land under irrigation. In addition, the inefficiencies of the present collective farming system will also make it difficult for the USSR to increase N consumption at rates anywhere near its rates of the past decade. The situation, of course, could change if irrigation facilities were developed and incentives put into place to increase food production.

B. Improvements in the Use of N

The rapid increase in energy prices and the threat of depletion of nonrenewable resources such as oil and natural gas have stimulated interest in finding ways to improve the use efficiency of commercial fertilizer N by using plant residues and organic manures more effectively and, where appropriate, taking advantage of biologically fixed N within the cropping system.

Technologies have already been developed in each of the areas mentioned. Recovery of commercial N fertilizer can be increased markedly by either deep placement, use of coatings, or addition of inhibitors (Mudahar & Hignett, 1981). Recent work at the International Fertilizer Development Center (IFDC) and the International Rice Research Institute (IRRI) has revealed that by using deep placement or coating of N it is possible to reduce N applications by 20 to 40% for rice (*Oryza sativa* L.) grown on flooded soils and still achieve the same yield (Flinn & O'Brien, 1982). Use of human and animal wastes as a fertilizer has been perfected in a number of countries with the most notable examples being in India and the PRC (Jin, 1982). The United States and western Europe have shown renewed interest in using these wastes in modern farming systems. Major advances are also being made in the use of biologically fixed N as a main means of supplying the crops' N needs. Chinese scientists have successfully used a system of narrow-wide row spacings of rice and azolla (*Azolla caroliniana* Willd.) (two rows of rice grown 6–9 cm apart followed by 50–70 cm where the azolla is grown) as a means of using biological N sources to meet 70 to 80% of the supplemental N needs of the rice crop. This is accomplished primarily through N release as the result of the rapid decomposition of azolla. Chinese scientists are attempting to select azolla species that excrete NH_3 directly from their roots and thereby serve as an immediate source of N to the rice crop (Liu, 1982).

Cost, relative to other sources of N, is the primary factor restricting the application of these technologies. If the price of commercial N increases substantially, one or more of these technologies will likely be used extensively by farmers. The significant application by farmers of technologies in any of these areas will likely reduce the rate of growth of commercial fertilizer N.

C. Future Supplies of N Fertilizer

The location and quantities of future supplies of N will be determined not only by demand but also by the selling price, the factors affecting its cost of production (particularly the cost of feedstock and capital investment), ownership, and development of technologies that will improve the efficiency of production.

D. Price of N

Currently the world N market is depressed because of the combined effect of high costs of feedstock, depressed prices of food crops in key fertilizer-consuming areas (United States, Canada, western Europe, Brazil, and India), and oversupply due to the large-scale entry into the international market of N products from eastern Europe, the USSR, and the Middle East. In spite of depressed prices and the existence of substantial idle NH_3 capacity, major expansions in N-production units are being continued, mainly in the USSR, Canada, the Middle East, Mexico, India, Malaysia, and Indonesia. With the exception of India and the PRC, a major portion of this capacity will be aimed at the export market. A number of these countries (eastern Europe and the USSR) are facing serious economic problems and need these exports to earn foreign exchange. As one means of earning the needed foreign exchange, these countries are aggressively selling large quantities of N on the international market at very low prices. This trend is likely to continue and may actually intensify as certain N-producing OPEC countries (densely populated) face the need to earn foreign exchange lost as a result of the drop in oil revenues. As a result, prices of N on the international market are not likely to increase over the next 12 to 24 months and significant increases are not likely to occur before 1985.

This will put even greater pressure on the already beleaguered industries of Japan, the United States, and western Europe. Japan, which has already scrapped more than 2 million tonnes of N capacity since 1975, is likely to announce permanent closure and eventual dismantling of another 1 million tonnes of N capacity. The U.S. industry is likely to go through an equally great reorganization of its N sector. As deregulation of natural gas becomes complete by 1985, the majority of the small-size (< 600 t/d NH_3) units (total 5 million tonnes of N) are likely to be closed and eventually dismantled. In addition, an unknown number of larger N units are likely to be permanently closed and possibly dismantled.

The net results of this move will be near balance in N supply and demand within the next 36 months. If the world comes out of its current economic recession and the demand for and price of food crops improve, the demand for N could increase sharply and thereby signal a gradual but substantial increase in N prices shortly thereafter. The speed of the price increase will depend to some degree on the amount of new capacity being built in the USSR and certain developing countries. These increases in price may result in the reactivation of some of the capacity currently idle, particularly in the United States.

The export price of urea as of May 1982 was about $130/t bagged, f.o.b. U.S. Gulf Coast (Anon., 1982f). Anhydrous ammonia from the same area was about $150/t.

E. Location of N Production

The makeup of the world N industry through the remainder of this century is likely to be determined by investments made in new capacity over the next 3 to 6 yr. This in turn will be highly influenced by three factors: (i) the international cost of feedstock (oil and natural gas), (ii) the international price of N, and (iii) the willingness of the developed countries with a food surplus to rely on N imports to meet their needs.

1. PRICE OF FEEDSTOCKS

The main feedstocks to be used to produce NH_3 in the future will be natural gas and oil. The price of each will be a major factor in determining the price of N. Inventories of crude oil by mid-1982 were the highest on record (Anon., 1982g). As a result, prices of oil have already dropped from a peak of $40/barrel to a spot price reported to be slightly less than $28/barrel, well below the official OPEC price set at $32/barrel for Saudi light crude. OPEC, in an effort to prevent further erosion of prices, has slashed production to nearly 50% of its 1979 production of 29 million barrels per day and is now reviewing its present price structure. All indicators point to a drop in the official price by 1983 to somewhere between $28 and $30/barrel (Anon., 1982h).

Production of oil from non-OPEC sources has not declined and, as a result of record levels of drilling and exploration during the past 3 yr, non-OPEC oil production should become an increasingly important factor in the immediate years ahead in supplying the world's energy needs. Therefore, it would appear that the world will not experience the large increases in energy prices that have characterized the energy market since 1972. Neither should one expect a sharp drop in energy prices. It is generally believed that OPEC will not collapse but will discipline its member countries in supporting the present pricing system. It is generally accepted that, because of the growth in non-OPEC oil supplies and major reduction in energy consumption through conservation, OPEC will no longer be able to manipulate energy prices at will. What can be expected is a stabilization of energy prices slightly below their present official levels, and starting in 1984 or 1985, a gradual increase per year (3–4%) in real energy costs (Fesharaki & Johnson, 1982).

In addition, the price of natural gas in many developed countries will be close to, but probably below, the international price of oil. The degree to which this is true will vary by country and depend on the actual reserves of natural gas and oil and the pressures that exist for their development. At May 1982 official oil prices ($32/barrel), this would be about $5.50 to $6.00/M scf. Stabilization of energy costs is only one factor that will affect the future structure of the N industry.

2. CAPITAL INVESTMENT COSTS

The current world price of N is too low to attract investment of private capital in financing new plants (FAO, 1981b). According to the World Bank (private communication), the current price of urea is below the realization price.[2] Using estimated investment costs of $220 million, $300 million, and $400 million for sites in a (i) developed country, (ii) developing country with some infrastructure in place, and (iii) grass roots location in a developing country, the realization prices for bagged urea were $304.93, $242.78, and $272.16/t, respectively (Table 15).

The price of bagged urea in May 1982 ($160/t f.o.b. U.S. Gulf Coast) was well below the realization price of bagged urea everywhere, including

[2] Price at which new capacity, if efficiently operated, will provide a reasonable return on investment (15–20% return on capital).

Table 15—Estimated investment, production cost, and realization price (1980 U.S. $/t) for urea (FAO, 1981c).†

Parameters	Developed site		Developing site: Some infra-structure	Developing site: Grassroots remote area
	dollars			
Total investment (millions)	220.00	220.00	300.00	400.00
Raw material costs				
Gas price (M scf)	3.00	4.50	1.50	0.80
Gas cost	105.00	157.50	52.50	28.00
Other variable costs/tonne	16.00	16.00	16.00	16.00
Fixed costs/tonne	64.09	64.09	82.45	105.72
Production costs/tonne	185.09	237.59	150.95	149.72
Capital charge (15%)/tonne	67.34	67.34	91.83	122.44
Realization price/tonne (ex-factory)	252.43	304.93	242.78	272.16

† Basis: 1000 t/d of NH_3, 1660 t/d bagged urea, operating at 90% capacity based on 330 d/yr.

Table 16—World N supply/demand balance for 1981–1986 (FAO, 1982).

Region	1981	1982	1983	1984	1985	1986
	t × 10^3 N					
Developed market economies	1.35	0.45	−0.22	−0.22	−0.53	−1.16
North America†	1.27	0.74	0.43	0.19	−0.07	−0.58
Western Europe	−0.13	−0.40	−0.75	−0.43	−0.46	−0.49
Oceania	−0.07	−0.06	−0.06	−0.07	−0.07	−0.09
Other developed market economies	0.28	0.17	0.16	0.09	0.07	0.00
Developing market economies	−2.57	−2.23	−1.94	−1.17	−0.30	0.00
Africa	−0.63	−0.64	−0.31	−0.32	−0.35	−0.32
Latin America	−0.11	−0.03	0.12	0.52	0.68	0.71
Near East	0.12	0.29	0.19	0.42	0.75	0.70
Far East	−1.95	−1.85	−1.94	−1.79	−1.38	−1.09
Centrally planned economies	2.34	3.14	3.99	4.41	3.85	2.99
Asia	−1.72	−1.74	−1.85	−1.96	−2.09	−2.20
Eastern Europe and USSR	4.06	4.88	5.84	6.37	5.94	5.19
World Total	1.12	1.36	1.83	3.02	3.02	1.83

† Takes into account nearly 5 million tonnes of NH_3 capacity now idle in North America, which is assumed will be reactivated by 1985.

sites where cheap feedstock prevails. Investors (public or private) in fertilizer factories will not operate at a loss over extended periods, and they desire some return on their investment. Therefore, urea prices must eventually rise sustantially. This is likely to happen as soon as the present energy price situation stabilizes and the supply of N becomes more in balance with demand, the demand for fertilizer increases, and present idle NH_3 capacity is either dismantled or reactivated. When this occurs, N prices, in 1982 dollars, can be expected to rise and will likely be at least $250/t for bagged urea. The actual rate of increase, of course, will be dictated by the market forces in play at that time.

F. Supply/Demand Balance—1981–1982 to 1986–1987

Recent plant closures and slackening of new plant expansions, particularly in developed market economies, coupled with the rapid increase in N consumption in Asia, will be instrumental in very nearly balancing the supply-demand situation by 1983. Only eastern Europe and the USSR will be in net surplus situations (Table 16). From 1983 onward, N demand is likely to increase and shortages will develop regionally; this situation should signal the beginning of a price increase. The speed at which prices increase will be determined by the cost of feedstock and the possibility of reactivating NH_3 plants now idle in the United States. These events have important implications for most developing countries, particularly the PRC and India—both major importers of N and facing a constraint of foreign exchange necessary to finance needed imports. Obviously, new plant expansions in these countries should be announced soon. However, the current outlook is such that new capacity will be built in time and price increases will be orderly. It is very doubtful that the rapid price increase of the early 1970's will be repeated in the foreseeable future.

REFERENCES

Anon. 1982a. Market Report. European Chemical News 38(1015):8–10.

Anon. 1982b. Market Report. European Chemical News 38(1016):13.

Anon. 1983c. Market Report. European Chemical News 38(1023):10.

Anon. 1982d. Japan subsidizes domestic naphtha. Chemical Week 130(16):25.

Anon. 1982e. World trends: is the light at the end of the tunnel? Nitrogen 135:3–4.

Anon. 1982f. Green markets price scan. Green Markets 6(18):5.

Anon. 1982g. More U.S. crude oil prices trimmed by $1–$2/bbl. Oil and Gas Journal 80(11): 40.

Anon. 1982h. Demand slump restrains global oil, condensate output in 1981. Oil and Gas Journal 80(10):99–103.

Beck, R. J. 1982. Forecast and review. Oil and Gas Journal 80(4):127–144.

Curtis, C. E. 1982. When the back door is better than the front. Forbes. 12 April, p. 57–60.

Feng, Y. 1982. The production and application of chemical fertilizer in China. Int. Workshop and Training Course on Nitrogen Management, Fuzhou, People's Republic of China. 26 Apr.–15 May 1982. Int. Fertilizer Development Center, Muscle Shoals, Ala.

Fesharaki, F., and T. M. Johnson. 1982. Short term and medium term outlook for oil: a review and analysis of recent studies. Working Paper WP-82-2, East-West Resource Systems Inst., The East-West Center, Honolulu, Hawaii.

Flinn, J. C., and D. T. O'Brien. 1982. Economic considerations in the evaluation of urea fertilizers in wetland rice farming. Int. Workshop and Training Course on Nitrogen Management, Fuzhou, People's Republic of China. 26 Apr.–15 May 1982. Int. Fertilizer Development Center, Muscle Shoals, Ala.

Food and Agriculture Organization. 1975. Annual fertilizer review 1974. Food and Agric. Org. of the United Nations, Rome, Italy.

Food and Agriculture Organization. 1981a. Fertilizer yearbook, Vol. 30. Food and Agric. Org. of the United Nations, Rome, Italy.

Food and Agriculture Organization. 1981b. Current fertilizer situation and outlook. Commission on Fertilizers, Seventh Session, 7–10 Sept. 1981, Food and Agric. Org. of the United Nations, Rome, Italy.

Food and Agriculture Organization. 1981c. Major factors affecting supply, demand, and prices. Commission on Fertilizers, Seventh Session, 7–10 Sept. 1981, Food and Agric. Org. of the United Nations, Rome, Italy.

Food and Agriculture Organization. 1982. Current fertilizer situation and outlook. Commission on Fertilizers, Seventh Session, 7–10 September, Agriculture Organization of the United Nations, Rome, Italy.

Food and Agriculture Organization/FIAC. 1982. FAO/FIAC working party on fertilizer statistics report. FAO/FIAC Meetings, 1–4 Mar. 1982, Rome, Italy.

Harris, G. T., and E. A. Harre. 1979. World fertilizer situation and outlook—1978–85. Tech. Bull. T-13, Int. Fertilizer Development Center, Muscle Shoals, Ala.

International Fertilizer Development Center. 1979. Fertilizer manual. Reference Manual R-1, Int. Fertilizer Development Center, Muscle Shoals, Ala.

Jin, W.-X. 1982. Recycling of organic materials—use of organic wastes and bio-nitrogen as fertilizers in China. Int. Workshop and Training Course on Nitrogen Management, Fuzhou, People's Republic of China. 26 Apr.–15 May 1982. Int. Fertilizer Development Center, Muscle Shoals, Ala.

Lastigzon, J. 1981. World fertilizer progress into the 1980's. Tech. Bull. T-22, Int. Fertilizer Development Center, Muscle Shoals, Ala.

Liu, C.-C. 1982. Biological nitrogen—azolla. Int. Workshop and Training Course on Nitrogen Management, Fuzhou, People's Republic of China. 26 Apr.–15 May 1982. Int. Fertilizer Development Center, Muscle Shoals, Ala.

Mudahar, M. S., and T. P. Hignett. 1981. Energy and fertilizer: policy implications and options for developing countries—executive brief. Tech. Bull. T-19, Int. Fertilizer Development Center, Muscle Shoals, Ala.

Patel, A. S., and R. K. Pandey. 1982. Demand for nitrogenous fertilizers in Indian agriculture. Fertiliser News 27(3):13–16.

Reidinger, R. 1976. World fertilizer review and prospects to 1980/81. Foreign Agric. Rep. no. 115. U.S. Department of Agricultural Economics Research Service, Washington, D.C.

Stangel, P. J. 1976. The impact of the energy crisis and other factors on the fertilizer industry of Asia. p. 81–152. *In* Impact of fertilizer shortage: focus on Asia. Asian Productivity Organization, Tokyo, Japan.

Stangel, P. J. 1980. The world fertilizer sector at the crossroads. p. 17–118. *In* Food situation and potential in the Asian and Pacific Region. ASPAC-FFTC Series no. 17, Food and Fertilizer Technology Center, Taipei, Taiwan.

Tennessee Valley Authority. 1982. World fertilizer capacity file. Tennessee Valley Authority, Muscle Shoals, Ala.

United Nations Industrial Development Organization. 1978. World-wide study of the fertilizer industry 1975–2000. United Nations Industrial Dev. Org., Vienna, Austria.

United Nations Industrial Development Organization. 1980. World-wide study of the fertilizer industry 1975–2000. Revised Document 81, United Nations Industrial Dev. Org., Vienna, Austria.

Universal Corporation (UNICO). 1977. Long-term forecast for world nitrogen fertilizer demand and supply. UNICO, Genowan, Japan.

Verghese, M. C. 1978. Issues facing the world fertilizer industry. p. PS/1-1–PS/1-41. *In* R. S. Giroti et al. (ed.) FAI–IFDC fertilizer seminar 1977 trends in consumption and production proc. The Fertiliser Assoc. of India, New Delhi, India.

3

L. E. Schrader
University of Wisconsin
Madison, Wisconsin

Functions and Transformations of Nitrogen in Higher Plants

Dry matter in plants is generally comprised of approximately 2 and 40% N and C, respectively (Beevers & Hageman, 1980), although certain species and plant parts (e.g., soybean seeds [*Glycine max* (L.) Merr.]) have much higher requirements for N. A requirement for N exists throughout the development of a plant to maintain growth, as N is a constituent of both structural (e.g., cell walls) and nonstructural (e.g., enzymes, chlorophyll, and nucleic acids) components of the cell. Most N for vegetative growth is supplied either by the assimilation of (i) N absorbed from the soil and/or (ii) N fixed from atmospheric N_2 in the case of leguminous crop species. Some N may be reassimilated several times during the growth cycle. For example, photorespiration causes the release of NH_3 from glycine, and turnover of proteins and nucleic acids may release N for reassimilation. During reproductive growth, much N is remobilized from vegetative tissues by hydrolysis of proteins to amino acids. Whether these amino acids are transported directly to developing seeds or first converted to selected compounds for transport is not known at present.

I. ASSIMILATION OF NITRATE-N INTO AMINO ACIDS

A. Biochemical Reactions

Because NO_3^--N is the predominant form of inorganic N available to plants under normal field conditions, and because N_2 fixation will be discussed elsewhere (Chapt. 7 and 8, this book), major emphasis will be given to the assimilation of NO_3^--N into amino acids and proteins. Assimilation of NO_3^--N includes three reductive processes and one nonreductive process in converting NO_3^--N to amino-N (Schrader & Thomas, 1981). The reac-

Nitrogen in Crop Production.

tions of this pathway have been discussed in several recent reviews (Beevers & Hageman, 1980; Hewitt, 1975; Hewitt & Cutting, 1979; Miflin, 1980; Miflin & Lea, 1976, 1977; Schrader & Thomas, 1981; Vennesland & Guerrero, 1979). The four enzymes involved are nitrate reductase (NR), nitrite reductase (NiR), glutamine synthetase (GS), and glutamate synthase (GOGAT). As shown below, the three reductive processes require 10 electrons for the reduction of NO_3^--N to glutamate (ATP = adenosine triphosphate; α-kg = α-ketoglutarate):

$$NO_3^- \xrightarrow[2e^-]{NR} NO_2^- \xrightarrow[6e^-]{NiR} NH_4^+ \xrightarrow[ATP]{GS} \text{Glutamine} \xrightarrow[\alpha\text{-kg} \;\; 2e^-]{GOGAT} \text{Glutamate} + \text{Glutamate}$$

(one Glutamate returns to the GS step)

In chlorophyllous tissues, NADH (reduced nicotinamide adenine dinucleotide) is usually the electron donor for NR, whereas reduced ferredoxin is the electron donor for both NiR and GOGAT. The involvement of the light reactions of photosynthesis in supplying these reductants was recently reviewed (Schrader & Thomas, 1981). Lee (1980) reviewed the sources of reductant for nitrate assimilation in nonphotosynthetic tissues. Although pyridine nucleotides serve as reductant for NR, additional research is needed to establish the natural reductant for NiR in nongreen tissues, as ferredoxin has not yet been isolated from nongreen plant tissue. In nongreen tissues, GOGAT generally uses NADH as a reductant more effectively than NADPH (reduced nicotinamide adenine dinucleotide phosphate) (Beevers & Storey, 1976; Chiu & Shargool, 1979).

If photochemical reactions supply reduced ferredoxin directly to NiR and GOGAT in illuminated green tissue, the equivalent of 20 photons are required for reduction of each NO_3^--N to glutamate (Schrader & Thomas, 1981). For nonchlorophyllous tissues or dark reduction in chlorophyllous tissues, the reductants must all come from oxidation of carbohydrates or organic acids; Schrader and Thomas (1981) estimated that the equivalent of 35 to 38 photons are required for each NO_3^- converted to glutamate. The quantum requirement for NO_3^- reduction in nongreen tissue or green tissue in the dark appears to be almost double that in green tissue under high levels of light. In order to minimize energy requirements and maximize crop productivity, it seems desirable to assimilate a large portion of the NO_3^--N in illuminated green tissue.

B. Localization of Enzymes Involved

Research on intracellular and intercellular location of the enzymes of NO_3^- assimilation was reviewed recently (Beevers & Hageman, 1980; Schrader & Thomas, 1981). Controversy still exists about the intracellular localization of NR, which appears to be in the leaf cytoplasm, but could be loosely associated with the outer envelope of the chloroplast. Chloroplasts contain the enzyme required for reduction of NO_2^- (Dalling et al., 1972a; Ritenour et al., 1967; Wallsgrove et al., 1979), and evidence has been presented (Dalling et al., 1972b; Emes & Fowler, 1979) that NiR in nongreen tissues is associated with proplastids. Lee (1980) recently reviewed the evidence that GS is localized both within the plastids and cytosol in green and nongreen tissues. This suggests that GS is present to detoxify NH_3 formed by NO_2^- reduction in chloroplasts and proplastids; the NH_3 released into the cytoplasm during photorespiration and other deamination reactions is assimilated by GS in the cytoplasm (Keys et al., 1978), as GS is not present in mitochondria (Wallsgrove et al., 1980). Chloroplasts contain GOGAT in green tissues (Anderson & Done, 1977, 1978; Lea & Miflin, 1974; Wallsgrove et al., 1979), whereas proplastids contain GOGAT in nongreen tissues (Emes & Fowler, 1979; Miflin & Lea, 1977).

The localization studies are supported by experiments in which NO_2^- was reduced to NH_3 and amino acids by isolated chloroplasts in a light-dependent, CO_2-independent reaction without other additives (Magalhaes et al., 1974; Miflin, 1974). Later studies showed NO_2^--dependent and glutamine plus α-ketoglutarate–dependent O_2 evolution (Anderson & Done, 1977, 1978). These results indicate that NO_2^- reduction and glutamate synthesis in green tissues are dependent on reductants generated by the chloroplast electron transport chain.

II. TRANSPORT OF N

Both the xylem and phloem participate in transporting N in plants, but the composition of the two saps varies widely from species to species (Pate, 1973). According to Pate (1976, 1980), N solutes frequently comprise the major component of dry matter in xylem sap, and are second only to carbohydrate in the phloem sap. The C/N weight ratio for xylem and phloem sap varies from 1.5 to 6 and from 15 to 200, respectively.

A. Xylem Transport

The xylem is the principal path for long-distance transport of nitrogenous solutes from roots to organs that transpire (Pate, 1973, 1975, 1980; Weissman, 1964; Hill-Cottingham & Lloyd-Jones, 1979). The xylem, therefore, transports NO_3^- from roots to shoots in addition to N reduced in roots, and the xylem sap composition will reflect the amount of NO_3^- reduced in the roots. Over 95% of the xylem N may consist of NO_3^--N in some species that have low NR activity in their roots (e.g., *Cucumis, Xanthium,* and *Gossypium*), whereas species with high NR activity in roots (e.g., *Pisum, Raphanus*, and *Lupinus*) have $<$ 20% of their xylem N as NO_3^--N (Pate, 1973, 1976). It is not clear why the relative proportion of NO_3^--N reduced in roots differs among genotypes and species, but Jackson (1978) discussed the potential competition between reduction and xylem deposition components, and suggested that NR in roots appears to have priority over xylem deposition for incoming NO_3^-.

Reduced N is incorporated into a limited number of amino acids, amides, and other N solutes for transport from roots to shoots, but each plant species appears to have a characteristic spectrum of these N compounds; the most common N solutes in plants grown on NO_3^--N are aspartate and glutamate and their amides (Pate, 1973, 1976, 1980). Ureides and other solutes are important in xylem transport in certain species grown on NO_3^- (Thomas et al., 1979), but ureides (allantoin and allantoic acid) are most important as a transport solute in certain tropical legumes—e.g., soybeans and cowpeas (*Vigna*)—that are dependent on N_2 fixation for much of their reduced N (McClure & Israel, 1979).

B. Phloem Transport

The phloem is the principal transporter of N assimilated in one part of the shoot and transported to another (e.g., from leaf to seed). The composition of the phloem sap varies from species to species, but the phloem contents of only a limited number of species have been examined because of difficulties in collecting phloem sap from most species (Pate, 1976, 1980). In contrast to the xylem, N solutes in the phloem are organic solutes (Pate, 1976, 1980). Nitrate is usually absent from, or present in only trace amounts in the phloem (MacRobbie, 1971; Pate et al., 1975). In *Lupinus,* asparagine is highest in the phloem (Pate et al., 1975), but glutamine/glutamate and asparagine/aspartate fractions can both be high, as in *Ricinus* phloem exudate (Hall & Baker, 1972). Nelson (1962) reported serine to be exported in highest quantity from young soybean leaves, but soybean petioles also transport significant amounts of aspartate, alanine, glutamate, and γ-aminobutyrate (Housley et al., 1979; Schrader et al., 1980).

The transport of amino acids via the phloem from soybean leaves was verified by Housley et al. (1977). When a petiole was heat-girdled, phloem but not xylem functioning was disrupted. The attached leaf was fed $^{14}CO_2$ for 2 h, but neither ^{14}C–sugars nor ^{14}C–amino acids moved out of the fed leaf, thus demonstrating the importance of the phloem in exporting amino N from leaves. When no heat girdling was applied, only five amino acids in the transport pathway were substantially labeled with ^{14}C after a 2-h exposure to $^{14}CO_2$ (Housley et al., 1979). The distribution of ^{14}C in aspartate and γ-aminobutyrate was much higher in the path than in the source leaf, whereas serine, alanine, and glutamate were substantially labeled in both the path and source leaf. The total percentage of ^{14}C recovered in amino acids also differed, with 8 to 17% in leaf amino acids, but only 2 to 6% in amino acids in the petiole. This suggests some selectivity in the efflux processes from mesophyll cells into the apoplast and/or in the loading processes. Recent work (K. E. Koch & L. E. Schrader, unpublished) indicates a higher percentage of current photosynthate partitioned to amino acids during the peak periods of demand for N by the developing seeds.

Much remains to be learned about the mechanisms of phloem loading and transport of N solutes. For example, are specific carriers involved in loading each amino acid into the phloem? Is active transport involved? What is the capacity of the phloem for transporting N solutes?

Our earlier work (Housley et al., 1977; Servaites et al., 1979) suggested that phloem transport velocities for amino acids are similar to those for sugars and other photosynthetic products. The possible involvement of carriers for amino acids was reviewed earlier (Schrader & Thomas, 1981). The concentration dependence of sucrose and leucine loading and transport in soybean leaves suggests the involvement of carriers (Servaites et al., 1979). Inhibition of amino acid and sucrose loading by metabolic inhibitors suggests that metabolism-dependent reactions are involved in phloem loading, but further research is needed to establish whether loading is an active process (Servaites et al., 1979).

C. Transport of N as a Possible Limitation on Productivity

Crop species such as soybeans require large amounts of reduced N during seed development (Sinclair & deWit, 1975). As discussed earlier by Schrader and Thomas (1981), the demand for N during seed fill may exceed the soybean plant's capacity to provide and/or transport reduced N to the seeds. To briefly summarize these calculations (Schrader & Thomas, 1981), 1.74 g of photosynthate are required to produce 1 g of soybean seed with 42% protein, 20% lipid, 31% carbohydrate, 4% lignin, and 3% ash. If amino acids and/or other N solutes provide the C skeletons required for protein synthesis in the seed, then 0.42 g of dry matter (42% protein by dry wt) will come from N transport, and the remainder (1.74 g − 0.42 g = 1.32 g) will come from sucrose transport. Sucrose will provide C for energy production and for synthesis of carbohydrates, lipids, and lignin. Therefore, sucrose will provide about three times more C than do N solutes. However, sucrose has 12 C per molecule, whereas an average amino acid has about four C atoms, so that one can predict that one amino acid or other N compound must be transported for every sucrose in the phloem during seed development. A C/N ratio of 16:1 will be provided to the seed if these assumptions are valid. If amides and/or ureides with a C/N ratio lower than 4:1 are important transport compounds to the seed, the C/N ratio of the phloem sap would be modified.

Because chloroplasts are the major site of amino acid biosynthesis in green leaves (Mills et al., 1980), it seems reasonable to expect that ^{14}C–amino acids and ^{14}C–sucrose synthesized in a leaf during a 2-h period of steady-state exposure to $^{14}CO_2$ would incorporate similar proportions of ^{14}C and ^{12}C. On the basis of the calculations and assumptions presented above, about 25% of the radioactivity in the petiole of the fed leaf should be recovered in amino acids and 75% in ^{14}C–sugars. Recent data (K. E. Koch & L. E. Schrader, unpublished) indicate that $\leq$ 12% of the radioactivity in the soluble constituents of the petiole was in amino acids with the remainder in sucrose. The peak of 12% was observed during mid-podfill when the demand for reduced N is very high in the seed. A maximum of 17% of the radioactivity in soluble constituents of the fed leaf was observed in amino acids at the same stage of development. These data indicate that the photosynthetic production and transport of amino acids are lower than the apparent requirements of the developing soybean seeds. The lower proportion of recovered radioactivity in amino acids of the petiole vs. the fed leaf supports a suggestion that transport rather than availability of N limits productivity of soybeans. The failure of foliar-applied nutrients to increase soybean yields consistently (Garcia & Hanway, 1976; Boote et al., 1978) may be due to a limited capacity for transport of N to the seeds from the leaves and other vegetative tissues. More research is needed to better understand the mechanisms and capacity for partitioning and transport of N in soybeans and to examine the role current photosynthate plays in providing amino N for the seed at various stages of seed fill.

III. SITE OF PROTEIN SYNTHESIS AND DEGRADATION

A. Vegetative Tissues

As noted earlier, the enzymes for assimilation of NO_2^- to glutamate are in the chloroplast. Mills et al. (1980) reported the photosynthetic formation of amino acids belonging to the aspartate family in isolated chloroplasts. More recently, Miflin et al. (1980) reported evidence that nearly all amino acids are synthesized in the chloroplasts. In those plant species that reduce most of their NO_3^- in the shoots, the chloroplasts become a major source of amino acids for protein synthesis in the chloroplasts and other plant parts. Much of the protein synthesized in leaves is synthesized or at least accumulated in the chloroplast. For example, ribulose bisphosphate (RuBP) carboxylase (Fraction I protein) varies in concentration from species to species, but can comprise up to 65% of the total soluble protein of grass or legume leaves (Huffaker & Miller, 1978). Although the small subunit of this protein is synthesized on cytoplasm ribosomes (80S), the large and small subunits are assembled in the chloroplast to form a multifunctional protein. Not only does the protein exhibit carboxylase and oxygenase activities important in photosynthesis and photorespiration, respectively, but its high concentration and slow turnover qualify it as a major storage protein in the leaf (Huffaker & Miller, 1978). Friedrich and Huffaker (1980) noted that RuBP carboxylase accounted for 74 to 81% of the soluble protein in an attached primary barley (*Hordeum vulgare* L.) leaf at 10 to 14 d after planting. The concentration of soluble protein began to decrease at 8 d after planting, and the loss of RuBP carboxylase accounted for about 85% of the decrease in soluble protein that was observed during senescence of that leaf.

Waters et al. (1980) used wheat (*Triticum aestivum* L.) leaves to examine a large number of enzymes representative of both exo- and endopeptidases during the period from anthesis to grain maturity. The soluble protein in the flag leaf declined rapidly, beginning with the onset of grain growth. Glumes continued to accumulate N until 5 d after anthesis, and then remobilization to the grain began. Stems lost relatively less N. The flag leaf appeared to be the major contributor of grain N, as it lost about 20 mg of protein compared with 9 mg for glumes between anthesis and maturity. Nearly all of the soluble protein from the leaves and glumes was mobilized. These authors concluded that a diverse group of peptide hydrolases including both exo- and endopeptidases are involved in mobilization of protein from vegetative organs.

In a subsequent study, Peoples et al. (1980) observed that RuBP carboxylase constituted about 60% of the total soluble protein of wheat leaves. During early grain fill, nonselective hydrolysis of soluble protein in leaves occurred until about 35% of the total protein had been degraded. Then, RuBP carboxylase loss increased more rapidly beginning about 20 d after anthesis. This rapid loss of RuBP carboxylase coincided closely with the threefold increase in acid peptide-hydrolase activity. These authors previously concluded that the only peptide-hydrolase enzymes that could be detected using RuBP carboxylase as substrate are characterized by a pH optimum near 5. No acid peptide-hydrolase activity could be detected within the vacuole, although up to 15% of the activity was found associated with chloroplasts; the major portion was in the cytoplasm. Whether RuBP carboxylase degradation is due to the chloroplast enzyme is not known, but as they noted, compartmentation of protein hydrolysis within a cell would provide a mechanism for achieving control of senescence. Electron micrographs of these cells at various stages of development showed that chloroplasts were surrounded by an intact double-membrane envelope until at least 20 d after anthesis, and that the inner membrane appeared intact until at least 35 d after anthesis. Mesophyll cells of senesced leaves had fewer intact organelles and membranes at 35 d after anthesis than did bundle sheath or guard cells of that leaf. This is particularly interesting in that it suggests that leaves can protect certain cell types during senescence to ensure that gas exchange and movement of amino acids and sugars within the vascular tissues are maintained as long as possible.

Thus, it is evident that N is remobilized or redistributed to other plant parts during ontogeny. It seems likely that an atom of N may be used several times for protein synthesis during that plant's ontogeny. Mobilization of N from vegetative tissues is quite important during grain fill, as a large portion of the grain protein appears to be provided by proteins mobilized from leaves (Friedrich & Schrader, 1979; Peoples et al., 1980; Waters et al., 1980). Further research is needed to elucidate the mechanisms by which these processes are regulated.

B. Storage Proteins in Seeds

Due to space limitations, this topic will not be reviewed. The interested reader is referred to recent reviews (e.g., Boulter, 1979; Hall et al., 1980, Larkins et al., 1980, Miflin et al., 1980; Miflin & Shewry, 1981) discussing synthesis of storage proteins in both cereals and legumes. For a general review of protein synthesis, Szekely (1980) has reviewed the processes from DNA to protein.

IV. DISTRIBUTION OF NITROGENOUS ASSIMILATES IN VARIOUS PLANT PARTS AT DIFFERENT STAGES OF GROWTH

In an earlier review (Schrader, 1978), seasonal patterns of accumulation of various nitrogenous assimilates in maize (*Zea mays* L.) were reported. Nitrate accumulated principally in the lower internodes of the stalk with maximal accumulation about 85 d after planting. Accumulation of free amino N was higher in leaves than in stalks, and was maximal in leaves between 50 and 70 d after planting, with a decrease observed thereafter. Total N per plant increased until 100 d after planting; increases in grain N occurred largely at the expense of leaf N during later stages of ear filling, suggesting remobilization of N from leaves to ears.

Friedrich and Schrader (1979) reported that $^{15}NO_3$ accumulated in maize stalks and roots prior to silking could be remobilized during grain fill if N deprivation to roots were imposed. However, if high levels of $^{15}NO_3^-$-N were provided during ear fill, newly absorbed N was assimilated, and much of the NO_3^--N that accumulated during earlier stages of development was not remobilized. In spite of this observation, N absorbed prior to silking accounted for most (84%) of the N found in the grain at maturity, even though high levels of NO_3^--N were provided to some plants throughout development. This suggests the importance of supplying adequate N to maize prior to silking. It also supports the idea that N stored in vegetative tissues earlier in development is the principal source of N for grain protein.

The site of accumulation of NO_3^--N in barley leaves was recently shown to be the vacuoles (Martinoia et al., 1981). Vacuoles isolated from mesophyll protoplasts were shown to contain 99% of the NO_3^--N, whereas 52 and 5%, respectively, of the amino acids and soluble proteins were in the vacuoles. Additional research is needed to elucidate the transport mechanisms responsible for NO_3^- accumulation in vacuoles, and the relationship between this NO_3^- pool and NR activity in leaves.

In summary, the type and quantity of N accumulated in different plant parts vary at different stages of development and in different plant species. Most of the NO_3^--N that is transported to leaves is rapidly assimilated to form amino N, but NO_3^--N accumulates in other plant parts (e.g., roots and stems). This amino N may be used to synthesize protein in the leaf in which the amino N is formed, or the amino N may be transported directly to other plant parts for protein synthesis. Regardless of its fate, most of the amino N is stored in the form of protein in both vegetative and reproductive tissues. Thus, the free amino acid concentration remains low in healthy plants that are not under environmental stress. Protein may be accumulated to high levels in vegetative tissues, as discussed earlier, in both forage and grain crops, but in the grain crops, much of the protein in vegetative tissues is eventually hydrolyzed to amino acids and transported to the developing seeds for incorporation into seed storage proteins.

REFERENCES

Anderson, J. W., and J. Done. 1977. A polarographic study of glutamate synthase activity in isolated chloroplasts. Plant Physiol. 60:345-359.

Anderson, J. W., and J. Done. 1978. Light-dependent assimilation of nitrite by isolated pea. Plant Physiol. 61:692-697.

Beevers, L., and R. H. Hageman. 1980. Nitrate and nitrite reduction. p. 115-168. *In* B. J. Miflin (ed.) The biochemistry of plants. A comprehensive treatise. Volume 5, Amino acids and derivatives. Academic Press, New York.

Beevers, L., and R. Storey. 1976. Glutamate synthetase in developing cotyledons of *Pisum sativum*. Plant Physiol. 57:862-866.

Boote, K. J., R. N. Gallaher, W. K. Robertson, K. Hinson, and L. C. Hammond. 1978. Effect of foliar fertilization on photosynthesis, leaf nutrition, and yield of soybeans. Agron. J. 70:787-791.

Boulter, D. 1979. Structure and biosynthesis of legume storage proteins. p. 125-134. *In* Seed protein improvement in cereals and grain legumes. Vol. 1. Int. Atomic Energy Agency, Vienna.

Chiu, J. Y., and P. D. Shargool. 1979. Importance of glutamate synthase in glutamate synthesis by soybean cell suspension cultures. Plant Physiol. 63:409-415.

Dalling, M. J., N. E. Tolbert, and R. H. Hageman. 1972a. Intracellular location of nitrate reductase and nitrite reductase. I. Spinach and tobacco leaves. Biochim. Biophys. Acta 283:505-512.

Dalling, M. J., N. E. Tolbert, and R. H. Hageman. 1972b. Intracellular location of nitrate reductase and nitrite reductase. II. Wheat roots. Biochim. Biophys. Acta 283:513-519.

Emes, M. J., and M. W. Fowler. 1979. The intracellular location of the enzymes of nitrate assimilation in the apices of seedling pea roots. Planta 144:249-253.

Friedrich, J. W., and R. C. Huffaker. 1980. Photosynthesis, leaf resistances, and ribulose-1,5-bisphosphate carboxylase degradation in senescing barley leaves. Plant Physiol. 65:1103-1107.

Friedrich, J. W., and L. E. Schrader. 1979. N deprivation in maize during grain filling. II. Remobilization of ^{15}N and ^{35}S and the relationship between N and S accumulation. Agron. J. 71:466-472.

Garcia, R. L., and J. J. Hanway. 1976. Foliar fertilization of soybeans during the seed-filling period. Agron. J. 68:653-657.

Hall, S. M., and D. A. Baker. 1972. The chemical composition of *Ricinus* phloem exudate. Planta 106:131-140.

Hall, T. C., S. M. Sun, B. U. Buchbinder, J. W. Pyne, F. A. Bliss, and J. D. Kemp. 1980. Bean seed globulin in mRNA: translation, characterization and its use as a probe towards genetic engineering of crop plants. p. 259-272. *In* C. J. Leaver (ed.) Genome organization and expression in plants. Plenum Publ. Corp., New York.

Hewitt, E. J. 1975. Assimilatory nitrate-nitrite reduction. Annu. Rev. Plant Physiol. 26:73-100.

Hewitt, E. J., and C. V. Cutting (ed.). 1979. Nitrogen assimilation of plants. (Sixth Long Ashton Symposium, 1977). Academic Press, London.

Hill-Cottingham, D. G., and C. P. Lloyd-Jones. 1979. Translocation of nitrogenous compounds in plants. p. 397-405. *In* E. J. Hewitt and C. V. Cutting (ed.) Nitrogen assimilation of plants. Academic Press, London.

Housley, T. L., D. M. Peterson, and L. E. Schrader. 1977. Long distance translocation of sucrose, serine, leucine, lysine, and CO_2 assimilates. I. Soybean. Plant Physiol. 59:217-220.

Housley, T. L., L. E. Schrader, M. Miller, and T. L. Setter. 1979. Partitioning of ^{14}C-photosynthate, and long distance translocation of amino acids in preflowering and flowering, nodulated and nonnodulated soybeans. Plant Physiol. 64:94-98.

Huffaker, R. C., and B. L. Miller. 1978. Reutilization of ribulose bisphosphate carboxylase. p. 139-152. *In* H. W. Siegelman and G. Hind (ed.) Photosynthetic carbon assimilation. Plenum Publ. Corp., New York.

Jackson, W. A. 1978. Nitrate acquisition and assimilation by higher plants: processes in the root system. p. 45-88. *In* D. R. Nielsen and J. G. MacDonald (ed.) Nitrogen in the environment. Vol. 2. Academic Press, London.

Keys, A. J., I. F. Bird, M. J. Cornelius, P. J. Lea, R. M. Wallsgrove, and B. J. Miflin. 1978. Photorespiratory nitrogen cycle. Nature 275:741–743.

Larkins, B. A., K. Pederson, W. J. Hurkman, A. K. Handa, A. C. Mason, C. Y. Tsai, and M. A. Hermodson. 1980. Maize storage proteins: characterization and biosynthesis. p. 203–217. *In* C. J. Leaver (ed.) Genome organization and expression in plants. Plenum Publ. Corp., New York.

Lea, P. J., and B. J. Miflin. 1974. Alternative route for nitrogen assimilation in higher plants. Nature 251:614–616.

Lee, R. B. 1980. Sources of reductant for nitrate assimilation in non-photosynthetic tissue: a review. Plant Cell Environ. 3:65–90.

MacRobbie, E. 1971. Phloem translocation. Facts and mechanisms. A comparative survey. Biol. Rev. 46:429–481.

Magalhaes, A. C., C. A. Neyra, and R. H. Hageman. 1974. Nitrite assimilation and amino nitrogen synthesis in isolated spinach chloroplasts. Plant Physiol. 53:411–415.

Martinoia, E., U. Heck, and A. Wiemken. 1981. Vacuoles as storage compartments for nitrate in barley leaves. Nature 289:292–294.

McClure, P. R., and D. W. Israel. 1979. Transport of nitrogen in the xylem sap of soybean plants. Plant Physiol. 65:411–416.

Miflin, B. J. 1974. Nitrite reduction in leaves: studies on isolated chloroplasts. Planta 116: 187–196.

Miflin, B. J. 1980. Nitrogen metabolism and amino acid biosynthesis in crop plants. p. 255–296. *In* P. S. Carlson (ed.) The biology of crop productivity. Academic Press, New York.

Miflin, B. J., and P. J. Lea. 1976. The pathway of nitrogen assimilation in plants. Phytochemistry 15:873–885.

Miflin, B. J., and P. J. Lea. 1977. Amino acid metabolism. Annu. Rev. Plant Physiol. 28:299–329.

Miflin, B. J., P. J. Lea, W. R. Mills, J. K. Sainis, R. M. Wallsgrove, and M. Mazelis. 1981. Amino acid biosynthesis in chloroplasts. p. 731–740. *In* G. Akoyunoglou (ed.) Photosynthesis. Volume 4, Regulation of carbon metabolism. Balaban Int. Sciences Services, Philadelphia.

Miflin, B. J., J. A. Matthews, S. R. Burgess, A. J. Faulks, and P. R. Shewry. 1980. The synthesis of barley storage proteins. p. 233–243. *In* C. J. Leaver (ed.) Genome organization and expression in plants. Plenum Publ. Corp., New York.

Miflin, B. J., and P. R. Shewry. 1981. Seed storage proteins: genetics, synthesis, accumulation and protein quality. p. 195–248. *In* J. D. Bewley (ed.) Nitrogen and carbon metabolism. Martinus Nijhoff/Dr. Junk Publ., The Hague.

Mills, W. R., P. J. Lea, and B. J. Miflin. 1980. Photosynthetic formation of the aspartate family of amino acids in isolated chloroplasts. Plant Physiol. 65:1166–1172.

Nelson, C. D. 1962. Translocation of organic compounds in plants. Can. J. Bot. 40:757–770.

Pate, J. S. 1973. Uptake, assimilation and transport of nitrogen compounds by plants. Soil Biol. Biochem. 5:109–119.

Pate, J. S. 1975. Exchange of solutes between phloem and xylem and circulation in the whole plant. p. 451–473. *In* M. H. Zimmerman and J. A. Milburn (ed.) Encyclopedia of plant physiology. Volume 1, Transport in plants. I. Phloem transport. Springer-Verlag, Berlin.

Pate, J. S. 1976. Nutrients and metabolites of fluids recovered from xylem and phloem: significance in relation to long-distance transport in plants. p. 253–281. *In* I. F. Wardlaw and J. B. Passioura (ed.) Transport and transfer processes in plants. Academic Press, New York.

Pate, J. S. 1980. Transport and partitioning of nitrogenous solutes. Annu. Rev. Plant Physiol. 31:313–340.

Pate, J. S., P. J. Sharkey, and O. A. M. Lewis. 1975. Xylem to phloem transfer of solutes in fruiting shoots of a legume, studied by a phloem bleeding technique. Planta 122:11–26.

Peoples, M. B., V. C. Beilherz, S. P. Waters, R. J. Simpson, and M. J. Dalling. 1980. Nitrogen redistribution during grain growth in wheat (*Triticum aestivum* L.). II. Chloroplast senescence and the degradation of ribulose-1,5-bisphosphate carboxylase. Planta 149: 241–251.

Ritenour, G. L., K. W. Joy, J. Bunning, and R. H. Hageman. 1967. Intracellular localization of nitrate reductase, nitrite reductase and glutamic acid dehydrogenase in green leaf tissue. Plant Physiol. 42:233–237.

Schrader, L. E. 1978. Uptake, accumulation, assimilation, and transport of nitrogen in higher plants. p. 101–141. *In* D. R. Nielsen and J. G. MacDonald (ed.) Nitrogen in the environment. Vol. 2. Academic Press, London.

Schrader, L. E., T. L. Housley, and J. C. Servaites. 1980. Amino acid loading and transport in phloem. p. 101–109. *In* F. T. Corbin (ed.) Proc. World Soybean Res. Conf. II. Westview Press, Boulder, Colo.

Schrader, L. E., and R. J. Thomas. 1981. Nitrate uptake, reduction and transport in the whole plant. p. 49–93. *In* J. D. Bewley (ed.) Nitrogen and carbon metabolism. Martinus Nijhoff/Dr. Junk Publ., The Hague.

Servaites, J. C., L. E. Schrader, and D. M. Jung. 1979. Energy-dependent loading of amino acids and sucrose into the phloem of soybean. Plant Physiol. 64:546–550.

Sinclair, T. R., and C. T. deWit. 1975. Photosynthate and nitrogen requirements for seed production by various crops. Science 189:565–567.

Szekely, M. 1980. From DNA to protein. John Wiley & Sons, Inc., New York.

Thomas, R. J., U. Feller, and K. H. Erismann. 1979. The effect of different inorganic nitrogen sources and plant age on the composition of bleeding sap of *Phaseolus vulgaris* (L.). New Phytol. 82:657–669.

Vennesland, B., and M. G. Guerrero. 1979. Reduction of nitrate and nitrite. p. 425–444. *In* M. Gibbs and E. Latzko (ed.) Encyclopedia of plant physiology. Photosynthesis II. Vol. 6. Springer-Verlag, Berlin.

Wallsgrove, R. M., A. J. Keys, I. F. Bird, M. J. Cornelius, P. J. Lea, and B. J. Miflin. 1980. The location of glutamine synthetase in leaf cells and its role in the reassimilation of ammonia released in photorespiration. J. Exp. Bot. 31:1005–1017.

Wallsgrove, R. M., P. J. Lea, and B. J. Miflin. 1979. The distribution of the enzymes of nitrogen assimilation within the pea leaf cell. Plant Physiol. 63:232–236.

Waters, S. P., M. B. Peoples, R. J. Simpson, and M. J. Dalling. 1980. Nitrogen redistribution during grain growth in wheat (*Triticum aestivum* L.). I. Peptide hydrolase activity and protein breakdown in the flag leaf, glumes and stem. Planta 148:422–428.

Weissman, G. S. 1964. Effect of ammonium and nitrate nutrition on protein level and exudate composition. Plant Physiol. 39:947–952.

4

R. H. Hageman
University of Illinois
Urbana, Illinois

Ammonium Versus Nitrate Nutrition of Higher Plants

The answer to the question "which form of N, NO_3^- or NH_4^+ is superior for obtaining maximum crop productivity?" remains unanswered. Because of the technical difficulties in evaluating this problem with field-grown crops, most studies have been done under growth chamber or greenhouse conditions. The results reported from such studies vary with plant species and environmental conditions. Most plant species when grown under appropriate conditions, can effectively utilize either form of N; however, growth responses over a wide array of environments, especially low light intensity, are usually superior for NO_3^- than for NH_4^+.

Because NH_3 is rapidly converted to NO_3^- by microorganisms in most soils when aeration and temperature are optimal for plant growth, NO_3^- is considered to be the primary form of N available to rain-fed crop plants. Under such conditions, the direct role of NH_4^+ in crop production becomes insignificant. The development of nitrification inhibitors that block the conversion of NH_3 to NO_3^- in the soil provides a reason for reexamination of the relative merits of NH_4^+ and NO_3^-. As the effectiveness and soil half-life of the nitrification inhibitors are increased the relative amount and maintenance of NH_3 in the soil will be enhanced. These developments could improve crop productivity, as several workers have found that mixtures of nitrate and ammonium salts support better plant growth than either form supplied separately.

I. ASPECTS OF AMMONIUM AND NITRATE NUTRITION

The effect of forms of inorganic N on plant growth and productivity has been under investigation and dispute for a long time (Miller, 1938). There are many reasons why it has been difficult to resolve the effects of NH_4^+ and NO_3^- per se on plant growth.

Nitrogen in Crop Production.

First, the inherent characteristics and properties of the two ions are different; NH_4^+ is a cation and NO_3^- is an anion, and in a negatively charged medium (i.e., soil), NH_4^+ is bound, while NO_3^- remains mobile. Hence, NO_3^- can move with the soil solution to the root or be more readily leached from the soil.

Second, in fertilizer salts, the two forms of N are associated with a different companion ion. In some instances the companion ion can directly affect plant growth (Allred & Ohlrogge, 1974), while in other cases the effects are indirect. Successive, annual applications of equal and high amounts of N as $(NH_4)_2SO_4$ or $NaNO_3$ to soils of low exchange capacity can, in time, cause the soil to become more acidic or more alkaline, respectively. As the pH of the soil changes, micronutrients or P may become unavailable and thus restrict plant growth.

Third, when NH_4^+ is taken up from the medium by plants, the medium tends to become acidic, while the absorption of NO_3^- results in an alkaline medium. As previously mentioned, these changes in pH can affect the availability of other nutrients.

Fourth, in well-aerated soils above 5°C, NH_4^+ is converted by soil organisms (Nitrobacteriaceae) to NO_3^- and results in acidification of the soil. The addition of anhydrous or liquid NH_4^+ initially causes the soil in and around the application zone to become alkaline. As NH_4^+ is converted to NO_3^-, the soil becomes acidic. In vitro studies of such systems have shown that these pH shifts are associated with changes in the amounts of extractable Mn; availability of other ions could also be affected.

Fifth, NH_4^+ may displace other cations that are bound to the negatively charged soil particles, thus affecting their availability to the plant.

Experiments designed to compare NH_4^+ with NO_3^- nutrition in soil or nonsterile media have sometimes ignored the fact that NH_4^+ can be converted to NO_3^-. Calcium carbonate, which is sometimes added to regulate pH of NH_4^+-type nutrient solutions, has been shown to enhance nitrification (Hewitt, 1966). In contrast, only small amounts of NH_4^+ have been found in NO_3^- nutrient solutions (Greidanus et al., 1972). The origin of this NH_4^+ was not determined.

Uptake of NH_4^+ and NO_3^- are affected differently by temperature and pH of the rooting medium (Hewitt, 1966; 1970). Absorption of NH_4^+ is highly temperature dependent (optimum 27°C) when the pH of the nutrient solution is varied from 4.0 to 6.5. Between pH 6.5 and 8.5, NH_4^+ absorption was independent of temperature (Lycklama, 1963). Absorption of NO_3^- is also temperature dependent, with an optimum of 35°C in some cases. According to one report (Williams & Vlamis, 1962), little NO_3^- is absorbed when temperatures are below 13°C. For NO_3^- uptake, a nutrient medium pH of 4.5 to 6.0 is considered optimal. While a pH of 6.0 to 7.0 is best for NH_4^+ (Hewitt, 1966; 1970), when NO_3^- concentrations were low (2 meq/L), pH had little effect on uptake (Wallace & Mueller, 1957). When NO_3^- levels were 8 meq/L, absorption decreased with increased pH. For these and other reasons, Munn and Jackson (1978) have cautioned against generalizing about the effects of ambient pH on uptake of NH_4^+ and NO_3^-.

Studies with some species (Hewitt, 1966; Michael et al., 1970) have indicated that young plants absorb NH_4^+ more readily than NO_3^-; however, as the plant ages, the converse is observed.

In experiments designed to minimize pH effects. Arnon (1937), found that barley (*Hordeum vulgaris* L.) plants grown on NO_3^- produced more dry weight of roots and shoots than plants cultured on NH_4^+ salts when solutions were not aerated. This was true for comparisons made at pH 4.0, 5.0, 6.0, and 6.7 during every season of the year. Plants grown on aerated NH_4^+ medium at pH 6.0 produced slightly more dry matter than plants grown on NO_3^- only in June, when light intensity was high and the photoperiod long. Dry matter production of plants grown on aerated NH_4^+ nutrient solution and supplied with additional Cu^{2+} or Mn^{2+}—especially the latter—nearly equaled the dry matter production of plants grown on aerated NO_3^- medium under all environmental conditions tested. It was suggested that NO_3^-, in addition to providing N, was also serving as an oxidizing agent (N of NO_3^- can accept eight electrons; in contrast, N of NH_4^+ undergoes no change in valence in its assimilation to amino acids), which could be beneficial to growth-related metabolism. It was implied that the aeration and extra Mn^{2+} and Cu^{2+} enabled the NH_4^+ plants to carry out these necessary oxidation-reduction reactions. From this and other studies, it was concluded that either NO_3^- or NH_4^+ salts can serve as an adequate source of N for plant growth, but, in general, NO_3^- salts are considered the "safer" fertilizer for plant production.

As with all generalizations, there are exceptions to this statement. Calcifuge (acid-loving) plants that grow naturally in acid soils where little nitrification occurs utilize NH_4^+ in preference to NO_3^- (Haynes & Goh, 1978). For example, cranberry [*Oxycoccus macrocarpus* (Ait.) Parsh] plants were not able to absorb or metabolize NO_3^- (Greidanus et al., 1972). In contrast, calcioles (plants with a wide pH tolerance) preferably utlize NO_3^- (Haynes & Goh, 1978). The differential capability to utilize NO_3^- or NH_4^+ in some instances can be related to variations in potential to produce the enzyme nitrate reductase (Stewart et al., 1974); calcifuge plants have limited ability for production of nitrate reductase.

Although paddy rice (*Oriza sativa* L.) is more productive when supplied with ammonium than nitrate salts (Martin et al., 1976), data were not found that clearly distinguish between the efficiency with which the plant utilizes the two forms of N and the denitrification of NO_3^- in the paddy. Corn (*Zea mays* L.) hybrids grown on sandy soil and provided with supplemental urea or NH_4NO_3 produced significantly more grain and stover of higher N content than plants supplied with KNO_3 (Jung et al., 1972). Some NO_3^- could have been leached from these sandy soils because of its mobility; conversely, some gaseous N could have been evolved during denitrification. The failure to determine actual levels of NH_4^+ and NO_3^- in the soil complicates the interpretation of these data. Because inhibitors of nitrification were not added, it is likely that some of the NH_4^+ ions added were converted to NO_3^-. This raises the possibility that a mixture of NO_3^- and NH_4^+ may be more favorable for plant growth than either ion supplied separately.

This concept is supported by the following reports. Mixtures of NH_4^+ and NO_3^- salts provide the most suitable source of inorganic N for asceptically cultured cells (Eriksson, 1965; Filner, 1966; Gamborg & Shyluk, 1970; Gamborg et al., 1968). Filner (1966) was unable to culture tobacco (*Nicotiana tabacum* L.) cells on ammonium salts regardless of the associated anion, while rice cells could be cultured on ammonium citrate but not diammonium sulfate (Yatazawa & Furuhoshi, 1968). Although soybean [*Glycine max* (L.) Merr.] cells could be cultured on 25 m*M* NO_3^- plus 2 m*M* $(NH_4)_2SO_4$ (B5 medium), they would not grow on NO_3^- alone (Gamborg & Shyluk, 1970). Cell growth of soybeans, wheat (*Triticum aestivum* L.), and flax (*Linum usitatissimum* L.) was satisfactory only when cells were supplied with ammonium salts of Krebs cycle acids. However, cell growth on these ammonium salts was slightly less than growth obtained with the B5 medium. Higher yields of wheat and dry matter and protein production by sunflower (*Helianthus annuus* L.) were obtained with NH_4^+ plus NO_3^- than with either source alone (Cox & Reisenauer, 1971). Schrader et al., (1972) found that 47-day-old corn plants grown on a nutrient media containing NO_3^-/NH_4^+ mixtures (ratios of 1:3, 1:1, and 3:1) produced more dry matter than plants grown on either NO_3^- or NH_4^+ as the sole source of N. Significant differences were noted in dry matter production of roots and stems, but not leaves, when plants were grown on equal proportions of NO_3^- and NH_4^+ compared with NO_3^- alone.

With respect to the question "which form of N, NO_3^- or NH_4^+, is superior for obtaining maximum crop productivity?", the opinions advanced by Arnon (1937) and Hewitt and Smith (1975) that most plant species grow better when supplied with nitrate than with ammonium salts, is considered valid. This is especially true when culture conditions are optimized for each N source. Under most field conditions, the preceding question is difficult to resolve because of the problems of precise control of pH, temperature, and the effect of the form of N on the other nutrient ions. Under conditions favorable for plant growth, the soil organisms rapidly convert NH_4^+ to NO_3^-. Considering the dynamic nature of the soil N complex, it is difficult to visualize a well-aerated soil devoid of NH_4^+, other forms of reduced N, or NO_3^- during the entire course of the growing season.

II. EFFECTS OF AMMONIUM AND NITRATE IONS ON PLANT METABOLITES AND METABOLISM

A. Cations and Anions

Tomato [*Lycopersicon esculentum* (Mill.)] seedlings grown for 20 d on a nutrient medium containing 5 meq L^{-1} NO_3^- produced more (2- to 3-fold) dry matter per plant part than seedlings grown on NH_4^+ (Kirkby & Mengel, 1967). The NO_3^--grown plants contained from 1- to 3-fold (depending on ion and plant part) more cations (Ca^{2+}, Mg^{2+}, and K^+), and from 2- to 22-fold more nonvolatile organic acids (primarily malate and citrate) than NH_4^+-grown plants. Plant parts from NH_4^+-grown plants contained more SO_4^- (1.5- to 2.4-fold) and $H_2PO_4^-$ (1.2-fold, roots excepted) than the NO_3^-

counterparts. The cation/anion ratio was approximately 1.0, regardless of the forms of N supplied. Pill and Lambert (1977) also found that tomato plants grown on NO_3^- 20 meq L^{-1}) produced more dry matter and accumulated more Ca^{2+}, Mg^{2+}, and K^+ than when grown on NH_4^+.

Similar effects of the two forms of N on growth, cation, and organic acid accumulation by leaves of white mustard (*Brassica hirta* Moench.) plants were obtained by Kirkby (1968). His plants were grown on a flowing dilute nutrient media that contained 0.004 meq L^{-1} NO_3^-, or 0.003 meq L^{-1} NH_4^+ plus 0.001 meq L^{-1} NO_3^-, with controlled pH. White mustard grows poorly on NH_4^+ as the sole source of N. In contrast to the previous study, the NH_4^+ plants did not accumulate any more inorganic anions than the NO_3^- plants. Barker and Maynard (1972) found that pea (*Pisum sativum* L.) or cucumber (*Cucumis sativus* L.) plants grown on NO_3^- (5–20 meq L^{-1}) accumulated significantly more dry weight than plants cultured on comparable levels of NH_4^+. While NH_4^+ treatment, relative to NO_3^-, suppressed the accumulation of cations (expressed as percent of dry matter) in cucumber shoots, it did not affect cation accumulation in pea shoots. Based on work of others (Coic et al., 1962; 1970), they suggested that most of the NO_3^- was reduced in the pea root; consequently, NO_3^- could not serve as a companion ion for transport of cations to the shoot. However, their assays showed that nitrate reductase activity was 10-fold higher in cucumber roots than in pea roots.

Blair et al. (1970) found no difference in dry matter production by corn seedlings grown for 14 and 28 d on flowing medium (pH 6.8) that contained either 2 m*M* NO_3^- or 2 m*M* NH_4^+. The NH_4^+-grown plants accumulated more $H_2PO_4^-$ and SO_4^{2-} than the NO_3^- plants. More Ca^{2+} was found in the shoots and roots, and more Mg^{2+} in the roots of NO_3^-, than in NH_4^+ plants. In contrast to previous work, the form of N had no effect on K^+ accumulation. Blair et al. (1970) state that the use of dilute nutrient medium eliminates growth reduction and injury frequently observed with NH_4^+-grown plants, and that comparisons made between effects of NH_4^+ and NO_3^- when growth is unequal is not valid. Although the flowing dilute solutions used may more closely approximate the composition of the soil solution throughout the season, availability of nutrients under field conditions vary as a function of water supply and season. The concentration of salts (i.e., 15 m*M* NO_3^-) in Hoagland's solution was based on the analysis of solution displaced from a fertile soil. By midseason, under cropping conditions, NO_3^- concentration in the soil solution would be much lower. While the point made on unequal plant size is well-taken, it is suggested that these experimental conditions (Blair et al., 1970) are biased in favor of the NH_4^+-grown plants. At pH 6.8, NH_4^+ uptake would be enhanced and NO_3^- uptake repressed, and corn frequently exhibits Fe deficiency at this pH (Hageman et al., 1961). An alternate approach would be to utilize conditions that would maximize plant growth rates for both forms of N. As indicated previously, it is difficult to compare the effects of NH_4^+ and NO_3^- per se on plant growth.

The reason for the decreased concentration of cations found in plants grown on NH_4^+, relative to NO_3^-, is considered to be due to increased competition in the uptake process. Wadleigh and Shive (1939) found that absorption of NH_4^+ by corn plants reduced the uptake of cations, especially Ca^{2+}. Although cation accumulation in the pea shoot was unaffected by the forms of N supplied, growth on NH_4^+ relative to NO_3^- was reduced almost as much in the pea as in the cucumber (Barker & Manyard, 1972). In contrast, NH_4^+-grown cucumber plants had much lower levels of cations. This raises the possibility that a considerable degree of variability in cation concentration can be tolerated without affecting plant productivity. It is not implied that under certain conditions the level of a given cation could not fall below a "critical level." Support for this may be indicated by the favorable effects on plant growth obtained by addition of extra K^+ to NH_4^+-grown plants. This extra K^+ could also decrease the rate of uptake of NH_4^+, thus helping to keep the internal NH_4^+ below toxic levels.

Kirkby and Mengel (1967) reported that macerated tissues from NO_3^--grown plants were less acidic (pH 5.5) than plants grown on NH_4^+ (4.8). The consequences, if any, of a differential in internal pH per se derived from NO_3^- or NH_4^+ nutrition on growth and development of plants are not known. Information is not available as to the inter- and intracellular sites of acid production or oxidation, compartmentalization, and effect of such pH changes on the various enzymes in situ. However, corn seedlings rapidly became chlorotic and stunted when grown on a medium containing NO_3^- if the pH of the external medium is not continuously adjusted (Hageman et al., 1961).

The cause of the differential in internal pH may be related to a differential level of cations and the metabolic reactions associated with the assimilation of NH_4^+ and NO_3^-. Kirkby (1968), Kirkby and Mengel (1967) and others indicate that NO_3^- reduction creates internal alkalinity according to the following equation:

$$NO_3^- + 8H^+ + 8e \rightarrow NH_3 + 2H_2O + OH^-. \qquad [1]$$

To maintain electrical neutrality and provide some measure of pH control, the plant synthesizes organic acids equivalent to the OH^- produced. However, Dijkshoorn (1962) showed that this was valid only for the assimilated NO_3^- that was absorbed as the companion ion of the metallic cations (Ca^{2+}, Mg^{2+}, K^+, or Na^+). In a nutrient solution where NO_3^- is the sole source of N, anion (largely NO_3^-) uptake exceeds cation uptake. To maintain internal electrical neutrality, he suggests that H^+ is absorbed or OH^- excreted to balance the excess of NO_3^- absorbed. This movement of H^+ or OH^- would create an alkaline external medium. In either event, there would be an internal cation present to balance the NO_3^- charge. The following equations illustrate why the companion uptake cation affects the internal alkalinity upon NO_3^- reduction:

$$H^+ + NO_3^- + 8H^+ + 8e \rightarrow NH_3 + 3H_2O \qquad [2]$$

$$K^+ + NO_3^- + 8H^+ + 8e \rightarrow NH_3 + 2H_2O + KOH \qquad [3]$$

(Other metallic cations can substitute for K^+.) Although NH_3 reacts with water as shown in Eq. [4], electrical balance is not altered.

$$H_2O + NH_3(aq.) \leftrightharpoons NH_4^+ + OH^- \quad [4]$$

[the hydrolysis constant (K) of ammonium is 5.52×10^{-10}].

The same rationale can be applied to the uptake of NH_4^+, when NH_4^+ is the sole source of N. Assuming that NH_4^+ is the ionic species that is absorbed and that electrical neutrality must be maintained, the uptake of NH_4^+ would be accompanied by the uptake of an anion (e.g., $H_2PO_4^-$ or OH^-) or the efflux of H^+ into the medium. The uptake of OH^- or the efflux of H^+ would render the external medium acidic. When OH^- was the companion uptake ion (or equivalent efflux of H^+), the assimilation of NH_3 would result in the production of H_2O. However, when $H_2PO_4^-$ or a similar anion was the uptake companion ion, internal acidity would result after NH_3 assimilation. In analogy to NO_3^- assimilation, an equivalent amount of organic acids would be catabolized to maintain electrical neutrality. When NH_3(aq) is adsorbed instead of NH_4^+ (Bennett, 1971), the following events seem probable. Assuming that $(NH_4)_2SO_4$ was added as the nutrient salt, the NH_4^+ ion would donate a H^+ to the medium to form NH_3; NH_3, after entry into or absorption by the roots, could be directly assimilated or react with H_2O—see Eq. [4].

Some of the consequences of change in external medium pH are more fully documented (Hewitt, 1966). For example, corn seedlings grown on a NO_3^--nutrient solution become chlorotic and stunted because of unavailability of Fe unless the pH of the external medium is continuously acidified (Hageman et al., 1961). The resulting acidification of an NH_4^+-nutrient solution may be equally injurious to other species. The buffer resin column developed by Harper and Nicholas (1976) provides a convenient method for regulating pH of the external solution for small pot cultures. While these same changes occur in the external medium under field conditions, the pH fluctuations are minimized by the buffering capacity of the clay colloids in most soils. However, in soils—at least at certain times—the microenvironment around the root should exhibit fluctuations in pH.

B. Site of Assimilation

Based on the proportion of NO_3^- to reduced N found in the xylem exudate of detopped seedlings, Pate (1973) found that the amounts of NO_3^- reduced in the root relative to the shoot varied widely among plant species. In his experiments, white lupine (*Lupinus alba*) reduced most (90–95%) of the NO_3^- in the root while cocklebur (*Xanthium pensylvanicum*) reduced all of the NO_3^- in the shoot. With cereal species, the amount of root NO_3^- reduction varied from 25 to 60% of the total. Using a similar experimental procedure, Raghuveer (1977) found that maize (*Zea mays* L.) roots reduced about one-third of the NO_3^- (range 22–54% for 11 genotypes). Preliminary evidence indicated that the percent of NO_3^- reduced by the root decreased with plant age.

Although it has been stated that all cells have the potential to assimilate NO_3^- in certain organs or tissues (e.g., cocklebur root and bundle sheath cells of plants with C_4 type photosynthesis), this genetic potential for the synthesis of enzymes of NO_3^- reduction, especially nitrate reductase, is not expressed.

The reduction and assimilation of NO_3^- that occurs in the roots yields amino acids. These amino acids may be used for synthesis of root protein or transported via the xylem elements to the shoot. Nitrate as such is also transported via the xylem elements to the shoot, where it is partitioned between the stalk and the leaves. The prime destination is the leaves, where the bulk of the enzymes for NO_3^- assimilation are located. In some crops (e.g., maize), the pith cells of the stalk serve as a storage site for NO_3^-. Upon entering the mesophyll cells of the leaves, the NO_3^- is partitioned between a "readily available" pool (presumed to be in the cytoplasm) and a "storage" pool (presumed to be in the vacuole). Because leaf NO_3^- content can be depleted, NO_3^- in the "storage" pool must be made available, over time, for assimilation. The currently accepted intracellular localization of the enzymes and pathway for assimilation in green leaves is as follows:

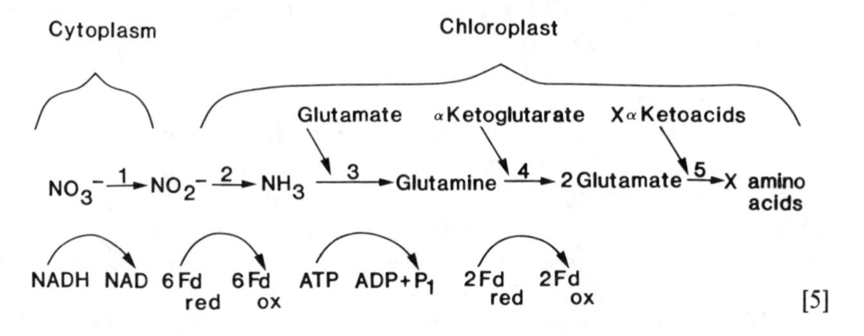

[5]

The metabolites, glutamate and various α ketoacids (such as α ketoglutarate, oxaloacetate, and pyruvate) shown in the top row are required for the assimilation of NH_3. These metabolites are supplied by other systems. The assimilation of NO_3^- (middle row) requires the following enzymes: (i) nitrate reductase, (ii) nitrite reductase, (iii) glutamine synthetase, (iv) glutamate synthetase (GOGAT), and (v) transaminases. The energy carriers required for the assimilation of NO_3^-, shown on the bottom row are: NADH and NAD—the reduced and oxidized forms of nicotiniamide adenine dinucleotide, respectively ($2H^+$ and 2e per molecule); Fe_{red} and Fd_{ox}—the reduced and oxidized forms of ferredoxin, respectively (1e per molecule, found only in the chloroplast); and ATP and ADP—adenosine tri- and diphosphate, respectively (1 high-energy phosphate per molecule). The components that supply the electrons and energy (bottom row) for the assimilation of NO_3^- are supplied by associated systems. The oxidation of glyceraldehyde-3-phosphate (3PGAld) and/or malate is the likely source of the reductant, NADH. The conversion of light energy by photosystems I and II to chemical energy (i.e., ATP and reduced ferredoxin) provides the energy for the final stages of assimilation. Thus, NO_3^- assimilation competes with CO_2 for light-derived chemical energy. In roots, it has been suggested that nitrite reductase is localized in a "plastid type" organelle. The source of reductant for nitrite reductase and glutamate synthetase in the root is not known; however, some evidence indicates that reduced nicotinamide adenine dinucleotide phosphate (NADPH—an energy carrier—$2H^+$ and 2e) may be involved. In any event, the energy for NO_3^- assimilation in roots is derived solely from the oxidation of organic compounds. A more complete description of these systems has been published (Miflin, 1980).

In contrast to NO_3^-, the bulk of the NH_4^+ absorbed from the soil is converted by the metabolic systems of the root into amino acids, amides, ureides, and other organic N compounds (McClure & Israel, 1978; Pate & Wallace, 1964). When pea (*Pisum arvense*) plants were supplied 10 mg N/plant as NH_4Cl, NH_4^+ accounted for $< 1\%$ of the N compounds found in the xylem exudate of detopped plants (Pate & Wallace, 1964). Photosynthate must be transported to the root to provide the organic compounds needed to detoxify (conversion of NH_3 into organic N compounds) the absorbed NH_4^+. Thus, there is need to balance the flux of NH_4^+ and photosynthate into the root to avoid root damage. Limited root growth is commonly associated with plants grown in NH_4^+ relative to those from NO_3^--grown plants (Kirkby & Mengel, 1967).

C. Organic Acids and Carbohydrates

McKee (1962) states that the main effect of supplying plants with high levels of NO_3^- is an increase in their organic acid levels. In contrast, plants supplied with high levels of NH_4^+, especially when grown under low light intensities, show a depletion of both organic acids and carbohydrate reserves (McKee, 1962). This depletion in carbohydrates is primarily in starch and not sugars. However, sugars are also affected, as Kirkby (1968) found that NH_4^+-grown plants had approximately 30% more reducing sugars than NO_3^- plants. Total carbohydrates were not reported. Maynard and Barker (1969) and Takashashi and co-workers published a series of reports (Matsumato et al., 1971b, 1968, 1969) in which they compared the changes in metabolites and enzymes of cucumber plants grown on noninjurious (1.5 m*M*) and injurious (15 m*M*) levels of NH_4^+. After 9 d of treatment, plants grown with high NH_4^+ had 1700% more glucose, 50% more sucrose, and 30% less starch. Although $^{14}CO_2$ was incorporated into glucose at the same rate, the incorporation of glucose into di- and trisaccharides was hindered by the high levels of NH_4^+. They concluded that the high levels of glucose of plants grown on high NH_4^+ was derived from photosynthate and not degradation of starch. High NH_4^+ in relation to low NH_4^+ treatments reduced $^{14}CO_2$ incorporation into starch by 50%, while incorporation into sugars was markedly increased by the high levels of NH_4^+. Injured plants contained 250% more uridine diphosphoglucose (UDPG), but less UDPX (where X was an unidentified sugar or sugar derivative) than the noninjured plants. They found that plants grown on 15 m*M* NH_4^+ for 5 d had lost all of their UDPG-dependent and 40% of their adenine diphosphoglucose-dependent starch synthetase activity. Of the numerous glycolytic and respiratory enzymes assayed, higher activities were found in the tissue from plants cultured on high levels of NH_4^+. Only ATPase and aldolase activities were higher in plants grown on low levels of NH_4^+. Although the authors implicated the inhibition of starch synthesis as the main cause of glucose accumulation, a reduction in aldolase activity could reduce the flow of glucose to pyruvate, which in turn could reduce the rate of operation of the Krebs cycle.

These findings are consistent with the concept that plants supplied with high levels of NH_4^+ have low levels of organic acids and carbohydrate reserves. Consequently, growth and yield of NH_4^+-grown plants should be more dependent upon factors that alter the continuous input of photosynthate.

The reason why NH_4^+, relative to NO_3^--grown plants, have low levels of organic acids is related to differences in metabolic reactions associated with the assimilation of NH_4^+ and NO_3^-. When absorbed NH_4^+ is accompanied by phosphate or a similar anion, a negative charge is generated upon NH_4^+ assimilation. The plant catabolizes an organic acid to achieve electrical neutrality and regulate pH. The converse is true for NO_3^- assimilation, as was previously detailed.

In the reduction of NO_3^- to NH_3, the N (not the three oxygens) accepts eight electrons, thereby serving as an oxidizing agent (Arnon, 1937). In the initial NO_3^- reduction step, the electrons are supplied by NADH, which may be generated by the oxidation of glyceraldehyde-3-phosphate (Klepper et al., 1971). The coupling of NO_3^- reduction to the oxidation of glyceraldehyde-3-phosphate—Eq. [6]—may enhance the flow of metabolites through the glycolytic system, which could enhance the level of precursors for organic acid production. This is analogous to alcohol fermentation where the reduction of acetaldehyde to ethanol provides the NAD for the oxidation of glyceraldehyde-3-phosphate, thereby increasing the flow of metabolites through the glycolytic system.

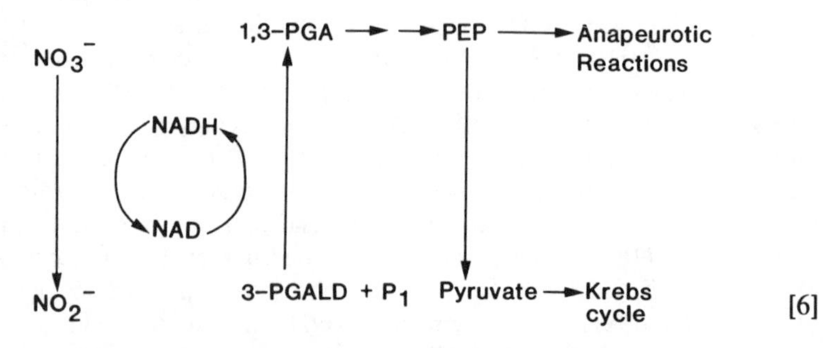

[6]

where 1,3-PGA = 3-phosphoglycerolphosphate and PEP = phosphoenolpyruvate.

In contrast, some workers (Wakiuchi et al., 1971) have reported that toxic levels of NH_4^+ depressed aldolase (one of the enzymes of glycolysis) activity. This could suppress the flow of metabolites through the glycolytic pathway, thereby directly or indirectly decreasing the production of organic acids.

When external NH_4^+ is readily available, NH_4^+ absorption can be quite rapid. The detoxification of NH_4^+ competes for α ketoacids of the Krebs cycle. Kirkby and Mengel (1967) found that roots of tomato plants grown on NH_4^+ had extremely low levels of citrate and malate. Root growth was also stunted. Extensive depletion of such organic acids could diminish the rate of operation of the Krebs cycle, and hence the availability of α ketoacids for amino acid formation (Lehninger, 1964). Anaplerotic reactions should also generate organic acids for metabolic use, and evidence that roots can fix CO_2 into organic acids has been observed (Hiatt & Hendricks, 1967; Jacobson & Ordin, 1954; Torii & Laties, 1966). Neither the mitochondria nor the anaplerotic reaction appear capable of maintaining high levels of organic acids in NH_4^+-grown plants.

Little applied work has been done on the effects of differential levels of internal organic acids on the growth and development of plants. de Wit et al. (1963) and Noggle (1966) have observed a positive correlation between organic acid content in the vegetation and yield.

D. N Compounds

In general, plants supplied with NH_4^+ have higher concentrations of total N and N components—especially free NH_4^+, amino acids, and amides—than do plants grown on comparable levels of NO_3^- (Coic et al., 1970; Hoff et al., 1974; Kirkby, 1968; Maynard & Barker, 1969; Naftel, 1931; Pill & Lambert, 1977; Weissman, 1964). Most work indicates that absorbed NH_4^+ is rapidly detoxified by conversion to amino acids and amides (Barash et al., 1974; McParland et al., 1976; Nightingale, 1948; Vickery et al., 1936; Weissman, 1972a; Yemm & Willis, 1956). Detoxification is considered to occur in the roots, because relatively low amounts of NH_4^+ have been detected in xylem exudates. The work of Vickery et al. (1936) suggests that the synthesis of amino acids and amides by the roots is a function of carbohydrate supply rather than any special enzymic mechanism.

Regardless of origin, NH_3 is rapidly converted (detoxified) into glutamine by glutamine synthetase (Eq. [5]). Glutamine synthetase is found in both chloroplasts and cytoplasm (Miflin, 1980). Previous to the discovery of glutamine synthetase, NH_3 was thought to be detoxified by the following reaction:

$$NH_3 + \underset{\alpha\ \text{Ketoglutarate}}{O{=}C(OH){-}C{=}O{-}HCH{-}HCH{-}HC(OH){=}O \ \ O^-} \xrightarrow[\text{Glutamate Dehydrogenase}]{NADH + H^+ \;\rightarrow\; NAD} \underset{\text{Glutamate}}{O{=}C(OH){-}C{-}NH_2{-}HCH{-}HCH{-}C(OH){=}O \ \ O^-} + H_2O \qquad [7]$$

Because glutamate dehydrogenase has a "low affinity" for NH_3, the reaction would not serve as an efficient detoxifying system. A current view is that under normal in situ conditions, the reaction (Eq. [7]) proceeds in the reverse direction and that the role of glutamate dehydrogenase is in the deamination of glutamate. However, a 1979 review (Givan, 1979) indicates that the following enzyme systems or enzymes are the ones primarily involved in detoxification of NH_4^+: (i) glutamine synthetase–glutamate synthetase; (ii) glutamate dehydrogenase; (iii) glutamate–oxaloacetate transferase; and (iv) aspargine synthetase. These enzymatic mechanisms are also involved in detoxifying NH_4^+ arising from NO_3^- reduction. Givan (1979) states that the amides play a key role in the detoxification process. Amides also serve as N-storage forms. They can be transported and utilized as needed. Recent work (McClure & Israel, 1978) has indicated that in certain legumes, ureides are produced from nodular fixed N_2. No work was found that implicates ureides as the product of NH_4^+ detoxification. With respect to detoxification and storage of N, ureides with their C/N composition ratio of 1.0 would be ideal.

E. Toxicity of NH_4^+ and NO_3^-

Evidence that NH_4^+ is toxic to plants has been recently summarized by Givan (1979). In contrast, within reasonable limits, absorbed NO_3^- is not toxic, and the stored NO_3^- can be assimilated as required.

While NH_4^+ per se may be toxic to plant tissue (Barker et al., 1966; Naftel, 1931), many workers consider NH_3(aq) to be the toxic component of aqueous ammonical N (Allred & Ohlrogge, 1974; Bennett, 1971; Blanchar, 1967; Megie et al., 1967; Naftel, 1931; Vines & Wedding, 1960; Warren, 1962). Vines and Wedding (1960) found that equivalent concentrations of NH_3 gas and undissociated NH_3(aq) (derived from ammonium salts) were equally effective in inhibiting barley root respiration. When roots were treated at pH 7.0 with $(NH_4)_2SO_4$ concentrations high enough to supply 3 m*M* NH_3(aq), respiration decreased to 40% of control during the first 2 h, and remained at that level for an additional 2 h. Calculations made with their data indicated that the roots were treated with 280 m*M* $(NH_4)_2SO_4$. Had these studies been conducted at pH 6.0, 2.8*M* $(NH_4)_2SO_4$ would have been required to provide 3 m*M* NH_3(aq). Matsumato et al. (1968) found cucumber leaves from plants grown on toxic (15 m*M*) levels of NH_4^+ salts had 3.8 m*M* NH_4^+. At pH 6.0, the concentration of NH_3(aq) was estimated to be $2 \times 10^{-6}M$. This is 0.0007 the concentration of NH_3(aq) necessary to inhibit barley root respiration by 60% (Vines & Wedding, 1960). These observations indicate either that NH_3(aq) is toxic at extremely low concentrations or that NH_4^+ also contributes to the toxicity.

In assimilation reactions (Eq. [5] and [7]), NH_3 is the form utilized. Because the N of NH_3 has an unshared pair of electrons, the N is a nucleophillic atom, and can react with an electrophillic C atom of oxo or carboxy groups, thereby forming amino or amido-N via Shiff's base formation. The nucleophillic nature of NH_3 permits this compound to react with other electrophillic atoms with or without an enzyme catalyst. The uncharged NH_3

can readily penetrate membranes, thus raising the concentration of NH_3 and NH_4^+ around enzyme systems that may normally be protected from NH_4^+ by membranes. Vines and Wedding (1960) reported that NH_3 (supplied as ammonium salts) caused slight (10%) damage to membranes as measured by anthocyanin leakage from beet (*Beta vulgaris* L.) discs.

Vines and Wedding (1960) suggested that the probable site of NH_3 toxicity was NADH–NAD-dependent mitochondrial reactions. Additional support for this concept has been provided by Katsunuma et al. (1965). They found that NH_4^+ caused a rapid (5 min) destruction of NAD(P)H, but not the oxidized form. They isolated two enzymes, both activated by NH_4^+, that were responsible for cofactor destruction. The following scheme was proposed:

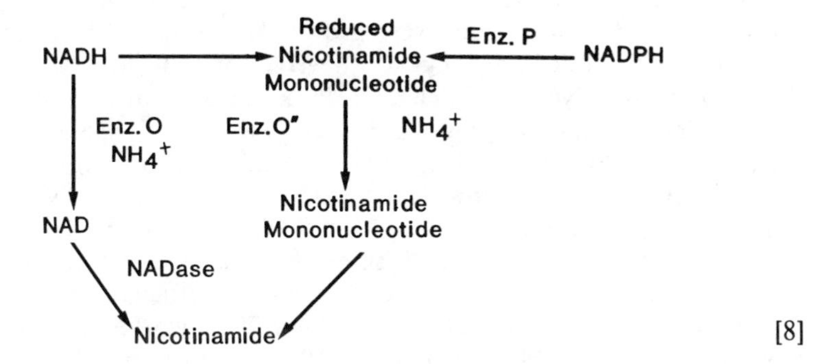

[8]

where enzyme O = an oxidase of NADH, O′ = an oxidase of reduced nicotinamide mononucleotide, P = pyrophosphate NAD(P)H splitting, and NADase = nucleotide diphosphate nuclease.

The authors state that enzyme O and O′ were activated by NH_4^+. However, at the concentration of ammonium salt applied and the assay pH used, the amount of NH_3 present would be approximately $5 \times 10^{-4} M$.

Although Weissman (1972b) found that NO_3^--grown sunflower and soybean plants had more (10–60%) NAD plus NADH, and less nicotinamide adenine dinucleotide phosphate (NADP) plus NADPH than NH_3^+-grown plants, he found no marked depletion of reduced pyridine nucleotides, as might be expected from the work of Katsunuma et al. (1965). However, with the culture techniques used (Weissman, 1972), they reported no adverse effects of NH_4^+ on plant growth.

The destruction of pyridine nucleotides would be catastrophic and adequate to explain the "scalded" appearance of leaves given toxic levels of NH_4^+ (or NH_3). This "acute" type of toxicity would be in contrast to "chronic" toxicity. Chronic toxicity would be more directly related to the gradual depletion of starch and organic acids, the accumulation of glucose, and diminished mitochondrial respiration arising from the depletion of organic acids by NH_4^+ detoxification (Lehninger, 1964). In practice, it would be difficult to separate the two types of toxicity, as toxicity would be a function of the rate of NH_4^+ accumulation in a given tissue and the ability of that tissue to convert NH_3 to innocuous forms. The detoxification process would be a function of enviornmental factors that affect photosynthate supply.

The ammonium ion uncouples photophosphorylation (Avron, 1960) in vitro, and at 6 m*M* it inhibits ferricyanide reduction in the Hill reaction (Barker et al., 1966; Hind & Wittingham, 1963). In vitro, fixation of CO_2 is also impaired by NH_4^+ in the $10^{-3}M$ range (Trebst et al., 1960). The photosynthetic O_2 evolution leaves from tomato plants supplied with toxic levels of NH_4^+ (15 m*M*) were 70 and 10% that of comparable leaves from NO_3^-- (15 m*M*) grown plants after 2 and 19 d of treatment, respectively (Purtich & Barker, 1967). The NH_4^+ concentrations in the leaves from the NH_4^+-grown plants were 0.13 and 0.27 mg N g^{-1} fresh weight after 4 and 19 d of treatment, respectively. Assuming 80% moisture and pH 6.0, these values equate to 11.6 and 24.0 m*M* NH_4^+ and 6.0×10^{-6} and $1.3 \times 10^{-5}M$ NH_3, respectively. The leaves from the NO_3^--grown plants sampled at the same time contained 0.05 and 0.03 mg N g^{-1} fresh weight of NH_4^+ after 4 and 19 d of treatment, respectively. Purtich and Barker (1967) also found that as NH_4^+ toxicity symptoms progressed, there were gross changes in chloroplast morphology, plastid degradation, and loss of chlorophyll. Although cucumber plants grown on 15 m*M* NH_4^+ also showed NH_4^+ toxicity symptoms, the concentration of NH_4^+ in the leaves was reported to be 0.38 m*M* (Matsumato et al., 1968). With similar cucumber leaves, the rate of CO_2 fixation into glucose was not different from that of leaves supplied with 1.5 m*M* NH_4^+ (Matsumato et al., 1969). These workers failed to detect any decrease in ATP (Matsumato et al., 1968) or any uncoupling of respiration (Matsumato et al., 1971a) from high levels of NH_4^+. The different results obtained in various experiments (Matsumato et al., 1971; Purtich & Barker, 1967) illustrates the differences in NH_4^+ tolerance among species and raises the question as to difference in tolerance within a species.

Gauch (1972) indicates that NO_3^- toxicity is rare, but occasionally results from excessive application of fertilizer.

F. Energy Requirements

Warburg and Negelein (1920) computed that the oxidation of NO_3^- to NH_3 by a unicellular green algae (*Chlorella pyrenoidsa* L.) required the input of 162 000 cal/mole. This represents 23.4% of the energy available from the combustion of a mole of glucose. If energy (photosynthate) is the limiting factor in plant productivity, then NH_4^+-grown plants should always be more productive than NO_3^--grown plants. However, this is contingent upon the coordination of NH_4^+ uptake with environment-induced variations in photosynthetic supply, so that toxic levels of NH_4^+ are never accumulated throughout the plant's life cycle.

While these energy considerations are irrefutable, the mechanism of NO_3^- reduction and assimilation that occurs in illuminated green leaves (Eq. [5]) eliminates the need for extensive carbohydrate catabolism.

The NADH for NO_3^- reduction to NO_2^- would be derived via carbohydrate oxidation pathways. Because ferredoxin (Fd) is reduced by light energy trapped by the chloroplast, 75% of the 162 000 cal/mole required for NO_3^- reduction would not be channeled through carbohydrate metabolism. Although NO_3^- reduction in the chloroplasts competes with C reduction for light energy (via Fd), Radner and Kok (1977) state that available light energy exceeds the energy needs of C reduction. Under certain conditions, and with C_4 species, this may not be completely valid; however, such conditions would be even more unfavorable for NH_4^+-grown plants. There are some indications that photosynthate supply may not be as limiting to crop yields as previously believed (Below et al., 1981).

The major route of NH_3 assimilation in the chloroplast (Miflin & Lea, 1976) is shown in Eq. [5]. The ATP and Fd_{red} required by glutamine synthesis and glutamate synthesis would be produced from light energy, and therefore bypass carbohydrate metabolism.

For plants grown on NH_4^+, it has been shown that most of it is assimilated (detoxified) in the root. Thus, ATP and equivalent energy (possibly NADPH as a substitute for Fd_{red}) required for glutamate formation or NAD(P)H required in the glutamate dehydrogenase reaction (Eq. [7]) would be derived from carbohydrate catabolism.

From these observations, it can be concluded that reduction and assimilation of NO_3^- by the leaf requires no more carbohydrate catabolism than assimilation of NH_3 by the root. In NH_4^+-grown plants, photosynthate for assimilation energy and carbohydrate skeletons for amino acids and amides must be transported to the root. Portions of the N-compounds must then be transported to the leaf. The energy cost for transport and the efficiency of distribution of photosynthate and N compounds to the various organs is not known. Reduction of NO_3^- in the root is dependent upon carbohydrate metabolism and would require 162 000 cal/mole to produce NH_3. In terms of carbohydrate catabolism, the energy cost of reducing NO_3^- to NH_3 is a function of the proportion of NO_3^- reduced in the root relative to that reduced in the leaves.

G. Plant Disease

The review of Huber andWatson (1974) shows that the form of N influences the severity of disease infestation of many plant species. In view of the effects of form of N on both external (root medium) and internal pH, enzyme activities, and level of metabolites in the plant, the variability in susceptibility to various diseases is not surprising. The effects of form of N on disease susceptibility of plants is complex and influenced by a multitude of factors, especially in the field where it is most difficult to maintain the form, concentration, and availability of both NO_3^- and NH_4^+.

Tabular data (Huber & Watson, 1974) show that NO_3^- treatment decreased the severity of corn root rot induced by the pathogen *Aphanomyces*, and increased the severity when rot was induced by *Pythium*. The severity of stalk rot induced by the pathogens *Fusarium* and *Diplodia* were decreased and increased, respectively, when the plant was supplied with NO_3^-. The severity of Northern leaf blight (*Helmin thosporium*) of corn was decreased when the plants were supplied with NO_3^-. For root and stalk rot and leaf blight, the effect of NH_4^+ on the severity of the disease was the converse of the NO_3^- effect.

Unfortunately, no generalizations of the effects of form of N on physiological parameters and susceptibility to disease are evident.

H. Conclusion

Relative to NO_3^- nutrition, NH_4^+ nutrition requires less energy (162 000 cal/mole) only in nonchlorophyllous tissue or organs. However, in chlorophyllous organs (primarily leaves), NO_3^- assimilation costs no more in terms of carbohydrate reserves. In contrast, four advantages of NO_3^- nutrition over NH_4^+ nutrition can be listed. First, NO_3^- per se is not toxic and can be stored for future use in various organs (e.g., stalk of maize) and cell vacuoles without apparent injury. Second, NO_3^- assimilation is coordinated and regulated by C metabolism. Reduction of NO_3^- requires NADH, which is produced by oxidation of glyceraldehyde-3-phosphate (Klepper et al., 1971) or alternatively by the oxidation of malate. If malate is produced by mitochondrial activity, as proposed (Reed & Hageman, 1977; Sawhaney et al., 1978a; 1978b), this system would also ensure an adequate supply of carbohydrate skeletons for amino acid production. Third, the initiation of NO_3^- assimilation is catalyzed by the enzyme nitrate reductase. This enzyme is more sensitive to heat and drought stresses than most plant enzymes, including those of the photosynthetic apparatus (Hageman, 1979). Fourth, in leaves, NO_3^- is assimilated more slowly in the dark than in the light, and in some species, NO_3^- is not assimilated in dark under certain conditions (Canvin & Atkins, 1974). These various regulatory mechanisms ensure that NH_4^+ will not be produced from NO_3^- when the plant is deficient in carbohydrate skeletons or has minimal requirements for N.

It would appear that "there is nothing new under the sun." The conclusion reached by Arnon in 1937 that either NO_3^- or NH_4^+ can serve as an adequate source of N for plant growth and productivity, but, in general, NO_3^- salts are considered the "safer" fertilizer, appears equally valid today.

REFERENCES

Allred, S. E., and A. J. Ohlrogge. 1974. Principles of nutrient uptake from fertilizer bands: III. Germination and emergence of corn as affected by ammonia and ammonium phosphate. Agron. J. 56:309–313.

Arnon, D. I. 1937. Ammonium and nitrate nitrogen nutrition of barley at different seasons in relation to hydrogen-ion concentration, manganese, copper, and oxygen supply. Soil Sci. 44:91–120.

Avron, M. 1960. Photophosphorylation by swiss chard chloroplasts. Biochim. Biophys. Acta 40:257–272.

Barash, I., T. Sadon, and H. Mor. 1974. Relationship of glutamate dehydrogenase levels of free amino acids, amides, and ammonia in excised oat leaves. Plant Cell Physiol. 15: 563–566.

Barker, A. V., and D. A. Manyard. 1972. Cation and nitrate accumulation in pea and cucumber plants as influenced by N nutrition. J. Am. Soc. Hortic. Sci. 97:27–30.

Barker, A. V., R. J. Volk, and W. A. Jackson. 1966. Growth and N distribution patterns in bean plants (*Phaseolus vulgaris* L.) subjected to ammonium nutrition: I. Effects of carbonates and acidity control. Soil Sci. Soc. Am. Proc. 30:228–232.

Below, F. E., L. E. Christensen, A. J. Reed, and R. H. Hageman. 1981. Availability of reduced N and carbohydrates for ear development of maize (*Zea mays* L.). Plant Physiol. 68:1186–1190.

Bennett, A. C. 1971. Toxic effects of aqueous ammonia, copper, zinc, lead, boron, and manganese on root growth. p. 669–683. *In* E. W. Carson (ed.) The plant root and its environment. Univ. of Virginia Press, Charlottesville.

Blair, G. J., M. H. Miller, and W. A. Mitchell. 1970. Nitrate and ammonium as sources of N for corn and their influence on uptake of other ions. Agron. J. 62:530–532.

Blanchar, R. W. 1967. Determination of the partial pressure of NH_3 in soil air. Soil Sci. Soc. Am. Proc. 31:791–795.

Canvin, D. T., and C. A. Atkins. 1974. Nitrate, nitrite, and ammonia assimilation by leaves: effect of light, carbon dioxide, and oxygen. Planta 116:207–224.

Coic, Y., C. Lesaint, and F. LeRoux. 1962. Comparison du mais de la tomato quant al effet de la nature nitrique on ammoniacale la nutrition azote su absorption et let metabolisme des anions-cations. C. R. Acad. Sci. 254:549–551.

Coic, Y., C. Lesaint, M. T. Piollat, and M. Lelandis. 1970. Effect of nitrate N nutrition on changes in concentrations of cations, K, Ca, Mg in the aerial parts of tobacco plants. Potash Rev. 3:1–8.

Cox, W. J., and H. M. Reisenauer. 1971. Critical external nitrate or ammonium concentrations for plant growth. Agron. Abstr. p. 87.

de Wit, C. T., W. Dijkshoorn, and J. C. Noggle. 1963. Ionic balance and growth of plants. Versl. Landbouwkd. Onderz. 69:15–21.

Dijkshoorn, W. 1962. Metabolic regulation of the alkaline effects of nitrate utilization in plants. Nature 194:165–167.

Eriksson, T. 1965. Studies on the growth requirements and growth measurements of cell cultures of *Haplopappus gracilis.* Physiol. Plant. 18:976–993.

Filner, P. 1966. Regulation of nitrate reductase in cultured tobacco cells. Biochim. Biophys. Acta 44:791–799.

Gamborg, O. L., and J. P. Shyluk. 1970. The culture of plant cells with ammonium salts as the sole sources of nitrogen. Plant Physiol. 45:598–600.

Gamborg, O. L., R. A. Miller, and K. Ojima. 1968. Nutrient requirements of suspension cultures of soybean root cells. Exp. Cell Res. 50:151–158.

Gauch, H. G. 1972. Inorganic plant nutrition. p. 378–394. Drawden, Hutchinson, and Ross, Inc., Stroudsburg, Pa.

Givan, C. V. 1979. Metabolic detoxification of ammonia in tissues of higher plants. Phytochemistry 18:375–381.

Greidanus, T., L. A. Peterson, L. E. Schrader, and M. N. Dana. 1972. Essentiallity of ammonia for cranberry nutrition. J. Am. Soc. Hortic. Sci. 97:272–277.

Hageman, R. H. 1979. Integration of N assimilation in relation to yield. p. 591–611. *In* E. J. Hewitt and C. V. Cutting (ed.) Nitrogen assimilation of plants. Academic Press, New York.

Hageman, R. H., D. Flesher, J. J. Wabol, and D. H. Storck. 1961. An improved nutrient culture technique for growing corn under greenhouse conditions. Agron. J. 53:175–180.

Harper, J. E., and J. C. Nicholas. 1976. Control of nutrient solution pH with ion exchange: effect on soybean nodulation. Physiol. Plant. 38:24–28.

Haynes, R. J., and K. M. Goh. 1978. Ammonium and nitrate nutrition of plants. Biol. Rev. 53:465–510.

Hewitt, E. J. 1966. Sand and water culture methods used in the study of plant nutrition. 2nd Ed. Commonwealth Agric. Bureau, Farnham Royal Bucks, England.

Hewitt, E. J. 1970. Physiological and biochemical factors which control the assimilation of inorganic N supplies by plants. p. 68–82. *In* E. A. Kirkby (ed.) Nitrogen nutrition of plants. Univ. of Leeds Press, England.

Hewitt, E. J., and T. A. Smith. 1975. Plant mineral nutrition. p. 176–222. English Univ. Press, London.

Hiatt, A. J., and S. B. Hendricks. 1967. The role of CO_2 fixation in accumulation of ions by barley roots. Z. Pflanzenphysiol. 56:1528–1532.

Hind, G., and C. P. Wittingham. 1963. Reduction of ferricyanide by chloroplasts in the presence of N bases. Biochim. Biophys. Acta 75:194–202.

Hoff, J. E., G. E. Wilcox, and C. M. Jones. 1974. The effect of nitrate and ammonium N on the free amino acid composition of tomato plants and tomato fruits. J. Am. Soc. Hortic. Sci. 99:27–30.

Huber, D. M., and R. D. Watson. 1974. Nitrogen form and plant disease. Annu. Rev. Phytopathol. 12:139–165.

Jacobson, L., and L. Ordin. 1954. Organic acid metabolism and ion absorption in roots. Plant Physiol. 29:70–75.

Jung, P. E., L. A. Peterson, and L. E. Schrader. 1972. Response of irrigated corn to time, rate, and source of applied N on sandy soils. Agron. J. 64:618–670.

Katsunuma, N., M. Okada, and Y. Nishi. 1965. Regulation of urea cycle and TCA cycle by ammonia. Adv. Enzyme Regul. 4:317–335.

Kirkby, E. A. 1968. Influence of ammonium and nitrate nutrition on the cation-anion balance and N and carbohydrate metabolism of white mustard plants grown in dilute nutrient solutions. Soil Sci. 105:133–141.

Kirkby, E. A., and K. Mengel. 1967. Ionic balance in different tissues of the tomato plant in relation to nitrate, urea, or ammonium nutrition. Plant Physiol. 42:6–14.

Klepper, L. E., D. Flesher, and R. H. Hageman. 1971. Generation of reduced nicotinamide adenine dinucleotide for nitrate reduction in green leaves. Plant Physiol. 48:580–590.

Lehninger, A. L. 1964. Control and integration of the citric acid cycle. p. 132–156. *In* A. L. Lehninger (ed.) The mitochondrion. W. A. Benjamin, Inc., New York.

Lycklama, J. C. 1963. The absorption of ammonium and nitrate by perennial rye-grass. Acta Bot. Nerrl. 12:361–423.

Martin, J. W., W. H. Leonard, and D. L. Stamp. 1976. Principles of field crop production. p. 547–548. Macmillan Publ. Co., New York.

Matsumato, H., N. Wakiuchi, and E. Takahaski. 1968. Changes in sugar levels in cucumber leaves during ammonium toxicity. Physiol. Plant. 21:1210–1216.

Matsumato, H., N. Wakiuchi, and E. Takahashi. 1969. The suppression of starch synthesis and the accumulation of uridine diphosphoglucose in cucumber leaves due to ammonium toxicity. Physiol. Plant. 22:537–545.

Matsumato, H., N. Wakiuchi, and E. Takahashi. 1971a. Changes in some mitochondrial enzyme activities of cucumber leaves during ammonium toxicity. Physiol. Plant. 25:353–357.

Matsumato, H., N. Wakiuchi, and E. Takahashi. 1971b. Changes in starch synthetase activity of cucumber leaves during ammonium toxicity. Physiol. Plant. 24:102–110.

Maynard, P. A., and A. V. Barker. 1969. Studies on the tolerance of plants to ammonium nutrition. J. Am. Soc. Hortic. Sci. 94:235–239.

McClure, P. R., and D. W. Israel. 1978. Transport of N in the xylem of soybeans. Agron. Abstr. p. 81.

McKee, H. S. 1962. Nitrogen metabolism in plants. p. 1–38. Clarendon Press, Oxford, England.

McParland, R. H., J. G. Guevara, R. R. Beeker, and H. J. Evans. 1976. The purification and properties of glutamine synthetase from the cytosol of soya-bean root nodules. Biochem. J. 153:597–601.

Megie, C. A., R. W. Pearson, and A. E. Hiltbold. 1967. Toxicity of decomposing crop residues to cotton germination and seedling growth. Agron. J. 59:197–199.

Michael, G., P. Martin, and I. Owassia. 1970. The uptake of ammonium and nitrate from labelled ammonium nitrate in relation to the carbohydrate supply of the root. p. 92–110. *In* E. A. Kirkby (ed.) Nitrogen nutrition of plants. Univ. of Leeds Press, England.

Miflin, B. J. (ed.). 1980. Biochemistry of plants: a comprehensive treatise. Vol. 5, Amino acids and derivatives. Academic Press, New York.

Miflin, B. J., and P. J. Lea. 1976. The pathway of N assimilation in plants. Phytochemistry 15:873–885.

Miller, E. C. 1938. Plant physiology. p. 642–725. McGraw-Hill Book Co., New York.

Munn, D. A., and W. A. Jackson. 1978. Nitrate and ammonium uptake by rooted cuttings of sweet potato. Agron. J. 70:312–316.

Naftel, J. A. 1931. The absorption of ammonium and nitrate N by various plants at different stages of growth. J. Am. Soc. Agron. 23:142–158.

Nightingale, G. T. 1948. The N nutrition of green plants. Bot. Rev. 14:185–198.

Noggle, J. C. 1966. Ionic balance and growth of 16 plant species. Soil Sci. Soc. Am. Proc. 30: 763–766.

Pate, J. S. 1973. Uptake, assimilation, and transport of N compounds by plants. Soil Biol. Biochem. 5:109–119.

Pate, J. S., and W. Wallace. 1964. Movement of assimilated nitrogen from the root system of the field pea (*Pisum arvense* L.). Ann. Bot. London 28:83–99.

Pill, W. G., and V. N. Lambert. 1977. Effects of NH_4 and NO_3 nutrition with and without pH adjustments on tomato growth, ion composition, and water relations. J. Am. Soc. Hortic. Sci. 102:78–81.

Purtich, G. S., and A. V. Barker. 1967. Structure and function of tomato leaf chloroplasts during ammonium toxicity. Plant Physiol. 42:1229–1238.

Radner, R., and B. Kok. 1977. Photosynthesis: limited yields, unlimited dreams. BioScience 27:599–605.

Raghuveer, P. 1977. Characteristics of nitrate uptake from nutrient solution and root nitrate reductase activity among corn genotypes. Ph.D. Thesis. Univ. of Illinois, Urbana-Champaign.

Reed, A. J., and R. H. Hageman. 1977. Relationship between nitrate reduction and dark respiration in wheat (*Triticum aestivum* L.). Plant Physiol. 59S:127.

Sawhaney, S. K., M. S. Naik, and D. J. D. Nicholas. 1978a. Regulation of NADH supply for nitrate reduction in green plants via photosynthesis and mitochondrial respiration. Biochem. Biophys. Res. Commun. 81:1209–1216.

Sawhaney, S. K., M. S. Naik, and D. J. D. Nicholas. 1978b. Regulation of nitrate reduction, light, ATP, mitochondrial respiration in leaves. Nature 272:647–648.

Schrader, L. E., D. Domska, P. E. Jung, Jr., and L. A. Peterson. 1972. Uptake and assimilation of ammonium-N and nitrate-N and their influence on the growth of corn (*Zea mays* L.). Agron. J. 64:690–695.

Stewart, G. E., J. A. Lee, T. O. Orebamjo, and D. C. Havill. 1974. Ecological aspects of nitrogen metabolism. p. 41–47. *In* R. L. Bielski et al. (ed.) Mechanisms of regulation of plant growth. Bull. no. 12, Royal Soc. of New Zealand, Wellington.

Torii, K., and G. G. Laties. 1966. Dual mechanism of ion uptake in relation to vacuolization in corn roots. Plant Physiol. 41:515–518.

Trebst, A., V. M. Losada, and D. I. Arnon. 1960. XII. Inhibitors of CO_2 assimilation in a reconstituted chloroplast system. J. Biol. Chem. 235:840–844.

Vickery, H. B., G. W. Pucher, and H. E. Clark. 1936. Glutamine metabolism of the beet. Plant Physiol. 11:413–420.

Vines, H. M., and R. T. Wedding. 1960. Some effects of ammonia on plant metabolism and a possible mechanism for ammonia toxicity. Plant Physiol. 35:820–825.

Wadleigh, C. H., and J. W. Shive. 1939. Base content of corn plants as influenced by pH of substrate and form of N supply. Soil Sci. 47:273–283.

Wakiuchi, N., H. Matsumato, and E. Takahashi. 1971. Changes in some enzyme activities of cucumber leaves during ammonium toxicity. Physiol. Plant. 24:248–253.

Wallace, A., and R. T. Mueller. 1957. Ammonium and nitrate absorption from sand by rough lemon seedlings. Proc. Am. Soc. Hortic. Sci. 69:183–188.

Warburg, O., and E. Negelein. 1920. Uber the reduktion des solpetersoure in gruen zellen. Biochem. Z. 110:66–78.

Warren, W. S. 1962. Ammonia toxicity and pH. Nature 195:47–49.

Weissman, G. S. 1964. Effect of ammonium and nitrate nutrition on protein level and exudate composition. Plant Physiol. 39:947–952.

Weissman, G. S. 1972a. Influence of ammonia and nitrate nutrition on enzymatic activity in soybean and sunflower. Plant Physiol. 49:138–141.

Weissman, G. S. 1972b. Influence of ammonium and nitrate nutrition on the pyridine and adenine nucleotides of soybeans and sunflower. Plant Physiol. 49:142–145.

Williams, D. E., and J. Vlamis. 1962. Differential cation and anion absorption as affected by climate. Plant Physiol. 37:198–202.

Yatazawa, M., and K. Furuhoshi. 1968. Nitrogen sources for the growth of rice callus tissue. Soil Sci. Plant Nutr. 14:73–79.

Yemm, E. W., and A. J. Willis. 1956. The respiration of barley plants. IX. The metabolism of roots during the assimilation of N. New Phytol. 55:229–237.

5

Stanley A. Barber
Purdue University
West Lafayette, Indiana

Nutrient Balance and Nitrogen Use

Producing high field crop yields requires both sufficient amounts and balance of the nutrients required for plant growth. When one nutrient is present in large amounts, it may depress uptake of some other nutrients and reduce yield. If nutrients were absorbed by the plant in the same ratio as found in the soil, it would be difficult to adjust the amounts of each nutrient to the appropriate level, especially since much larger amounts of nutrients such as N and K are needed by the plant. Fortunately, nutrient uptake by plant roots is selective, so that plants can tolerate a range in the relative levels of nutrients in the soil and still maintain an appropriate balance of the nutrients absorbed and translocated to the shoot. Also, soil acts as a buffer for added nutrients so that when larger than necessary amounts of nutrients are added, some may be adsorbed on the soil surfaces and only a portion remains in soil solution at the root surface. Then, the effect on uptake is much less than if no soil adsorption occurred. Balance of nutrients also needs to be maintained throughout the growth cycle of the plant. If quantities of a nutrient such as N are less than needed for the growth cycle, sufficient N may be available during early growth, but not during later stages of growth. Hence, nutrient balance would be satisfactory during early growth but not later on.

Nutrient balance as it relates to N use involves consideration of the principles involved in N uptake by plant roots growing in soil. Nitrogen may be present in soil solution either as NH_4^+ (a cation) or NO_3^- (an anion). Their root absorption characteristics and the adsorption and transport of these ions in the soil will affect nutrient balance considerations. The relation of NH_4^+ and NO_3^- uptake to other nutrient uptake has usually been investigated in solution culture where levels of the ions can be controlled. Results from solution culture research will be used in the discussion of these nutrient interactions. In soil systems where crops ordinarily are produced

Nitrogen in Crop Production.

transport of NH_4^+, NO_3^-, and the other nutrients to the root before they can be absorbed becomes an important factor in discussing nutrient balance. Hence, following discussions of solution data, the balance in soil systems will be discussed.

I. NUTRIENT BALANCE IN SOLUTION CULTURE

Sufficiency of a nutrient in solution for obtaining maximum growth depends on the level of the nutrient in solution. Nutrient uptake rate per unit of root surface, influx, depends on the level maintained at the root surface. The relation between P concentration in solution and P influx for 18-day-old corn (*Zea mays* L.) is shown in Fig. 1. The curvillinear relation can be described by the Michaelis-Menten equation for enzyme kinetics. This equation is

$$I = \frac{I_{max} C}{K_m + C}, \quad [1]$$

where I is influx, I_{max} is maximum influx with high values of C, C is concentration of the nutrient in solution (mol/L), and K_m is the Michaelis-Menten constant, which is the concentration where $I = 1/2\, I_{max}$.

When plants are allowed to deplete a solution, the concentration is reduced to a minimum value below which no net influx, I_n, occurs. Net influx ceases before C reaches zero. The C where $I_n = 0$ is C_{min}. The relation in Eq. [1] is modified to include C_{min} in Eq. [2], so that the experimental data are accurately described by the equation

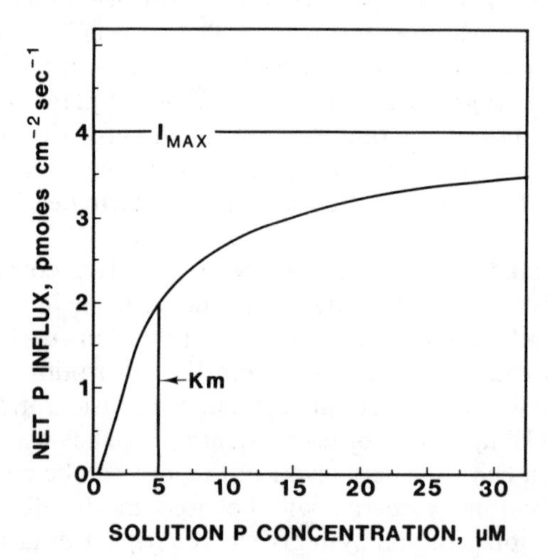

Fig. 1—Relation between P influx by plant roots and P concentration in stirred solution (Barber, 1976).

$$I_n = \frac{I_{max}(C\text{-}C_{min})}{K_m + C\text{-}C_{min}} \cdot \qquad [2]$$

For N uptake by 18- to 58-day-old corn grown in the greenhouse, C_{min} values were 3 to 9 μM, K_m values were 10 to 20 μM, and I_{max} values were 0.05 to 1.0 pmoles/cm of root length per second (Edwards & Barber, 1976). Results for NO_3^- and NH_4^+ were similar.

As nutrient levels are increased beyond the level needed for I_{max}, plant growth may decrease, because C becomes so high that the high uptake of one nutrient depresses uptake of other nutrients, and the nutrient balance required for maximum yield is no longer present. In addition, maximum yield can be obtained without maximum I_n. As C is increased, the nutrient level in the plant increases because I_n increases. This nutrient level increases yield; however, yield of the plant usually reaches a maximum value before nutrient uptake reaches it maximum. Hence, for nutrient balance and maximum yield, a nutrient concentration outside the root that is somewhat less than that giving maximum uptake of the particular nutrient but still high enough to give maximum growth is the desired level.

The level of other nutrients in solution will often affect the influx of the one investigated. Increasing N level in solution will increase P influx. Miller (1974) reviewed the research on the effect of NH_4^+ and NO_3^- in solution on P influx. Presence of either form of N increased P absorption as well as translocation of P within the plant. Hence, presence of N increased the efficiency of P use.

II. FORM OF N IN SOLUTION

Form of N, whether NH_4^+ or NO_3^-, affects uptake of Ca, Mg, and K so that level of Ca, Mg, and K required in solution to get the needed uptake rate may be influenced by the form of N (Table 1). Barber (1974) grew ryegrass (*Lolium perenne*) in solution culture using NH_4^+ in one treatment and NO_3^- in a second treatment, with the other nutrients at the same level in

Table 1—Effect of NH_4^+ and NO_3^- form of N on nutrient uptake from soil by corn seedlings.†

	Plant composition of:				
	N	P	K	Ca	Mg
			%		
		Princeton sand			
NO_3^-	2.4	0.55	4.99	0.71	0.24
NH_4^+	2.8	0.70	3.66	0.69	0.18
		Fincastle silt loam			
NO_3^-	2.5	0.34	3.48	0.62	0.44
NH_4^+	2.2	0.41	3.13	0.47	0.30

† Claassen and Wilcox, 1974.

Table 2—Effect of form of N on P, Ca, Mg, and K uptake from solution culture by ryegrass.†

	NH_4^+	NO_3^-	$H_2PO_4^-$	K^+	Ca^{2+}	Mg^{2+}
	meq/pot					
NO_3^-		4.72	0.34	1.87	0.46	0.52
NH_4^+	5.59		0.37	1.76	0.20	0.30

† Barber, 1974.

both treatments. Solution pH was adjusted daily to 6.5. With NH_4^+ nutrition the solution gradually becomes more acid; with NO_3^- nutrition the solution gradually becomes basic. The growth of the ryegrass was not affected by source of N; however, the nutrient composition was affected. Table 2 shows that P was higher and Ca, Mg, and K were lower with NH_4 than with NO_3. Increasing NH_4^+ usually depresses Ca and Mg uptake. Hence, it is important to have adequate levels of Ca and Mg present when NH_4 is the only or major source of N present. Decreasing solution pH also tends to depress Ca and Mg uptake.

Potassium will increase NO_3^- uptake. In a split-root experiment where part of the root system was in nutrient solution containing K and the remainder was in nutrient solution lacking K, Claassen and Barber (1977) obtained the results shown in Table 3. Where K was present, the NO_3^- uptake rate was from 1.6 to 2.2 times that of the roots that were in solution lacking K. The increased NO_3^- uptake where K was present indicates that K plays some role in the uptake and/or translocation of NO_3^- within the plant. The level of K in the plant influences the assimilation of NO_3^- into protein (Koch & Mengel, 1974), and this in turn may influence the uptake rate of NO_3^-. It may also be that these ions cooperate in each other's transport to the shoot. In field experiments, S. A. Barber (unpublished data) has shown increasing N fertilization increases K uptake by corn plants. Increasing N from 62 to 212 kg/ha caused an increase in ear leaf K concentration from 1.53 to 1.94% K. This occurred even though root length with 212 kg N/ha was much less than with 62 kg N/ha.

Table 3—Influence of presence of K on uptake of NO_3^- in a split-root experiment.†

Proportion of root	Presence of K	Average NO_3^- uptake rate μmoles $g_{F.W.}^{-1}$ $sec^{-1} \times 10^4$‡
%		
50	+	32.3
50	+	35.2
50	+	39.6
50	–	25.2
25	+	50.0
75	–	25.5
15	+	39.2
85	–	18.4

† Claassen and Barber, 1977.
‡ $g_{F.W.}$ = grams fresh weight.

III. FORM OF N IN SOIL

Nitrogen in the soil may be either NH_4^+ or NO_3^-. Usually nitrification converts most of the NH_4^+ to NO_3^- and uptake is principally as NO_3^-. However, nitrification inhibitors that minimize the conversion of NH_4^+ are now in use. Also, in acid soils or under cool conditions, fertilizer N applied as NH_4^+ may remain as NH_4^+ for a considerable length of time.

When all of the N is absorbed as NH_4^+, more cations are absorbed than anions, and H^+ is released by the root to balance the charge. When all the N is absorbed as NO_3^-, more anions are absorbed than cations, and OH^- or HCO_3^- is released to balance the charge. Hence, the pH of the root environment will change depending on the form of N absorbed. Barber (1974) found that when NH_4^+ was the source of N for ryegrass growing in solution culture, cation uptake was 7.88 meq/pot, while anion uptake was only 1.79 meq/pot—there was a difference of 7.09 meq/pot, which was balanced by H^+ release into solution. When NO_3^- was the form of N, cation uptake was 2.88 meq/pot and anion uptake was 5.56 meq/pot; hence, 2.68 meq/pot of OH^- was released and the pH increased.

When plants are growing in soil, the release of H^+ or OH^- changes the pH of the rhizosphere soil. This change in pH can change the availability of nutrients as well as elements such as Al, which may be detrimental to crop growth. Reducing soil pH increases Al solubility. Riley and Barber (1971) grew soybeans [*Glycine max* (L.) Merr.] in soil fertilized with either NH_4^+ or NO_3^- and found that the change in soil pH near the root caused by NH_4^+ or NO_3^- uptake affected P uptake. There was a correlation (R^2 = 0.94) between P concentration of the shoot and pH of the rhizocylinder (root plus rhizophere soil). A similar relation was also found for B uptake (Barber, 1971) and pH of the rhizocylinder. Decreasing rhizocylinder pH increased B uptake. In both of these cases, the plant root changed the pH in its soil environment, and this changed the level of the B and P in the soil solution, and hence their uptake.

Changing the pH of the rhizocylinder could also influence the uptake of ions such as Al and Fe. If NH_4^+ were the main source of N, reduction of the pH of an acid soil that contains Al to pH levels < 5.2 would increase the availability of Al, and this could be deterious to plant growth. Under this situation, the soil would need to be limed to a higher pH so that after pH reduction in the rhizocylinder soil, pH would still be high enough that no Al problems would occur. One can also assume that if NO_3^- fertilizer was used on a soil of pH ≤ 5.2, the soil pH in the rhizocylinder may be increased sufficiently so that Al would not be a problem.

Some soils do not supply sufficient Fe for maximum growth of some crops. Using NH_4^+ as the N source on these soils would cause a pH reduction in the rhizocylinder and bring more Fe into solution. This could alleviate Fe deficiency. Using the same reasoning, NO_3^- would aggravate Fe deficiency.

IV. SOIL SUPPLY MECHANISMS AND NUTRIENT INTERACTION

Level of nutrient at the absorbing surface of the root differs from the average level in the soil because nutrients must move to the plant root to be absorbed. The movement to the root is by mass flow and diffusion. Mass flow is the transport of nutrients to the root in the convective flow of water caused by water uptake by the plant. If mass flow plus the nutrients in the soil initially at the root surface do not supply the plant requirement, nutrient upake by the root will reduce the concentration at the root. The reduced concentration at the root creates a concentration gradient, along which nutrients diffuse to the root. However, if uptake is less than the rate supplied by mass flow, nutrient concentration will increase at the root.

The significance of mass flow and diffusion varies with nutrient. Nutrients such as Ca, Mg, and NO_3^- may be mainly supplied by mass flow. The ammonium ion and nutrients such as P and K are usually supplied mainly by diffusion. When nutrients mainly reach the root by diffusion, the concentration at the root is less than that initially in the soil. The degree of reduction will vary with nutrient. Rates of diffusion vary with nutrient—P is the slowest, then K^+ and NH_4^+; NO_3^- would diffuse the fastest, since it is not adsorbed by the soil. Figure 2 from Barber (1976) illustrates the gradients of NO_3^-, P, and K that might occur about a root growing in the soil.

Since the levels at the root differ from those initially in solution, this difference may affect the balance of nutrients required for maximum plant growth.

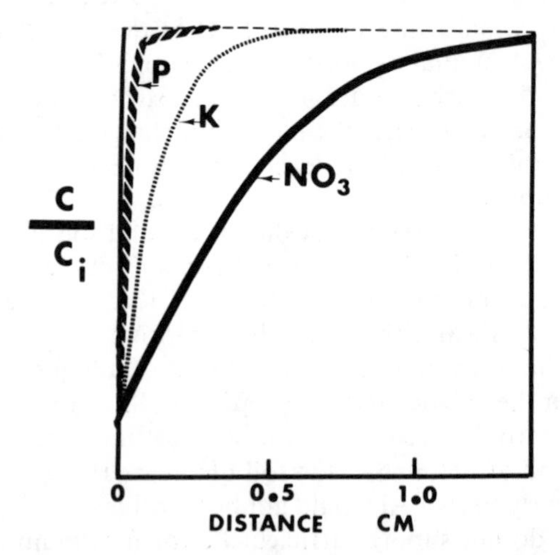

Fig. 2—Distribution of P, K, and NO_3^- radially from a plant root growing in silt loam soil after uptake of 5 d where diffusion controlled the nutrient flux to the root. C is the concentration as a fraction of C_i, the initial concentration of available nutrients (Barber, 1976).

Amount of nutrients transported to the root by mass flow may be greater than plant demand, and the level at the root absorbing surface may be two or more times greater than that initially in the soil solution. Accumulation or depletion at the root would not occur in solution culture; hence, results from solution culture compared with results from plants growing in soil must consider variation in levels due to differences in rates of supply to the root. Balance of nutrients in the soil before cropping is different than that at the root surface after uptake has been in progress for several days.

V. NUTRIENT BALANCE FOR PLANT ANALYSIS

Two approaches have been used in using plant analysis in determining nutrient balance for crops growing in the field. These are: (i) the critical value approach and (ii) the Diagnosis and Recommendation Integrated System (DRIS) approach proposed by Sumner (1977).

The critical value approach has been widely used. Plants or plant parts are analyzed for a single nutrient such as N in experiments where yield varies due to N fertilization. A plot is made of the N concentration vs. yield. A critical value is obtained, which is the value above which yield does not increase significantly. There is a nutrient range that indicates the nutrient level is adequate. Separate studies are made for each nutrient. Plant analysis for a particular plant growth stage will indicate (i) which nutrients are at higher levels than needed, (ii) which nutrients are below the critical level, and (iii) that lack of sufficient amounts may be reducing yield. When nutrient concentration increases above a certain level, yield depression may occur because of excess nutrients (Fig. 3). One of the problems in using this method to balance nutrients is that when a nutrient such as N restricts yield, the level of other nutrients may be higher than would occur if N level was not limiting yield.

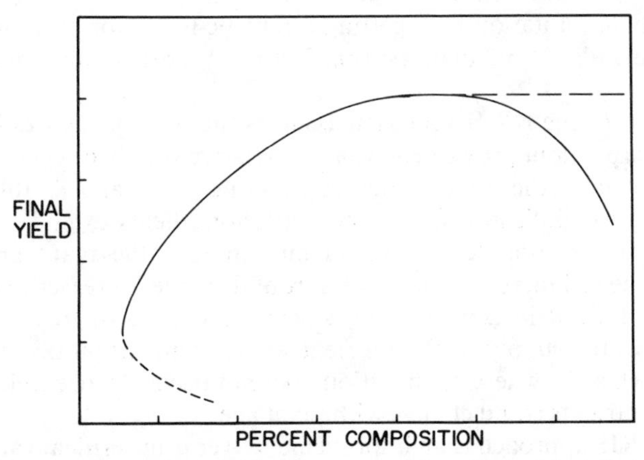

Fig. 3—Relation between plant yield and concentration of a nutrient that causes yield depression at high levels.

In the DRIS approach, all factors that can be measured and that have an effect on yield are measured. It is assumed that high yields can only be obtained when all factors are favorable or optimal. As one or more of the yield factors become unfavorable, crop yield will decrease. The factors most amendable to measurement are nutrient concentration in the leaf, and most of the relationships are based on the relation between leaf composition and yield. The DRIS system characterizes the components of yield in terms of indices. The indices allow one to classify yield factors in order of limiting importance, and this incorporates the concept of balance of nutrients into the system. An important advantage of the DRIS approach is that it diagnoses from plant analysis irrespective of age, while the critical value approach is usually restricted to a particular plant age since composition changes with age. The DRIS approach also indicates the most limiting nutrient.

The indices for each nutrient are calculated from surveying a large amount of data. An example is given in the calculation of the N index:

$$\text{N index} = \frac{f(\text{N/P}) + f(\text{N/K})}{2}, \qquad [3]$$

where

$$f(\text{N/P}) = 100\left(\frac{\text{N/P}}{n/p} - 1\right)\frac{10}{\text{CV}} \text{ when N/P} > n/p,$$

in which N/P is the actual ratio of nitrogen and phosphorus in the leaf under consideration, and n/p is the ratio for the population of high-yielding plants; CV is the coefficient of variation for the high-yielding plant. Similar indices are calculated for P, K, etc. When all indices are compared, the nutrient with the most negative indice is the most limiting, the next most negative is the next most limiting, etc. Hence, the DRIS approach is an approach to give the optimum balance of nutrients.

The main limitation is in getting a reliable index for each nutrient for high-yielding plants and in measuring factors other than nutrients that limit yield.

The use of the DRIS approach assumes that the nutrient ratios remain the same throughout growth and that the relative supply of each nutrient to the plant remains the same. When considering N, P, and K, this may not occur. Figure 2 indicates that concentration gradients extend varying distances from the root depending on nutrient. As the plant grows, root density in the soil increases and root to root distance decreases. Roots begin to compete for nutrients when the gradient extends out to greater distances. This reduces the supply of the nutrient where competition occurs, but not the one, such as P, where competition does not occur. Hence, relative availability of nutrients could change with plant age.

The DRIS approach is an improvement over using critical values, and it does consider optimum nutrient balance, which is important in some instances for obtaining high yields.

In order to maximize the yield increase from N application, all other nutrients need to be at a level that does not restrict yield either because they are too high or too low. Fortunately, because plant roots are selective in uptake, there is a range in concentration of each nutrient that can be tolerated. Unfortunately, we do not have adequate information on the exact values for these ranges.

REFERENCES

Barber, S. A. 1974. The influence of the plant root on ion movement in soil. p. 525–564. *In* E. W.Carson (ed.) The plant root and its environment. Univ. Press of Virginia, Blacksburg.

Barber, S. A. 1976. Efficient fertilizer use. p. 13–29. *In* F. L. Patterson (ed.) Agronomic research for food. Am. Soc. of Agron. Spec. Pub. no. 26. Am. Soc. of Agron., Madison, Wis.

Barber, S. A. 1971. The influence of the plant root system in the evaluation of soil fertility. Int. Symp. Soil Fert. Eval. Proc. 1971 1:249–256.

Claassen, N., and S. A. Barber. 1977. Potassium influx characteristics of corn roots and interaction with N, P, Ca, and Mg influx. Agron. J. 69:860–864.

Claassen, M. E., and G. E. Wilcox. 1974. Comparative reduction of calcium and magnesium composition of corn tissue by NH_4^--N and K fertilization. Agron. J. 66:521–522.

Edwards, J. H., and S. A. Barber. 1976. Nitrogen uptake characteristics of corn roots at low N concentrations as influenced by plant age. Agron. J. 68:17–19.

Koch, K., and K. Mengel. 1974. The influence of potassium nutritional status on the absorption and incorporation of nitrate nitrogen. p. 209–218. *In* Plant analysis and fertilizer problems. Proc. 7th Int. Colloq., Hannover, Germany.

Miller, M. H. 1974. Effects of nitrogen on phosphorus absorption by plants. *In* E. Carson (ed.) The plant root and its environment. Univ. of Virginia Press, Charlottesville.

Riley, D., and S. A. Barber. 1971. Effect of ammonium and nitrate fertilization on phosphorus uptake as related to root-induced pH changes at the root-soil interface. Soil Sci. Soc. Am. Proc. 35:301–306.

Sumner, M. E. 1977. Use of the DRIS system in foliar diagnosis of crops at high yield levels. Commun. Soil Sci. Plant Anal. 8:251–268.

6

Sham S. Goyal and Ray C. Huffaker
University of California
Davis, California

Nitrogen Toxicity in Plants

Nitrogen is probably the single most important factor limiting crop yields, as most plants require large quantities. Organic matter mineralization and N fixation by living organisms are the major sources of available N in most soils. According to estimates, only 1 to 4% of the N in organic matter may become available each year (Thompson & Troeh, 1973). Dinitrogen fixation by living organisms (except for some symbiotic N_2 fixers) is generally quite limited (Tisdale & Nelson, 1967), so N must be added to soils of most cropping systems to achieve high yields demanded by today's commercial farming.

The utilization of fertilizer N by crop plants (see Table 1 for Latin names of plants used in this chapter) is influenced by many factors, some of which are still beyond our control. Moreover, the bulk of available N in soils is prone to losses through such processes as leaching, denitrification, and volatilization. To ensure adequate N supply for crops, quantities of fertilizer N larger than actually required are often added to the soil. This may at times result in inefficient use of N fertilizer and have toxic effects on plant growth.

This paper describes possible toxic effects of N on plant growth that may be caused by a particular form of N, excessively high levels of N, or both. Although the word "toxic" means "poisonous," it has been used by plant scientists to designate any state in which the yield of a desired plant part is adversely affected. That definition is followed in this paper.

I. TOXICITY OF DIFFERENT FORMS OF N

A. NH_4^+ Toxicity

Of all the forms of inorganic N that plant roots might encounter in a cropping system, NH_4^+ is probably the most toxic to many plants. The literature on the effect of NH_4^+ nutrition on plant growth and processes is

Nitrogen in Crop Production.

Table 1—List of plants and their Latin names used in this chapter.

Plant name	Latin name
Alfalfa	*Medicago sativa* L.
Apple	*Malus sylvestris* Mill.
Avocado	*Persea americana* Mill.
Azalea	*Rhododendron calendulaceum* (Michx.) Torr.
Barley	*Hordeum vulgare* L.
Bean	*Phaseolus vulgaris* L.
Blueberry	*Vaccinium* sp.
Buckwheat	*Fagopyrum esculentum* Moench
Cauliflower	*Brassica oleracea* (Botrytis Group)
Coastal bermudagrass	*Cynodon dactylon* (L.) Pers.
Coffee	*Coffea arabica* L.
Corn	*Zea mays* L.
Cotton	*Gossypium hirsutum* L.
Cranberry	*Vaccinium macrocarpon* Ait.
Cucumber	*Cucumis sativus* L.
Grain sorghum	*Sorghum bicolor* (L.) Moench
Groundsel	*Senecio vulgaris* L.
Guineagrass	*Panicum maximum* Jacq.
Horse gram	*Cicer arietinum*
Kidney bean	*Phaseolus vulgaris* L.
Lime bean	*Phaseolus lunatus* L.
Lime	*Citrus aurantiifolia* (Christm.) Swingle
Maize	*Zea mays* L.
Mustard	*Brassica* sp.
Pea	*Pisum sativum* L.
Pine	*Pinus sylvestris*
Potato	*Solanum tuberosum* L.
Radish	*Raphanus sativus* L.
Red beet	*Beta vulgaris* L.
Rice	*Oryza sativa* L.
Rose	*Rosa* sp.
Ryegrass	*Lolium* sp.
Spinach	*Spinacia oleracea* L.
Spruce	*Picea sitchensis*
Soybean	*Glycine max* (L.) Merr.
Strawberry	*Fragaria* sp.
Sugarbeet	*Beta vulgaris* L.
Sunflower	*Helianthus annus* L.
Sweet corn	*Zea mays* L.
Tobacco	*Nicotiana tabacum* L.
Tomato	*Lycopersicon esculentum* Mill.
Wheat	*Triticum aestivum* L.

voluminous. Although NH_4^+ is an intermediate in the assimilation of mineral N, it is toxic to many plants when it is the sole exogenous source. Ammonium supplied exogenously as a substrate presents a different situation than NH_4^+ being generated internally from NO_3^- reduction and photorespiration. With NO_3^- as substrate, the steady-state concentration of NH_4^+ in the cells is quite low (Goyal, 1974; Goyal & Huffaker, 1981), perhaps because NH_4^+ is utilized almost as rapidly as it is formed. In contrast, when NH_4^+ is applied exogenously, large amounts can be taken up, flooding the cells with concentrations usually not present (Goyal, 1974). Moreover, with

NO_3^- nutrition, NH_4^+ is formed in chloroplasts (Neyra & Hageman, 1978), whereas with NH_4^+ nutrition, the cytoplasm would be exposed to absorbed NH_4^+ in addition to that generated from photorespiration. The causes of NH_4^+ toxicity are unclear. Many physiological and biochemical effects of NH_4^+ toxicity have been reported, as described below.

1. NH_4^+ EFFECTS ON PLANT GROWTH

Growth probably best indicates overall performance of a plant in its own environment. Plant species differ greatly in growth in response to NH_4^+ nutrition. The effect of NH_4^+ on growth generally may be characterized as "very injurious" to "essential," depending on the plant species. However, most crop species react adversely to NH_4^+ nutrition by exhibiting reduced growth rate and foliar damage; NH_4^+, as the sole source of N, was injurious to the growth of bean, sweet corn, cucumber, pea (Maynard & Barker, 1969), and radish plants (Goyal, 1974). Dry weights of all plant parts of lima beans grown in solution culture were consistently lower when $\geq$ 25% of N was supplied as NH_4^+ (McElhannon & Mills, 1978). In addition, the dry weights of southern pea roots, stems, leaves, pods, seeds, and total plants were significantly lower at three different stages of development when grown in solution culture in which NH_4^+ was the sole source of N; $\geq$ 75% of the N had to be supplied as NO_3^- to obtain maximum growth (Sasseville & Mills, 1979). Dirr et al. (1973) reported that the number and length of *Leucothoe catesbaei* shoots were, respectively, lower and shorter when cultured on NH_4^+ than when cultured on NO_3^-. Ammoniacal-N decreased growth of tomato tops and roots, compared with NO_3^--N (Torres de Claassen & Wilcox, 1974). Fresh and dry weights of pea and cucumber shoots grown in NH_4^+-N were significantly lower than those of shoots grown in equivalent concentrations of NO_3^--N at four concentrations of N (Barker & Maynard, 1972). Furthermore, Hohlt et al. (1970), in tests of several cultivated species of the family Solanaceae for NH_4^+ tolerance, found that tobacco was the most tolerant to applications of $(NH_4)_2SO_4$, and the tomato was the most susceptible. Barker et al. (1970) reported that the germination percentage of cucumber seeds was lower in 0.1N $(NH_4)_2SO_4$ than in the same concentration of K_2SO_4, even though the percentage of emergence of primary roots from seeds was not affected. This indicates that NH_4^+ was toxic only after it had been absorbed physiologically by roots.

The release of gaseous NH_3 from fertilizer bands can also result in NH_3 toxicity (Court et al., 1964a, b; Adams, 1966; Bennett & Adams, 1970b). Continuous exposure of corn seeds during the initial stages of germination (3 d) to partial vapor pressures of NH_3 of as low as 0.063 mm Hg significantly injured the early physiological development of corn; NH_3 was toxic at even lower levels when the exposure time was increased (Allred & Ohlrogge, 1964).

When NH_4^+ replaced NO_3^- at equivalent N concentration, the medium did not support the growth of excised tomato roots (Robbins & Schmidt, 1938) or groundsel roots (Skinner & Street, 1954). In general, NH_4^+ seems more deleterious to growth of the root than to that of the shoot (Bennett et al., 1964; Haynes & Goh, 1977). Ammonium ions were ineffective as a N source for the growth of callus tissue (Riker & Gutsche, 1948; Burkholder & Nickell, 1949; Heller, 1953, 1954).

From a practical standpoint, NH_4^+ toxicity under field conditions may not be as serious a problem as it might seem from solution culture experiments. There are many possible reasons for this: (i) soil can act as a buffer to pH changes; therefore, acidification of the root medium due to NH_4^+ uptake may be minimal; (ii) nitrification of NH_4^+ occurs, which constantly decreases the NH_4^+ concentrations; (iii) most soils have some NO_3^- at all times (even before the nitrification of recently applied NH_4^+ begins), which can modify NH_4^+ assimilation; and (iv) roots have the option of avoiding the zones where NH_4^+ might be too concentrated, and can approach it at a later time when nitrification has begun.

Although NH_4^+ nutrition is toxic to many higher plants, it nevertheless is a better source of N than NO_3^- for many other plants. Ammonium is superior to NO_3^- as a N source of rice (Wahhab & Bhatti, 1957), azalea (Stuwart, 1947; Colgrove & Roberts, 1956), and blueberries (Cain, 1952). Greidanus et al. (1972) reported that NH_4^+ was "essential" for growth of cranberries; the plant apparently was unable to absorb NO_3^- from soil or nutrient cultures and lacked the genes required for induction of NR (see Table 2 for explanation of abbreviations, used in this chapter). Many plants in the family Ericaceae (e.g., blueberries, cranberries) growing predominantly in acid soils, and known as "Calcifuges" ("acid-loving"), grow better in NH_4^+ than in NO_3^- (Haynes & Goh, 1978). Another group of

Table 2—Explanation of abbreviations used in this chapter.

Abbreviation	Meaning
ADP:	adenosine diphosphate
ADPG:	adenosine diphosphoglucose
AMP	adenosine monophosphate
ATP:	adenosine triphosphate
F-6-P:	fructose-6-phosphate
G-1-P:	glucose-1-phosphate
G-3-PDH:	glyceraldehyde-3-phosphate dehydrogenase
G-6-P:	glucose-6-phosphate
G-6-PDH:	glucose-6-phosphate dehydrogenase
GDH:	glutamate dehydrogenase
Gln:	glutamine
Glu:	glutamate
ICDH:	isocitrate dehydrogenase
α-Kg	α-ketoglutarate
NADH:	reduced nicotinamide adenine dinucleotide
NADP:	nicotinamide adenine dinucleotide phosphate
NR:	nitrate reductase
PEP:	phosphoenol pyruvate
RuBP:	ribulose-1,5-bisphosphate
UDPG:	uridine diphosphoglucose
UDPX:	uridine compound containing sugar or sugar derivative

plants, apparently capable of growing under waterlogged conditions for extended periods (e.g., rice), also seem to prefer NH_4^+ to NO_3^-. Wheat and ryegrass initially grew better and absorbed more N when fertilized with NH_4^+ than when fertilized with NO_3^- (Spratt & Gasser, 1970). When growth and N uptake were fastest, wheat grew more rapidly when fertilized with NH_4^+ than when fertilized with NO_3^-.

In general, the plants that grow naturally where little or no nitrification takes place seem to prefer (or at least tolerate) NH_4^+ to NO_3^-. Since these plants naturally grow under acid or waterlogged conditions, it is logical to postulate that the plants evolved under "no NO_3^-" conditions, and are therefore well adapted to the form of N that would predominate in these soils (NH_4^+). Since inhibition of NH_4^+ nitrification is now feasible in many areas, future genetic manipulation of species presently not well adapted to NH_4^+ nutrition, for improved capacity to assimilate NH_4^+, could be an important means to utilize fertilizer N more efficiently.

Reports indicating that NH_4^+ is as good a source of N as NO_3^- are also not uncommon. Blair et al. (1970) found no difference in yield of tops or roots of corn grown in solutions of low (2 m*M*) concentrations of NO_3^- or NH_4^+. The fresh and dry weights of sugarbeet plants cultured on NO_3^- or NH_4^+ were quite similar (Breteler, 1973). Morris andGiddens (1963) reported no significant differences in plant weights due to NH_4^+ or NO_3^- nutrition for cotton, corn, grain sorghum and coastal bermudagrass grown in soil. The fronds of *Spirodela oligorrhiza* grew equally well whether NH_4^+ or NO_3^- was the sole source of N (Ferguson & Bollard, 1969).

Many reports show that NH_4^+ provided along with NO_3^- gives more beneficial effects than either form alone. Mohanty and Fletcher (1976) reported that 'Paul's Scarlet' rose cells in suspension culture grew twice as much when 0.91 m*M* NH_4^+ was included in medium containing 25 m*M* NO_3^-. Yields of wheat were higher when grown in NH_4^+ plus NO_3^- than when grown in either source alone in a continuous-flow culture system (Cox & Reisenauer, 1973). Weissman (1964) reported that dry weights, total protein content, protein concentration, and percent protein of total N were all higher in leaves from sunflower grown on NH_4^+ plus NO_3^- than on either source alone. Schrader et al. (1972) reported fresh weights of corn plant parts were higher when plants were supplied 100 mg/kg of N as a NO_3^-/NH_4^+ mixture (25:75, 50:50, or 75:25) than when 100 mg/kg of either NO_3^- (100:0) or NH_4^+ (0:100) were provided. Burdine (1976) concluded that 15 to 30% of total N supplied as NH_4^+ was better for growth of radish plants than 100% NO_3^-. Gamborg (1970) reported that soybean cells in suspension culture did not grow on NO_3^- unless supplemented with NH_4^+ or glutamine.

The current literature indicates that more monocot species may tolerate NH_4^+ than dicots, even though it may not be universally true. We have not seen a report of a dicot plant (other than Calcifuges) cultured successfully with NH_4^+ as the sole source of N. The degree to which a dicot may be susceptible to NH_4^+ also differs greatly. The reasons for higher tolerance of monocots to NH_4^+ than dicots are unclear.

2. NH_4^+ TOXICITY AND pH

In most biological fluids, ammonical-N exists in two forms, i.e., NH_4^+ and NH_3 (aqueous—will be referred to as "aq" hereafter), and the relative proportions of each are determined primarily by the pH of the solution (Sienko & Plane, 1957). According to Bennett and Adams (1970b), the toxicity of NH_4^+ to plants depends on the substrate-solution concentration of NH_3 (aq). Since the NH_3 (aq) concentration increases as pH increases (Weir et al., 1972), NH_4^+ should be more toxic when pH is higher. Weir et al. (1972), attempting to explain the superiority of $(NH_4)_2SO_4$ to urea in neutral-to-alkaline soil (Lorenz et al., 1954; Tyler et al., 1962), considered the equilibrium reactions between NH_3 and NH_4^+ in aqueous solutions. They concluded that the toxic effects of urea, under field conditions, on potatoes could be due to the formation of higher levels of NH_3 resulting from high pH in the fertilizer band. Toxic effects on growth of maize seedlings during the first week after urea was added to a sandy loam soil were attributed to high levels of free NH_3, because soil pH rose to 8.0 (Court et al., 1964b).

In most cases, relatively low and high pH of the nutrient solution, respectively, favored higher rates of absorption of NO_3^- and NH_4^+ (Clark & Shive, 1934; Wallace & Mueller, 1957). These reports agree with the theoretical aspects of ammoniacal-N forms occurring in solutions. When pH is higher, so is the concentration of NH_3 (aq), the form to which cell membranes are relatively more permeable (Milne et al., 1958).

The above reports indicate that NH_4^+ is more toxic when solution pH is higher, but numerous reports show that a decrease in solution pH, resulting from NH_4^+ absorption, limits NH_4^+ utilization. Inhibition of plant growth can be overcome by using $CaCO_3$ in the root medium as a buffer (Barker et al., 1966a, b; Barker, 1967). Maynard and Barker (1969) showed that bean, sweet corn, cucumber, and pea plants, which are inhibited by NH_4^+, grew normally when the pH of the sand culture was maintained near neutrality by adding 1% $CaCO_3$. The inability of NH_4^+ to serve as a source of N for the growth of excised roots was altered when the medium was buffered at pH 7.0 using $CaCO_3$ (Hanny et al., 1959; Sheat et al., 1959). The enhanced conversion of absorbed NH_4^+ to organic nitrogenous compounds in root tissue under controlled conditions of acidity has also been reported (Barker et al., 1966a, b). Movement of NH_4^+ to shoots was restricted and thus decreased the detrimental effects of high NH_4^+ concentrations in the leaves.

The literature, therefore, suggests that NH_4^+ can inhibit plant growth at all pH's except at or near neutrality, but the reasons for that are unclear.

3. NH_4^+ TOXICITY AND OTHER INORGANIC IONS

Numerous reports have shown that NH_4^+ nutrition reduces the uptake of other cations; Ca^{2+}, Mg^+, and K^+ are affected most (Holley et al., 1931; Arnon, 1939; Wander & Sites, 1956; Barker et al., 1967; Kirkby, 1968). Arnon (1939) postulated that K^+ and NH_4^+ competed for uptake since NH_4^+

solutions contained two cations (NH_4^+ and K^+), whereas NO_3^- solutions contain K^+ as the only rapidly absorbed cation. Tromp (1962) reported that K^+ uptake by wheat roots was reduced by half or more by the concurrent uptake of NH_4^+. Plants grown in both soil and solution culture with NH_4^+ contained lower levels of K^+ than those grown in NO_3^-, even though the amount of K^+ supplied was the same in both cases (Maynard et al., 1968; Ajayi et al., 1970).

Cox and Reisenauer (1973) reported that cation uptake decreased as NH_4^+ levels increased in the root medium, and explained it as ionic competition at the site of intake, either with NH_4^+ per se or with H^+ produced during NH_4^+ uptake. Smith and Epstein (1964) reported that NH_4^+ uptake was strictly competitive with K^+ and Rb^+. Epstein (1972) found that the uptake of NH_4^+ resembles that of K^+, indicating that the two ions are transported by the same carrier.

These reports suggest that NH_4^+ toxicity to plants may be due to reduced uptake of K^+. Ajayi et al. (1970) and Maynard et al. (1968) were able to correct the NH_4^+ toxicity symptoms by using excessive amounts of K^+. However, a greatly increased K^+/NH_4^+ and Ca^{2+}/NH_4^+ ratio in solution culture did not help alleviate NH_4^+ toxicity in radish plants (Goyal, 1974). Apparently the effects of NH_4^+ nutrition are manifold.

Another line of evidence explaining higher levels of cations in NO_3^--fed plants indicates that NO_3^- actually stimulates both cation uptake and translocation. Jackson and Williams (1968) reported that cation uptake increased as NO_3^- concentration in the root medium increased. They also showed that uptake of both monovalent (Cs^+, Na^+, and K^+) and divalent (Sr^{2+}, Mn^{2+}, and Mg^{2+}) cations was stimulated by NO_3^- in N-depleted wheat seedlings. Jackson and Williams (1968) theorized that the rise in the rhizosphere pH as a consequence of rapid NO_3^- uptake favored cation uptake. Jackson et al. (1974b) observed that the onset of the accelerated NO_3^- uptake was associated with the increased K^+ uptake. Blevins et al. (1974) reported higher rates of K^+ uptake from KNO_3 than from KCl by barley seedlings, which were accompanied by higher pH of the growing medium and cell sap.

Stimulation of cation translocation by NO_3^- has also been shown. Upward movement of Ca^{2+} and K^+ in tobacco plants treated with NO_3^- was greater than movement in those treated with NH_4^+ (Yoshida, 1969). Wallace et al. (1966) observed that external K^+ and Ca^{2+} appeared more rapidly in root exudates when supplied as nitrates than when supplied as chlorides.

The reduced uptake of Ca^{2+} due to NH_4^+ has also been studied. Adams (1966) suggested Ca^{2+} deficiency was the mechanism of $(NH_4)_2HPO_4$ injury to cotton seedlings, based on root growth experiments in soil. Hoff et al. (1974) also suggested that NH_4^+ toxicity was a manifestation of intracellular Ca^{2+} deficiency. Wilcox et al. (1973) reported that topical applications of Ca^{2+} prevented blossom end rot of tomatoes caused by NH_4^+. However, Bennett and Adams (1970a), using sterilized soil cultures, demonstrated that Ca^{2+} deficiency and NH_4^+ toxicity were separate causal factors of $(NH_4)_2HPO_4$ injury to seedlings. The addition of limestone, or $CaCO_3$, or use of NH_4HCO_3 permitted plants otherwise highly susceptible to injury to have near-normal growth in NH_4^+ (Prianishnikov, 1951; Gouny, 1955; Barker et al., 1966b). Reported causes lack unanimity. Increased Ca^{2+} supply, general increased cation absorption by roots, and assimilation of carbonates and bicarbonates into organic acids as extra C skeletons have been proposed (Chouteau, 1963; Gouny, 1955; Prianishnikov, 1951). By contrast, Barker et al. (1966b), using ^{13}C-labeled carbonate, showed that carbonate was not absorbed or incorporated into plant tissue.

Phosphorus uptake increased when NH_4^+ replaced NO_3^- as the N source (Mattson, 1966; Klemm, 1967). The effect of N source on P uptake has been attributed to the changes in pH of the rhizosphere, which was lowered during NH_4^+ absorption (Riley & Barber, 1971; Soon & Miller, 1977). Lowering of growth medium pH, resulting from NH_4^+ absorption, increased the concentration of P in the $H_2PO_4^-$, which is absorbed several times faster than HPO_4^{2-} (Miller et al., 1970; Soon & Miller, 1977). Moreover, HPO_4^{2-} was precipitated as $CaHPO_4 \cdot 2H_2O$ at the root surface in the absence of NH_4^+ (Miller et al., 1970).

It seems clear that NH_4^+ lowers the uptake of other cations, whereas NO_3^- can stimulate their uptake. There is insufficient evidence for concluding that reduced uptake of cations (K^+, Ca^{2+}, Mg^{2+}) is the sole cause of NH_4^+ toxicity to plants.

4. BIOCHEMICAL AND PHYSIOLOGICAL CHANGES DURING NH_4^+ TOXICITY

a. Photosynthesis—Numerous reports have shown that effects of NH_4^+ on the physiology and biochemistry of plants contrast with effects caused by NO_3^-. Increased concentration of NH_4^+ in plant tissue is one of the first and basic elements of NH_4^+ toxicity (Wall, 1940; Barker et al., 1966a, b; Maynard & Barker, 1969; Goyal, 1974). The consequences of this can be manifold.

Ammonium ions acted as an uncoupler of photophosphorylation in isolated chloroplasts (Gibbs & Calo, 1959; Krogmann et al., 1959; Avron, 1960), resulting in increases in lower energy forms of adenine nucleotides (AMP and ADP), and concurrent decreases in the amounts of ATP (Krogmann et al., 1959; Trebst et al., 1960; Losada et al., 1973). Concentrations as low as 0.6 m*M* NH_4^+ inhibited ATP formation by > 50% (Krogmann et al., 1959).

Gibbs and Calo (1959) reported that NH_4^+ inhibited CO_2 fixation in isolated spinach chloroplasts by 81% in a concentration of 1.0 m*M* and by 99% in 10.0 m*M*. Photosynthesis of intact radish leaves on NH_4^+-fed plants was 50% of leaves on those fed NO_3^- (Goyal, 1974). No distinct grana in tomato leaf chloroplasts were detected after 4 weeks of NH_4^+ nutrition, even though the mitochondria in the same cells appeared normal (Puritch & Barker, 1967).

Recent findings, however, show that NH_4^+ increased photosynthetic CO_2 fixation in isolated cells of spinach (Woo & Canvin, 1980a, b) and *Papaver somniferum* (Paul et al., 1978) and in intact spinach chloroplasts

(Bennedetti et al., 1976; Slovacek & Hind, 1977; Heath & Leech, 1978). The stimulation of photosynthesis by NH_4^+ was described as an activation of RuBP carboxylase (Benedetti et al., 1976). Heath and Leech (1978), however, explained this was caused by altered stromal pH, since raising the medium pH from 7.6 to 8.2 lowered the NH_4^+-induced stimulation of photosynthesis. As external pH was increased further, NH_4^+ inhibited CO_2 fixation.

In some studies, NH_4^+ did not stimulate photosynthetic CO_2 fixation; for example, in isolated cotton cells, NH_4^+ in the presence of 5 m*M* $Ca(NO_3)_2$ for up to 7.5 h did not affect total photosynthetic CO_2 fixation (Rehfeld & Jensen, 1973). Furthermore, Platt et al. (1977) found that NH_4^+ did not affect total CO_2 fixation in leaf discs from alfalfa plants grown on Hoagland's solution (which should have NO_3^-). In studies where NH_4^+ stimulated CO_2 fixation, no other source of reduced or reducible N, such as glutamine, was made available. Stimulated CO_2 fixation may be a response to reduced N because of additional sinks created for C, since stimulation occurred only when no other source of reduced N was available. However, Heath and Leech (1978) showed that NH_4^+-induced stimulation of CO_2 fixation in isolated chloroplasts can be duplicated by substituting methylamine chloride for NH_4Cl.

b. Respiration—If photosynthesis is adversely affected by NH_4^+, respiration can be expected to change. In general, N addition (via NH_4^+ or NO_3^- salts) to N-starved algae or higher plants reportedly caused an acceleration of respiratory rates (Folkes et al., 1952; Willis & Yemm, 1955; Syrett, 1956a, b; Yemm & Willis, 1956; Hattori, 1958). Reports comparing respiration rates during NH_4^+ or NO_3^- assimilation are contradictory and often confusing. Both gaseous NH_3 and undissociated NH_3 inhibited respiration (Vines & Wedding, 1960) in barley roots and were believed to be the effective inhibitory forms rather than NH_4^+. Ammonia inhibited NADH oxidation in red beet root homogenates (Wedding & Vines, 1959). Furthermore, the ability of NH_3 to specifically inhibit the oxidation of NADH, thus blocking the respiratory electron transport chain, was suggested as a partial cause of NH_3 toxicity (Vines & Wedding, 1960).

Some authors have disputed the findings that NH_4^+ inhibits respiration. Burkhart (1938) concluded that etiolated seedlings of several species absorb and utilize NH_4^+ until available carbohydrates are exhausted. Breakdown of carbohydrates due to enhanced respiration during rapid NH_4^+ assimilation was accelerated (Willis & Yemm, 1955; Syrett, 1956a). Wakiuchi et al. (1971) reported respiration of cucumber leaves was higher in those grown in 14.3 m*M* NH_4Cl than in those grown in 1.43 m*M* NH_4^+Cl, which was not due to uncoupling in mitochondria (Matsumoto et al., 1971b). Ammonium also slightly increased respiration of detached barley leaves (Berner, 1971). The accelerated respiration and carbohydrate catabolism during NH_4^+ toxicity may be required for the supply of organic acids, especially α-Kg, to counteract NH_4^+ (Givan, 1979; Matsumoto et al., 1971b). By contrast, the dark respiration rates of bean (Barker et al., 1965) and radish leaves (Goyal, 1974) from plants grown in NO_3^- or NH_4^+ were the same.

c. N-Containing Compounds—Plants grown in NH_4^+ invariably contained higher levels of free NH_4^+ and amide N than those grown in NO_3^- (Weissman, 1964; Maynard et al., 1966; Barker, 1967; Hoff et al., 1974; Tromp & Ovaa, 1979; Kato, 1980). Amide N contributed 71% of total N in exudates of sunflower plants grown in NH_4^+ and 19% in NO_3^- (Weissman, 1964). Weissman also reported that NH_4^+ nutrition favored alanine, arginine, leucine, serine, and valine in exudates, whereas NO_3^- nutrition favored γ-aminobutyric acid, aspartic acid, glutamic acid, and lysine. Tobacco cells grown in suspension cultures and supplied with NO_3^- plus NH_4^+ contained 50 to 100 times more glutamine and alanine than those supplied with NO_3^- alone (Bergmann et al., 1976). De Kock and Kirkby (1969) reported that NH_4^+-fed buckwheat and mustard plants contained much more free amino acids than those fed with NO_3^-, even though the different N forms did not create differences in total N content. Hoff et al. (1974) also found NH_4^+ nutrition elevated levels of most free amino acids in tomato plant parts; particularly high were glutamic and aspartic acids and their amides. Soluble organic N in embryos of wheat seedlings cultured on NH_4^+ was about five times higher than N in seedlings cultured on NO_3^- (Weissman, 1951). Barley plants grown in NH_4^+ had higher levels of non-protein N, free amino acids (particularly aspartic), and inorganic N than those grown on NO_3^- (Richter et al., 1975). *Spirodela oligorrhiza* plants utilizing NH_4^+ contained higher levels of NH_4^+, total N, arginine, asparagine, and glutamine than did those utilizing NO_3^-, whereas concentrations of other amino acids were similar (Ferguson & Bollard, 1969). Ammonium nutrition enhanced production of asparagine and retarded its conversion to arginine in young apple trees (Tromp & Ovaa, 1979). However, Schrader et al. (1972) did not find any differences in amino acid composition and total protein content of corn leaf blades due to NH_4^+ or NO_3^- nutrition. Mohanty and Fletcher (1980) reported that rose cell suspension cultures grown on media containing 1920 μmoles NO_3^- + 72.8 μmoles NH_4^+ had higher levels of soluble amino acids, total N, N in protein, amino acids, and total soluble and membrane-bound proteins than those grown on media containing 1920 μmoles NO_3^- alone. Alfalfa discs floating on buffer containing NH_4Cl and photosynthesizing with $^{14}CO_2$ produced more label in glutamine, glutamate, aspartate, alanine, glycine, serine, and UDPG than did discs floating on buffer that did not contain NH_4Cl (Platt et al., 1977). To explain the elevated levels of free amino acids with NH_4^+ nutrition, it has been suggested that NH_4^+ plays a regulatory role in diverting C traffic from carbohydrate biosynthesis into amino acid synthesis, presumably by activating pyruvate kinase and PEP carboxylase in the anaplerotic pathway (Kanazawa et al., 1972; Platt et al., 1977; Paul et al., 1978).

d. Organic Acids—Kirkby and Mengel (1967), who prepared a balance sheet of inorganic and organic ions, showed that organic acid anions (fumaric, succinic, malonic, malic, citric, oxalate, and total) in leaves, petioles, stems, and roots of tomato plants grown in NO_3^- were several hundred percent higher than in those grown in NH_4^+. Differences in malic, citric, and oxalate were particularly large. Breteler (1973) reported a similar observation for sugarbeets; the effect was more pronounced in tops than in roots. Cells of tobacco, grown in suspension culture on media containing NO_3^- as sole source of N, contained 10 to 15 times more malate than did those grown on NO_3^- + NH_4^+ (Bergmann et al., 1976). Accumulation of organic acids with NO_3^- nutrition has been explained as a cellular response to counteract OH^- produced during NO_3^- reduction, in order to maintain intracellular pH (Davies, 1973; Raven & Smith, 1976).

e. **Carbohydrates**—Ammonium may also affect carbohydrate synthesis. Platt et al. (1977) showed that NH_4^+ increased transfer of photosynthetically incorporated C to synthesis of amino acid skeletons at the expense of sucrose synthesis by activating pyruvate kinase. Kanazawa et al. (1972) and Paul et al. (1978) have made similar suggestions. Cucumber leaves had higher levels of glucose, fructose, UDPG, G-1-P, and F-6-P and lower levels of starch, UDPX, and G-6-P during NH_4^+ toxicity (Matsumoto et al., 1968, 1969). The accumulated glucose and fructose were the products of newly fixed C and not a degradative product of a glucose polymer. Glucose translocation after its synthesis, at least up to starch synthesis, may have been inhibited, resulting in a disorder of carbohydrate metabolism during NH_4^+ toxicity.

f. **Activity of Enzymes**—Since the biochemical composition of plants is altered distinctly when NH_4^+ is the source of N, enzyme activities may be expected to change. The respective accumulation and depletion of free sugars (glucose and UDPG) and starch in cucumber leaves during NH_4^+ toxicity were related to the decreased in vitro activity of granule-bound starch synthetase when either UDPG or ADPG was a glucose donor (Matsumoto et al. 1968, 1971a). Ammonium addition to the reaction medium also inhibited starch synthetase activity when UDPG was a glucose donor. In vitro activities of enzymes GDH and all those operating in glycolysis, TCA cycle, and mitochondrial respiratory chain were higher (except aldolase and ATPase) in NH_4^+-injured cucumber plants (Wakiuchi et al., 1971; Matsumoto et al., 1971b). The highest increase (fivefold) was found in phosphofructokinase. These workers concluded that when NH_4^+ is toxic to cucumber plants, starch synthesis was impaired but carbohydrate catabolism was accelerated to meet higher demand of C skeleton (particularly α-Kg) in order to detoxify NH_4^+. However, they did not explain how NH_4^+-injured plants would expectedly have extra glucose for conversion to starch under increased respiratory activity unless photosynthesis was increased accordingly. Although NH_4^+ inhibited starch synthesis, that does not seem to explain the NH_4^+ toxicity of cucumber plants.

Activity of malic enzyme in cells of tobacco grown in suspension culture with media containing both NO_3^- and NH_4^+ was 3 to 4 times higher than activity of the enzyme in cells fed NO_3^- alone (Bergmann et al., 1976). This may explain the reported lower levels of malate due to NH_4^+ nutrition.

Activities of GDH and Glu synthase in 'Paul's Scarlet' rose cells grown in suspension culture containing both NO_3^- and NH_4^+ were higher than activity in those grown with NO_3^- alone. However, the reverse was true for Gln synthetase (Mohanty & Fletcher, 1980). The enzymatic potential of Gln synthetase and GDH greatly exceeded the actual in vivo rate of N assimilation. Therefore, it was concluded that these enzymes did not limit N assimilation and that Glu synthase may have limited N assimilation instead, thus diverting N to GDH.

Activities of particulate and soluble ICDH in rose cells grown in suspension culture on medium containing 24 m*M* NO_3^- was similar to activity in 24 m*M* NO_3^- and 0.91 m*M* NH_4^+. The maximum activity of G-3-PDH attained was also the same in both cell types, even though it peaked 1 d earlier in NH_4^+ cells (Mohanty & Fletcher, 1976).

Activities of GDH in plants grown in NH_4^+ nutrition have been higher than those in plants grown in NO_3^- (Baley et al., 1972; Gamborg & Shyluk, 1970; Weissman, 1972; Goyal, 1974). Gamborg and Shyluk (1972) reported that activities of ICDH and Gln synthetase of soybean cells grown in NH_4^+ were lower than those in cells grown in NO_3^-, but considering other observations, they concluded, lower levels of these enzymes did not limit NH_4^+ utilization. Results of studies of ICDH in radish plants were similar (Goyal, 1974). Sunflower and soybean roots grown in NH_4^+ had higher and lower activity, respectively, of Gln synthetase than those grown in NO_3^- (Weissman, 1972). Both soybean and sunflower roots grown in NO_3^- had higher activity of G-6-PDH than those grown in NH_4^+. However, the leaves of both species showed the same activity of Gln synthetase and G-6-PDH in either form of N (Weissman, 1972).

In conclusion, NH_4^+ nutrition alters the activities of many enzymes, and the effect varies greatly from one species to another. There is no conclusive evidence, however, that the activity of a particular enzyme limits NH_4^+ utilization when it is toxic.

g. Water Uptake—Tomato (Quebedaux & Ozbun, 1973) and sugarbeet (Stuart & Haddock, 1968) plants grown in NH_4^+ had a lower water uptake than those grown in NO_3^-. Radish plants absorbed similar amounts of water whether grown in NH_4^+ or NO_3^- (Goyal, 1974).

B. NO_2^- toxicity

Nitrite, an intermediate product in the conversion of NH_4^+ to NO_3^- and NO_3^- to N_2 in soils and of NO_3^- to NH_4^+ in plants, usually does not accumulate appreciably in either soils or plants. Some soils accumulate considerable amounts of NO_2^-, especially if they have a pH of neutral-to-alkaline and are fertilized heavily with urea or ammoniacal fertilizers (Chapman & Liebig, 1952) or if they are amended with sewage sludge (Yoneyama & Yoshida, 1978). Some NO_2^- may also accumulate in soils during anaerobiosis. Thus, exposure of plants to ambient NO_2^- is not totally theoretical.

Even though NO_2^- is generally considered harmful to higher plants, its toxicity is a function of many factors. A pH and concentration-dependent NO_2^- toxicity to avocado and citrus seedlings (Curtis, 1949) and to growth of tomato and barley plants (Bingham et al., 1954) was reported. According to Bingham et al. (1954), NO_2^--N concentrations of ≥ 50 mg/kg must prevail in the root zone to damage plants. In tomato, the concentrations of NO_2^- required to produce toxicity symptoms and the severity of toxicity symptoms were closely linked to root supply of Mg^{2+}, Fe^{2+}, and air (Phipps & Cornforth, 1970). Rice plants were very susceptible to NO_2^-, whereas cucumber and wheat were tolerant (Oji, 1974; Quoted by Yoneyama et al., 1980). Similarly, avocado was more susceptible than citrus (Curtis, 1949). In the absence of lime, $NaNO_2$ (40 mg/kg) was as good as $(NH_4)_2SO_4$ for the growth of guineagrass (Oke, 1966). Sahulka (1973) reported that average length and fresh weights of excised pea roots were significantly lower when grown on 2 m*M* KNO_2 than when grown on 2 m*M* KNO_3. Toxicity arising from heavy application of urea to soil was due to accumulation of NO_2^- (Court et al., 1962).

In general, conditions of increased acidity, low O_2 supply to roots, and higher ambient NO_2^- concentrations appear to favor NO_2^- toxicity. Recently, Lee (1979) reported that low O_2 supply to roots increased the sensitivity of barley and maize plants to NO_2^-, but the changes produced by NO_2^- were not significant when enough O_2 was present. The undissociated nitrous acid (HNO_2) (the concentration of which increases with decreasing pH) rather than the dissociated NO_2^- may be related to growth inhibition (Bingham et al., 1954; Lee, 1979).

Jackson et al. (1974b) showed that NO_2^- inhibits NO_3^- uptake in wheat seedlings (also unpublished results from our lab). In N-starved wheat seedlings, the uptake of K^+ and NO_2^- from KNO_2 was significantly lower than the uptake of K^+ and NO_3^- from equimolar KNO_3 (Jackson et al., 1974b). However, NO_2^- was absorbed and converted to amino N more rapidly than NO_3^- by roots of kidney beans, corn, and sunflower (Yoneyama et al., 1980).

Numerous authors have shown that plants can absorb and assimilate NO_2^-. A complete reduction and assimilation pathway for NO_2^- seems to be active in the roots of most plants (Yemm & Willis, 1956; Sheat et al., 1959; Yoneyama et al., 1980). Even though little or no NO_2^- may be exported from the roots to shoots (Jackson et al., 1974a; Oji, 1974; Yoneyama et al., 1980), leaves of many species effectively reduce and metabolize exogenously provided NO_2^- in both light and darkness (Vanecko & Varner, 1955; Jones & Sheard, 1978; Aslam et al., 1979; Yoneyama et al., 1980). This suggests that metabolites can move into the chloroplasts and reduce NADP, which in turn may reduce ferredoxin via NADP reductase, thus supplying electrons to nitrite reductase. However, light dependence of NO_2^- reduction in barley leaves (Canvin & Atkins, 1974) and sunflower (Ito & Kumazawa, 1978) has also been reported. Vanecko and Varner (1955) showed that NO_2^- reduction during light depended on chlorophyll.

C. Urea and Biuret Toxicity

Many studies have shown that crops respond poorly and even adversely to soil applications of urea. Urea application to soils of neutral-to-alkaline reaction lowered yields of potato as compared with $(NH_4)_2SO_4$ at an equal rate of N (Lorenz et al., 1954; Tyler et al., 1962). Urea was only 66 to 80% effective as calcium ammonium nitrate in promoting yields of potato (Singh et al., 1979a). It also inhibited root development and foliage yield of maize in pot experiments as compared with no N controls (Court et al., 1964b). For several crops tested, yields were lower with urea than with $(NH_4)_2SO_4$ or NH_4NO_3 (Stephen & Waid, 1963a).

Reduced yields and phytotoxicity resulting from soil application of urea may be caused by urea itself or by products of its transformation in soil (Court et al., 1964a). Unfortunately, most reports dealing with the toxicity of urea come from field and soil culture experiments. Furthermore, most urea applied to soil is rapidly hydrolyzed (Court et al., 1964b). Therefore, determining whether urea itself has been toxic is generally difficult. Urea in the presence of biuret inhibited protein synthesis but had no effect on proteolysis. Neither urea nor biuret alone had any effect (Webster et al., 1957).

The inefficient and toxic effects of urea seem to be associated with high soil pH. Yields of potato in acid soils were the same with both urea and $(NH_4)_2SO_4$ (Tyler et al., 1962). The addition of H^+, via sulfuric or phosphoric acid, alleviated the toxic effects of urea on maize plants (Stephen & Waid, 1963b; Court et al., 1964b). Soil pH increased following urea addition and was attributed to hydrolysis of urea to ammonium carbonate (Court et al., 1964b). Urea toxicity seems to occur in two phases. The first phase (which occurs early and damages young seedlings) seems to be due to accumulation of NH_4^+ from hydrolysis of urea, which under conditions of increased pH (≈ 8.00) becomes toxic because of high vapor pressure of NH_3 (Cooke, 1962; Court et al., 1964b). In the second phase, established plants may be damaged by accumulation of NO_2^-, because high pH inhibits nitrification of NO_2^- (Bingham et al., 1954; Court et al., 1962, 1964b; Stephen & Waid, 1963c). Reduced uptake and utilization of P, due to pH being increased by urea fertilization, was believed to cause the potato to use urea inefficiently (Tyler et al., 1962).

Biuret, which is formed while urea is manufactured, has been considered potentially phytotoxic. Soil application of urea containing biuret $\leq 1\%$ does not cause any toxicity (Singh et al., 1979a; Tyler et al., 1962; Sharma et al., 1975). Therefore, the risk of biuret phytotoxicity due to the use of commercial urea is almost negligible. However, urea for foliar sprays that contain $> 0.5\%$ biuret may affect yields adversely (Singh et al., 1979b; Sharma et al., 1975; Bhargava, 1978).

D. NO_3^- Toxicity

Most plants tolerate high levels of NO_3^- without any physiological disorder, but excess NO_3^- nutrition can be toxic; why it can be toxic is not known (Barker & Mills, 1980). Nitrate is toxic mainly to plants in which NH_4^+ proved to be a superior source of N compared with NO_3^-, such as members of family Ericaceae, some species of genus *Pinus,* and rice.

High levels of NO_3^- may cause Fe deficiency and produce symptoms of well-known "lime-induced" chlorosis. As early as 1923, Willis and Carrero (1923) showed NO_3^--associated chlorosis of rice plants that would be corrected using NH_4^+. Blueberry (Cain, 1952) and azalea (Colgrove & Roberts, 1956) plants grew poorly and developed leaf chlorosis, similar to that of Fe deficiency when NO_3^- was the sole source of N. Since blueberry plants receiving NO_3^- and showing chlorosis contained as much Fe as did those receiving NH_4^+, Cain (1952) concluded that the toxicity of NO_3^- did not affect uptake of Fe, but did affect internal functions. Colgrove and Roberts (1956) suggested that Fe was inactivated in the plant tissue by higher tissue pH promoted by NO_3^-. However, in 'Sitka' spruce (*Picea sitchensis*) and 'Scots' pine (*Pinus sylvestris*), chlorosis due to NO_3^- nutrition was associated with higher organic anions (Nelson & Selby, 1974). The chlorosis caused by NO_3^- seems to be best explained by a competitive chelation theory proposed by Wallace (1971), which suggests that Fe activity in plant tissue was reduced by excess organic anions. Higher levels of organic acid anions due to NO_3^- nutrition were discussed earlier in the present paper.

Greidanus et al. (1972) reported that cranberry plants grown in NO_3^- were N-deficient, since the plant apparently lacked NO_3^- absorption and reduction mechanisms. Maynard and Barker (1971) showed that the dry matter yield of spinach decreased significantly when NO_3^--N supplied in sand culture was 24 and 48 mg/L instead of 12 mg/L. They concluded that the toxic effects of $Ca(NO_3)_2$ at the high concentration can be attributed to a specific NO_3^- effect, since spinach is classified as a halophyte. Nitrate toxicity in strawberry plants was characterized as color change of leaf margins to dark brown with a purplish tinge (Jackson, 1972). In severe cases, interveinal tissue was also affected. Whiptail of cauliflower, a Mo deficiency symptom, can apparently be caused by accumulation of large quantities of NO_3^- in leaves (Agarwala, 1952).

Nitrate nutrition favors accumulation of higher levels of NO_3^- and organic acids such as malate and oxalate. Excessive consumption of NO_3^- and oxalate can be toxic to humans and animals. Therefore, NO_3^- nutrition may lower the food quality of some plants, especially those whose foliage is consumed fresh.

II. TOXICITY OF EXCESS N

The effects of excess N on most plants are well known. Excess N causes many plants to grow vigorously and be dark green, but have some characteristic changes in developmental patterns such as elongation of vegetative phase, delayed maturity, elongation of the entire plant life cycle, and increased succulence. In grain crops, the plants may grow unusually tall, which increases the danger of grain loss through lodging. Plants may become more susceptible to disease and pest attack because of increased succulence. In fruit plants, vegetative growth may be stimulated in lieu of flower or fruit set and development (Mills & Jones, 1979).

Excess N may also change the biochemistry of the plants. Sugarbeets grown on high levels of N accumulate less sugar (Mills & Jones, 1979). Excess N may induce S deficiency in certain plants (Beaton et al., 1971). In coffee plants, excess N may cause imbalances of S and Mn, which in turn adversely affect protein synthesis (Muller, 1966).

The level at which N may become excessive depends largely on the species and, in some cases, on variety. The levels of N in plant tissue beyond which it may be in excess have been defined for some crops. Total N level of 2.2% in apple can reduce fruit quality and yield (Boynton, 1966). Embleton and Jones (1966) found that the most productive range of avocado was when leaf N content varied from 1.6 to 2.0%. The relative growth of spinach plants decreased sharply when NO_3^--N content of leaves rose beyond 0.9% on a dry-weight basis (Maynard & Barker, 1971). Some varieties of potato tolerate less N because of their tendency to produce more haulm and relatively less tuber growth under the influence of high N (de Geus, 1973). An application of 40 to 45 kg/ha of N to horse gram and certain other tropical legumes would cause vegetation to grow very vigorously but significantly reduce flower and pod setting (personal observation, S. Goyal).

The "critical concentration concept" (Ulrich, 1952) is a guide to N fertilization that will help ensure maximum yields and avoid excessive fertilization.

III. CONCLUSIONS

Ammonium and nitrite may be toxic to plant growth, but the actual causes of toxicity are not well understood. Root medium pH, aeration, and their concentrations all play an important role. Moreover, plants differ greatly in their responses to these forms of N.

There is little evidence to indicate that urea itself may be toxic; instead, urea toxicity may be inflicted indirectly through NH_4^+ and NO_2^-. High biuret content of urea can be potentially phototoxic. Excessively high level of NO_3^- may be toxic in a few cases, but in general they do not seem to affect plant growth adversely.

REFERENCES

Adams, F. 1966. Calcium deficiency as a causal agent of ammonium phosphate injury to cotton seedlings. Soil Sci. Soc. Am. Proc. 30:485–488.

Agarwala, S. C. 1952. Relation of nitrogen supply to the molybdenum requirement of cauliflower grown in sand culture. Nature 169:1099.

Ajayi, O., D. N. Maynard, and A. V. Barker. 1970. The effect of potassium on ammonium nutrition of tomato (*Lycopersicon esculentum* Mill). Agron. J. 62:818–821.

Allred, S. E., and A. J. Ohlrogge. 1964. Principles of nutrient uptake from fertilizer bands. VI. Germination and emergence of corn as affected by ammonia and ammonium phosphate. Agron. J. 56:309–313.

Arnon, D. I. 1939. Effect of ammonium and nitrate nitrogen on the mineral composition and sap characteristics of barley. Soil Sci. 48:295–307.

Aslam, M., R. C. Huffaker, D. W. Rains, and K. P. Rao. 1979. Influence of light and ambient carbon dioxide concentrations on nitrate assimilation by intact barley seedlings. Plant Physiol. 63:1205–1209.

Avron, M. 1960. Photophosphorylation by swisschard chloroplasts. Biochim. Biophys. Acta 40:257–272.

Baley, L. M., J. King, and O. L. Gramborg. 1972. The effects of the source of inorganic nitrogen on growth and enzymes of nitrogen assimilation in soybean and wheat cells in suspension cultures. Planta 105:15–24.

Barker, A. V. 1967. Growth and nitrogen distribution patterns in bean (*Phaseolus vulgaris* L.) plants subjected to ammonium nutrition. II. Effects of potassium in a calcium carbonate buffered system. Adv. Front. Plant Sci. 18:7–22.

Barker, A. V., and D. N. Maynard. 1972. Cation and nitrate accumulation in pea and cucumber as influenced by nitrogen nutrition. J. Am. Soc. Hortic. Sci. 97:27–30.

Barker, A. V., D. N. Maynard, B. Mioduchawska, and A. Buch. 1970. Ammonium and salt inhibitions of some physiological processes associated with seed germination. Physiol. Plant. 23:898–907.

Barker, A. V., D. N. Maynard, and W. H. Lachman. 1967. Induction of tomato stem and leaf lesions and potassium deficiency by excessive ammonium nutrition. Soil Sci. 103:319–327.

Barker, A. V., and H. A. Mills. 1980. Ammonium and nitrate nutrition of horticultural crops. Hortic. Rev. 2:395–423.

Barker, A. V., R. J. Volk, and W. A. Jackson. 1965. Effects of ammonium and nitrate nutrition on dark respiration of excised bean leaves. Crop Sci. 5:439–444.

Barker, A. V., R. J. Volk, and W. A. Jackson. 1966a. Root environment acidity as a regulatory factor in ammonium assimilation by the bean plant. Plant Physiol. 41:1193–1199.

Barker, A. V., R. J. Volk, and W. A. Jackson. 1966b. Growth and nitrogen distribution patterns in bean plants (*Phaseolus vulgaris* L.), subjected to ammonium nutrition. I. Effect of carbonates and acidity control. Soil Sci. Soc. Am. Proc. 30:228–232.

Beaton, J. D., S. L. Tisdale, and J. Platore. 1971. Crop responses to sulphur in North America. Sulfur Inst. Tech. Bull. 18:1–18.

Bennedetti, E. D., G. Forti, F. M. Garlaschi, and L. Rosa. 1976. On the mechanism of ammonium stimulation of photosynthesis in isolated chloroplasts. Plant Sci. Lett. 7:85–90.

Bennett, A. C., and F. Adams. 1970a. Calcium deficiency and ammonia toxicity as separate causal factors of $(NH_4)_2HPO_4$-injury to seedlings. Soil Sci. Soc. Am. Proc. 34:255–259.

Bennett, A. C., and F. Adams. 1970b. Concentrations of NH_3 (aq.) required for incipient toxicity to seedlings. Soil Sci. Soc. Am. Proc. 34:259–263.

Bennett, W. F., J. Pesak, and J. J. Hanway. 1964. Effect of NO_3^- and NH_4^+ on growth of corn in nutrient solution sand culture. Agron. J. 56:342–345.

Bergmann, L., W. Grosse, and P. Koth. 1976. Influences of ammonium and nitrate on N-metabolism, malate accumulation and malic enzyme activity in suspension cultures of *Nicotiana tabacum* var. "samsun." Z. Pflanzenphysiol. 80:60–70.

Berner, E., Jr. 1971. Studies on the nitrogen metabolism of barley leaves. II. The effect of nitrate and ammonium on respiration and photosynthesis. Physiol. Plant. Suppl. 6:46–56.

Bhargava, B. S. 1978. Behaviour of rice towards large doses of biuret applied to soil and foliage. Mysore J. Agric. Sci. 12:559–565.

Bingham, F. T., H. D. Chapman, and A. L. Pugh. 1954. Solution-culture studies of nitrite toxicity to plants. Soil Sci. Soc. Am. Proc. 18:305–308.

Blair, G. J., M. H. Miller, and W. A. Mitchell. 1970. Nitrate and ammonium as sources of nitrogen for corn and their influence on the uptake of other ions. Agron. J. 62:530–532.

Blevins, D.G., A. J. Hiatt, and R. H. Lowe. 1974. The influence of nitrate and chloride uptake on expressed sap pH, organic acid synthesis and potassium accumulation in higher plants. Plant Physiol. 54:83–87.

Boynoton, D. 1966. Apple nutrition. p. 1–50. *In* N. F. Childers (ed.) Nutrition of fruit crops. Somerset Press, Inc., Somerville, N.J.

Breteler, H. 1973. A comparison between ammonium and nitrate nutrition of young sugar-beet plants grown in constant acidity. I. Production of dry matter, ionic balance and chemical composition. Neth. J. Agric. Sci. 21:227–244.

Burdine, H. W. 1976. Radish responses to nitroen source. Soil Crop Sci. Soc. Fla. Proc. 35: 59–63.

Burkhart, L. 1938. Ammonium nutrition and metabolism of etiolated seedlings. Plant Physiol. 13:265–293.

Burkholder, P. R., and L. G. Nickell. 1949. Atypical growth of plants. I. Cultivation of virus tumors of Rumex on nutrient agar. Bot. Gaz. Chicago 110:426–432.

Cain, J. C. 1952. A comparison of ammonium and nitrate nitrogen for blueberries. Am. Soc. Hortic. Sci. Proc. 59:161–166.

Canvin, D. T., and C. A. Atkins. 1974. Nitrate, nitrite and ammonia assimilation by leaves: effect of light, carbon dioxide and oxygen. Planta 116:207–224.

Chapman, H. D., and G. F. Liebig, Jr. 1952. Field and laboratory studies of nitrite accumulation in soils. Soil Sci. Soc. Am. Proc. 16:276–282.

Chouteau, J. 1963. E tude de la nutrition nitrique et ammoniacale de la plante de tabac en presence des doses croissantes de bicarbonate dans de milieu nutritif. Ann. Inst. Exp. Tab. Bergerac. 4:319–322.

Clark, H. E., and L. W. Shive. 1934. The influence of the pH of a culture solution on the rates of absorption of ammonium and nitrate nitrogen by the tomato plant. Soil Sci. 37:03–225.

Colgrove, M. S., Jr., and A. N. Roberts. 1956. Growth of azalea as influenced by ammonium and nitrate nitrogen. Am. Soc. Hortic. Sci. Proc. 68:522–536.

Cooke, I. J. 1962. Damage to plant roots caused by urea and anhydrous ammonia. Nature 194: 1262–1263.

Court, M. N., R. C. Stephen, and J. S. Ward. 1962. Nitrite toxicity arising from the use of urea as a fertilizer. Nature 194:1263–1265.

Court, M. N., R. C. Stephen, and J. S. Ward. 1964a. Toxicity as a cause of the inefficiency of urea as a fertilizer. I. Review. J. Soil Sci. 15:42–48.

Court, M. N., R. C. Stephen, and J. S. Ward. 1964b. Toxicity as a cause of the inefficiency of urea as a fertilizer. II. Experimental. J. Soil Sci. 15:49–65.

Cox, W. J., and H. M. Reisenauer. 1973. Growth and ion uptake by wheat supplied nitrogen as nitrate, or ammonium, or both. Plant Soil 38:363–380.

Curtis, D. S. 1949. Nitrite injury on avocado and citrus seedlings in nutrient solution. Soil Sci. 68:441–450.

Davies, D. S. 1973. Control of and by pH. Symp. Soc. Exp. Biol. 27:513–529.

de Geus, J. G. 1973. Fertilizer guide for the tropics and subtropics. Centre d'Etude de l'Azote, Zurich.

De Kock, P. C., and E. A. Kirkby. 1969. Uptake by plants of various forms of nitrogen and effects on plant composition. Tech. Bull. Minist. Agric. Fish. Food (G.B.) 15:7–14.

Dirr, M. A., A. V. Barker, and D. N. Maynard. 1973. Growth and development of leucothoe and rhododendron under different nitrogen and pH regimes. Hortic. Sci. 8:131–132.

Embleton, T. W., and W. W. Jones. 1966. Avocado and mango nutrition. p. 51–76. *In* N. F. Childers (ed.) Nutrition of fruit crops. Somerset Press, Inc., Somerville, N.J.

Epstein, E. 1972. Mineral nutrition of plants. Principles and perspectives. John Wiley & Sons, Inc., New York.

Ferguson, A. R., and E. G. Bollard. 1969. Nitrogen metabolism of *Spirodela oligorrhiza*. I. Utilization of ammonium nitrate, and nitrite. Planta 88:344–352.

Folkes, B. F., A. J. Willis, and E. W. Yemm. 1952. The respiration of barley plants. VII. The metabolism of nitrogen and respiration in seedlings. New Phytol. 51:317–341.

Gamborg, O. L. 1970. The effects of amino acids and ammonium on the growth of plant cells in suspension culture. Plant Physiol. 45:372–375.

Gamborg, O. L., and J. P. Shyluk. 1970. The culture of plant cells with ammonium salts as the sole nitrogen source. Plant Physiol. 45:598–600.

Gibbs, M., and N. Calo. 1959. Factors affecting light induced fixation of carbon dioxide by isolated spinach chloroplasts. Plant Physiol. 34:318–323.

Givan, C. V. 1979. Review. Metabolic detoxification of ammonia in tissues of higher plants. Phytochemistry 18:375–382.

Gouny, P. 1955. Role du calcaire dans l'assimilation de lazote ammoniacale. C. R. Acad. Sci. 241:95–97.

Goyal, S. S. 1974. Studies on the inhibitory effects of ammoniacal nitrogen on growth of radish plants (*Raphanus sativus*) and its reversal by nitrate. Ph.D. Dissertation. Univ. of California, Davis.

Goyal, S. S., and R. C. Huffaker. 1981. Interaction between NO_3^-, NO_2^- and NH_4^+ during assimilation in detached barley leaves. p. 561–568. *In* J. M. Lyons et al. (ed.) Genetic engineering of symbiotic N_2 fixation and conservation of fixed nitrogen. Plenum Press, New York.

Greidanus, T., L. A. Peterson, L. E. Schrader, and M. N. Dana. 1972. Essentiality of ammonium for cranberry nutrition. J. Am. Soc. Hortic. Sci. 97:272–277.

Hanny, J. W., B. L. Fletcher, and H. F. Street. 1959. Studies on the growth of excised roots: IX. The effects of other nutrient ions upon the growth of excised tomato roots supplied with various nitrogen sources. New Phytol. 58:142–154.

Hattori, A. 1958. Studies on the metabolism of urea and other nitrogenous compounds in *Chlorella ellipsoidea*. II. Changes in levels of amino acids and amides during the assimilation of ammonia and urea by nitrogen-starved cells. J. Biochem. 45:57–64.

Haynes, R. J., and K. M. Goh. 1977. Evaluation of potting media for commercial plants. II. Effects of media, fertilizer nitrogen and a nitrification inhibitor on yield and nitrogen uptake of *Callistephus chinensis* (L.) Nees 'Pink Princess.' N.Z. J. Agric. Res. 20:371-381.

Haynes, R. J., and K. M. Goh. 1978. Ammonium and nitrate nutrition of plants. Biol. Rev. Cambridge Philos. Soc. 53:465-510.

Heath, R. L., and R. M. Leech. 1978. The stimulation of CO_2-supported O_2 evolution in intact spinach chloroplasts by ammonium ion. Arch. Biochem. Biophys. 190:221-226.

Heller, R. 1953. Recherches sur la nutrition minerale des tissus vegetaux cultives *in vitro.* Ann. Sci. Nat. Bot. Biol. Veg., 11th ser. 14:1-223.

Heller, R. 1954. Les besoius mineraux des tissus en culture. Annee Biol. 30:361-374.

Hoff, J. E., G. E. Wilcox, and C. M. Jones. 1974. The effect of nitrate and ammonium nitrogen on the free amino acid composition of tomato plants and tomato fruits. J. Am. Soc. Hortic. Sci. 99:27-30.

Hohlt, M. E., D. N. Maynard, and A. V. Barker. 1970. Studies on the ammonium tolerance of some cultivated Solanaceae. J. Am. Soc. Hortic. Sci. 95:345-348.

Holley, K. T., T. A. Pickett, and T. G. Dulin. 1931. A study of ammonia and nitrate nitrogen for cotton: I. Influence on absorption of other elements. Georgia Exp. Stn. Bull. 169.

Ito, O., and K. Kumazawa. 1978. Amino acid metabolism in plant leaf. III. The effect of light on the exchange of ^{15}N-labelled nitrogen among several amino acids in sunflower discs. Soil Sci. Plant Nutr. 24:327-336.

Jackson, D. C. 1972. Nitrate toxicity in strawberries. Agrochemophysica 4:45.

Jackson, W. A., R. E. Johnson, and R. J. Volk. 1974a. Nitrite uptake by nitrogen-depleted wheat seedlings. Physiol. Plant. 32:37-42.

Jackson, W. A., R. E. Johnson, and R. J. Volk. 1974b. Nitrite uptake patterns in wheat seedlings as influenced by nitrate and ammonium. Physiol. Plant. 32:108-114.

Jackson, W. A., and D. C. Williams. 1968. Nitrate-stimulated uptake and transport of strontium and other cations. Soil Sci. Soc. Am. Proc. 32:689-704.

Jones, R. W., and R. W. Sheard. 1978. Accumulation and ability of nitrite in intact aerial leaves. Plant Sci. Lett. 11:285-291.

Kanazawa, T., K. Kanazawa, M. R. Kirk, and J. A. Bassham. 1972. Regulatory effects of ammonia on carbon metabolism in *Chlorella pyrenoidosa* during photosynthesis and respiration. Biochim. Biophys. Acta 265:656-669.

Kato, T. 1980. Nitrogen assimilation in citrus trees. 1. Ammonium and nitrate assimilation by intact roots, leaves, and fruits. Physiol. Plant. 48:416-420.

Kirkby, E. A. 1968. Influence of ammonium and nitrate nutrition on the cation-anion balance and nitrogen and carbohydrate metabolism of white mustard plants grown in dilute nutrient solution. Soil Sci. 105:133-141.

Kirkby, E. A., and K. Mengel. 1967. Ionic balance in different tissues of the tomato plant in relation to nitrate, urea, or ammonium nutrition. Plant Physiol. 42:6-14.

Klemm, K. 1967. The effect of nitrogen form on the nutrient uptake by plants. Bodenkultur 18:210-228.

Krogmann, D. W., A. T. Jagendorf, and M. Avron. 1959. Uncouplers of spinach chloroplasts photosynthetic phosphorylation. Plant Physiol. 34:272-277.

Lee, R. B. 1979. The effect of nitrite on root growth of barley and maize. New Phytol. 83: 615-622.

Lorenz, O. A., J. C. Bishop, B. J. Hoyle, M. P. Zobel, P. A. Minges, L. D. Doneen, and A. Ulrich. 1954. Potato fertilizer experiments in California. Calif. Agr. Exp. Stn. Bull. 744.

Losada, M., J. Herrera, J. M. Maldonado, and A. Paneque. 1973. Mechanism of nitrate reductase reversible inactivation by ammonia in *Chlamydomonas.* Plant Sci. Lett. 1:31-37.

Matsumoto, H., N. Wakiuchi, and E. Takahashi. 1968. Changes of sugar levels in cucumber leaves during ammonium toxicity. Physiol. Plant. 21:1210-1216.

Matsumoto, H., N. Wakiuchi, and E. Takahashi. 1969. The suppression of starch synthesis and the accumulation of uridine diphosphoglucose in cucumber leaves due to ammonium toxicity. Physiol. Plant. 22:537-545.

Matsumoto, H., N. Wakiuchi, and E. Takahashi. 1971a. Changes of starch synthetase activity of cucumber leaves during ammonium toxicity. Physiol. Plant. 24:102-105.

Matsumoto, H., N. Wakiuchi, and E. Takahashi. 1971b. Changes of some mitochondrial enzyme activity of cucumber leaves during ammonium toxicity. Physiol. Plant. 25:353-357.

Mattson, S. 1966. The ionic relationship of soil and plant. Acta Agric. Scand. 16:135-143.

Maynard, D. N., and A. V. Barker. 1969. Studies on the tolerance of plants to ammonium nutrition. J. Am. Soc. Hortic. Sci. 94:235–239.

Maynard, D. N., and A. V. Barker. 1971. Critical nitrate levels for leaf lettuce, radish, and spinach plants. Commun. Soil Sci. Plant Anal. 2:461–470.

Maynard, D. N., A. V. Barker, and W. H. Lachman. 1966. Ammonium-induced stem and leaf lesions of tomato plants. Proc. Am. Soc. Hortic. Sci. 88:561–520.

Maynard, D. N., A. V. Barker, and W. H. Lachman. 1968. Influence of potassium on the utilization of ammonium by tomato plants. Proc. Am. Soc. Hortic. Sci. 92:537–542.

McElhannon, W. S., and H. A. Mills. 1978. Influence of percent NO_3^-/NH_4^+ on growth, N absorption, and assimilation by lima beans in solution culture. Agron. J. 70:1027–1032.

Miller, M. H., C. P. Mamaril, and G. J. Blair. 1970. Ammonium effects on phosphorus absorption through pH changes and phosphorus precipitation at the soil-root interface. Agron. J. 62:524–527.

Mills, H. A., and J. B. Jones, Jr. 1979. Nutrient deficiencies and toxicities in plants: Nitrogen. J. Plant Nutr. 1:101–122.

Milne, M. D., B. H. Schribner, and M. A. Crawford. 1958. Non-ionic diffusion and the excretion of weak acids and bases. Am. J. Med. 24:709–729.

Mohanty, B., and J. S. Fletcher. 1976. Ammonium influence on the growth and nitrate reductase activity of 'Paul's Scarlet' rose suspension cultures. Plant Physiol. 58:152–155.

Mohanty, B., and J. S. Fletcher. 1980. Ammonium influence on nitrogen assimilating enzymes and protein accumulation in suspension cultures of 'Paul's Scarlet' rose. Physiol. Plant. 48:453–459.

Morris, M. D., and J. Giddens. 1963. Response of several crops to ammonium and nitrate forms of nitrogen as influenced by soil fumigation and liming. Agron. J. 55:372–374.

Muller, L. 1966. Coffee nutrition. p. 685–776. *In* N. F. Childers (ed.) Nutrition of fruit crops. Somerset Press, Inc., Somerville, N.J.

Nelson, L. E., and R. Selby. 1974. The effect of nitrogen sources and iron levels on the growth and composition of sitka spruce and scots pine. Plant Soil 41:573–588.

Neyra, C. A., and R. H. Hageman. 1978. Pathway of nitrate assimilation in corn (*Zea mays* L.) leaves: cellular distribution of enzymes and energy sources for nitrate reduction. Plant Physiol. 62:618–621.

Oji, Y. 1974. Comparative studies on the nutriophysiology of inorganic nitrogen assimilation by plants. Ph.D. Thesis. Kyoto Univ., Japan.

Oke, O. L. 1966. Nitrite toxicity to plants. Nature 212:528.

Paul, J. S., K. L. Cornwell, and J. A. Bassham. 1978. Effect of ammonia on carbon metabolism in photosynthesizing isolated cells from *Papaver somniferum* L. Planta 142: 49–54.

Phipps, R. H., and I. S. Cornforth. 1970. Factors affecting the toxicity of nitrite nitrogen to tomatoes. Plant Soil 33:457–466.

Platt, S. G., Z. Plaut, and J. A. Bassham. 1977. Ammonia regulation of carbon metabolism in photosynthesizing leaf discs. Plant Physiol. 60:739–742.

Prianishnikov, D. N. 1951. Nitrogen in the life of plants. Translated by S. A. Wilde. Kramer Business Service Inc., Madison, Wis.

Puritch, G. S., and A. V. Barker. 1967. Structure and function of tomato leaf chloroplasts during ammonium toxicity. Plant Physiol. 42:1229–1238.

Quebedaux, B., and J. L. Ozbun. 1973. Effects on ammonium nutrition on water stress, water uptake and root pressure in *Lycopersicon esculentum* Mill. Plant Physiol. 52:677–679.

Raven, J. A., and F. A. Smith. 1976. Nitrogen assimilation and transport in vascular land plants in relation to intracellular pH regulation. New Phytol. 76:415–431.

Rehfeld, D. W., and R. G. Jensen. 1973. Metabolism of separated leaf cells. III. Effect of calcium and ammonium on product distribution during photosynthesis with cotton cells. Plant Physiol. 52:17–22.

Richter, R., W. Dijkshoorn, and C. R. Vonk. 1975. Amino acids of barley plants in relation to nitrate, urea or ammonium nutrition. Plant Soil 42:601–618.

Riker, A. J., and A. E. Gutsche. 1948. The growth of sunflower tissue *in vitro* on synthetic media with various organic and inorganic sources of nitrogen. Am. J. Bot. 35:227–232.

Riley, D., and S. A. Barber. 1971. Effect of ammonium fertilization on phosphorus uptake as related to root-induced pH changes at the root-soil interface. Soil Sci. Soc. Am. Proc. 35: 301–306.

Robbins, W. J., and M. B. Schmidt. 1938. Growth of excised roots of tomato. Bot. Gaz. Chicago 99:671.

Sahulka, J. 1973. The regulation of glutamate dehydrogenase, nitrite reductase, and nitrate reductase in excised pea roots by nitrite. Biologia Plant. 15:298–301.

Sasseville, D. N., and H. A. Mills. 1979. N form and concentration: effects on N absorption growth, and total N accumulation with southern peas. J. Am. Soc. Hortic. Sci. 104:586–591.

Schrader, L. E., D. Domska, P. E. Jung, Jr., and L. A. Peterson. 1972. Uptake and assimilation of ammonium-N and nitrate-N and their influence on the growth of corn (*Zea mays* L.). Agron. J. 64:690–695.

Sharma, R. C., J. S. Grewal, K. C. Sud, and S. S. Saini. 1975. Effect of biuret content in urea and nitrogen source on the yield and composition of potato tubers. Plant Soil 43:701–705.

Sheat, D. E. G., B. H. Fletcher, and H. E. Street. 1959. Studies on the growth of excised roots. VIII. The growth of excised tomato roots supplied with various inorganic sources of nitrogen. New Phytol. 58:128–141.

Sienko, J. E., and R. A. Plane. 1957. Chemistry. McGraw-Hill Book Co., New York.

Singh, D., M. Singh, and H. S. Sandhu. 1979a. Effect of different nitrogen sources and of biuret in urea on the growth and yield of potato and its nutrient uptake. Indian J. Agric. Sci. 49:641–648.

Singh, D., M. Singh, and H. S. Sandhu. 1979b. Effect of biuret levels in urea applied as foliar spray on the growth and yield of potato and its nutrient uptake. Indian J. Agric. Sci. 48: 649–654.

Skinner, J. C., and H. E. Street. 1954. Studies on the growth of excised roots. II. Observations on the growth of excised groundsel roots. New Phytol. 53:44–67.

Slovacek, R. E., and G. Hind. 1977. Influence of antimycin A and uncouplers on anaerobic photosynthesis in isolated chloroplasts. Plant Physiol. 60:538–542.

Smith, R. C., and E. Epstein. 1964. Ion absorption by shoot tissue: kinetics of potassium and rubidium absorption by corn leaf tissue. Plant Physiol. 39:992–996.

Soon, Y. K., and M. H. Miller. 1977. Changes in rhizosphere due to ammonium and nitrate fertilization and phosphorus uptake by corn seedlings (*Zea mays* L.). Soil Sci. Soc. Am. Proc. 41:77–80.

Spratt, E. D., and J. K. R. Grasser. 1970. The effect of ammonium sulphate treated with a nitrification inhibitor, and calcium nitrate, on growth and N-uptake of spring wheat, rye grass and kale. J. Agric. Sci. 74:111–117.

Stephen, R. C., and J. S. Waid. 1963a. Pot experiments on urea as a fertilizer. I. A comparison of responses by various plants. Plant Soil 18:309–316.

Stephen, R. C., and J. S. Waid. 1963b. Pot experiments on urea as a fertilizer. II. The influence of other fertilizer constituents on the response of maize to urea. Plant Soil 19:97–104.

Stephen, R. C., and J. S. Waid. 1963c. Pot experiments on urea as a fertilizer. III. The influence of rate, form, time and placement. Plant Soil 19:184–192.

Stuart, N. W. 1947. Some studies in azalea nutrition. Natl. Hortic. Mag. 26:210–214.

Stuart, D. M., and J. L. Haddock. 1968. Inhibition of water uptake in sugar beet roots by ammonia. Plant Physiol. 43:345–350.

Syrett, P. J. 1956a. The assimilation of ammonia and nitrate by nitrogen-starved cells of *Chlorella vulgaris.* II. The assimilation of large quantities of nitrogen. Physiol. Plant. 9: 19–27.

Syrett, P. J. 1956b. The assimilation of ammonia and nitrate by nitrogen-starved cells of *Chlorella vulgaris.* III. Differences of metabolism dependent on the nature of the nitrogen source. Physiol. Plant. 9:28–37.

Thompson, L. M., and F. R. Troeh. 1973. Soils and soil fertility. McGraw-Hill Book Co., New York.

Tisdale, S. L., and W. L. Nelson. 1967. Soil fertility and fertilizers. The Macmillan Co., New York.

Torres de Claassen, M. E., and G. E. Wilcox. 1974. Effect of nitrogen form on growth and composition of tomato and pea tissue. J. Am. Soc. Hortic. Sci. 99:171–174.

Trebst, A. V., M. Losada, and D. I. Arnon. 1960. XII. Inhibition of CO_2 assimilation in a reconstituted chloroplast system. J. Biol. Chem. 235:840–844.

Tromp, J. 1962. Interactions in the absorption of ammonium, potassium and sodium ions by wheat roots. Acta Bot. Neerl. 11:147–192.

Tromp, J., and J. C. Ovaa. 1979. Uptake and distribution of nitrogen in young apple trees after application of nitrate or ammonium, with special reference to asparagine and arginine. Physiol. Plant. 45:23–28.

Tyler, K. B., O. A. Lorenz, F. H. Takatori, and J. C. Bishop. 1962. Urea nitrogen for potatoes. Am. Potato J. 36:88–89.

Ulrich, A. 1952. Physiological bases for assessing the nutritional requirements of plants. Annu. Rev. Plant Physiol. 3:207–228.

Vanecko, S., and J. E. Varner. 1955. Studies on nitrite metabolism in higher plants. Plant Physiol. 30:388–390.

Vines, H. M., and R. T. Wedding. 1960. Some effects of ammonia on plant metabolism and a possible mechanism for ammonia toxicity. Plant Physiol. 35:820–825.

Wahhab, A., and H. M. Bhatti. 1957. Effect of various sources of nitrogen on rice paddy yields. Agron. J. 49:114–116.

Wakiuchi, N., H. Matsumoto, and E. Takahashi. 1971. Changes of some enzyme activities of cucumber during ammonium toxicity. Physiol. Plant. 24:248–253.

Wall, M. E. 1940. The role of potassium in plants: III. Nitrogen and carbohydrate metabolism in potassium-deficient plants supplied with either nitrate or ammonium nitrogen. Soil Sci. 49:393–409.

Wallace, A. 1971. The competitive chelation hypothesis of lime induced chlorosis. p. 230–239. *In* A. Wallace (ed.) Regulation of the micronutrient status of plants by chelating agents and other factors. A. Wallace, Los Angeles.

Wallace, A., R. T. Ashcroft, O. R. Lunt, C. B. Joven, and J. A. Clark. 1966. Root pressure exudation in tobacco. I. Some general characteristics. p. 127–135. *In* A. Wallace (ed.) Current topics in plant nutrition. Edwards Brothers, Ann Arbor, Mich.

Wallace, A., and R. T. Mueller. 1957. Ammonium and nitrate absorption from sand culture by rough lemon cuttings. Proc. Am. Soc. Hortic. Sci. 69:183–188.

Wander, I. W., and J. W. Sites. 1956. The effects of ammonium and nitrate nitrogen with and without pH control on the growth of rough lemon seedlings. Proc. Am. Soc. Hortic. Sci. 68:211–226.

Webster, G. C., R. A. Berner, and A. N. Gansa. 1957. The effect of biuret on protein synthesis in plants. Plant Physiol. 32:60–61.

Wedding, R. T., and H. M. Vines. 1959. Inhibition of reduced diphosphopyridine nucleotide oxidation by ammonia. Nature 185:1226–1227.

Weir, B. L., K. N. Paulson, and O. A. Lorenz. 1972. The effect of ammoniacal nitrogen on lettuce (*Latuca sativa*) and radish (*Raphanus sativus*) plants. Soil Sci. Soc. Am. Proc. 36: 462–465.

Weissman, G. S. 1951. Nitrogen metabolism of wheat seedlings as influenced by the ammonium:nitrate ratio and the hydrogen ion concentration. Am. J. Bot. 38:162–174.

Weissman, G. S. 1964. Effects of ammonium and nitrate nutrition on protein level and exudate composition. Plant Physiol. 39:947–952.

Weissman, G. S. 1972. Influence of ammonium and nitrate nutrition on enzymatic activity in soybean and sunflower. Plant Physiol. 49:138–141.

Wilcox, G. E., J. E. Hoff, and C. M. Jones. 1973. Ammonium reduction of calcium and magnesium content of tomato and sweet corn and influence on incidence of blossom end rot of tomato fruits. J. Am. Soc. Hortic. Sci. 98:86–89.

Willis, A. J., and E. W. Yemm. 1955. The respiration of barley plants. VIII. Nitrogen assimilation and the respiration of the root system. New Phytol. 54:163–181.

Willis, L. G., and J. O. Carrero. 1923. Influence of some nitrogenous fertilizers on the development of chlorosis in rice. J. Agric. Res. 24:620–640.

Woo, K. C., and D. T. Canvin. 1980a. Effect of ammonia on photosynthetic carbon fixation in isolated spinach leaf cells. Can. J. Bot. 58:505–510.

Woo, K. C., and D. T. Canvin. 1980b. Effect of ammonia, nitrite, glutamate, and inhibitors of N metabolism on photosynthetic carbon fixation in isolated spinach leaf cells. Can. J. Bot. 58:511–516.

Yemm, E. W., and A. J. Willis. 1956. The respiration of barley plants. IX. The metabolism of roots during the assimilation of nitrogen. New Phytol. 55:229–252.

Yoneyama, T., and T. Yoshida. 1978. Nitrogen mineralization of sewage sludges in soil. Soil Sci. Plant Nutr. 24:139–144.

Yoneyama, T., E. Iwata, and J. Yazaki. 1980. Nitrite utilization in the roots of higher plants. Soil Sci. Plant Nutr. 26:9–23.

Yoshida, D. 1969. Effects of forms of nitrogen supplied on the distribution of nutrients in the tobacco plant. Soil Sci. Plant Nutr. 15:113–117.

Sources of Nitrogen

section II

7

Donald A. Phillips and Theodore M. DeJong
University of California
Davis, California

Dinitrogen Fixation in Leguminous Crop Plants

Associations between certain microorganisms and higher plants result in the reduction of N_2 to NH_3. Roots of legumes often are infected by *Rhizobium* bacteria, and nodules (the structures where N_2 reduction occurs) are formed. Specificities occur between *Rhizobium* and legume species. Classically, *Rhizobium* species have been defined by the host plant they nodulate (Vincent, 1974), but recent developments in microbial genetics have shown that genetic factors determining nodulating ability can be transferred between *Rhizobium* strains and "species" by conjugation (Johnston et al., 1978). The present chapter provides an overview of the basic process of N_2 fixation by the *Rhizobium*-legume symbiosis and suggests how available information and techniques can be applied to enhance N_2 fixation in agronomic legumes.

I. AGRONOMIC SIGNIFICANCE OF SYMBIOTIC N_2 FIXATION IN LEGUMES

The ancient agricultural practice of using leguminous plants to maintain soil fertility was placed in proper perspective in 1934 by careful N balance studies performed over 10 yr with various legumes and cereals (Lyon & Bizzell, 1934). Those data and others obtained under controlled conditions have been used to estimate the role of legumes in the global N economy (Burns & Hardy, 1975). The data in Tables 1 and 2 assess the significance of symbiotic N_2 fixation by legumes in specific agricultural systems. Only grain legumes were considered on a global basis (Table 1), because few data are available on forage legumes harvested in situ by grazing animals. For that same reason, estimates of N_2 fixation by forage legumes in the United States (Table 2) should be considered as a conservative approximation. The

Nitrogen in Crop Production.

Table 1—Magnitude of N_2 fixation in grain legumes.

Continent	Area harvested†	N_2 fixed‡
	ha × 10^3	t × 10^6/yr
Asia	86 718	5.17
North America	29 965	2.42
Africa	18 269	0.96
South America	15 878	0.81
Europe	3 549	0.20
Oceania	269	0.02
World	154 648	9.58

† *FAO Production Yearbook,* 1978. Estimates include oil source legumes (soybean, peanut) and major pulse legumes (edible dry beans, edible dry peas, lentils (*Lens esculenta* Moench.), chickpea (*Cicer arietinum* L.), cowpea [*Vigna unguiculata* (L.) Walp.], pigeon pea [*Cajanus cajan* (L.) Millsp.]).

‡ Calculated by using mean estimates for amount of N_2 fixed by various legumes according to Burns and Hardy (1975).

primary conclusion from Tables 1 and 2 is that symbiotic N_2 fixation by legumes contributes significantly to agricultural production.

II. ROOT NODULES: THE SITE OF N_2 REDUCTION IN LEGUMES

Root nodules represent a structure that has evolved in legumes for the purpose of restricting the pathogenic spread of *Rhizobium* cells, while at the same time maintaining their capacity for N_2 fixation (Fig. 1). Nodule morphology differs among legume species, but the major characteristics include (i) *Rhizobium* cells that have differentiated into bacteroids, a sometimes pleiomorphic form capable of N_2 reduction; (ii) the bacteroid-containing plant cells, which provide energy and O_2 protection for N_2 fixation as they assimilate NH_3; and (iii) vascular tissue for importing photosynthate and exporting fixed-N compounds.

Nitrogenase, the enzyme responsible for reducing N_2 to NH_3, and its specific reducing agent nitrogenase reductase (Hageman & Burris, 1978) are found in bacteroids. Both proteins are rather uniform physically and functionally when compared among diverse N_2-fixing organisms (Burns & Hardy, 1975). Sources of ATP and reductant are required for nitrogenase activity. The ATP probably comes from oxidative phosphorylation because *Rhizobium* does not function under a strictly anaerobic condition, but the source of reductant in *Rhizobium* bacteroids is unclear (Evans & Phillips, 1975). The host plant presumably provides a C source to *Rhizobium* bacteroids, but the identity of the predominant compound is unknown. Whether the plant affects *Rhizobium* metabolism through any hormonal signals is speculative at this point. Obviously, a more thorough understanding of those aspects of the symbiosis is desirable for long-term optimization of the association through genetic modification of the two organisms.

One important point about the biochemical characteristics of nitrogenase is that it shows a lack of specificity in the substrates that can be re-

Table 2—Estimates of area harvested and N_2 fixed by grain and forage legumes in the United States.

Legume crop	Area harvested†	N_2 fixed‡
	ha × 10^3	t × 10^3/yr
Grain legumes		
Soybean [*Glycine max* (L.) Merr.]	28 000	2 352
Dry edible beans (*Phaseolus* sp.)	716	36
Peanut (*Arachis hypogaea* L.)	613	29
Dry edible peas (*Pisum sativum* L.)	53	4
Others	33	2
	29 415	2 423
Forage legumes		
Alfalfa (*Medicago sativa* L.)	10 873	2 196
White clover (*Trifolium repens* L.)	5 431	755
Red clover (*Trifolium pratense* L.)	4 075	566
Sweet clover (*Melilotus* sp.)	1 128	157
Vetch (*Vicia* sp.)	760	65
Trefoil (*Lotus corniculatus* L.)	686	81
Crimson clover (*Trifolium incarnatum* L.)	328	46
	23 281	3 866
Total	52 696	6 289

† Estimates of area harvested for individual grain legumes and alfalfa are from U.S. Crop Reporting Board, 1980. The area of other grain legumes (identified in Table 1) is from *FAO Production Yearbook*, 1978. Estimates for forage legumes other than alfalfa are from Paulling (1969).

‡ Estimates of N_2 fixation were made using mean values for individual crops according to Lyon and Bizzell (1934) and Burns and Hardy (1975).

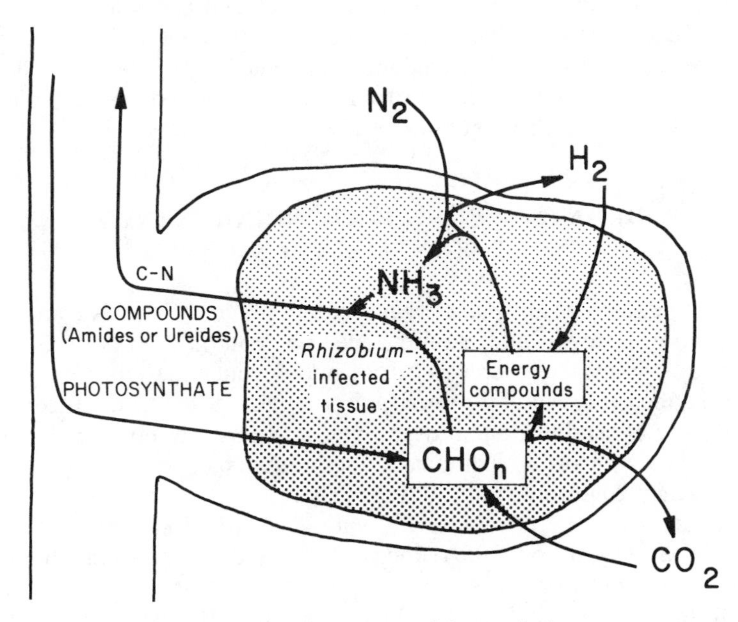

Fig. 1—Schematic of a leguminous root nodule attached to a root segment.

duced (Burns & Hardy, 1975). Many of those substrates are seldom encountered by *Rhizobium* cells. One molecule of that type, C_2H_2, provides the basis for an important indirect assay of nitrogenase activity. Protons, another substrate, are reduced to H_2 by nitrogenase under normal biological conditions. The significant amount of ATP and reductant used to produce H_2 must be considered in any estimate of the energy cost of N_2 fixation. The exact relationship between H_2 evolution by nitrogenase and N_2 reduction is not known, but the two reactions are coupled closely (Mortenson, 1978). During the past 40 yr, many workers have considered the energy lost in producing H_2, and have shown that uptake hydrogenase systems may function to recycle H_2 and recover some energy (Phillips, 1980). The potential agronomic significance of uptake hydrogenase activity and its occurrence in certain *Rhizobium* strains is discussed in Chapter 8 of this book.

The plant cell containing *Rhizobium* bacteroids functions in several roles important for symbiotic N_2 fixation. First, the presence of leghemoglobin prevents O_2 inhibition of nitrogenase activity at the same time it facilitates transport of O_2 to bacteroids, where it presumably permits the operation of oxidase systems that generate ATP (Appleby, 1974).

Second, the plant cell cytoplasm probably functions in metabolizing ammonia excreted by *Rhizobium* bacteroids. The biochemical pathways active in legumes that transport amides have been described (Scott et al., 1976), and reactions occurring in plants that transport ureides are being elucidated (Schubert & Boland, 1984). The importance of the ammonia assimilation pathway in the root nodule is that inhibition of nitrogenase by NH_3 is precluded at the same time that the N-containing compounds formed are being transported to other parts of the legume.

A third important function of the plant cell cytoplasm in root nodules is the reduction of CO_2 respired by bacteroids (Lawrie & Wheeler, 1975; Christeller et al., 1977). That reaction, presumably mediated by PEP carboxylase, could have the effect of providing C compounds needed to transport fixed N out of the root nodule.

III. INTEGRATING ROOT NODULES INTO THE INTACT PLANT

Root nodules, as any plant organ, require reduced C and N compounds for initiation, growth, and maintenance. The complex flows of C and N compounds necessary to integrate root nodules into the intact plant have been described for several legumes (Atkins et al., 1978). Undoubtedly, it is through the allocation of photosynthate, and possibly hormonal signals, that the plant controls N_2 fixation and maintains a balance between C and N metabolism as environmental conditions change.

Legumes growing symbiotically with *Rhizobium* can be limited primarily by either reduced C or reduced N. Early demonstrations that long-term CO_2 enrichment of air surrounding legumes promoted N_2 fixation and growth in red clover and alfalfa (Wilson et al., 1933) were supported by more recent experiments with soybean (Hardy & Havelka, 1976). Although those data strengthen the concept that such plants are C-limited, other studies clearly show that growth can be N-limited in both grain and forage legume seedlings dependent on *Rhizobium* (Fred et al., 1938; Fishbeck & Phillips, 1981). One hypothesis that reconciles those observations is that symbiotically grown legumes endure a period of N-limited growth before achieving N sufficiency and becoming C-limited. Presumably, such plants have insufficient reduced N available to construct an optimum photosynthetic apparatus (Bethlenfalvay et al., 1978; DeJong & Phillips, 1981). If one accepts that hypothesis, then several implications must be considered.

First, *Rhizobium*-dependent plants may never achieve the productivity of those that have adequate combined N available during the period of N-limited growth. Apparently, that possibility is more likely in annual grain legumes than in perennial forage plants. Limited data from soybean indicate that seed yields are greater when both NO_3^- and N_2 are available to the plant than when plants are totally dependent on N_2 (Harper, 1974). Alfalfa plants dependent on *Rhizobium* had lower herbage yield and crude protein content than plants supplied with 8.0 m*M* NH_4NO_3 for the first three harvests, but the differences declined with each harvest; at the fourth harvest, dry matter production and protein yield did not differ significantly between treatments (Fishbeck & Phillips, 1981).

A second implication of a N-limited growth period preceding the C-limited phase is that studies with applied objectives should encompass both periods. Thus, an increase in N_2 fixation and growth provided by a *Rhizobium* mutant or a host genotype during seedling development (Wacek & Brill, 1976; Maier & Brill, 1978) cannot necessarily be extrapolated into the period when plant growth is C-limited. For that reason, long-term growth chamber studies or, preferably, field trials are desirable.

IV. EFFECTS OF SOIL N ON SYMBIOTIC N_2 FIXATION

Soil N is one of many environmental factors that affect symbiotic N_2 fixation. Legumes that form symbioses with *Rhizobium* can use either soil N or N_2, but concentrations of available N above 2.0 m*M* generally decrease N_2 fixation. The presence of excess soil N during seedling development can inhibit root nodule formation, and the addition of combined N to a nodulated legume can decrease N_2 fixation by causing nodule senescence. The molecular mechanisms underlying both types of inhibitory effects are poorly understood. Increasing available combined N from 0.0 to 2.0 m*M* often promotes growth and N_2 fixation of symbiotically grown legumes. The latter phenomenon indicates that growth of legume seedlings entirely dependent upon *Rhizobium* for a source of N is primarily N-limited.

Little information is available on the factors that control the use of soil N by legumes as opposed to N_2. Most studies indicate that soil N is a primary resource and symbiotic N_2 fixation is a secondary alternative used when combined N available in the soil solution is inadequate for optimum growth. In general, the capacity of legumes to use both soil N and N_2 is an agronomic advantage, because most soil NO_3^- that is not used by plants will be lost from the system by leaching or denitrification. When both soil N and N_2 are available to a plant, their relative use may be affected by the energy costs associated with their assimilation. One recent study reported that root systems of nodulated soybeans respired more CO_2 than those of plants receiving abundant NO_3^- (Ryle et al., 1979). Work with subterranean clover supports the concept that N_2 fixation requires more energy than NO_3^- utilization; but in that system, the greater energy cost of N_2 fixation was reflected by both a decrease in crude protein (reduced N) content and a change in the growth coefficient (Silsbury, 1977).

The agronomic implications of interactions between soil N and symbiotic N_2 fixation are complex, but one controlling factor in all agricultural situations is the economic cost of a particular N source. Recent data show a distinct fiscal advantage for using N_2 fixation in beef-forage systems (Jacobs & Stricker, 1976), but fertilizer N presently is used on prime agricultural land for monocultures of nonlegumes. If adequate data are collected on long-term economic factors, including those associated with soil erosion in various legume/nonlegume cropping systems, more N_2 fixation might be justified on prime agricultural land (Heichel, 1978). Alternatively, fertilizer N use may increase if inexpensive methods, such as inorganic catalysts, are discovered for reducing N_2 under atmospheric pressure at field sites.

V. MEASURING SYMBIOTIC N_2 FIXATION IN FIELD-GROWN LEGUMES

Most agronomic investigators use measurements of N_2 fixation to identify biological materials or management strategies that will enhance the symbiotic process. They need an accurate, rapid estimate of N_2 fixation to guide their research. Unfortunately, techniques available for measuring N_2 fixation by field-grown legumes tend to be more qualitative than quantitative. The quantitative techniques that can be used in controlled environments are not entirely suitable for field applications. A discussion of the major methods for estimating N_2 fixation is available (Hardy & Holsten, 1977), and several more detailed reviews on the use of ^{15}N for measuring N_2 fixation in field environments have been presented (Hauck & Bremner, 1976; Rennie et al., 1978).

Direct measures of N_2 fixation are based on either N accumulation in the absence of combined N or $^{15}N_2$ incorporation. The first alternative is not suitable for precise field studies, because most agricultural soils contain significant amounts of N. Methods employing $^{15}N_2$ have the advantage of permitting short-term measurements of N_2 reduction, but exposing legumes to the expensive, gaseous $^{15}N_2$ requires special equipment and potentially abnormal physiological conditions.

Indirect methods generally have been used for estimating N_2 fixation in field environments. One classical method calculates N_2 incorporation from a total N balance sheet measured over 5 to 10 yr in soil frames containing legumes or nonlegumes (Lyon & Bizzell, 1934). The length of time required for such studies has discouraged its use. Indirect studies based on the total N difference between a legume and a grass or a nonnodulated legume provide information in single-season trials, but grass and legume root systems undoubtedly differ in their efficiency of soil N utilization. Furthermore, limited observations show that nonnodulating and nodulating soybean isolines have much different root morphologies. The nonnodulating lines, which are totally dependent on soil N, have larger root systems and presumably explore a different volume of soil than nodulated plants (L. E. Williams & D. A. Phillips, unpublished data).

Acetylene-dependent C_2H_4 production by nodulated legumes, less rigorously termed C_2H_2 reduction, is another indirect measure of symbiotic N_2 fixation (Hardy et al., 1968). That assay, which is based on the fact that nitrogenase reduces C_2H_2 to C_2H_4, is rapid, inexpensive, and more sensitive than other methods available for estimating N_2 fixation. It has been a valuable tool in many laboratory studies. The disadvantage of using the C_2H_2-reduction assay in the field is that the technique is so sensitive and rapid that many environmental changes produce significant fluctuations in the estimated rate of N_2 fixation. For that reason, assays must be repeated frequently in order to obtain valid estimates of seasonal N_2 fixation (Hardy et al., 1968). The destructive nature of the assay obviously necessitates separate subplots for N_2 fixation and yield measurements.

One reasonable method for measuring N_2 fixation by field-grown legumes is the indirect ^{15}N-dilution assay, because yield and N_2 fixation estimates can be made on the same plants. The theoretical concept of analyzing use of soil nutrients from two sources was presented by Fried and Dean (1953); in the same year, a specific example of estimating N_2 fixation by dilution of available $^{15}NH_4^+$ was reported (Newton et al., 1953). The technique is based on the fact that a N_2-fixing organism uses available combined ^{15}N and dilutes it with atmospheric $^{14}N_2$ in proportion to the amount of $^{14}N_2$ fixed. One control in a field trial must be a reference plant, such as a nonnodulating legume isoline or a grass, that uses no N_2. The fraction of legume N derived from N_2 is calculated by the following formula:

$$\text{Fraction of legume N from } N_2 = 1 - \frac{\text{Atom \% excess } ^{15}\text{N in legume}}{\text{Atom \% excess } ^{15}\text{N in reference crop}} .$$

Concerns about root morphology of a reference crop suggest that the ^{15}N-dilution technique is only a qualitative measure of N_2 fixation. Nevertheless, treatments such as plant density and legume fraction apparently alter N_2 fixation significantly in clover/grass plots studied with this assay (Phillips & Bennett, 1978). The major advantage of the ^{15}N-dilution assay is that a single sample at harvest integrates all N_2 fixation throughout the season. In grain legumes, concerns about isotopic fractionation among plant parts probably are minimal, because it is known that for soybean, the

^{15}N enrichment of seeds is similar to ^{15}N enrichment of the entire plant (Shearer et al., 1980).

In order to use the ^{15}N-dilution assay, the ^{15}N content of available soil N must differ significantly from that of N_2. Normal processes within the rhizosphere increase the natural abundance of ^{15}N in soil N over that in N_2, and this fact has been used to measure N_2 fixation in soybean (Kohl et al., 1980). Such differences in natural abundance, however, are extremely small, and if one is studying a small effect of the plant or *Rhizobium* genotype on N_2 fixation, it may be advantageous to increase that difference by applying a small amount of N highly enriched with ^{15}N. Theoretically, the ^{15}N-dilution assay can be applied in a reverse manner with ^{15}N-depleted materials. Such an approach would require that more total N be used to produce an equivalent change in soil ^{15}N. Additional problems of isotope discrimination and variations in natural abundance of ^{15}N would demand more attention than in studies with ^{15}N-enriched materials (Bremner, 1977).

VI. APPLYING BASIC PRINCIPLES TO ENHANCE SYMBIOTIC N_2 FIXATION IN LEGUMES

Symbiotic N_2 fixation by legumes already contributes large quantities of N to agricultural systems (Tables 1 and 2). Although the cost of fertilizer N may increase in the future, greater dependence on symbiotic N_2 fixation is a viable alternative only if agronomic yields are not decreased.

Symbiotic N_2 fixation in legumes can be increased by manipulating either the *Rhizobium* bacteria or the host plant. Classical improvements of commercial *Rhizobium* strains have resulted from screening for maximum promotion of plant growth in the absence of combined N. A modification of that technique examines the effect of a particular physiological trait, such as an uptake hydrogenase, by comparing plant growth in the presence of strains possessing or lacking the character (Albrecht et al., 1979). Such methods identify superior *Rhizobium* strains, but they do not create strains better than those that have survived in nature.

Various techniques can be used to create potentially superior *Rhizobium* strains that are not readily available in nature. Maier and Brill (1978) produced *Rhizobium* mutants with high C_2H_2-reduction activity. The strains enhanced N_2 fixation by soybeans grown under controlled conditions but had no beneficial effect in the field (Thomas et al., 1983). Another mutant strain of *R. japonicum* enhanced N_2 fixation and dry-matter accumulation of soybeans during the first 6 weeks of growth in a controlled environment and increased seed yield significantly in 2 yr of field trials (Williams & Phillips, 1983). Those successful results were obtained in a California soil which contained very few indigenous soybean rhizobia, and it seems doubtful that the strain would overcome the severe competition from other rhizobia in a normal soybean production area. Another promising technique for *Rhizobium* improvement involves genetic transfer of traits through bacterial conjugation. Genetic determinants for hydrogenase activity have been transferred between *R. leguminosarum* strains (Brewin et al., 1980), and the rhizobial plasmid pIJ1008 is quite effective at improving symbiotic performance of *R. leguminosarum* strains (DeJong et al., 1982).

Traditional plant-breeding programs in legumes have not been directed toward enhancing symbiotic N_2 fixation. Phillips (1980) summarized the few efforts that have been made in that direction. Several groups recently have screened alfalfa populations for C_2H_2-reduction capability with a commercial mixture of *Rhizobium* strains in the absence of soil N (Seetin & Barnes, 1977; Duhigg et al., 1978). Alfalfa populations produced by phenotypic recurrent selection for high C_2H_2 reduction showed the desired characteristics under glasshouse but not field conditions (Heichel et al., 1981). The plants selected in those experiments tended to form associations with the most effective rhizobial genotype present in the mixed inoculum (Hardarson et al., 1982). The disappointing performance of the selected plants under field conditions may have been associated with an inadequate number of the effective rhizobial genotypes in the field or with the fact that alfalfa genotypes selected for N_2 fixation often perform quite differently on soil N (Phillips et al., 1982).

An alternative method of breeding alfalfa for increased N_2 fixation may produce agronomically important results. That method combines two concepts (Phillips et al., 1982). First, the plants are selected for physiological performance under both N_2- and NH_4NO_3-dependent growth conditions. Second, under N_2-dependent conditions a single genotype of *Rhizobium* is present. This strategy has produced populations of plants which yielded more dry matter with higher reduced N concentration under glasshouse conditions (Teuber et al., 1984) and during the seeding year in the field (L. R. Teuber & D. A. Phillips, unpublished data). An important basis of those results apparently is the fact that the selected plants can induce a better symbiotic performance from diverse rhizobial genotypes (D. A. Phillips & L. R. Teuber, unpublished data). Several years of additional data from other field sites will be required to conclude that the technique is worthy of wider application.

One alternative to broad screening of diverse genotypes is a direct testing of host plant lines that have traits presumed to favor N_2 fixation. No successful examples of this strategy are available, but several related attempts have been made. In soybean, N_2 fixation declines with leaf senescence while seeds are filling (Harper, 1974; Thibodeau & Jaworski, 1975). Comparison of C_2H_2-reduction activities in male-sterile and male-fertile soybean isolines indicate that sterile lines can fix more N_2 than fertile lines in some cases (Imsande & Ralston, 1982; Riggle et al., 1984) but not in others (Wilson et al., 1978). The discovery of fertile soybean genotypes that maintained photosynthetically active green leaves until after seeds matured prompted the suggestion that such a phenotype might promote N_2 fixation and increase yield (Abu-Shakra et al., 1978). Subsequent work has not supported that concept (Phillips et al., 1984), but the results have shown that the $dt_1dt_1e_1e_1$ genotype present in some high-yielding midwestern cultivars is associated with the delayed leaf senescence phenotype under certain field conditions (Pierce et al., 1984).

Some basic principles, such as decreasing soil N to prevent inhibition of symbiotic N_2 fixation and the selection of superior, naturally occurring *Rhizobium* strains, are already being used to increase symbiotic N_2 fixation in commercial agriculture. The future for further application of basic principles to increase symbiotic N_2 fixation is bright. The development of new genetic techniques may facilitate an increase in *Rhizobium* efficiency through genetic manipulation. If such bacterial strains can be introduced successfully to the soil, they may contribute significantly to increasing agricultural production. At the same time, very promising results have been obtained in attempts to breed the host legume for increased N_2 fixation and soil N assimilation. It is possible that the leguminous crop plant itself may be useful as a tool for manipulating and using the unmanageable populations of *Rhizobium* found in most agricultural environments. As more agronomists become active in studying such problems, the true potential for enhancing symbiotic N_2 fixation will become known and, hopefully, realized.

REFERENCES

Abu-Shakra, S. S., D. A. Phillips, and R. C. Huffaker. 1978. Nitrogen fixation and delayed leaf senescence in soybeans. Science 199:973–975.

Albrecht, S. L., R. J. Maier, F. J. Hanus, S. A. Russell, D. W. Emerich, and H. J. Evans. 1979. Hydrogenase in *Rhizobium japonicum* increases nitrogen fixation by nodulated soybeans. Science 203:1255–1257.

Appleby, C. A. 1974. Leghemoglobin. p. 521–554. *In* A. Quispel (ed.) The biology of nitrogen fixation. North-Holland Publ. Co., Amsterdam.

Atkins, C. A., D. F. Herridge, and J. S. Pate. 1978. The economy of carbon and nitrogen in nitrogen-fixing annual legumes: experimental observations and theoretical considerations. p. 211–240. *In* Isotopes in biological dinitrogen fixation. Int. Atomic Energy Agency, Vienna.

Bethlenfalvay, G. J., S. S. Abu-Shakra, and D. A. Phillips. 1978. Interdependence of nitrogen nutrition and photosynthesis in *Pisum sativum* L. II. Host plant response to nitrogen fixation by *Rhizobium* strains. Plant Physiol. 62:131–133.

Bremner, J. M. 1977. Use of nitrogen-tracer techniques for research on nitrogen fixation. p. 335–352. *In* A. Ayanaba and P. J. Dart (ed.) Biological nitrogen fixation in farming systems of the tropics. John Wiley & Sons, Inc., New York.

Brewin, N. J., T. M. DeJong, D. A. Phillips, and A. W. B. Johnston. 1980. Co-transfer of determinants for hydrogenase activity and nodulation ability in *Rhizobium leguminosarum*. Nature 288:77–79.

Burns, R. C., and R. W. F. Hardy. 1975. Nitrogen fixation in bacteria and higher plants. Springer-Verlag, New York.

Christeller, J. T., W. A. Laing, and W. D. Sutton. 1977. Carbon dioxide fixation by lupin root nodules. I. Characterization, association with phosphoenolpyruvate carboxylase, and correlation with nitrogen fixation during nodule development. Plant Physiol. 60:47–50.

DeJong, T. M., N. J. Brewin, A. W. B. Johnston, and D. A. Phillips. 1982. Improvement of symbiotic properties in *Rhizobium leguminosarum* by plasmid transfer. J. Gen. Microbiol. 128:1829–1838.

DeJong, T. M., and D. A. Phillips. 1981. Nitrogen stress and apparent photosynthesis in symbiotically-grown *Pisum sativum* L. Plant Physiol. 68:309–313.

Duhigg, P., B. Melton, and A. Baltensperger. 1978. Selection for acetylene reduction rates in 'Mesilla' alfalfa. Crop Sci. 18:813–816.

Evans, H. J., and D. A. Phillips. 1975. Reductants for nitrogenase and relationships to cellular electron transport. p. 389–420. *In* W. D. P. Stewart (ed.) Nitrogen fixation by free-living micro-organisms. Cambridge Univ. Press, Cambridge, England.

Fishbeck, K. A., and D. A. Phillips. 1981. Combined nitrogen and vegetative regrowth of symbiotically-grown alfalfa. Agron. J. 73:975–978.

Food and Agriculture Organization. 1979. FAO production yearbook 1978. Vol. 32. United Nations FAO, Rome.

Fred, E. B., P. W. Wilson, and O. Wyss. 1938. Light intensity and the nitrogen hunger period in the Manchu soybean. Proc. Natl. Acad. Sci. USA 24:46–52.

Fried, M., and L. A. Dean. 1953. A concept concerning the measurement of available soil nutrients. Soil Sci. 73:263–271.

Hageman, R. V., and R. H. Burris. 1978. Nitrogenase and nitrogenase reductase associate and dissociate with each catalytic cycle. Proc. Natl. Acad. Sci. USA 75:2699–2702.

Hardarson, G., G. H. Heichel, D. K. Barnes, and C. P. Vance. 1982. Rhizobial strain preference of alfalfa populations selected for characteristics associated with N_2 fixation. Crop Sci. 22:55–58.

Hardy, R. W. F., and U. D. Havelka. 1976. Photosynthate as a major factor limiting nitrogen fixation by field-grown legumes with emphasis on soybeans. p. 421–439. *In* P. S. Nutman (ed.) Symbiotic nitrogen fixation in plants. Cambridge Univ. Press, Cambridge, England.

Hardy, R. W. F., and R. D. Holsten. 1977. Methods for measurement of dinitrogen fixation. p. 451–486. *In* R. W. F. Hardy (ed.) A treatise on dinitrogen fixation. Section IV. J. Wiley & Sons, Inc., New York.

Hardy, R. W. F., R. D. Holsten, E. K. Jackson, and R. C. Burns. 1968. The acetylene-ethylene assay for N_2 fixation: laboratory and field evaluation. Plant Physiol. 43:1185–1207.

Harper, J. E. 1974. Soil and symbiotic nitrogen requirements for optimum soybean production. Crop Sci. 14:255–260.

Hauck, R. D., and J. M. Bremner. 1976. Use of tracers for soil and fertilizer nitrogen research. Adv. Agron. 28:219–266.

Heichel, G. H. 1978. Stabilizing agricultural energy needs: role of forages, rotations, and nitrogen fixation. J. Soil Water Conserv. 33:279–282.

Heichel, G. H., C. P. Vance, and D. K. Barnes. 1981. Evaluating elite alfalfa lines for N_2-fixation under field conditions. p. 217–232. *In* J. M. Lyons et al. (ed.) Genetic engineering of symbiotic nitrogen fixation and conservation of fixed nitrogen. Plenum Press, New York.

Imsande, J., and E. J. Ralston. 1982. Dinitrogen fixation in male-sterile soybeans. Plant Physiol. 69:745–746.

Jacobs, V. E., and J. A. Stricker. 1976. Economic comparisons of legume nitrogen and fertilizer nitrogen in pastures. p. 109–127. *In* C. S. Hoveland (ed.) Biological N fixation in forage-livestock systems. Am. Soc. of Agron. Spec. Pub. no. 28. Am. Soc. of Agron., Madison, Wis.

Johnston, A. W. B., J. L. Beynon, A. V. Buchanan-Wollaston, S. M. Setchell, P. R. Hirsch, and J. E. Beringer. 1978. High frequency transfer of nodulating ability between strains and species of *Rhizobium.* Nature 276:635–636.

Kohl, D. H., G. Shearer, and J. E. Harper. 1980. Estimates of N_2 fixation based on differences in the natural abundance of ^{15}N in nodulating and nonnodulating isolines of soybeans. Plant Physiol. 66:61–65.

Lawrie, A. C., and C. T. Wheeler. 1975. Nitrogen fixation in the root nodules of *Vicia faba* L. in relation to the assimilation of carbon. II. The dark fixation of carbon dioxide. New Phytol. 74:437–445.

Lyon, T. L., and J. A. Bizzell. 1934. A comparison of several legumes with respect to nitrogen accretion. J. Am. Soc. Agron. 26:651–656.

Maier, R. J., and W. J. Brill. 1978. Mutant strains of *Rhizobium japonicum* with increased ability to fix nitrogen for soybean. Science 201:448–450.

Mortenson, L. E. 1978. The role of dihydrogen and hydrogenase in nitrogen fixation. Biochimie 60:219–223.

Newton, J. W., P. W. Wilson, and R. H. Burris. 1953. Direct demonstration of ammonia as an intermediate in nitrogen fixation by *Azotobacter.* J. Biol. Chem. 204:445–451.

Paulling, J. R. 1969. Trends in forage crops varieties. U.S. Fed. Ext. Ser. USDA. U.S. Government Printing Office, Washington, D.C.

Phillips, D. A. 1980. Efficiency of symbiotic nitrogen fixation in legumes. Annu. Rev. Plant Physiol. 31:29–49.

Phillips, D. A., and J. P. Bennett. 1978. Measuring symbiotic nitrogen fixation in rangeland plots of *Trifolium subterraneum* L. and *Bromus mollis* L. Agron. J. 70:671–674.

Phillips, D. A., R. O. Pierce, S. A. Edie, K. W. Foster, and P. F. Knowles. 1984. Delayed leaf senescence in soybean. Crop Sci. 24:518–522.

Phillips, D. A., L. R. Teuber, and S. S. Jue. 1982. Variation among alfalfa genotypes for reduced nitrogen concentration. Crop Sci. 22:606–610.

Pierce, R. O., P. F. Knowles, and D. A. Phillips. 1984. Inheritance of delayed leaf senescence in soybean. Crop Sci. 24:515–517.

Rennie, R. J., D. A. Rennie, and M. Fried. 1978. Concepts of ^{15}N usage in dinitrogen fixation studies. p. 107–130. *In* Isotopes in biological dinitrogen fixation. Int. Atomic Energy Agency, Vienna.

Riggle, B. D., W. J. Wiebold, and W. J. Kenworthy. 1984. Effect of photosynthate source-sink manipulation on dinitrogen fixation of male-fertile and male-sterile soybean isolines. Crop Sci. 24:5–8.

Ryle, G. J. A., C. E. Powell, and A. J. Gordon. 1979. The respiratory costs of nitrogen fixation in soybean, cowpea, and white clover. J. Exp. Bot. 30:145–153.

Schubert, K. R., and M. J. Boland. 1984. The cellular and intracellular organization of the reactions of ureide biogenesis in nodules of tropical legumes. p. 445–451. *In* C. Veeger and W. E. Newton (ed.) Advances in nitrogen fixation research. Martinus Nijhoff/Dr. Junk Publ., The Hague.

Scott, D. B., K. J. F. Farnden, and J. G. Robertson. 1976. Ammonia assimilation in lupine nodules. Nature 263:703–705.

Seetin, M. W., and D. K. Barnes. 1977. Variation among alfalfa genotypes for rate of acetylene reduction. Crop Sci. 17:783–787.

Shearer, G., D. H. Kohl, and J. E. Harper. 1980. Distribution of ^{15}N among plant parts of nodulating and nonnodulating isolines of soybeans. Plant Physiol. 66:57–60.

Silsbury, J. H. 1977. Energy requirement for symbiotic nitrogen fixation. Nature 267:149–150.

Teuber, L. R., R. P. Levin, T. C. Sweeney, and D. A. Phillips. 1984. Selection for N concentration and forage yield in alfalfa. Crop Sci. 24:553–558.

Thibodeau, P. S., and E. G. Jaworski. 1975. Patterns of nitrogen utilization in the soybean. Planta 127:133–147.

Thomas, R. J., D. Jokinen, and L. E. Schrader. 1983. Effect of *Rhizobium japonicum* mutants with enhanced N_2 fixation activity on N transport and photosynthesis of soybeans during vegetative growth. Crop Sci. 23:453–456.

United States Crop Reporting Board. 1980. Acreage. USDA, Washington, D.C.

Vincent, J. M. 1974. Root-nodule symbiosis with *Rhizobium.* p. 265–341. *In* A. Quispel (ed.) The biology of nitrogen fixation. North-Holland Publ. Co., Amsterdam.

Wacek, T. J., and W. J. Brill. 1976. Simple, rapid assay for screening nitrogen-fixing ability in soybean. Crop Sci. 16:519–522.

Williams, L. E., and D. A. Phillips. 1983. Increased soybean productivity with a *Rhizobium japonicum* mutant. Crop Sci. 23:246–250.

Wilson, R. F., J. W. Burton, J. A. Buck, and C. A. Brim. 1978. Studies on genetic male-sterile soybeans. I. Distribution of plant carbohydrate and nitrogen during development. Plant Physiol. 61:838–841.

Wilson, P. W., E. B. Fred, and M. R. Salmon. 1933. Relation between carbon dioxide and elemental nitrogen assimilation in leguminous plants. Soil Sci. 35:145–163.

8

David W. Emerich
University of Missouri
Columbia, Missouri

Harold J. Evans
Oregon State University
Corvallis, Oregon

Enhancing Biological Dinitrogen Fixation in Crop Plants

Earlier chapters in this volume have emphasized that N is the most common limiting nutrient in agriculture. Atmospheric N (N_2) is potentially available to all plants, but relatively few species are able to assimilate N from this vast pool. Crop plants that are capable of fixing atmospheric N_2 have a distinct advantage in agriculture. The legumes are the only agriculturally important crop plants capable of fixing atmospheric N_2 in significant amounts. While the cereals have been improved significantly by breeding programs and improved fertilizer practices, the legumes have not. The possibility exists for increased crop yields and improved seed quality of legumes via enhanced N_2 fixation through three general approaches: (i) plant breeding to select superior host plants, (ii) optimizing the symbiotic interaction between host and bacteria, and (iii) genetic and biochemical selection of *Rhizobium* strains for improved N_2 fixation potential. We will discuss each of these approaches briefly, and then describe in more detail two promising near-term approaches for enhancing N_2 fixation under item (iii). Postgate (1977) and Phillips (1980) have recently reviewed both long- and short-term possibilities for improving N_2 fixation.

Nitrogen in Crop Production.

I. APPROACHES

A. Plant Breeding

Progress toward improving yields of legumes through plant-breeding programs or improved fertilization practices has not approached that obtained with the cereals. Recent advances, particularly in the genetics and biochemistry of symbiotic N_2 fixation, have provided plant breeders with a new set of characteristics for which to screen. Quebedeaux et al. (1975) and Havelka and Hardy (1976) reported that photosynthate may be the primary factor limiting N_2 fixation potential. Thus, the selection of plant varieties that produce and/or transport more photosynthate to the root nodules would presumably increase N_2 fixation capacity. Bach et al. (1958) and Pate (1962) determined that sucrose and glucose with minor amounts of organic acids are the primary photosynthetic products transported to the root nodules. However, organic acids are more effective substrates for support of N_2 fixation than sucrose or glucose (Bergersen, 1958; Bergersen & Turner, 1967; Burris & Wilson, 1939). Selection of plant cultivars with capacities for supplying greater quantities of effective substrates to nodules would be expected to increase the rate of N_2 fixation.

Abu-Shakra et al. (1978) have described a soybean cultivar [*Glycine max* (L.) Merr.] that exhibits enhanced fixation. Most commercial soybean varieties senesce, and N_2 fixation declines during the pod-filling stage of growth, but the cultivar described by Abu-Shakra et al. (1978) delays the senescence process and increased yields are observed (Phillips, 1980).

B. Symbiotic Interactions

Traditionally, attempts to improve N_2 fixation have focused on either the plant cultivar or the bacterial symbiont. However, the formation and maintenance of an effective symbiosis requires an effective interaction of both symbionts. Matched selections of plant cultivar and *Rhizobium* strain may produce a more efficient symbiosis; but the broad, double screening required by this approach has been rarely attempted. Seetin and Barnes (1977) screened alfalfa cultivars (*Medicago sativa* L.) with commercial inoculum for enhanced fixation and found combinations that increased the annual N_2 fixation by 41%. The complexity of the symbiosis has retarded elucidation of the specific physiological changes that are associated with the increases.

C. *Rhizobium* Biochemistry and Genetics

Developments that have the greatest potential for enhancing N_2 fixation are those from the biochemistry and genetics of the N_2 fixation process. Recent research has provided considerable insight into the factors that limit the fixation process at the molecular level. Hageman and Burris (1978a, b, 1979) have shown that the activity of nitrogenase, the enzyme system catalyzing the reduction of N_2 to NH_4^+, is greatly influenced by the rate of electron flux. The electron flux is greatly influenced by the supply of adenosine triphosphate (ATP) and the generation of low-potential electrons, both of which are dependent on the C metabolism of the nodule cytosol and the bacteroids. These limiting metabolic steps are being identified, and the possibility exists that genetic and biochemical alterations may be made that would prevent such limitations. Maier and Brill (1978), for example, have selected mutant strains of *R. japonicum* that display greater N_2 fixation rates than the parent strains. Obviously, expression of traits such as this under field conditions would be beneficial.

For a more thorough discussion of the many possibilities of improving N_2 fixation potential, the reader is referred to recent reviews by Postgate (1977) and Phillips (1980). We will now focus on two near-term improvements that are related to the metabolism of soybean root nodule bacteroids.

II. H_2 METABOLISM

A. H_2 Evolution

During the reduction of N_2 to NH_4^+ by nitrogenase, H_2 is evolved. The evolution of H_2 is a constitutive catalytic function of the enzyme complex (Rivera-Ortiz & Burris, 1975), and under optimal conditions, no less than 1 mole of H_2 is evolved for each mole of N_2 reduced. The energy requirements for H_2 evolution are the same as for N_2 reduction (4–5 ATP molecules hydrolyzed for each two electrons transferred to substrate); thus, H_2 evolution represents a theoretical minimum loss of 25% of the total energy input into the nitrogenase reaction. The loss of energy in terms of electron flux through nitrogenase for crop legume species usually ranges between 25 and 35% (Evans et al., 1980a; Bethlenfalvay & Phillips, 1977). Thus, a considerable loss in legume productivity may be due to the inherent inefficiency of the fixation process (Fig. 1).

B. H_2 Oxidation

In 1941, Phelps and Wilson noted that a strain of N_2-fixing *R. leguminosarum* bacteroids did not evolve H_2. The nitrogenase complex was not more efficient from this organism, but rather the organism had evolved

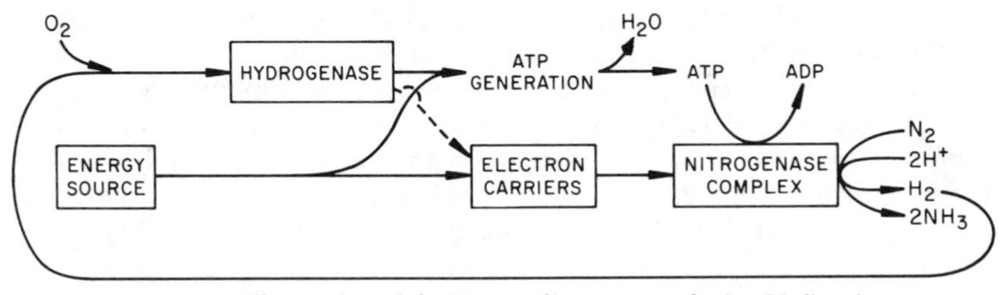

Fig. 1—An illustration of the H_2 recycling process during N_2 fixation.

a complementary enzyme system that consumed the H_2 produced by nitrogenase. Dixon (1967, 1968, 1972) confirmed this observation and demonstrated that the H_2 was consumed via an oxidation reaction with the generation of ATP. The H_2 oxidation reaction was postulated to conserve energy of the N_2-fixing system by: (i) alleviating H_2 inhibition of nitrogenase by reducing the partial pressure of H_2 in the vicinity of nitrogenase, (ii) protecting the O_2-labile nitrogenase proteins from O_2 damage by removal of O_2 during H_2 oxidation, and (iii) regaining part of the energy lost via nitrogenase-dependent H_2 evolution through ATP synthesis associated with H_2 oxidation. In regard to Dixon's postulate (i) above, Yates and Walker (1980) reported that under normal conditions H_2 produced by nitrogenase apparently does not accumulate to levels in vivo that are sufficient to inhibit N_2 reduction. Evidence supporting Dixon's proposals (ii) and (iii), however, have been well documented (Evans et al., 1980a; Phillips, 1980; Yates & Walker, 1980).

The contribution of respiratory protection has been difficult to quantitate. Respiratory protection would presumably contribute most significantly where the availability of photosynthate is limited by inadequate light and during the very early stages and the late stages of nodule development, when the content of leghemolgobin may be insufficient to maintain a microaerobic environment. In greenhouse trials, soybean plants inoculated with *R. japonicum* strains possessing H_2 oxidation activity reached their full developmental height earlier, and during their senescent phase, degraded their leghemoglobin more slowly than plants inoculated with strains of *R. japonicum* that did not possess H_2 oxidation activity (Zablotowicz et al., 1980). Ruiz-Argueso et al. (1979) have demonstrated respiratory protection of nitrogenase activity in washed suspensions of *R. japonicum* bacteroids.

Particulate preparations of *R. leguminosarum* bacteroids (Dixon, 1972) and washed suspensions of *R. japonicum* bacteroids (Emerich et al., 1979) demonstrated H_2-dependent ATP synthesis. Moreover, the rate of respiration as measured by CO_2 evolution decreases during the oxidation of H_2 (Emerich et al., 1980). The sparing of respiratory C substrates is particularly significant since photosynthate supply is believed to limit the N_2 fixation potential (Havelka & Hardy, 1976; Quebedeaux et al., 1975). Thus, H_2 oxidation regains energy through direct formation of ATP and conservation of energy-rich, respiratory C substrates.

Evans et al. (1980b) have tabulated the results of several surveys on the effect of H_2 oxidation on crop legumes. The mean loss of nitrogenase electron flux due to H_2 evolution in 91 replicates samples of soybeans, peas (*Pisum sativum* L.), alfalfa, and clover (*Trifolium* sp.) was 32%. Approximately 28 moles of ATP are required to reduce 1 mole of N_2 to 2 moles of NH_4^+, including the energy requirement for the concommitant reduction of two protons to H_2 (Evans et al., 1980a). This estimate consists of a minimum of 16 moles of ATP used directly by the nitrogenase complex, and in addition, 12 moles of ATP considered equivalent to the eight low-potential electrons required for the enzymatic reduction processes. A 32% inefficiency of N_2 reduction would produce 1.28 molecules of H_2 during each catalytic cycle (32% × 8 electrons = 2.56 electrons = 1.28 molecules of H_2). Assuming 2 moles of ATP are generated during the oxidation of each mole of H_2, 1.28 moles of H_2 would form 2.56 moles of ATP or a theoretical energy recovery of 9.1% [(2.56/28) × 100]. In comparisons of groups of hydrogen-uptake-positive (Hup^+) and hydrogen-uptake-negative (Hup^-) inoculants for soybeans, yield increases of this magnitude have been reported in both greenhouse and field trials (Evans et al., 1980b).

C. Greenhouse Trials

Schubert et al. (1977) found differences in yield of soybeans when a single strain of *Rhizobium japonicum* possessing H_2 oxidation activity was compared with a strain lacking this trait. Single-strain comparisons, however, may not be reliable, because genetic differences not related to differences in hydrogenase undoubtedly are present in strains that are not isogenic. In an effort to eliminate the effects of genetic variation, Albrecht et al. (1979) randomly selected five wild-type *R. japonicum* strains possessing H_2-oxidizing activity and five that lacked this characteristic. As a group, soybeans inoculated with the five strains possessing H_2-oxidizing capacity produced an average of 16% more dry matter and 26% higher N content than the plants inoculated with strains lacking H_2-oxidizing capacity. More recently, however, Cantrell et al. (1982) have reported that the Hup^+ strains used in these experiments showed no plasmids in the molecular weight range $<$ 170 MDa, whereas all but one of the Hup^- strains showed plasmids of this size. From these results, it might be argued that genetic variation other than the hydrogenase characteristic may not be random among the Hup^+ and Hup^- strains used in the comparisons.

D. Field Trials

Hanus et al. (1981) reported the results of four different field experiments conducted in soils containing relatively few indigenous *R. japonicum.* Each trial consisted of three to five randomly chosen strains lacking H_2-oxidizing activity and an equal number of strains possessing H_2-oxidizing activity. The average increase in crude protein in seed from plants inoculated with the hydrogenase-positive strains in the four field trials was 8.4% (Table 1). In these experiments, it was also assumed that genetic variation other than hydrogenase was random among the strains compared.

E. Isogenic Strains

The greenhouse and field trials described above were conducted with wild-type *Rhizobium* of different genetic makeups. Recently, Lepo et al. (1980) have isolated mutants (PJ 17 and PJ 18) of *R. japonicum* that apparently are identical to the parent strain with the exception of the H_2-oxidizing characteristic. On the basis of spontaneous reversion of the mutants to the parent at a rate of about 1 to 10^9, it is presumed that the mutants are iso-

Table 1—Crude protein content of grain from field-grown soybean plants nodulated by *R. japonicum* strains with and without H_2-oxidizing capacity.

Rhizobium, phenotype	Crude protein in dry grain†			
	Experiment designation (location and year)			
	Corvallis, Ore. 1977	Hermiston, Ore. 1979	Corvallis, Ore. 1979a	Corvallis, Ore. 1979b
	%			
Possessing H_2-oxidizing activity (H_{up+})	37.6**	39.9*	44.0**	41.2**
Lacking H_2-oxidizing activity (H_{up-})	35.6**	35.9*	41.6**	37.0**

*,** Significant at the 0.05 and 0.01 levels, respectively.

† Values represent means of groups—treatments in which *Rhizobium* inoculants were used possessing or lacking H_2-oxidizing activity, each replicated 4, 5, 7, and 6 times, respectively. For details, see Hanus et al., 1981.

genic with the parent strain. In initial growth cabinet trials, plants inoculated with mutants PJ 17 and PJ 18 yielded means of 59 and 71 mg N/plant, respectively, compared with 86 mg N for plants inoculated with the parent strain. Further testing of isogenic strains and their revertants in replicated field and greenhouse experiments should provide a reliable evaluation of the benefits of the H_2-oxidizing system.

F. Effect of H_2 Oxidiation on Dry Matter Accumulation

In considering the potential benefits of H_2 recycling to growth of legumes, it is assumed that N_2 fixation limits growth and that energy provided to nodules as products of photosynthesis limits N_2 fixation. If nodules are supplied with adequate energy in the form of products of photosynthesis to meet their needs, then losses of H_2 during N_2 fixation would not be expected to be significant.

Any benefits of the H_2 oxidation system would be expected to accrue over the course of the growth cycle of the legume. The early phases of growth of a legume dependent on N_2 can be described by Eq. [1], where m is the plant mass in grams dry matter, m_n is the plant mass at the onset of N_2 fixation, r is the specific growth rate in grams increase in plant mass per gram plant material per day, and t is the length of the growth period in days:

$$m = m_n e^{rt}. \quad [1]$$

F. J. Hanus and H. J. Evans (1980, unpublished data) have derived a specific growth rate of 0.0862 from an experiment of Schubert et al. (1978). Assuming a 9.1% energy recovery from oxidation of 1.28 moles of H_2 associated with each 8 electron turnover of nitrogenase, the specific growth

Table 2—Dry weight accumulation in soybeans fixing N_2 when nodulated by *R. japonicum* with and without H_2-oxidizing capacity.†

Growth period	Dry matter accumulation in plants nodulated by *R. japonicum*: Without H_2-oxidizing capacity	With H_2-oxidizing capacity	Difference	Increase
days	grams/culture			%
15	0.87	0.87	0	0
20	1.34	1.39	0.05	3.7
25	2.06	2.23	0.17	8.1
30	3.17	3.56	0.39	12.4
35	4.88	5.70	0.82	16.8
40	7.51	9.12	1.61	21.5

† Values were calculated by Eq. [1]. The specific growth rate for plants nodulated by H_2-oxidizing *R. japonicum* was 0.0940, and that for *R. japonicum* without this trait was 0.0862. Dinitrogen fixation activity commenced on day 15.

rate of plants inoculated with H_2-oxidizing strains would increase to 0.0940. After 40 d of growth, the equation predicts a 21% increase in dry weight of plants inoculated with H_2-oxidizing strains of *Rhizobium* (Table 2). No estimation of the benefits of respiratory protection were included in these calculations. In the experiment from which the initial specific growth rate was derived (Schubert et al., 1978), the dry weight increase after 40 d growth was 24%. Table 2 shows that the extent of increase depends upon the growth period at a particular specific growth rate. Although many other parameters need to be taken into account to describe the differences in yield expected from H_2 oxidation, Eq. [1], with appropriate increases in the specific growth rate for advantages of H_2 recycling, may be used to estimate benefits of the H_2 oxidation process.

G. Application

The use as inoculants of strains of *Rhizobium* capable of oxidizing all the H_2 evolved during N_2 fixation provides a near-term approach for increasing grain legume yields. Utilization of these strains requires inoculation of seed by established methods. The problem of obtaining nodulation by applied strains to soils that contain high populations of indigenous rhizobia remains to be solved. There is no reason to expect that the Hup^+ strains would be any less competitive than the indigenous strains or strains that are present in commercial inoculants; however, more research on these questions needs to be carried out. It is known that the H_2-oxidizing strains are capable of chemolithotrophic growth requiring H_2 and CO_2 (Hanus et al., 1979). It seems reasonable to expect that the metabolic flexibility of the Hup^+ strains might provide improved changes for survival and competition in the soil.

The limited number of surveys to date have revealed that approximately 30% of *R. japonicum* strains, 85% of cowpea strains [*Vigna unguiculata*

(L.) Walp. subsp. unguiculata], 5% of *R. leguminosarum* strains, and no *R. trifolii* or *R. meliloti* strains possess sufficient H_2 oxidation activity to recycle a significant proportion of the H_2 evolved during N_2 fixation (Evans et al., 1980a, b). Developments in *Rhizobium* genetics are occurring at a rapid rate, and it has already been reported by Brewin et al. (1980) that the determinants for nodulation and hydrogenase activity in *R. leguminosarum* 128C53 could be transferred to other strains of *R. leguminosarum* after recombination of a nontransmissable plasmid with one that is transmissable. Thus, it should soon be possible to combine H_2 oxidation with other desirable metabolic and/or physiological traits in several *Rhizobium* species. Improved strains could be incorporated into commercial inoculants; however, the problem of forming nodules by applied inoculants in soils containing enormous populations of indigenous *Rhizobium* strains remains to be solved.

III. POLY-β-HYDROXYBUTYRATE METABOLISM

Photosynthate supply has been reported to limit the N_2 fixation potential (Quebedeaux et al., 1975; Havelka & Hardy, 1976). In soybeans, at the peak of N_2 fixation activity, when the demand for photosynthate is greater, a C storage material, poly-β-hydroxybutyrate (PHB), begins to accumulate. The possibility exists that the accumulation restricts N_2 fixation since photosynthate needs for support of N_2 reduction is sequestered in the form of a storage material and apparently is unavailable for further metabolism and energy generation. Poly-β-hydroxybutyrate can accumulate to levels of up to 10% of the nodule dry weight, or 50% of the bacteroid dry weight (Klucas, 1974; Klucas & Evans, 1968; Wong & Evans, 1971). Assuming that PHB accumulation utilizes a significant proportion of the photosynthate supply provided to nodules, the elimination of PHB synthesis and the release of energy available to support N_2 fixation might result in significant increases in yield.

Many researchers have proposed that if the peak period of N_2 fixation could be prolonged by several days, the N content of the grain could theoretically double. We have made calculations to estimate the minimum yield increase expected in soybeans, assuming PHB accumulation could be eliminated and the spared photosynthate metabolized to supply ATP for N_2 fixation. These estimates indicate a minimum expected increase of 5%. The upper limit cannot be determined, because any additional effects of deleting the PHB synthetic pathway are unknown. However, any increase due to reduction of PHB accumulation should be additive or synergistic with those of H_2 oxidation since the two metabolic processes are independent. It would be desirable, therefore, to isolate mutants of H_2-oxidizing strains of *R. japonicum* that have lost their ability to accumulate PHB. This selection process is currently underway, and the preliminary results are encouraging (J. Waters & D. W. Emerich, unpublished results).

As described above for H_2-oxidizing strains of *Rhizobium,* application of PHB-deficient strains might be accomplished by incorporation of them into the inoculation medium. Research is underway to enhance nodulation by desired strains through the use of chemical additions to the inoculation medium (D. G. Blevins & D. W. Emerich, personal communication). Hopefully, this approach will retard nodulation by indigenous soil *Rhizobium* but not those applied to the seed, thus augmenting the effects of the improved *Rhizobium* strains.

IV. CONCLUSIONS

Genetically or biochemically improved strains of *Rhizobium* have potential for near-term enhancement of N_2 fixation. Once these strains are developed, they can rapidly be provided to the farmer by incorporation into inoculation medium; H_2-oxidizing strains in some *Rhizobium* species are available now. The problem of competition with the indigenous population in some soils still must be overcome. Other improved strains of *Rhizobium* undoubtedly will be available in the near future.

The aspects of biochemical and genetic alterations of *Rhizobium* that make this approach attractive also determine its limitations. By alleviating one rate-limiting step, a new process becomes rate limiting. The new rate-limiting step must be investigated, analyzed, and then, if possible, biochemically or genetically altered. Although research of this type requires time, it does appear that progress is being made in understanding limitations of N_2 fixation by legumes, and certainly more research seems justified.

ACKNOWLEDGMENT

The authors express appreciation of Miss Sheri Woods for typing the manuscript. The research program of Harold J. Evans has been supported by NSF Grant 77-08784, USDA/SEA Grant 78-59-2411-0-1-099-1, USDA/CR Grant 901-15-37, and the Oregon Agricultural Experiment Station from which this is Technical Paper no. 5728.

REFERENCES

Abu-Shakra, S. S., D.A. Phillips, and R. C. Huffaker. 1978. Nitrogen fixation and delayed leaf senescence in soybeans. Science 199:973–975.

Albrecht, S. L., R. J. Maier, F. J. Hanus, S.A. Russell, D. W. Emerich, and H. J. Evans. 1979. Hydrogenase in *Rhizobium japonicum* increases nitrogen fixation by nodulated soybeans. Science 203:1255–1257.

Bach, M. K., W.E. Magee, and R. H. Burris. 1958. Translocation of photosynthetic products to soybean nodules and their role in nitrogen fixation. Plant Physiol. 33:118–124.

Bergersen, F. J. 1958. The bacterial component of soybean root nodules changes in respiratory activity, cell dry weight and nucleic acid content with increasing nodule age. J. Gen. Microbiol. 19:312–323.

Bergersen, F. J., and G. L. Turner. 1967. Nitrogen fixation by the bacteroid fraction of breis of soybean root nodules. Biochim. Biophys. Acta 141:507–515.

Bethlenfalvay, G. J., and D. A. Phillips. 1977. Effect of light intensity on efficiency on carbon dioxide and nitrogen reduction in *Pisum sativum* L. Plant Physiol. 60:419–421.

Brewin, M. H., T. M. DeJong, D. A. Phillips, and A. W. B. Johnson. 1980. Co-transfer of determinants for hydrogenase activity and nodulation ability in *Rhizobium leguminosarum.* Nature 288:77–79.

Burris, R. H., and P. W. Wilson. 1939. Respiratory enzyme systems in symbiotic nitrogen fixation. Cold Spring Harbor Symp. Quant. Biol. 7:349–361.

Cantrell, M. A., R. E. Hickok, and H. J. Evans. 1982. Identification and characterization of plasmids in hydrogen uptake positive and hydrogen uptake negative strains of *Rhizobium japonicum.* Arch. Microbiol. 131:102–106.

Dixon, R. O. D. 1967. Hydrogen uptake and exchange by pea root nodules. Ann Bot. (London) 31:179–188.

Dixon, R. O. D. 1968. Hydrogenase in pea root nodule bacteroids. Arch. Microbiol. 62:272–283.

Dixon, R. O. D. 1972. Hydrogenase in legume root nodule bacteroids: occurrence and properties. Arch. Microbiol. 85:193–201.

Emerich, D. W., T. Ruiz-Argueso, T. M. Ching, and H. J. Evans. 1979. Hydrogen dependent nitrogenase activity and ATP formation in *Rhizobium japonicum* bacteroids. J. Bacteriol. 137:153–160.

Emerich, D. W., T. Ruiz-Argueso, S. A. Russell, and H. J. Evans. 1980. Investigation of the H_2 oxidation system in *Rhizobium japonicum* 122 DES bacteroids. Plant Physiol. 66: 1061–1066.

Evans, H. J., D. W. Emerich, T. Ruiz-Argueso, R. J. Maier, and S. L. Albrecht. 1980a. Hydrogen metabolism in the legume-rhizobium symbiosis. p. 67–86. *In* W. H. Orme-Johnson and W. E. Newton (ed.) Nitrogen fixation. Vol. 2, Symbiotic associations and cyanobacteria.

Evans, H. J., K. Purohit, M. A. Cantrell, G. Eisbrenner, S. A. Russell, F. J. Hanus, and J. E. Lepo. 1980b. Hydrogen losses and hydrogenases in nitrogen-fixing organisms. p. 84–96. *In* A. H. Gibson and W. E. Newton (ed.) Current perspectives in nitrogen fixation. Australian Academy of Science, Canberra.

Hageman, R. V., and R. H. Burris. 1978a. Kinetic studies on electron transfer and interactions between nitrogenase components from *Azotobacter vinelandii.* Biochemistry 14:4117–4124.

Hageman, R. V., and R. H. Burris. 1978b. Nitrogenase and nitrogenase reductase associate and dissociate with each catalytic cycle. Proc. Nat. Acad. Sci. USDA 75:2699–2702.

Hageman, R. V., and R. H. Burris. 1979. Changes in the EPR signal of dinitrogenase from *Azotobacter vinelandii* during the lag period before hydrogen evolution begins. J. Biol. Chem. 254:11189–11192.

Hanus, F. J., S. L. Albrecht, R. M. Zablotowicz, D. W. Emerich, S. A. Russell, and H. J. Evans. 1981. Yield and N content of soybean seed as influenced by *Rhizobium japonicum* inoculants possessing the hydrogenase characteristic. Agron. J. 73:368–372.

Hanus, F. J., R. J. Maier, H. J. Evans. 1979. Autotropic growth of H_2-uptake-positive strains of *Rhizobium japonicum* in an atmosphere supplied with hydrogen gas. Proc. Nat. Acad. Sci. USA 76:1788–1792.

Havelka, U. D., and R. W. F. Hardy. 1976. Legume N_2 fixation as a problem in carbon nutrition. p. 456–475. *In* W. E. Newton and C. J. Nyman (ed.) Proc. First Int. Symp. on Nitrogen Fixation. Washington State Univ. Press, Pullman.

Klucas, R. V. 1974. Studies on soybean nodule senescence. Plant Physiol. 54:612–616.

Klucas, R. V., and H. J. Evans. 1968. An electron donor system for nitrogenase-dependent acetylene reduction by extracts of soybean nodules. Plant Physiol. 43:1458–1460.

Lepo, J. E., F. J. Hanus, and H. J. Evans. 1980. Chemoautotrophic growth of hydrogen-uptake-positive strains of *Rhizobium japonicum.* J. Bacteriol. 141:664–670.

Maier, R. J., and W. J. Brill. 1978. Mutant studies of *Rhizobium japonicum* with increased ability to fix nitrogen for soybean. Science 201:448–450.

Pate, J. S. 1962. Root-exudation studies on the exchange of C^{14}-labelled organic substances between the roots and shoots of the nodulated legume. Plant Soil 17:333–355.

Phelps, A. S., and P. W. Wilson. 1941. Occurrence of hydrogenase in nitrogen-fixing organisms. Proc. Soc. Exp. Biol. Med. 47:473–476.

Phillips, D. A. 1980. Efficiency of symbiotic nitrogen fixation in legumes. Annu. Rev. Plant Physiol. 31:29–50.

Postgate, J.R. 1977. Possibilities for the enhancement of biological nitrogen fixation. Philos. Trans. R. Soc. London 281:249–261.

Quebedeaux, B., U. D. Havelka, K. L. Livak, and R. W. F. Hardy. 1975. Effect of altered pO_2 in the aerial part of soybean on symbiotic N_2 fixation. Plant Physiol. 56:761–764.

Rivera-Ortiz, J. M., and R. H. Burris. 1975. Interactions among substrates and inhibitors of nitrogenase. J. Bacteriol. 123:537–545.

Ruiz-Argueso, T., D. W. Emerich, and H. J. Evans. 1979. Hydrogenase system in legume nodules: a mechanism of providing nitrogenese with energy and protection from oxygen damage. Biochem. Biophys. Res. Commun. 86:259–264.

Schubert, K. R., J. A. Engelke, S. A. Russell, and H. J. Evans. 1977. Hydrogen reactions of nodulated leguminous plants. I. Effect of rhizobial strains and plant age. Plant Physiol. 60:651–654.

Schubert, K. R., N. T. Jennings, and H. J. Evans. 1978. Hydrogen reactions of nodulated leguminous plants. II. Effects on dry matter accumulation and nitrogen fixation. Plant Physiol. 61:398–401.

Seetin, M. W., and D. K. Barnes. 1977. Variation among alfalfa genotypes for rate of acetylene reduction. Crop Sci. 17:783–787.

Wong, P. P., and H. J. Evans. 1971. Poly-β-hydroxybutyrate utilization by soybean (*Glycine max* Merr.) nodules and assessment of its role in maintenance of nitrogenase activity. Plant Physiol. 47:750–755.

Yates, M. G., and C. C. Walker. 1980. Hydrogenase activity and hydrogen evolution by nitrogenase in nitrogen-fixing *Azotobacter chroococcum*. p. 95–110. *In* W.E. Newton and W. H. Orme-Johnson (ed.) Nitrogen fixation. Vol. I. Univ. Park Press, Baltimore.

Zablotowicz, R. M., S. A. Russell, and H. J. Evans. 1980. The effect of the hydrogenase system on *Rhizobium japonicum* on the nitrogen fixation and growth of soybeans at different stages of development. Agron. J. 72:555–559.

9

Peter van Berkum
Nitrogen Fixation and Soybean Genetics Laboratory
Agricultural Research Service, USDA
Beltsville, Maryland

Potential for Nonsymbiotic and Associative Dinitrogen Fixation[1]

Biological N_2 fixation is generally regarded to be a mechanism by which applications of fertilizer N to crops may be supplemented or in part replaced to enhance economic return. Several authorities have also advocated that the development of management techniques, which encourage N input into agriculture through biological N_2 fixation may reduce pollution of natural ecosystem. The legume-*Rhizobium* symbioses are the most well-known examples where biological N_2 fixation has been widely applied to agriculture for the production of grain and forage crops (see Chapt. 7, 8, and 25, this book). Although the legume-*Rhizobium* symbioses are obviously most important to agriculture, nonsymbiotic and associative N_2 fixation are considered to occur at magnitudes that are probably of agronomic significance (Dobereiner, 1978; Knowles, 1977; Mague, 1977; Moore, 1966). The diverse range of microorganisms capable of N_2 fixation include photosynthetic bacteria, blue-green algae, and heterotrophic bacteria. Some of these prokaryotes have gained ecological advantage by associating with organisms not capable of N_2 fixation. The *Azolla-Anabaena* symbioses are considered to be the most important N_2-fixing associations for crop production, but N_2-fixing endorhizosphere associations between *Azospirillum* and the roots of tropical grasses have been proposed and have received much publicity and attention.

[1] Supported in part by USDA Competitive Grant no. 5901-0410-9-0254-0 and Cooperative Agreement 12-14-1001-1387 between the USDA and the Agronomy Dep., Univ. of Maryland, College Park, MD 20742.

I. SOME ESTIMATES OF N CONTRIBUTED BY NONSYMBIOTIC AND ASSOCIATIVE N_2 FIXATION

The hypotheses that nonnodulated plants derive significant benefit from biological N_2 fixation originate from reported N balance studies in both temperate and tropical environments (van Berkum & Bohlool, 1980). It has been estimated that nonsymbiotic N_2 fixation may contribute significantly to the N economy of dryland legume-free soil-plant systems under temperate climatic conditions (Dart & Day, 1975). The long-term N balance studies at Rothamsted, England have provided the most convincing evidence for N accumulation in temperate soils through nonsymbiotic N_2 fixation (Dart & Day, 1975). Estimates of the contribution of N_2 fixation to the N economy of non-N fertilized wheat (*Triticum aestivum* L.) plots have been 18–23 kg N/ha annually, and to uncultivated legume-free plant-soil systems up to 49 kg N/ha annually. Free-living blue-green algae forming N_2-fixing crusts on the soil surface of the arable experiment and root-associated N_2 fixation by heterotrophic bacteria in the uncultivated area are considered to be responsible for gains of N on these sites (Day et al., 1975; Witty et al., 1979). Heterotrophic bacteria belonging to the family Enterobacteriaceae are believed to form N_2-fixing associations with the roots of the plants growing on these experimental sites (Day et al., 1975).

Nitrogen balance studies in the tropics involving forest and plant fallows have also provided evidence that N accumulates because of nonsymbiotic N_2 fixation. Greenland (1977) concluded that microorganisms probably contribute as much as 100 kg N/ha annually to soil-plant systems. Dinitrogen fixation by heterotrophic bacteria in the rhizosphere of the roots of grasses has been suggested as a mechanism to explain the substantial N gains observed in savanna fallows (Dobereiner, 1978; Greenland, 1977; Neyra & Dobereiner, 1977). Day (1977) noted that economically important forage grasses appear to support the highest rates of N_2 fixation. *Paspalum notatum* has been reported to stimulate N_2 fixation by *Azotobacter paspali* at rates to satisfy the N requirement for normal growth (Neyra & Dobereiner, 1977). Sugarcane (*Saccharum officinarum* L.) has been grown in monoculture in many tropical regions without the application of fertilizer N, and it has been suggested that this crop benefits from stimulating N_2 fixation by *Beijerinckia* (Neyra & Dobereiner, 1977). Most of the other tropical forage and grain grasses have been suggested to stimulate N_2 fixation by associated *Azospirillum* (Dobereiner, 1978; Neyra & Dobereiner, 1977). Free-living heterotrophic bacteria also occur in the phyllosphere of plants, but N_2 fixation in this environment is sporadic or negligible (Knowles, 1977).

The possibility that nonsymbiotic and associative N_2 fixation benefit aquatic ecosystems under temperate and tropical climatic conditions has also been suggested. Saltmarshes have been recognized to be highly productive aquatic ecosystems and contribute dissolved organic and inorganic N to receding tidal waters. Saltmarshes are believed to be major sites for N_2

fixation (Patriquin & McClung, 1978), and saltmarsh grasses do have some agronomic value as forage crops (Reimold et al., 1975). Total N accretion by a short stand of *Spartina alterniflora* in Nova Scotia has been estimated to be 149 kg N/ha during one growing season (Patriquin & McClung, 1978). The heterotrophic N_2-fixing bacterium *Campylobacter* has been reported to associate with the roots of *S. alterniflora* (McClung & Patriquin, 1980), but this grass and other saltmarsh plants may also benefit from blue-green algal N_2 fixation estimated to range from 4.3 to 462 kg N/ha annually (Jones, 1974).

Saltmarshes and rice paddies (*Oryza sativa* L.) are similar aquatic systems because both are cropped regularly (Patriquin & McClung, 1978). Paddy rice has been grown for centuries in the Far East without applications of N fertilizers, and a long-term N balance study at the International Rice Research Institute (IRRI) in the Philippines has revealed constant yields of rice without N fertilizer (Watanabe & Lee, 1977). Nonsymbiotic N_2 fixation by photosynthetic bacteria and blue-green algae are considered to benefit subsequent growth of rice by providing 30 kg N/ha (Knowles, 1977; Watanabe & Lee, 1977). The *Azolla-Anabaena* N_2-fixing symbioses are important when introduced as green manure to paddies, or benefit crop production directly when inoculated to paddies at the time rice plants are sown. The *Azolla-Anabaena* symbioses contribute N for the production of rice at rates estimated to be from 0.4 to 3.3 kg N/ha daily (Lumpkin & Plucknett, 1980; Moore, 1969). Dinitrogen-fixing associations between rice roots and a diverse group of heterotrophic bacteria is also believed to contribute N to the growing crop.

II. METHODOLOGY FOR THE MEASUREMENT OF N_2 FIXATION BY FIELD SAMPLES

Nonsymbiotic and associative N_2 fixation should be measured under field conditions with intact, in situ methods by quantifying the accumulated N from N_2 fixation or by following N_2-ase activity. The measurement of N_2 fixation with field samples in situ is more laborious than with isolated tissue in the laboratory. An intractable problem with field measurements is that erroneous estimates of N_2 fixation may result from the use of methods that disturb the natural environment.

A. Total N by Kjeldahl Analysis

Total N measurements with the Kjeldahl method with small subsamples of a particular system permit N accretion to be determined with minimal disturbance. However, the Kjeldahl analysis does not distinguish sources of N as fractions within the total; therefore, N balance sheets must be constructed to estimate N input from N_2 fixation. The construction of N balance sheets to determine N_2 fixation requires assiduous measurements of

N input and loss over long periods of time. The lysimeter, although a disturbed system, enables measurements of N accretion and loss to be made with great precision. However, it is difficult to regulate the water content of lysimeters, because the soil within the construction is detached from the water table. Nitrogen-balance studies have also been made for successive cropping of agricultural crops or for natural vegetation growing on soil depleted of N through agricultural exploitation. Dart and Day (1975) have cautioned that N accretions of < 100 kg N/ha per annum are difficult to measure with short-term field trials, but some N gains have been estimated to be in excess of this figure (Jensen, 1965).

B. The Use of Isotopes of N

The use of isotopes of N (^{13}N or ^{15}N) allows products of N_2 fixation to be distinguished within the total N of tissues. The restrictions imposed by the short half-life of the radioactive isotope ^{13}N (11 min) has discouraged its use as a method to measure N_2 fixation. The use of the stable isotope ^{15}N is preferred, and N_2 fixation is measured by deriving the ratio of ^{15}N to ^{14}N in the tissues of the test plants and comparing the increased abundance above the natural abundance of ^{15}N. It is most desirable to use the isotope of N in gaseous form when measuring N_2 fixation, but major limitations are enclosure of the test material and changes in environmental conditions with the necessary long-term incubations requiring complex control equipment. These limitations may be overcome if N_2 fixation is determined indirectly by enriching the soil with ^{15}N and measuring the dilution of the label in the tissues of the N_2-fixing system. This technique is more appropriate for the measurement of N_2 fixation associated with agricultural crops, rather than plants in an undisturbed natural environment. In order to measure N_2 fixation accurately with the enrichment of the soil N pool with ^{15}N: (i) no volatile losses of ^{15}N should occur; (ii) test and control plants should utilize ^{15}N equally well; (iii) the soil should be uniformly labeled; and (iv) the application of ^{15}N to the soil should not affect the rate of N_2 fixation by the test plants. The inhibition of N_2 fixation by applications of inorganic ^{15}N may be overcome by adding organic C to the soil, which incorporates the applied N into the organic N fraction of the soil. However, inadequate growth of test plants under severe N-limiting conditions may seriously restrict the use of this method to determine N_2 fixation in cereal and forage grasses. An alternative method to measure N_2 fixation would be to determine the differences in natural abundance of ^{15}N arising from mass discrimination effects resulting from N_2 fixation, NH_4^+ assimilation, and N transport ($\delta^{15}N$). The $\delta^{15}N$ has been reported to vary considerably with soil depth (Karamanos & Rennie, 1980), which may restrict the use of this method to determine N_2 fixation. Also, it might not be feasible to extend the $\delta^{15}N$ method to the measurement of N_2 fixation in grasses because of low rates, the probability that N transfer from fixation to these plants is restricted, and the limited choice of suitable control plants.

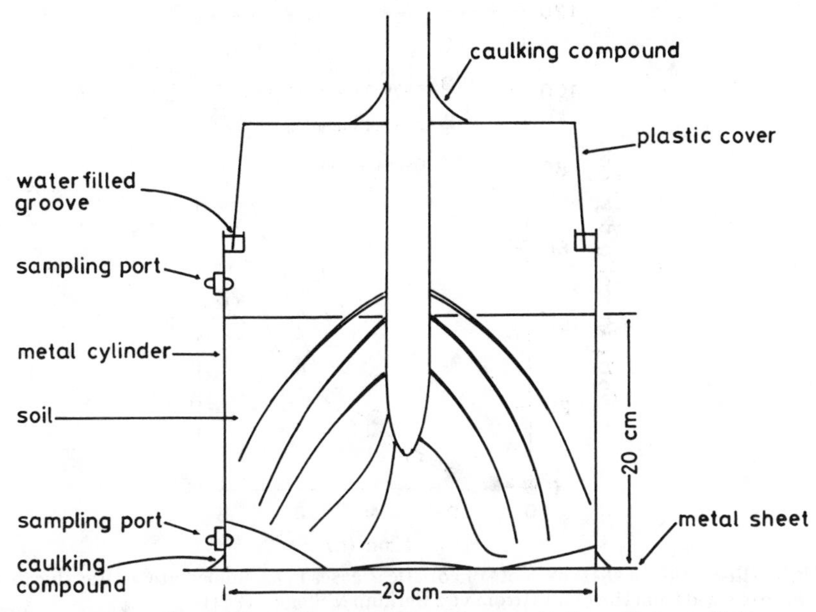

Fig. 1—In situ soil core device for measuring N_2-ase activity with large field-grown grasses (from Tjepkema & van Berkum, 1977).

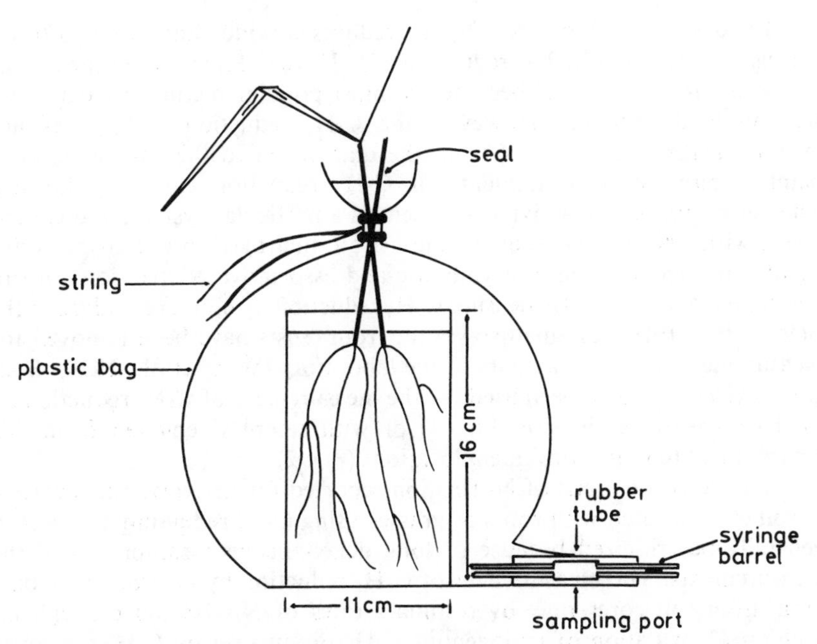

Fig. 2—The small soil core method for measuring N_2-ase activity with grasses (from van Berkum & Day, 1980; Dobereiner, 1980). Day et al. (1975) used similar soil cores of grasses with the plant tops cut off or folded into the 3.7-L confectionary jars for measurement of C_2H_2 reduction.

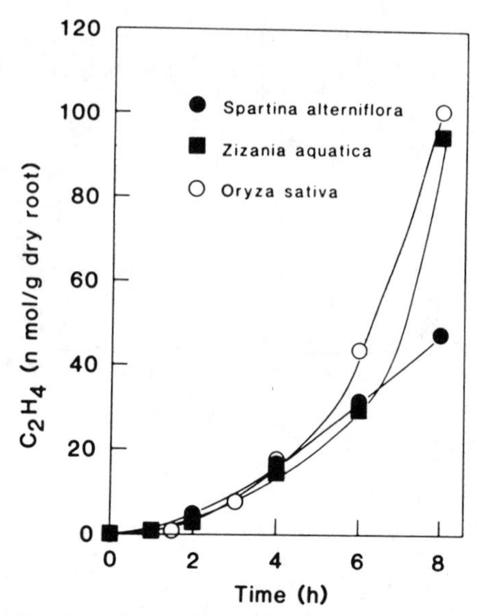

Fig. 3—Reduction of C_2H_2 by roots of aquatic grasses in air immediately after they had been excised from the plants (from van Berkum & Sloger, 1981).

C. Acetylene Reduction for the Measurement of N_2-ase Activity

Besides atmospheric N_2, N_2-ase reduces a wide range of substrates (Burns & Hardy, 1975). The reduction of C_2H_2 to C_2H_4 measures the rate of N_2-ase indirectly, but has become the most popular method by which N_2 fixation is determined. However, the C_2H_2 reduction assay does not measure transfer of fixed N from the diazotroph to the associated crop plant. Therefore, measurements of C_2H_2 reduction can only identify whether or not N_2-ase activity is present in a particular system, and experiments with ^{15}N are required to demonstrate that agricultural crops derive significant benefit from nonsymbiotic and associative N_2 fixation. In situ devices have been used to measure C_2H_2 reduction by blue-green algae in the field (Witty, 1979), or small soil cores from crusts have been removed for measurement of N_2-ase activity in the laboratory (Witty et al., 1979). Soil-core devices have also been used for the measurement of C_2H_2 reduction by field-grown grasses in situ (Fig. 1) or with samples removed from this environment to more convenient locations (Fig. 2).

The determinations of N_2 fixation reported for heterotrophic bacteria in soil or associated with roots of grasses using C_2H_2 reduction or $^{15}N_2$ can generally be criticized because of long delays between sampling and the measurements. The measurement of C_2H_2 reduction by heterotrophic bacteria using soil cores may overestimate rates of N_2 fixation through inhibition of oxidation of endogenous C_2H_4 production by C_2H_2 and poor penetration of C_2H_2 and C_2H_4 through soil, leading to prolonged incubations causing further induction of N_2-ase activity (van Berkum & Bohlool, 1980). Evidence that N_2-ase activity is associated with roots of grasses is based on measurements of C_2H_2 reduction by excised roots, because it is not

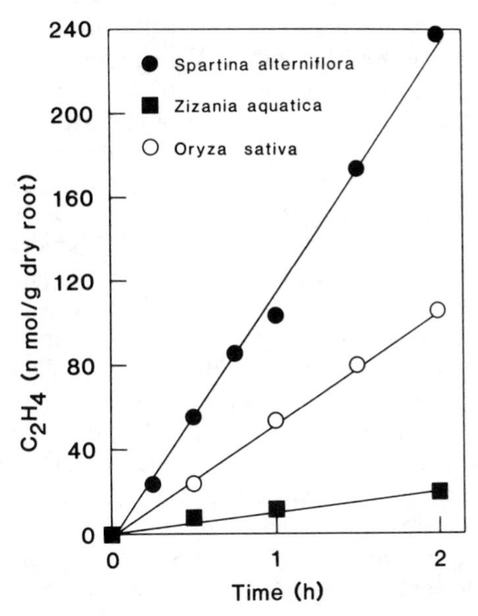

Fig. 4—Immediately linear rates of C_2H_2 reduction by roots of intact aquatic grasses protected from air during preparation of plants for assay (from van Berkum & Sloger, 1981).

possible to distinguish microsites of N_2 fixation with soil cores. However, an 8- to 18-h delay before C_2H_2 reduction can be detected with excised roots from most grasses indicates that N_2 fixation in situ is not root-associated (van Berkum & Bohlool, 1980). Dinitrogen fixation by heterotrophic bacteria in situ is associated with the roots and grasses when excised roots reduce C_2H_2 immediately at nonlinear rates in air (Fig. 3) and at linear rates when intact assays are used, which are devised to protect the roots from O_2 (Fig. 4).

III. OUR CURRENT UNDERSTANDING OF NONSYMBIOTIC AND ASSOCIATIVE N_2 FIXATION

A. N_2 Fixation, Carbon Supply, and Photosynthesis

Nitrogenase-catalyzed reduction of N_2 to NH_3 requires adenosine triphosphate (ATP) and electrons via reduced ferredoxin (Burns & Hardy, 1975). Stewart (1974) concluded that although blue-green algae are capable of limited heterotrophic growth with N_2 as the only N source, one of the roles of photosynthesis in these diazatrophs is to supply ATP and reductant to drive N_2-ase. A similar relationship between photosynthesis and N_2 fixation in photosynthetic bacteria has been proposed (Arnon & Yoch, 1974).

Although ATP and reductant as prerequisites for N_2 fixation are readily available in photoautotrophic microorganisms, links between photosynthesis and N_2-ase activity by heterotrophic bacteria are not so obvious.

Dinitrogen fixation in legumes is dependent on plant photosynthetic products to the bacteroids in the nodules for the production of reductants and synthesis of ATP through respiration, and the provision of C skeletons for the assimilation of NH_4^+. Similar direct links between photosynthesis and nonsymbiotic heterotrophic N_2-fixing bacteria do not exist, which suggests that in agricultural soil N_2 fixation occurs sporadically at low rates, even though these diazotrophs are widespread. However, measurements of C_2H_2 reduction and $^{15}N_2$ incorporation into carbohydrate or cellulose-amended soils have demonstrated significant potential gains of N through N_2 fixation by nonsymbiotic heterotrophic bacteria when C substrates are available. Under field conditions, the decay of plant material in soil probably is significant for nonsymbiotic N_2 fixation.

Studies of plant root exudation (Barber & Martin, 1976; Rovira, 1969) suggest that organic material excreted or sloughed off by roots may stimulate N_2 fixation in the rhizosphere. Dinitrogen fixation in the rhizosphere of plant roots is further favored because the exudates are low in N and N uptake by the plants lowers the N levels in the root zone (Jurgensen and Davey, 1970). Measurements of C_2H_2 reduction by soil cores of grasses and excised roots have generally been accepted as evidence that the root zone of grasses favors and stimulates N_2 fixation by free-living heterotrophic bacteria. Pronounced and rapid responses of N_2-ase activity with changing light incidence during diurnal cycle studies prompted Dobereiner and Day (1975) to suggest close relationships between plant photosynthesis and N_2 fixation by the root-associated bacteria. The rapid responses of N_2-ase to changing light intensity during the day-night cycles have been explained by root-cell symbioses with *Azospirillum,* which, because of their suggested internal location, have ready access to plant photosynthetic products (Dobereiner & Day, 1976). Observations of peak N_2-ase activities at night in some grasses was attributed to the solubilization of starch grains in the chloroplasts (Dobereiner & Day, 1975).

The evidence interrelating plant photosynthesis and root-associated N_2 fixation in grasses with diurnal cycle studies of C_2H_2 reduction rates is obscured because fluctuation of soil temperature also coincides with the cycle of N_2-ase activity during the day-night cycle (van Berkum & Bohlool, 1980). The reported diurnal cycle studies of the variation in the rate of C_2H_2 reduction by grasses do not indicate that plant photosynthesis directly drives N_2 fixation associated with the roots, because these experiments have not been attempted with controlled temperatures. Recently, van Berkum and Sloger (1981) reported that at 35°C, root-associated N_2-ase activity in rice in sunlight is fivefold higher than in the dark. Significant changes in the rate of root-associated C_2H_2 reduction within 15 min of transferring plants from sunlight into the dark, or vice versa, suggests a strong link between N_2-ase activity and light incidence in rice.

It has been suggested that the C_4 dicarboxylic acid pathway of photosynthesis of tropical grasses favors the establishment of root-associated N_2 fixation (Dobereiner et al., 1972; Neyra & Dobereiner, 1977). Potential high rates of photosynthesis in tropical environments, minimal losses of C through photorespiration, and the accumulation of photosynthetic products in the roots have been cited as evidence that C_4 grasses are more efficient at harvesting light energy for N_2 fixation than C_3 grasses (Neyra & Dobereiner, 1977). Most of the reports indicating a link between the C_4 photosynthetic pathway of photosynthesis and efficient N_2 fixation in grasses emanate from studies with preincubated excised roots in which N_2-ase activity is induced during the assay time. Measurements with intact C_4 grasses have indicated low rates or no associated N_2-ase activity (van Berkum & Bohlool, 1980); N_2-ase activity has also been recorded with grasses having the C_3 pathway of photosynthesis. Therefore, no significant information is available to indicate that: (i) photosynthesis by grasses directly provides root-associated bacteria with substrates for N_2 fixation; and (ii) the C_4 photosynthetic pathway confers plants with higher rates of more efficient N_2 fixation.

B. The Relationship between Soil Moisture and Nonsymbiotic N_2 Fixation

Microorganisms, like all other forms of life, require water as well as other nutrients for growth, reproduction, and maintainance. Therefore, it would not be difficult to imagine that water availability could significantly affect N_2 fixation in soils. It has become evident that unlike the *Rhizobium*-legume symbioses, nonsymbiotic N_2 fixation is favored by high moisture content or flooded conditions.

1. N_2 FIXATION BY HETEROTROPHIC BACTERIA

During their studies of N gains on Broadbalk at Rothamsted, Day et al. (1975) noticed that high rates of C_2H_2 reduction were always associated with soil cores of grasses and plants from wet areas. They reported that the rate of N_2-ase activity was positively correlated with soil moisture, and that the rate of C_2H_2 reduction increased exponentially with linear increases in soil moisture. Vlassak et al. (1973) also have reported correlations between soil moisture and the rate of N_2 fixation associated with grasslands. Day et al. (1975) hypothesized that the level of anaerobiosis in soil crumbs and the rhizosphere increases with higher soil moisture, and that the rate of N_2-ase activity is affected by the pO_2. This hypothesis is supported by observed higher rates of C_2H_2 reduction by soil cores of plants incubated under N_2 than under air (Tjepkema & Evans, 1976). Marginal C_2H_2 reduction was observed to be associated with the roots of wheat on Broadbalk arable, and it was suggested that under normal agricultural practice, soil is rarely wet enough to support significant N_2 fixation by heterotrophic bacteria (Day et al., 1975).

The positive correlation between soil moisture and N_2-ase activity is in support of the hypothesis that N_2 fixation by nonsymbiotic heterotrophic bacteria in soil or assocaited with roots of plants is enhanced in wet areas (Tjepkema & Evans, 1976). The growth habitat of grasses is probably an important factor for the development of root-associated N_2 fixation. Tjepkema and Evans (1976) reported N_2-ase activity associated with soil cores of wetland plants to be widespread, while C_2H_2 reduction associated with soil cores of dryland plants was observed to be much lower (Kana & Tjepkema, 1978). Similarly, the immediate rate of C_2H_2 reduction by excised roots of *S. alterniflora* was observed to be higher than with corn (*Zea mays*) roots (van Berkum & Sloger, 1979). It has been suggested that aquatic environments and high-moisture soils are low in available N due to denitrification and leaching of NO_3^- (Tjepkema & Burris, 1976). The depletion of N in aquatic environments may induce N_2 fixation by heterotrophic bacteria associated with the roots of grasses common to these sites and enable C_2H_2 reduction with excised samples to be detected without delay. This conclusion also supports the suggestion that washing excised roots from grasses of dry soil regions favors bacterial multiplication and the induction of N_2-ase activity during prolonged incubations by depleting the source of available combined N (van Berkum, 1980).

2. N_2 FIXATION BY BLUE-GREEN ALGAE

Dessication has been shown to limit N_2 fixation by blue-green algae in a variety of habitats (Fogg et al., 1973). Blue-green algae that had recovered from dessication were reported to fix N at significantly reduced rates, when compared with the rate of N_2-ase activity before moisture stress (Witty et al., 1979). Dessication was suggested to lead to extensive lysis of vegetative cell chains, and recovery of N_2-ase activity on rewetting was attributed to reactivation of existing cell systems. The canopy of crops and weeds on plants of Broadbalk not treated with herbicide was shown to favor N_2 fixation by blue-green algae (Witty et al., 1979). Herbicide-treated areas and plots with scarce canopy were observed to be drier, suggesting that dense vegetation helps to retain moisture on the soil surface, which prevents crusts of blue-green algae from dessicating. Witty et al. (1979) concluded that the importance of algal N_2 fixation in temperate agriculture is small compared with paddy soils because of limitations imposed by dessication.

C. Ecology of Free-Living N_2 Fixing Microorganisms

The ecology of free living N_2 fixing microorganisms has been extensively investigated, and several recent reviews have covered this subject exhaustively (Fogg et al., 1973; Knowles, 1977; Mague, 1977; Mulder & Brontonegoro, 1974; Stewart, 1974). My aim in this chapter is to focus specifically on the heterotrophic bacteria, which have been suggested to form N_2 fixing associations with the roots of grasses.

1. DISTRIBUTION OF DIAZOTROPHS WITH PARTICULAR REFERENCE TO THE ROOT REGION

Heterotrophic N_2 fixing bacteria and blue-green algae are widely distributed in nature. Representatives of these two groups of microorganisms have been found and have been isolated from a variety of terrestrial and aquatic habitats. The photosynthetic bacteria are ecologically more restricted than the heterotrophic N_2-fixing bacteria and blue-green algae.

The rhizosphere of the roots of plants appears to be an ideal habitat for heterotrophic bacteria, within which the N_2 fixers are well represented. Dinitrogen-fixing bacteria of many diverse genera occur in high numbers in the rhizosphere of roots of a wide variety of grasses. *Azotobacter paspali* is present in the rhizosphere of the roots of some tetraploid ecotypes of *P. notatum* (Neyra & Dobereiner, 1977). *Azospirillum* has been isolated from the roots of a wide variety of plants growing both in temperate as well as tropical environments. Nevertheless, *Azospirillum* is considered to be of special significance only in the tropics.

The organic materials available from root-cell debris and soluble root exudates are perhaps sufficient in quantity to satisfy the energy and C requirements of associated N_2-fixing bacteria. However, to be of much significance to the plants, these bacteria must reach high population densities. Although free-living heterotrophic bacteria are competent rhizosphere microorganisms, nonsymbiotic N_2 fixation is probably closely tied to their growth. The extent to which nonsymbiotic N_2 fixation in the rhizosphere of the roots of grasses benefits the plants has not been determined.

There would be a great competitive advantage for both the N_2-fixing bacteria and the plants if a more intimate internal association could be established. The ideal relationship is that between rhizobia and their leguminous hosts. Other microorganisms, such as mycorrhizal fungi and the symbiotic N_2-fixing actinomycete *Frankia,* also occupy inter- and intracellular spaces inside roots establishing highly integrated mutualistic associations with their plant hosts.

Evidence for the occurrence of nonsymbiotic N_2-fixing bacteria inside roots has been indirect. *Azospirillum* and *Campylobacter* have been isolated from "surface-sterilized" roots of field-grown grasses. However, in none of the descriptions reporting the isolation of N_2-fixing bacteria from inside grass roots has the surface sterility of the roots been demonstrated unequivocally. Dobereiner and Day (1976) observed that root pieces of *D. decumbens* exhibiting high rates of C_2H_2 reduction after they had been preincubated to induce N_2-ase activity had cells that were packed with tetrazolium-reducing bacteria-like particles. Tetrazolium-reducing bacteria-like structures have also been found in the inter- and intracellular spaces of some plant cells in the cortex, endodermis, xylem, and stele of the roots from other field and test tube-grown plants (Patriquin & Dobereiner, 1978). Tetrazolium staining of root tissue is not an adequate technique to identify bacteria in situ, but some information on the serology of *Azospirillum* already available could be valuable for experiments that are ecologically meaningful. Light and electron microscopy may also prove to be useful tools to establish whether, and to what extent, soil bacteria are capable of occupying specialized niches within the roots of plants. Figure 5 shows almost "bacteroid-like structures within living parenchyma cells of the roots of oats (*Avena sativa* L.), which appeared to be free from disease. Old et al. (1975) also reported unidentified "helically lobed" bacteria inside collapsed epidermal cells of *Ammophila aranaria* and within roots of *Z. mays.* It has been suggested that streptomycin resistance in *Azospirillum* confers advantages of infectivity of roots (Dobereiner & Baldani, 1979), and that

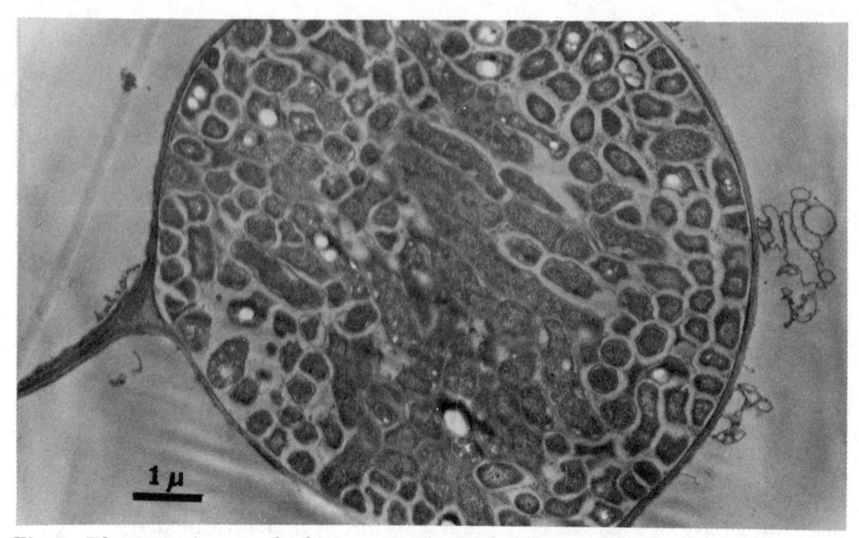

Fig. 5—Electron micrograph of a cross-section of healthy seedling root of California red oats (*Avena sativa* L.), showing a modified cortical parenchyma cell filled with bacteria. A series of similar "bacteroid-like" cells were observed encircling the stele outside the endodermis. Seedlings had been growing in vermiculite for 1 month prior to sampling with nutrient supplied by watering with one-half strength Hoaglands solution. By courtesy of Harold E. Moline, USDA.

entry of bacteria into epidermal and cortical cells occurs by intercellular penetration of the epidermal cells (Old & Nicolson, 1975).

2. HOST-PLANT SPECIFICITY

With the exception of the *Rhizobium*-legume symbioses, very little is known about possible specificity of the associations between N_2-fixing bacteria and plants. *Azotobacter paspali* appears to be restricted to the rhizosphere of roots of five tetraploid ecotypes of *P. notatum.* It has been suggested that other microorganisms (e.g., mycorrhizal fungi) may be involved in the establishment of this association (Neyra & Dobereiner, 1977). On the other hand, *Beijerinckia* associates with the roots of a variety of tropical plants, which include sugarcane, rice, and *Cyperus. Azospirillum* colonizes the roots of a wide variety of tropical and temperate region plants. It has been suggested that only specific species of *Azospirillum* are capable of endorhizosphere associations with particular host plants (Baldani & Dobereiner, 1980). Field-grown corn was suggested to form associations with *A. lipoferum,* and *A. brasilense* nir$^-$ (nitrite reductase negative) was mainly isolated from roots of wheat and rice. However, there remain serious doubts whether the roots of these grasses are capable of supporting N_2-ase activity by *Azospirillum*, and about the procedures used to surface sterilize roots of grasses. These are essential prerequisites to be able to define an endorhizosphere N_2-fixing association. Nevertheless, knowledge of the specificity of associations between N_2-fixing bacteria and various grasses and cereals may have significant implications in future research establishing N_2-fixing associations with selected or genetically manipulated strains of bacteria.

D. Transfer of Fixed N from Diazotroph to Crop Plants

The importance of nonsymbiotic and associative N_2 fixation in agricultural soils can only be realized if the N from N_2 fixation ultimately becomes available to plants under cultivation. Nonsymbiotic N_2 fixers usually efficiently assimilate the N they fix from the atmosphere; therefore, it is essential to agriculture that these microorganisms are induced to release N if it is to become available for crop production. The two mechanisms by which N from nonsymbiotic N_2 fixation may be released from the diazotrophs is through excretion and/or through cell lysis after death.

1. N FROM N_2 FIXATION BY HETEROTROPHIC BACTERIA

Dinitrogen fixation by nonsymbiotic bacteria in soil may occur significantly in flushes as C and energy substrates become available to them. The result of N_2 fixation would be to increase the population of nonsymbiotic N_2-fixing bacteria probably without a significant release of fixed N to the surrounding environment. As the substrates supporting nonsymbiotic N_2 fixation and growth of the bacteria expend, the increase in the population probably slows. Eventually, the population of the N_2-fixing bacteria probably decreases gradually as cells die and lyse through the absence of additional substrates. Nitrogen gains through lysis of nonsymbiotic N_2-fixing bacteria after their death may be of great significance in ecosystems with high rates of turnover of C and energy-rich materials.

Carbon and energy substrates are periodically available to N_2-fixing bacteria in the root environment of plants. The size of the population of N_2-fixing bacteria in the root zone is probably related to the quantity of substrates released by the plant, which is probably inversely related to the quantity of N available to this environment from N_2 fixation. However, the N released from N_2-fixing bacteria in the rhizosphere of plant roots need not necessarily be exclusively available to the plant providing the substrates for N_2 fixation. The roots of plants must compete for substrates in the rhizosphere with a diverse range of soil microorganisms.

It would be of great competitive advantage to the plant if the bacteria were provided with an exclusive niche inside the root. An intimate association of this nature would allow close biochemical interaction, and possibly stimulate the release of N from the N_2-fixing bacteria inside the root in exchange for a supply of photosynthate. In the legume systems, a highly integrated association between the partners has evolved, whereby the microsymbionts occupy a specialized niche inside the plant-derived nodules. The bacteroids in the nodules of legumes generally do not grow while N from N_2 fixation is released to the plant host. Considerable increases in N_2 fixation occur as the life-cycles of legume symbioses progress. This implies that high rates of N_2 fixation in legumes are possible with predominantly nongrowing bacteroids in the nodules. A similar highly integrated association has been -

suggested to exist between *Azospirillum* and the roots of grasses, but its intimacy is not so obvious. Mechanisms by which grasses stimulate the release of fixed N from associated *Azospirillum* have not been proposed, but efficiency of N_2 fixation has been coupled to O_2 depletion in this environment (Dobereiner & Day, 1976). Neyra and Dobereiner (1977) suggested that O_2 deficiency increases the rate of N_2-ase activity per bacterial cell with a concomitant decrease in growth rate. However, these inferences are based on pure culture studies; therefore, the relationships between O_2 deficiency, growth, efficient N_2 fixation, and N excretion by *Azospirillum* associated with roots of grasses remains to be determined.

2. $^{15}N_2$ INCORPORATION INTO THE TISSUES OF FIELD-GROWN GRASSES

The hypotheses that grass crops benefit from nonsymbiotic and associative N_2 fixation are predominantly based on indirect determinations. Only measurements with the incorporation of $^{15}N_2$ through N_2 fixation should be considered as evidence that grasses have access to N from the atmosphere, but few measurements have been attempted. *Paspalum notatum* and *Digitaria decumbens* have been reported to incorporate N from N_2 fixation, but this data would suggest only at low rates (DePolli et al., 1977; Dobereiner & Day, 1976). Ruschel et al. (1975) calculated that sugarcane seedlings derived between 0.2 and 23.8 g N/ha per day from N_2, but undisturbed, mature, field-grown plants showed no ^{15}N enrichment in the leaves and stems (Matsui et al., 1981). Measurements of $^{15}N_2$ incorporation into rice has indicated that products of N_2-fixation by root-associated heterotrophic bacteria are not readily available to the growing plants (Ito et al., 1980).

3. N FROM N_2 FIXATION BY BLUE-GREEN ALGAE

Pure culture studies with N_2-fixing blue-green algae have demonstrated that large quantities of ammonia and a variety of nitrogenous organic substances may be excreted by these microorganisms. The transfer of fixed N from the blue-green algae is of particular relevance when these diazotrophs have formed symbioses and associations with non-N_2-fixing organisms. It has been suggested that the hosts' need for combined N stimulates both the rate of N_2 fixation and the release of fixed N (Stewart, 1974).

Higher plants in natural ecosystems probably benefit from nonsymbiotic N_2 fixation by blue-green algae, because grasses and mosses of a dune-slack community became readily enriched with ^{15}N after the algal mat had been exposed to $^{15}N_2$ (Stewart, 1967). Evidence that N fixed by nonsymbiotic blue-green algae benefit associated crop plants has been inferred mainly from determinations of yield response to inoculation. The presence of blue-green algae in pot and field experiments with rice has been demonstrated to result in significant increases in yield. The production of plant growth–promoting substances, as well as the transfer of fixed N to the crop, has been suggested as a mechanism to explain yield responses to inoculation with blue-green algae (Stewart, 1974). Little information about transfer of ^{15}N to associated crop plants is available; therefore, the relative contribution made by fixed N transfer and hormone production to observed yield increases through inoculation remains to be determined. Fixed N may be transferred via extracellular products directly released by the living cells, through cell lysis after death caused by freezing and drought, or by rapid N cycling through the participation of grazing animals.

IV. IMPROVING N INPUT FROM ASSOCIATIVE N_2 FIXATION BY HETEROTROPHIC BACTERIA FOR AGRICULTURE

The aim of identifying associative N_2 fixation by heterotrophic bacteria in grasses is ultimately to use this characteristic for the production of grain and forage crops. Many fundamental aspects of associative N_2 fixation need to be established before it will be possible to predict whether or not this trait can be subjected to plant breeding, bacterial inoculation, and genetic engineering for eventual application to mechanized agriculture. Although this area of scientific investigation has suffered from highly optimistic inferences based on poor techniques, more recent research offers a glimmer of hope that some cereal and forage grasses may one day rely in part on N_2 fixation for a supply of N. Many of the programs to date endeavoring to improve associative N_2 fixation in grasses have relied on existing techniques and confidence that this trait could either be established or enhanced. The lack of rapid, simple, and meaningful techniques to quantify associative N_2 fixation in field-grown grasses has hindered any real progress in these programs.

A. Plant Breeding

The possibility of breeding plants with the ability to stimulate the growth of N_2-fixing bacteria in the root region is clearly indicated. Investigations with *P. notatum* have revealed that the N_2-fixing bacterium *A. paspali* specifically associates with the roots of five tetraploid ecotypes (Neyra & Dobereiner, 1977). Only these 5 of 33 cultivars tested stimulated *A. paspali* in the rhizosphere of the roots. Unfortunately, further studies of the microbial ecology of this association have not been reported. Larson and Neal (1978) have presented evidence that a facultative N_2-fixing *Bacillus* colonizes selectively the roots of a genetically defined line of wheat. Hopefully, more examples of specific associations between heterotrophic N_2-fixing bacteria and grass genotypes will be described in the future. These reports will be significant contributions in trying to understand the mechanisms by which desirable bacteria are stimulated in the rhizosphere of roots.

By measuring C_2H_2 reduction with preincubated, excised roots, differences in the ability between lines and genotypes of corn, *P. notatum, Pennisetum purpureum,* and wheat have also been described. However, values of C_2H_2 reduction by excised roots after the preincubation period reflects the relative ability of roots to support growth and subsequent N_2-ase activity by N_2-fixing bacteria during the preparation of samples for assay (van Berkum & Bohlool, 1980).

B. Inoculation with N_2-Fixing Bacteria

The legume-*Rhizobium* symbiosis is a good example of accumulated knowledge being applied to agriculture. Research has led to the development of techniques for inoculating plants with N_2-fixing bacteria and to evaluate the performance of the resulting symbioses. The inoculation of legumes with *Rhizobium* is practiced widely to nodulate crops with desirable microsymbionts to establish or enhance N_2 fixation (see Chapt. 25, this book).

The N_2-fixing bacteria *Azospirillum* are considered by several athorities to form endorhizosphere associations with the roots of grasses, and, therefore, have been the subject of most of the inoculation studies in recent years. *Azospirillum* is common in tropical soils, but its incidence in temperate countries is thought to be much lower. Dobereiner et al. (1976) suggested that inoculation of *Azospirillum* in these areas would help to establish these bacteria on the roots of grasses. Inoculating grain and forage grasses with *Azospirillum* was anticipated to result in significant increases in yield through biological N_2 fixation. This prospect originated from reports of high rates of C_2H_2 reduction with preincubated, excised roots, coupled with the suggestion that *Azospirillum* are primarily responsible for N_2 fixation in grasses. In practice, the inoculation of grasses with *Azospirillum* has at best only resulted in slight improvement of yield, but results have not been repeatable, neither in the same nor other locations (van Berkum & Bohlool, 1980). The inoculation of cereal grasses in Israel with a specific strain of *A. brasilense*, isolated in California from the roots of *Cynodon dactylon* [*A. brasilense* (Sp Cd)] (Eskew et al., 1977), has been reported to result in 10 to 35% yield increases over noninoculated control plants at intermediate levels of initial N fertilizer (Okon et al., 1981). However, the mechanisms by which Sp Cd caused increases in yield of these grass crops remains to be determined, because no satisfactory measurement for N_2 fixation was used. It has been suggested that the production of plant-growth hormones and N_2 fixation may be synergetic or alternative mechanisms by which inoculated bacteria produce yield increases in grasses (van Berkum & Bohlool, 1980). Host-plant specificity in the infection of cereal grasses with *Azospirillum* sp. have been suggested (Baldani & Dobereiner, 1980), which could explain why many of the inoculation experiments have failed. In view of the suggested host-plant specificity, the reported yield increases with the inoculation of Sp Cd are curious, because a wide range of grass crops responded and included *Zea mays, Panicum miliaceum, Sorghum bicolor, Setaria italica, Triticum turgidum* var. 'durum', and *T. aestivum.* Furthermore, Sp Cd is an *A. brasilense* nir$^+$ (nitrate reductase positive) isolate (Tarrand et al., 1978), which reportedly does not form endorhizosphere associations with corn or wheat grown in the field (Baldani & Dobereiner, 1980). I wish to emphasize that *Azospirillum* sp. have been isolated from the roots of grasses that reduce C_2H_2 only after a long delay, during which time the N_2-fixing bacteria proliferate 8- to 665-fold and induce N_2-ase activity (van Berkum &

Bohlool, 1980). *Azospirillum* sp. have not been demonstrated to be responsible for reported measurements of N_2-ase activity when in situ methods to determine C_2H_2 reduction were used. Also, there is no evidence that N_2-ase activity by *Azospirillum* is closely coupled to the assimilation of fixed N in grasses. Therefore, inoculating grasses with *Azospirillum* to obtain agriculturally significant yield increases through associative N_2 fixation for the present does not seem justifiable because, in most instances, this practice has little or no effect on crop yield.

V. CONCLUSIONS

It is clear that grasses may derive N from the atmosphere through the participation of nonsymbiotic or associative diazotrophs. Whether or not this natural phenomenon can be put to use in agriculture for the production of crops that rely in part on N_2 fixation for their N requirement remains an open question. Carbohydrates released by the decay of plant material may support the growth on N_2 by free-living heterotrophic bacteria. Blue-green algae may proliferate substantially on soil surfaces and in paddy water. The N retained in this biomass of these fractions of the microflora may have long-term soil enrichment properties, especially in undisturbed ecosystems. Long-term enrichment of soil that is able to support labor-intensive and subsistence farming may result from N_2 fixation by heterotrophic bacteria in association with roots of crops and by blue-green algal blooms in paddies and crusts on soil surfaces. However, the highly mechanized farming practices of developed countries probably would have to rely on crop production where root-associated N_2 fixation by bacteria is closely coupled to plant N metabolism similar to the legume-*Rhizobium* symbioses for maximal economic return. The *Azospirillum*-grass associative symbioses have been suggested to offer the greatest prospect of fulfilling this role, but N_2 fixation by this combination and fixed N transfer to the crop under field conditions still remains to be demonstrated. Nevertheless, this area of scientific investigation should not be dismissed, because the interrelationships between nonsymbiotic and associative N_2 fixation and the maintainance of soil fertility and the full potential for crop production have not been explored in detail.

REFERENCES

Arnon, D. I., and D. C. Yoch. 1974. Photosynthetic bacteria. p. 168–201. *In* A. Quispel (ed.) The biology of nitrogen fixation. American Elsevier Publ., Co., Inc., New York.

Baldani, V. L. D., and J. Dobereiner. 1980. Host-plant specificity in the infection of cereals with *Azospirillum* spp. Soil Biol. Biochem. 12:433–439.

Barber, D. A., and J. K. Martin. 1976. The release of organic substances by cereal roots into soil. New Phytol. 76:69–80.

Burns, R. C., and R. W. F. Hardy. 1975. Nitrogen fixation in bacteria and higher plants. p. 189. *In* A. Kleinzeller et al. (ed.) Molecular biology biochemistry and biophysics. Vol. 21. Springer-Verlag, New York.

Dart, P. J., and J. M. Day. 1975. Non-symbiotic nitrogen fixation in soils. p. 225–252. *In* N. Walker (ed.) Soil microbiology—a critical review. Butterworths, London.

Day, J. M. 1977. Nitrogen-fixing associations between bacteria and tropical grass roots. p. 271–288. *In* A. Ayanaba and P. J. Dart (ed.) Biological nitrogen fixation in farming systems of the tropics. John Wiley & Sons, Inc., New York.

Day, J. M., D. Harris, P. J. Dart, and P. van Berkum. 1975. The Broadbalk experiment. An investigation of nitrogen gains from non-symbiotic fixation. p. 71–84. *In* W. D. P. Stewart (ed.) Nitrogen fixation by free-living microorganisms. Int. Biological Programme Series. Vol. 6. Cambridge Univ. Press, Cambridge, England.

DePolli, H., E. Matsui, J. Dobereiner, and E. Salati. 1977. Confirmation of nitrogen fixation in two tropical grasses by $^{15}N_2$ incorporation. Soil Biol. Biochem. 9:119–123.

Dobereiner, J. 1978. Potential for nitrogen fixation in tropical legumes and grasses. p. 13–34. *In* J. Dobereiner et al. (ed.) Limitations and potentials for biological nitrogen fixation in the tropics. Plenum Publishing Corp., New York.

Dobereiner, J. 1980. Forage grasses and grain crops. p. 535–555. *In* F. J. Bergersen (ed.) Methods for evaluating biological nitrogen fixation. John Wiley & Sons, Inc., New York.

Dobereiner, J., and V. L. D. Baldani. 1979. Selective infection of maize roots by streptomycin-resistant *Azospirillum lipoferum* and other bacteria. Can. J. Microbiol. 25:1264–1269.

Dobereiner, J., and J. M. Day. 1975. Nitrogen fixation in the rhizosphere of tropical grasses. p. 39–56. *In* W. D. P. Stewart (ed.) Nitrogen fixation by free-living microorganisms. Int. Biological Programme Series. Vol. 6. Cambridge Univ. Press, Cambridge, England.

Dobereiner, J., and J. M. Day. 1976. Associative symbioses in tropical grasses: characterization of microorganisms and dinitrogen fixing sites. p. 518–538. *In* W. E. Newton and C. J. Nyman (ed.) Proc. of the First Int. Symp. on Nitrogen Fixation. Washington State Univ. Press, Pullman.

Dobereiner, J., J. M. Day, and P. J. Dart. 1972. Nitrogenase activity and oxygen sensitivity of the *Paspalum notatum-Azotobacter paspali* association. J. Gen. Microbiol. 71:103–116.

Dobereiner, J., I. E. Marriel, and M. Nery. 1976. Ecological distribution of *Spirillum lipoferum* Beijerinck. Can. J. Microbiol. 22:1464–1473.

Eskew, D. L., D. D. Focht, and I. P. Ting. 1977. Nitrogen fixation, denitrification, and pleomorphic growth in a highly pigmented *Spirillum lipoferum.* Appl. Environ. Microbiol. 34:582–585.

Fogg, G. E., W. D. P. Stewart, P. Fay, and A. E. Walsby. 1973. The blue-green algae. p. 459. Academic Press, New York.

Greenland, D. J. 1977. Contribution of microorganisms to the nitrogen status of tropical soils. p. 13–26. *In* A. Ayanabe and P. J. Dart (ed.) Biological nitrogen fixation in farming systems of the tropics. John Wiley & Sons, Inc., New York.

Ito, O., D. Cabrera, and I. Watanabe. 1980. Fixation of dinitrogen-15 associated with rice plants. Appl. Environ. Microbiol. 39:554–558.

Jensen, H. L. 1965. Nonsymbiotic nitrogen fixation. *In* W. V. Bartholomew and F. E. Clark (ed.) Soil nitrogen. Agronomy 10:436–480. Am. Soc. of Agron., Madison, Wis.

Jones, K. 1974. Nitrogen fixation in a salt water marsh. J. Ecol. 62:583–595.

Jurgensen, M. F., and C. B. Davey. 1970. Nonsymbiotic nitrogen-fixing microorganisms in acid soils and the rhizosphere. Soils Fert. 33:435–446.

Kana, T. M., and J. D. Tjepkema. 1978. Nitrogen fixation associated with *Scirpus atrovirens* and other nonnodulated plants in Massachusetts. Can. J. Bot. 56:2636–2640.

Karamanos, R. E., and D. A. Rennie. 1980. Changes in natural ^{15}N abundance associated with pedogenic processes in soil. II. Changes on different slope positions. Can. J. Soil Sci. 60: 365–372.

Knowles, R. 1977. The significance of asymbiotic dinitrogen fixation by bacteria. p. 33–83. *In* R. W. F. Hardy and A. H. Gibson (ed.) A treatise on dinitrogen fixation, section 4. John Wiley & Sons, Inc., New York.

Larson, R. I., and J. L. Neal. 1978. Selective colonization of the rhizosphere of wheat by nitrogen-fixing bacteria. *In* U. Granhall (ed.) Environmental role of nitrogen-fixing blue-green algae and asymbiotic bacteria. Ecol. Bull. 26:331–342.

Lumpkin, T. A., and D. L. Plucknett. 1980. Azolla: botany, physiology, and use as a green manure. Econ. Bot. 34:111–153.

Mague, T. H. 1977. Ecological aspects of dinitrogen fixation by blue-green algae. p. 85–140. *In* R. W. F. Hardy and A. H. Gibson (ed.) A treatise on dinitrogen fixation, section 4. John Wiley & Sons, Inc., New York.

Matsui, E., P. B. Vose, N. S. Rodrigues, and A. P. Ruschel. 1981. Use of ^{15}N enriched gas to determine N_2 fixation by undisturbed sugar-cane plant in the field. p. 153–161. *In* P. B. Vose and A. P. Ruschel (ed.) Associative N_2-fixation Vol. II. CRC Press, Inc., Boca Raton, Florida.

McClung, C. R., and D. G. Patriquin. 1980. Isolation of a nitrogen-fixing *Campylobacter* species from the roots of *Spartina alterniflora* Loisel. Can. J. Microbiol. 26:881-886.

Moore, A. W. 1966. Non-symbiotic nitrogen fixation in soil and soil-plant systems. Soils Fert. 29:1185-1207.

Moore, A. W. 1969. Azolla: biology and agronomic significance. Bot. Rev. 35:17-34.

Mulder, E.G., and S. Brotonegoro. 1974. Free-living heterotrophic nitrogen-fixing bacteria. p. 37-85. *In* A. Quispel (ed.) The biology of nitrogen-fixation. American Elsevier Publ. Co., Inc., New York.

Neyra, C. A., and J. Dobereiner. 1977. Nitrogen fixation in grasses. Adv. Agron. 29:1-38.

Okon, Y., Y. Kapulnik, S. Sarig, I. Nur, J. Kigel, and Y. Henis. 1981. *Azospirillum* increases crop yields in fields of Israel. p. 492. *In* A. H. Gibson and W. E. Newton (ed.) Current perspectives in nitrogen fixation. Australian Academy of Science, Canberra.

Old, K. M., S. Hallam, and T. N. Nicolson. 1975. Helically-lobed bacteria in plant roots. Soil Biol. Biochem. 7:73-75.

Old, K. M., and T. H. Nicolson. 1975. Electron microscopical studies of the microflora of roots of sand dune grasses. New Phytol. 74:51-58.

Patriquin, D. G., and J. Dobereiner. 1978. Light microscopy observations of tetrazolium-reducing bacteria in the endorhizosphere of maize and other grasses in Brazil. Can. J. Microbiol. 24:734-742.

Patriquin, D. G., and C. R. McClung. 1978. Nitrogen accretion and the nature and possible significance of N_2-fixation (acetylene reduction) in a Nova Scotian *Spartina alterniflora* stand. Mar. Biol. 47:227-242.

Reimold, R. J., R.A. Linthurst, and P. L. Wolf. 1975. Effects of grazing on a salt marsh. Biol. Conserv. 8:105-125.

Rovira, A. D. 1969. Plant root exudates. Bot. Rev. 35:35-57.

Ruschel, A. P., Y. Henis, and E. Salati. 1975. Nitrogen-15 tracing of N-fixation with soil-grown sugarcane seedlings. Soil Biol. Biochem. 7:181-182.

Stewart, W. D. P. 1967. Transfer of biologically fixed nitrogen in a sand dune slack region. Nature (London) 214:603-604.

Stewart, W. D. P. 1974. Blue-green algae. p. 202-237. *In* A. Quispel (ed.) The biology of nitrogen fixation. American Elsevier Publ. Co., Inc., New York.

Tarrand, J. J., N. R. Krieg, and J. Dobereiner. 1978. A taxonomic study of the *Spirillum lipoferum* group, with descriptions of a new genus, *Azospirillum* gen. nov. and two species, *Azospirillum lipoferum* (Beijerinck) comb, nov. and *Azospirillum brasilense* sp. nov. Can. J. Microbiol. 24:967-980.

Tjepkema, J. D., and R. H. Burris. 1976. Nitrogenase activity associated with some Wisconsin prairie grasses. Plant Soil 45:81-94.

Tjepkema, J. D., and H. J. Evans. 1976. Nitrogen fixation associated with *Juncus balticus* and other plants of Oregon wetlands. Soil Biol. Biochem. 8:505-509.

Tjepkema, J. D., and P. van Berkum. 1977. Acetylene reduction by soil cores of maize and sorghum in Brazil. Appl. Environ. Microbiol. 33:626-629.

van Berkum, P. 1980. Evaluation of acetylene reduction by excised roots for the determination of nitrogen fixation in grasses. Soil Biol. Biochem. 12:141-145.

van Berkum, P., and B.B. Bohlool. 1980. Evaluation of nitrogen fixation by bacteria in association with roots of tropical grasses. Microbiol. Rev. 44:491-517.

van Berkum, P., and J. M. Day. 1980. Nitrogenase activity associated with soil cores of grasses in Brazil. Soil Biol. Biochem. 12:137-140.

van Berkum, P., and C. Sloger. 1979. Immediate acetylene reduction by excised grass roots not previously preincubated at low oxygen tensions. Plant Physiol. 64:739-745.

van Berkum, P., and C. Sloger. 1981. Physiology of root-associated nitrogenase activity in *Oryza sativa* L. p. 490. *In* A. H. Gibson and W. E. Newton (ed.) Current perspectives in nitrogen fixation. Australian Academy of Science, Canberra.

Vlassak, K., E. A. Paul, and R. E. Harris. 1973. Assessment of biological nitrogen fixation in grassland and associted sites. Plant Soil 38:637-649.

Watanabe, I., and K. K. Lee. 1977. Non-symbiotic nitrogen fixation in rice and rice fields. p. 289-306. *In* A. Ayanabe and P. J. Dart (ed.) Biological nitrogen fixation in farming systems of the tropics. John Wiley & Sons, Inc., New York.

Witty, J. F. 1979. Algal nitrogen fixation on temperate arable fields. Algal inoculation experiments. Plant Soil 52:165-183.

Witty, J. F., P. J. Keay, P. J. Frogatt, and P. J. Dart. 1979. Algal nitrogen fixation on temperate arable fields. The Broadbalk experiment. Plant Soil 52:151-164.

10

James E. Harper
Agricultural Research Service, USDA
University of Illinois
Urbana, Illinois

Uptake of Organic Nitrogen Forms by Roots and Leaves

The major fraction of organic N in soils is relatively stable and not directly available to plants. However, a portion of it becomes available to plants during the growing season through mineralization to inorganic forms by soil microorganisms. The mineralization process generally converts from 1 to 3% of the total organic N per year to inorganic N. The stability of organic N in soil is largely due to (i) the absence of adequate sources of energy to support vigorous microbial activity (ii) inclusion in other stable soil organic constituents, and (iii) adsorption by clay minerals (Bremner, 1965). Although soil organic N serves as an important reservoir of N for crop use, little of it appears to be available to plants in the organic form.

One form of organic N that can be taken up directly by plants is urea. Although urea is not thought to occur naturally in soils in significant quantities, it is the primary dry N fertilizer source in use today (Harris & Harre, 1979). Since urea is an important N source in agricultural practice, this chapter will cover its uptake and metabolism by the roots and foliage of plants in some detail. Evidence that other organic N compounds, such as amino acids, may be taken up by plants is also presented. Finally, a perspective as to the overall importance of organic N to crop production is presented.

I. UREA UPTAKE BY ROOTS

A. Available Form and Growth Response

Soil application of urea generally results in rapid hydrolysis to NH_4^+ by soil urease, with subsequent conversion of NH_4^+ to NO_3^-. However, under conditions where urea is not hydrolyzed, such as sterile media, urea is

Nitrogen in Crop Production.

utilized directly by many plants. Bollard (1959) listed several plant species that have the ability to utilize urea as a sole N source. Furthermore, several of the plant species produced dry matter yields as good or better than those obtained with NO_3^-. Notable exceptions were tobacco (*Nicotiana tabacum*) and *Baeria chrysostoma,* with growth on urea being less than half that on NO_3^-.

B. Mechanism of Uptake

Although it has sometimes been suggested that urea may be degraded on the surface of roots, the preponderance of evidence suggests that urea is taken up as an intact molecule by most plants. Following uptake, it may either be metabolized in the root or translocated through the xylem as an intact molecule and eventually metabolized in the aerial portion of the plant. There are conflicting views on the mechanism of urea uptake by plant roots. Both passive and active uptake have been reported. Hentschel (1970) noted that urea uptake by roots of beans (*Phaseolus vulgaris*) was inhibited by anaerobic conditions (N_2 bubbling) and by dinitrophenol, suggesting active uptake. The lack of linearity in uptake with increasing urea concentration in nutrient solution, further supported an involvement of a metabolic process. However, Olszanska (1968) found no influence of potassium cyanide, dinitrophenol, or sodium azide on urea uptake by roots of corn (*Zea mays*), suggesting that it was not an active process. Rice (*Oryza sativa* L.) seedlings absorbed increasing amounts of urea with increasing urea concentration in the nutrient solution without any sign of a stationary uptake phase (Hirose & Goto, 1961). Thus, urea uptake by rice appears to be a simple diffusion process. Apparently, both active and passive urea uptake processes exist and appear to be species-dependent.

C. Rate of Uptake

The rate or urea uptake is generally less than that of NH_4^+ (Hentschel, 1970) or NO_3^- (Vigue et al., 1977; Kirkby & Mengel, 1970). Even at one-fourth the concentration of urea, the uptake of NO_3^- still exceeded uptake of urea by sunflower (*Helianthus* sp.) (Kirkby & Mengel, 1970). With soybean [*Glycine max* (L.) Merr.], the uptake of urea by young plants was initially slower than that of NO_3^-, but with older plants, equivalent uptake of the two N sources was obtained (Vigue et al., 1977). Urea uptake was inhibited by NH_4^+ in bush bean when both were supplied at equivalent concentrations (Hentschel, 1970), while NO_3^- and urea were taken up by soybean independent of the concentration of the two N sources (Vigue et al., 1977). The uptake of urea by soybean increased with increased concentration of urea, in contrast to NO_3^- uptake, which was similar at 2 and 6 m*M* N levels. Thus, urea uptake appears independent of NO_3^- uptake, but may be competitively inhibited by NH_4^+. Presumably, the same uptake mechanism and/or site is involved in urea and NH_4^+ uptake.

D. Metabolism and Translocation

Although urea is generally believed to be absorbed as an intact molecule, some breakdown at the root surface with subsequent intake of NH_4^+ may occur. Following uptake of the intact molecule, urea may be hydrolyzed to NH_3 and CO_2 by the urease enzyme in the root, or it may be translocated intact to the aerial portion of the plant. Whether urea is metabolized in the root or translocated intact is in part dependent on the presence or absence of urease in the root—an apparent species-dependent trait. In certain plant species—cucumber (*Cucumis sativus*), bean, tomato (*Lycopersicum esculentum*), corn, potato (*Solanum tuberosum*), celery (*Apium graveolens*), and *Chlorella*—urease activity has not been detected (Bollard, 1959). Whether these plant species do in fact lack the enzyme, or whether the proper techniques for detection of the enzyme have not been developed, is still open to question. Mothes (1961) reviewed the metabolism of urea and has suggested that urea may be directly incorporated into allantoin via allantoic acid, the latter being synthesized from condensation of glyoxylate and urea. Another possible metabolic pathway for urea, although not confirmed, is a reversal of the classical urea-ornithine cycle, which could allow assimilation of urea without prior hydrolysis to NH_3 and CO_2.

II. UREA UPTAKE BY FOLIAGE

A. Plant Tolerance

Absorption of urea readily takes place in the foliage of many plants (see also Chapt. 39, this book). Urea has been successfully applied to foliage of apple (*Malus sylvestris*), tomato, celery, lima bean (*Phaseolus lunatus*), potato, cantaloupe (*Cucucumis melo*), cucumber, and sugarcane (*Seccarum officinarum*) (Donahue et al., 1971). In some instances, foliar application of urea causes visible damage. Plants showing the least tolerance to urea generally respond most favorably to physiologically tolerable amounts. The level of urease activity in the leaf is inversely correlated with tolerance to urea application; higher urease activity may cause rapid accumulation of NH_3, which is conducive to foliar injury. Plants are also damaged by urea fertilizers containing biuret; generally, levels in excess of 0.25% should be avoided when utilizing urea for foliar application, while as much as 2.5% can be tolerated when urea is applied to soil (Kilmer & Webb, 1968).

B. Uptake Process in Leaves

The process of urea uptake by leaves differs from that by roots. The leaf cell wall is covered by a cuticular layer and urea must first diffuse across the cuticle and cell wall where it is then absorbed to the surface of the plasma membrane. Movement across the plasma membrane into the cytoplasm appears to be an active step requiring metabolic energy. Foliarly applied urea has been observed to penetrate the cuticular membrane at a rate of 10- to 20-fold higher than other ions, and the rate of penetration was independent of concentration (Franke, 1967). Other ions penetrate leaf cells more rapidly when applied in conjunction with urea than when applied singly, suggesting that urea alters membrane permeability. Yamada et al. (1965) concluded that the effect of urea on cuticular membrane permeability was based on changing the ester, ether, and diether bonds between the macromolecules of the cutin.

C. Entry and Metabolism

Urea is rapidly taken up following foliar application, with 80% or more generally being absorbed within 48 h of application. Absorption of urea is more rapid from lower- than upper-leaf surfaces of fruit trees, whereas, in tobacco, uptake is similar from the two leaf surfaces (younger leaves are generally more active in uptake, regardless of plant species). Lower leaf surfaces of fruit trees also tend to have a greater number of stomates, suggesting that uptake may be partially through the stomates. However, with tobacco, urea absorption was increased 10-fold by artificially damaging epidermal hairs, which indicates that stomates may not be the major pathway of foliar absorption (Volk & McAuliffe, 1954). Urea uptake by tobacco was also greater in the dark than in the light, which would support that uptake was not through the stomata.

As with root uptake, foliar absorption of urea may be followed by subsequent hydrolysis by urease, likely the predominant mechanism of incorporation, or it may be incorporated through more direct mechanisms, as suggested for root tissue. Although urease is reported to be present in numerous plant species and tissue types, the primary role of the enzyme may not be a hydrolyze urea, since free urea is seldom available to the plant unless specifically applied to the foliage. Changes in urease and arginase activities during germination of soybean are parallel, and it has been suggested that urease is involved in arginine metabolism (Varner, 1960). Regardless of primary function of urease, the hydrolysis of urea by urease to NH_3 and CO_2 can occur with subsequent incorporation into amino acids and amides.

III. OTHER ORGANIC N SOURCES

There is evidence that several amino compounds are absorbed quite readily by higher plant cells (Luttge & Higinbotham, 1979). Different uptake systems, with some overlap, have been identified for basic, acidic, and neutral amino acids using soybean root cells (King & Hirji, 1975). Evidence for both carrier-mediated and H^+ cotransport uptake mechanisms for amino acids have been presented for various plant cell types and species

(Kirk & Kirk, 1978; Jung & Luttge, 1980). In addition to uptake, release of amino acids followed by reabsorption has been reported for plant cells in liquid culture. Plants also absorb applied herbicides, many of which are organic N compounds, which provides further evidence that organic N compounds can be taken up. Although mechanisms apparently exist for uptake of various organic N compounds, the contribution of these organic N compounds to overall N requirements of plants in field environments is not well-defined, but would appear to be of minimal consequence.

IV. IMPORTANCE OF ORGANIC N TO CROP PRODUCTION

It is generally assumed that NH_4^+ and NO_3^- are the predominant forms of N taken up from the soil by most higher plants. Even when organic N (e.g., urea) is applied to the soil, it is expected that plants utilize this N following conversion of NH_4^+ to NO_3^--N. However, the evidence that plants can utilize urea as a sole N source raises the possibility that under certain conditions, such as green manuring or grazing animals, plant roots may be exposed to and take up organic N forms directly from the soil. Under most agricultural practices, however, the actual availability of organic N compounds to plant roots is generally minimal, and hence, cannot be considered a major pathway of N utilization by higher plants.

The efficiency of N utilization from soil-incorporated urea is not significantly different from that of NH_4^+ and NO_3^- fertilizers. However, under conditions involving surface application of urea, such as to turf grass, considerable volatilization of NH_3 may occur and the efficiency of N utilization is decreased. The recovery of foliarly applied urea by the plant does generally exceed that of soil-applied N sources. Thus, the practical potential of foliar fertilization must be considered in view of the ever-increasing costs of producing N fertilizers and the environmental impact of their inefficient utilization through soil application. A renewed interest in foliar fertilization of field crop plants was enhanced by reports of positive responses to foliar feeding of soybeans with N, P, K, and S fertilizers (Garcia & Hanway, 1976). Efforts to reproduce these responses have unfortunately been somewhat disappointing with numerous no-response results intermingled with a few positive responses (see also Chapt. 39, this book). Nitrogen from foliar feeding of soybean with urea does enter the soybean plant, and primarily ends up in the seed fraction (Vasilas et al., 1980). Both rice and wheat (*Triticum aestivum* L.) have shown increases in protein content of the seed following foliar urea application, indicating responsiveness to foliar feedings. However, the mixed results, with respect to yield response to foliar N fertilization, still makes this a questionable practice in field crop plants such as soybeans. The more positive responses to foliar application of urea in certain citrus crops has made this method of N fertilization feasible, especially if the N is applied in conjunction with the pesticide applications routinely made in the citrus industry.

The significance of organic N to crop production is obvious when one considers the vast soil N reserves that are primarily in organic forms. However, in terms of direct plant use of organic N without some prior conversion to an inorganic form, the only apparent source of any importance is that of urea applied foliarly. Additional efforts are needed to sort out the requisite conditions necessary to realizing consistent responses by field crop plants to foliar urea application.

REFERENCES

Bollard, E. E. 1959. Urease, urea, and ureides in plants. p. 304–329. *In* Utilization of nitrogen and its compounds by plants. Symposia of the Soc. Exp. Biol. XIII. Academic Press, New York.

Bremner, J. M. 1965. Organic nitrogen in soils. *In* W. V. Bartholomew and F. E. Clark (ed.) Soil nitrogen. Agronomy 10:93–149. Am. Soc. of Agron., Madison, Wis.

Donahue, R. L., J. C. Shickluna, and L. S. Robertson. 1971. Soils—an introduction to soils and plant growth. 3rd Ed. Prentice-Hall, Inc., New Jersey.

Franke, W. 1967. Mechanisms of foliar penetration of solutions. Annu. Rev. Plant Physiol. 18:281–300.

Garcia, L. R., and J. J. Hanway. 1976. Foliar fertilization of soybeans during the seed-filling period. Agron. J. 68:653–657.

Harris, G. T., and E. H. Harre. 1979. World fertilizer situation and outlook—1978–1985. Tech. Bull. IFDC-T-13. Int. Fert. Dev. Center, Muscle Shoals, Ala.

Hentschel, G. 1970. The uptake of ^{15}N-labelled urea by bush bean. p. 30–34. *In* E. A. Kirkby (ed.) Nitrogen nutrition of the plant. The Univ. of Leeds, Leeds, England.

Hirose, S., and Y. Goto. 1961. Mode of absorption of urea by rice seedlings. Soil Sci. Plant Nutr. 7:85.

Jung, K. D., and U. Luttge. 1980. Amino acid uptake by *Lemna gibba* by a mechanism with affinity to neutral L- and D-amino acids. Planta 150:230–235.

Kilmer, V. J., and J. Webb. 1968. Agronomic effectiveness of different fertilizers. p. 33–65. *In* L. B. Nelson (ed.) Changing patterns in fertilizer use. Soil Sci. Soc. of Am., Madison, Wis.

King, J., and R. Hirji. 1975. Amino acid transport system of cultured soybean root cells. Can. J. Bot. 53:2088–2091.

Kirk, M. M., and D. L. Kirk. 1978. Carrier-mediated uptake of arginine and urea by *Volvox carteri* f *nagariensis.* Plant Physiol. 61:549–555.

Kirkby, E. A., and K. Mengel. 1970. Preliminary observations on the effect of urea nutrition on the growth and nitrogen metabolism of sunflower plants. p. 35–38. *In* E. A. Kirkby (ed.) Nitrogen nutrition of the plant. The Univ. of Leeds, Leeds, England.

Luttge, U., and N. Higinbotham. 1979. Transport in plants. Springer-Verlag, New York.

Mothes, K. 1961. The metabolism of urea and ureides. Can. J. Bot. 39:1785–1807.

Olszanska, B. 1968. Studies on the mechanism of urea uptake by plant roots. I. Kinetics of uptake, effect of concentration and inhibitors. Acta Soc. Bot. Pol. 37:39–49.

Varner, J. E. 1960. Urease. p. 247–256. *In* P. D. Boyer et al. (ed.) Vol IV, The enzymes. 2nd Ed. Academic Press, New York.

Vasilas, B. L., J. O. Legg, and D. C. Wolf. 1980. Foliar fertilization of soybeans: absorption and translocation of ^{15}N-labeled urea. Agron. J. 72:271–275.

Vigue, J. T., J. E. Harper, R. H. Hageman, and D. B. Peters. 1977. Nodulation of soybeans grown hydroponically on urea. Crop Sci. 17:169–172.

Volk, R., and C. McAuliffe. 1954. Factors affecting the foliar absorption of ^{15}N-labelled urea by tobacco. Soil Sci. Soc. Am. Proc. 18:308–312.

Yamada, Y., S. H. Wittwer, and M. J. Bukovac. 1965. Penetration of organic compounds through isolated cuticular membranes with special reference to C^{14} urea. Plant Physiol. 40:170–175.

11

F. E. Broadbent
University of California
Davis, California

Plant Use of Soil Nitrogen

The evaluation of plant uptake of soil N in quantitative terms is a difficult matter since such uptake varies with the nature of the plant, depth and distribution of roots, kind and amount of fertilizer N applied, relative amounts and distribution of inorganic and organic forms of N in the soil profile, climatic factors such as temperature and rainfall, and a number of management factors as well. Whereas it is possible to distinguish between soil N and recently applied fertilizer N by labeling the latter and determining the quantity of labeled N in the plant, it is much more difficult to measure the relative contributions of the various organic and inorganic components of soil N. Moreover, since interchange occurs between fertilizer N and N already present in the soil, beginning as soon as the fertilizer is applied and continuing for a long time (months or years) thereafter, it is not possible to provide a clear-cut answer to the question of just what constitutes soil N.

Obviously, the N-supplying capacity is not the same in all soils. Estimates based upon laboratory incubations or greenhouse pot experiments often give values somewhat different from measurements based upon plant uptake under field conditions. For that reason, the data cited in this chapter and the conclusions derived from them are primarily, though not entirely, from field experiments, many of them utilizing isotopically labeled N.

I. METHODS OF MEASUREMENT

Since virtually all nonleguminous crops in commercial production are fertilized with N, it is appropriate to consider the methods used to estimate the relative contributions of soil and fertilizer N to crop uptake. By far, the most common method is the so-called difference method, which assumes that uptake of soil N by a fertilized crop is the same as that in an unfertilized crop, the difference in total N uptake being attributed to the fertilizer. However, a fertilized plant will often have a larger root system, enabling it to exploit a larger volume of soil than an unfertilized plant, and since other differences may also exist, the assumption underlying use of the difference

Nitrogen in Crop Production.

method is frequently invalid. Numerous investigators have published evidence of the so-called priming effect, or stimulating effect, of N fertilizers on uptake of soil N (Allison, 1966; Sapozhnikov et al., 1968; Westerman & Kurtz, 1973; Hauck & Bremner, 1976). In some cases, uptake of soil N is increased two- or threefold by addition of fertilizer (Kissel & Smith, 1978). Frequently, estimation of soil N uptake by an unfertilized control will underestimate actual uptake of soil N, but in some cases it may give an overestimate, particularly in the presence of large additions of fertilizer N.

A second method of estimating the uptake of soil N is the regression method proposed by Terman and Brown (1968). This method is applicable where three or more fertilizer rates are applied under otherwise similar conditions. A response curve of N uptake vs. N applied is drawn and extrapolated to the ordinate, the intercept on which gives a value of soil N uptake by the crop. Quite often in greenhouse experiments and in some field experiments, a linear relationship is obtained that makes it easy to determine the intercept. Examples of this method are found in the work of Terman and Brown (1968) and of Broadbent and Reyes (1971). In field trials, the response curve is more commonly curvilinear unless a rather narrow range of fertilizer application rates is selected and determination of the intercept becomes less obvious. If the tangent to the curve at a particular point is taken, the intercept will increase with increasing N rate, whereas plant uptake may actually decrease, particularly at N rates that exceed the optimum.

A third, and by most criteria the best, method for estimating plant uptake of soil N is by use of isotopically labeled fertilizer N. Although some interchange between applied inorganic N and soil organic N may occur, the isotopic composition of the plant still provides the most accurate measure of soil N utilized, which is available at present. Most of the data cited in this chapter are based upon experimental work utilizing isotopically labeled fertilizer materials in which plant uptake of unlabeled N is used as a measure of the contribution of soil N.

II. QUANTITIES OF SOIL N UTILIZED BY PLANTS

Various estimates of the quantity of soil N supplied to plants have been made, such as Scarsbrook's (1965) estimate of 2 to 4% per year of total soil N. Values expressed as a percentage of total N in the surface soil may be misleading, since plants may obtain N from a considerable depth below the surface, and since subsoil N may be mineralized at rates different from that which occurs at the surface. Among surface soils, there is some indication that the rate of N mineralization is inversely related to the total soil N. In a greenhouse pot experiment with 12 surface soils ranging from 0.041 to 0.34% total N, Legg and Stanford (1967) found uptake of soil N by a crop of oats (*Avena sativa* L.) fertilized with NO_3^- at rates ranging from 0 to 200 mg/kg N to vary from 1.8 to 9.7% of the total soil N. Tabatabai and Al-

Table 1—Uptake of soil N by some fertilized crops.

Crop	N rate	Plant N from soil	Soil N in crop	Reference
	kg/ha	%	kg/ha	
Corn	50–168	68–77	137–177	Bigeriego et al., 1979
Corn	112–560	22–47	55–79	Broadbent & Carlton, 1978
Corn	90–360	40–76	84–148	Broadbent & Carlton, 1979
Corn	100	32–85	39–114	IAEA, 1970
Wheat	50–100	56–79	64–78	Olson et al., 1979
Wheat	20–164	35–75	32–108	Campbell & Paul, 1978
Wheat	60–120	41–62	--	Smirnov et al., 1976
Wheat	60	52–82	--	Koritskaya, 1969
Rice	37–75	64–88	45–58	Koyama et al., 1973
Rice	50	79–93	94–113	IRRI, 1975
Rice	100	64–72	63–64	IAEA, 1978
Rice	100	56–62	50–57	Patrick & Reddy, 1976
Barley (*Hordeum vulgare* L.)	60	75–86	--	Sapozhnikov et al., 1974
Oats	90	46–61	--	Zamyatina et al., 1971
Sudangrass [*Sorghum sudanense* (Piper) Stapf.]	56–168	61–84	136–169	Westerman et al., 1972
Bermudagrass (*Cynadon dactylon*)	560	38	173	Kissel & Smith, 1978
Sugarbeet	56–280	60–88	164–169	Hills et al., 1978
Tomato (*Lycopersicon esculentum* Mill.)	56–224	57–90	196–216	Broadbent et al., 1980

Khafaji (1980) reported percentages of organic N mineralized at 20°C in a laboratory incubation experiment with 12 soils ranging from 0.115 to 0.418% total N to vary from 1.8 to 6.7% of organic N. In both of these investigations, there was a negative correlation between total N and percent of total N mineralized. Broadbent and Nakashima (1968) reported uptake of 5.1 to 13.1% of soil N by a succession of four crops grown in greenhouse pots over a period of 314 d.

Estimates of rates of N mineralization based on slopes of curves of N content of the surface soil vs. time were obtained from unfertilized, continuous corn (*Zea mays* L.) plots of long-term experiments at Sanborn field in Missouri and the Morrow plots in Illinois. The values ranged from 0.5 to 1.1% of total N mineralized per year (Broadbent, 1971). Rauhe and Seiberlich (1976) reported that mineralized N constituted 2.3% of soil N in experiments with cereals fertilized with urea. In two field trials with labeled fertilizer, where the depth of rooting of corn was about 180 cm, plant uptake was 0.69% of total soil N in this 180-cm layer in a fertile soil and 3.69% in a less fertile soil containing about one-fifth of the total N (Broadbent & Carlton, 1978).

Plant uptake of soil N expressed as percent of total plant N recovered and as kilograms N per hectare for a number of important crops is given in Table 1. These data were all obtained under field conditions with adequately fertilized crops. A broad spectrum of soils and climatic conditions are represented. In most cases, well over half of the N utilized by plants was soil N, with quantities ranging from < 40 to > 200 kg N/ha.

Table 2—Rates of soil N uptake for several crops.

Crop	Days after planting 30	60	90	120	Reference
	kg N ha^{-1} d^{-1}				
Corn	1.11	1.22	0.84	--	F. J. Hills et al. (unpublished)
Corn	1.55	2.30	1.77	--	Bigeriego et al., 1979
Rice	0.29	0.49	0.58	0.56	Patrick & Reddy, 1976
Rice	0.96	0.93	0.47	--	IRRI, 1974
Tomato	0.99	1.81	0.73	--	Hills et al., 1983
Sugarbeet	1.44	1.67	1.53	1.01	Hills et al., 1983
Sudangrass	1.62	1.22	0.78	0.31	Westerman & Kurtz, 1973
Bermudagrass	1.11	0.85	0.66	0.53	Kissel & Smith, 1978

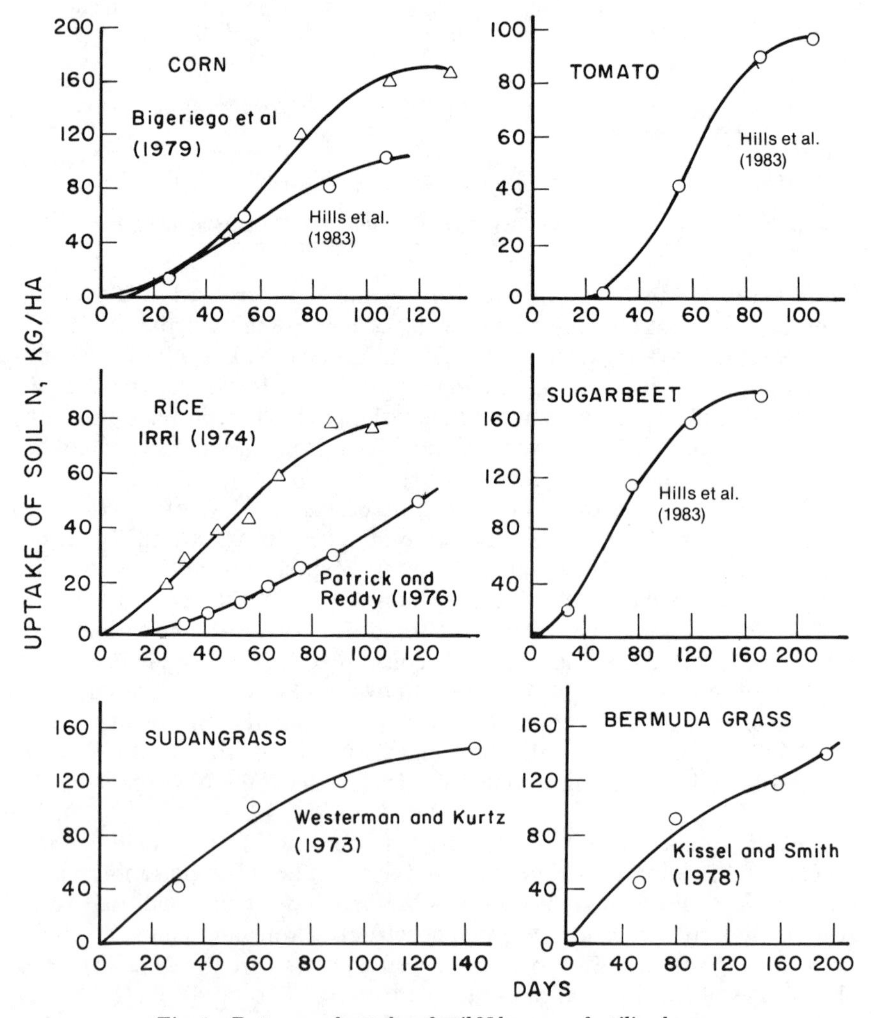

Fig. 1—Patterns of uptake of soil N by some fertilized crops.

III. UPTAKE PATTERNS OF SOIL N

In general, the mineralization of organic N in soils does not proceed at a constant rate, but varies according to soil temperature and other factors. Early in the growth period of an annual crop, the need for N is low, increasing during the period of most rapid growth and tapering off at maturity. The pattern of uptake of soil N in the usual situation where additional N has been supplied as fertilizer may be followed in experiments with labeled fertilizers where intermediate harvests have been made to follow the time course of plant development. In Fig. 1, soil N uptake patterns are presented for several crops. Smooth curves have been fitted to these values calculated from some published and some unpublished data. The curves are usually sigmoidal in shape and are described by third-order polynomial equations. There is a need for additional data of this type for a greater variety of crops and soils, which could be used as a guide for development of more efficient fertilizer practices. However, it should be recognized that uptake patterns of soil N by a given crop may be influenced by rate and timing of fertilizer applications, by fertilizer characteristics, and by soil environmental conditions.

Rates at which soil N is assimilated by crops can be calculated from the slopes of curves such as those shown in Fig. 1. Values calculated at 30-d intervals for the crops of Fig. 1 are given in Table 2. Nearly all values fall in the range 0.5 to 2.0 kg N/ha per day in spite of the different crops, soils, and locations represented. Values for rice (*Oryza sativa* L.) tend to be lower than for other crops, although values for a given crop may vary considerably, as is illustrated by the two sets of values for corn, where uptake of soil N varied by a factor of 2, even though grain yields were comparable.

IV. DEPTH OF UTILIZATION OF SOIL N

The depth of soil from which various crops are able to extract N and the relative proportions of total N derived from different profile depths can be estimated from changes in the pattern of distribution of inorganic N in the profile as a function of depth, as observed by comparing patterns before and after cropping. This procedure provides an approximate quantitative estimate of N uptake from horizons where the mineralization of organic N contributes little to the N supply during the period of plant growth, and where leaching and denitrification losses are negligible. An example of this procedure is illustrated in Fig. 2, which shows utilization of soil N by corn down to a depth of 150 cm. The data of Bauer and Conlon (1974) for wheat (*Triticum aestivum* L.) alternating with fallow showed relatively little change below 60 cm. Another approach to this problem is that employed by Gass et al. (1971), in which the response of the plant to isotopically labeled N placed at various depths was measured. These workers found no uptake of N from below 120 cm by corn early in the growing season, but consider-

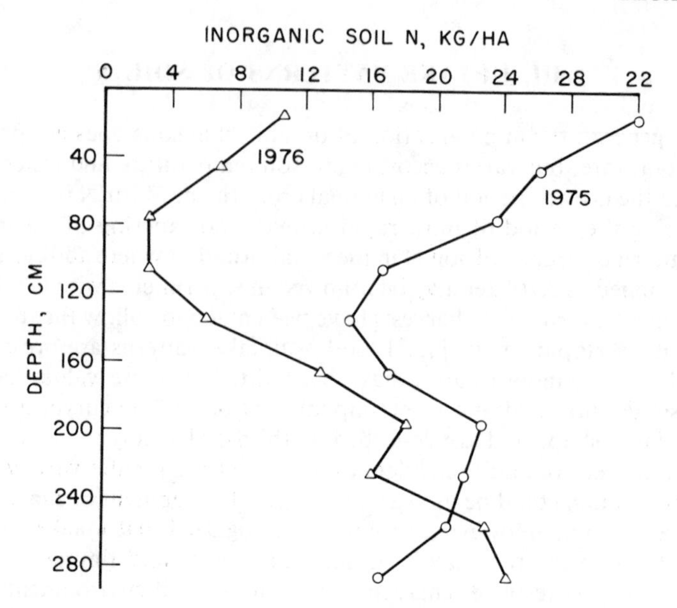

Fig. 2—Changes in distribution of inorganic N in a soil profile as a result of growing a corn crop, showing depth of N utilization.

able utilization of N placed as low as 180 cm occurred during the later stages of growth, as evidenced by the ^{15}N content of the grain. This method gives a more precise estimate of the depth from which inorganic N is obtained by plants and permits an approximate calculation of the contribution from a particular depth at a given level of inorganic N, but does not measure mineralization of organic N at a particular depth.

Craswell and Strong (1976) used the technique of applying labeled N at various depths to show that wheat recovered more applied N from the 45-cm depth than from 15 cm. Kudeyarov et al. (1974) found that N uptake from the 0- to 10- and 20- to 30-cm horizons by buckwheat (*Fagooyrum sagittatum* Gilib.) was greatest at time of budding, whereas uptake from layers between 40 and 90 cm was greatest at the flowering stage. Peterson et al. (1979) reported that sugarbeets (*Beta vulgaris* L.) use N from layers as deep as 210 cm, with slight absorption of N even at 240 cm.

V. IMMOBILIZATION AND REMINERALIZATION OF N

A part of fertilizer N applied to soils is converted to organic N as a component of the microbial or higher plant biomass. This organic N undergoes further transformation during immobilization and mineralization, involving interchange among various pools of soil N. It is well-known that net immobilization of N occurs with the incorporation of mature crop residues of low N content, but some degree of turnover occurs without residue incorporation. In addition to microbiological activities, uptake of fertilizer N by plant roots that leave residues in the soil also contributes to the conversion of inorganic to organic N, which may be regarded as a way that fertilizer N is changed to soil N. The quantities of fertilizer N thus fixed in the soil may be considerable. Smirnov (1977) reported that 30 to 40% of applied fertilizer N was fixed in soil organic matter in pot and field experiments with cereals. Values of 11 to 18% have been reported with rice (Tsukada et al., 1968); 20% with sorghum [*Sorghum bicolor* (L.) Moench.] (Kissel et al., 1976); 21 to 39% with oats (Ikonomova, 1971); and 33% with wheat and corn (Broadbent & Krauter, 1974).

Jansson (1958) introduced the concept of an active pool or labile fraction of soil organic N that equilibrates with a somewhat larger pool of passive organic N and with a smaller one of inorganic or available N. The purpose of soil-testing procedures is to attempt to sample the active pool and to determine its size, which represents potentially available N. Although much of the soil N made available to plants undoubtedly is derived from some such labile fraction, no practical way of determining the relative amounts of plant N obtained from active and passive pools has yet been devised. The transition from available to active to passive forms of N is a progressive and continuous one, probably involving intermediate stages between the conceptual active and passive pools. The supply of soil N that is potentially available to plants is more likely represented by a continuum than by discrete pools.

An indication of changing availability of soil N over time can be obtained by following plant uptake of recently immobilized N, which is assumed to first enter the active pool. Stewart et al. (1963) found that about

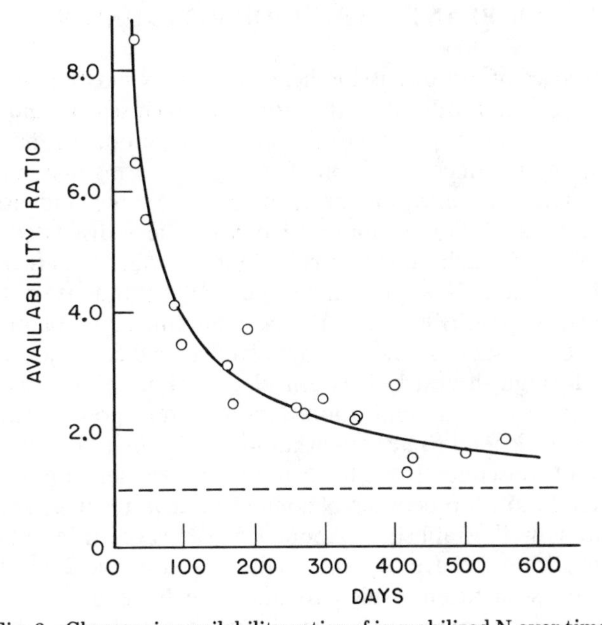

Fig. 3—Changes in availability ratios of immobilized N over time.

75% of the N taken up by plants after a period of initial immobilization was derived from the nondistillable acid-soluble N fraction, which consists primarily of amino-acid N. Recently immobilized ^{15}N-labeled N in this fraction was mineralized more rapidly than the soil N initially present. However, one-half of the immobilized N remained in the soil after four croppings. Broadbent (1967) used a value called the "availability ratio" to compare plant uptake of recently immobilized, labeled N with that of unlabeled N already present in the soil organic fraction. A value of 1.0 indicates equal availability of the two sources, while values greater than unity indicate higher availability of recently immobilized N. Figure 3 is based on values obtained in experiments with three different soils that were cropped continuously for periods ranging from 498 to 599 d following the initial immobilization of labeled N. The availability ratio decreased progressively from a value of 7.14 at 30 d to 1.49 at 600 d. Extrapolation of the curve shows the immobilized and soil N becoming equal in availability in 3.5 yr. Similar results were reported by Smirnov and Sukov (1970), who found that plants utilized only 5 to 9% of the n immobilized in the previous year.

The implication of these findings is that the so-called active pool becomes progressively less active over time, eventually becoming indistinguishable from the passive pool. One might speculate that the familiar amino acid and other biologically active forms of N undergo reaction with polyphenols or quinones to form stable polymers, which are eventually built into the complex, high molecular weight structure of soil organic matter with an attendant reduction in biological and chemical reactivity.

VI. PLANT UPTAKE OF RESIDUAL N

For purposes of our discussion here, residual N may be defined as any fertilizer N that is not utilized by the crop in a given season and which carries over to the period of growth of the succeeding crop. It is partly organic and partly inorganic in composition. The organic forms may have resulted from microbial immobilization or from plant uptake and assimilation, usually both. Availability of immobilized N has been discussed in the preceding section, and much of what was said there applies to N tied up in crop residues. Plant residue N is somewhat more available than is the bulk of soil organic N, but its contribution to the N requirements of plants decreases over time, as does that of immobilized N. Under field conditions, it is rarely possible to distinguish clearly between plant residue N and that resulting from microbiological immobilization. An important point of difference is that microbial cells tend to be somewhat higher in their N content than are plant cells, and consequently are likely to mineralize their contained N more readily. The relationship between N content of crop residues and N mineralization rate is well-established. About 1.5% N is the threshold value for net mineralization of N from plant residues, and above 2.5% the rate of mineralization is sufficiently high to meet the N requirements of many crops, depending, of course, on the total quantity of plant residues present.

Table 3—Residual inorganic N in relation to total N upake and grain yield in two unfertilized soils cropped to corn.

Year	Residual N 0–180 cm	Plant uptake	Yield	Residual N 0–180 cm	Plant uptake	Yield
	kg/ha					
	Yolo fine sandy loam†			Hanford sandy loam†		
1973	61.0	155	7920	20.3	33.1	870
1974	81.4	82.8	5900	12.0	37.3	1220
1975	81.0	89.5	5980	25.1	38.7	1520
1976	68.9	62.9	4170	11.2	32.9	1060
1977	54.3	66.7	4140	51.6	30.6	960

† Typic Xerorthents.

In green manures, one-half or more of the total N may be mineralized in a growing season. Tokunaga (1975) obtained 15 to 20% recovery of the N from labeled corn stover by a crop of wheat. Giddens et al. (1971) reported 72% of the total N uptake by sudax was from soil N in plots that had been in fescuegrass (*Festuca arundinacea*) for the previous 5 yr, compared with 44% from plots on which fescuegrass had not been grown, an increase of 28% due to residual effects of the sod. Moore (1974) investigated the availability to Rhodesgrass (*Chloris gayana*) of N in tops and roots of this plant added to soil and found that after 16 months about one-third was utilized by plants and two-thirds remained in the soil. Watanabe et al. (1977) found that about 75% of the total N in dry *Azolla–Anabaena* residues was mineralized in 6 to 8 weeks in a submerged soil. This is an unusually high value for residual organic N.

A part of every fertilizer application of N remains unused in the inorganic form after the crop is harvested. This is a reflection of the fact that fertilizer uptake efficiencies are usually in the vicinity of 50% or less, so that it is necessary to add twice as much N as is desired for the crop. Stated another way, plants lack the ability to reduce the concentration of inorganic N in the soil profile to zero. Nitrate-N concentrations in the soil solution in the root zone of N-deficient corn have been reported at 14 mg/kg (Broadbent & Rauschkolb, 1977). Table 3 gives values for residual inorganic N in the 0- to 180-cm profile of two unfertilized soils cropped to corn over a

Table 4—Some reported values for plant uptake of residual N.

Crop	Residual N uptake	Residual N utilized	Reference
	kg/ha	%	
Rice	2.7–3.1	5.0–6.8	Reddy & Patrick, 1978
Bermudagrass	37	17	Kissel & Smith, 1978
Sudangrass	2.5–10	13–18	Westerman & Kurtz, 1972
Oats	--	2.0–3.9	Jansson, 1963
Oats	--	3.1–14	Koritskaya, 1969
Barley	--	1–4	Koyama et al., 1977
Corn	3.1–49	2.0–6.8	Broadbent, 1980
Sorghum	10.3	3.2	Broadbent, 1980

period of 5 yr. Yields and total N uptake were low as a result of severe N deficiency, but in spite of that fact, considerable amounts of inorganic N were found in the soil immediately after harvest, illustrating the inability of the crop to utilize N below a threshold value that varies with the soil and nature of the crop.

In a practical situation, residual N that is available to plants is a composite of unused inorganic N from the previous season and N derived from the mineralization of plant residues and microbial cells or their degradation products. Most published data show that the value of this residual N to a following crop is quite low, either in terms of kg N/ha supplied or expressed as a fraction of the residual N utilized. Table 4 presents values for several different crops. In those cases where values greater than a few kilograms per hectare have been reported, there has usually been application of fertilizer N considerably in excess of plant needs, leaving residual N predominantly in the inorganic form.

REFERENCES

Allison, F. E. 1966. The fate of nitrogen applied to soils. Adv. Agron. 18:219–258.

Bauer, A., and T. J. Conlon. 1974. Effect of tillage interval of fallow on available soil nitrogen, some soil physical properties and wheat yields. North Dakota Res. Rep. no. 51. North Dakota State Univ., Fargo.

Bigeriego, R., R. D. Hauck, and R. A. Olson. 1979. Uptake, translocation and utilization of ^{15}nitrogen-depleted fertilizer in irrigated corn. Soil Sci. Soc. Am. J. 43:528–533.

Broadbent, F. E. 1967. Reversion of fertilizer nitrogen insoils. Soil Sci. Soc. Am. Proc. 31: 648–652.

Broadbent, F. E. 1971. Nitrogen in soil and water. Proc. of the Symp. on Nitrogen in Soil and Water. Dep. of Soil Sci., Univ. of Guelph, Ontario.

Broadbent, F. E. 1980. Residual effects of labeled N in field trials. Agron. J. 72:325–329.

Broadbent, F. E., and A. B. Carlton. 1978. Field trials with isotopically labeled nitrogen fertilizer. p. 1–44. *In* D. R. Nielsen and J. G. MacDonald (ed.) Nitrogen in the environment. Vol. I, Nitrogen behavior in field soil. Academic Press, New York.

Broadbent, F. E., and A. B. Carlton. 1979. Field trials with isotopes—plant and soil data for Davis and Kearney sites. p. 433–465. *In* Final report to the National Science Foundation for grant nos. GI34733X, GI43664, AEN74-11136 A01, ENV76-10283 and PFR7610283, Nitrate in effluents from irrigated lands, Univ. of California.

Broadbent, F. E., and C. Krauter. 1974. Nitrate movement and plant uptake of N in a field soil receiving ^{15}N-enriched fertilizer. Proc. 2nd Annu. Natl. Sci. Foundation–Res. Applied to Natl. Needs Trace Contaminants Conf., Asilomar, Pacific Grove, Calif. no. LBL-3217. p. 236–239. The Lawrence Berkeley Lab., Univ. of California, Berkeley.

Broadbent, F. E., and T. Nakashima. 1968. Reversion of fertilizer nitrogen in soils. Soil Sci. Soc. Am. Proc. 31:648–652.

Broadbent, F. E., and R. S. Rauschkolb. 1977. Nitrogen fertilization and water pollution. Calif. Agric. 31(5):24–25.

Broadbent, F. E., and O. C. Reyes. 1971. Uptake of soil and fertilizer nitrogen by rice in some Philippine soils. Soil Sci. 112:200–205.

Broadbent, F. E., K. B. Tyler, and D. M. May. 1980. Tomatoes make efficient use of applied nitrogen. Calif. Agric. 34(11 & 12):24–25.

Campbell, C. A., and E. A. Paul. 1978. Effects of fertilizer N and soil moisture on mineralization, N recovery and A-values under spring wheat grown in small lysimeters. Can. J. Soil Sci. 58:39–51.

Craswell, E. T., and W. M. Strong. 1976. Isotopic studies of the nitrogen balance in a cracking clay. III. Nitrogen recovery in plant and soil in relation to the depth of fertilizer addition and rainfall. Aust. J. Soil Res. 14:75–83.

Gass, W. B., G. A. Peterson, R. D. Hauck, and R. A. Olson. 1971. Recovery of residual nitrogen by corn (*Zea mays* L.) from various soil depths as measured by ^{15}N tracer techniques. Soil Sci. Soc. Am. Proc. 35:290–294.

Giddens, J., R. D. Hauck, W. E. Adams, and R. N. Dawson. 1971. Forms of nitrogen and nitrogen availability in fescuegrass sod. Agron. J. 63:458–460.

Hauck, R. D., and J. M. Bremner. 1976. Use of tracers for soil and fertilizer nitrogen research. Adv. Agron. 28:219-266.

Hills, F. J., F. E. Broadbent, and M. Fried. 1978. Timing and rate of fertilizer nitrogen for sugarbeets related to nitrogen uptake and pollution potential. J. Environ. Qual. 7:368–372.

Hills, F. J., F. E. Broadbent, and O. A. Lorenz. 1983. Fertilizer nitrogen utilization by corn, tomato and sugarbeet. Agron. J. 75:423–426.

Ikonomova, E. 1971. A study of fertilizer nitrogen uptake by plants and of the nitrogen balance with the aid of ^{15}nitrogen. Soil Sci. Agrochem. 6:47–54.

International Atomic Energy Agency. 1970. Fertilizer management practices for maize: results of experiments with isotopes. Joint FAO/IAEA Div. of Atomic Energy in Food & Agric., tech. Rep. Series no. 121. IAEA, Vienna.

International Atomic Energy Agency. 1978. Isotope studies on rice fertilization. Joint FAO/IAEA Div. of Atomic Energy in Food & Agric., Tech. Rep. Series no. 181. IAEA, Vienna.

International Rice Research Institute. 1974. Annual Report 1973. IRRI, Los Banos, Philippines.

International Rice Research Institute. 1975. Annual Report 1974. IRRI, Los Banos, Philippines.

Jansson, S. L. 1958. Tracer studies on nitrogen transformations in soil with special attention to mineralization-immobilization relationships. Ann. R. Agric. Coll. Swed. 24:101-361.

Jansson, S. L. 1963. Balance sheet and residual effects of fertilizer N in a 6-year study with N^{15}. Soil Sci. 95:31-37.

Kissel, D. E., and S. J. Smith. 1978. Fate of fertilizer nitrate applied to coastal bermudagrass on a swelling clay soil. Soil Sci. Soc. Am. J. 42:77–80.

Kissel, D. E., S. J. Smith, and D. W. Dillow. 1976. Disposition of fertilizer nitrate applied to a swelling clay soil in the field. J. Environ. Qual. 5:66–71.

Koritskaya, I. A. 1969. Use of calcium nitrate-^{15}N and ammonium-^{15}N sulfate by spring wheat fertilized at various growth stages. Agrokhimiya 1969(9):3–7.

Koyama, T., C. Chammek, and N. Niamsrichand. 1973. Nitrogen application technology for tropical rice as determined by field experiments using ^{15}nitrogen tracer technique. Trop. Agric. Res. Center Tech. Bull. TARC 3, Tokyo, Japan.

Koyama, T., M. Shibuya, M. Tokuyasu, T. Shimomura, and K. Ide. 1977. Balance sheet and residual effects of fertilizer nitrogen in Saga paddy field. p. 289–296. *In* Proc. Int. Seminar Soil Environ. and Fertility Management in Intensive Agriculture, Tokyo. Soc. of the Sci. of Soil and Manure, Tokyo.

Kudeyarov, V. N., O. A. Sokolov, and A. N. Bochkarev. 1974. Availability of nitrate nitrogen from different horizons of grey forest soil to plants. Int. Congr. Soil Sci., Trans. 10th (Moscow) 1974 9:117-123.

Legg, J. O., and G. Stanford. 1967. Utilization of soil and fertilizer N by oats in relation to the available N status of soils. Soil Sci. Soc. Am. Proc. 31:215–219.

Moore, A. W. 1974. Availability to rhodesgrass (*Chloris gayana*) of nitrogen in tops and roots added to soil. Soil Biol. Biochem. 6:249–255.

Olson, R. V., L. S. Murphy, H.C. Moser, and C. W. Swallow. 1979. Fate of tagged fertilizer nitrogen applied to winter wheat. Soil Sci. Soc. Am. J. 43:973–975.

Patrick, W. H., Jr., and K. R. Reddy. 1976. Fate of fertilizer nitrogen in a flooded rice soil. Soil Sci. Soc. Am. J. 40:678–681.

Peterson, G. A., F. N. Anderson, G. E. Varvel, and R. A. Olson. 1979. Uptake of ^{15}N-labeled nitrate by sugarbeets from depths greater than 180 cm. Agron. J. 71:371–372.

Rauhe, K., and R. Seiberlich. 1976. Role of inorganic nitrogen fertilizers in the regeneration of organic substances in relation to the carbon and nitrogen supply in the soil (nitrogen-15-labeled urea studies). Pochvovedenie 9:57–64.

Reddy, K. R., and W. H. Patrick, Jr. 1978. Residual fertilizer nitrogen in a flooded rice soil. soil Sci. Soc. Am. J. 42:316–318.

Sapozhnikov, N. A., N. A. Ivanova, T. K. Livanova, I. P. Rusinova, V. V. Siderova, L. B. Sirota, and T. Tarvis. 1974. Transformation of nitrogen in soil and nitrogen nutrition of plants studied according to the data from research using nitrogen-15. Int. Congr. Soil Sci., Trans. 10th (Moscow) 1974 9:30–38.

Sapozhnikov, N. A., E. I. Nesterova, I. P. Rusinova, L. B. Sirota, and T. K. Livanova. 1968. The effect of fertilizer nitrogen on plant uptake of nitrogen from different podzolic soils. Int. Congr. Soil Sci., Trans. 9th (Adelaide) 1968 2:467–474.

Scarsbrook, C. E. 1965. Nitrogen availability. *In* W. V. Bartholomew and F. E. Clark (ed.) Soil nitrogen. Agronomy 10:481–502. Am. Soc. of Agron., Madison, Wis.

Smirnov, P. M. 1977. Soil nitrogen problems and results of nitrogen-15 studies. Agrokhimiya 1977(1):3–25.

Smirnov, P. M., and A. A. Sukov. 1970. Availability to plants and transformation in soil of residual immobilized fertilizer nitrogen. Agrokhimiya 1970(12):3–15.

Smirnov, P. M., M. V. Tekhina, and S. D. Bazilevich. 1976. The effect of irrigation and different rates of ammonium sulphate on the yield and quality of winter wheat, the transformation of fertilizer nitrogen in the soil, and its utilization by the plants. Izv. Timiryazevsk. Skh. Akad. 4:90–99.

Stewart, B. A., D. D. Johnson, and L. K. Porter. 1963. The availability of fertilizer nitrogen immobilized during decomposition of straw. Soil Sci. Soc. Am. Proc. 27:656–659.

Tabatabai, M. A., and A. A. Al-Khafaji. 1980. Comparison of nitrogen and sulfur mineralization in soils. Soil Sci. Soc. Am. J. 44:1000–1006.

Terman, G. L., and M. A. Brown. 1968. Crop recovery of applied fertilizer nitrogen. Plant Soil 29:48–65.

Tokunaga, Y. 1975. The effects of cereal crop residues on upland soils. ASPAC Food & Fert. Technol. Center, Ext. Bull. 48. Asian & Pacific Council Food & Fertilizer Technology Center, Taiwan.

Tsukada, T., K. Ogawa, J. Kinbara, and M. Deguchi. 1968. Utilization by upland rice plants of ^{15}N-labeled nitrogen fertilizer and the effect of rice straw compost on the fate of ^{15}N in soil. J. Sci. Soil Manure Jpn. 39:520–525.

Watanabe, I., C. R. Espinas, N. S. Berja, and B. V. Alimagno. 1977. Utilization of the *Azolla–Anabaena* complex as a nitrogen fertilizer for rice. Int. Rice Res. Inst. Res. Paper Series no. 11. IRRI, Los Banos, Philippines.

Westerman, R. L., and L. T. Kurtz. 1972. Residual effects of ^{15}N-labeled fertilizers in a field study. Soil Sci. Soc. Am. Proc. 36:91–94.

Westerman, R. L., and L. T. Kurtz. 1973. Priming effect of nitrogen-15-labeled fertilizers on soil nitrogen in field experiments. Soil Sci. Soc. Am. Proc. 37:725–727.

Westerman, R. L., L. T. Kurtz, and R. D. Hauck. 1972. Recovery of ^{15}N-labeled fertilizers in field experiments. Soil Sci. Soc. Am. Proc. 36:82–86.

Zamyatina, V. B., N. I. Borisova, N. M. Varyushina, and S. V. Burtseva. 1971. Transformations of nitrogen compounds of the soil and fertilizers based on data from studies using nitrogen-15. Primen. Izot. Yad. Izluch. Selsk. Khoz. Dokl. Vses Nauchno. Tekh. Konf. 1968:142–151.

12

Darrell A. Russel
National Fertilizer Development Center
Tennessee Valley Authority
Muscle Shoals, Alabama

Conventional Nitrogen Fertilizers

In 1974, some exporting countries embargoed petroleum shipments to the United States. Fertilizer prices, especially those for N, increased sharply and the American farmer paused in his drive to use more and more N as a production input. For the first time since World War II, there was a significant decrease in the quantity of N used. Although another decrease occurred in 1978 because of weather and government agricultural policy, N use has increased steadily year after year (Fig. 1). Rightly or wrongly, farmers seem convinced that conventional N fertilizers are essential.

In contrast, agronomists, chemists, economists, and chemical engineers are in a quandary about N fertilizers. They are concerned about the efficiency of different N sources; energy needed to manufacture, transport, and apply them; granule size and hardness; time and rate of application; effects on the environment; and other physical and chemical characteristics. Several research and development programs are under way to moderate some of these concerns (NFDC, 1980; Sheldon et al., 1981).

I. TYPES OF MATERIALS AND PRODUCTION TECHNOLOGY

Until 1900, the N sources used in fertilizing materials were largely animal and vegetable wastes from other industries, plus Chilean nitrate (sodium nitrate). These were augmented after 1900 with by-product ammonium sulfate (AS) from the coking industry. In the late 1950's, synthetic NH_3 became the dominant source and it likely will retain this position well into the next century. Several materials, such as ammonium nitrate (AN), N solutions, and urea, vie for second place.

Commonly, N fertilizers are classed as either NH_4^+- or NO_3^--types. Commercially available NH_4^+-types include anhydrous ammonia (AA),

Nitrogen in Crop Production.

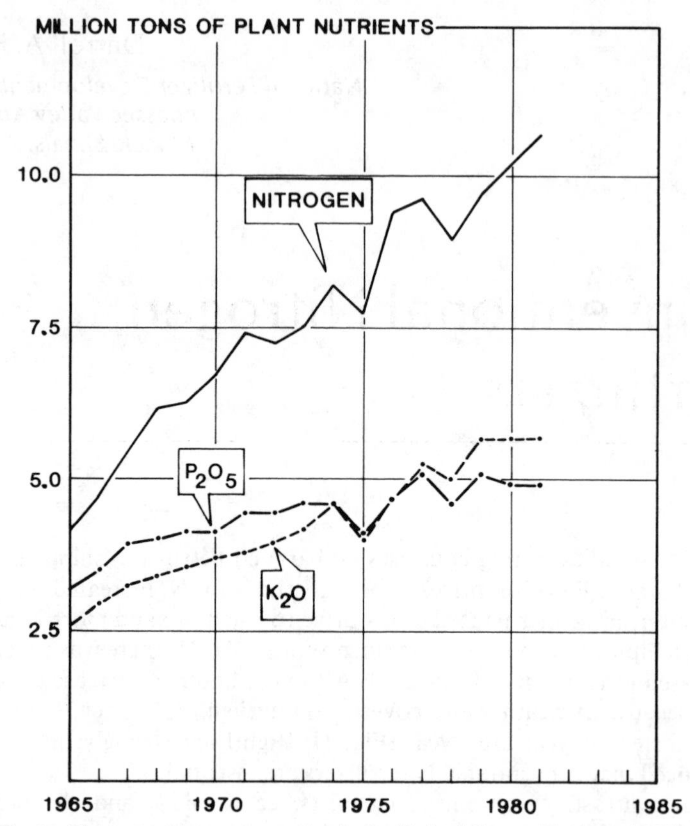

Fig. 1—Plant nutrient consumption in the United States (Bridges, 1980).

aqua ammonia (AQA), AN, AS, N solutions, urea, NH_4^+-phosphates, and mixed fertilizers. The NO_3^--type fertilizers include AN, N solutions, and some mixed fertilizers. Outside the United States, calcium nitrate (CN) and sodium nitrate are important sources of NO_3^-; small quantities of these are imported into the United States. A few slow- or controlled-release N fertilizers also are available, but they are considered in other chapters of this book.

A. Anhydrous and Aqua Ammonia

In fiscal year 1979, 38.5% of all N fertilizer was applied directly to the soil as either AA or AQA. Although some AA is used for nonfertilizer industrial processes, most is used in the fertilizer industry, either for direct application or to make other N fertilizers. Not all this AA is produced in the United States. Since 1974, the United States has been a net importer of N fertilizers. Imported materials are AA, AN, and urea. In 1978, the net imports included 721 000 t of N as AA and 109 000 t as AN; this was partially offset by exports (net) of 153 000 t of N as AS and 35 000 t as urea (Bridges, 1980).

Fixation of atmospheric N by synthetic means is an energy-intensive process. There are continued investigations of direct fixation of atmospheric N by processes that resemble biological fixation and by catalysis. Such systems are presumed to require less energy than the engineering process systems now in use or to obtain energy from sources that ordinarily are not available to engineers. All of these biotechnical methods are still in the research and development stage.

Synthetic AA is manufactured by high-pressure, high-temperature processes originally developed by Haber and Bosch in Germany. Many of these processes have been improved during the last 50 yr. Economies to size, at least in the developed countries, have led to larger and larger plants. Development of the steam-driven centrifugal compressors in the 1960's caused production from a single train to jump from 340 t N/d to $\geq$ 500 t N/d.

In the United States, the H_2 that is combined with N_2 to produce AA is obtained from natural gas. Availability and cost of this gas are becoming the major constraints to production. To remove these constraints, alternate sources of H_2 are being examined. One possibility is electrolysis. Future improvements in electrolytic-cell design coupled with more efficient means of power generation—such as fuel cells, magnetohydrodynamics, or the breeder reactor—could make this the preferred route.

Another route is coal gasification, in which thermal energy from coal is used to break the O–H bond in steam. The basis for this approach is that coal is more abundant than natural gas. Thus, coal gasification is projected to provide H_2 at a lower cost than natural gas as gas supplies shrink. TVA is retrofitting a coal gasification plant to an existing AA plant (NFDC, 1979). Energy constraints also can be reduced by increasing efficiency of the production process (such as improved catalysts) and by recovering heat and energy normally lost to the process.

In the production processes, there have been some shifts in the relative importance of various components of cost. For many years, the cost of natural gas—for both raw material and energy requirements—was relatively unimportant. Capital and labor costs were $\geq$ 80% of the production costs (UNIDO, 1967). Capital costs have increased alarmingly in the last 10 yr, but natural gas costs have escalated even faster—to the point where they now make up 40 to 50% of the cost of production. They are projected to become even more significant (IFDC, 1979; TVA/IFDC, unpublished data).

Production statistics for AQA are not precise, because some AA is converted to AQA without being reported. Aqua ammonia usually contains about 20% N. It is easier and safer to use than AA and usually is not injected as deeply into the soil. The cost for producing and distributing AQA is higher than for AA, but the cost of application is lower.

B. Ammonium Nitrate

Ammonium nitrate has been the solid N material of choice since World War II to both the American and the European farmer. When first adopted, it had the advantage of containing about twice as much N as AS and

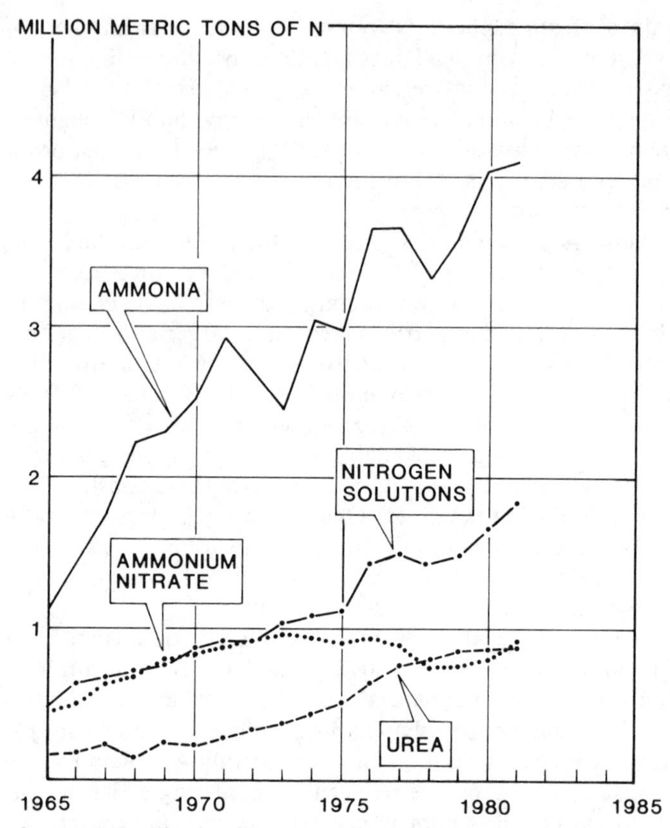

Fig. 2—Consumption of N fertilizer materials in the United States (Bridges, 1980).

sodium nitrate. It contained both NH_4^+- and NO_3^--N, so it was versatile. Losses of N from surface applications were minimal. In the United States, AN was compatible with most other materials used in bulk blending. Improvements in production processes reduced its hygroscopicity and minimized disintegration caused by temperature fluctuations.

Use of AN for direct application to the soil peaked in 1973 and has declined slightly since then (Fig. 2). Trends in use of AN in bulk blends cannot be quantified. Bulk application of fertilizer mixtures still appears to be increasing, and so is the amount of N applied in bulk (Fig. 3); precise information about the source of the N is not available. Urea is replacing AN in many of its traditional uses. It seems likely that use of AN in bulk blends will soon start to decrease if it has not already started. The one area where AN use still is increasing is in the production of urea–ammonium nitrate (UAN) solution. Use of UAN solution is increasing (Fig. 2); because of this, production of AN in the United States is fairly stable. Since 1977, six new plants have been built to produce AN. About one-third of the production of these plants is used for direct application; the other two-thirds is used to produce UAN solution (NFDC, 1980).

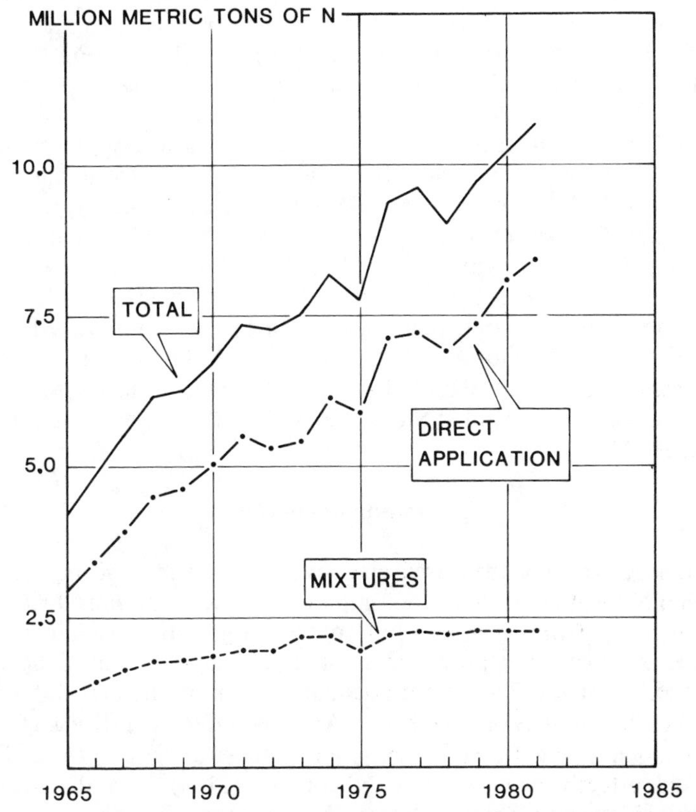

Fig. 3—Nitrogen fertilizer consumption in the United States (Bridges, 1980).

C. Urea

For many years, bulk blenders resisted using urea. Prilled urea did not blend satisfactorily with granular diammonium phosphate (DAP) and granular KCl because the prills were smaller than the granules and the blends segregated. Urea formed undesirable chemical compounds with superphosphates and was not compatible with AN. A mixture of urea and AN had a lower critical humidity than either material alone and was excessively hygroscopic. Even mixtures of dust from the two materials created problems in a plant. Also, prills broke when they impinged on spinner blades during bulk application. Farmers, too, resisted urea because surface applications sometimes resulted in substantial losses of N by volatilization. Incorporation within hours after application was recommended but did not always fit a farmer's schedule.

Nevertheless, the high analysis and the associated savings in long-distance transportation promoted urea as a world commodity. From 1968 to 1973, many new manufacturing facilities were built and urea exports grew rapidly. Sales within the United States also were promoted so that plants could operate at near capacity. By 1978, urea was competing with AN as the main solid source of N applied directly to the soil (Fig. 2). Development of granular urea also made this material suitable for use by bulk blenders.

Concurrent production of urea and AA is customary in both developed and developing countries. Manufacture of AA from natural gas yields CO_2 as a by-product—but CO_2 and AA are primary ingredients for urea manufacture. Farming systems in most developing countries cannot be adapted to direct application of AA or AQA. Farms are too small, and farmers do not have the specialized application equipment. Urea, however, can be applied without sophisticated equipment. Consequently, developing countries with natural gas supplies choose the AA–urea combination. Surplus production of urea is sold on the world market to generate foreign exchange. The United States generally has been a net importer of urea since 1972.

D. Ammonium Sulfate

This fertilizer has been used extensively worldwide and has been the traditional N fertilizer for most developing countries. Early production was as a by-product from town gas and coke-oven gas. In the United States, about 400 000 t of AS is produced annually as a by-product of the coking and textile industries. This is not adequate to supply the demand and another 2.0 million tonnes of "synthetic" AS is manufactured. It has many advantages, such as low hygroscopicity, good physical properties, and good agronomic effectiveness. Its low N content (21%) is a disadvantage, particularly if the material must be transported a long distance.

The S content (24%), however, makes the material a two-nutrient fertilizer, particularly adapted to S-deficient areas. Many areas in the United States, western Canada, and tropical countries show S deficiencies when crop production is accelerated. Thus, there is a continued worldwide interest in AS as a source of both N and S. While American farmers generally do not buy S fertilizers, they are willing to pay more for the N in AS than that in AN.

Production processes are based on adsorption of AA from by-product gases into H_2SO_4, absorption of synthetic AA into by-product H_2SO_4, treating gypsum or anhydrite with ammonium carbonate, and reacting synthetic AA with pure H_2SO_4. Studies of the conversion of stack-gas SO_2 into AS show that enormous tonnages could be produced. Economic feasibility studies usually pass this point successfully but are stopped by distribution costs. The initial product is a solution of AS containing many impurities and about 8% N. The competitive distribution area for this material is limited to a radius of about 10 km around the production plant. Concentrating the material from 8% N to 18 or 19% N requires energy and does not solve the distribution problem. Long-distance transportation is needed to move all the material away from the point of production. The fertilizer, now low in value to the producer because of supply/demand imbalance, is also low in N content, compared with other N fertilizers. Distribution costs, per unit of N, make AS from stack-gas scrubbers too expensive for farmers to use. Regardless of final decisions on how clean air must be, the supply of AS fertilizer is not likely to be affected greatly.

E. Fluid N Fertilizers

For many years, UAN solution has been the main material in this classification. Occasionally, urea solutions (19–0–0) and other minor types were made. Aqua ammonia is included in this class by some statisticians but generally is treated separately or included with AA.

Use of fluid N fertilizers is increasing steadily in the United States (Fig. 2). The UAN solutions (28–32% N) contain nearly as much N as AN. Being fluid, however, they are easier to handle mechanically than are the solid N fertilizers. Thus, in situations where labor is in short supply or timeliness is critical, fluid N fertilizers are advantageous. New N facilities have been justified primarily by the expanding need for UAN solutions.

F. Mixed Fertilizers and Ammonium Phosphates

Mixed fertilizers supply about 25% of the N used in the Unied States. Use of mixtures, however, is not increasing nearly as fast as direct application of N fertilizers. Mixed fertilizers can be chemically mixed or physically mixed (bulk blended) and either dry or fluid. Of the many possible N sources, the most common are NH_4^+-phosphates, urea, AN, AA, AQA, and UAN solutions.

About 50% of the fertilizer used in the United States is applied as a mixed fertilizer; the other 50% is applied directly to the soil as straight materials (Fig. 4). Nearly 75% of the mixtures is applied as bulk blends and fluid fertilizers; use of bagged material is decreasing steadily (Fig. 5). Similarly, the NH_4^+-phosphates are used for direct application, in bulk blends, and in fluid fertilizers.

The NH_4^+-phosphates are included in this class because they contain N. They usually are considered to be P fertilizers because, in the oxide system, they contain more P than N. For example, DAP is 18–46–0 on an oxide basis. On an elemental basis it is 18–20–0 and the N suddenly seems just as important as the P.

The NH_4^+-phosphates include both the ortho- and the polyphosphate systems. Common orthophosphates are DAP and monoammonium phosphate (MAP), with DAP the predominant fertilizer on the world market. Countries needing P order DAP instead of triple superphosphate. The P contents of both materials are the same, and so are the freight costs. With DAP, however, 18 units of N are transported at no cost. Because U.S. fertilizer production is largely adapted to the world market, the supply of fertilizer within the United States also is largely controlled by the world demand.

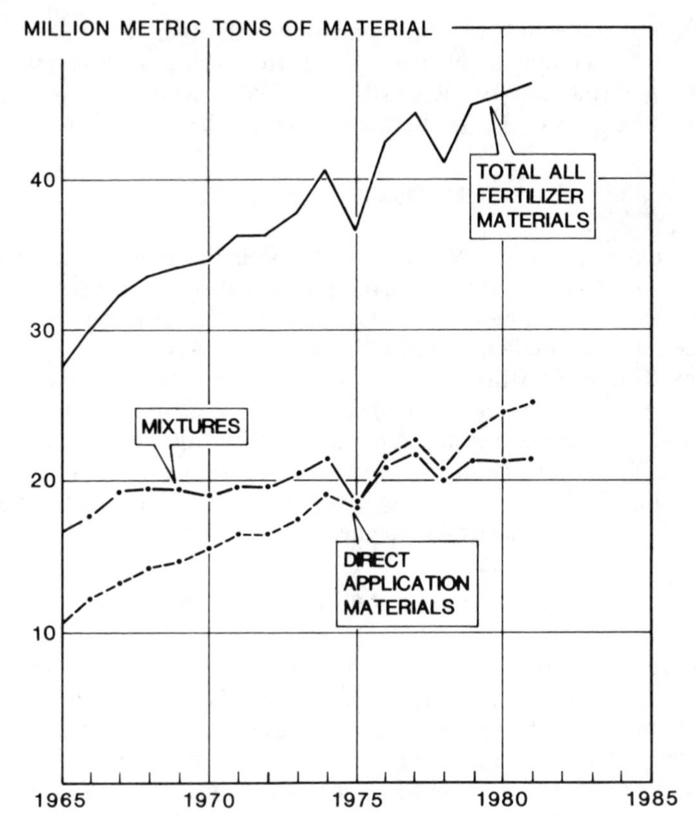

Fig. 4—Consumption of fertilizer materials in the United States (Bridges, 1980).

Monoammonium phosphate, however, is peculiarly adapted to the United States. One reason is the high cost of transporting phosphoric acid from the southeastern United States to the Midwest, where P use is highest. By adding AA, the phosphoric acid is transformed into a dry material that is easier to handle, cheaper to transport, and about as versatile as the original phosphoric acid. The Midwest also is a major user of N. Much N is moved by pipeline into the area as AA from the Gulf Coast. Transportation costs are minimal. Any diversion of AA to the Florida–North Carolina area to produce NH_4^+-phosphate adds measurably to the cost of the final product. Consequently, there are economic advantages for producing MAP instead of DAP for the U.S. market. The relative importance of the U.S. market vs. the world market to phosphate manufacturers will determine how fast MAP replaces DAP in the United States.

Production of NH_4^+-phosphates through new plants and expansions of present capacity has outstripped demand. Manufacturers now are keenly interested in counterbalancing a sagging demand for their product and in reducing their production costs so their product can be priced low enough for farmers to continue to purchase it.

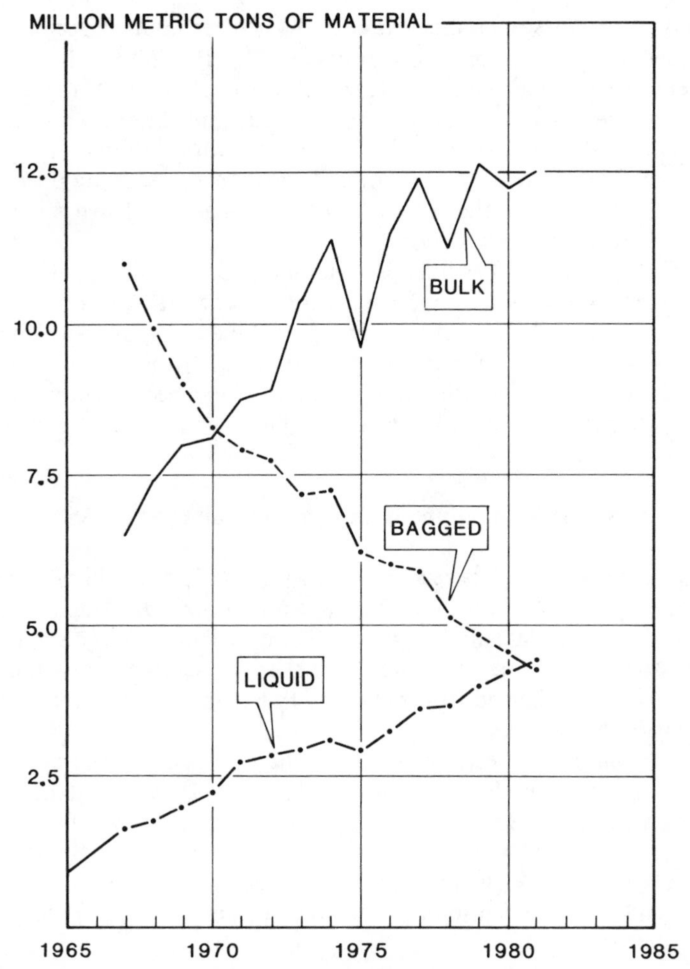

Fig. 5—Consumption of mixed fertilizers by class in the United States (Bridges, 1980).

II. NEW GENERATION USES FOR CONVENTIONAL N FERTILIZERS

Trends in N fertilizer production are fogged by the dominance of urea in the world market, the continued supply of by-product AS and potential supply from emission controls on fossil-fuel power plants, traditional preferences for AN, and the economics of AA. The fertilizer industry has not shown any clearcut preference for one type of material over another. Industry's position is understandable. The present high cost of money does not lead to many new manufacturing plants and certainly not to innovations. The main concern is to stay in business.

Farmers, too, are looking for ways to stay in business. Their money costs also are high and commodity prices are low. Initial reactions are to cut crop production costs, including the cost of fertilizer, in the hope that yields will not decline before commodity prices rise and the soil-fertility program can be reinstated. An alternative is to increase the efficiency of the fertilizer, perhaps by switching from broadcast to band application. This seems to apply more to P and K than to N. The nitrification inhibitors, however, have attracted much interest. Still, most farmers are specifying urea, NO_3^-, N solution, or AA when they order N fertilizer—or they leave the decision mostly to their fertilizer supplier.

At TVA's National Fertilizer Development Center, a different attitude prevails. Here, we also are concerned with the escalating costs of energy, labor, and capital. We seek ways to minimize these costs, but we also recognize the inevitability of urea, AA, AN, and AS. Most of our work on formamide, oxamide, and other new forms of N has been greatly curtailed. We have concentrated on new ways to combine urea with other conventional N fertilizers.

A. Urea-Ammonium Sulfate Suspension (29-0-0-5S)

In a pilot plant, TVA is reacting AA with H_2SO_4 and adding urea melt to produce 29-0-0-5S. This material has a dual potential. The 6:1 ratio of N/S makes the material suitable for many S-deficient soils. The fluid fertilizer industry already has established a market for S. Urea-ammonium sulfate suspension is expected to supply S to this market at a lower cost than the present fluid S sources.

The N content is almost 40% higher than that of AS and application costs, per unit of N, are decreased accordingly. The 29% N content is about the same as for UAN solution (28-32% N). In those areas of the United States where 28% UAN solution prevails, the new urea-AS suspension should have a slight advantage, even without regard to the S content.

A granular urea-AS also has been studied. One possible grade for this material is 40-0-0-4S.

B. Urea-Ammonium Nitrate Suspension (31-0-0)

This material is intended primarily for use by the fluid fertilizer dealer to produce other high analysis suspension fertilizers and for direct application. The N content is of the same order as for UAN clear solutions, but its 1.5% clay content makes the material more suitable for the N source in NPK suspensions. It also is used for direct application with herbicides. About 40 000 t of UAN suspension has been produced for TVA's industry demonstration program in the last 3 or 4 yr.

TVA also is researching a 36-0-0 UAN suspension. The N content is 12.5 to 25% higher than for commercial UAN solutions and is about the maximum that can be made.

C. Urea-Ammonium Phosphate (28-12-0 and 35-7-0)

The NH_4^+-phosphates are another family of conventional N materials that can be combined satisfactorily with urea. Both NH_4^+-orthophosphate and NH_4^+-polyphosphate can be enhanced in N content by adding urea melt to the process just ahead of the granulation step. In the conventional oxide system, N/P_2O_5 ratios of 1:1 and 2:1 are possible. These grades are 28-28-0 and 35-17-0 or 28-12-0 and 35-7-0 on an elemental basis. Adding urea to the NH_4^+-phosphate granule decreases the resistance of the granule to humidity effects somewhat, compared with AN, but the product still is satisfactory.

d. Urea Phosphate and Urea-Urea Phosphate

Urea phosphate is a crystalline adduct $[CO(NH_2)_2 \cdot H_3PO_4]$ that forms readily by the reaction of urea with orthophosphoric acid. TVA has been interested in this product as another way to convert phosphoric acid into a solid form for shipping, and then reconverting it into fluid fertilizers. Greenhouse tests of this material showed some surprising characteristics. In the soil, the phosphoric acid creates an extremely acidic environment for the urea, thereby greatly reducing NH_3 volatilization. When made from relatively pure phosphoric acid, the grade is 17-19-0 (17-44-0, oxide). At a N/P_2O_5 ratio up to 2:1, the acidic effect is still strong enough to keep NH_3 volatilization to nearly zero for a few days. Thus, urea-urea phosphate grades of 28-28-0 and 35-17-0 (oxide) or 28-12-0 and 35-7-0 (elemental) are feasible.

More recently, TVA has been investigating urea phosphate made from wet-process acid without first removing the impurities in the acid. This urea phosphate has a grade of about 12-40-0 (oxide) or 12-18-0 (elemental). Retention of the impurities may improve granule quality, especially for the urea-urea phosphates.

III. SUMMARY

Economic conditions are forcing reexamination of conventional N fertilizers rather than development of new kinds of N fertilizers. As a result, urea is proving to be a versatile material. It can be added to other conventional N sources such as AS, AN, and NH_4^+-phosphate. The increased N content extends the usefulness of these fertilizers; thus, farmers can continue to use familiar products. At the same time, agronomists, chemists, economists, and chemical engineers have achieved some of their goals to increase the efficiency of N fertilizers and to reduce the energy required for producing, transporting, and applying N fertilizers.

REFERENCES

Bridges, J. D. 1980. Fertilizer trends 1979. Bull. Y-150. NFDC, TVA, Muscle Shoals, Ala.

International Fertilizer Development Center. 1979. Fertilizer manual. Reference manual IFDC-R-1. IFDC, Muscle Shoals, Ala.

National Fertilizer Development Center. 1979. Symposium: ammonia from coal. Bull. Y-143. NFDC, TVA, Muscle Shoals, Ala.

National Fertilizer Development Center. 1980. New developments in fertilizer technology, 13th demonstration, 7–8 Oct. 1980. Bull. Y-158. NFDC, TVA, Muscle Shoals, Ala.

Sheldon, V. L., D. A. Russel, and J. W. Aldridge. 1981. Directory, fertilizer research in the U.S.: agronomic, economic, marketing. Bull. Y-164. NFDC, TVA, Muscle Shoals, Ala.

United Nations Industrial Development Organization. 1967. Fertilizer manual. Publication ST/CID/15. United Nations, New York.

13

Seward E. Allen
National Fertilizer Development Center
Tennessee Valley Authority
Muscle Shoals, Alabama

Slow-Release Nitrogen Fertilizers

Hauck and Koshino (1971) reviewed the chemistry of slow-release N (SRN) fertilizers and their use in crop production, while Hauck (1972) discussed their reactions in the soil environment. More recent papers by Allen and Mays (1974), Allen (1977), and Davies (1976), emphasized S-coated urea (SCU). Murray and Horn (1979) reviewed organic SRN fertilizers other than SCU. Young (1974) was concerned with the engineering aspects of SCU development.

As regards SRN fertilizers in crop production, Allen (1980) and Staib and Hays (1980) were concerned with turfgrasses. Allen and Mays (1974) restricted their paper to forage crops, while Engelstad et al. (1972) reported on development of more efficient fertilizers for rice (*Oryza sativa* L.). This paper: (i) discusses theoretical benefits from SRN fertilizers, (ii) reviews the chemistry of selected SRN products and their mode of action in soil, and (iii) relates this information to the response of selected crops to SRN fertilizers.

I. POTENTIAL BENEFITS FROM SLOW-RELEASE N FERTILIZERS

Potential benefits from SRN fertilizers include: (i) more efficient use of N by the crop, (ii) less leaching of N, (iii) lower toxicity, (iv) longer lasting N supply, (v) reduced volatilization losses of N, and (vi) lower application cost.

A. Efficiency of N Use

Crops often take up more N than is required for optimum yield, a process referred to as "luxury consumption." With annual crops harvested for seed or fiber [corn (*Zea mays* L.), small grains, cotton (*Gossypium hirsu-*

Nitrogen in Crop Production.

tum L.)], this seldom results in reduced efficiency of N use, because excess N stored in leaves is transferred to the seed during maturation. With forage and turfgrasses, excess N accumulated in early growth is removed in clippings, and less N remains available for late-season production. Similarly, leafy vegetables may accumulate high levels of NO_3^--N during early growth, a process resulting in food products of inferior quality. Luxury consumption or accumulation of NO_3^--N often may be controlled by split application of soluble fertilizers or with a single application of SRN fertilizers.

B. Decreased Leaching of N

All conventional water-soluble N fertilizers contain NO_3^--N, urea, or NH_4^+-N. The latter two sources are readily converted to NO_3^--N by soil organisms. Thus, the dominant source of N in well-aerated soils is NO_3^--N. Nitrates leach readily when large amounts of water pass through the soil profile. Movement of NO_3^--N beyond the rooting zone results in less N available for crop response and may contribute to pollution of drainage water. All SRN fertilizers eventually convert to NO_3^--N by processes to be discussed later. However, since their conversion is slow, less NO_3^--N may be present in the soil when leaching incidents occur (heavy rains or excess irrigation). Thus, nutrients "saved" may become available later and be utilized in crop growth.

C. Lower Toxicity

When soluble N fertilizers are surface-placed, especially on turfgrasses, or row-placed with seeds, damage to foliage or germinating seedlings may occur. Toxicity results from salt effects (competition between plants and dissolving fertilizer granules for available water) or from fumigation of tender plant parts with NH_3 derived from urea or fertilizer salts containing NH_4^+-N. In either case, toxicity may be reduced with SRN fertilizers, because the amount solubilized at any given time usually is small.

D. Longer Lasting N Supply

Soluble fertilizers may be wasted through luxury consumption or leaching. In either case, the supply from a single application may not last all season. Slow-release N fertilizers often last longer, because the insoluble portion is released slowly and in better relationship with the actual crop uptake pattern. One application of SRN fertilizers may last all season, and there is evidence that residual effects can be measured in later years.

E. Reduced Volatilization Losses

Nitrogen may be lost to the atmosphere by release of NH_3 from surface-applied urea or NH_4^+ salts, or by release of N_2 or N_2O through denitrification of NO_3^--N in flooded soils. Ammonia loss is greatest from heavy rates where the rate of release exceeds the capacity of soil or crops to absorb it. Thus, losses may be reduced with SRN fertilizers that dissolve or solubilize slowly. Similarly, denitrification loss may be less from SRN fertilizers, because the supply of NO_3^--N may be less than from soluble fertilizers.

F. Lower Application Cost

Some crops require frequent application of soluble N. With SRN fertilizers, the number of applications may be reduced. Savings in application cost are substantial, particularly when aerial application is eliminated.

II. SLOW-RELEASE N FERTILIZERS AND THEIR MODE OF ACTION

Figure 1 shows structural formulas for some SRN fertilizers (Hauck & Koshino, 1971). The same materials still are of practical and research interest. However, major changes in emphasis on specific products have occurred. Urea-formaldehyde (UF) remains a popular source of SRN. Emphasis on coated products, particularly SCU, has increased dramatically since 1970, and about 50 000 t were produced at three locations in the United States and Canada during 1980. Formolene®[1] (a 1:1 mixture of urea and methylolurea) was introduced recently by Ashland Chemical Company, and is being marketed as a fluid SRN fertilizer for turfgrasses. Interest in oxamide has waned since 1970, but recently improved methods of synthesis offer promise of future development. Thiourea and dicyandiamide were dropped as SRN fertilizers, but they may be useful as additives to urea for control of nitrification and/or urea hydrolysis. These possibilities are covered in more detail elsewhere (Chapt. 37, this book). In this paper, only UF, isobutylidene diurea (IBDU), Formolene®, and coated products (SCU and Osmocote®) will receive further consideration.

A. Ureaformaldehyde

Urea-formaldehyde is prepared by reacting urea with formaldehyde. Linear and cross-linked polymers of varying complexity result from varying the ratio of reactants. Commercial UF (38% N) is a mixture of unreacted urea and short-chain polymers with one, two, or three methylene linkages. More complex polymers are too unreactive for use as SRN fertilizers. Biodegradation of UF apparently occurs at the C-bond rather than by hydrolysis (formaldehyde has not been detected in soil treated with UF). Once the initial degradation takes place, ammonification and nitrification proceed rapidly and crops utilize NO_3^--N. Urea-formaldehyde is available in granular form for direct application and in powder suitable for suspension fertilizers. Although fluid UF (small particle size) is more rapidly available than is granular UF, the particle size effect is much smaller than for IBDU.

[1] Any mention of a trade name does not constitute endorsement of the product by TVA nor exclusion of a similar product.

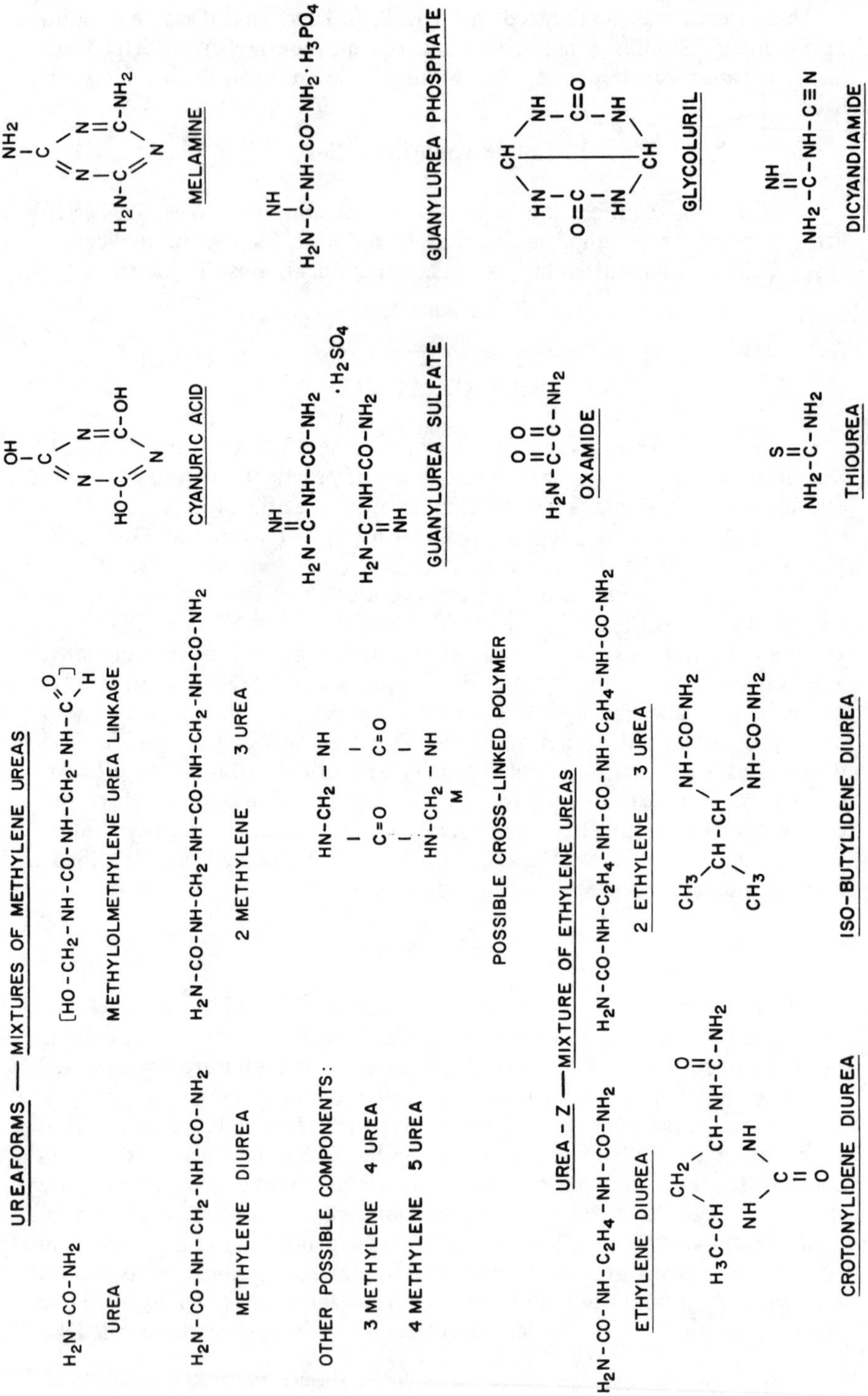

UREAFORMS — MIXTURES OF METHYLENE UREAS

$H_2N-CO-NH_2$

UREA

$[HO-CH_2-NH-CO-NH-CH_2-NH-CHO]$

METHYLOLMETHYLENE UREA LINKAGE

$H_2N-CO-NH-CH_2-NH-CO-NH_2$

METHYLENE DIUREA

$H_2N-CO-NH-CH_2-NH-CO-NH-CH_2-NH-CO-NH_2$

2 METHYLENE 3 UREA

OTHER POSSIBLE COMPONENTS:

3 METHYLENE 4 UREA

4 METHYLENE 5 UREA

HN–CH2 – NH, C=O, C=O, HN–CH2 – NH, M

POSSIBLE CROSS-LINKED POLYMER

UREA - Z — MIXTURE OF ETHYLENE UREAS

$H_2N-CO-NH-C_2H_4-NH-CO-NH_2$

ETHYLENE DIUREA

$H_2N-CO-NH-C_2H_4-NH-CO-NH-C_2H_4-NH-CO-NH_2$

2 ETHYLENE 3 UREA

CH2, H3C–CH, CH–NH–C(=O)–NH2, NH, NH, C=O

CROTONYLIDENE DIUREA

$(CH_3)_2CH-CH(NH-CO-NH_2)_2$

ISO-BUTYLIDENE DIUREA

OH, C, N, N, HO–C, C–OH, N

CYANURIC ACID

NH2, C, N, N, H2N–C, C–NH2, N

MELAMINE

$H_2N-C(=NH)-NH-CO-NH_2$ $\cdot H_2SO_4$
$H_2N-C(=NH)-NH-CO-NH_2$

GUANYLUREA SULFATE

$H_2N-C(NH)-NH-CO-NH_2 \cdot H_3PO_4$

GUANYLUREA PHOSPHATE

$H_2N-C(=O)-C(=O)-NH_2$

OXAMIDE

CH, HN, NH, O=C, C=O, HN, NH, CH

GLYCOLURIL

$NH_2-C(=S)-NH_2$

THIOUREA

$NH_2-C(=NH)-NH-C\equiv N$

DICYANDIAMIDE

Fig. 1—Structural formulas of some slow-release N materials.

Since initial release of N from UF is through biodegradation, the process is temperature sensitive and most rapid during summer months. The primary manufacturer of UF is Boots-Hercules Co. (Wilmington, Del.).

B. Isobutylidene Diurea

Isobutylidene diurea is prepared by reaction of urea and isobutyraldehyde. The product formed (IBDU, 31% N) is very insoluble in water. But once dissolution takes place, hydrolysis proceeds rapidly with regeneration of the original reactants. Consequently, SRN effects result only from granular IBDU, which dissolves slowly. The primary product of IBDU hydrolysis is urea, and soil transformations are identical to those for urea from any other source. Fine IBDU (rapid dissolution) is not a useful SRN fertilizer. Because hydrolysis is the primary step in release of N from IBDU, the process is less temperature sensitive than for UF. On this basis, IBDU is more effective than UF in cold soil; hydrolysis of IBDU is also more rapid in acidic than in calcareous soils. Estech General Chemical Co. (Chicago, Ill.) markets IBDU in the United States.

C. Formolene®

Formolene® (30% N) is marketed by Ashland Chemical Co. (Dublin, Ohio) as a liquid fertilizer containing equal parts of urea and methylolurea. Little information is available on crop response or soil transformations. However, it is reasonable to predict that it is the most soluble of all SRN fertilizers.

D. Coated Products

After nearly 20 yr of research and development by TVA, SCU is now marketed by TVA, Agricultural Industries Manufacturing Corp. (Columbia, Ala.), and Canadian Industries, Ltd. (Ingersoll, Ontario). All suppliers prepare SCU by spraying substrate urea granules or prills with molten S, followed by a light coat of sealant wax and conditioner. Typical products contain about 36% N, 16% S, and 5% sealant plus conditioner. Since a coating of finite thickness is required, large and small granules often contain less or more S, respectively, than the average value reported. For quality control purposes, products are often characterized by determining the amount of urea released in 7 d in water at 38°C. Thus, SCU-30 refers to a product that releases 30% of total N under the prescribed conditions. The 7-d dissolution value is mostly a measure of the relative number of imperfectly coated granules.

Much of the dissolution of the SRN fraction of SCU proceeds through physical processes. Once water gains access to the substrate, osmotic forces dominate the reaction and individual granules empty in a few days. On this basis, the SRN pattern results from many granules releasing urea at different times rather than from gradual release of urea from all granules at the same time. Once urea is released, soil transformations proceed as for all fertilizers containing urea. Several soil and plant factors modify the dissolution reaction (Allen, 1980). Some of the more important variables follow.

1) Release is accelerated in warm soil, which suggests that biodegradation is also involved in coating failure.
2) Moisture tension accelerates release. In controlled-environment tests in a silt loam soil at 25°C, the relative order was: dry (10% H_2O > alternating moist/dry > continuously moist. Lowering soil temperature to 5°C slowed dissolution, but did not change the ranking of soil moisture regimes.
3) Dissolution is more rapid with surface placement than with mixed or layering in the soil. This effect is believed to be related to much wider ranges in soil moisture and temperature at the soil surface.
4) Root action accelerates dissolution of SCU.
5) Dissolution is not affected by soil pH in the range of 5 to 8.
6) Biodegradation of the coating residue (elemental S) proceeds slowly with formation of H_2SO_4. Thus, SCU is an effective source of both SRN and SRS. Sulfuric acid formed may also contribute to soil acidity in direct relation to rate of applied SCU and CEC of the soil.

Osmocote®, developed by Archer Daniels Midland Co. (Minneapolis, Minn.) is marketed by Sierra Chemical Co. (Milpita, Calif.). In all formulations, soluble substrate granules are coated with a semipermeable plastic membrane. Release of nutrient proceeds by diffusion of solutes through the membrane. Osmocote®, available in formulations with varying NPK content, is used largely for container-grown ornamentals.

III. SLOW-RELEASE N FERTILIZERS IN CROP PRODUCTION

It is obvious that SRN fertilizers are entirely different from conventional fertilizers containing water-soluble N. To use them effectively, one must understand their mode of action and crop uptake pattern. This section relates basic information (sections I and II) to actual situations with several types of crops.

A. Turfgrass

The main objective in turfgrass fertilization is to maintain uniform growth of high-quality turf (good color and vigor) throughout the season without danger of burning grass and with as few applications as possible. Turf is a high-value crop, and relative cost of fertilizer N may be less important than with agronomic crops. On this basis, the primary benefits from SRN fertilizers for turf result from their low burn potential and long-lasting N supply. This conclusion is supported by Waddington et al. (1976), who fertilized Kentucky bluegrass (*Poa pratense* L.) for 7 yr with UF, IBDU, activated sludge, and soluble N sources. Soluble fertilizers produced excessive growth after fertilization, UF and IBDU supported much more uniform growth, and UF had the most residual effects. Wilkinson (1977) reported similar results, but also noted that IBDU was much more effective than UF in early spring when soil temperature was low. Waddington and Duich (1976) concluded that the SRN effect from SCU was intermediate between IBDU and UF. Leaching is seldom a problem in well-established turf, except on very coarse-textured soils under heavy rainfall or excessive irrigation. However, Allen et al. (1978) were able to measure large benefits from SCU with grass grown in cylinders under accelerated leaching conditions. In their experiment, 30% of applied urea or ammonium nitrate (AN) was lost to the drainage water, as compared with 5% for SCU. This resulted in 40% more forage from SCU than from soluble N sources. Golf greens are usually established on artificial media designed more for good drainage than for retention of nutrients. Thus, UF and IBDU are popular for greens use; SCU is not recommended for greens because close mowing may fracture granules and destroy the SRN effect. The net result from the above is that UF has long been a popular turf fertilizer, while use of IBDU, and especially SCU, has increased rapidly in the last 5 yr.

B. Forage Grasses

The chief objective of forage fertilization is to produce maximum yield of acceptable quality forage with least cost. For this reason, recent research has been concentrated on SCU (the least expensive SRN fertilizer). Allen and Mays (1974) concluded that the major benefit in normal cropping situations may result from greater efficiency of N use (control of luxury uptake). On this basis, one might—under ideal conditions—achieve the same result with a single application of SRN fertilizer as with several split applications of AN. Davies (1973) achieved this goal with SCU on clipped ryegrass (*Lolium perenne* L.), but UF was inferior in all parameters measured. In certain situations where leaching or volatilization losses reduce the effective rate of soluble fertilizers, benefits from SRN fertilizers may be large. Maschmedt and Cocks (1976) fertilized annual ryegrass (*Lolium rigidum* Gaud.) with urea, AN, and SCU. Total yield of forage was similar from AN and SCU, but seasonal distribution was much more uniform with SCU. Yield from urea was lower and only 44% of applied N was recovered in forage, as compared with 78% from SCU and 88% from AN. In this case, volatilization losses from urea were large, and SCU reduced but did not entirely eliminate losses. Lopez (1980) fertilized guineagrass (*Panicum maximum*) with urea and SCU and measured yield and uptake of N for 2 yr. Recovery of N was 91% with SCU and 52% with urea. Leaching losses were much reduced with SCU in this experiment (conducted in eastern Peru), where annual rainfall exceeded 2 m.

C. Sugarcane and Pineapple

Sugarcane (*Saccharum officinarum* L.) and pineapple [*Ananas comosus* (L.) Merr.] are long-term crops requiring up to 2 yr of growth before harvest. They are routinely fertilized several times (often by air) with soluble N sources; thus, they may benefit from the long-lasting properties, as well as protection from leaching, offered by SRN fertilizers. El Wali et al. (1980) conducted two greenhouse experiments involving fertilization of sugarcane with SCU and urea under two leaching regimes. Losses of NO_3^--N were least from SCU, but differences in growth and uptake of N could not be measured. Gascho and Snyder (1976) compared SCU applied in November with four split applications of ammonium sulfate (AS). Excellent growth was obtained from SCU in spring and early summer, but a single fall application of SCU failed to maintain optimum levels of N late in the season. Yield from SCU approached, but did not equal, that from split AS. They concluded that SCU with a lower solubility and/or a later date of application might eliminate the need for split application of soluble N sources.

D. Corn and Small Grains

As indicated previously, these crops absorb large amounts of N early in their growth cycle, store it in leaves, and transfer it to the developing seed. This type of growth results in very efficient use of soluble N. Thus, there may be little benefit from SRN fertilizers, except where there are large leaching losses soon after application.

E. Horticultural Crops

Osmocote®, IBDU, UF, and SCU have proven useful for ornamentals grown in containers. These are high-value crops where fertilizer is a minor input in production cost. Also, the growth medium is often designed for good drainage and aeration rather than for retention of nutrients. In this type of soil, soluble fertilizers may be toxic (salt effects) and it is not possible to supply enough N at planting to mature the crop. Effective SRN fertilizers offer solutions to both problems. Interest in SRN fertilizers for field-grown horticultural crops has been confined largely to SCU on soils where leaching may occur or with crops that benefit from split applications of soluble N sources. Locascio and Fiskell (1970), Locascio et al. (1974), and Locascio et al. (1978) conducted several experiments with watermelon (*Citrullus lanatus*) fertilized with SCU and soluble N sources on sandy soil under severe leaching conditions. At several sites, yields were greater with SCU than with a single application of urea. However, mulching the soil with plastic was more effective than SCU in reducing leaching losses. Liegel and Walsh (1976) grew potatoes (*Solanum tuberosum* L.) for 3 yr in Wisconsin sandy soil; SCU was a superior source in 1 yr when there were leaching losses. Forage trials with oats (*Avena sativa* L.) following potatoes showed greater residual N in SCU plots, a result attributed to incomplete dissolution of SCU in the previous year. Shelton (1976) fertilized trellised tomatoes (*Lycopersicon esculentum* L.) with SCU and AN at very high rates (392, 560, or 729 kg N/ha). Yields from the lower two rates of SCU were superior to those from single or split applications of AN. At the highest rate, AN was slightly toxic and SCU was not. Most of the benefits from SCU were attributed to uniform regulation of N supply without excessive amounts during early fruit set and development.

F. Rice

Rice is the major commercial crop customarily grown in flooded soil. The soil-fertilizer regime is entirely different from other crops, particularly in regard to the fate of fertilizer N. If fertilizers containing NO_3^--N are used, or if NH_4^+-N nitrifies prior to flooding, losses through denitrification may be large. For this reason, AS has long been the favored N source for flooded rice. Urea or AS may be applied in floodwater with little loss, except for a small amount of NO_3^--N that may form in the oxidized layer at the soil-water interface. If this NO_3^--N leaches into the reduced layer below, it may be lost through denitrification. Small amounts of NH_3-N also may be lost to the atmosphere when floodwater becomes alkaline during daylight hours, as algae consume all available carbonate in photosynthesis. Urea and NH_4^+-N flooded soon after application remain effective N sources, provided flooding is maintained until the crop is nearly mature. In areas dependent on natural rainfall, or where the supply of irrigation water is inadequate, soils may dry enough to permit nitrification. When paddies are reflooded, large losses of N may occur. With alternate flooding and drying, SRN fertilizers may be more effective, since that portion of applied N not yet dissolved or nitrified is protected from denitrification and may be used later by the crop. In recent years, there has been much interest in SCU and IBDU for rice. There has been little interest in UF, because release of N from this source results from aerobic biodegradation. There is also evidence that SRN fertilizers supply less N during early growth and more during maturation, a release pattern that often results in less lodging and more grain.

Engelstad et al. (1972) summarized a large number of rice experiments conducted in Asia and South America under conditions of intermittent flooding. When applied as a basal treatment, SCU was superior to urea in 46 of 56 comparisons. In one experiment in Peru, SCU was 4 times as effective as urea. In this case, fertilizers were surface-applied on coarse-textured calcareous soil several weeks before flooding, and there were frequent wet-dry cycles during growth. Under these conditions, it was possible to lose NH_3-N to the atmosphere, leach NO_3^--N and/or NH_4^+-N, and denitrify NO_3^--N. All of these losses were controlled in varying degrees by SCU, and the overall superiority of SCU over urea was exceptionally large. Sanchez et al. (1973) also reported on rice experiments in Peru where yield was increased an average of 59% more by pretransplant-incorporated SCU than by urea or AS applied in the same manner. Similarly, SCU produced 20% more grain than did top-dressed urea or AS. Sharma (1973) compared SCU at transplanting with urea at transplanting or in several split combinations. Grain yield was 15 to 20% higher with SCU than with any treatment with urea. Wells and Shockley (1975) compared preplant or top-dressed urea, SCU, IBDU, and AS for direct-seeded rice flooded 2 weeks after seedling emergence. Growth was slowed early in the season by SRN sources, but grain yield was similar for all sources. On clay soils with intermittent flooding, SCU resulted in increased yield, as compared with preplant urea, and was superior to split, top-dressed urea in 1 of 5 yr. Both SCU and IBDU were suitable SRN fertilizers under the water management systems used in these studies.

G. Forestry and Tree Crops

Forest fertilization has become common in the Pacific Northwest and Southeast; about 500 000 ha has been fertilized in the past 10 yr. There has been considerable interest in SRN fertilizers in forestry, but few experiments have been conducted long enough to measure long-term benefits. Forest trees use N very efficiently through recycling of nutrients from leaf residue on the forest floor. For this reason, response to a single application of N often lasts several years. Thus, the major benefit from SRN fertilizers may result from reduction of leaching in coarse-textured forest soils. This conclusion is supported by Ballard (1979), who reported less leaching from SCU and IBDU than from soluble fertilizers. Interest also is developing in SRN fertilizers for citrus and pecans (*Carya illinoiensis* L.), crops often grown on soils subject to leaching.

IV. ENVIRONMENTAL CONSIDERATIONS

Effects of N management on environmental quality are considered elsewhere (Chapt. 45, this book). Comments relate only to the possible role of SRN fertilizers in reducing the flow of NO_3^--N to drainage water.

When soluble fertilizers are applied before planting on coarse-textured soils, there may be leaching losses if heavy rains occur before the crop is well established. Obviously, SRN fertilizers may reduce the danger of groundwater pollution during this critical period. After crop establishment, there may be little danger of leaching during summer when evapotranspiration is high and plant roots are active. Lysimeter experiments usually show maximum loss of NO_3^--N during late winter and early spring, when there is no crop and water moves freely through many soils. With soluble fertilizers, crops may remove most of the available N during growth and leave little residual NO_3^--N for winter leaching. When SRN fertilizers leave more available soil N after cropping, or when dissolution of SRN granules occurs during winter, they may actually increase leaching losses in winter or early spring. On this basis, winter leaching could be less from UF (slow biodegradation in cold soil) than from SCU or IBDU. If the above assumptions (less leaching from SRN fertilizers during cropping and more leaching during winter) are correct, the average annual leaching loss may be similar for conventional and SRN fertilizers.

V. CONCLUSIONS

The preceding discussion has emphasized certain cropping situations where SRN fertilizers have been used to advantage. Little mention was made of the many situations where they were not superior to soluble fertilizers. It should be emphasized that SRN fertilizers are only one of several better soil and crop management techniques (pest control, irrigation, improved cultivars, etc.) available for increasing or modifying crop yield. One would expect maximum benefit in situations where serious problems with soluble fertilizers may be controlled with SRN products and where other potential limiting factors are supplied at optimum level. Unfortunately, there are few reliable criteria for forecasting or identifying these problem areas. The final decision to use SRN fertilizers and select appropriate sources depends entirely on the anticipated cost/benefit ratio. In other words, does the potential benefit justify the extra cost? In 1980, the relative cost per unit of N was 1.0, 2.0, 2.5, and 3.0 for urea, SCU, Formolene®, and IBDU or UF, respectively. In the past, cost factors have severely limited acceptance of SRN fertilizers for most field crops. In contrast, use for high-value crops (turfgrass and horticultural crops) continues to increase and may rise even more rapidly with recent development of cheaper, more effective SRN fertilizers.

REFERENCES

Allen, S. E. 1977. Controlled-release fertilizers: theory and practice. p. 289–298. *In* R. L. Goulding (ed.) Proc. Controlled Release Pesticide Symposium, Oregon State Univ., Corvallis.

Allen, S. E. 1980. Sulfur-coated urea: cover controls nitrogen release. Weeds Trees & Turf Mag. June 1980, p. 54–55.

Allen, S. E., and D. A. Mays. 1974. Coated and other slow-release fertilizers for forages. p. 559–582. *In* D. A. Mays (ed.) Forage fertilization. Am. Soc. of Agron., Madison, Wis.

Allen, S. E., G. L. Terman, and H. G. Kennedy. 1978. Nutrient uptake by grass and leaching losses from soluble and S-coated urea and KCl. Agron. J. 70:264–268.

Ballard, R. 1979. Transformations of nitrogen fertilizers and movement of nutrients from the surface of a rhyolitic pumice forest soil. N.Z. J. For. Sci. 9:53–67.

Davies, L. H. 1973. Two grass field trials with a sulfur-coated urea to examine its potential as a slow-release nitrogen fertilizer in the United Kingdom. J. Sci. Food Agric. 24:63–67.

Davies, L. H. 1976. Slow-release fertilizers, particularly sulphur-coated urea. Proc. no. 53. Fertilizer Society, London.

El Wali, A. M. O., F. Le Grand, and J. G. Gascho. 1980. Nitrogen leaching from soil and uptake by sugarcane from various urea-based fertilizers. Soil Sci. Soc. Am. J. 44:119–122.

Engelstad, O. P., J. G. Getsinger, and P. J. Stangel. 1972. Tailoring of fertilizers for rice. Bull. Y-52. NFDC, Muscle Shoals, Ala.

Gascho, J. G., and G. H. Snyder. 1976. Sulfur-coated fertilizers for sugarcane. Soil Sci. Soc. Am. J. 40:119–122.

Hauck, R. D. 1972. Synthetic slow-release fertilizers and fertilizer amendments. p. 633–690. *In* C. A. I. Goring and J. W. Hamaker (ed.) Organic chemicals in the soil environment. Marcel Dekker, Inc., New York.

Hauck, R. D., and M. Koshino. 1971. Slow-release and amended fertilizers. p. 455–495. *In* R. A. Olson et al. (ed.) Fertilizer technology and use. 2nd ed. Soil Sci. Soc. of Am., Madison, Wis.

Liegel, E. A., and L. M. Walsh. 1976. Evaluation of sulfur-coated urea applied to irrigated potatoes and corn. Agron. J. 68:457–463.

Locascio, S. J., and J. G. A. Fiskell. 1970. Effects of sulfur-coated urea and potassium chloride on watermelon production and soluble salts. Soil Crop Sci. Soc. Fla. Proc. 30: 113–122.

Locascio, S. J., J. G. A. Fiskell, and G. W. Elstrom. 1978. Comparison of sulfur-coated urea and uncoated urea for watermelons. Soil Crop Sci. Soc. Fla. Proc. 37:197–200.

Locascio, S. J., J. G. A. Fiskell, and F. G. Martin. 1974. Evaluation of sulfur-coated urea in laboratory and field studies. Soil Crop Sci. Soc. Fla. Proc. 33:191–196.

Lopez, C. E. 1980. Nutritional requirements for *Panicum maximum* production in the Amazon jungle of Peru. Diss. Abstr. B41:766.

Maschedt, D. J., and P. S. Cocks. 1976. Sulfur-coated urea compared with urea and ammonium nitrate as a source of nitrogen for swards of annual ryegrass. Agric. Rec. (South Australia) 3:4–7.

Murray, T. P., and R. C. Horn. 1979. Organic nitrogen compounds for use as fertilizers. Tech. Bull. IFDC-T-14. IFDC, Muscle Shoals, Ala.

Sanchez, P. A., A. Gavidia, G. E. Ramirez, R. Vergara, and F. Minguillo. 1973. Performance of sulfur-coated urea under intermittently flooded rice culture in Peru. Soil Sci. Soc. Am. Proc. 37:789–792.

Sharma, S. N. 1973. Relative efficiency of sulfur-coated urea and urea in rice soils. Indian J. Agron. 18:399–400.

Shelton, J. E. 1976. Evaluation of sulfur-coated urea as a preplant total season nitrogen supply for trellised tomatoes. Soil Sci. Soc. Am. J. 40:126–129.

Staib, C. R., and J. T. Hays. 1980. Fertilizer basics. Lawn Care Industry Mag. February 1980, p. 18.

Waddington, D. V., and J. M. Duich. 1976. Evaluation of slow-release nitrogen fertilizers on Penpar creeping fescue. Agron. J. 68:812–815.

Waddington, D. V., E. L. Moberg, J. M. Duich, and T. L. Watschke. 1976. Long-term evaluation of slow-release nitrogen sources on turfgrasses. Soil Sci. Soc. Am. J. 40:593–597.

Wells, B. R., and P. A. Shockley. 1975. Conventional and controlled-release nitrogen sources for rice. Soil Sci. Soc. Am. Proc. 39:549–551.

Wilkinson, J. F. 1977. Effect of isobutylidene diurea and ureaformaldehyde rate, date, and frequency of application on 'Merion' Kentucky bluegrass. Agron. J. 69:657–661.

Young, R. D. 1974. TVA's development of sulfur-coated urea. Bull. Y-79. NFDC, Muscle Shoals, Ala.

14

Lee E. Sommers
Purdue University
West Lafayette, Indiana

Paul M. Giordano
National Fertilizer Development Center
Tennessee Valley Authority
Muscle Shoals, Alabama

Use of Nitrogen from Agricultural, Industrial, and Municipal Wastes[1]

A wide variety of sludges and effluents are generated during the treatment of domestic, industrial, and agricultural wastes by physical, chemical, and biological treatment processes. Other waste materials include municipal refuse and garbage, wood processing wastes, and so forth. In contrast with conventional fertilizer materials that contain fixed concentrations of plant nutrients, most waste materials contain variable concentrations of plant nutrients, as well as numerous nonnutrient constituents (e.g., heavy metals or persistent organics). The plant nutrient content of waste materials depends on temporal variations in inputs to and properties of the waste treatment process. This results in a need for accurate characterization of the processed waste.

The use of wastes on agricultural cropland can be viewed from either of two standpoints. Firstly, the waste material is used as a substitute for conventional fertilizers. The amount applied equals the fertilizer requirement (usually N or P) for the crop grown or some proportional part of the nutrient requirement. Similarly, application rates of some metal-bearing wastes are dictated by restrictions on loading rates of certain metals such as Cd (CAST, 1980). The second approach is based on the soil being used as

[1] Joint contribution from Agric. Exp. Stn. (Journal Paper no. 8706), Purdue University, W. Lafayette, Ind. and Agricultural Research Branch, National Fertilizer Development Center, Tennessee Valley Authority, Muscle Shoals, Ala.

Nitrogen in Crop Production.

merely a waste disposal site with the rate of application being limited by a noncrop growth parameter, such as infiltration rate or trafficability. The former approach will be emphasized in this discussion because disposal sites are normally not used for crop production, and because waste application rates consistent with nutrient needs of the crop grown tend to minimize the potential for adverse environmental impacts.

There are some constraints associated with applying waste materials on soils that do not apply to conventional fertilizer materials. Depending on the waste material, these constraints may include (i) pathogens (e.g., bacteria, viruses, or parasitic worms); (ii) soluble salts; (iii) heavy metals (e.g., Pb, Zn, Cu, Ni, or Cd); (iv) other trace elements (e.g., Se, Mo, or B) and; (v) persistent organics [e.g., polychlorinated biphenyls (PCB's) or chlorinated hydrocarbon pesticides]. Obviously, the chemical composition in conjunction with the application rate determines which one of the above constraints applies to a given waste material. Pathogens are present in wastes from municipal sewage treatment plants, whereas they are a minimal concern for many industrial and agricultural processing wastes. Soluble salts tend to be a greater potential problem in liquid than in solid wastes, especially in irrigated regions. Heavy metals or other trace elements are present in all sewage sludges, some industrial wastes, and relatively few agricultural processing wastes. Either effective waste-treatment processes (e.g., pathogen reduction) or appropriate application rates (e.g., metals limitation), along with best management practices (e.g., soil pH control), are needed to minimize the above constraints and to enable the use of waste materials as a nutrient source for crop production. The reader should consult the following sources for additional information concerning application of wastes on land: pathogens in municipal wastewaters and sludges (Sagik & Sorber, 1978; Bitton et al., 1980); heavy metals in sewage sludge (Chaney & Giordano, 1977; CAST, 1976, 1980; USEPA, 1979; Bitton et al., 1980); determining application rates for sewage sludge (Knezek & Miller, 1978; USEPA, 1979, 1983; Sommers & Nelson, 1978); fate of wastewater constituents in soils (CRREL, 1978; Bouwer & Chaney, 1974; Sopper & Kardos, 1973); and determining application rates for municipal wastewater (USEPA, 1981). It is beyond the scope of this chapter to discuss in detail the constraints associated with using the diverse wastes referred to above.

I. COMPOSITION OF WASTE MATERIALS

Some representative N, P, and K data for a variety of solid or semisolid (e.g., refuse, tankage, and sewage sludge) and liquid (e.g., municipal wastewater, whey, and agricultural processing) wastes are shown in Table 1. These data clearly show that wastes contain appreciably lower concentrations of N, P, and K than commercial fertilizer materials. The majority of solid or semisolid wastes contain $< 5\%$ total N, while liquid wastes vary from 19 to 1500 mg total N/L. It should be realized that many wastes (e.g., sewage sludge) are commonly a suspension containing from < 1 to 15% solids. Because the N/P/K ratio required by many cereal crops is approxi-

Table 1—Total N, P, and K concentrations in selected waste materials.

Waste material	N	P	K	Reference
	Concentration of			
	%†			
Solid or semisolid:				
Composted or shredded refuse	0.57–1.30	0.08–0.26	0.27–0.98	McCalla et al., 1977
Leather tankage	12.80	0.01	0.36	Sims & Boswell, 1980
Waste food fiber	2.00	0.01	0.10	King, 1979
Paper mill sludge	0.15–2.33	0.16–0.50	0.44–0.85	Dolar et al., 1972
Citric acid production wastes	0.51–4.13	0.06–0.29	0.01–0.19	King & Vick, 1978
Antibiotic production mycelium waste	3.29	0.71	0.34	Nelson & Sommers, 1979
Tomato processing wastes	2.33	0.29	0.28	Timm et al., 1980
Municipal sewage sludge	<0.1–17.6	<0.10–14.30	0.02–2.64	Sommers, 1977
	mg/L‡			
Liquids:				
Municipal wastewater	16–37	7–13	14–22	Sommers & Sutton, 1980
Potato starch waste	420	62	580	de Haan et al., 1973
Whey	1500	500	1820	Watson et al., 1977
Antibiotic production broth	936	690	604	Nelson & Sommers, 1979
Vegetable and fruit processing wastes	19–318	4–91	--	Sommers & Sutton, 1980

† Expressed on a dry-weight basis.
‡ Expressed on a wet-weight basis (commonly contain suspended solids).

mately 4:1:2 and the ratio in many wastes is 6:4:1, it is obvious that few wastes constitute a balanced fertilizer material. For many solid wastes, application to soils at a rate selected to satisfy the N needs of the crop will result in excess additions of P and insufficient levels of K (e.g., sewage sludge). In contrast, similar concentrations of N and K are found in many liquid wastes. The data presented in Table 1 are the total concentrations of N, P, and K, with varying proportions of the total N and P being contributed by inorganic and organic forms.

The majority of wastes contain both organic and inorganic forms of N. Inorganic N is typically present as either NH_4^+ or NO_3^-; the concentrations are a function of the waste-treatment process employed. For example, aerobic digestion of a waste results in both mineralization of organic N to NH_4^+ and nitrification of NH_4^+ to NO_3^-, whereas an anaerobically digested waste will only contain NH_4^+. For wastes that are not subjected to a treatment process, the NH_4^+ and NO_3^- content will be controlled by the composition of the raw materials collected (e.g., refuse).

Handling procedures have a marked effect on the inorganic N content of wastes. Sewage sludges and other waste suspensions can be dewatered by vacuum filters, centrifuges, or sand-drying beds, resulting in significant decreases in the levels of NH_4^+ and NO_3^-. In contrast, dewatering by air- or heat-drying decreases only the NH_4^+ content through NH_3 volatilization (Ryan & Keeney, 1975). If liquid or suspension wastes are stored in a lagoon prior to application, inorganic N can be lost through NH_3 volatilization and denitrification. In summary, the inorganic N content of wastes can constitute from < 10 to > 90% of the total N present, but it is strongly influenced by methods of waste processing, handling, and storage.

A variety of organic N compounds can be present in wastes. If a biological treatment process is employed, or if a natural product is processed (e.g., antibiotics, fermentations, vegetables), the majority of the organic N will be in the form of proteins, peptides, or intermediate degradation products. Small amounts of purines, pyridines, glucosamines, and other heterocyclic N compounds could also be present. Fractionation of the organic N in sewage sludges indicates that 30 to 40% of the organic N is found in amino acids, and that < 10% of the organic N is present in hexoseamines (Ryan et al., 1973; Sommers et al., 1972). The wastes from most fermentation processes consist of microbial cells and their degradation products, and, as such, likely contain significant levels of proteinaceous and cell-wall components. Wastes from agricultural processing plants include appreciable concentrations of readily degradable organic C in addition to organic N.

Some waste materials may contain inorganic N, organic N, or varying proportions of both. It is typically assumed that all the inorganic N applied to soils will be available for plant uptake, whereas only 5 to 75% of the organic N will be mineralized within 1 yr after application. It must be strongly emphasized that waste materials are variable in composition, not only from one treatment or processing plant to another, but also with time for a specific plant. Prior to applying wastes as a source of N, a sound waste sampling and analysis program is needed to minimize the potential for adverse environmental impacts, and to maximize the benefit attainable from plant nutrients contained in the waste. The relatively low concentrations of plant-available N in most wastes also necessitates application of from < 2 to 20 t/ha to meet the needs of many crops. This has obvious unfavorable economic implications, because excessive distance for transport may negate the benefits derived from the nutrients contained in the waste product.

II. FATE OF WASTE-DERIVED N IN SOILS

Even though N transformations are similar in soils treated with conventional fertilizer materials and wastes, the rate or extent of NH_3 volatilization, nitrification, and denitrification are altered by the presence of waste-borne organic matter and other components (e.g., salts). Concurrent mineralization-immobilizations processes can also occur after waste application to soils. In the following discussion, data from studies with sewage sludge are emphasized to illustrate the importance of N transformations in soils amended with wastes.

A. NH_3 Volatilization

Many wastes are surface applied to soils as a liquid or a suspension. Following surface application, the water containing dissolved NH_4^+ can percolate into the soil or evaporate. Laboratory studies with sewage sludge indicate that up to 50% of the NH_4^+-N applied to the soil surface can be volatilized (Ryan & Keeney, 1975). Ammonia volatilization was decreased by increasing the cation exchange capacity of the soil or decreasing the amount of sludge applied. Subsequent laboratory studies (Terry et al., 1978) showed that the percentage of NH_3-N volatilized after surface application of sewage sludge on a Tracy sandy loam soil (Ultic Hapludalfs) was increased from < 10% at an air flow rate of 20 mL/min to 35% at 500 mL/min (100 mL/min equalled 1 volume change per minute). This study also demonstrated that NH_3 volatilization was increased by increasing sludge application rates and by decreasing the cation exchange capacity of the soil and the relative humidity of the air. Ammonia volatilization was < 10% of the applied NH_4^+-N when sludge was incorporated into the soil.

Experiments have also been conducted under field conditions to determine the extent of NH_3 volatilization following application of sewage sludge. One experimental approach involves the periodic measurement of NH_3 volatilized from soil amended with sludge. This has indicated that either incorporation of sludge or surface application of an acidic sludge (pH 5–5.5) to a neutral silt loam results in < 5% of the NH_4^+-N being volatilized (Sommers et al., 1981). The NH_4^+-N lost after application of an alkaline sludge (pH 7.5) was doubled if sludge was applied to a water-saturated soil. The NH_3 losses ranged from 10 to 20% of the NH_4^+-N applied. Another study found that from 56 to 60% of the NH_4^+-N applied was volatilized following application of sludge to the soil surface (Beauchamp et al., 1978). From 70 to 80% of the NH_3 loss occurred during the first 3 d after sludge application. Both of the above studies showed that diurnal variations occur for rates of NH_3 volatilization and losses of NH_3 were greatest at midday when air temperatures approached the maximum.

Ammonia volatilization is a significant mechanism for N losses following the application of many liquid and suspension wastes to the soil surface. This mechanism is minimal following surface application of solid or semisolid wastes (i.e., dried sludges, refuse). The available data suggest that up to 60% of the NH_4^+-N in liquid sewage sludges can be lost following surface applications. However, it must be realized that factors such as waste and soil pH, initial soil conditions (permeability, water content), and climate will alter NH_3 losses to a great extent. The data available also strongly indicate that a direct measurement of NH_3 loss is essential and that losses cannot be estimated from soil analysis. Current approaches to developing recommendations for sludge application rates on agricultural cropland assume that 0 and 50% of the NH_4^+-N is volatilized for incorporated and surface applications, respectively.

B. Mineralization and Immobilization

For many waste materials, a significant percentage of the N applied to soils is in various organic forms. Soil microorganisms will initiate various decomposition processes after waste application, resulting in either mineralization or immobilization of N. Because many wastes contain organic compounds with a range of C/N ratios, it is likely that mineralization and immobilization occur simultaneously.

The mineralization of organic N added to soils in sewage sludge has been evaluated primarily in laboratory studies. The experimental technique involves either (i) static incubation of waste-treated soils with periodic NH_4^+ and NO_3^- analysis on a subsample, or (ii) periodic leaching of a soil-waste mixture with a dilute salt solution followed by analysis of NH_4^+ and NO_3^- in the leachate. The latter method was widely used by Stanford and coworkers to evaluate N mineralization in diverse soils (Stanford & Smith, 1972). Data for the mineralization of organic N in 24 sludges by the above two approaches are presented in Table 2. The sludges ranged in organic N from 0.5 to 6.81%. Primary and waste-activated sludges contained the largest concentrations of organic N. The results indicate that the percentage of organic N mineralized was quite variable, not only for sludges treated by different processes, but also for sludges treated by the same process. For example, N mineralization in composted sludges ranged from 3 to 9% of the organic N, and anaerobically digested sludges varied from 2 to 25%. In

Table 2—Mineralization of organic N in a Fincastle silt loam soil (Aeric Ochraqualfs) amended with different sewage sludges and incubated for 16 weeks under laboratory conditions (from Sommers et al., 1981).

		Experimental procedure used to determine N mineralization§	
Type of sludge†	Organic N	Periodic leaching	Periodic subsampling
	%‡	— % organic N mineralized —	
Primary (1)	4.21	11.2	16.0
Primary + waste-activated (1)	4.43	43.9	42.1
Primary + CaO (1)	2.32	21.6	27.9
Primary + wet-air oxidized (2)	1.32–1.74	0–3.3	0–5.5
Waste-activated (2)	3.89–6.81	30.8–57.8	33.1–55.1
Aerobically digested (1)	2.52	31.9	25.0
Anaerobically digested (13)	0.50–3.03	0.1–31.2	2.1–24.9
Primary + CaO + composted (1)	1.31	5.3	12.0
Anaerobically digested + composted (2)	0.72–1.54	2.7–3.3	4.5–9.4
Range	0.50–6.81	0–57.8	0–55.1
Mean	2.17	15.9	17.5

† Number of sludges studied shown in parenthesis.
‡ Air-dry weight basis.
§ Leaching method based on Stanford & Smith (1972); periodic subsampling involved removal of soil sample from incubation unit for N analyses.

general, the amount of N mineralized from sewage sludge decreases with increased degree of treatment or processing within the treatment plant. Other studies have shown that composting either primary or anaerobically digested sludges decreases N mineralization from approximately 40 to 6% of the organic N present (Epstein et al., 1978). Even though the amount of N mineralized increases with increasing rate of sludge application to soils, the percentage of sludge organic N released as inorganic N tends to decrease, especially at N additions of $> 1000\ \mu g$ N/g (Ryan et al., 1973).

Laboratory incubations have been conducted to determine the amount of residual mineralizable N present several years after sludge application. Stark and Clapp (1980) collected soil samples from field plots treated with several sludge types (aerobic, anaerobic, and mixed primary–waste–activated) and evaluated N mineralization with the leaching procedure of Stanford and Smith (1972). The amounts of N mineralized indicated that the rate of sludge application was of greater importance than sludge type. Chemical indices of N availability (organic C; organic N; indigenous, inorganic N) were correlated with plant uptake of N and were as reliable as N mineralized during laboratory incubations in estimating release of N for plant uptake.

The mineralization of N has been evaluated for a variety of industrial and agricultural processing wastes. From 81 to 93% of the organic N in a microbial residue generated during citric acid production was mineralized in soil within 32 weeks. Mineralization of N in fermentation wastes can range from 56 to 96% of the organic N (Wright, 1978; Volz & Heichel, 1979). Evaluation of wastes associated with antibiotic production indicate that the amount of N mineralized ranged from 31 to 43, 10 to 56, and 39 to 52% of the organic N in fungal mycelium (grown to synthesize the antibiotic), "spent" broth (culture solution for growing the fungus), and activated sludge (treated "spent" broth), respectively (Sommers et al., 1979). The amounts of N mineralized following application of tomato (*Lycopersicon esculentum* Mill.) wastes to soil ranged from 44 to 58% of the organic N applied (Timm et al., 1980). In contrast, virtually no N was mineralized from finely shredded leather scrap residue in soil when subjected to long-term leaching in a perfusion apparatus (P. M. Giordano, unpublished data). In many cases, there is surprisingly good agreement between N mineralized in a laboratory incubation and that estimated from soil and plant N data obtained under field conditions.

Immobilization of N may occur following addition of some wastes to soil. As shown in Table 2, net immobilization of N occurred in soils amended with a primary sludge treated by a wet-air oxidation (elevated temperature and pressure) process. The addition of paper mill sludges with a C/N ratio $> 20/1$ to soil decreased oat (*Avena sativa* L.) yields because of N immobilization (Dolar et al., 1972). The addition of a primary sewage sludge amended with $K^{15}NO_3$ to soils revealed that 33% of the added ^{15}N was converted into organic N (i.e., immobilization) during the initial 7 d (Epstein et al., 1978). Interestingly, the primary sludge had a C/N ratio of 12:1, and also exhibited *net* mineralization of N. Because the sludge contained 39% organic C, and because only 11% of ^{15}N was recovered after 5 weeks of incubation, the authors suggested that the sludge addition caused rapid growth of microorganisms and the development of anaerobic regions where denitrification occurred. Nitrogen-15 has also been used to demonstrate that mineralization and immobilization of N occur simultaneously in soils amended with a synthetic sewage sludge prepared by anaerobic digestion of yeast cells (Terry et al., 1981).

Mineralization of organic N must be considered when determining the appropriate rate for waste application on agricultural cropland. For sewage sludges, it is not possible to predict the extent of N immobilization or mineralization from the C/N ratio. In addition, the methods of waste treatment and handling may have a marked effect on the amount of mineralizable N present. Undoubtedly, the percentage of organic N released will decrease with each succeeding year after application and will ultimately approach the N mineralization rate of soil organic N (i.e., 1–4%/yr). For example, it has been suggested that the percentage of organic N mineralized will be 15, 6, 4, and 2% for the first, second, third, and fourth year after sewage sludge applications, respectively (Kelling et al., 1977). A general decay series such as this is needed, but recent information shows that the initial rate will be dependent on waste-treatment processes. The initial rate of N mineralization will also vary in different regions of the United States because it is a function of soil temperature and water potential.

C. Nitrification and Denitrification

The NH_4^+ applied to soils in wastes is readily nitrified to NO_3^-, although several studies have shown that nitrification is somewhat delayed at elevated waste application rates. In contrast with conventional N fertilizers, waste applications result in substantial additions of available C along with inorganic N. This results in the potential for concurrent nitrification and denitrification. For example, a laboratory column study indicated that NO_3^- leached from intact soil cores was less for sludge-amended soils than for nontreated soils (Sommers et al., 1979). Since the soils were maintained at −0.33 bar water potential, the data suggest that nitrification was occurring on the surface of soil aggregates followed by diffusion of the NO_3^- into the anaerobic microenvironments. Similarly, it was found that 15 to 20% of the N applied to soil columns was denitrified during a 22-week incubation (King, 1973). Additional studies have shown that N recovery from field plots treated with wastes was significantly less than anticipated, with denitrification being suggested as the most likely explanation for the low N recoveries.

III. UTILIZATION OF WASTE-DERIVED N BY CROPS

Both greenhouse and field studies have been conducted to evaluate the utilization of N by crops grown on waste-amended soils. The interpretation of such data is difficult because both organic and inorganic forms of N are added, and because N losses, such as denitrification and NH_3 volatilization,

Table 3—Effect of sewage sludge application on the yield and uptake of N by successive crops of sorghum-sudan and corn (from Kelling et al., 1977).

	Sewage sludge applied, t/ha					
Cropping sequence	0	3.75	7.5†	15	30	60
			Yield			
			t/ha			
Sorghum-sudan	1.93	3.82	5.12	5.06	4.98	4.27
Corn	3.14	3.84	4.82	5.68	6.38	7.13
Corn	2.00	2.13	2.93	3.55	5.34	5.91
Corn	1.35	1.07	1.35	3.14	3.14	4.78
			Uptake of N (grain + stover)			
			kg/ha			
Sorghum-sudan	54	99	133	121	117	116
Corn	54	76	92	111	123	145
Corn	34	32	48	63	98	116
Corn	27	20	28	33	64	101
			Cumulative recovery of N			
			% of N applied			
(All crops)	--	17(48)‡	19(55)	12(33)	9(24)	6(16)

† This rate supplied 433 kg organic N/ha, 249 kg NH_4^+-N/ha, and 9 kg NO_3^--N/ha.
‡ Figures in parentheses represent percentage of applied N that was estimated to be available for plant uptake. Estimated available N was 50% NH_4^+-N, 100% NO_3^--N, and 15% organic N for sorghum-sudan. Organic N mineralized was 6, 4, and 2% of the remaining organic N, respectively, for the succeeding corn crops.

occur after waste additions to soils. Numerous studies have shown that comparable crop yields can be obtained by substituting wastes for conventional N fertilizers, but not necessarily at the same level of total N addition. In contrast with conventional N fertilizers, yield depressions can occur when excessive rates of some wastes are applied to soils. Decreases in yield are typically caused by soluble salts, metal (Cu, Zn, Ni), or other trace element (B) toxicities. Yield depressions are rarely encountered if the rate of waste application is consistent with the N requirements of the crop grown. In fact, yields may increase beyond that expected from the response to N alone as a result of supplying other essential plant nutrients and improving soil physical conditions (water-holding capacity, aeration).

Representative data on the response of sorghum-sudan [*Sorghum bicolor* (L.) Moench × *Sorghum sudanense* (Piper) Staph] and corn (*Zea mays* L.) to a single application of sewage sludge are shown in Table 3. Yields of sorghum-sudan were increased by sludge applications at 7.5 to 30 t/ha. Dry-matter yields of corn were enhanced at all sludge application rates in the three cropping seasons following sorghum-sudan. Data for both yield and N uptake indicate that mineralization of sludge organic N was supplying a significant amount of N for plant uptake, and that N mineralization was decreasing with time. The amount of N recovered in the four crops ranged from 6 to 19% of the total N applied, values considerably lower than those found for conventional N fertilizers. Because the sludges were sur-

Table 4—Silage corn and hay yields on soil treated with NH_4NO_3 and sewage sludge (from Magdoff & Amadon, 1980).

Treatment†	N applied‡		Yield of:		
			Silage corn	Hay	
	Inorganic	Organic	(1977)	1977	1978 (1st cutting)
	kg/ha		t/ha		
None	--	--	10.8	1.0	1.0
NH_4NO_3	50	--	11.5	2.1	0.3
Sewage sludge	12–20	91–99	13.0	2.5	0.5
NH_4NO_3	100	--	12.5	3.1	0.5
Sewage sludge	21–40	184–198	13.2	3.9	1.2
NH_4NO_3	150	--	12.8	3.8	0.8
Sewage sludge	36–59	272–298	13.8	4.3	1.7
NH_4NO_3	200	--	13.2	3.9	0.8
Sewage sludge	48–79	363–397	14.1	4.8	1.7

† Sewage sludge was applied for corn prior to planting and to hay after first and second cuttings. Hay was also harvested in 1978 before additional sludge applications.
‡ Range shown for inorganic and organic N reflects differences in sludge composition applied to corn and hay.

face-applied and allowed to dry before incorporation into the soil, a significant percentage of the applied NH_4^+ was undoubtedly volatilized, and thus not present for plant uptake. When NH_3 volatilization and expected N mineralization are employed to calculate the amount of estimated available N added to the soil, the N recoveries range from 16 to 55%. These recovery values are more typical of those obtained with conventional N fertilizers. For the sludge used, the authors concluded that application rates on cropland should be between 7.5 and 15 t/ha (190–380 kg estimated available N per hectare).

A comparison of silage corn and hay yields obtained by adding NH_4NO_3 and sewage sludge are shown in Table 4. For both crops, yields were somewhat greater when sewage sludge was applied. Because the majority of the N added in the sewage sludge was present in organic forms, N mineralization was contributing a significant proportion of the N for crop growth. It was calculated that 55% of the sludge organic N was mineralized during the growing season. Similar N mineralization results were obtained in a laboratory incubation study. The continued release of plant-available N is shown by the first cutting hay yields in 1978 for plots treated with sewage sludge (0.5–1.7 t/ha) as compared with NH_4NO_3 (0.3–0.8 t/ha). In a similar study, yields of bromegrass (*Bromus inermis* Leyss.) were comparable when fertilized with either 800 kg N/ha from sewage sludge (organic N was 65% of total N), or 400 kg NH_4NO_3-N/ha (Soon et al., 1978). Maximum corn yields were obtained with 200 and 100 kg N/ha from sewage sludge and NH_4NO_3, respectively.

Many industrial wastes also offer potential as sources of N for crop production. For example, application of presscake from citric acid production to rye (*Secale cereale* L.) and bermudagrass (*Cynodon dactylon* L. Pers.) resulted in yields of 15.6 to 26.1 t/ha as compared with 18.5 t/ha for conventional N (NH_4NO_3), P, and K sources (King, 1980). Recovery of N in the soil and crop was 67% for NH_4NO_3 and from 26 to 32% for several rates of presscake application. Essentially all of the N applied in the presscake was present in organic forms. Corn and orchardgrass (*Dactylis glomerata* L.) yields were similar when fertilizer N or antibiotic production wastes (mycelium, "spent" broth, and activated sludge) were applied to soils (Nelson & Sommers, 1979). For all three types of wastes, crop yields tended to increase as the amount of total N applied increased from 168 to 773 kg/ha. Crop yields were also increased the second year after application because of mineralization of residual organic N. One interesting aspect of this study was that marked yield reductions resulted from spring, but not fall, applications of activated sludge produced by treatment of the "spent" broth. For fall applications, the phytotoxic substance was degraded, volatilized, or leached between the time of application (October) and planting (May). No phytoxicity was encountered with the untreated "spent" broth. This illustrates one of the potential problems that can be encountered in applying waste materials on cropland.

Studies have been conducted in several locations to determine yields of crops irrigated with municipal wastewater. Clapp et al. (1978) summarized data obtained from field experiments in Minnesota, New Hampshire, Pennsylvania, Florida, Massachusetts, and Alberta, Canada. Uptake of N by corn varied from 110 to 145 kg/ha (67% of N applied) for wastewater irrigation, as compared with 175 kg N/ha (58% of N applied) for NH_4NO_3 application. The amount of N assimilated by corn reached a maximum at 180 kg N/ha. Nitrogen uptake by a variety of grasses was similar whether the N was applied in wastewater or conventional fertilizer materials. From 40 to 100% of the N applied in wastewater was recovered in the forage.

IV. FUTURE PROSPECTS

The anticipated increased costs of petroleum-based fertilizers could result in greater emphasis on utilizing wastes in crop production. The amounts of sewage sludge generated in the United States will increase because of federal legislation requiring secondary treament of wastewater discharged in natural waters. Likewise, municipal refuse and industrial and agricultural processing wastes will also be produced in greater amounts. Although comparable crop yields can be obtained by substituting many wastes for conventional fertilizers, there is not a sufficient amount of some wastes to significantly reduce agriculture's need for fertilizer N. For example, $< 2\%$ of the cropland in the United States could be fertilized with sewage sludge. An additional factor is the increased cost of transporting and applying low-analysis waste materials. Typical waste application rates on soils are from 5 to 20 t/ha, while conventional N fertilizers are nearly always applied at < 1 t/ha.

Future state and federal regulations may have a dramatic impact on limiting the use of wastes in crop production. For wastes that contain "significant quantities" of toxic metals (e.g., Cd, Pb) or persistent organics (e.g., PCB's), regulations could preclude their application to soils used for growing crops entering the human diet. Current regulations limit the amount of Cd that can be applied to cropland on both an annual and cumulative basis (USEPA, 1979). It is possible that cumulative metal limitations will be established for other metals (e.g., Pb, Zn, Cu, Ni) or organic compounds. Guidelines have been set in several states concerning metal additions to cropland. Since metal availability to plants usually increases with decreasing soil pH, soil pH must be maintained at $\geq$ 6.5 if soils are treated with Cd-containing wastes. The use of wastes may therefore increase production costs as a result of greater amounts or more frequent applications of agricultural limestone. In general, a more sophisticated level of management is needed when using waste materials as compared with conventional fertilizer materials. In spite of these limitations, it has been well established that waste materials can be applied to cropland in an environmentally sound manner, and they result in yields of many crops that are comparable to those obtained with conventional fertilizer materials.

The N transformations occurring after waste application to soils are similar to those of fertilizer N. Additional studies are needed to determine the extent of NH_3 volatilization, N mineralization, and denitrification in soils treated with wastes. Current guidelines consider these factors in determining the appropriate waste application rate for a specific crop, but additional data, preferably under field conditions, is required to optimize use of waste-borne N in crop production.

REFERENCES

Beauchamp, E. G., G. E. Kidd, and G. Thurtell. 1978. Ammonia volatilization from sewage sludge applied in the field. J. Environ. Qual. 7:141–146.

Bitton, G., B. L. Damron, G. T. Edds, and J. M. Davidson. 1980. Sludge—health risks of land application. Ann Arbor Science Publ., Inc., Ann Arbor, Mich.

Bouwer, H., and R. L. Chaney. 1974. Land treatment of wastewater. Adv. Agron. 26:133–176.

Chaney, R. L., and P. M. Giordano. 1977. Microelements as related to plant deficiencies and toxicities. p. 235–284. *In* L. F. Elliott et al., (ed.) Soils for management of organic wastes and waste waters. Soil Sci. Soc. of Am., Madison, Wis.

Clapp, C. E., A. J. Palazzo, W. E. Larson, G. C. Marten, and D. R. Linden. 1978. Uptake of nutrients by plants irrigated with municipal wastewater effluent. p. 395–404. *In* State of knowledge in land treatment of wastewater. Vol. 1. U.S. Army Corps of Engineers, CRREL, Hanover, N.H.

Cold Regions Research and Engineering Laboratory. 1978. State of knowledge in land treatment of wastewater, Vol. 1 and 2. U.S. Army Corps of Engineers, CRREL, Hanover, N.H.

Council for Agricultural Science and Technology. 1976. Application of sewage lsudge to cropland: appraisal of potential hazards of the heavy metals to plants and animals. Report no. 64. CAST, Ames, Iowa. Reprinted as MCD-33 by the USEPA, Washington, D.C.

Council for Agricultural Science and Technology. 1980. Effects of sewage sludge on the cadmium and zinc content of crops. Report no. 83. CAST, Ames, Iowa.

de Haan, F. A. M., G. J. Hoogeveen, and F. RiemVis. 1973. Aspects of agricultural use of potato starch wastewater. Neth. J. Agric. Sci. 21:85–94.

Dolar, S. G., J. R. Boyle, and D. R. Keeney. 1972. Paper mill sludge disposal on soils: effects on the yield and mineral nutrition of oats (*Avena sativa* L.). J. Environ. Qual. 1:405–409.

Epstein, E., D. B. Keane, J. J. Meisinger, and J. O. Legg. 1978. Mineralization of nitrogen from sewage sludge and sludge compost. J. Environ. Qual. 7:217–221.

Kelling, K. A., A. E. Peterson, L. M. Walsh, J. A. Ryan, and D. R. Keeney. 1977. A field study of the agricultural use of sewage sludge: I. Effect on crop yield and uptake of N and P. J. Environ. Qual. 6:339–345.

King, L. D. 1973. Mineralization and gaseous loss of nitrogen and soil-applied liquid sewage sludge. J. Environ. Qual. 3:356–358.

King, L. D. 1979. Waste wood fiber as a soil amendment. J. Environ. Qual. 8:91–95.

King, L. D. 1980. Use of fermentation residue from citric acid production as a soil amendment. J. Environ. Qual. 9:443–447.

King, L. D., and R. L. Vick, Jr. 1978. Mineralization of nitrogen in fermentation residue from citric acid production. J. Environ. Qual. 7:315–318.

Knezek, B. D., and R. H. Miller. 1978. Application of sludges and wastewaters on agricultural land: a planning and educational guide. North Central Regional Research Pub. 235. Reprinted by USEPA, (MCD-35), Washington, D.C.

Magdoff, F. R., and J. F. Amadon. 1980. Nitrogen availability from sewage sludge. J. Environ. Qual. 9:451–455.

McCalla, T. M., J. R. Peterson, and C. Lue-Hing. 1977. Properties of agricultural and municipal wastes. p. 11–43. *In* L. F. Elliott et al. (ed.) Soils for management of organic wastes and waste waters. Soil Sci. Soc. of Am., Madison, Wis.

Nelson, D. W., and L. E. Sommers. 1979. Recycling antibiotic production wastes on cropland. p. 533–548. *In* Proc. 2nd Annual Madison Conf. of Appl. Res. and Practice on Municipal and Industrial Waste, Univ. of Wisconsin-Extension, Madison.

Ryan, J. A., and D. R. Keeney. 1975. Ammonia volatilization from surface applied wastewater sludge. J. Water Pollut. Control Fed. 47:386–393.

Ryan, J. A., D. R. Keeney, and L. M. Walsh. 1973. Nitrogen transformations and availability of an anaerobically digested sewage sludge in soil. J. Environ. Qual. 4:489–492.

Sagik, B. P., and C. A. Sorber. 1978. Risk assessment and health effects of land application of municipal wastewater and sludges. Center Appl. Res. and Technol., Univ. of Texas, San Antonio.

Sims, J. T., and F. C. Boswell. 1980. The influence of organic waste and inorganic nitrogen sources on soil nitrogen, yield, and elemental composition of corn. J. Environ. Qual. 9: 512–518.

Sommers, L. E. 1977. Chemical composition of sewage sludges and analysis of their potential use as fertilizers. J. Environ. Qual. 6:225–232.

Sommers, L. E., and D. W. Nelson. 1978. A model for application of sewage sludge on cropland. Proc. of the First Annual Conf. of Appl. Res. and Practice on Municipal and Industrial Waste, Univ. of Wisconsin-Extension, Madison.

Sommers, L. E., D. W. Nelson, and D. J. Silviera. 1979. Transformations of carbon, nitrogen, and metals in soils treated with waste materials. J. Environ. Qual. 8:287–294.

Sommers, L. E., D. W. Nelson, J. E. Yahner, and J. V. Mannering. 1972. Chemical composition of sewage sludge in selected Indiana cities. Proc. Indiana Acad. Sci. 82:424–432.

Sommers, L. E., C. F. Parker, and G. J. Myers. 1981. Volatilization, plant uptake and mineralization of nitrogen in soils treated with sewage sludge. Tech. Rep. no. 133. Purdue Univ. Water Resour. Res. Center, Purdue Univ., West Lafayette, Ind.

Sommers, L. E., and A. L. Sutton. 1980. Use of waste materials as sources of phosphorus. p. 515–544. *In* F. E. Khasawneh et al. (ed.) The role of phosphorus in agriculture. Am. Soc. of Agron., Madison, Wis.

Soon, Y. K., T. E. Bates, E. G. Beauchamp, and J. R. Moyer. 1978. Land application of chemically treated sewage sludge: I. Effects on crop yield and nitrogen availability. J. Environ. Qual. 7:264–269.

Sopper, W. E., and L. T. Kardos (ed.). 1973. Recycling treated municipal wastewater and sludge through forest and cropland. Penn. State Univ. Press, State College.

Stanford, G., and S. J. Smith. 1972. Nitrogen mineralization potential for soils. Soil Sci. Soc. Am. Proc. 36:465–472.

Stark, S. A., and C. E. Clapp. 1980. Residual nitrogen availability from soils treated with sewage sludge in a field experiment. J. Environ. Qual. 9:505–512.

Terry, R. E., D. W. Nelson, and L. E. Sommers. 1981. Nitrogen transformations in sewage sludge-amended soils as affected by soil environmental factors. Soil Sci. Soc. Am. J. 45: 506–513.

Terry, R. E., D. W. Nelson, L. E. Sommers, and G. J. Myers. 1978. Ammonia volatilization from wastewater sludge applied to soil. J. Water Pollut. Control Fed. 50:2657–2665.

Timm, H., W. J. Flocker, M. D. Akesson, and M. O. Bryan. 1980. Mineralization of soil-incorporated tomato solid waste. J. Environ. Qual. 9:211–214.

U.S. Environmental Protection Agency. 1981. Process design manual for land treatment of municipal wastewater. EPA 625/1-81-013. Center Environ. Res. Inf., Cincinnati, Ohio.

U.S. Environmental Protection Agency. 1983. Process design manual—land application of municipal sludge. EPA 625/1-83-016. Center Environ. Res. Inf., Cincinnati, Ohio.

U.S. Environmental Protection Agency. 1979. Criteria for classification of solid waste disposal facilities and practices: final, interim final, and proposed regulations. Fed. Regist. 44: 53438–53468.

Volz, M. G., and G. H. Heichel. 1979. Nitrogen transformations and microbial population dynamics in soil amended with fermentation residue. J. Environ. Qual. 8:434–439.

Watson, K. S., A. E. Peterson, and R. D. Powell. 1977. Benefits of spreading whey on agricultural land. J. Water Pollut. Control Fed. 49:24–35.

Wright, W. R. 1978. Laboratory and field mineralization of fermentation residues. J. Environ. Qual. 7:343–346.

15

D. R. Bouldin, S. D. Klausner, and W. S. Reid
Cornell University
Ithaca, New York

Use of Nitrogen from Manure[1]

In reviewing a subject that has been important for centuries, one supposes the reviewers must "reinvent the wheel." Lest this opening statement drive potential readers away, let us assure you that at present the "manure wheel" is far from round; in fact, it is four-sided and the hub is off-center. One of our objectives is to demonstrate that this is true. A second objective is to demonstrate the tremendous potential for improving manure management and that the incentives for improvement are developing rapidly.

As excreted, manure is a mixture of metabolic products such as urea and uric acid, living and dead organisms, and residues from the original feed (Salter & Schollenberger, 1939; Azevedo & Stout, 1974; Wilkinson, 1979). In most cases, 40 to 60% of the total N is in compounds such as urea and uric acid, which are readily hydrolyzed and/or decomposed to yield ammoniacal compounds. During these processes, the equilibrium partial pressure of NH_3 increases. Furthermore, when temperature conditions are favorable for microbial activity, N in some of the more complex organic compounds is converted to ammoniacal N.

The net effect is that within a few days, only about one-half of the total excreted N remains in insoluble organic forms. The ammoniacal forms are in solution and are easily lost in drainage water and by NH_3 volatilization. The losses can be prevented by rapid collection (within 1 h or less) of urine and feces and storage under conditions such that drainage of liquid and drying are prevented.

Ammonia volatilization continues during handling and storage and only ceases when the ammoniacal N is leached or incorporated into the soil. In the soil, the ammoniacal fraction will be converted to NO_3^- if temperatures are favorable for microbial activity. Subsequently, the NO_3^- may be lost by leaching and by denitrification associated with alternate wetting and drying. The potential for denitrification is enhanced because the organic matter in the manure furnishes energy for microbial activity.

[1] Journal Article no. 1419, Department of Agronomy, New York State College of Agriculture and Life Sciences, Cornell University, Ithaca, NY 14853.

Nitrogen in Crop Production.

In the soil, the remaining *organic* fraction decomposes and ammoniacal N is released. This mineralization is favored by the same moisture and temperature conditions as crop growth, and the rate of mineralization is approximately proportional to crop growth rates. The organic fraction of manure has many of the properties of the ideal N fertilizer—it is not subject to leaching or denitrification losses, it is not toxic to plants, and it mineralizes N at a rate dependent on the same climatic conditions as plant growth.

The foregoing discussion points out the many management variables determining the value of manure to crops. We will discuss: (i) the total N in manures produced in the United States, (ii) N balance sheets for animal production systems, watersheds, farms, and fields, (iii) management factors that reduce N losses, and (iv) manure management in the future.

I. TOTAL N IN ANIMAL MANURE IN THE UNITED STATES

Several methods of estimating the total N in animal manures in the United States have been used. Differences among the estimates result from choice of methods and data inputs. Examples are illustrated in Table 1 with laying hens and egg production.

The discrepancies in Table 1 arise from some or all of the following: (i) variation in loss of N from manure between time of excretion and analysis, (ii) inaccuracies in *Agricultural Statistics* (USDA, 1977), (iii) inclusion of more or less N from wasted feed and from bedding materials, (iv) inappropriate constant for conversion of feed N to manure N, (v) inappropriate constant for percent protein in feed, and (vi) inclusion or exclusion of feed N required to produce a laying hen.

Two common objectives of making estimates of N in manure follow. One objective is to construct a N balance sheet for the animal industry; in

Table 1—Estimates of N in manure from laying hens in the United States, 1975.

Basis of calculation	Reference	N, t $\times 10^6$
1. 0.43 kg N/layer per year	Gilbertson et al. (1979b)	
277×10^6 layers	USDA (1977)	0.12
2. 27 kg feed per 100 eggs	USDA (1977)	
17% protein in feed	Hashimoto (1974)	
64.8×10^9 eggs/yr	USDA (1977)	
80% of feed N in manure	Hashimoto (1974)	0.38
3. 0.7 g N in eggs/layer per day	Hashimoto (1974)	
277×10^6 layers	USDA (1977)	
20% feed N in eggs		
80% feed N in manure	Hashimoto (1974)	0.28
4. 45 kg feed/layer per year	USDA (1977)	
17% protein in feed	Hashimoto (1974)	
277×10^6 layers/yr	USDA (1977)	
80% of feed N in manure	Hashimoto (1974)	0.27
5. 0.57 kg N/layer per year	Azevedo & Stout (1974)	
277×10^6 layer per year	USDA (1977)	0.16

Table 2—Estimated amount of N in animal feed in the United States, and fertilizer N use, 1975.

Source	t × 10⁶	% N	N, t × 10⁶
High protein†	18.8	7	1.3
Concentrates‡	157	1.5	2.4
Hay	120	1.5	1.8
Silage§	112	1.0	0.2
Pasture‡	220	1.5	3.3
Total in feed			9.0
Total N in animal products¶			1.3
N in manure (by difference)#			7.7
Fertilizer N (1976)			9.4

† Expressed as 44% protein soybean meal equivalents.
‡ Expressed as corn equivalents.
§ 25% dry matter.
¶ N in liveweight of cattle, swine, and poultry slaughtered, and milk and eggs produced (Ensminger & Olentine, 1978; USDA, 1977).
# Based on *requirements* and maximum production rates, the fraction of feed N recovered in manure should be 0.63 for broilers, 0.75 for eggs, 0.78 for swine, and 0.76 for milk (Ensminger & Olentine, 1978). In actual production units, ratios observed are 0.8 for eggs (Hashimoto, 1974) 0.80–0.89 for cattle in feedlots (Gilbertson et al., 1971), 0.52 for broilers (Jensen, 1977), and 0.76 for dairy cows in New York (see following section). Azevedo and Stout (1974), on the basis of literature review, estimate the ratios to range from 0.7 to 0.8 and average about 0.75. In most production units, feed is spilled, and hence ends in manure without benefit to animals, animals and product are lost by various accidents, feed is used for raising production animals, over feeding, slowed rate of production, and improper formulation all increase feed inputs relative to requirements. Based on N in feed and N in product as shown above, the fraction is 0.85, which means about 15% of feed N is recovered in the animal products.

this case, N loss between excretion and analysis should be eliminated. A second objective is to estimate the amount of N that is available for recycling for crop-animal systems; in this case, typical losses should be included.

Table 2 contains estimates of total N in manure in the United States, based on estimates of N in animal feeds. Several approximations were used; the major uncertainties are associated with the N content of pasture and hay, two very large components of the balance. Note that total N in animal feed is almost equal to total N in fertilizers.

Several estimates of N in animal manures in the United States are presented in Table 3. The estimates based on Gilbertson et al. (1979b) include some losses. The estimate based on Taiganides and Stroshine (1971) may be higher than the others because a very large fraction (70%) is associated with beef production; the others estimate less total N produced on rangeland by brood cows and calves.

Table 3—Estimates of N in animal manure.

Reference	N, t × 10⁶
Taiganides & Stroshine (1971)	10.5
Lauer (1975)	7.4
Table 2	7.7
Gilbertson et al. (1979b) & USDA (1977)	6.3
Yeck et al. (1975)	5.3

The major point in Table 3 is that the N in animal manures is approximately 75% of the fertilizer N used annually. As we will demonstrate in following sections, much of the manure N is not recycled through farming; we will discuss some problems and opportunities relative to substitution of more of the manure N for fertilizer N.

II. N BALANCE FOR FARM UNITS

In the following sections, selected data are examined with the objective of estimating recycling of manure N in farming units. In most cases the data are incomplete and we arrive at ambiguous answers. Even with these uncertainties, the preponderance of evidence supports the conclusion that large amounts of manure N are lost from farming in the United States each year.

A. N Balance of Food Production in United States

The national N balance for agriculture illustrated in Fig. 1 is an approximation based on *Agricultural Statistics* (USDA, 1977). A portion of the balance is based on the assumption that 50% of the N in soybeans [*Glycine max* (L.) Merr.] is obtained from N_2 fixation, the remainder coming from soil N and residual fertilizer N applied to corn (*Zea mays* L.) in a

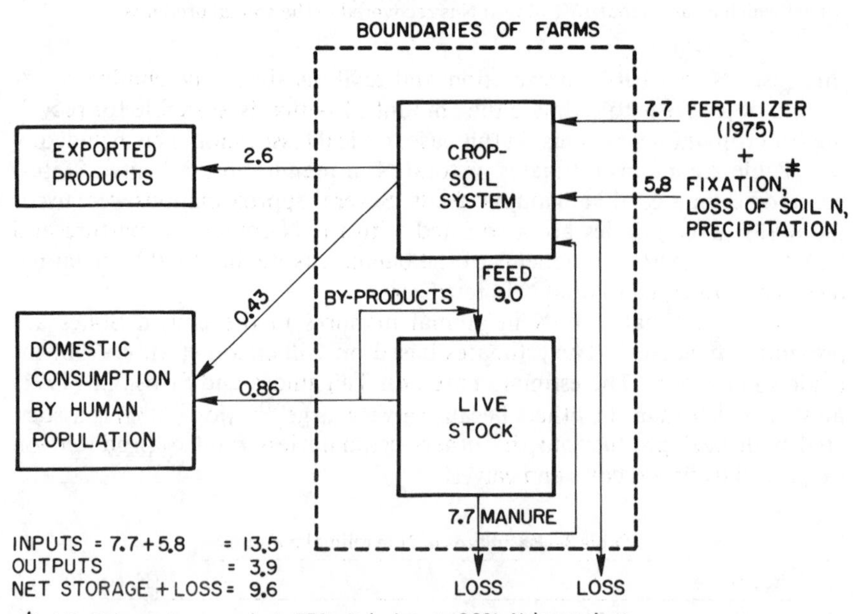

Fig. 1—Nitrogen balance for farming in the United States. Numbers are million tonnes N/yr.

corn-soybean rotation; 75% of the N in hay, and all of the N in pasture, is obtained from N_2 fixation and the soil. The most interesting statistic is the difference between annual input of 13.5 million tonnes of N and output of 3.9 million tonnes of N—on the order of 9.6 million tonnes of N is lost as NO_3 to groundwater; lost to the atmosphere as N_2, N_2O, and NH_3; or stored in the soil as organic N. There is no particularly convincing evidence on the net change in organic N in farmed soils; based on a variety of evidence and our judgements, we hypothesize that most of the 9.6 million tonnes is not stored in the soil, but is lost from farmland as NO_3^-, N_2, N_2O, and NH_3.

We consider now the amount of N from manure that is recycled, since this is a very large component of the overall balance and approximately equals 75% of the N from fertilizer. We will not try to derive a national figure for recycling of manure N; rather, we will illustrate the situation with specific examples.

B. N Balance for Animal Production Units

1. N BALANCE OF THE FALL CREEK WATERSHED

The Fall Creek Watershed lies in central New York, and it is a major tributary of Cayuga Lake. It is representative of central New York with respect to climate, geology, and land use. The total area is 330 km^2 (80 000 acres), and approximately 40% is farmed. Most of the farming is devoted to corn, hay, small grains, and pasture for 6800 dairy cows in the watershed. Other details on the watershed can be found in Porter (1975).

The estimated inputs and outputs of N are summarized in Table 4 for the watershed as a whole, and in Table 5 for the animal portion. The combined inputs are 1200 t/yr, and measured outputs are 480 t/yr. Thus, 720 t/yr must be accounted for by storage, NH_3 volatilization, and denitrification (Table 4).

In the 60% of the land not farmed, the inputs in precipitation (Table 4) are about 9 kg N/ha per year, and outputs as NO_3^--N are about 1 kg N/ha per year; the unaccounted-for N on nonfarmed land is equal to 150 t/yr. We hypothesize that 50% of the N from the human food input is denitrified, and the remainder is leached into streams as NO_3^-. Thus, 720–35–150 leaves about 535 t N/yr, which is the unaccounted-for N associated with farming. Most of this unaccounted-for N must be associated with about 7000 ha of intensively farmed land, since the remainder of the farmed land received very small amounts of fertilizer and manure.

With respect to storage in organic forms, in farmed soils, the following considerations are pertinent:

1) Most farmers have about 1 ha/cow of intensively farmed land that is rotated between corn and alfalfa-grass hay (*Medicago sativa* L.). The bulk of the manure and fertilizer is applied to this land. Thus, if storage is a major component of the unaccounted-for N, it must occur on this land.

Table 4—Nitrogen balance, Fall Creek Watershed.

Source	N, t
Inputs from outside watershed + legume fixation inside watershed	
Precipitation	300
Estimated as 8.8 kg NO_3^--N + NH_4^+-N in precipitation (Likens, 1972).	
Fertilizer N	270
Unpublished data, based on survey of 20% of farms in the watershed. (J. J. Jacobs and G. L. Casler. Dep. of Agricultural Economics, Cornell Univ., Ithaca, N.Y.)	
Purchased feed	320
Unpublished data, based on survey of 20% of farms in the watershed. (J. J. Jacobs and G. L. Casler. Dep. of Agricultural Economics, Cornell Univ., Ithaca, N.Y.)	
Purchased feed (human diet)	70
Estimates from average per capita consumption of protein, given in *Agricultural Statistics*, 1972.	
Legume fixation	260
Estimated as 50% of the N content of the legume and grass hay harvested in the watershed.	
Total	1200
Outputs from the watershed	
Milk	190
Estimated from milk sales figures.	
Meat	50
Estimated as 2500 animals (about 33% of the dairy cows) at 675 kg (1500 pounds) each, 17% protein.	
NO_3^--N in Fall Creek	180
Measured.	
Particulate N in suspended solids, Fall Creek	60
Estimated from 14 000 t of suspended solids as 0.4% N.	
NH_3 volatilization + denitrification + storage	720
Calculated as difference between inputs and known outputs.	
Total	1200

2) Most of the intensively farmed land has been farmed for at least a century, and it has received most of the manure and grown most of the legumes. It is probably approaching a steady-state level of organic N.
3) The unaccounted-for N is 75 kg/ha per year, averaged over the intensively farmed land. Over a period of 25 yr, this amounts to about 0.1% N for the surface 15 cm of soil; since the soil contains 0.2 to 0.3% total N presently, it seems unreasonable to suppose that the organic N content has doubled over the last 25 to 40 yr. Therefore, storage in soil is unlikely to amount to an appreciable fraction of the unaccounted-for N.

Available evidence supports the hypothesis that about 50% of the total N in the manure is lost by NH_3 volatilization (Gracey, 1979; Hermanson et al., 1980; Hoff et al., 1981; Kolenbrander, 1981b; Lauer et al., 1976; Luebs et al., 1973, 1974; Reddy et al., 1979; Steenhuis et al., 1979). Because 960 t N is contained in the manure, the loss by NH_3 volatilization accounts for approximately 480 t N (Table 5).

Table 5—Nitrogen balance for farm animals in Fall Creek Watershed.

Source	N, t
N in feed	
Purchased	320
Unpublished data, based on survey of 20% of farms in the watershed. (J. J. Jacobs and G. L. Casler. Dep. of Agricultural Economics, Cornell Univ., Ithaca, N.Y.)	
Feed grown in watershed	
Unpublished data, based on survey of 20% of farms in the watershed. (J. J. Jacobs and G. L. Casler, De. of Agricultural Economics, Cornell Univ., Ithaca, N.Y.)	
Corn silage	200
Corn grain	70
Legume hay	340
Grass hay	190
Pasture	80
Total	1200
Animal products	
Milk	190
Estimated from milk sales figures.	
Meat	50
Estimated as 2500 animals (about 33% of the dairy cows) at 675 kg (1500 pounds) each, 17% protein.	
Manure	960
Manure as total N in feed minus N in milk and meat.	
Total	1200

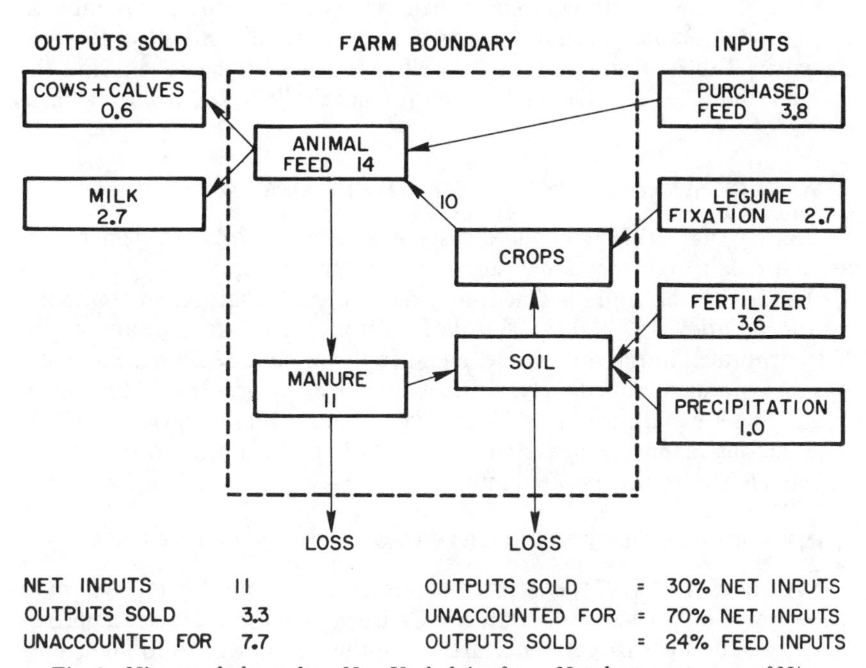

Fig. 2—Nitrogen balance for a New York dairy farm. Numbers are tonnes of N/yr.

Table 6—Fraction of the excreted N in dairy cow manure that is applied to the field, incorporated into the soil, and costs. Based on Safley et al. (1976) and Safley (1977).

Management system	Excreted N		Capital costs†	Operating costs†
	To field	Incorporated in soil		
	%	%	$ per cow	$ per cow per year
1) Alley flush, aerobic lagoon, liquid spreading, incorporated in soil 1-3 weeks after spreading	40	20	900	181
2) Free stall or stantion, daily cleaning and hauling, spring incorporation	60	40	180	88
3) Slotted floor with concrete pit below flood, yearly cleaning, incorporation in soil 1-3 weeks after spreading	60	40	930	180
4) Alley scraper, manure pump, pit storage, yearly cleaning, immediate incorporation	80	80	675	220

† 100-cow dairy, 1 yr storage.

2. N BALANCE FOR A NEW YORK DAIRY FARM

As a further confirmation of the results of the Fall Creek study presented above, a N balance for a "typical" New York dairy farm is presented in Fig. 2 (75 cows + 50 replacement heifers). The results are comparable to the national picture presented in Fig. 1 and the results in Fall Creek presented in Tables 4 and 5; that is, well over 50% of the N entering the farming unit(s) is unaccounted for and it probably is lost from the farm units.

3. MANURE MANAGEMENT SYSTEMS FOR DAIRY COWS

Safley et al. (1976) compared costs of several manure management systems for dairy cows and measured the amount of excreted N that was applied in the field and the fraction of the applied N that was in ammoniacal form (Safley, 1977). Table 6 is derived from these data. The amount of N incorporated in the soil was set equal to the organic N applied, unless it was incorporated immediately; if the manure was incorporated immediately, it was set equal to total N. These data demonstrate remarkable differences among management systems in terms of costs and the fraction of the excreted N that is incorporated into the soil.

3. INTENSIVE DAIRY PRODUCTION IN SOUTHERN CALIFORNIA

Adriano et al. (1971) describe the intensive dairy production system in the Chino-Corona basin in southern California where there are approximately 20 cows per irrigated hectare. With this ratio of cows to land, they estimate only about 20% of the manure N is recycled through crops.

5. BEEF CATTLE FEED LOTS IN NEBRASKA

Gilbertson et al. (1971) reported results from unpaved beef feed lots in Nebraska, in which they measured N inputs in feed and outputs of N in animal products and manure. The lots were cleaned at intervals ranging from 112 to 203 d. On the average, the manure N excreted was 75% of feed input, and the amount of manure N at cleaning relative to the amount excreted was about 45%, and ranged from 15 to about 70%, depending upon weather and management variables. About 5% of the N was lost in runoff. Probably most loss was the result of NH_3 volatilization.

6. GRAZING ANIMALS

Whitehead (1970) summarized N losses from grazed lands. He assembled probable values for the various components in N balance sheets for grazing animals on grass swards fertilized with N and in grass/legume swards. For a fertilized grass sward, he estimated that when 134 kg/ha of fertilizer N was applied, the output of N in milk was 27 kg N/ha, the net increase in soil N storage was about 40 kg N/ha, and losses by NH_3 volatilization, denitrification, and leaching totaled 75 kg N/ha. Somewhat smaller losses were projected for the grass/clover sward.

7. N BALANCE FOR CAGE LAYERS IN "HIGH RISE" HOUSES

In the "high rise" systsem, the manure from cage layers drops directly into a pit. Fans exhaust air over the pits, which dries the manure and dissipates NH_3, which volatilizes. Usually, the manure is removed from the pits once each year. Sobel (1976) studied the N balance over a 3-yr period and found that 36% of the manure N was recovered at the end of the year; presumably, 64% of the N was lost by NH_3 volatilization during the year.

8. GENERALIZATIONS

Gilbertson and colleagues (Gilbertson et al., 1979a, b) and Vanderholm (1979) have generalized the various losses under a wide range of conditions and management systems between time of excretion and hauling to the field. In addition, Van Dyne and Gilbertson (1978) and Lauer (1975) have estimated the amounts of N that they judged to be available for recycling through the crop-production systems.

C. N Balance in Field Experiments where Manure has been Applied

Summarized in Table 7 are several experiments where a reasonable N balance can be made and where manured treatments can be compared with a no-manure and no-fertilizer N treatment. The numbers in the table represent the net effect of manure relative to a no-manure, no-fertilizer N treatment on soil organic N and crop removal. The striking aspects of the data are (i) appreciable, cumulative effects of manure on soil organic N are evi-

Table 7—Summary of N balances in selected field experiments with manure.

Reference	Period	N added	Applied N		
			Accumulated in soil	In crop	Unaccounted for
			%		
Jenkinson & Johnston, 1976, 'Hoosfield' continuous barley	1852–1972	27.2 t/ha	15	25	60
Johnston, 1975, 'Stackyard Field' continuous wheat	1877–1926	5.5 t/ha	30	30	40
Jenkinson, 1976, 'Broadbalk' continuous wheat	1852–1972	25.7 t/ha	13	26	60
Pratt et al., 1976, dry manure	1970–1974	4 t/ha	45	15	40
Pratt et al., 1976, liquid manure	1970–1974	4 t/ha	20	20	60
Magdoff, 1978, continuous corn, three manure rates, Panton clay†	1965–1970	0, 350, 700, 1050 kg/ha	76	10	14
Magdoff, 1978, continuous corn, three manure rates, Calais loam‡	1970–1975	0, 460, 920, 1380 kg/ha	30	14	56
Oveson, 1966, wheat-fallow, 10 t "strawy" manure/acre per 2 years	1931–1964	1500 kg N/ha	40	15	35
Klausner et al., 1979, continuous corn, liquid manure (Unpublished, Cornell Univ.)	1975–1977	400 kg N/ha	0	40	60
Mattingly, 1973, arable crops	1965–1971	1580 kg N/ha	40	10	50
Webb et al., 1980, wheat (manure applied ≅ every 4 yr)	1899–1979	1240 kg N/ha	0	24	76
Herron & Erhart, 1965, irrigated grain sorghum§	1960–1963	0, 350, 700 kg N/ha	55	35	10

† Typic Ochraqualfs.
‡ Typic Fragiorthods.
§ Grain sorghum, *Sorghum bicolor* (L.) Moench.

dent where large amounts of manure are added over long periods of time, and (ii) large amounts (50%) of the manure N cannot be accounted for. Thus, even though one is tempted to hypothesize that the organic N in manure mineralizes more or less in concert with plant needs for N, there is still evidence of large losses of N. In some cases, the amount of inorganic N available to the crop exceeded the ability of the crop to utilize N and the excess was lost by leaching and denitrification. In many cases, the manure was managed such that NH_3 was volatilized before incorporation into the soil and/or the manure was applied in advance of active crop uptake; hence, the leaching and denitrification losses of the inorganic N were excessive.

D. Summary of N Balance

The examples all illustrate a large loss of N from the animal production systems. They illustrate large losses between time of excretion and incorporation into the soil, and they illustrate large losses of the N actually applied to the soil. In Fig. 1, the national N balance indicates that about 9.6 million tonnes of N cannot be accounted for each year. Probably about 75% of the manure N is lost each year, which means that about 50% of the unaccounted-for N in the national balance sheets can be attributed to losses in the manure portion of the N balance sheet.

Considering the overall balances for food production in the United States, we cannot account for 70% of the annual N inputs. For the dairy production system in Fall Creek (with some feed imported from outside the watershed), the corresponding figure is about 45%. For dairy farms in New York, the value is about 70%.

Considering the recovery of feed N in product and the manure removed, about 45% of feed N cannot be accounted for in beef feedlots, and for cage layers in high-rise houses with yearly cleaning, the value is about 45%. For dairy manure management systems in New York, the percent loss of excreted N ranges from 20 to 60% before field spreading, and after field spreading losses continue. In southern California, 20% or less of the excreted N from dairy cows is recycled through crops. With these dismal figures as a background, we now proceed to discuss manure management and recycling of manure N through farming systems.

III. CROP RESPONSE

A. Factors Influencing Crop Response

Manure is a source of all plant nutrients and it is a soil conditioner. Under many experimental conditions, as well as on-farm situations, the crop response to manure cannot be attributed entirely to N. However, very often the major effects are associated with N. With these reservations in mind, we illustrate some generally accepted practices, which enhance crop response to manure, probably in large part because of enhancement of N supply. We rely heavily on the summary of Salter and Schollenberger (1939), and the interested reader should consult their paper for more details.

Table 8—Nitrogen content of greenhouse crops as influenced by solid and solid plus liquid portions of manure (Heck, 1931).

Manure source	pp2m N added	Increase in N content of crops, pp2m of soil: First crop	Next three crops	Total
	Fresh cow manure			
Solid only	31	0.6	2.0	2.6
Solid and liquid	66	14	3.5	17.5
	Fermented cow manure			
Solid only	31	1.2	5.0	6.2
Solid and liquid	66	10	3.2	13.2

Table 9—Effect of date of application of manure on next crop of mangels. Average of 16 yr of data from Denmark, quoted by Salter & Schollenberger (1939, p. 36).

Application date	Relative effectiveness for next crop
15 October	58
15 December	84
1 February	83
1 March	75
15 April	100

The data in Table 8 illustrate that conservation of the urine or liquid fraction of manure clearly enhances plant response to manure because (i) it increases the N content of the manure, and (ii) the liquid fraction contains inorganic N, which is immediately available, while the organic fraction alone becomes available over many cropping periods.

Conservation of inorganic N between time of excretion and application to the soil will be of little benefit if the manure is applied so far in advance of crop demand for N that leaching and denitrification losses have dissipated most of the inorganic N that was conserved. This is illustrated by the data in Table 9, which shows considerable increase in plant response from manure as the time of application was moved closer to the period of most rapid plant uptake. More recently, Klausner and Guest (1981) compared topdressing of manure with sidedress injection during the early stages of corn growth. The advantage of injection was equivalent to 60 kg of fertilizer N for 40 t/ha of manure.

Some manure storage conditions are so poor that equivalent losses of N take place whether the manure is stored or spread in the field. Under this situation, it would not make any important difference when the manure was spread. Similarly, conservation of N between time of excretion and application in the field is of little benefit if the manure is left on the surface long enough so that the inorganic N in the manure is depleted by NH_3 volatilization. This is illustrated by Table 10 and the several references cited earlier (section IIB).

The residual effects would probably not be changed appreciably by methods of collection, handling, and storage. These effects are dependent upon the decomposition over many years of the more stable organic fractions, and the behavior of these fractions is not influenced to a great extent by storage, season of application, or delays in incorporation of manure into the soil.

Table 10—Relative effectiveness of manure in increasing crop yields as influenced by time between spreading and plowing under [Salter & Schollenberger (1939, p. 35)].

Treatment	Relative value in increasing crop yields (average of 34 experiments)
Manure plowed under immediately	100
Manure plowed under 6 h after spreading	85
Manure plowed under 24 h after spreading	73
Manure plowed under 4 d after spreading	56

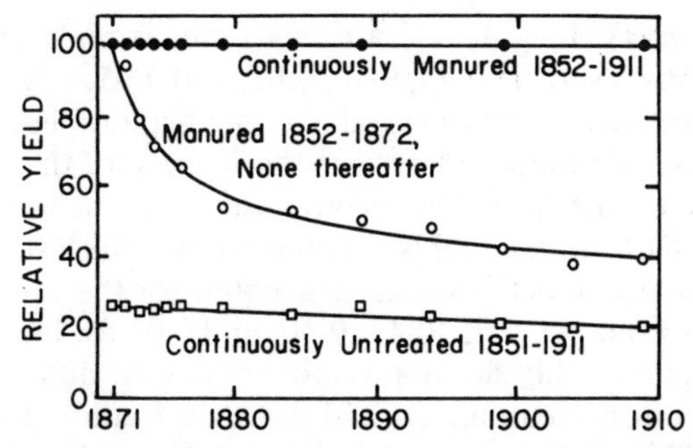

Fig. 3—Relative yields of wheat over time illustrating the long residual effects of manure (Salter & Schollenberger, 1939).

The residual effects of manure are appreciable, and they last for many years when large amounts of manure have been added. Perhaps the best illustration of this is Fig. 3, which is reproduced from Salter and Schollenberger (1939).

B. Interpretation of Data

Since the review by Salter and Schollenberger in 1939, the following factors have materially changed, and these changes have had important implications for manure management.

1) Fertilizer N has become readily available and during the last 20 yr it has been inexpensive in relation to costs of manure management.
2) Yields have increased greatly and there have been corresponding increases in the amount of N that must be supplied in order to achieve these yields.
3) Some livestock operations have become very large. They use purchased feed, and the source of feed and manure are separated by large distances, which makes recycling impractical. The manure from many large livestock operations is considered a disposal problem rather than a valuable source of nutrients. Particularly troublesome is the potential for water pollution.

The major effect of the foregoing developments has been to focus attention on the amount of manure N mineralized each year, on the amount of N that manure furnishes to crops, and on the amount of fertilizer N that manure can replace. These are somewhat different questions than those posed by previous generations, and the following developments reflect these changing concerns.

1. DECAY SERIES

The concept of a "decay" series has been developed by Pratt and his colleagues (Pratt et al., 1973, 1976). This concept recognizes two major factors: (i) the variable amount of inorganic N in different manures that is available to the first crop, and (ii) the gradual mineralization of the organic

N over several years. In practice, a decay series is represented by several consecutive numbers such as 0.45, 0.10, 0.05, and 0.05, where the first number represents the fraction of the total N in the manure that mineralizes the first year, the second number represents the fraction of the N residual from the first year that mineralizes the second year, etc. The last number represents the fraction of the residual N that mineralizes the last and all succeeding years. The first number is varied to account for the variable amount of inorganic N; for example, 0.75, 0.15, 0.10, and 0.05 for liquid manure high in inorganic N and readily decomposable organic N, and 0.45, 0.10, 0.05, and 0.05 for manure accumulated and dryed in feedlots before spreading (Pratt et al., 1976). This concept has been elaborated on or modified by Powers et al. (1975); Gilbertson et al. (1979a); Mathers and Goss (1979); Magdoff (1978); and Castellanos and Pratt (1981).

Kolenbrander (1981a) and Sluijsmans and Kolenbrander (1977) describe a simpler system based on three fractions in manure, which can be measured chemically, namely (i) an inorganic fraction, (ii) a fraction that decomposes the first year, and (iii) a fraction that will only mineralize slowly (over several years).

There is no entirely satisfactory procedure for measuring the parameters in the decay series directly; therefore, parameters have been hypothesized, and then the accumulation of N in soil based on the hypothesized parameters is compared with the measured accumulation (Pratt et al., 1976; Mathers & Goss, 1979), or else the amount of fertilizer N that is replaced by the manure is estimated (Powers et al., 1975). In neither case is there any large body of data that can be used to document any given set of parameters. Several sets of parameters are listed by Pratt et al. (1973) and by Gilbertson et al. (1979a). However, the parameters of these series are based on judgments, a minimum of experimental data, and probably only apply to the fairly specific situations for which they were developed (Sluijsmans & Kolenbrander, 1977). Basically, the decay series concept and its elaborations far exceed the experimental data available for their quantitative verification. Surely, no one would argue seriously with the validity of the concept (or some modification); however, the actual numbers to be used in the decay series and their variability among years, soils, cropping systems, and sources of manure still does not advance the art of management on the farm as much as one might hope. However, they are a major development because they provide a logical framework for more experimental work and for summarizing available data.

2. FERTILIZER N EQUIVALENTS

The mineralization series as formulated by Pratt et al. (1973) does not take into account leaching, denitrification, and NH_3 volatilization losses once mineralization has occurred. Hence, plant uptake, the ultimate objective of manure applications, is one step removed from the mineralization series. This could be accomplished by adjusting the coefficients in the decay series to reflect losses or to estimate losses as proposed by Gilbertson et al. (1979a) and Powers et al. (1975).

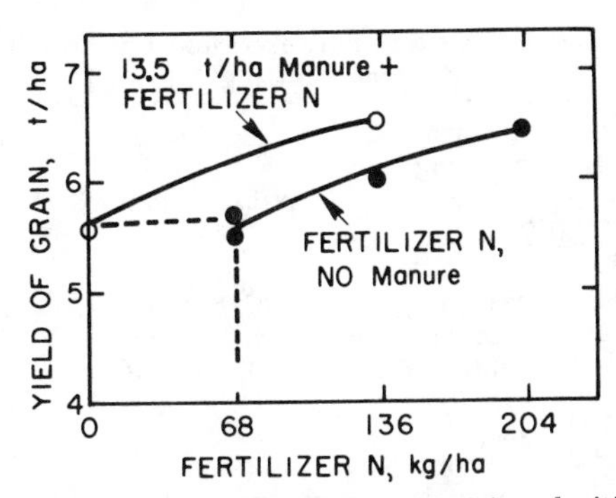

Fig. 4—Response of grain corn to fertilizer N with and without manure.

An appealing hypothesis is that mineralization rates are primarily a function of the properties of the manure and that losses are primarily the result of management variables such as timing of application, length of time the manure remains on the surface, etc. According to this hypothesis, the behavior of the sources is independent of soil properties. Careful review of Table 7 illustrates that there are large differences among soils with respect to mineralization rates, and with respect to unaccounted-for losses. Therefore, the mineralization series and losses are very much dependent upon an array of soil, climatic, and cultural variables, as well as properties of the manure and management variables such as timing, methods of application, etc.

Another method of evaluating manure is to compare it in terms of fertilizer N equivalents (Herron & Erhart, 1965; Klausner & Guest, 1981).

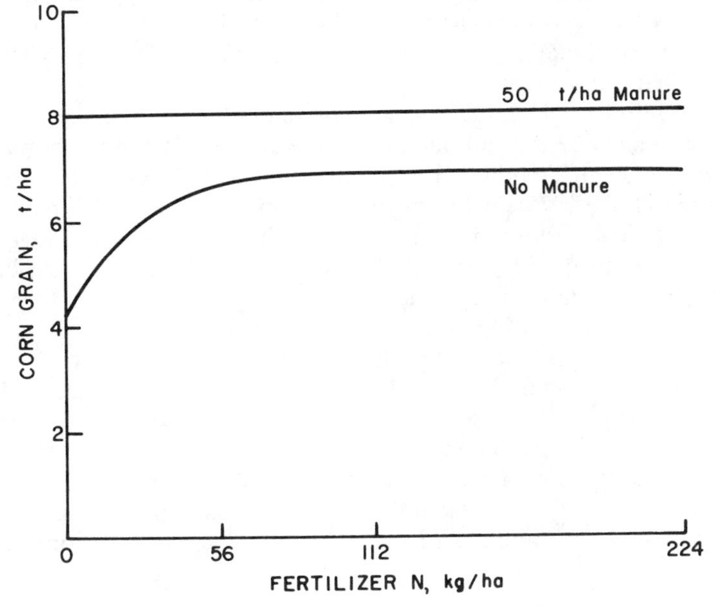

Fig. 5—Response of corn to fertilizer N with and without manure.

Table 11—Summary of data from a manure experiment at Rothamsted, Harpenden, Herts, England (Mattingly, 1973).†

Manure added in autumn of	Cumulative manure N added	Crop in following summer‡	$\Delta N_P/\Delta N_F$§ Fertilizer	Manure	ΔN_P from manure¶	Fertilizer N equivalents of manure
	kg/ha					kg/ha
1965	293	Barley	0.31	0.27	6	22
1966	640	Potatoes	0.32	0.31	16	52
1967	826	Winter wheat	0.48	0.37	34	92
1968	1128	Sugarbeets	0.67	0.78	48	62
1969	1379	Beans	--	--	22	--
1970	1582	Rye	--	--	29	--

† From autumn of 1965 to autumn of 1971 the soil in plots receiving only fertilizers decreased 70 kg N/ha and the soil receiving manure increased 585 kg N/ha. Approximately 40% of the applied manure N accumulated in the soil, approximately 10% was recovered in increased plant N (in aboveground dry matter) and approximately 50% cannot be accounted for. Approximately 43% of the fertilizer N was recovered in the aboveground dry matter over the initial 4 yr. Approximately 16% of the N mineralized from the manure was recovered in aboveground dry matter.

‡ Potatoes, *Solanum tuberosum* L.; sugarbeets, *Beta vulgaris* L.; beans, *Phaseolus* sp.; rye, *Secale cereale* L. Other plants identified in the text.

§ Increase in N content of aboveground dry matter per unit of fertilizer N.

¶ Increase in N conent of aboveground dry matter in manured plots relative to that in plots receiving only inorganic fertilizer.

That is, estimate the amount of fertilizer N that 1 t of manure replaces in terms of crop production. An example is afforded by the results in Fig. 4, where N response curves for fertilizer N alone and in combination with manure are illustrated (Baldock & Musgrave, 1980). Application of 13.5 t/ha of manure was equivalent to 68 kg/ha of fertilizer N. Evidently, the only property of manure in these experiments that is influencing yield is the N furnished by the manure.

The fertilizer N equivalents of manure cannot be calculated when the limiting yields with and without manure are different. Figure 5 shows such a case where the differential in maximum yield was 1100 kg/ha of corn grain (Klausner et al., 1979). Additionally, the yield at zero N with manure exceeded the yield without manure and 224 kg/ha of fertilizer N. Adequate P and K were added with fertilizer. Manure is contributing to yield in ways other than a N contribution. The results of Herron and Erhart (1965) suggest the same type of behavior in their experiments.

Some of the complexities of data analysis are illustrated by a study at Rothamsted, England (Mattingly, 1973). A summary of the results are presented in Table 11. In this study, main plots received 50 t/ha per year of manure ($\approx$10 t/ha per year dry matter) and a companion series of main plots received an equivalent amount of P, K, and Mg in the form of inorganic fertilizers. Each of the main plots was split, and one of four quantities of fertilizer N was applied to the subplots. The same treatments were applied to the same plots in succeeding years. The sequence of crops grown is shown in Table 11.

The N contents of the aboveground dry matter of crops were approximately linear functions of amount of fertilizer N added (as "nitrochalk" in the spring), and the increase of N in aboveground dry matter (ΔN_P) as a function of fertilizer N added (ΔN_F) is shown in Table 11 under the heading $\Delta N_P/\Delta N_F$. Note that there was much more variation in $\Delta N_P/\Delta N_F$ among years than between main plots receiving either inorganic sources of P, K, Mg (column headed "Fertilizer"), or these sources of nutrients from manure (column headed "Manure"). In the next column (ΔN_P from manure), the differences in N content of the aboveground dry matter between "Fertilizer" and "Manure" main plots are shown to vary considerably among years. Finally, in the last column, the "fertilizer N equivalents" of the manure is shown (calculated as the ratio of ΔN_P from manure to $\Delta N_P/\Delta N_F$ for manure plots).

Some of the conclusions that can be drawn are the following:

1) The N mineralized from the manure averaged about 60% (950 kg N/ha) of the applied N, since only 40% remains in the soil. Of this, approximately 16% (155 kg N/ha) was recovered in the aboveground dry matter. In the first 4 yr of the experiment, the N in the aboveground dry matter of the crops was increased by an amount equivalent to about 45% of the fertilizer N applied. Hence, the inorganic N mineralized from the manure was only about one-third as effective as N supplied as "nitrochalk." Since the differences in $\Delta N_P/\Delta N_F$ between "fertilizer" and "manure" plots in any given year was small relative to this difference, the effect was probably not differential denitrification during the cropping season. The low recovery of mineral N from manure may result from fall application; probably the inorganic N in the manure in the autumn and any that mineralized during the winter was lost (recall that the fertilizer N was applied in the spring and did not undergo the overwinter losses).

2) The effectiveness of the N fertilizer varied widely among years (from 0.31 in 1965 to 0.67 in 1968). Presumably, this is a consequence of differential leaching and denitrification among years. The fertilizer N equivalents is calculated as the quotient of ΔN_P from manure and $\Delta N_P/\Delta N_F$. In a practical sense, this is the amount of fertilizer N replaced by the manure and it is one measure of the economic value of manure. As an average over the 4 yr, application of 50 t/ha wet manure per year increased plant uptake by an amount equivalent to about 50 kg fertilizer N/ha, or approximately 1 kg fertilizer N/t of manure. Conceptually, one might argue that the mineralized N from manure and fertilizer N undergo similar losses and that the "fertilizer N equivalents" represents a reasonable approximation to the actual amount of N mineralized during the year. This seems unlikely for reasons outlined above. However, there may be some situations where such an argument would be an acceptable approximation.

3) Examination of the ΔN_P from manure or the fertilizer N equivalents indicates the difficulty of finding a decay series that will be consistent with these two parameters, although it would be quite easy to find a series consistent with the fraction of the added N that has been accumulated in the soil.

IV. SUMMARY OF PRESENT SITUATION

The evidence is overwhelming that at least 50% of the manure N is not recycled through the farming system, and there is reasonable evidence that no more than 25% of the manure N from most feedlot dairy and poultry operations is recycled (50% is lost between time of excretion and spreading in the field, and 50% of what is spread in the field is lost).

Management variables such as storage conditions, timing of application, incorporation following spreading, and quantity of manure are major factors determining the amount of manure N that will be recovered in crops. These management variables have an important influence on the response of the first crop following application, but they are expected to have a small influence on the residual effects.

Many livestock production systems (e.g., open feedlots, high-rise cage laying operations) do not lend themselves to easy conservation of N between excretion and spreading. Very high concentrations of livestock operations make efficient recycling difficult, because the manure produced exceeds the cropland within reasonable hauling distances.

In the period prior to about 1940, many experiments with manure were concerned with how to obtain maximum benefit in terms of crop production. However, the effects of nutrients other than N often make interpretation difficult. Since 1940, many experiments have been primarily concerned with disposal rather than effective recycling of nutrients. Most recent experiments have been primarily concerned with evaluating what farmers are doing rather than demonstrating (or evaluating) the impact of improved management systems on the recycling of manure N. The experimental treatments are hardly ever innovative. This should not be interpreted to mean the experiments were useless; most actually achieved the objective of giving the farmer good information about the value of the manure as he is presently managing it. They also furnish good evidence about how effectively farmers are recycling nutrients, since experimental procedures simulate farmer operations closely.

Characterization of the manure actually spread, and details on how the cultural operations were managed once the manure was spread are almost universally missing from descriptions of the experimental methods. This makes useful comparisons among experiments impossible.

V. SOME FUTURE CONSIDERATIONS

A. Incentives for Substitution of Manure N for Fertilizer N

1. ENERGY CONSERVATION

Production and distribution of fertilizer N requires about 18 000 kcal/kg (Pimental et al., 1973). This is roughly the amount of energy contained in 1.9 L of gasoline. In 1979, farmers used about 9.6×10^6 t fertilizer N,

Table 12—Energy required for producing corn by conventional and no-tillage systems (Phillips et al., 1980).

	Energy required per hectare per year	
	Conventional tillage	No tillage
	kcal × 10^3	
Machinery	237	195
Supplies other than N	323	358
Cultural operations	2210	1890
Fertilizer N	1950	1950
Total	4720	4390

which required about 173 × 10^{12} kcal of fossil fuel. White (1978) estimates that total on-farm energy use was about 3% of the national total and that < 1% of the national total was used to produce all fertilizers.

Comparison of the amount of manure N lost from farms and fertilizer N use reveals that lost N is in the range of 25 to 75% of fertilizer N. Therefore, simultaneous reduction of manure N loss and substitution of this N for fertilizer N may conserve energy.

In the following example, we compare the possible impact of this substitution with tillage reduction in corn production. As shown in Table 12, no-till required 93% as much energy as conventional tillage. However the fertilizer N required 44% of the total fossil energy input into the no-tillage system. The point we wish to make is that reduction of tillage is a small percent of total energy use; if energetically efficient ways can be found to substitute manure N for fertilizer N, then there are possibilities for reductions of energy requirements.

Another aspect of energy conservation through recycling of manure N lies in reduced dependence on imports. In the future, the energy conserved through substitution of manure N for fertilizer N reduces our dependence on imported fossil fuel. If NH_3 is purchased from areas that now have plentiful natural gas or excess NH_3 capacity, then NH_3 production capacity within the United States will be reduced, and we will become dependent upon another import.

2. ECONOMIC

Energy is the major cost item in fertilizer N production. Natural gas is the usual source, and thus irrespective of increases in costs of capital, labor, transportation, storage, etc., cost of NH_3 will increase with cost of natural gas (Douglas, 1981). Each increase in cost of fertilizer provides more incentive for substitution of manure N for fertilizer N. Enhanced utilization of manure N will likely increase farm operating costs and energy inputs, but perhaps not as much as the costs and inputs of fertilizer N.

3. ENVIRONMENTAL

Based on NO_3^--N in water, one judges that only a small fraction of the unaccounted-for N from manure ends up as NO_3^--N in surface water. Most appears to be lost as NH_3 to the atmosphere. Presumably, the NH_3 is dispersed in the atmosphere and is effectively diluted. Perhaps it even serves a very useful function of neutralizing some of the acidity in rain.

Despite the fact that most unaccounted-for manure N does not end in groundwater as NO_3^-, it should not be taken to mean that NO_3^--N from manure is never a problem on a local scale. Such a potential problem is illustrated by the data of Pratt et al. (1976).

Conservation of more manure N would undoubtedly entail immediate incorporation of manure in the soil just prior to active crop uptake. This means less P, organic matter, and other constituents will be carried to streams and lakes by surface runoff.

B. Constraints

In the preceding three subsections, some incentives toward increasing substitution of manure N for fertilizer N were listed. In this section we will deal with some general constraints.

1. CONSERVATION OF NH_3 BETWEEN EXCRETION AND INCORPORATION IN SOIL

There is no known method of conserving manure N when animals are pastured. There is no obvious feasible technology for conserving NH_3 in unpaved, open feedlots—the alternatives are slotted floors; periodic (daily or more frequent) cleaning; and either storage under conditions that conserve NH_3, or immediate spreading and incorporation in the soil. As illustrated in Table 6, some of the alternatives for dairy cows are expensive, and energy costs may also be high relative to energy costs of synthetic fertilizer.

2. CONSERVATION OF INORGANIC N AFTER INCORPORATION IN SOIL

As pointed out earlier (section III), conservation of NH_3 and timing of application have a major impact on crop uptake for the first crop subsequent to the addition of manure. Timing of application would have minimum impact if the manure did not contain appreciable amounts of inorganic N.

To obtain any benefit from conserving NH_3 between excretion and time of application, the following two additional conditions must be met: (i) the manure must be incorporated into the soil almost immediately after spreading in order to prevent NH_3 volatilization; and (ii) the manure must be applied only a short period of time before rapid crop uptake in most climates, or else denitrification and leaching will dissipate the NH_3 that has been conserved and incorporated in the soil.

For example, with corn in the humid regions, a number of experiments have demonstrated that an appreciable amount of fertilizer N applied in the fall is lost to the succeeding corn crop. Thus, we would expect the same thing to happen with ammoniacal N in fall-applied manure. Similar restrictions would apply to fall applications of manure on small grains such as wheat (*Triticum aestivum* L.), oats (*Avena sativa* L.), and barley (*Hordeum vulgare* L.). This means: (i) storage facilities have to be large enough for several months to 1 yr of storage; (ii) all of the manure has to be spread in a very short period of time in the spring; (iii) expensive machinery is used for only a short period of time; and (iv) labor requirements are high during an already busy season. Furthermore, incorporation of manure in the spring on small grain crops planted the previous fall is not feasible. In Table 7, Pratt et al. (1976) and Herron and Erhart (1965) report unaccounted-for N of 50% and 10%, respectively, under irrigated conditions. Thus, under some conditions at least (e.g., Pratt et al., 1976), the limitations for humid regions also apply under irrigation.

3. THE SUPPLY OF MANURE MAY EXCEED CROP NEEDS

Since manure is a bulky commodity, it is expensive to transport, and the energetically and economically feasible transport distance is limited. With large animal units, the cropland within this "feasible" distance may not need all of the N if NH_3 conservation is practiced. For example, in New York, much of the intensively cropped land on dairy farms appears to be in a 6-yr rotation, with 3 yr of alfalfa followed by 3 yr of corn for silage. Usually there will be about 1 ha of such land per cow. Approximately 50% of the manure N plus the legume residues supply the necessary N for the corn crops, and the "average" dairy farmer in New York does not need to use over 50% of his manure N under present circumstances. In this case, the incentive for N conservation can only come from the sale of manure to neighbors, or else expension of the area devoted to nonlegumes.

In many cases, the animal-feeding operations are very large and they are not associated with a feed production enterprise. In these cases, disposal of the manure is a major problem, and dissipation of N in the feedlot is a much more important objective than conservation of N. The amount of manure N available if it were all conserved would expand the hauling radius such that recycling becomes energetically and economically impractical.

C. Research and Development Needs

1. SETTING PRIORITIES

During the last 30 yr, manure research has been mainly aimed at disposal in the cheapest and most environmentally sound manner. Almost no emphasis has been given to improving nutrient conservation. Based on what we have read, and our own observations in the northeast, we conclude that not much emphasis has been placed on maximizing nutrient recycling when animal production units/farms are planned/built. Some manure management/storage systems have been advocated on the basis of N conservation, but often, no consideration was given to how the farmer could utilize the N on his limited area of nonlegume crops, how he could manage to spread and incorporate the manure in the spring before corn planting time, etc. The only design criteria was conservation of NH_3 from excretion to spreading without considering the other essential aspects of manure management. This is probably pretty typical of recent manure management; the management schemes ignore some of the basic requirements for recycling of the N.

On the present typical New York dairy farm, the milk per cow produces an income of $1500 to $2000; an amount of fertilizer N equivalent to the amount of N in the manure will cost about $40 to $60 per cow per year. Examination of the capital and operating costs of various manure management systems listed in Table 6 illustrate that manure management is indeed a "losing" proportion; it costs more to do something with it than an equivalent amount of fertilizer N costs. Similar figures would apply to other livestock enterprises and to dairy farms in other regions (Vanderholm, 1979).

The question that must be considered carefully is whether or not this will continue through the next 10 to 30 yr. New animal production units are continually being designed as old units are replaced and as shifts in production occur. Since "retro fitting" is often an expensive process, we will likely have to use the manure management systems we are now building for a long time—Are they adequate for the future? Will they be obsolete soon? Regardless of how we answer the question, it will likely be a self-fulfilling prophecy, because the research, demonstration, and extension programs will be tied to the answer. As pointed out earlier, most crop production experiments with manure in the recent past have studied what farmers are doing; there were few if any innovative treatments. This probably results from concluding that farmers would have no incentive to change to some hypothetical, innovative system.

The economics of manure management systems intertwine with many other economic questions: Are animal production units that are small enough for efficient manure distribution on cropland economically viable? Will the same person manage the animal production and crop production, or will the complexities of animal and crop management require separate management? If separate managements, how can they be intergrated?

There are no general or universal answers to these complex questions. The question of how much emphasis should be placed on recycling of manure N (and the other nutrients in manure) needs careful study in each region of the United States for each enterprise. However, remember the answer is likely to be a self-fulfilling prophecy.

2. AGRONOMIC RESEARCH NEEDS

In the previous sections, emphasis was placed on simultaneously (i) conserving NH_3 between time of excretion and spreading, (ii) immediate incorporation in the soil, and (iii) application as close to time of plant demand as feasible. These place very difficult demands on farm management, labor availability, and efficient utilization of machinery. Some means of alleviating these requirements must be found. Some possibilities follow.

a. Inhibition of Nitrification—With fertilizer N, timing of application is a reasonable means to ensure efficient utilization. However, it is not so easy to properly time manure applications because of the bulky nature of the product. Hence, nitrification inhibition appears to be an extremely important tool. Particularly interesting in this respect is the inhibition of nitrification by C_2H_2 (Walter et al., 1979), dicyandiamide (Amberger & Vilsmeier, 1979), and N-Serve® (Gorlitz & Hecht, 1980).

b. Methane Generation from Manure is a way to Produce Energy—Perhaps this can be combined with NH_3 stripping so that the inorganic N is concentrated and the organic and inorganic N fractions are separated and can be managed in an appropriate fashion. Perhaps NH_3 recovery from exhaust air in closed barns would be feasible.

c. Considerable Study on Denitrification in Manure–Soil Mixtures is Needed—Perhaps denitrification is so serious in some situations that NH_3 conservation will be of no benefit to subsequent crops.

REFERENCES

Adriano, D. C., P. F. Pratt, and J. E. Bishop. 1971. Fate of inorganic forms of N and salt from land disposed manure from dairies. p. 243–246. *In* Livestock waste management and pollution abatement. Proc. Int. Symp. on Livestock Wastes. Am. Soc. Agric. Eng. Publ. Proc.-271. ASAE, St. Joseph, Mich.

Amberger, A., and K. Vilsmeier. 1979. The inhibition of the nitrification of slurry nitrogen by decyandiamide. Z. Acker Pflanzenbau 148:239–246.

Azevedo, J., and P. R. Stout. 1974. Farm animal manures. California Agric. Exp. Stn. Manual 44.

Baldock, J. O., and R. B. Musgrave. 1980. Manure and mineral fertilizer effects in continuous and rotational crop sequences in Central N.Y. Agron. J. 72:511–518.

Castellanos, J. Z., and P. F. Pratt. 1981. Mineralization of manure nitrogen. Correlation with laboratory indexes. Soil Sci. Soc. Am. J. 45:354–357.

Douglas, J. 1981. Fertilizer Costs—1985—Can farmers afford them? Fertilizer Progress 12(5): 14–20.

Ensminger, M. E., and C. G. Olentine. 1978. Feeds and nutrition. Ensminger Publ. Co., Clovis, Calif.

Gilbertson, C. B., T. M. McCalla, J. R. Ellis, and W. R. Woods. 1971. Characteristics of manure accumulations removed from outdoor, unpaved beef cattle feedlots. p. 56–59. *In* Livestock waste management and pollution. Proc. Int. Symp. on Livestock Wastes. Am. Soc. Agric. Eng. Publ. PROC-271. ASAE, St. Joseph, Mich.

Gilbertson, C. B., F. A. Norstadt, A. C. Mather, R. F. Holt, A. R. Bennett, T. M. McCalla, C. A. Onatad, and R. A. Young. 1979a. Animal waste utilization on cropland and pastureland. USDA-EPA-60012-79-059. National Technical Information Service, Springfield, Va.

Gilbertson, C. B., D. L. Van Dyne, C. J. Clanton, and R. K. White. 1979b. Estimating quantity and constituents in livestock and poultry manure residue as reflected by management systems. Trans. ASAE 22:602–611.

Gorlitz, H., and W. Hect. 1980. Reducing the nitrification of slurry nitrogen and nitrogen translocation to the subsoil by the addition of active principles. Arch. Aker Pflanzenbau Bodenkd 24:151–159.

Gracey, J. F. 1979. Nutrient content of cattle slurry and losses of N during storage. Exp. Husb. 35:47–51.

Hashimoto, A. G. 1974. Characterization of white leghorn manure. *In* Processing and management of agricultural wastes. Cornell Univ., Ithaca, N.Y.

Heck, A. F. 1931. Conservation and availability of the nitrogen in farm manure. Soil Sci. 31: 335–359.

Hermanson, R. E., M. H. Ehlers, and D. Bezdicek. 1980. A mass balance analysis of a surface-aerated dairy manure storage lagoon. J. Agric. Eng. Res. 25:299–310.

Herron, G. M., and A. B. Erhart. 1965. Value of manure on an irrigated calcareous soil. Soil Sci. Soc. Am. Proc. 29:278–281.

Hoff, J. D., D. W. Nelson, and A. L. Sutton. 1981. Ammonia volatilization from liquid swine manure applied to cropland. J. Environ. Qual. 10:90–95.

Jenkinson, D. S. 1976. The nitrogen economy of the Broadbalk Experiment 1. Nitrogen balance in the experiment. Rothamsted Experiment Station Report for 1972. Part 2. p. 103–109. Rothamsted Exp. Stn., Harpenden, Herts, England.

Jenkinson, D. S., and A. E. Johnston. 1976. Soil organic matter in the Hoosfield continuous barley experiment. Rothamsted Experiment Station Report for 1976. Part 2. p. 87–101. Rothamsted Exp. Stn., Harpenden, Herts, England.

Jensen, L. S. 1977. Recent developments in applied broiler nutrition. p. 123–148. *In* W. Haresign and D. Levis (ed.) Recent advances in animal nutrition. Butterworths, London.

Johnson, A. E. 1974. Experiments made on Stackyard Field, Woburn, 1879–1974. Rothamsted Experiment Station Report for 1974. Part 2. p. 45–78. Rothamsted Exp. Stn., Harpenden, Herts, England.

Klausner, S. D., D. R. Bouldin, W. S. Reid, and D. Wilson. 1979. Influence of manure on the yield of corn. Mimeo 80-25, Department of Agronomy, Cornell Univ., Ithaca, N.Y.

Klausner, S. D., and R. W. Guest. 1981. Influence of NH_3 conservation from dairy manure on the yield of corn. Agron. J. 73:720–723.

Kolenbrander, G. J. 1981a. Limits to the spreading of animal excrement on agricultural land. *In* J. C. Brogan (ed.) Nitrogen losses and surface runoff from landspreading of manures. Martinus Nijhoff, The Hague/Boston/London.

Kolenbrander, G. J. 1981b. Effect of injection of animal waste on ammonia losses by volatilization of arable land and grassland. *In* J. D. Brogan (ed.) Nitrogen losses and surface runoff from landspreading of manures. Martinus Nijhoff, The Hague/Boston/London.

Lauer, D. A. 1975. Limitations of animal waste replacement for inorganic fertilizers. *In* W. J. Jewell (ed.) Energy, agriculture and waste management. Ann Arbor Science Publ., Inc., Ann Arbor, Mich.

Lauer, D. A., D. R. Bouldin, and S. D. Klausner. 1976. Ammonia volatilization from dairy manure spread on the soil surface. J. Environ. Qual. 5:134–141.

Likens, G. E. 1972. The chemistry of precipitation in the Finger Lakes Region. Tech. rep. no. 50. Cornell Univ. Water Resources & Marine Sciences Center, Ithaca, N.Y.

Luebs, R. E., K. R. Davis, and A. E. Laag. 1973. Enrichment of the atmosphere with ammonia compounds volatilized from a large dairy area. J. Environ. Qual. 1:37–41.

Luebs, R. E., K. R. Davis, and A. E. Laag. 1973. Diurnal fluctuation and movement of atmospheric ammonia and related gases from dairies. J. Environ. Qual. 3:265–269.

Magdoff, F. 1978. Influence of manure application rates and continuous corn on soil N. Agron. J. 70:629–632.

Mathers, A. C., and D. W. Goss. 1979. Estimating animal waste applications to supply crop nitrogen requirements. Soil Sci. Soc. Am. J. 43:364–366.

Mattingly, G. E. G. 1973. The Woburn organic manuring experiment. Rothamsted Experiment Station Annual Report for 1973. Part 2. p. 98–151. Rothamsted Exp. Stn., Harpenden, Harts, England.

Oveson, M. M. 1966. Conservation of soil nitrogen in a wheat summer fallow farming practice. Agron. J. 58:444–447.

Phillips, R. E., R. L. Blevins, G. N. Thomas, W. W. Frye, and S. H. Phillips. 1980. No-tillage agriculture. Science 208:1108–1113.

Pimental, D., L. E. Hurd, A. C. Bellotti, M. J. Forster, I. N. Oka, O. D. Sholes, and R. J. Whitman. 1973. Food production and the energy crisis. Science 182:443–449.

Porter, K. S. 1975. Nitrogen and phosphorus. Food production, waste and the environment. Ann Arbor Science Publ., Inc., Ann Arbor, Mich.

Powers, W. L., G. W. Wallingford, and L. S. Murphy. 1975. Formulas for applying organic wastes to land. J. Soil Water Conserv. 30:286–289.

Pratt, P. F., F. E. Broodbent, and J. P. Martin. 1973. Using organic wastes as nitrogen fertilizers. Calif. Agric. 27:10–13.

Pratt, P. F., S. Davis, and R. G. Sharpless. 1976. A four-year field trial with animal manures. Hilgardia 44:99–125.

Reddy, K. R., R. Khaliel, M. R. Overcash, and P. W. Westerman. 1979. A non-point source model for land areas receiving animal wastes. II. Ammonia volatilization. Trans. ASAE 22:1398–1405.

Safley, L. M. 1977. System selection and optimization models for dairy manure handling systems. Ph.D. Thesis. Cornell Univ. Library, Ithaca, N.Y.

Safley, L. M., D. R. Price, and D. C. Ludington. 1976. Network analysis for dairy waste management alternatives. Trans. ASAE 19:920–924.

Salter, R. M., and C. J. Schollenberger. 1939. Farm manure. Ohio Agric. Exp. Stn. Bull. no. 605.

Sluijsmans, C. M. J., and G. J. Kolenbrander. 1977. The significance of animal manure as a source of nitrogen in soils. *In* Proc. Int. Seminar on Soil Fertility and Fertility Management in Intensive Agriculture. The Soc. of the Sci. of Soil and Manure, Tokyo, Japan.

Sobel, A. T. 1976. High rise system of manure management. AWM-76-01, Dep. of Agric. Eng., Cornell Univ., Ithaca, N.Y.

Steenhuis, T. S., G. D. Bubinger, and J. C. Converse. 1979. Ammonia volatilization of winter spread manure. Trans. ASAE 22:152–157.

Taiganides, E. P., and R. L. Stroshine. 1971. Impact of farm animal production and processing on the total environment. p. 95–98. *In* Livestock waste management and pollution abatement. Proc. Int. Symp. on Livestock Wastes. Am. Soc. Agric. Eng. Publ. PROC-271. ASAE, St. Joseph, Mich.

U.S. Department of Agriculture. 1977. Agricultural statistics. U.S. Government Printing Office, Washington, D.C.

Vanderholm, D. H. 1979. Handling of manure from different livestock and managements systems. J. Am. Sci. 48:113–120.

Van Dyne, D. L., and C. B. Gilbertson. 1978. Estimated inventory of livestock and poultry manure resources in the U.S. Agric. Wastes 1:259–266.

Walter, H. M., D. R. Keeney, and I. R. Fillery. 1979. Inhibition of nitrification by acetylene. Soil Sci. Soc. Am. J. 43:195–196.

Webb, B. B., B. B. Tucker, and R. L. Westerman. 1980. The Magruder plots: taming the prairies through research. Okla. Agric. Exp. Stn. Bull. B-750.

White, W. C. 1978. Energy, food and fertilizers. Fertilizer Progress (July–August):14–18.

Whitehead, D. C. 1970. The role of nitrogen in grassland productivity. Bull. 48. Commonwealth Bureau of Pastures and Field Crops. Hurley, England. p. 56–58.

Wilkinson, S. R. 1979. Plant nutrient and economic value of animal manures. J. Am. Sic. 48: 121–133.

Yeck, R. G., L. W. Smith, and C. C. Calvert. 1975. Recovery of nutrients from animal wastes —an overview of existing options and potentials for use in feed. *In* Managing livestock wastes. Proc. 3rd Int. Symp. on Livestock Wastes, ASAE, St. Joseph, Mich.

Management of Crops for Nitrogen Utilization

section III

16

T. C. Tucker
University of Arizona
Tucson, Arizona

Diagnosis of Nitrogen Deficiency in Plants

A half century ago, and for more than the decade that followed, corn (*Zea mays* L.) and other nonleguminous crops received little fertilizer. A typical application of fertilizer was 2–4.4–3.3 or 2–5.3–5 (2–10–4 or 2–12–6) at rates up to 100 kg/ha. When a light green color appeared, or later a "firing" of the lower leaves, it was assumed that rainfall was inadequate or the soil was droughty. Long after agricultural scientists associated these symptoms with N deficiency, many farmers continued to attribute the problem to drought (Hambidge, 1941). Specific visual N deficiency symptoms were confirmed for many crops, and to a degree quantitatized by field tissue tests and laboratory analyses. Research has continued to improve the knowledge of the quantitative relationship between plant tissue N and the nutritional status of various crops. Recent activity has focused attention again on "visual" symptoms by use of infrared imagery of crops from aircraft and satellites. Thus, diagnosing N deficiency in crops is a dynamic, everchanging endeavor.

Nitrogen occurs in plants both in inorganic form and as a functional constituent in a number of organic compounds. The role of N in plant metabolism is discussed in a previous chapter. Thus, it is sufficient here to point out that N contained in the inorganic form, in the reserve proteins, and in chlorophyll serves as the best indicator of the N status of the plant. Inorganic N in the plant is indicative of the active supply available for metabolism at a given time. Organic reserves reflect what the N supply for the plant has been prior to that time. The N used in chlorophyll synthesis aids in visual diagnosis of a N deficiency by influencing the plant color. These facts serve as a rational basis for diagnosis of N deficiency in plants.

Nitrogen in Crop Production.

I. VISUAL DEFICIENCY SYMPTOMS

Nitrogen deficiency in plants can vary in severity, from slight with no visual symptoms, to acute with very obvious changes in appearance. The acute deficiency is an obvious signal that plant productive capacity has been affected. At the opposite end of the scale, some reduction in productivity can occur despite the absence of visual symptoms.

Generally, with N deficiency, plant appearance is that of sparse growth with small leaves, thin stems, and fewer lateral branches, tillers, or shoots. In early growth stages, leaves are pale and yellowish-green in color due to limited chlorophyll. The symptoms are more apparent on older leaves because N is mobile in the plant. Both inorganic forms and degradation products are translocated from the older tissue for reuse in the younger leaves. The leaves may develop yellow, red, or purple colors at later growth stages as pigments other than chlorophyll have a predominating effect. As a N shortage becomes more acute, the older leaves turn brown, starting at the tip and progressing over the leaf until the entire leaf is dead. The younger leaves remain green until the stage of deficiency is severe. Although N deficiency symptoms are among the most reliable of those caused by nutrient shortage, other complexities can be confused with N deficiency. For example, S deficiency symptoms are easily confused with early symptoms of N deficiency, in that the leaves have a general pale-green-to-yellowish appearance. Plant damage caused by disease, insects, or other environmental factors can be confused with N deficiency, as can other nutrient deficiencies at late stages of growth and development.

A detailed description of specific N deficiency symptoms on a crop-by-crop basis is almost a never-ending endeavor. There are several reviews, textbooks, and standard references that provide this type of information (Hambidge, 1941; Sprague, 1964; Jones, 1966; Black, 1968; English & Maynard, 1978).

In plants of the grass family, advanced stages of N deficiency symptoms are rather characteristic. The tips of the older leaves turn yellow, followed by dying or "firing" of the leaves. In corn, the typical V-shaped pattern of dead tissue proceeds upward along the midrib. In other grasses, the symptoms are less specific with general yellowing and firing of the lower leaves. In sorghum [*Sorghum bicolor* (L.) Moench], the symptoms sometimes resemble K deficiency, in that the leaf tissue along the margins may deteriorate.

With plants such as cotton (*Gossypium hirsutum* L.), N deficiency symptoms are less pronounced. With early N deficiency, the plants show a lack of vigor and have small leaves, short petioles, and reduced internode elongation. The entire older leaves are pale-to-yellowish-green in color. As severity of N shortage intensifies, the lower leaves turn brown, die, and eventually drop. Shedding of leaves is similar to natural leaf-drop at maturity. A more specific effect of N deficiency is found in tobacco (*Nicotiana tabacum* L.), with a yellowing of lower leaves, followed by drying or firing of the yellowed leaves. The more acute the N shortage, the more leaves that are affected.

Visual symptoms of N in vegetable crops are, in general, characterized by chlorosis progressing from light green to yellow. Under prolonged stress, the entire plant becomes yellow, growth is severely restricted, and some plants drop older leaves. Marketability of green, leafy vegetables is severely reduced by spindly appearance or lack of green color.

In fruit and nut tree crops, the symptoms of N deficiency vary somewhat with specific crop. However, the leaves are generally small, pale-green-to-yellowish-green in color, the foliage appears sparse and stunted, and twigs may die back. In various species, colors other than yellow appear, such as brown, orange, red, and purple. Older leaves fall, and fruit is sparse and small.

II. EVALUATION OF PLANT N STATUS

A. Theory

For plant analysis to be useful in diagnosing the N status of a plant, a point of reference must first be established. This point differs with kind of plant, plant part, position on the plant, stage of growth, or plant product specific to a particular plant.

1. CRITICAL CONCENTRATION

The concept of critical concentration was first proposed by Macy (1936). It was defined as the concentration of a nutrient at the point that separates the zone of deficiency from the zone of adequacy. According to Ulrich (1952), the critical concentration is a point slightly below the concentration at maximum yield and in the transition zone between deficiency and adequacy. These two viewpoints differ only slightly and are indistinguishable in practical application. Figure 1 is a graphical illustration of

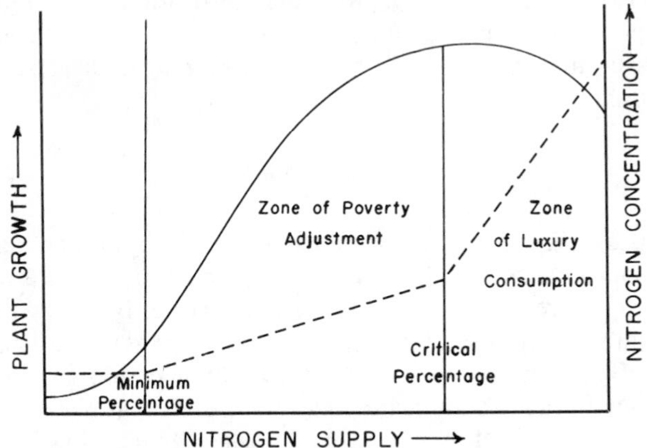

Fig. 1—Illustration of Macy's concept in relation to growth response to increasing N supply.

Macy's concept in relation to plant growth. While this model provides a rationale for the gross relationship, it fails to provide quantitative functions allowing adequate considerations of different plants, parts, and positions of sampling at various stages of growth, nor is it specific for chemical form of the nutrient to be measured.

2. OTHER NUTRIENT DEFICIENCIES

The critical concentration for N is generally determined by growing plants in soil or culture medium well supplied with nutrients other than N, and adding N in increments until an adequate or excessive level is attained. This approach presupposes that other nutrient deficiencies affect the accumulation of N and its meaning in relation to plant growth. Well-known examples of higher N concentrations associated with other nutrient deficiencies are with P and S. Deficiencies of several nutrients are known to interfere with NO_3^- reduction, protein synthesis, and other metabolic involvement of N in the plant. Use within the plant is usually more severely affected than absorption; thus, higher N concentrations occur in the plant than is indicative of N supply.

Environmental factors other than nutrients also influence N accumulation in plants. In these cases, growth or yield is usually not a function of N supply.

3. NONNUTRITIONAL FACTORS

Soil moisture stress can cause lower concentrations of plant N by its effect on N mineralization, movement, and absorption by the plant or it can result in higher N accumulation by reducing plant growth and inhibition of specific metabolic processes. Both cases were cited by Munson and Nelson (1973). Salinity stress can act similarly to moisture stress and usually causes higher N concentrations associated with limited growth and inhibition of protein synthesis (Frota & Tucker, 1978a, b).

Temperature affects crop growth rate, absorption of N, and microbial activity responsible for N transformation in the soil. Thus, N concentration in plant tissue may reflect both temperature effect and N supply in the growth medium.

B. Plant N Forms

For diagnostic purposes, the most common approach has been to measure total tissue N by a Kjeldahl procedure or to measure the NO_3^- in the cell sap from fresh tissue or water-soluble extract from dry tissue. Each has its advantages and disadvantages. Total analysis better reflects the summation of effects with respect to N status for a period of time prior to sampling, but it fails to indicate clearly the status at the exact time of sampling. This value is often more a postmortem than evaluation of current condition, although it is commonly used for corn and sorghum (Jones & Eck, 1973). On the other hand, NO_3^- in the tissue is indicative of current N within the plant at time of sampling. This interpretation presumes that NO_3^- is the principal N form absorbed by the plant in question. Nitrate in the tissue at a given time fails to indicate a shortage that may have occurred at an earlier stage of growth unless fertilizer was not applied and growing conditions have not appreciably changed.

Thus, the measurement of the organic N accumulation best reflects the cumulative effect of N supply to the plant; the inorganic form (NO_3^-) best reflects the current and most recent supply. Deficiencies of other nutrients or nonnutritional factors can cause undue accumulation of either organic N or NO_3^- by limiting growth.

C. Plant Part

1. WHOLE PLANT

The choice of plant part for sampling and analysis depends upon the kind of plant and purpose of the assay. Whole plant samples of small plants are easy to take and analyze. However, such a sample often includes tissue differing in age and function, and the analysis value represents an average. This value also depends upon the rate of plant growth and proportion of older tissue to young tissue, as well as the N supply in the soil. Values from the analysis can be interpreted if levels of deficiency and adequacy are known for that particular plant at the same stage of growth. Whole plant samples are necessary in N research when total uptake is part of a study of fertilizer use efficiency or N balance. Older and larger plants are more difficult to sample, dry, and grind for analysis. Tree crops are certainly the least suitable for whole plant analysis.

2. LEAF AND STEM TISSUE

Leaf samples, often including the petiole, are most commonly used for Kjeldahl N determinations, either with or without the inclusion of NO_3^-. Again, these values better indicate the cumulative N status until the time of sampling and are not necessarily an accurate representation of current status. Nitrate transported to the leaf blade is rapidly reduced in most plants with normal growth. Stem tissues contain much higher concentrations of NO_3^- than leaf tissues; NO_3^- also is stored in associated parenchyma cells. For example, with an adequate N supply, a cotton petiole may contain 20 or more times as much NO_3^- as the leaf blade. Furthermore, the change in NO_3^- in the petiole is much more sensitive to change in NO_3^- supply to the plant. Stem tissue thus appears the most suitable plant part to sample and analyze for evaluation of current N status of actively growing annual crops.

3. POSITION ON PLANT FOR SAMPLING

Nitrogen compounds in plants are continuously degraded, retranslocated, and resynthesized into new compounds. The older leaves are poor competitors for receiving the breakdown products because of internal competition for N in a plant. New leaves are stronger competitors and fruits are even stronger sinks for metabolites. Thus, leaf age is a more important consideration than leaf location. On annual plants, the lower leaves are commonly the older leaves, but on trees, the old and new leaves occur over the entire plant.

Most important in diagnosis of N status is that plant part and position on the plant be standardized and interpretation be based on known values of adequacy or deficiency for a specific plant. For current plant N status, this means a soluble N determination in petioles or stem tissue from recently, fully developed leaves or main stem tissues above the soil level.

D. Stage of Growth

The concept of critical nutrient concentration is useful in theory for separating the zone of adequacy from the zone of deficiency in terms of nutrient concentration within the plant. However, it fails to serve as a practical guide for N fertilization of most growing crops throughout the season. As an example, consider a crop that has a "critical" NO_3^--N value of 1000 mg/kg. A plant analysis showing this critical level at an early growth stage essentially indicates a deficient condition. Fertilizer cannot be applied and N absorbed before a longer period of deficiency ensues. With some crops, even a short period of early deficiency reduces yield. With an actively growing crop and an adequate supply of N, young plants may normally contain $\geq$ 15 000 mg/kg NO_3^--N, or 15× the critical concentration. Upon initiation of fruiting, the NO_3^--N will normally drop to 8 to 10× the critical concentration—although the plant is still well supplied with N from the soil. As

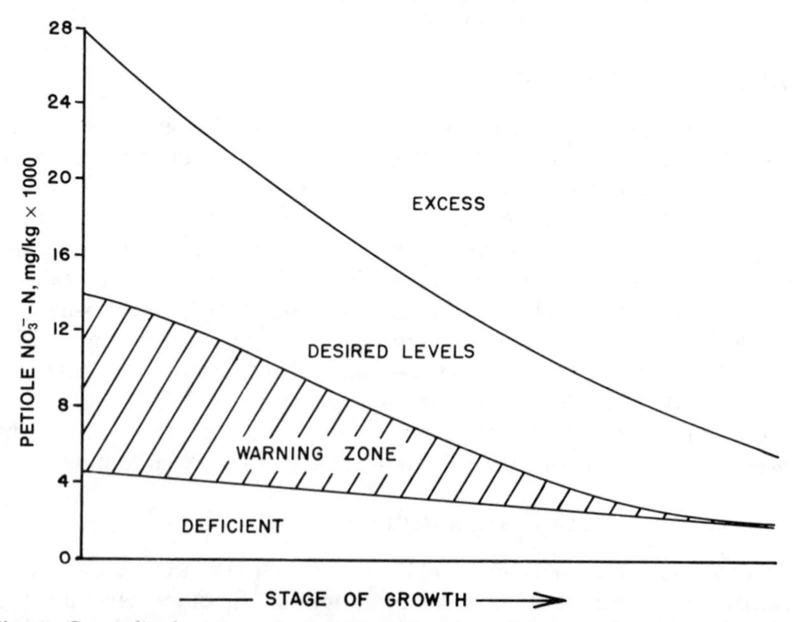

Fig. 2—Generalized pattern for petiole NO_3^- at various growth stages of an indeterminate crop such as cotton (adapted from Tucker & Tucker, 1968).

the physiological age of the crop approaches maturity, the NO_3^- level further declines to near or even below the "critical" concentration, regardless of the adequacy of the supply of soil N.

Thus, the NO_3^- level in a plant tissue sample must be interpreted differently at different physiological ages of the plant. When using these values as a guide to fertilization, the decline in NO_3^- to a deficient level must be anticipated by the rate of decline before a deficiency occurs. Fertilizer can be applied before the plant NO_3^- closely approaches the critical value. This is illustrated in Fig. 2 by the "warning" zone, which lies between the deficiency level and the desirable range. It can be observed also that the desirable levels of NO_3^- decline as the plant increases in physiological age.

Periodic sampling and analysis of the plant during the season can serve as a postmortem evaluation. Such use of this diagnostic technique is valuable in assessing the N soil fertility program, and, to some extent, in planning for the next year's crop. Values from samples at late stages of growth usually fall into this category, even if earlier vlaues are used to guide N fertilization.

III. PRACTICAL APPLICATION

A. Diagnostic Techniques for Postmortem Evaluation

Visual symptoms of N deficiency are probably the most definitive of nutrient deficiency symptoms. Yet, these seldom can be used as the sole basis for corrective action without a reduction in growth and final yield; some irreparable damage has resulted by the time the effect is visible. On the other hand, deficiency symptoms are the easiest of the diagnostic techniques for evaluation of a N fertilizer program. The intensity of shortage is qualitatively indicated by the severity of the symptoms. Color photography can be useful in detecting small differences in hue, but infrared imagery has greater potential for locating incipient deficiencies with the added advantage of large-scale rapid detection. A chemical analysis of designated plant parts or products can serve in the detection of incipient deficiency not evident by visual appearance. Multiple nutrient analyses can also confirm the visual diagnosis.

The common goal of maximum food and fiber production with minimum inputs can best be approached—insofar as the N fertilizer input is concerned—by using techniques that anticipate N deficiency before it actually occurs. This includes sequential sampling and analysis during the period of plant growth that yield or quality can be influenced by N fertilizer. Thus, timely fertilization can maintain an adequate N supply.

B. Guide for Fertilization

In a practical program using diagnostic techniques as a guide to N fertilization, chemical determinations serve as the primary basis for decisions of fertilizer application or evaluation of fertilizer practices. The values must be interpreted for each crop, specific plant part, specific N fraction or determination, and specific stage of growth. In addition, consideration must be given to a number of other environmental and plant factors in evaluating the chemical analysis values. Standard or reference values must be based on reasonably normal growing conditions of soil, temperature, moisture, adequacy of other nutrients, freedom from disease and insect pests, toxic factors including excess salinity, cultivars of similar behavior, and past cropping history.

Some of the factors have been discussed that can cause plants to accumulate NO_3^- in the tissue. Such accumulations do indicate the plant has adequate N under existing conditions, but they cannot be used to forecast the changes that will occur if the problem restricting plant growth and metabolism is corrected. For example, correcting S deficiency or removing salinity or water stress can result in rapid growth and utilization of accumulated plant NO_3^-, creating a shortage, unless the soil supply is continuously adequate. In either case, the concentration of N will be lower by "dilution" from rapid growth. Some conditions that cause low values that inaccurately reflect soil N supply are prolonged drought, diseases that restrict absorption and translocation, unusually heavy fruiting, and NH_4^+ fertilizer application (assuming NO_3^- is measured in the tissue sample). A previous legume crop often results in continued slow release of N and generally lower NO_3^- in the crop than when fertilized with N. Thus, for the most effective use of these diagnostic values, one must know as much as possible about the physiological development and health of the plant and other current conditions at the time of sampling. It can be ascertained that the value is low, adequate, or high under whatever conditions prevail, but accurate predictions cannot be made unless it is known what conditions can be changed or are likely to change.

Nitrate in a specific plant tissue at different stages of growth appears to be the common approach to tissue analysis as a guide to fertilization of most annual cultivated crops. Total or "Kjeldahl N" is most commonly used for tree crops and rice (*Oryza sativa* L.) and in some areas of corn and other grain crops. Examples will be discussed for two crops—cotton and head lettuce (*Lactuca sativa* L.)—in which NO_3^- is measured in tissue samples at advancing stages of growth. These crops are somewhat extreme representatives of growth habits.

Cotton has an indeterminate growth pattern, initially vegetative, followed by initiation of early fruiting forms or "squares," then flowers and bolls. After initiation of flowers and bolls, all growth phases continue simultaneously. When a large number of bolls are developing, vegetative growth and fruit initiation decline, and sometimes growth ceases. Tissue analysis samples are taken biweekly for NO_3^- determination by collecting the petioles from the most-recently fully expanded leaf blade from square stage to early boll maturation. Additional samples are desirable during early to midseason when NO_3^- is marginal. The values follow the generalized pattern illustrated in Fig. 2. With adequate N, petioles are high in NO_3^- early in

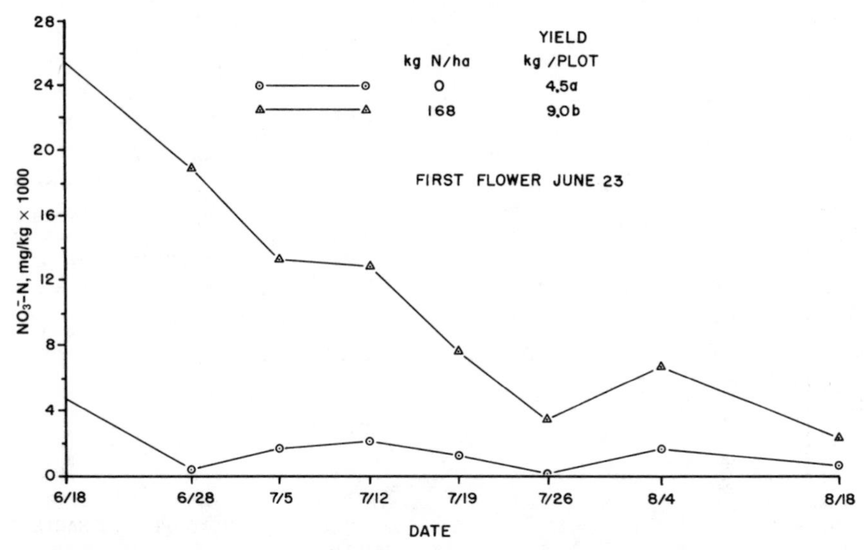

Fig. 3—Nitrate in cotton petioles of N deficient and sufficient plants (Gardner & Tucker, 1967).

the season (during vegetative growth), and this level declines with advancing growth stage and fruiting. The time from vegetative to late square or early flowering is probably the most important period during which N fertilization is the most effective. The rate and extent of decline is the key to N status of the plant and anticipation of need for N fertilization. An example of petiole NO_3^- patterns with adequate and deficient NO_3^- levels is shown in Fig. 3. Nitrate values appreciably below those from the N-treated plot would represent opportunities for fertilizer application until approximately mid-July at this location. The length of the growing season determines how late fertilizer applications can be made. Obviously, the form of N and method of application, as well as the time, amount, and method of irrigation or rainfall, will influence the effectiveness of the fertilizer and the lag period for utilization.

In contrast with cotton, lettuce is vegetative for the entire growth period. The characteristic NO_3^- decline does not occur, nor is it desirable. A typical pattern for fall lettuce is shown in Fig. 4. These NO_3^- values were determined on the midribs of the outer wrapper leaf.

When NO_3^--N values of $\geq$ 8000 mg/kg are maintained throughout most of the season, head size is larger and maturity is hastened. The quality factors, color, and size are favorably influenced, as well as yield. If NO_3^--N levels fall below 5000 mg/kg, deficiency symptoms appear. Short periods of N deficiency during early vegetative stages can be corrected by N applications with only a very slight loss in yield; there also is a delay in maturity and harvest date (Gardner & Pew, 1972). It is desirable to have NO_3^--N levels $>$ 8000 mg/kg before a cold period when mean temperatures

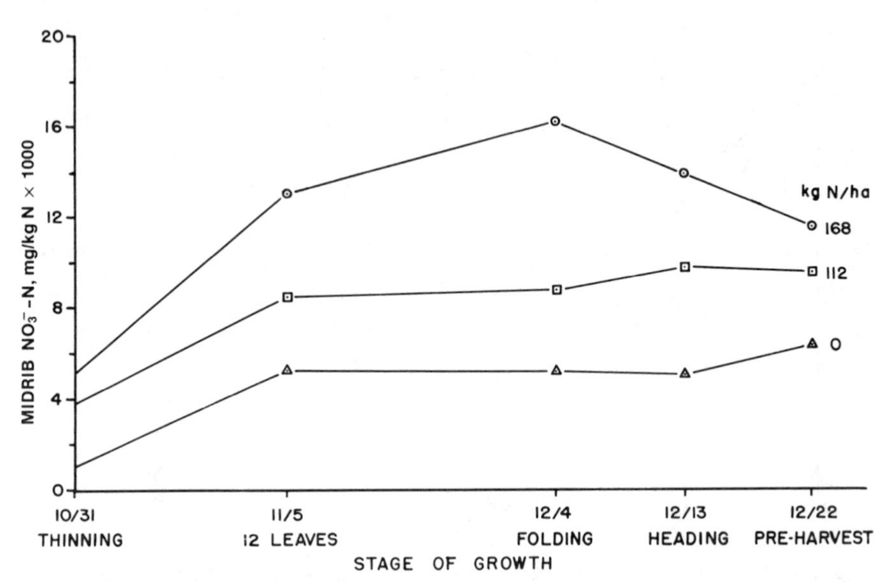

Fig. 4—Nitrate in lettuce midribs as related to N supply (Gardner & Pew, 1972).

fall below 13°C (Gardner & Pew, 1974). At low temperatures, the NO_3^- absorption rate is severely reduced (Frota & Tucker, 1972).

The contrast between these two crops is summarized in Table 1. It is much easier to establish and interpret critical levels of NO_3^- in a vegetative crop such as lettuce than in an indeterminate fruiting crop such as cotton. In cotton, a critical level is not meaningful between the early vegetative and early boll maturation stages of growth; however, this is the period during which N fertilization should be used as a corrective measure to avoid a deficiency. The N management objective should be to prevent a deficiency without use of excess N. Therefore, a deficiency must be anticipated in time for N application to effectively avoid NO_3^- declining to deficient levels. It is more appropriate to think in terms of a desirable range of NO_3^- values declining with stage of cotton growth but remaining more or less constant with lettuce. Values slightly lower than those in the desirable range do not indicate an incipient deficiency as such, but should be interpreted as an opportunity to take corrective action—namely, fertilize with N before a deficiency occurs. This zone below the desirable levels and above the deficient concentrations can be regarded as a warning zone.

A similar approach can be taken with a number of other crops. The values in Tables 2 and 3 are offered as guides for selected crops for which tissue analysis has been useful in modifying the N fertilizer program during the season. In many cases, an impending deficiency can be predicted by tissue NO_3^- analysis in time to apply N fertilizer and prevent loss in yield and/or quality. Often tissue NO_3^- values are found to be adequate, indicating fertilizer is not needed, which is an important advantage to the grower. This aspect of the diagnostic program is energy conserving and cost effective.

Table 1—Diagnosing N deficiencies in cotton and lettuce.

Crop growth stage	Cotton	Lettuce
	Visual deficiency symptoms	
Thinning stage	Stunted growth, small stems and leaves, pale-green-to-yellowish color.	
Midlife	Short tems, light-green-to-yellow color.	Light-green color, small leaves
Maturity	Leaf shed, purple color.	Light-green color, small heads.
	Concentration of NO_3^--N per mg/kg (desirable range)	
Thinning stage	>15 000	>8 000
Midlife	>6 000	>8 000
Maturity	<2 000	6 000
	Plant part to sample	
Thinning stage	Petiole	Leaf midrib
Weekly intervals during midlife	Petiole	Leaf midrib
Maturity	Petiole	Leaf midrib
	Responsiveness to N fertilizer after N deficiency has been detected	
Seedling stage	High	High
Midlife	Moderate	High
Maturity	None	Low to none

Table 2—Guide for tissue analysis of selected field crops.

Crop growth stage	Plant part	NO_3^- concentration, mg/kg Deficient	Intermediate	Sufficient	Reference
		Cantaloupe (*Cucumis melo* L.)			
Early, four leaves	Petiole from most-recent, full-size leaf	4 000	8 000	12 000	Pew & Gardner, 1972
Runner		4 000	8 000	12 000	
Small melon		4 000	6 000	10 000	
Full size melon		2 000	5 000	7 000	
		Celery [*Apium graveolens* L. var. dulce (Mill.) Pers.]			
Midgrowth	Petiole of youngest, fully elongated leaf	5 000	7 000	9 000	Geraldson et al., 1973
Near maturity		3 000	4 500	6 000	
		Lettuce (*Lactuca sativa* L.)			
Thinning	Midrib of wrapper leaf	4 000	6 000	8 000	Gardner & Pew, 1972
Folding to heading		4 000	6 000	8 000†	
		Potatoes (*Solanum tuberosum* L.)			
Early, six to eight leaves	Petiole from most-recent, full-sized leaf	8 000	16 000	20 000	Adapted from Gardner & Jones, 1975; Geraldson et al., 1973
Tubers, 1–2.5 cm		6 000	10 000	16 000	
Tubers, 3–5 cm		4 000	6 000	10 000	
		Spinach (*Spinacia oleracea*)			
Midgrowth	Petiole of young, mature leaf	4 000	6 000	8 000	Geraldson et al., 1973

(continued on next page)

Table 2—Continued.

Crop growth stage	Plant part	NO_3^- concentration, mg/kg: Deficient	Intermediate	Sufficient	Reference
		Sweet Corn (Zea mays)			
Tasseling	Midrib and first leaf above primary ear	500	1 000	1 500	Geraldson et al., 1973
		Tomato (Lycopersicon esculentum Mill.)			
Early bloom	Petiole, fourth leaf from growing tip	8 000	10 000	12 000	Geraldson et al., 1973
Fruit, 2–5 cm		6 000	8 000	10 000	
First color		2 000	3 000	4 000	
		Watermelon [Citrullus lanatus (Thunb.)]			
Early fruit	Petiole from sixth leaf	5 000	7 000	9 000	Geraldson et al., 1973
		Cotton (Gossypium hirsutum L.)			
Square	Petiole from most-recent, full-size leaf	4 000	10 000	18 000	Adapted from Gardner & Tucker, 1967; Maples et al., 1977
Early bloom		3 000	8 000	12 000	
Peak bloom		2 000	5 000	8 000	
First full-size boll		1 000	3 000	5 000	
		Grapes (Vitis labrusca L.)			
Full bloom	Petiole opposite the cluster	350	600	1 200	*Western Fertilizer Handbook*, 1980
		Sugar Beets (Beta vulgaris L.)			
Thinning	Petiole from most-recent, full-size leaf	1 000	3 000	5 000	Adapted from Johnson et al., 1978
Midseason		1 000	1 500	2 000	
Late season		400	600	1 000	
		Wheat (Triticum aestivum L.)			
Jointing	Stem tissue aboveground level	2 000	3 000	6 000	Adapted from Gardner & Jackson, 1976
Boot		1 500	2 000	3 000	
Milk stage		500	1 000	2 000	

† Higher value desired when temperatures are below 13°C.

Table 3—Guide for tissue analysis of fruit and nut crops.

Crop†	Total N: Deficient	Intermediate	Sufficient	Reference
	%			
Almond	1.8	2.0	2.5	*Western Fertilizer Handbook*, 1980; Beutel et al., 1978
Apple, pear	1.7	1.9	2.4	
Apricot, prune	2.0	2.2	2.5	
Filbert, walnut	2.0	2.2	2.8	
Peach	2.5	2.6	3.5	
Sweet cherries	2.0	2.5	3.0	
Oranges	2.1	2.3	2.5	

† Almond (*Prunus amygdalus* Batsch)
Apple (*Malus sylvestris* Mill.)
Pear (*Pyrus communis* L.)
Apricot (*Prunus armeniaca* L.)
Prune (*Prunus domestica* L.)
Filbert (*Corylus avellana* L.)
Walnut (*Juglans regia* L.)
Peach (*Prunus persica* Batsch)
Sweet cherries (*Prunus avium* L.)
Oranges (*Citrus sinensis* Osbeck)

IV. SUMMARY

Any thought that a diagnostic program will eliminate all nutrient problems, shortages and excesses, and result in maximum production and quality will cause disappointment. Plant performance is affected by too many variables other than tissue N concentration. Several of these have been mentioned in this chapter. Practical experience and knowledge of the specific crop and behavior of different cultivars on different soils under various climatic conditions are essential prerequisites to interpretation and application of test values. The more that is known about the normal behavior of the crop at all physiological stages of development, the better the chances of proper evaluation of diagnostic values. Soil-test values for NO_3^- before planting or early in the season often give an indication of the crop's early N status, and can frequently help reconcile later tissue analysis values. Knowledge of the previous cropping history is especially important in understanding a continued gradual release of soil N. Soil analyses for other nutrients, salinity, and possible toxic effects provide information needed to evaluate NO_3^- in the proper tissue. Many irrigation waters contain appreciable amounts of N in NO_3^- form, and this quantity should be known. Sampling variation due to plant and soil variability, as well as human error in consistently collecting the identical plant part, is to be recognized.

In spite of all shortcomings of tissue analysis, the technique provides the most exact procedure now available for effectively maximizing N fertilizer use efficiency. The values given as guides can be further refined through research and experience for local areas and new crops and cultivars. Visual symptoms of N deficiency should not be used as a guide for N fertilization because they should not appear with an effective tissue analysis program. The potential use of infrared or other imagery outside the visible spectrum offers intriguing possibilities for the future.

REFERENCES

Beutel, J., K. Uriv, and O. Lilleland. 1978. Leaf analysis for California deciduous fruits. p. 11–14. *In* H. M. Reisenauer (ed.) Soil and plant-tissue testing in California. Univ. of California Bull. 1879. Univ. of California, Berkeley.

Black, C. A. 1968. Soil-plant relationships. 2nd ed. John Wiley & Sons, Inc., New York.

English, J. E., and D. N. Maynard. 1978. A key to nutrient disorders of vegetable plants. Hortic. Sci. 13:28–29.

Frota, J. N. E., and T. C. Tucker. 1972. Temperature influence on ammonium and nitrate absorption by lettuce. Soil Sci. Soc. Am. Proc. 36:97–100.

Frota, J. N. E., and T. C. Tucker. 1978a. Absorption rates of ammonium and nitrate by red kidney beans under salt and water stress. Soil Sci. Soc. Am. J. 42:753–756.

Frota, J. N. E., and T. C. Tucker. 1978b. Salt and water stress influences nitrogen metabolism in red kidney beans. Soil Sci. Soc. Am. J. 42:743–746.

Gardner, B. R., and E. B. Jackson. 1976. Fertilization, nutrient composition, and yield relationships in irrigated spring wheat. Agron. J. 68:75–78.

Gardner, B. R., and J. P. Jones. 1975. Petiole analysis and the nitrogen fertilization of 'Russet Burbank' potatoes. Am. Potato J. 52:195–200.

Gardner, B. R., and W. D. Pew. 1972. Response of fall grown head lettuce to nitrogen fertilization. Ariz. Agric. Exp. Stn. Tech. Bull. 199.

Gardner, B. R., and W. D. Pew. 1974. Response of spring grown head lettuce to nitrogen fertilizer. Ariz. Agric. Exp. Stn. Tech. Bull. 210.

Gardner, B. R., and T. C. Tucker. 1967. Nitrogen effects on cotton: II. Soil and petiole analyses. Soil Sci. Soc. Am. Proc. 31:785–791.

Geraldson, C. M., G. R. Klacan, and O. A. Lorenz. 1973. Plant analysis as an aid in fertilizing vegetable crops. p. 365–379. *In* L. M. Walsh and J. D. Beaton (ed.) Soil testing and plant analysis. Soil Sci. Soc. of Am., Inc., Madison, Wis.

Hambidge, G. 1941. Hunger signs in crops. Am. Soc. of Agron. and Natl. Fert. Assoc., Washington, D.C.

Johnson, G. V., J. L. Stroehlein, and J. L. Abbott. 1978. Critical tissue levels for predicting nitrogen needs of sugarbeets at Mesa, Arizona. J. Am. Soc. Sugar Beet Technol. 20:65–72.

Jones, J. B., Jr., and H. V. Eck. 1973. Plant analysis as an aid in fertilizing corn and grain sorghum. p. 348–364. *In* L. M. Walsh and J. D. Beaton (ed.) Soil testing and plant analysis. Soil Sci. Soc. of Am., Madison, Wis.

Jones, W. W. 1966. Nitrogen. p. 310–323. *In* H. D. Chapman (ed.) Diagnostic criteria for plants and soils. Univ. of California, Div. of Agric. Sci., Riverside.

Macy, P. 1936. The quantitative mineral nutrient requirements of plants. Plant Physiol. 11: 749–764.

Maples, R., J. G. Keogh, and W. E. Sabbe. 1977. Nitrate monitoring for cotton production in Loring-Calloway silt loam. Ark. Agric. Exp. Stn. Bull. 825.

Munson, R. D., and W. L. Nelson. 1973. Principles and practices in plant analysis. p. 223–248. *In* L. M. Walsh and J. D. Beaton (ed.) Soil testing and plant analysis. Soil Sci. Soc. of Am., Madison, Wis.

Pew, W. D., and B. R. Gardner. 1972. Nitrogen effects on cantaloupes. Commun. Soil Sci. Plant Anal. 3:467–476.

Sprague, H. B. 1964. Hunger signs in crops. 3rd ed. David McKay Co., Inc., New York.

Tucker, T. C., and B. B. Tucker. 1968. Nitrogen nutrition. p. 183–211. *In* F. C. Elliot et al. (ed.) Advances in prediction and utilization of quality cotton: principles and practices. Iowa State Univ. Press, Ames.

Ulrich, A. 1952. Physiological basis for assessing the nutritional requirements of plants. Annu. Rev. Plant Physiol. 3:207–228.

Western Fertilizer Handbook. 1980. Soil and tissue testing. p. 138–153. The Interstate Printers & Publishers, Inc., Danville, Ill.

17

George Stanford[1]
Agricultural Research Service, USDA
Beltsville, Maryland

J. O. Legg
University of Arkansas
Fayetteville, Arkansas

Nitrogen and Yield Potential

There is continued interest in increasing the yield potential of crop plants in order to meet world needs of food and fiber. In most cases, major crop plants have greater genetic yield potential than is being realized through current production practices. For the purposes of this chapter, "potential yield" is defined as the ultimate yield attainable in any geographic region when all controllable production factors are at an optimum and when the uncontrollable factors (mainly weather conditions) are generally favorable. Although some soil conditions are not easily modified, they will be generally classified as controllable factors. From a soil fertility standpoint, N is the element that is of primary importance in the determination of plant yields, especially in relation to the attainment of maximum yield potentials.

Since the beginning of plant domestication, attempts have been made to increase crop yields. Early efforts were directed toward selecting plants with large, numerous seeds that were retained to maturity. Over the years, gradual improvements in yield were also attained through fallowing, use of organic manures, growing legumes, etc.; however, it was not until the latter part of the 19th century that dramatic yield increases began to occur. This initial upsurge in plant yields was based chiefly on increasing knowledge of the chemical elements required in plant nutrition and supplementation of soil sources of these elements with chemical fertilizers. Evans (1980) illustrated graphically the historical trends in yields of wheat (*Triticum aestivum* L.) in England and rice (*Oryza sativa* L.) in Japan from about 1200 A.D. to

[1] Deceased 28 January 1981.

Nitrogen in Crop Production.

the present. There is no indication that the increase in average yields that began at the end of the 19th century is slowing down. It was also shown by Evans (1980) that yields for all cereals in several countries have increased dramatically with increased fertilizer applications. In Japan, where up to 500 kg/ha of fertilizer were applied, no leveling off of cereal yields was apparent. Such results indicate that average yields have not approached potential yields, but they do not show the maximum yields that can be attained or how crop yield potentials have changed in recent years. Data for current grain yields indicate that maximum potential yields are at least two to three times greater than average yields. In this chapter, our major objective is to examine the role of N as one of the most important factors in the attainment of high potential yields. This will be done without regard to economic considerations; however, this does not mean necessarily that the high inputs required for maximum potential yields are not economically feasible.

I. COMPONENTS OF HIGH YIELDS

Only under the most optimum conditions is it possible to attain the yield potential of crop plants. As agronomists, we concentrate on improvements in plant breeding, nutrient and moisture supply, soil physical conditions, and pest control to attain high yield potential. Other highly important factors are associated with weather conditions such as temperature, excess or deficient rainfall, storm patterns, and the amount of radiant energy reaching the plant canopy. Good management and cultural practices are required to promote the most favorable environment for plant growth and to harvest the plant products.

The factor that sets the upper limit on potential yield is the quantity of energy that crop tissues capture from the sun (Johnson, 1980). This factor is not generally the most limiting one, however, since Johnson (1980) calculated that at maximum photosynthesis, the energy received from the sun in the central Corn Belt on the average is sufficient to produce corn (*Zea mays* L.) yields of about 31 383 kg/ha (500 bu/acre). The current record yield of corn is about 70% of this upper limit. Nevertheless, it is possible to increase the capture of radiant energy to a certain degree by the selection of cultivars that have good leaf arrangement and form a closed canopy early in the growth period.

Under field conditions, it is impossible to control all of the factors influencing potential yield. The record yields attained in recent years have resulted mainly from improved cultivars, pest control, improved soil physical and chemical conditions, moisture control, and good management practices. Perhaps the most critical element in promoting extremely high yields is supplying nutrients in sequence with crop demand, without creating toxic conditions.

II. RECORD YIELDS REPORTED FOR MAJOR CROPS

Most of the reports of record high yields obtained on farms for different crops occur in farm and trade magazines, and all of the factors influencing the yields are not always reported and analyzed. Normally, the record yields have been independently verified by county extension agents or contest officials, and they can be considered to be reliable estimates. For this report, all of the individual sources of information for record farm yields will not be identified.[2] All of the yields may not be the highest that have been attained, but they will serve to illustrate the yields that can be realized under farm conditions with the technology that is currently available to farmers.

The yields given in Table 1 were generally obtained from measured areas of 0.405 ha (1 acre) or greater. According to Dennis and Owens (1980), such record yields were obtained by giving special attention to (i) selection of adapted varieties and use of quality seed; (ii) use of optimum planting dates and plant populations; and (iii) minimizing plant stress in relation to fertility, water, weeds, insects, or diseases. Proper timing of operations to achieve maximum plant growth and yields is also important.

It is obvious from Table 1 that record yields are at least twice as high as average yields for most crops. The tabulated values represent *attainable* potential yields and serve as goals for those lower on the production scale. In the case of corn yields, it is interesting to note that the first 18 830 kg/ha (300 bu/acre) yield was reported in 1955 (over 25 yr ago), and this level has been exceeded at least 12 times since then. It may be another 25 yr before the 25 106 kg/ha (400 bu/acre) yield barrier is broken, but such a yield is certainly within the realm of possibility.

[2] Grateful appreciation is expressed to Robert E. Dennis, Extension Agronomist, Univ. of Arizona, for much of the record yield information contained herein.

Table 1—Record yields of agronomic crops.

Crop	Year	State	Yield
Corn (*Zea mays* L.)	1977	Michigan	22 131 kg/ha (352.6 bu/acre)
Corn silage	1980	New Jersey	101.1 t/ha (45.1 tons/acre) (32% dry matter)
Wheat (*Triticum aestivum* L.)	--	Washington	14 055 kg/ha (209 bu/acre)
Grain sorghum [*Sorghum bicolor* (L.) Moench]	1976	Arizona	12 421 kg/ha (11 090 lb/acre)
Grain sorghum	1978	Arkansas	12 979 kg/ha (11 588 lb/acre)
Grain sorghum	1973	New Mexico	12 684 kg/ha (11 325 lb/acre)
Grain sorghum	--	Texas	12 632 kg/ha (11 279 lb/acre)
Soybeans (*Glycine max* L.)	1977	Nebraska	5 380 kg/ha (80 bu/acre)
Soybeans	1980	New Jersey	6 052 kg/ha (90+ bu/acre)
Sugarbeets (*Beta vulgaris* L.)	1972	California	139.9 t/ha (62.4 tons/acre)
Alfalfa (*Medicago sativa* L.)	--	California	37 t/ha (16.5 tons/acre)

III. N REQUIREMENT OF CROPS IN RELATION TO POTENTIAL YIELD

It is recognized that all nutrient elements are required in optimum amounts to achieve high potential yields. Compared with N, however, other nutrients can be adjusted more easily to optimum concentrations in the soil, because they are either held by the exchange complex or have low solubility and are not so easily lost from the system. On the other hand, N fertilizers are highly soluble and may be readily leached, volatilized as NH_3, or denitrified when in the NO_3^- form. Substantial quantities of N also may be immobilized in organic forms that are not readily available to plants. Furthermore, when N is supplied in the amounts required for maximum potential yield, plant toxicity may result if proper application procedures are not followed.

The goal in supplying N for a particular crop cannot be defined quantitatively without a knowledge of the crop's requirement for this element (Stanford, 1966). The N requirement is defined as the minimum amount of N in the aboveground portion of crops associated with maximum production. Although Viets (1965) concluded that the total N requirement of a crop cannot be accurately predicted, Stanford (1966, 1973) presented convincing evidence that reasonable estimates of internal N requirements can be derived for use in supplying N fertilizer needs for maximum crop production. Stanford (1966) determined the N requirements for a number of crops. Such determinations are easily made, using data from N rate experiments in which the total N content of the aboveground portion of the plant is known. Data for corn furnish a good example (Stanford, 1973). When the dry matter yield (grain plus stover) is plotted against N uptake (Fig. 1), yields begin leveling off when the N content of plants reaches 12 g/kg (1.2%), and there is essentially no further increase in yields with increasing N content. It is especially interesting to note that although attainable maximum yields varied greatly with years, the percentage N required for the plants to attain those yields did not vary perceptibly. This can be illustrated in another way (Fig. 2) in which the relation between percent attainable yield (corn grain plus stover) and percent N in the dry matter is depicted. Again, at least 12 g/kg (1.2%) N is required for 100% attainable yields under the existing conditions, while further increases in N content have no effect on yields. This relationship provides a means of determining the adequacy of N in a fertilization program, as well as providing information needed for an estimation of N requirements for any anticipated yield level.

The N content of corn grain has been used by Pierre et al. (1977a, b) as an indicator of N sufficiency in the plants. In their study, they hypothesized that the N percentage in corn grain should provide a measure of N sufficiency if the N percentage was related to yield expressed as a percentage of the maximum rather than to the absolute yield or yield increase. They found that the N content of corn grain at 100% of maximum yield in Iowa experi-

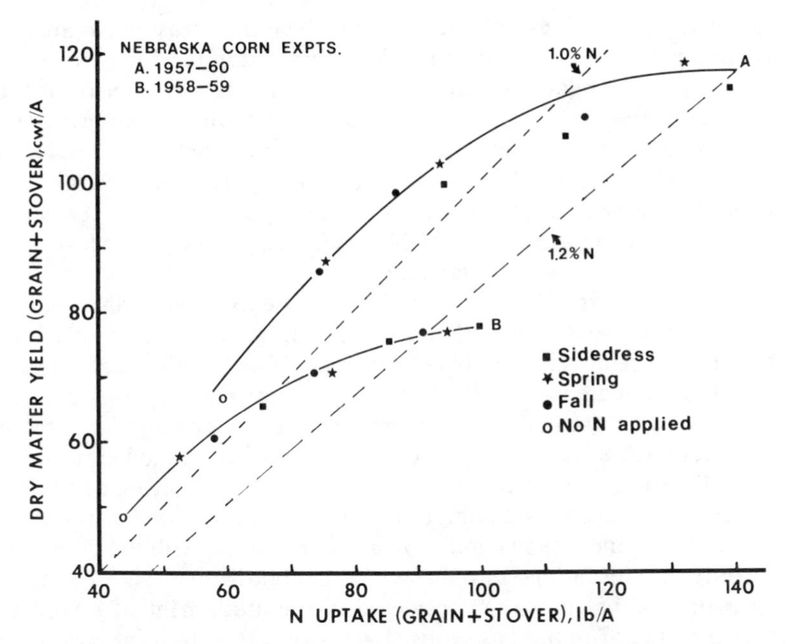

Fig. 1—Relation of dry matter yield (corn grain plus stover) to total N uptake in two groups of experiments with widely differing average attainable yields. Adapted from Stanford (1973).

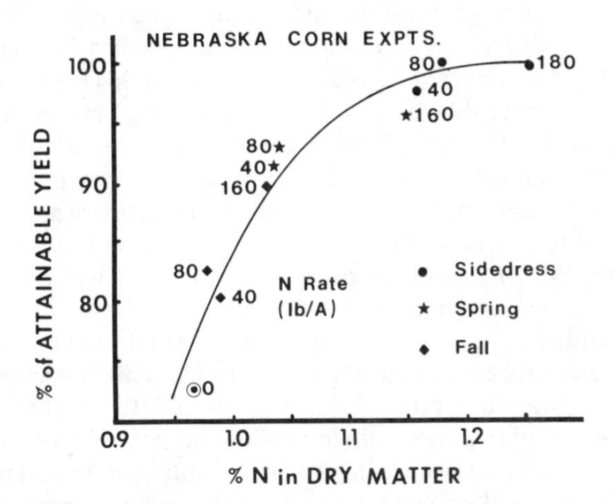

Fig. 2—Relation between percent of attainable total dry matter yield and percent N in total dry matter (corn grain plus stover) as influenced by timing and rate of N fertilizer application. Adapted from Stanford (1973).

ments varied from 14 to 16 g/kg (1.40–1.60%) [mean, 15.2 g/kg (1.52%)] as determined by a graphical method, and from 14.3 to 17.1 g/kg (1.43–1.71%) [mean, 15.4 g/kg (1.54%)] as determined by a regression method. This shows a rather wide variation of N percentage at maximum yield, although the mean values obtained appear to be in a reasonable area. The range in percentage N at maximum yield obtained in the Iowa experiments could possibly be related to differences in translocation of N from leaves and stems to the grain during maturation. It is obviously easier to obtain and prepare grain samples for N analysis, compared with whole plant samples, and this method of assessing the N requirements of grain plants deserves further study. At the present time, it appears advantageous to use whole plants in the assessment of N requirements; this procedure will be used in further treatment of this subject.

Other cereal crops have similar N requirements as corn, although there are generally less data available under varying growth conditions. Forage grasses have a relatively high N requirement [25 to 30 g/kg (2.5–3.0%) for 'Coastal' bermudagrass, *Cynodon dactylon* (L.) Pers.] since they are generally harvested before maturity. The importance of knowing the N requirement of grasses for maximum potential yield may be questioned since an increased digestible protein content can be achieved by applying additional N fertilizer. To estimate the optimum fertilizer N application, therefore, it is necessary to balance the maximum yield of dry matter with the desired protein content. This brings into focus the qualitative aspects of harvested plant products that must be considered in the attainment of quantitative potential yields of numerous crops. Excessive use of N fertilizers, for example, can result in decreased cotton (*Gossypium hirsutum* L.) fiber quality, as well as harvesting difficulties caused by excessive vegetative growth. The percentage sugar in sugar crops decreases with excessive N fertilization, although total dry matter production may increase. This does not mean that the internal N requirement of such crops cannot be determined. Stanford (1966, 1973) found that near-maximum sugar yields of 2-yr sugarcane (*Saccharum officinarum*) in Hawaii invariably were associated with approximately 2 g/kg (0.2%) N in the total dry matter (leaves and cane). This level corresponds to 1 kg of N/t of millable cane. Once the desired quality-quantity relationship has been determined, the N requirement can be calculated for maximum potential yields that will maintain the desired quality of the harvested plant product.

Considerable research effort has been directed toward improving the quality and quantity of protein in cereal grains. This has chiefly involved plant breeding and nutrition studies. For example, the study by Johnson et al. (1973) illustrated the high protein production possible with winter wheat as fertilizer N was increased, using a variety with genetic potential for high grain protein. It is well known that lodging of small grains can occur at high N rates, but this effect has been reduced by the development of dwarf or semidwarf varieties that resist lodging. This has made it possible to produce over 14 t of wheat/ha (Table 1). Since there is normally a desired increase in protein content with N applications sufficient to produce high yields, problems associated with small grains at high N levels are much different from the undesired effects in examples cited above.

IV. EFFECTIVE USE OF N FERTILIZERS TO PRODUCE MAXIMUM YIELDS

In this section, we shall be concerned with the effective use of high rates of N fertilizer to achieve maximum potential yield, primarily from the standpoint of supplying N in amounts sufficient to induce maximum production without causing any adverse effects. Although economic considerations are not involved here, it is deemed prudent to maintain the available N supply in such a way that only a small excess remains after the crop growth period.

The amount of fertilizer N that should be applied in any given situation depends on a number of variables within a given agricultural region. First of all, one must make a reasonable estimate of potential yield attainable under the expected prevailing conditions. Using corn as an example, assume that an adapted hybrid will yield 18 834 kg/ha (300 bu/acre) in a given area when all inputs are optimum. Total aboveground dry matter can be calculated if we assume that the grain comprises 55% of the plant weight at high yield levels, and the dry matter in a bushel of corn is approximately 22.68 kg (50 lb). Thus, 18 834 × (50/56) × (100/55) = 30 575 kg/ha of dry matter [or 300 × 50 × (100/55) = 27 273 lb/acre].

Since the critical N percentage for the total dry matter is 1.2%, the internal N requirement is 366 kg/ha (327 lb/acre). Allowing 70% efficiency in plant uptake, the total amount of N required is 524 kg/ha (468 lb/acre). This is the minimum amount needed; therefore, it should be increased to 560 kg/ha (500 lb/acre) to allow for a small excess of N.

In the above calculations, it is interesting to note that the internal N requirement of 366 kg/ha (327 lb/acre) for a yield of 18 830 kg/ha (300 bu/acre) is slightly more than 194 g/kg (1 lb/bu). This provides a simple approximation of the N requirement for corn, based on sound, scientific evidence that is similar to the "rule-of-thumb" method (1 lb N/bu corn) that has been used by farmers in the Corn Belt for supplying N fertilizer. It should be emphasized, however, that the calculations are based on actual N uptake, and allowances must be made for fertilizer efficiency and soil N availability. In this case, a fertilizer application of 336 kg N/ha (300 lb/acre) for a corn yield of 18 830 kg/ha (300 bu/acre) would be appropriate only if the soil were supplying about 200 kg N/ha (179 lb/acre) of available N.

With the high plant populations required for maximum yields, it is likely that with soils of low N content, most of the mineralized soil N would be required for root growth. In many cases, however, the residual NO_3^- from previous cropping and the mineralization of soil organic N should be taken into consideration. Ideally, an estimate of the N mineralization potential of the soil (Stanford & Smith, 1972) provides a sound basis for predicting fertilizer N requirements. In the absence of this information, it is possible to obtain a rough estimate of soil N availability from previous cropping experience. For example, if a 6277 kg/ha (100 bu/acre) corn crop requires only 84 kg of fertilizer N/ha (75 lb/acre), then the soil is supplying about 112 kg/ha (100 lb/acre), assuming equal efficiency of fertilizer and soil-derived N of about 70%. Since it is necessary to have almost ideal soil conditions with respect to moisture, temperature, and aeration for the attainment of maximum potential yields, mineralization of soil organic N would also be at a maximum during the cropping period.

Once the N requirement for maximum potential yield has been determined, it is necessary to apply the fertilizer N in an effective manner for crop utilization. In the above example, the 560 kg N/ha (500 lb/acre) rate represents about 1680 kg/ha (1500 lb/acre) of NH_4NO_3 fertilizer. Experimental data for such high rates of N on grain crops are rather limited. With forage crops, however, rates of 1008 to 1344 kg N/ha (900–1200 lb/acre) are not too uncommon. In such cases, the grass sod is already established, several harvests are removed annually that contain high percentages of N, and split applications of N can be broadcast easily without damaging the crop. The usual practice with high N applications to row crops, such as corn, is to apply part of the N at or near seeding time, and the remainder as a side-dressing while the plants are still young enough not to be damaged appreciably by the side-dressing treatment. This is not necessarily the most effective way of promoting maximum N utilization and growth by the crop.

An alternative method of applying N to crops is through irrigation water. This has been practiced to some extent in the major irrigation regions, but rarely in the more humid areas. It is being recognized, however, that even under normal rainfall conditions that produce good crops, water is usually limiting at certain periods for maximum yields. In order to achieve the goal of maximum potential yields, therefore, it is almost mandatory to provide supplementary irrigation. This provides the key to better N fertilizer management practices, since N can be applied in accordance with actual plant needs for both N and water. Sprinkler irrigation is often used as a supplementary water source for large areas, while trickle or drip irrigation is suitable for smaller areas. The latter probably offers the best means of applying water and N most uniformly at frequent intervals.

How does the hypothetical N requirement for corn given above compare with actual N fertilizer applications used to obtain record high yields? In Table 1, the 101.1 t/ha silage yield provides a good comparison, since it was obtained from New Jersey research plots, and the grain yield (19 583 kg/ha or 312 bu/acre) was measured as well. On a dry-matter basis, the grain comprised about 54% of the total aboveground production of 32 356 kg/ha (28 864 lb/acre). If this amount of dry matter contained 1.2% N, then 388 kg N/ha (346 lb/acre) were in the harvested crop. Assuming 70% efficiency in fertilizer N uptake, 554 kg N/ha (495 lb/acre) should have been applied as a minimum rate. The actual amount applied in this case was 560 kg N/ha (500 lb/acre). Without an analysis of the plant material, the adequacy of this N rate is unknown; however, it agrees fairly well with the theoretical N requirement.

The record high yield of 22 131 kg/ha (352.6 bu/acre) of corn given in

Table 1 presents a case that is not so easily analyzed. An internal N requirement for such a crop was calculated to be 407 kg N/ha (363 lb/acre), or 576 kg N/ha (514 lb/acre) as fertilizer N at 70% efficiency. The total amount of fertilizer N added was 426 kg/ha (380 lb/acre), indicating that a considerable amount of mineralized N from the soil was utilized by the crop or that fertilizer efficiency was > 70%. This points to the need for information concerning previous cropping history, organic amendments, effective rooting depth, and N mineralization rate to refine predictions of fertilizer N needs for maximum potential yields. In addition, it is useful, after the crop is grown, to analyze the plant material to determine the sufficiency of the fertilizer N application rate.

Space does not permit an analysis of all crops in terms of N requirements, but the procedure should be similar to that used for corn. It is more difficult to determine the N requirement of crops with indeterminate growth, but as more data become available from N rate experiments, it should be possible to determine realistic N fertilizer rates for maximum potential yields. Currently, there is a great need for data obtained at high fertilizer levels in which potential yields are being explored.

V. SUMMARY AND CONCLUSIONS

Average yields of major crops have increased steadily since the latter part of the 19th century, but it is evident that they are still far below maximum potential yields. All of the factors contributing to yield must be at an optimum to realize the potential yielding ability of crop plants. Record crop yields attained in recent years have resulted mainly from improved cultivars, pest control, improved soil physical and chemical conditions, moisture control, and good management practices. Perhaps the most critical element in promoting extremely high yields is supplying nutrients, especially N, in sequence with crop demand without creating toxic conditions.

An estimate of the fertilizer N needs of a particular crop cannot be determined quantitatively without a knowledge of the crop's requirement for this element. The N requirement is defined as the minimum amount of N in the aboveground portion of crops associated with maximum production. The N requirements for a number of crops have been previously determined (Stanford, 1966, 1973) using data from N rate experiments in which the total N content of the plants was known. Corn, with a N requirement of approximately 12 g/kg (1.2%) in the stover plus grain, furnishes a good example for calculations of fertilizer N needs. Total dry matter production can be estimated from grain yields. The calculated internal N requirement of the crop must be adjusted by an efficiency factor for fertilizer N uptake, and the actual fertilizer N application should be further refined by an estimate of the N derived from the soil.

Application of large amounts of N fertilizer necessary to attain maximum yields may present problems for certain crops (toxicity, lodging, etc.). Generally, some form of irrigation will be necessary to exempt plants from periods of water stress, even in humid areas where rainfall is sufficient for good plant growth. Additions of N fertilizers periodically throughout the growing season in irrigation water provides the best means of fertilization without injury to plants. Adequacy of N fertilizer applications can be determined best at the end of the growing season by an analysis of samples of the total dry matter produced to see if the internal N requirement has been met. In addition, an analysis of soil NO_3^- after plants are mature would provide an estimate of excess N not utilized by the crop.

Although the internal N requirement of most grain crops can be readily assessed, crops in which qualitative factors (such as protein or sugar content) are important require an assessment that balances the qualitative and quantitative features of maximum production. At the present time, there are insufficient data available to assess the N requirements of many crops, and further research is needed in this area. It can be concluded that the high-yield potentials of modern plant cultivars have not been fully exploited, and that continued research is required to elucidate the effects of high inputs of all factors, especially N, that contribute to the attainment of maximum yield potentials.

REFERENCES

Dennis, R. E., and H. I. Owens. 1980. The components of high yields. Cultural practices. Crops Soils 32(5):10-13.

Evans, L. T. 1980. The natural history of crop yield. Am. Sci. 68:388-397.

Johnson, R. R. 1980. The components of high yields. How high can yields go? Crops Soils 32(9):9-11.

Johnson, V. A., A. F. Dreier, and P. H. Grabouski. 1973. Yield and protein responses to nitrogen fertilizer of two winter wheat varieties differing in inherent protein content of their grain. Agron. J. 65:259-263.

Pierre, W. H., L. Dumenil, and J. Henao. 1977a. Relationship between corn yield, expressed as a percentage of the maximum, and the N percentage in the grain. II. Diagnostic use. Agron. J. 69:221-226.

Pierre, W. H., L. Dumenil, V. D. Jolley, J. R. Webb, and W. D. Shrader. 1977b. Relationship between corn yield, expressed as a percentage of the maximum, and the N percentage in the grain. I. Various N rate experiments. Agron. J. 69:215-220.

Stanford, G. 1966. Nitrogen requirements of crops for maximum yield. p. 237-257. *In* W. H. McVickar et al. (ed.) Agricultural anhydrous ammonia—technology and use. Soil Sci. Soc. of Am., Madison, Wis.

Stanford, G. 1973. Rationale for optimum nitrogen fertilization in corn production. J. Environ. Qual. 2:159-166.

Stanford, G., and S. J. Smith. 1972. Nitrogen mineralization potentials of soils. Soil Sci. Soc. Am. Proc. 36:465-472.

Viets, F. G., Jr. 1965. The plant's need for and use of nitrogen. *In* W. V. Bartholomew and F. E. Clark (ed.) Soil nitrogen. Agronomy 10:503-549. Am. Soc. of Agron., Madison, Wis.

18

B. R. Bock
National Fertilizer Development Center
Tennessee Valley Authority
Muscle Shoals, Alabama

Efficient Use of Nitrogen in Cropping Systems

Efficient use of N in cropping systems is often viewed from agronomic, economic, or environmental perspectives. A given N management system may provide highly efficient use of N from one perspective but be relatively inefficient from another. This chapter reviews basic factors affecting N use efficiency (NUE) from agronomic, economic, and environmental perspectives and discusses interrelationships among these perspectives. The agronomic section reviews ways of characterizing NUE based on yield and N recovery in relation to N inputs and discusses physical potentials and limitations for improving these relationships. Since Chapt. 17 (this book) deals with the role of N in attaining high yield potentials, the agronomic section in this chapter emphasizes factors affecting the slope of yield curves rather than factors affecting maximum yield levels with a nonlimiting N rate. The economic section reviews factors that determine optimum N rates and discusses economic considerations for optimizing inputs (e.g., application practices and nitrification inhibitors) that affect the slope of yield curves. The environmental section examines relationships between N fertilizer recovery and environmental impact of N fertilizer use and also environmental implications of maximizing return from N fertilizer use.

I. AGRONOMIC PERSPECTIVE

A. Characterizing Efficiency

The phrase *N use efficiency* (NUE) usually has referred to relationships between yield and N rate (yield efficiency), N recovered and N rate (N recovery efficiency), or yield and N recovered (physiological efficiency). For

Nitrogen in Crop Production.

each of these relationships, several definitions have been used to characterize NUE. This subsection characterizes NUE from an agronomic perspective and discusses limitations of definitions for comparing NUE among N management systems and cropping situations.

1. YIELD EFFICIENCY

Yield efficiency is defined here as the average yield increase per unit of applied N for a specified portion of a yield curve. Examples of yield efficiency values and methods of calculation are presented in Table 1, column 7 (incremental basis) and column 10 (cumulative basis). These efficiency values are based on N rate and yield data presented in columns 1 and 3 of Table 1 and illustrated in Fig. 1b.

When yield response curves are curvilinear, yield efficiency values vary not only with shape of the response curve but also with that portion of the curve used for calculation. For example, depending on the range of N rates used, yield efficiency values calculated on an incremental basis in column 7 of Table 1 ranged from 9.7 to 19.6 kg corn grain/kg applied N for spring application and ranged from 2.8 to 36.2 kg corn grain/kg applied N for side-dress application. Thus, yield efficiency values have little merit for comparison between N management systems and cropping situations unless accompanied with information concerning shape of the response curve and N rates used for calculation.

Yield efficiency indexes with respect to N rate have been used to characterize yield response curves for N rates up to that required to achieve maximum yield (Y_m). These indexes greatly expedite characterization of yield efficiency, particularly when yield response to N is curvilinear; one index characterizes yield efficiencies for an entire yield curve with N rates up to that required for Y_m. Also, yield efficiency indexes provide a basis for comparing N management systems and cropping situations when Y_m and/or yield with no N applied (Y_o) differ among the response curves to be compared.

The efficiency factor (C) in the Mitscherlich equation for exponential yield response curves is a well-known yield efficiency index defined as follows: $dY/dN = C(Y_m - Y)$ and $Y = Y_m(1 - e^{-CN})$ for $Y_o = 0$. Capurro and Voss (1981) defined a yield efficiency index (E_N) for quadratic yield responses to applied N, which relates the derivative of yield with respect to N rate to relative yield as follows: $dY/dN = 10\,E_N(1 - Y/Y_m)^{1/2}$. This yield efficiency index, which is similar to C in the Mitcherlich equation, ranged from 2.5 to 9.3 in 33 corn experiments conducted on well, moderately well, and somewhat poorly drained soils in Iowa. The yield efficiency index for quadratic yield responses (E_N) is used to characterize yield efficiencies over a range of N rates in Fig. 1 and in other examples in the remainder of this chapter. Bartholomew (1972) argued that cereal yield responses to N fertilizer are virtually rectilinear for a given site in a given year; therefore, the slope of the yield vs. N rate curve with $Y < Y_m$ was defined as a yield efficiency index. When yield response to N is rectilinear, the yield response slope with $Y < Y_m$ can be easily compared between N management systems and cropping situations regardless of the level of Y_m and Y_o.

Table 1—Examples of N use efficiency expressions and values for corn (*Zea mays* L.) as affected by N rate and application timing.†

				N use efficiency					
				Incremental basis			Cumulative basis		
N rate (N) (1)	N recovered (NR) (2)	Corn grain yield (Y) (3)	i (4)	Recovery efficiency $\left(\frac{NR_i - NR_{i-1}}{N_i - N_{i-1}}\right) \times 100$ (5)	Physiological efficiency $\frac{Y_i - Y_{i-1}}{NR_i - NR_{i-1}}$ (6)	Yield efficiency $\frac{Y_i - Y_{i-1}}{N_i - N_{i-1}}$ (7)	Recovery efficiency $\left(\frac{NR_i - NR_o}{N_i}\right) \times 100$ (8)	Physiological efficiency $\frac{Y_i - Y_o}{NR_i - NR_o}$ (9)	Yield efficiency $\frac{Y_i - Y_o}{N_i}$ (10)
kg/ha				%	kg corn/kg N		%	kg corn/kg N	
0	81	4520	0	--	--	--	--	--	--
				Spring application					
45	93	5400	1	26.7	73.3	19.6	26.7	73.3	19.6
90	108	6280	2	33.3	58.7	19.6	30.0	65.2	19.6
180	140	7150	3	35.6	27.2	9.7	32.8	44.6	14.6
				Side-dress application					
45	107	6150	1	57.7	62.7	36.2	57.7	62.7	36.2
90	130	7030	2	51.1	38.3	19.6	54.4	51.2	27.9
180	146	7280	3	17.8	15.6	2.8	36.1	42.5	15.3

† Yield and N recovered data from Olson et al., 1964.

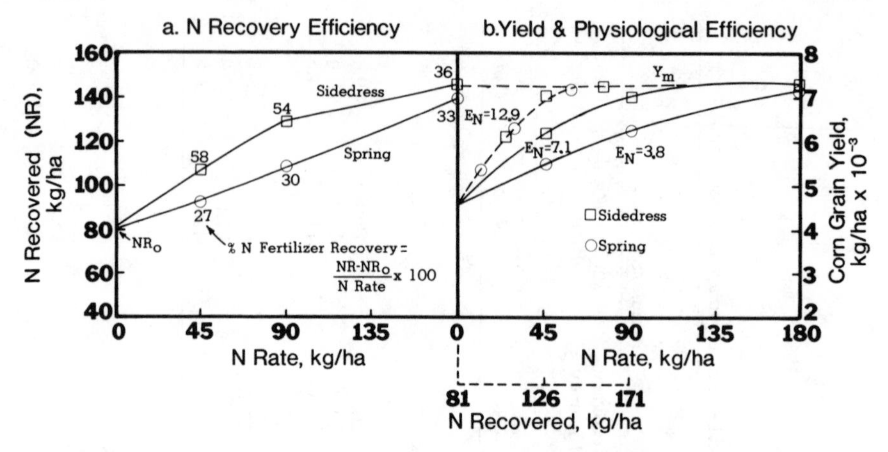

Fig. 1—Illustration of (*a*) N recovery efficiency and (*b*) physiological and yield efficiency data in Table 1. Numbers on yield curves are E_N values as defined by Capurro and Voss (1981) for $Y < Y_m$. $E_N = 1/5\ (B_2\ Y_m)^{1/2}$ with $N < -B_1/2B_2$, where $Y = Y_o + B_1 N + B_2 N^2$, $B_1 = 10\ E_N\ (1 - Y_o/Y_m)^{1/2}$, and $B_2 = -25\ E_N^2/Y_m$. $Y = Y_m$ and $E_N = 0$ with $N > -B_1/2B_2$. A segmented regression model (SAS Inst., 1979) was used to determine Y_o, B_1, and B_2.

2. N RECOVERY AND PHYSIOLOGICAL EFFICIENCIES

Formulae for calculating N recovery and physiological efficiencies on incremental and cumulative bases are presented in Table 1. As can be seen, yield efficiency equals the product of N recovery and physiological efficiencies. Nitrogen recovery efficiency also has important implications for the environmental impact of N fertilizers, as will be discussed in more detail later in this chapter.

Characterization of physiological efficiency is directly analogous to characterization of yield efficiency where N recovered is substituted for N rate as indicated in Table 1 and Fig. 1. Physiological efficiency, depicted by the dashed curve and dashed abscissa in Fig. 1 and columns 6 and 9 in Table 1, represents the theoretical attainable yield efficiency with N recovery efficiency equal to 100%. In this example, N application timing had little effect on physiological efficiency as indicated by the excellent fit of points on one line in Fig. 1 ($R^2 = 0.99$) for both spring and side-dress application times.

Percent N fertilizer recovery in the aboveground plant parts is probably the most commonly used definition of NUE. Perhaps this is because more emphasis has been placed on technology and practices for altering N recovery efficiency than for altering physiological efficiency as a means of improving yield efficiency and because of implications of N recovery efficiency for environmental impact.

The traditional approach for measuring percent recovery of applied N is by the difference method in which total N recovery in fertilized plots minus N recovery in unfertilized plots is divided by the quantity of applied N. This method indicates only the apparent percent recovery of applied N because (i) applied N can affect transformation of indigenous N in the soil and (ii) the larger and more vigorous root system in the fertilized plots usually recovers more N from the soil, excluding that added as fertilizer, than does the unfertilized crop. Calculation of percent N recovery by the difference method is shown in Table 1 on incremental and cumulative bases in columns 5 and 8, respectively. Tracer techniques provide measures of N recovered from both fertilizer and the composite of other N sources. Advantages and disadvantages of the difference and tracer methods of measuring percent recovery of applied N are discussed in detail by Hauck and Bremner (1976) and Broadbent (1981).

B. Physical Potentials and Limitations for Improved N Use Efficiency

From a physical or agronomic perspective, there appears to be considerable potential for improving NUE in cropping systems. With current understanding and technology, however, there are also some important limitations. Some of these potentials and limitations are reviewed below.

1. N RECOVERY

a. Maintaining N in Forms Available to Plants—Allison (1966) indicated that recovery of applied N under average field conditions is often no greater than 50 to 60%, even if immobilization is taken into account. Kundler (1970) reported a range of 30 to 70% recovery of applied N by crops during the year of application with 10 to 40% of applied N incorporated into organic matter, 5 to 10% lost by leaching, and 10 to 30% lost in gaseous form. Bartholomew (1972) referred to the commonly cited figure of 50% recovery of applied N by corn and wheat (*Triticum aestivum* L.) and indicated that recoveries of 70 to 80% by these crops are physically feasible in most situations with efficient N application timing and placement. A recent review of N budgets by Legg and Meisinger (1982) indicates that the above observations are still applicable today. Thus, from an agronomic perspective there is considerable opportunity for improving efficiency of N recovery by reducing the fraction of plant-available N introduced into cropping systems that is either lost from the root zone or immobilized in soil organic matter.

Leaching, denitrification, immobilization, and NH_3 volatilization are the processes known to be of practical significance in lowering availability of N to plants. Of these processes, leaching and denitrification usually are considered of greatest importance (Legg & Meisinger, 1982). However, many studies on which this conclusion is based involved precautions such as injection or mechanical incorporation for preventing NH_3 volatilization from fertilizers. These practices are not always practical in production agriculture. The full significance of immobilization for NUE has not been fully determined, because unknown quantities of immobilized N are remineralized in the long term (Tucker & Hauck, 1978). Although a theoretical basis for NH_3 volatilization from plant foliage (Wetselaar & Farquhar, 1980) and chemodenitrification (Bremner & Nelson, 1968) has been established, the practical significance of these pathways has not been determined.

General approaches for minimizing the impact of leaching, denitrification, and immobilization include: (i) optimum use of the crop's ability to compete with these processes for plant-available N and (ii) direct lowering of the rate and duration of the loss processes themselves. Two key elements of the first approach are assuring that N inputs do not exceed crop assimilation capacity, and applying N in phase with crop demand (Legg & Meisinger, 1982; Keeney, 1982; Scarsbrook, 1965). Examples of the second approach are use of altered N sources (Chapt. 13 and Chapt. 36, this book) and nitrification inhibitors (Kurtz, 1980) to control or partially control transformation of N to NO_3^--N, thereby directly lowering leaching and denitrification losses. Improved irrigation practices are also instrumental in directly lowering these N-loss processes (Chapt. 21, this book).

Although the full potential of the above approaches for lowering denitrification has not been realized in production agriculture, these approaches appear to have fundamental limitations for completely eliminating denitrification. Broadbent and Clark (1965) classified N loss by denitrification in two categories: rapid and extensive losses associated with heavy rains and irrigations, and continuing small losses over an extended period of time in anaerobic microsites. Nitrogen and irrigation management practices generally affect the former more than the latter. Hauck and Bock (1980) concluded that some control over the extent of denitrification is possible but, in general, at the present time, losses (presumably via denitrification) ranging from 10 to 45% of applied N are inevitable under most cropping conditions.

Pratt (1979) summarized what seems to be a consensus concerning the potential for lowering N leaching losses.

> The potentially leachable N in soils, assuming proper timing and placement of fertilizers and relatively efficient irrigation, should increase very little with increases in the amount of fertilizer N applied within the range of N rates where yields are increased with increased application rate. This is the result of increased uptake of available N as the N rate increases. However, when the point of maximum yield is exceeded, the potentially leachable NO_3^--N should increase substantially with N rate.

Barraclough (1979) cited the inability to accurately predict the N requirement of crops throughout the growing season as the major obstacle to efficient N timing and rate for lowering NO_3^--N leaching losses. More reliable methods are also needed for predicting the supply of available N from sources other than N fertilizer (Chapt. 26, this book). These operational constraints also limit the extent that denitrification can be minimized in production agriculture at the present time.

Ammonia volatilization from N fertilizers is associated primarily with application of ammoniacal or urea-containing fertilizers on the soil surface without immediate incorporation (Nelson, 1982). Crop recovery of N is often not a factor in lowering NH_3 volatilization from the soil surface because losses occur before surface-applied N reaches plant roots, or even before a crop is established. When immediate incorporation is physically impractical or uneconomical, as in no-till systems or most topdress applications, NH_3 losses from surface-applied fertilizers can be lowered by avoiding environmental conditions that provide a high potential for NH_3 loss (Terman, 1979). Another option is to modify fertilizers chemically or physically to reduce NH_3 loss (Chapt. 36, this book).

b. Recovery of Plant-Available Forms—Nitrogen recovery efficiency is sometimes limited by a crop's ability to recover N that remains in a chemically available form in the root zone. This may be due to either the distance between N and the roots combined with rate of N movement to roots (Barber, 1976; Bartholomew, 1972) or perhaps a lower-than-optimum rate of N absorption by roots with relatively low soil solution levels of plant-available N (Huffaker & Rains, 1978; Barber, 1976). There is some opportunity for manipulating these factors to improve N recovery efficiency.

Most N recovered by plants must move to root surfaces before being absorbed since roots usually occupy only about 1 to 2% of the soil volume in the root zone (Barber, 1976). Providing a relatively high soil moisture level is probably the most effective way of enhancing rate of NO_3^--N movement to roots, since NO_3^--N moves in soil largely by mass flow. Bartholomew (1972) indicated that soil moisture levels need to be lowered and replenished a number of times for high recovery of soluble N, because significant quantities of soluble N remain with water held at tensions above the wilting point.

Root density in the soil affects the distance N must travel to reach root surfaces. If NO_3^--N can be absorbed rapidly per unit of root, root density does not have to be high to recover NO_3^--N. However, plants with roots that have capability for a high rate of N absorption and that can reduce NO_3^--N and NH_4^+-N in soil solution to low levels may be needed to maximize N recovery efficiency (Barber, 1976). Root density may have a greater effect on recovery of NH_4^+-N than NO_3^--N since NH_4^+-N moves relatively slowly to roots. This may be an important consideration when a nitrification inhibitor is used or when NH_4^+-N fertilizer is applied in increments in phase with crop demand, resulting in a relatively large fraction of plant-available N being in the NH_4^+-N form.

2. PHYSIOLOGICAL EFFICIENCY

There are obvious limitations for increasing yield relative to quantity of N recovered by crops if a crop is to be used principally as a protein source. However, there is some potential for increasing the fraction of recovered N translocated from nonharvested to harvested components of these crops (Gerloff, 1976). There is greater opportunity for increasing yield relative to quantity of recovered N in crops not used principally as a protein source, because N content of the harvested products is less critical. Maranville et al. (1980), for example, observed a 20% difference in physiological efficiency among 12 grain sorghum [*Sorghum bicolor* (L.) Moench] hybrids.

Other opportunities for improved physiological efficiency relate to N fertilizer management. Nitrogen recovery as affected by time of application, placement, and soil moisture content can influence the relative production of harvested and nonharvested plant parts (Viets, 1965). Form of N recovered by plants can also affect physiological efficiency (Chapt. 4, this book). Optimum N rates are important from the standpoint of physiological as well as N recovery efficiency. With excessive levels of plant-available N, excessive vegetative growth and lodging may result and most crops tend to recover more N than required for optimum growth, resulting in decreased physiological efficiency. Yields or quality are sometimes sharply depressed with excessively high levels of recovered N in small grain crops and other crops such as cotton (*Gossypium hirsutum* L.) and sugarbeets (*Beta vulgaris* L.) (Chapt. 41, this book).

Improved N recovery efficiency may not always be totally compatible with improved physiological efficiency. For example, inhibiting nitrification may increase N recovery efficiency but lower physiological efficiency in some cropping systems. Timing of N application can affect N recovery efficiency and also physiological efficiency, particularly if timing is delayed too long. In such situations, optimizing both recovery and physiological efficiency rather than maximizing one or the other is required for maximum yield efficiency.

II. ECONOMIC PERSPECTIVE

The primary objective of this section is to review the fundamentals of maximizing farm income from N fertilizer use. Secondary objectives are to provide a basis for discussing (i) economic constraints on NUE from an agronomic perspective and (ii) environmental implications of maximizing return from N fertilizer use. These objectives are first pursued assuming that type of yield response, prices for N and related inputs, and crop prices can all be predicted in advance with reasonable certainty, since much of the literature is based on this assumption. Following this classical treatment, the above objectives are addressed assuming uncertain yield response and prices. These latter remarks are largely the author's ideas. They are included mainly to underscore information needs since the literature provides little information about maximizing return from N with uncertain yield response and prices. Finally, a third subsection touches on indirect effects of N fertilizer management on return from cropping systems. Again, this subsection is included primarily to call attention to information needs.

A. Return from N: Certain Yield Response and Prices

Major determinants of return from N fertilizer use are (i) the yield increase from applied N, (ii) crop price, (iii) N rate, (iv) N fertilizer price, and (v) costs of N fertilizer use, which are fixed relative to N rate. Nitrogen application and additive (e.g., nitrification inhibitor) costs are considered here as fixed relative to N rate. When these costs vary with N rate they should be included with N fertilizer as variable costs. Nitrogen fertilizer management affects each of the above factors at the farm level except crop price, assuming that selection of N source affects the price/unit of N paid by the farmer. This subsection discusses relationships among the above factors, assuming prior selection of cropping system inputs other than those directly associated with N fertilizer use.

1. OPTIMUM N RATE

a. Unlimited Capital—With unlimited capital and diminishing returns from a continuous variable input such as N, maximum return from N occurs at the point where marginal return equals zero (i.e., where the added cost from the last increment of N equals the value of the added yield from that increment). Thus, with unlimited capital, optimum N rate depends on shape of the yield curve (i.e., yield efficiency index and Y_m) and N/crop price ratio as discussed by Capurro and Voss (1981), but is independent of the vertical position of the yield response curve.

Engelstad and Parks (1971) showed that a fairly wide range in choice of N rate for corn is possible without seriously affecting either return from N or yield. This was defined as the "relevant range," which should be determined for cropping systems as a basis for N rate recommendations. Figure 2b, which is based on yield curves in Fig. 2a and a N/corn price ratio

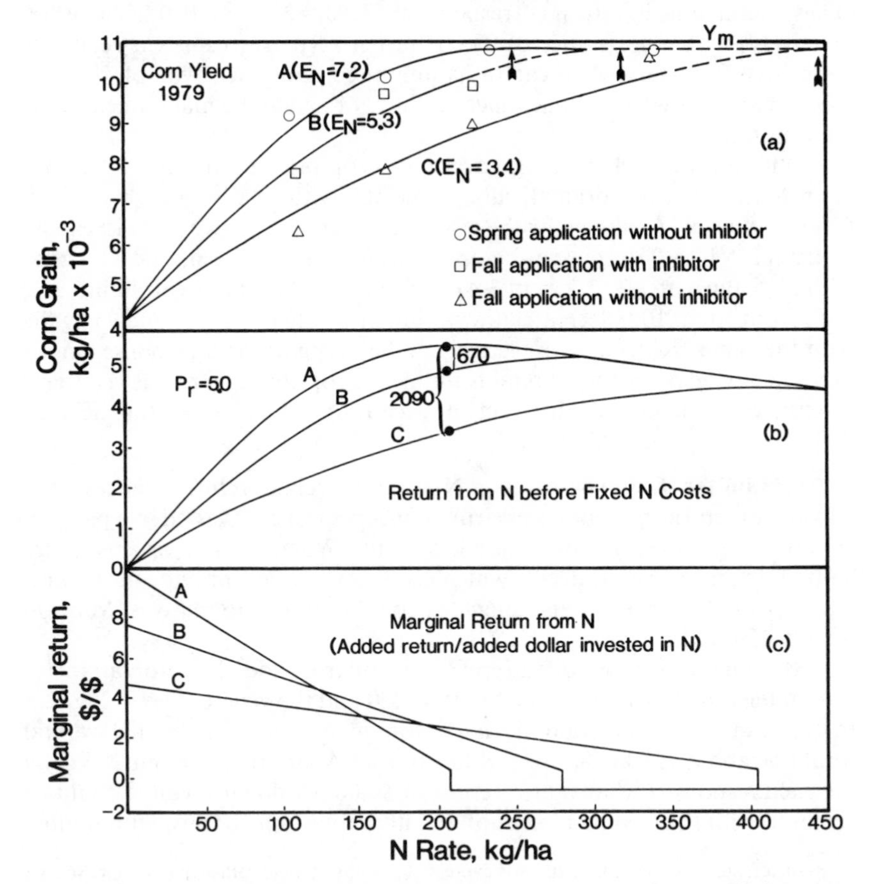

Fig. 2—Typical effects of N fertilizer management on (*a*) E_N values (yield data from Hoeft et al., 1981), (*b*) return from N before fixed N costs, and (*c*) marginal return from N. Arrows denote minimum N rate to achieve Y_m.

Table 2—Effect of yield efficiency index (E_N) and N/corn price ratio on the economically optimum N rate.†

E_N	Minimum N rate to achieve maximum yield	P_r ($ per kg N/$ per kg corn)‡ 2.5	5.0	10.0
		Economically optimum N rate§		
	kg/ha			
3.4	500	450	405	310
5.3	320	300	280	245
7.2	240	225	215	195

† Data based on yield response curves in Fig. 2a.
‡ P_r values of 2.5, 5.0, and 10.0 correspond to 1:20, 1:10, and 1:5 $ per pound N/$ per bushel corn, respectively.
§ N rate with zero marginal return.

(P_r) of 5.0 ($ per kg N/$ per kg corn), illustrates that return from N on a particular yield curve is rather insensitive to N rate.[1] For example, return from N is lowered by < 75, 48, and 18 kg corn/ha, respectively, for curves A, B, and C by adjusting N rate ± 25 kg/ha from the optimum level. (These reductions in return correspond to $3.60, $2.20, and $0.90/acre, respectively, with a corn price of $3.00/bushel.) Return from N is relatively insensitive to N rate when capital is unlimited and when type of yield response can be predicted in advance so that N rate can be maintained in the relevant range.

With a given yield response curve, N/crop price ratio affects the optimum N rate, but not dramatically, especially with relatively high yield efficiency indexes. As shown by the examples in Table 2, a twofold increase in P_r from 2.5 to 5.0 decreased the optimum N rate by 10, 20, and 45 kg/ha with E_N values of 7.2, 5.3, and 3.4, respectively. Another twofold increase in P_r from 5.0 to 10.0 decreased the optimum N rate by 20, 35, and 95 kg/ha with the same E_N values, respectively. When type of yield response can be predicted in advance and capital is unlimited, optimum N rate is relatively insensitive to changes in the N/crop price ratio, except when the yield efficiency index is relatively low.

b. Limited Capital—Adding N until marginal return reaches zero maximizes return, but not necessarily return per dollar invested in a particular cropping enterprise or farming system. When capital or credit are limited, the farmer must decide which production inputs provide the highest return per dollar invested and allocate capital to those alternatives (Munson & Doll, 1959).

Marginal return or added return per dollar of added cost on an incremental basis is shown in Fig. 2c for P_r = 5.0 and the yield curves in Fig. 2a. In this example, the economically optimum N rate with limited capital would be about 125 kg/ha for yield responses A and B and about 25 kg/ha for yield response *C* if an added return of $4.00 per dollar invested could be achieved with the next increment of an alternative input. Thus, with limited

[1] Return from N in Fig. 2b, 3, and 4 is expressed in units of yield rather than dollars so that stated N/corn price ratios will have more general applicability.

capital the optimum N rate is highly dependent on yield efficiency index, if other inputs provide a marginal return higher than or comparable to that from N fertilizer; in such cases, the optimum N rate is less than the N rate associated with zero marginal return.

With a given yield curve, marginal return from N is inversely proportional to the N/crop price ratio. Consequently, the N/crop price ratio may be expected to greatly affect optimum N rate when capital is limited, more so in most cases than when capital is unlimited.

Crop quality considerations may also result in the economically optimum N rate being significantly different from the minimum required for Y_m. For example, economically optimum N rates for barley (*Hordeum vulgare* L.) are often lower than required for Y_m to maintain protein levels acceptable for the brewing industry (Olson & Koehler, 1968). Alternatively, N rates higher than the minimum required for Y_m are sometimes used to achieve either desired crop uniformity as with lettuce (*Lactuca sativa* L.), for example (Pratt, 1979), or desired crop quality as in the case of wheat protein levels (Olson & Koehler, 1968).

2. OPTIMUM YIELD EFFICIENCY INDEX

Consideration also needs to be given to optimizing the yield efficiency index. Selection of N source, additives, and application practices affects return from N fertilizer use, primarily by affecting the yield efficiency index. Nitrogen source, additives, timing, and placement usually have relatively little direct affect on Y_m levels with N rate nonlimiting but can affect Y_m levels indirectly as discussed later in this chapter (subsection C).

When type of yield response can be predicted, N rate can be adjusted accordingly and yield reductions from a lower-than-expected yield efficiency index are not a consideration. In these situations, optimizing the yield efficiency index involves optimizing N management-N fertilizer substitutions. The upper limit on added investment in measures for achieving a higher yield efficiency index is in essence the value of the associated N rate reduction.

Typical relationships between return from N before fixed N costs (yield increase from N minus variable N costs) and E_N are presented in Fig. 3 for three P_r levels. These relationships were calculated using the optimum N rate for each E_N level and Y_o and Y_m values in Fig. 2a. Whereas yield efficiency at maximum yield [$(Y_m - Y_o)$/minimum N rate to achieve Y_m] is a linear function of E_N, return from N before fixed N costs follows diminishing returns with each unit increase in E_N. This means that each successive increase in E_N will justify a smaller added investment in fixed N costs to achieve the higher E_N.

Added return from N before fixed N costs is denoted in Fig. 3 for ΔE_N increments of 3.4 to 5.3 and 5.3 to 7.2. These E_N values correspond to the yield curves in Fig. 2a. Even with a P_r of 2.5 in metric units (e.g., $3.00/bushel corn and $0.15/pound N in English units), the added return ranges from $48.40/ha to $23.72/ha ($19.60 to $9.60/acre) for these ΔE_N increments. Added return from the same ΔE_N increments increases roughly

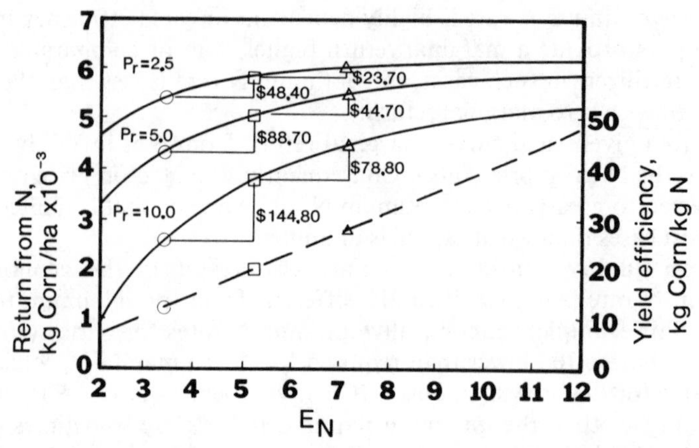

Fig. 3—Return from N before fixed N costs using optimum N rate (solid lines) and yield efficiency at maximum yield ($Y_m - Y_o$/minimum N rate to achieve Y_m) (dashed line) in relation to E_N. Values for spring application without inhibitor (Δ), fall application with inhibitor (□), and fall application without inhibitor (○) correspond to yield curves in Fig. 2a. Added return ($/ha) for ΔE_N increments is based on a corn price of $0.12/kg ($3.00/bushel).

proportionally with N price. These levels of added return would more than pay for use of a nitrification inhibitor or delaying N application from fall to spring in this example, even if delayed application required use of custom application services. (See Table 3 for typical costs of custom fertilizer application.)

The above example shows that relatively high yield efficiency indexes are generally economical when type of yield response can be predicted and N rates are adjusted accordingly. However, adverse effects of efficient N fertilizer management on overall labor and equipment efficiency may, in some cases, place limits on the yield efficiency index level that can be attained economically (see subsection C).

B. Return from N: Uncertain Yield Response and Prices

The previous discussion, based on the assumption of certain yield response and prices, discussed pertinent basic concepts concerning maximizing return from N fertilizer use. But it has serious limitations in many situations where the type of yield response with a particular N management system is highly unpredictable. Even with a particular N management system at a particular location, Y_m, soil N supply, and yield efficiency index often are highly unpredictable, primarily because of year-to-year changes in weather patterns. Crop and N prices are also unpredictable, but as noted by Iback and Williams (1971), several studies of yield response data have demonstrated that the gains to be made from ability to forecast prices and applying N accordingly would be substantially less than gains to be made from being able to foretell natural events that affect type of yield response.

1. OPTIMUM N RATE

Pratt (1979) indicated that with uncertain yield response, low N/crop price ratios, and low N costs relative to total crop production costs there is a tendency to add excess N to insure that there are no N deficiencies. The *insurance approach* to N fertilizer management is defined here as use of an N rate that is higher than the economic optimum in most crop years to minimize yield reductions in years with lower-than-normal yield efficiency index or lower-than-normal soil N supply and to achieve yield increases in years with higher-than-normal Y_m. Another prerequisite for use of the insurance approach is that neither yields nor crop quality, if of economic consequence, is adversely affected by receiving more N than required to achieve Y_m.

A hypothetical example is presented in Fig. 4 to show effects of N/crop price ratio on return from insurance N. This example assumes that soil N supply and Y_m remain stable from year to year, an E_N of 8.0 is obtained on an average of 2 out of 3 yr on a long-term basis, E_N is either 6.0 in each of the remaining years or 4.0 in each of the remaining years, and that the sequence of years with high and low E_N values cannot be predicted. With these assumptions, net return from insurance N with E_N alternating between 8.0 and 6.0 would be 35, −196, and −356 kg corn/ha per 3 yr with P_r of 2.5, 5.0, and 10.0, respectively. With E_N alternating between 8.0 and 4.0,

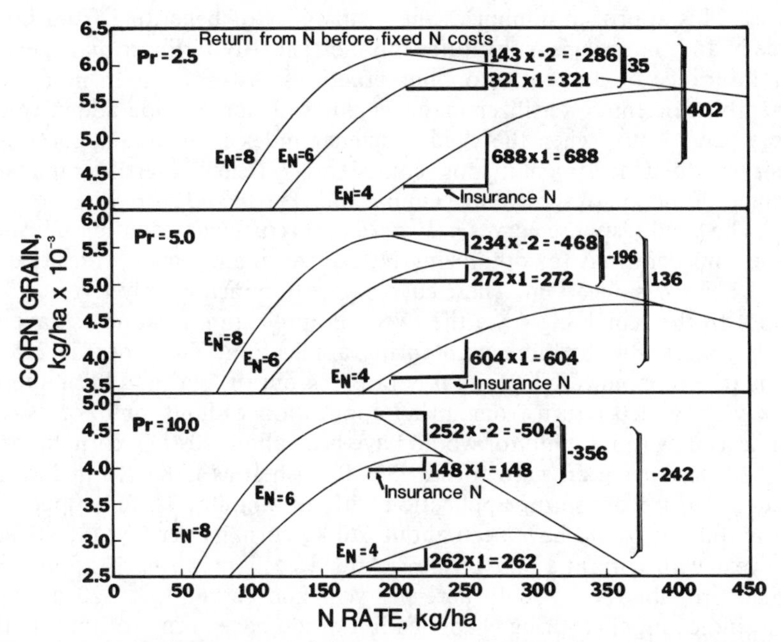

Fig. 4—Hypothetical example showing return from insurance N as affected by N/corn price ratio with the following assumptions: E_N of 8.0, 2 out of 3 yr; E_N of either 6.0 or 4.0 in the remaining years; and unpredictable sequence of E_N values.

net return from insurance N would be 402, 136, and −242 kg corn/ha per 3 yr with P_r of 2.5, 5.0, and 10.0, respectively.

The above example shows that long-term return from insurance N in a given cropping situation with uncertain yield response is highly dependent on N/crop price ratio. Thus, with uncertain yield response, N/crop price ratio markedly affects optimum N rate via the impact on return from insurance N. This contrasts to the relatively minor impact of N/crop price ratio on optimum N rate with a particular yield curve (i.e., with certain yield response) and unlimited capital.

When type of yield response is uncertain, the insurance approach will tend to be economical for high-value crops that do not give yield reductions with higher N rates than required for Y_m. Insurance N will tend to be economical for lower-value field crops such as corn and wheat only when N prices are relatively low—for example, when relatively low-priced anhydrous NH_3 is used. If N prices increase more rapidly than crop prices, the insurance approach may no longer be economical for lower-value field crops, even when relatively low-priced N sources are used.

2. OPTIMUM YIELD EFFICIENCY INDEX

A second basic approach for minimizing the adverse impact of uncertain yield efficiency index on return from N is use of N fertilizer management, which consistently provides relatively high yield efficiency indexes, even with weather potentially conducive to unusually low yield efficiency indexes. This approach minimizes the probability of benefiting from insurance N and, in essence, substitutes more efficient N fertilizer management for insurance N. In most cropping situations, investments in more consistently efficient N fertilizer management will not provide added return every year. Thus, when the yield efficiency index is uncertain, the upper limit on added investment in more consistently efficient N fertilizer management is a function of the average annual added yield and crop price.

The yield response curves in Fig. 2a and return curves in Fig. 2b show some considerations for optimizing N fertilizer management other than N rate when E_N is uncertain. These curves were obtained in only 1 yr in a 5-yr study. In the remaining 4 yr, the two fall application treatments gave E_N levels comparable to those with spring application. Therefore, if 215 kg N/ha (the economically optimum N rate in 4 out of 5 yr) had been applied each year, added return from spring application without inhibitor vs. fall application without inhibitor would have been about 2090 kg corn/ha per 5 yr ($20/acre per year, with corn at $3.00/bushel) as indicated in Fig. 2b. Added return from spring application without inhibitor vs. fall application with inhibitor would have been about 670 kg corn/ha per 5 yr ($6.40/acre per year with corn at $3.00/bushel). Added cost for spring N application would probably be < $6.40/acre per year and certainly < $20/acre per year, indicating that spring N application would have been profitable in this example, even though the E_N, and therefore yield, increased with spring N application in only 1 out of 5 yr.

The above example suggests that occasional marked reductions in yield efficiency index will justify considerable added investment in measures for

Table 3—Regional average charges for custom fertilizer application.†

Fertilizer type	Region				
	Northeast	North Central	South	Plains	Mountain & Pacific
	dollars/ha				
Bulk dry	9.40	6.20	7.15	4.70	8.40
Liquid	9.25	7.40	8.15	5.20	10.15
Side-dressing	10.40	9.90	9.00	7.90	14.80
Anhydrous NH_3	--	10.25	9.65	9.10	11.60
Aqua NH_3	--	8.75	8.15	--	13.35

† Doane's Agricultural Report Staff, 1981.

more consistently efficient N fertilizer management. As noted in Table 3, the cost per additional custom N application operation to delay part of the N application is generally about $12.50/ha per application ($5.00/acre per application) or less. The added cost of injecting N rather than broadcasting ranges approximately from $2.50 to $5.00/ha ($1.00 to $2.00/acre) as indicated by the difference between custom application costs for anhydrous NH_3 and bulk dry materials. Costs for use of a nitrification inhibitor are about $12.50/ha ($5.00/acre). Average annual added yields from improved N fertilizer management will frequently more than justify added costs such as these for more consistently efficient N fertilizer management.

More studies are needed to provide site-specific guidelines for maximizing return from aspects of N fertilizer management affecting the yield efficiency index. These studies should consider probabilities of achieving various yield efficiency indexes in relation to N fertilizer management, direct costs of N fertilizer management packages, and indirect effects of N fertilizer management on return from cropping systems as discussed in the next section (subsection C).

A third basic approach for minimizing the adverse impact of uncertain yield response on return from N is crop logging, monitoring the N status of the crop throughout the growing season, and applying N accordingly. This approach has been most common with crops such as cotton (Luckhardt & Ensminger, 1968) and sugarbeets (Humbert & Ulrich, 1968), which, as indicated earlier, give substantial reductions in yield or quality with excess N. As with the second approach, using more consistently efficient N fertilizer management, the upper limit on investments in monitoring crop N status is mainly a function of the average added yield and crop price over a number of years and the premium received for improved quality.

C. Indirect Effects of N Fertilizer Management on Return from Cropping Systems

Nitrogen fertilizer management can indirectly affect the Y_m level with N rate nonlimiting by influencing timeliness of field operations such as seedbed preparation and planting or herbicide application and cultivation. In the short term, farmers generally have little latitude in changing overall labor and equipment inputs in a given farming system. Thus, if N fertilizer application overlaps with other field operations, the effect of N timing and placement on timeliness of other field operations can be pronounced, particularly when weather severely limits number of work days in the field (Hoeft & Siemans, 1982; Tice, 1979). When available, custom application services can be used to alleviate adverse effects of N fertilizer management on timeliness of other operations. But Richards (1977) concluded that a farmer's concern for timely planting should not cause him to lose sight of profitable practices, such as banding, that improve the agronomic efficiency of fertilizer use. In the long term, farmers may choose to change parts of their overall labor and equipment input to accommodate N fertilizer management without adversely affecting timeliness of general field operations, a second possible indirect effect of N fertilizer management on return from cropping systems. These points are expanded below.

Consider, for example, possible effects of knifing vs. broadcasting N on timeliness of general field operations. Broadcasting is faster and can be achieved when soils are either too wet or too dry for knifing. Also, there generally are more options for custom application and joint herbicide-N application with broadcasting than knifing. For these reasons, broadcasting usually affects timeliness of general field operations much less than does knifing, particularly when weather is uncooperative. Compared with broadcasting, however, knifing can result in higher yield efficiency indexes.

Carrying the above example further, the probability of N application interfering with seedbed preparation and corn planting increases as application time is delayed from fall to early spring or from early spring to planting time. This is particularly true if N is knifed rather than broadcasted. Further delay in N application to side-dressing time can result in competition for labor and equipment for such operations as cultivation, herbicide application, and soybean [*Glycine max* (L.) Merr.] planting. But delaying N application from fall to side-dressing improves agronomic NUE by corn when weather conditions are potentially conducive to N loss by leaching and denitrification.

Selection of N source and use of a nitrification inhibitor both can affect N timing and placement decisions, and therefore also enter into timeliness considerations. For example, one use of nitrification inhibitors is to allow earlier N application without lowering the yield efficiency index. This allows labor and equipment workloads to be less compressed and spread over a longer period of time. Injection or immediate incorporation to achieve desired yield efficiency indexes is often more important for urea than ammonium nitrate, an example of N source selection impacting on N placement requirements. Thus, N source, additives, and associated timing and placement as a package not only directly affect the yield efficiency index, but also timeliness of general field operations. These, in turn, can indirectly affect Y_m levels.

More economic analyses are needed to evaluate effects of N fertilizer management on labor and equipment efficiency (i.e., land area farmed in a timely fashion with given labor and equipment inputs), Y_m levels as related to timeliness of general field operations, and ultimately, return from crop production. This need is especially critical because of evidence that high yield efficiency indexes and minimum adverse impact of N fertilizer management on timeliness and Y_m levels are not always economically compatible.

III. IMPLICATIONS OF AGRONOMIC AND ECONOMIC N USE EFFICIENCY FOR ENVIRONMENTAL IMPACT

Agronomic NUE is important from an environmental perspective, since N that is not recovered by crops can be lost from soil-plant systems and can adversely affect the environment. Keeney (1982) reviewed possible effects of N on environmental quality and concluded that NO_3^--N contamination of groundwater, eutrophication of surface waters, and possible ozone depletion by release of N_2O into the stratosphere are the primary environmental concerns related to N from soil-plant systems. Nitrate leaching and denitrification as a source of N_2O are considered here.

A. Environmental Impact Relative to N Recovery

As indicated earlier, incomplete recovery of applied N is partially due to N losses from soil-plant systems by NO_3^--N leaching and/or denitrification during the growing season. However, a significant portion of the unrecovered N usually remains in either immobilized or inorganic forms in the root zone at the end of the growing season and does not necessarily pose a hazard to the environment. In Fig. 5, for example, 36, 25, and 23% of the N applied during a 5-yr period was retained in the soil in either NO_3^--N or other forms (0- to 300-cm depth) with N rates of 112, 224, and 448 kg/ha, respectively.

Stanford (1973) concluded that with minimum N rates required for achievement of Y_m, 50 to 70% recovery of applied N is physically possible in most soil-plant systems and up to 30 to 50% of applied N can be immobilized during decomposition of corn stover and roots if that much inorganic N remains in the root zone at the end of the growing season. Similar

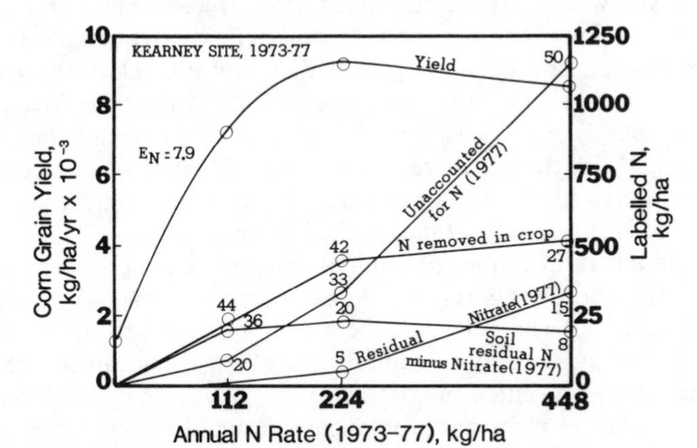

Fig. 5—Typical relationships among yield, N rate, and aspects of N economy in a cropping system with relatively high E_N and N recovery (Broadbent & Carlton, 1979). Numeric values on N curves are percentages of total N applied during a 5-yr period.

results may be expected from other cereal crops that also have relatively low N contents in their residues. Nitrogen immobilized during decay processes is remineralized very slowly (Bartholomew, 1965). Thus, immobilized N usually has little immediate impact on the environment. Timing of remineralization largely determines the environmental impact from immobilized N. The literature is inadequate for drawing generalizations concerning the long-term environmental impact of remineralized N.

The potential for adverse environmental impact with a given percent N recovery is relatively high when the N removed in the harvested portion of annual crops is a small fraction of the total N in the total weight of the crop (Pratt, 1979). Fraction of N removed in the harvested portions of annual crops is generally in the following order: forages > grains > vegetables > fruits.

Especially in drier climates, much of the residual inorganic N persists in the rooting zone until the following crop is established. Crop response to applied N is often highly correlated with residual inorganic N in these areas (Chapt. 22, this book), indicating crop recovery of residual N. In these situations, values for percent recovery of applied N for a single growing season do not indicate environmental impact (actual or potential) of cropping systems. Percent recovery of applied N over a period of years is a more reliable index of potential for adverse environmental impact but is still inferior to measurements of NO_3^--N leaching and N_2O evolution from denitrification or, at least, unaccounted-for N.

B. Return from N vs. Environmental Impact

Typical relationships among yield, unaccounted-for N, and N rate for cropping systems with moderately efficient irrigation (total application about 10% greater than evapotranspiration) and N application practices from an agronomic viewpoint (E_N = 7.9) are shown in Fig. 5. One-third of the ammonium sulfate fertilizer was applied preplant with the remainder side-dressed 3 weeks later. Both applications were incorporated to prevent NH_3 volatilization. With these conditions, type of yield response was relatively consistent from year to year and, therefore, relatively predictable. Residual NO_3^--N and average unaccounted-for N over the 5-yr period increased gradually with N rate until Y_m was achieved and then increased more sharply with further increases in N rate. Most of the unaccounted-for N was attributed by the authors to denitrification at the lower N rates with leaching increasing in importance at the higher N rates.

In the above example, percent of applied N unaccounted for at the economically optimum N rate would have been < 33%, which was associated with 224 kg N/ha per year, the actual amount depending on N/crop price ratio and availability of capital. Use of the economically optimum N rate would have prevented the dramatic increase in unaccounted-for N observed at higher N rates but would have resulted in limited N loss. Parr (1973) made a similar observation in a review of considerations for maximizing NUE, indicating that since the point of greatest economic return from N is somewhere below the point of Y_m, it should be possible to adjust

N rates for maximum return and minimum N loss to the environment. Economically optimum N rates may be less compatible with minimum N loss to the environment when crop quality considerations require higher N rates than the minimum required for Y_m or when insurance N is economical. In the latter case, maximum return from N becomes more compatible with minimum N loss as the N/crop price ratio increases.

Beyond use of economically optimum N rates, *long-term* improvements in N recovery efficiency are usually associated with decreases in N loss at a given N rate, at least with N rates less than or equal to those required for Y_m. As indicated earlier, return from N application practices, sources, and additives that improve the yield efficiency index is highly site- and farming-system–specific, making generalizations difficult. Kurtz (1980) concluded that even with judicious application, N losses due to leaching and denitrification probably cannot be avoided completely. However, physically attainable minimum levels of these losses are consistent with relatively high return from N in most cropping systems. Further lowering of leaching and denitrification losses will become profitable if crop and N fertilizer prices increase more rapidly than the cost of efficient methods of N use.

IV. SUMMARY AND CONCLUSIONS

Since NUE can be defined in many ways, care must be taken to specify which definition is used. From a physical or agronomic perspective, this involves specifying whether yield efficiency, physiological efficiency, or N recovery efficiency is intended and whether the type of efficiency of interest is characterized on an incremental basis, cumulative basis, or by use of a yield efficiency index. From an economic perspective, it is important to specify whether NUE is defined in terms of total return from N fertilizer use or in terms of marginal return (i.e., added return per added dollar invested in N fertilizer use).

Yield efficiency depends on both N recovery efficiency and physiological efficiency. There is considerable potential for improving N recovery efficiency since recovery is often no greater than 50 to 60%, even if immobilization is considered. With current understanding and technology, leaching losses can be held to relatively low levels by using optimum N management. The major limitation is not being able to predict accurately the N fertilizer requirment of crops throughout the growing season. Some control over the extent of denitrification is possible, but significant denitrification losses are apparently inevitable in anaerobic microsites under most cropping systems. Most NH_3 volatilization can be eliminated by proper subsurface application or by immediate incorporation. These measures, however, are not always physically compatible with cropping systems, especially with limited tillage and with multiple applications of N. The full significance of immobilization for long-term N recovery efficiency has not been fully determined. There is also opportunity for increasing N recovery efficiency by developing plants with greater root density and with higher rates of N absorption when soil solution levels of plant-available N are low.

Physiological efficiency can sometimes be improved by increasing the ratio of harvested to nonharvested plant parts and by increasing the fraction of recovered N translocated from nonharvested to harvested parts. Significant dilution of N recovered in the harvested portion is not generally an acceptable option for improving physiological efficiency. Genetic improvement and N management are both important in these regards. However, there is probably less opportunity for improving physiological efficiency than for improving N recovery efficiency.

With relatively certain yield responses and unlimited capital, the economically optimum N rate is usually slightly lower than the minimum required for Y_m. The main exception is when crop quality considerations require higher N rates. Economically optimum N rates are usually lower with limited than with unlimited capital. When yield response is uncertain there is a tendency to use an insurance approach (i.e., N rates that are higher than the eocnomic optimum in most crop years to minimize yield reductions in years with a lower-than-normal yield efficiency index or lower-than-normal soil N supply and to achieve yield increases in years with higher-than-normal Y_m). The insurance approach is favored by relatively low N/crop price ratios and low N costs relative to total crop production costs.

When type of yield response can be predicted, the upper limit on added investment in measures for achieving a higher yield efficiency index is in essence the value of the associated N rate reduction. Alternatively, when type of yield response is uncertain, the upper limit on added investment in measures for achieving consistently high yield efficiency indexes is a function of the average added yield achieved over a period of years and crop prices. In the latter case, occasional marked reductions in yield efficiency index will justify considerable added investment in either insurance N or measures to maintain consistently high yield efficiency indexes, even with conditions potentially conducive to high N losses. More studies are needed to provide site-specific guidelines for maximizing return from aspects of N fertilizer management affecting the yield efficiency index.

There is some evidence that N fertilizer management can affect timeliness of general field operations and thereby indirectly affect Y_m levels. More economic analyses are needed to evaluate effects of N fertilizer management on labor and equipment efficiency, Y_m as related to timeliness of general field operations, and, ultimately, return from crop reduction. This is particularly critical because of evidence that high yield efficiency indexes and minimum adverse impact of N fertilizer management on timeliness and Y_m are not always economically compatible goals.

A significant portion of unrecovered N in a given year usually remains in either immobilized or inorganic form in the root zone at the end of the growing season and does not necessarily pose a hazard to the environment. Percent recovery of applied N over a period of years is a more reliable index of potential for environmental impact but is still inferior to measurements of NO_3^--N leaching and N_2O evolution from denitrification or, at least, unaccounted-for N.

Since the point of greatest economic return from N is usually somewhere below the point of Y_m, it should be possible to adjust N rates for maximum return from N and minimum N loss to the environment, if type of yield response can be predicted in advance with some degree of certainty. Economically optimum N rates may be less compatible with minimum N loss to the environment when crop quality considerations require higher N rates than the minimum required for Y_m or when insurance N is economical. When insurance N is economical, maximum return from N becomes more compatible with minimum N loss as N/crop price ratio increases.

Yield efficiency index as a reflection of N recovery efficiency has important implications for environmental impact; however, return from measures that improve the yield efficiency index by reducing N losses is highly site- and farming-system-specific, making generalizations about economic return relative to environmental impact difficult. Even with judicious N fertilizer management, N losses due to leaching and denitrification probably cannot be avoided completely. However, physically attainable minimum levels of these losses are consistent with relatively high return from N in most cropping systems. Further lowering of leaching and denitrification losses will become profitable if crop and N fertilizer prices increase more rapidly than the cost of efficient N fertilizer management.

REFERENCES

Allison, F. E. 1966. The fate of nitrogen applied to soils. Adv. Agron. 18:219–258.

Barber, S. A. 1976. Efficient fertilizer use. p. 13–29. *In* F. L. Patterson (ed.) Agronomic research for food. ASA Spec. Pub. no. 36. Am. Soc. of Agron., Madison, Wis.

Barraclough, D. 1979. Factors influencing the efficiency of nitrogenous fertilizers. p. 35–55. *In* Chemistry and agriculture. The Chem. Soc. Spec. Pub. 36. Burlington House, London.

Bartholomew, W. V. 1965. Mineralization and immobilization of nitrogen in the decomposition of plant and animal residues. *In* W. V. Bartholomew and Francis E. Clark (ed.) Soil nitrogen. Agronomy 10:285–306.

Bartholomew, W. V. 1972. Soil nitrogen supply processes and crop requirements. Soil Fert. Eval. Improv. Tech. Bull. no. 6. Dep. of Soil Sci., North Carolina State Univ.

Bremner, J. M., and D. W. Nelson. 1968. Chemical decomposition of nitrite in soils. Int. Congr. Soil Sci., Trans. 9th (Adelaide) 1968 2:495–503.

Broadbent, F. E. 1981. Methodology for nitrogen transformation and balance in soil. Plant Soil 58:383–399.

Broadbent, F. E., and A. B. Carlton. 1979. Field trials with isotopes—plant and soil data for Davis and Kearney sites. p. 433–465. *In* Final report to the Natl. Sci. Found. for nitrate in effluents from irrigated lands. Univ. of California, Riverside.

Broadbent, F. E., and F. E. Clark. 1965. Denitrification. *In* W. V. Bartholomew and Francis E. Clark (ed.) Soil nitrogen. Agronomy 10:344–359.

Capurro, E., and R. Voss. 1981. An index of nutrient efficiency and its application to corn yield response to fertilizer N: I. Derivation, estimation, and application. Agron. J. 73: 128–135.

Doane's Agricultural Report Staff. 1981. 1981 Machinery custom rates guide. Doane's Agric. Report 44:23–25. Doane's Farm Marketing, St. Louis, Mo.

Engelstad, O. P., and W. L. Parks. 1971. Variability in optimum N rates for corn. Agron. J. 63:21–23.

Gerloff, G. C. 1976. Plant efficiencies in use of nitrogen, phosphorus, and potassium. p. 161–173. *In* M. J. Wright (ed.) Plant adaption to mineral stress in problem soils. Natl. Agric. Library, Beltsville, Md.

Hauck, R. D., and B. R. Bock. 1980. Comments on nitrogen loss from soil-plant systems. FAO seminar on maximizing fertilizer use efficiency, New Dehli, India. 15–19 Sept. 1980. FAO, Rome.

Hauck, R. D., and J. M. Bremner. 1976. Use of tracers for soil and fertilizer nitrogen research. Adv. Agron. 28:219–266.

Hoeft, R. G., and J. C. Siemans. 1982. Considerations for fall and spring applied N. p. 17–22. *In* Situation 82. TVA Fertilizer Conference, St. Louis, Mo. 18–19 Aug. 1982. TVA, Muscle Shoals, Ala.

Hoeft, R. G., L. F. Welch, and G. S. Brinkman. 1981. What affects nitrogen efficiency? Solutions Mag. May–June, p. 53–62.

Huffaker, R. C., and D. W. Rains. 1978. Factors influencing nitrate acquisition by plants: assimilation and fate of reduced nitrogen. p. 1–43. *In* D. R. Nielsen and J. G. MacDonald (ed.) Nitrogen in the environment. Vol. 2, Soil-plant-nitrogen relationships. Academic Press, New York.

Humbert, R. P., and A. Ulrich. 1968. Fertilizer use on sugar crops. p. 379–401. *In* L. B. Nelson et al. (ed.) Changing patterns in fertilizer use. Soil Sci. Soc. of Am., Madison, Wis.

Ibach, D. B., and M. S. Williams. 1971. Economics of fertilizer use. p. 81–103. *In* R. A. Olson et al. (ed.) Fertilizer technology and use. Soil Sci. Soc. of Am., Madison, Wis.

Keeney, D. R. 1982. Nitrogen management for maximum efficiency and minimum pollution. *In* F. J. Stevenson et al. (ed.) Nitrogen in agricultural soils. Agronomy 22:605–649.

Kundler, P. 1970. Utilization, fixation, and loss of fertilizer nitrogen (review of international results of the last ten years basic research). Albrecht–Thaer–Arch. 14:191–210.

Kurtz, L. T. 1980. Potential for nitrogen loss. p. 1–17. *In* J. J. Meisinger et al. (ed.) Nitrification inhibitors—potentials and limitations. ASA Spec. Pub. no. 38. Am. Soc. of Agron., Madison, Wis.

Legg, J. O., and J. J. Meisinger. 1982. Soil nitrogen budgets. *In* F. J. Stevenson et al. (ed.) Nitrogen in agriculture soils. Agronomy 22:503–557.

Luckhardt, R. L., and L. E. Ensminger. 1968. Fertilizer use on cotton. p. 221–252. *In* L. B. Nelson et al. (ed.) Changing patterns in fertilizer use. Soil Sci. Soc. of Am., Madison, Wis.

Maranville, J. W., R. B. Clark, and W. M. Ross. 1980. Nitrogen efficiency in grain sorghum. J. Plant Nutr. 2:577–589.

Munson, R. D., and J. P. Doll. 1959. The economics of fertilizer use in crop production. Adv. Agron. 11:133–168.

Nelson, D. W. 1982. Gaseous losses of nitrogen other than through denitrification. p. 327–358. *In* F. J. Stevenson et al. (ed.) Nitrogen in agricultural soils. Agronomy 22:327–358.

Olson, R. A., A. F. Drier, C. Thompson, K. Frank, and P. H. Grabouski. 1964. Using fertilizer nitrogen effectively on grain crops. Neb. Agric. Exp. Stn. Bull. SB479.

Olson, R. V., and F. E. Koehler. 1968. Fertilizer use on small grain. p. 253–271. *In* L. B. Nelson et al. (ed.) Changing patterns in fertilizer use. Soil Sci. Soc. of Am., Madison, Wis.

Parr, J. F. 1973. Chemical and biochemical considerations for maximizing the efficiency of fertilizer nitrogen. J. Environ. Qual. 2:75–84.

Pratt, P. F. 1979. Integration, discussion, and conclusions. p. 719–769. *In* Final report to the Natl. Sci. Found. for nitrate in effluents from irrigated lands. Univ. of California, Riverside.

Richards, G. E. 1977. Economics of band application. p. 120–128. *In* G. E. Richards (ed.) Band application of phosphatic fertilizers. Olin Corp., Little Rock, Ark.

Scarsbrook, C. E. 1965. Nitrate availability. *In* W. V. Bartholomew and F. E. Clark (ed.) Soil nitrogen. Agronomy 10:481–502.

Stanford, G. 1973. Rationale for optimum nitrogen fertilization for corn production. J. Environ. Qual. 2:159–166.

Statistical Analysis System Institute. 1979. NLIN procedure, segmented model. p. 325–328. *In* SAS user's guide. SAS Inst., Inc., Cary, N.C.

Terman, G. L. 1979. Volatilization losses of nitrogen as ammonia from surface-applied fertilizers, organic amendments, and crop residues. Adv. Agron. 31:189–223.

Tice, T. F. 1979. An investigation of nitrogen application under uncertainty. Ph.D. Thesis. Purdue Univ., West Lafayette, Ind.

Tucker, J. C., and R. D. Hauck. 1978. Removal of nitrogen by various irrigated crops. p. 135–167. *In* P. F. Pratt (ed.) Proc. National Conf. on Management of Nitrogen in Irrigated Agriculture, Sacramento, Calif. Dep. of Soil and Environ. Sci., Univ. of California, Riverside.

Viets, F. G., Jr. 1965. The plant's need for and use of nitrogen. *In* W. V. Bartholomew and F. E. Clark (ed.) Soil nitrogen. Agronomy 10:503–549.

Wetselaar, R., and G. D. Farquhar. 1980. Nitrogen losses from tops of plants. Adv. Agron. 33:263–302.

19

L. T. Kurtz, L. V. Boone, T. R. Peck,
and R. G. Hoeft
University of Illinois
Urbana, Illinois

Crop Rotations for Efficient Nitrogen Use[1]

This short title drags from the closet the skeletons of several paradoxes, contradictions, and unsolved mysteries. A few decades ago, crop rotations were widely considered the cornerstones of agronomic science, and every textbook dealing with soil fertility, crop production, or soil conservation devoted lengthy sections to the merits and benefits of crop rotations. At the same time, some crops [e.g., sugarcane (*Saccharum officinarum* L.)] were being produced in monoculture, and many others were grown on a year-by-year basis. Even in regions where rotations were "officially" accepted and vigorously recommended, annual fluctuations in land area of specific crops occurred in response to economic, weather, and other current conditions.

I. IMPORTANCE OF N IN CROP ROTATIONS

A. Rotations and Technological Developments

After World War II, when fertilizer N became readily available and generally accepted, the necessity for (and emphasis on) crop rotations declined greatly. From the standpoint of soil fertility and crop production, an "informed" opinion gradually developed that the benefits of rotations were largely, if not entirely, associated with N. Since N could be supplied conveniently and economically in fertilizer, rotations were gradually abbreviated or abandoned by the great majority of farm operators on the more

[1] Contribution from the Agronomy Department, University of Illinois and the Illinois Agricultural Experiment Station, Urbana-Champaign, IL 61801.

Nitrogen in Crop Production.

level lands. The value of some of the other recognized benefits from crop rotations—control of weeds, insects, and diseases—was diminished when chemical pesticides came into wide use. Additional advantages and disadvantages of crop rotations remain alleged but unproven, some are proven but unexplained, and still others have been rendered immaterial by changing conditions.

Crop rotations do receive attention in connection with soil erosion and water quality. The Universal Soil Loss Equation, which is likely to continue to be extensively utilized, contains ' "*C* factor" derived from consideration of the cropping system, crop residues, and the tillage practices (Walker & Pope, 1980). Continuous soil cover by crops and residues—a feature of desirable rotations—are known to reduce N losses from leaching and erosion. Benefits from these practices, however, are difficult to demonstrate when compared with production of row crops using soil-conserving tillage practices and generous inputs of N fertilizer (Phillips et al., 1980).

B. Rotations for Efficient N Use

Referring again to our title, we might begin and quickly end this discussion by pointing out that crop rotations are rarely selected to conserve N. In our industralized agriculture, we choose the crop we want to produce from those adapted to the region. While production practices will usually be designed to make effective use of fertilizer N, efficient N use is rarely a major consideration in the choice of the crops to be grown. If efficient use of N were the deciding factor in crop selection, we might decide to have a continuous grass sod receiving 20 kg/ha of fertilizer N early every summer. Such a cropping system would achieve very efficient N use.

Another conceivable cropping system that should attain efficient use of N might be one composed entirely of legume crops that produce their own symbiotically fixed N and provide continuous soil cover. Legumes and nonlegumes grown in close succession, or together and without N fertilizer, should also utilize N very efficiently.

We may also speculate that crop species or cultivars to be developed in the future will be more effective in the uptake of N from the soil or perhaps be more active in symbiotic or nonsymbiotic N fixation. Such developments might drastically change the choices of crops in rotations. Efficient use of N might again become important in selecting a crop rotation.

C. Rotations, Management Practices, and N

From the standpoint of a desirable, permanent agriculture, soil management practices that accompany crop production are equally or more important than the sequence of crops. In this context, rotations achieving efficient N use would be characterized by:

1) Nitrogen inputs approximating the amounts required for near-maximum yields,
2) Timing of application of N for maximum effectiveness to the current crop,

3) Provisions for soil cover for the greatest practical proportion of the time to minimize losses by erosion,
4) Provision for a growing crop for the greatest practical proportion of the time to minimize losses by leaching,
5) Utilization of crops that are capable of taking part in biological fixation of N,
6) Utilization of the by-products of the cropping system such as crop residues, animal manures, and other N-bearing wastes, and
7) Avoidance of crop-sequences or cultural practices in which important detrimental (e.g., allelopathic) effects among crops may occur.

Having summarized the topic in its generalities, we may discuss further some of the reasons for our lack of specific recommendations of rotations.

II. THE CHANGING ROLE OF ROTATIONS

A. Sources of N

Biological fixation of atmospheric N is the major known source of N for plants in natural systems. Biological fixations through legume-rhizobia associations are utilized extensively as N sources in agricultural systems. In some cases, the legume crop, itself, is a marketable product or can be used directly for food. Often, the legume crop is used as forage and marketed through animals. If only legume crops were grown, N management would be a simple program.

However, since the diet of the world's people depends heavily on cereal grains, the quantity and quality of yields depend on the presence of adequate N for crop production. Field research has demonstrated an almost universal need for supplemental N for nonleguminous crops. Cereal grain farming systems require major supplemental inputs of N. These inputs may come from the following sources with the choices heavily influenced by economic considerations:

1) Nitrogen-containing wastes and by-products, such as manure, sludges, and residues,
2) Green manure crops such as alfalfa (*Medicago sativa* L.), clover (*Trifolium* sp.), or other legumes grown during a season for the purpose of supplying N and other associated benefits for the following crop,
3) Leguminous "catch-crops" grown during the off season or as companion-crops, and
4) Fertilizer N carriers.

Since 1950, the use of fertilizer N has become an increasingly dominant crop-production practice. At present, it is generally considered that maximum yields of cereal crops cannot be attained in most instances without fertilizer N, even when grown immediately after a legume crop.

During a major proportion of the history of scientific agriculture in this country, N for crop growth came largely from fixation by the legume-rhizobia symbiosis. Production of the better yields of cereal crops was directly linked to the production of legumes and to cropping sequences that featured a legume crop immediately before the grain crop.

B. Advantages and Disadvantages of Rotations

Rather elaborate cropping systems were developed to fit the soils and the production patterns of different regions. Crop rotations were strongly recommended by agricultural colleges, demonstrated in field experiments, and followed by many leading farmers. An example might be a 3-yr rotation, in which the farm would be divided into three equal land areas, with one each in corn (*Zea mays* L.), oats (*Avena* sp.), and clover (*Trifolium* sp.). In such a rotation, corn would always follow the legume to benefit from the symbiotically fixed N. A given field would thus rotate in successive years through corn, oats, and clover, and these crops would also rotate about the farm into a different field each year (Tables 1 and 2).

During that period when rotations were in vogue, long lists of their advantages and benefits evolved (Pieters, 1927). These were expounded at length, but were seldom supported by quantitative information. The disadvantages or the relative importance of the different benefits were not often discussed (Table 3).

Acceptance of the concept of crop rotation, of course, varied from time to time and place to place. Although yields of cash grain crops were demonstrably greater, there was always some proportion of farmers who did not follow established, scientific rotations. Either they didn't understand the benefits or they were willing to accept lower yields in order to grow higher-value crops more frequently. Some were unable to grow good legume crops, and thus did not realize the full value of the cropping system.

Table 1—Example of a 3-yr rotation having crop management "*C* factors" of 0.073–0.11 under conventional tillage in northern Illinois.†

	Field		
Year	1	2	3
1	Corn	Small grain	Meadow
2	Small grain	Meadow	Corn
3	Meadow	Corn	Small grain

† Walker & Pope, 1980.

Table 2—Example of a 4-yr rotation having crop management "*C* factors" of 0.25–0.38 under conventional-tillage in northern Illinois.†

	Field			
Year	1	2	3	4
1	Corn	Small grain‡	Soybeans	Corn
2	Corn	Corn	Small grain‡	Soybeans
3	Soybeans	Corn	Corn	Small grain‡
4	Small grain‡	Soybeans	Corn	Corn

† Walker & Pope, 1980. ‡ Catch crop legume seeding.

Table 3—Advantages and disadvantages listed for rotations.

Advantages	Disadvantages
Chemical effects	
Source of N	Symbiotic N may be expensive
Solubilization of soil minerals by decay of residues	Symbiotic N supply may be variable and unreliable
Accumulation of nitrates during fallow	C/N, C/P, C/S ratios may be unfavorable
Increased soil organic matter	Reduced additions of residues
Reduced leaching of nutrients	Uneven fertility distribution
Physical effects	
Protection against soil erosion and impact of raindrops	Increased draft and compaction
Improved aeration and drainage from root action and earthworms (*Lumbricus* sp.)	Proper timing of operations sometimes difficult
Increased water-holding capacity and infiltration rates	
Improved seedling emergence	
Reduced weather risks	
Moisture conservation by residues, straw, fallow	
Biological effects	
Control of weeds, insects, diseases, nematodes, rootworms	Reduced yields of companion crops and interseedings
Elimination of autoallelopathy	
Increased microbial activity	
Nonagronomic effects	
Better distribution of labor, income, and products	Inflexible; does not permit maximum land area in highest value crop
Encourages diversified, animal agriculture	Requires more equipment
Helps satisfy the "Swiss-Family-Robinson" desire	May require production of crops of variable quality for undependable markets
	May require more labor

The expression "follow a rotation" was apt in more than one way. Not only should the crops follow each other in specific sequence in the different fields, but many farmers were "followers" almost in the sense of being disciples of a cause. Many seemed persuaded that having an acceptable rotation was a moral obligation and a prerequisite of good stewardship of the land. An acceptable rotation always had a legume–meadow crop preceding corn. Alternating between corn and soybeans [*Glycine max* (L.) Merr.] in a 2-yr sequence, for example, was definitely not "following a rotation."

C. Apportioning the Effects of Rotations

Since there was little practical reason to make separate evaluations, the benefits of crop rotations were usually lumped together. Field experiments of the period normally confounded the effect of crop sequences and management practices, and did not appraise the individual effects of fertility, tilth, or individual crop. There was little inclination to question the alleged benefits or to wonder which were the most important.

1. THE N EFFECT

The need for definitive information about the relative importance of the different effects of rotations assumed major proportions after World War II, as fertilizer N became generally available at nominal prices. Debates raged for years about the extent to which fertilizer N could or should be substituted for legumes in rotations. Field experiments and "farm trials" with fertilizer N suggested that many, if not all, of the benefits from rotations were due to N. Often, the effects of N were not clearly separated, but the "agricultural establishment" that had been committed for decades to diversified and animal agriculture was suspected of being biased and unwilling to change.

One of the more dramatic and effective demonstrations of the effect of N in different rotations was described as follows by Allaway (1957) in his discussion of "Cropping Systems and Soil" in the *USDA Yearbook of Agriculture*:

> Corn had been grown continuously on the Morrow Plots at the University of Illinois alongside a 3-year rotation of corn, oats, and meadow since 1876. The results of these comparisons formed one of the cornerstones of the belief that rotations were essential to the maintenance of soil productivity.
>
> But liberal amounts of N, P, and K fertilizers were applied in 1955 to part of each plot of this experiment. Where this fertilizer was applied, the 79th consecutive crop of corn yielded 86 bushels an acre. It made 36 bushels an acre without the fertilizer. —Thus on the Morrow Plots, differences in soil productivity built up by 79 years of cropping systems were reduced by one application of fertilizer.

2. THE PHYSICAL EFFECT

It is not reasonable to expect that the beneficial effects of rotations would be qualitatively or quantitatively equal on all soils and under all cropping conditions. Nevertheless, the relative effect of N is usually so large that other effects are often overlooked or ignored.

The Morrow Plot results (Table 4) gave no specific answers, and the title of one report, "All the Way Back in One Year?" (Russell, 1956) was

Table 4—Corn yields on the Morrow Plots in 1955 (Russell, 1956).

Treatment	Continuous corn	Two-year rotation: Corn–oats†	Three-year rotation: Corn–oats legume
	kg/ha (bu/acre)		
None	2300 (36)	2700 (43)	4000 (63)
1955 Treatment only	5400 (86)	6100 (97)	6400 (102)
Manure–Lime–Phosphate	5000 (79)	6200 (98)†	6300 (100)
Manure–Lime–Phosphate	6200 (98)	6700 (107)†	6300 (101)

† Legume "catch-crop" in oats.

correctly given as a question. The discussion stated that differences in corn yields associated with past management were largely eliminated by liberal application of plant food. The article concluded that "differences in the physical condition of the soil associated with previous management practices were not yet materially influencing corn yields on the Morrow Plots."

Those who wished to substantiate their faith in rotations could point out that the best yields on the Morrow Plots came in the rotations. Grain farmers pointed out that with adequate fertilizer, they could grow corn every year and obtain yields almost as good as in well-fertilized rotations.

For most soils, the physical benefits of crop rotations are not shown to be sufficiently evident to have great influence on crop selection. In the era of crop rotations, the effects of erosion often became apparent rather quickly. Loss of surface soil meant a loss of appreciable soil fertility and soon affected yields. The subsoil was more difficult to plow and the loss of tilth may have been noticeable. In contrast, if a little loss of the soil surface occurs in present day agriculture, giant plows bring up a compensating amount of subsoil. Fertilizers are generously applied and yield trends continue upward. Loss of tilth may be discussed, but it doesn't appreciably modify activities, since the power available in modern equipment is more than adequate for all operations. Thus, detrimental effects are likely to be obscured for long periods of time, except on very shallow soils.

3. THE CROP SEQUENCE EFFECT

Numerous field experiments were begun in the early and mid-1950's to study the interactions between N and other effects of rotations. More specifically, an important objective was to determine if a legume had other effects in a crop rotation besides providing N. There was a growing contention that continuous corn with plenty of N would yield as well as corn in rotations.

Commercial N came to be regarded as a soil management practice, which increased the farmers' options in crops planted. Land less subject to erosion could be used more frequently for row crops, while more sloping land would ideally be committed to meadow.

During the first 10 yr of such studies, data from many locations indicated that equal corn yields could be obtained whether the source of N was provided by legumes or by fertilizer and that the effect of meadow (legumes) in the rotation in many soils was only to furnish N. It was still appropriate to point out, however, (Shrader, 1968) that: "Where oat yields are 80 bu/A and on livestock farms where hay and straw are needed, rotations of corn, corn, oats, and hay or corn, soybeans, oats, and hay can be used with little or no sacrifice compared to continuous corn." A widely cited, contemporary discussion of the role of rotations was given by Aldrich (1964).

The conclusion that the effect of meadow (legumes) in the rotation on many soils was only to furnish N was apparently temporary (Anon., 1981). More recent results in many of the same locations (Voss & Shrader, 1979) show that corn following corn does not usually yield as well as after other

Table 5—Yields of continuous corn and corn following soybeans, Morrow Plots.†

Year	Corn yield	
	After corn	After soybeans
	kg/ha (bu/acre)	
1969	8 500 (136)	9 000 (145)
1971	9 200 (146)	10 600 (169)
1973	9 300 (149)	11 300 (180)
1975	10 100 (161)	12 000 (191)
1977	7 100 (113)	8 900 (142)
Average	8 800 (141)	10 300 (165)

† The corn crops were fertilized at the rate of 360 kg N/ha (320 lb N/A). Welch, 1979.

crops. This relationship seems to be consistent across the corn-growing region. All results appear to indicate that with adequate fertility, particularly N, equal corn yields are obtained after all crops except corn (Higgs et al., 1976). No amount of N fertilizer or other recognized treatment seems to eliminate this yield disadvantage of 5 to 10% for corn following corn (Table 5).

These lower yields of corn following corn, regardless of N level, is one of the current mysteries of crop production. Randall (1978) discussed this situation and offered speculations about seven possible factors that might be involved. These factors were soil moisture, fallow effect, allelopathy, microbial activity, nutrient availability, compaction, tilth, rootworms (*Diabrotica* sp.), and nematodes. Carbon-to-sulfur ratios of corn residues might be included in these speculations under the nutrient availability category. Auto allelopathy is being studied (Klein & Miller, 1980). Whatever the cause, it is apparently alleviated in one season, since there is no additional benefit from having two, rather than one, different crops preceding corn.

D. Compatibility of N Sources in Rotations

In contemporary crop production, legumes are usually grown for their value as crops rather than as a source of N. Furthermore, the importance of their contributions as N fixers is somewhat obscure now when larger and larger amounts of N fertilizer are being used in the cropping system.

Due to the combination of technological improvements, corn yields have been increasing steadily (Fig. 1) and the requirements for fertilizer N is directly dependent on crop yield (Anon., 1981; Brown et al., 1981). This requirement is ordinarily adjusted for other N inputs, including those from legumes. Accordingly, the amounts of fertilizer N recommended for a nonlegume is reduced when the preceding crop was a legume. These adjustments (Table 6) are intended to be a function of the net amount of N remaining in the soil and are usually derived by comparing N response curves in field experiments involving crop sequences with and without legumes. While these adjustments from different sources agree rather well, there is considerable uncertainity and "glancing-over-the-shoulder" on the part of those making the recommendations. Uncertainity arises from our inability

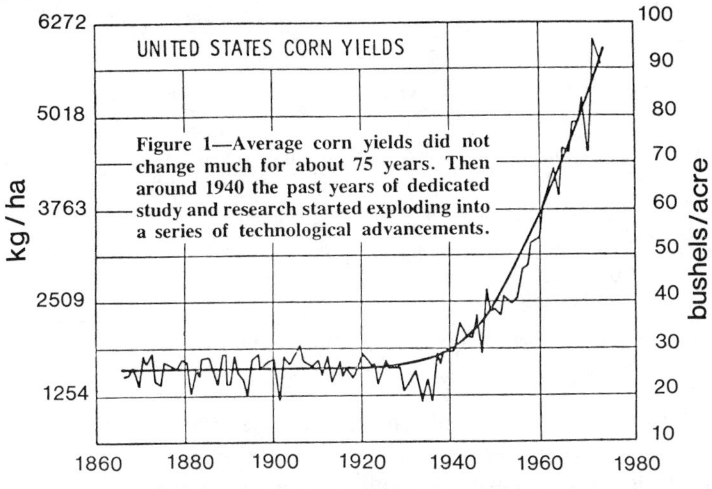

Fig. 1—Trend in corn yields (Newman, 1976).

to accurately estimate the amount of N fixed by legumes in different situations.

Many investigations have shown that the proportion of symbiotically fixed N in a legume is inversely related to the amount of N the legume obtains from the soil. Amounts of N fixed and amounts obtained from the soil by legumes have to be estimated. One procedure for making such estimates is to grow nodulating and nonnodulating isolines of a legume species in adjacent plots. It is assumed that both isolines take up the same amount of N from the soil and that additional N in the nodulated plants comes from symbiotic N fixation. Results from an early experiment utilizing this procedure are shown in Table 7 and they illustrate the reduction in symbiotic fixation as the N levels in the soil increase.

Table 6—Adjustments in N recommendations for corn after legumes.

	N reduction					
	First year after alfalfa† or sweet clover‡			Second year after alfalfa or sweet clover	After soybeans	
Location	Good	Average	Poor			Reference
	kg/ha					
Illinois	112	56	0	34	45	Anon., 1981
Indiana	78–90	68	34			Stivers, 1977
Iowa	157	112	22		45	Voss, 1980
Michigan		60				Warnke & Christenson, 1980
Minnesota	112	56		56	22	Anon., 1976
Missouri	56				34	Buchholz et al., 1981
Ohio	79–180	101	56	22		Anon., 1980
Wisconsin	90	45	22		1.12 kg/ha yield	Walsh et al., 1976

† *Medicago sativa* L. ‡ *Melilotus* sp.

Table 7—Effect of soil N status on N fixed by soybeans (Sears, 1953).

		Soybean grain				
		Nodulated		Nonnodulated		
Soil N status	Soil treatment†	Yield	N content	Yield	N content	Air-derived N
		kg/ha				%
High	337 kg N/ha	2460	223	2220	178	21
Medium	None	2420	216	1960	115	47
Low	Straw mulch, 11.25 t/ha	2380	199	1410	73	63

† Previous crop was an established timothy (*Phleum pratense* L.) sod.

Another more recent representation of the amounts of N fixed symbiotically at different levels of available N are given in Fig. 2. Again, N fixation declined as levels of soil and fertilizer N increased. At the highest level of added N, only about 10% of the N in the soybean crop came from symbiotic fixation. Current informed opinion is that maximum symbiotic fixation occurs only when the plants are under a mild N stress.

Field experiments have also shown that legumes are excellent scavengers for inorganic N in the soil (Olsen et al., 1970; Schertz & Miller, 1972). Residual fertilizer N and N released from soil organic matter are effectively taken from the soil by legumes, and symbiotic fixation is reduced correspondingly.

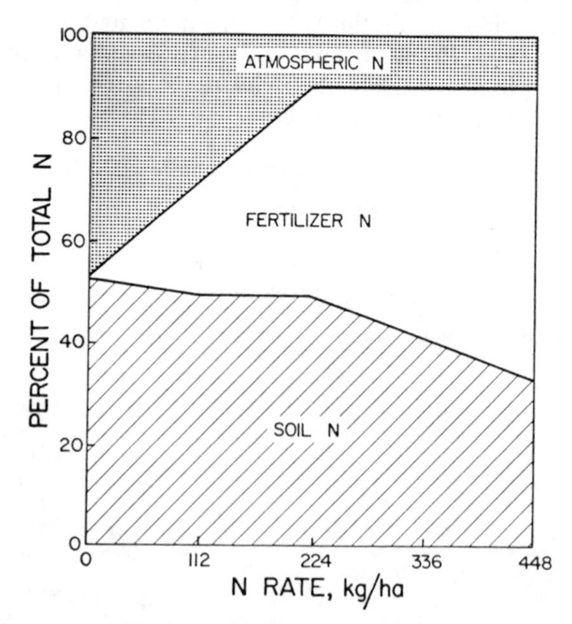

Fig. 2—Contribution of atmospheric, fertilizer, and soil N to soybeans (Johnson et al., 1975).

Thus, in some respects, the goals of obtaining maximum N fixation by legumes and maximum corn yields in the same crop rotation are incompatible. Potential corn yields are gradually increasing and larger amounts of N fertilizer are required if these potential yields are to be obtained. In one way or another, increased use of fertilizer N will result in higher levels of soil N so that the proportion of N fixed by a legume in the rotation will decline. If we pursue this line of reasoning, we might conclude that little gain in N efficiency would be obtained at high N levels by including a legume in the rotation.

Efficiency of N use can be greatly influenced by conditions external to the soil-crop system; for example, economic factors such as the price ratio between grain and fertilizer. Much discussion has also recently taken place about the possibilities of soil-crop management systems that would minimize the use of fertilizer N and chemical pesticides (Aldrich, 1980; Lockeretz et al., 1981). While maximum income or yield may not be attained in such systems, increased efficiency of certain inputs may be possible. Some comparisons indicate that farm size and land class influence the profitability of different systems (Johnson et al., 1977). Undoubtedly, new cultivars and management practices will also be developed that will influence crop rotations and N use.

This discussion about crop rotations and N efficiency reports the experiences in recent times in our region of the world. In other geographic areas and economic conditions, conclusions about the relation of N to crop combinations and sequences may be quite different. A rotation that was a necessity at one time and place may be an anachronism at another. More complete knowledge about plant-soil and plant-plant interactions will be helpful in predicting the behavior of N and crops in soil systems. However, crop selection will probably always involve numerous nonagronomic considerations.

REFERENCES

Aldrich, S. R. 1964. Are crop rotations out of date? Proc. of Hybrid Corn Industry-Research Conf. 1–7. 9–10 Dec. *In* W. Heckendorn and J. I. Sutherland (ed.) Publ. no. 19 Am. Seed Trade Assoc., Washington, D.C.

Aldrich, S. R. 1980. Nitrogen in relation to food, environment, and energy. Spec. Pub. 61, Ill. Agric. Exp. Stn., Univ. of Illinois, Champaign-Urbana.

Allaway, W. H. 1957. Cropping systems and soil. p. 386–396. *In* A. Stefferud (ed.) Soil—the 1957 yearbook of agriculture. U.S. Department of Agriculture. U.S. Government Printing Office, Washington, D.C.

Anon. 1976. Fact sheet soils no. 24 (Revised), Agric. Ext. Service, Univ. of Minnesota, St. Paul.

Anon. 1980. Agronomy guide. Bull. 472. Coop. Ext. Service, Ohio State Univ., Columbus.

Anon. 1981. Illinois agronomy handbook 1981–82. Circ. 1186, Univ. of Illinois, Urbana.

Brown, J. R., R. G. Hanson, and D. D. Buchholz. 1981. Interpretation of Missouri soil tests. Agron. Misc. Pub. 80-04. Univ. of Missouri, Columbia.

Buchholz, D. D., J. R. Brown, R. G. Hanson, H. W. Wheaton, and J. D. Garrett. 1981. Soil test interpretations and recommendations handbook. Dep. of Agronomy, Univ. of Missouri, Columbia.

Higgs, R. L., W. H. Paulson, J. W. Pendelton, A. F. Peterson, J. A. Jackobs, and W. D. Shrader. 1976. Crop rotations and nitrogen. Res. Bull. R2761. College of Agric. and Life Sciences, Univ. of Wisconsin, Madison.

Johnson, J. W., L. F. Welch, and L. T. Kurtz. 1975. Environmental implications of nitrogen fixation by soybeans. J. Environ. Qual. 4:303–306.

Johnson, W. A., V. Stoltzfus, and P. Craumer. 1977. Energy conservation in Amish agriculture. Science 198:373–378.

Klein, R. R., and D. A. Miller. 1980. Allelopathy and its role in agriculture. Commun. Soil Sci. Plant Anal. 11:43–56.

Lockeretz, W., G. Shearer, and D. Kohl. 1981. Organic farming in the Corn Belt. Science 211: 540–547.

Newman, J. E. 1976. Weather, technology, and corn yields. Better Crops Plant Food 40:25–27.

Olsen, R. J., R. F. Hensler, O. J. Attoe, S. A. Witzel, and A. L. Peterson. 1970. Fertilizer nitrogen and crop rotation in relation to movement of nitrate nitrogen through soil profiles. Soil Sci. Soc. Am. Proc. 34:448–452.

Phillips, R. E., R. L. Blevins, G. W. Thomas, W. W. Frye, and S. H. Phillips. 1980. No-tillage agriculture. Science 208:1108–1113.

Pieters, A. J. 1927. Benefits other than increase of nitrogen. p. 118–135. *In* Green manuring-principles and practices. John Wiley & Sons, New York.

Randall, G. W. 1978. Effect of preceding crop on corn yield. Proc. Soils and Fertilizers Short Course. p. 2–8. Inst. of Agriculture, Forestry, and Home Economics, Univ. of Minnesota, St. Paul.

Russell, M. B. 1956. All the way back in one year? Plant Food Rev. 2:18–19.

Schertz, D. L., and D. A. Miller. 1972. Nitrate-N accumulation in the soil profile under alfalfa. Agron. J. 64:660–664.

Sears, O. H. 1953. Legumes as nitrogen fixers. Agron. Facts. SR 8. Agronomy Dep., Univ. of Illinois, Urbana-Champaign.

Shrader, W. D. 1968. Crop rotations, 1968 viewpoint. Iowa Farm Sci. 23:3–6.

Stivers, R. 1977. Fertilizer recommendations for field crops, gardens, and lawns. S. T. no. 1. Soil and Plant Anal. Lab., Purdue Univ., West Lafayette, Ind.

Voss, R. D. 1980. Understanding your soil test report. Pm-429 (Revised), Coop. Ext. Service, Iowa State Univ., Ames.

Voss, R. D., and W. D. Shrader. 1979. Crop rotations—effect on yields and response to nitrogen. Pm-905, Coop. Ext. Service, Iowa State Univ., Ames.

Walker, R. D., and R. A. Pope. 1980. Estimating your soil erosion losses with the Universal Soil Loss Equation. Coop. Ext. Service, Univ. of Illinois, Urbana-Champaign.

Walsh, L. M., E. E. Schulte, J. J. Genson, and E. A. Liegel. 1976. Soil test recommendations. Coop. Ext. Program A2809. College of Agric. and Life Sciences, Univ. of Wisconsin, Madison.

Warncke, D. D., and D. R. Christenson. 1980. Fertilizer recommendations. Ext. Bull. E-550, Coop. Ext. Service, Michigan State Univ., East Lansing.

Welch, L. F. 1979. Nitrogen use and behavior in crop production. Ill. Agric. Exp. Stn. Bull. 761, Univ. of Illinois, Urbana-Champaign.

20

F. M. Rhoads
University of Florida
North Florida Research and Education Center
Quincy, Florida

Nitrogen or Water Stress: Their Interrelationships

Nitrogen stress in crop plants can result from either a deficiency or an excess. Deficiency symptoms are exhibited as pale green color and stunted growth. However, other elements such as S and Mo may show similar effects. A reduced number of kernel rows has been observed in sweet corn (*Zea mays* L.) as a result of N deficiency (Viets, 1965). Stress symptoms for excessive N include excessive height due to self competition for light, lodging, and delayed flowering (Viets, 1965). Nitrogen stress in crop plants may not be visually evident, but show up in reduced yield.

Shaw and Laing (1966) state: "Moisture stress is the result of an imbalance between the supply furnished by the soil-water and the amount needed by the plant as determined by the atmosphere, assuming a complete crop cover." Water stress is observed as wilting of leaves and loss of turgor. However, the plant experiences water stress long before wilting occurs because the leaf water potential becomes considerably negative near midday with full sunlight, even when soil-water level is high (Slatyer, 1969). Wilting of leaves is an expression of degree of water stress, not just an indication of stress vs. no stress. Crop plants can be stressed under flooded conditions, but the stress is most likely due to lack of O_2 rather than excessive water.

Interrelationships between N stress and water stress are extensively documented, but only a limited number of references can be cited here. The influence of N stress and water stress on crop yield will be treated separately, and then their interrelationships will be examined.

Nitrogen in Crop Production.

I. N FERTILIZATION AND CROP YIELD

The influence of residual N is important when evaluating crop response to N fertilization. Data cited by Viets (1965) from a furrow-irrigated corn experiment showed a yield of 1200 kg/ha for zero N and 6300 kg/ha for 224 kg N/ha as fertilizer on a virgin soil with no N applied the previous year. However, where 224 kg N/ha was applied to the corn crop in the previous year, yields were 4400 kg/ha with no N applied and 7000 kg/ha with 224 kg N/ha applied in the current year. The importance of residual N is illustrated in more recent data with sprinkler-irrigated corn in southeastern United States (Rhoads & Stanley, 1979). Grain yields ranged from 4100 to 6900 kg/ha with no N fertilizer and varied inversely with population (Table 1). Total dry matter production varied directly with population at all levels of N fertilization, but at the zero N rate, a larger portion of the dry matter was contained in the stalks and leaves at the high population (90 000 plants/ha). Uptake of N increased to maximum levels at the highest rate of fertilization for each plant population (Table 2). Near-maximum yields were obtained at each population with 3 g of N applied per plant, or 90 kg N/ha at 30 000 plants/ha, 180 kg N/ha at 60 000 plants/ha, and 270 kg N/ha at 90 000 plants/ha. Residual N taken up where no fertilizer was applied ranged from 74 to 92 kg/ha. The increase in N uptake between zero applied N and 3 g applied N/plant ranged from 50 to 72% of the applied N.

Response of barley (*Hordeum vulgare* L.) to N fertilization was reported by Stanberry and Lowrey (1965). The N rates in their experiment ranged from 0 to 270 kg/ha. Only the irrigated portion of the experiment

Table 1—Effect of population and fertilization on grain yield of irrigated corn (Rhoads & Stanley, 1979).

	N fertilizer rate (g/plant)					
Population	0	1.0	2.0	3.0	5.0	r
plants/ha	kg/ha					
30 000	6 900	7 400	8 700	9 700	10 000	**
60 000	6 300	9 500	11 200	12 700	12 300	*
90 000	4 100	10 300	13 200	14 500	14 500	*

**,* Correlation coefficient significant at the 1 and 5% levels, respectively.

Table 2—Effect of population and rate of fertilization on N uptake of irrigated corn plants 87 d after planting (Rhoads & Stanley, 1979).

	N fertilizer rate (g/plant)					
Population	0	1.0	2.0	3.0	5.0	r
plants/ha	kg N/ha					
30 000	92	99	112	137	162	**
60 000	74	152	167	204	246	**
90 000	88	153	206	239	296	**

** Correlation coefficient significant at the 1% level.

will be considered in this section. With no applied N, yield was 672 kg/ha. A rate of 67 kg N/ha increased the yield to 3456 kg/ha—an increase of 42 kg of grain/kg of applied N. The second 67 kg/ha increment of N increased the yield to 4368 kg/ha or an increase of 14 kg of grain/kg of additional N. Further increase in N fertilization did not result in a significant yield increase.

Winter wheat (*Triticum aestivum* L.) in Nebraska exhibited N response similar to barley (Black, 1966). Available water for the growing season was 51.6 cm. Rates of N fertilization were 0, 22, 45, and 90 kg/ha. Grain yield tended to level off at 45 kg/ha of applied N, but total dry matter increased substantially between 45 and 90 kg/ha of applied N. The first 22 kg/ha increment of N increased grain yield an average of 26 kg/kg N (from 1368 to 1941 kg/ha). Yield of grain increased from 2295 to 2603 where applied N was increased from 45 to 90 kg/ha.

Legumes such as soybeans [*Glycine max* (L.) Merr.], peanuts (*Arachis hypogaea* L.), alfalfa (*Medicago sativa* L.), etc. obtain their N from the atmosphere via a symbiotic relationship with bacteria living in nodules attached to the plant's roots. However, it is conceivable that yield could be increased by the application of fertilizer N to supplement the N obtained from the atmosphere. A yield response to N is not usually expected for legumes if the soil is adequately innoculated with the specific bacteria required by the species in question. Numerous researchers have studied this and found the anticipated lack of response. Smith et al. (1968) conducted extensive research in Florida to test the effect of fertilizer N on yield of soybeans. Rates of N varied from 0 to 672 kg/ha and number of applications from one to three per season. Yield with no applied N was 2730 kg/ha and with 672 kg N/ha the yield was 2580 kg/ha. None of the differences between treatments was statistically signfiicant. This is in agreement with data obtained by other researchers (Hardy, 1959; Weber, 1966). Peanuts did not respond to N in a 16-yr study in Florida (Lipscomb et al., 1965).

II. WATER STRESS AND CROP YIELD

Timing of water stress with respect to plant growth stage is extremely important in relationships between yield and water stress. According to Slatyer (1969): "Stress at anthesis can markedly reduce fertilization and grainset in most cereals." Corn suffered the greatest yield loss when wilted 6 to 8 d at tasseling time (Robins & Domingo, 1953). Their experiment included several soil-water regimes. Grain yield of corn was 5958 kg/ha with the last irrigation at tasseling time and 8675 kg/ha with three subsequent irrigations. Available soil-water was maintained at $\geq$ 60% in both treatments during the vegetative growth period. However, yield was 8650 kg/ha for corn with three irrigations, all subsequent to tasseling. Yield was $<$ 5000 kg/ha where corn was allowed to wilt 6 d or more at tassel. Water stress imposed on corn 30 to 60 d after planting reduced yield 25%, whereas 1 or 2 d of wilting during tasseling reduced yield 22%, and 6 to 8 d wilting during tasseling reduced yield by 51% (Shaw & Laing, 1966).

Table 3—Effect of maximum soil-water suction (SWS) and amount of irrigation on corn yield and irrigation water use efficiency (Rhoads & Stanley, 1973).

SWS	Days/ irrigation	Water per irrigation	Number of irrigations	Yield	IWUE†
cbar		cm		kg/ha	kg/ha-cm
20	3	2.0	11	11 900	214
40	5	3.0	6	11 000	211
60	7	3.0	4	10 000	233
--‡	--	0	0	7 200	--
LSD 0.05				1 200	

† Irrigation water use efficiency. ‡ Not irrigated.

Total dry matter production of cereal grains generally shows a greater response than grain yield to an increase in available water at suctions > 100 cbar. Sweet corn was irrigated when soil-water suction at the 15- or 30-cm depth reached 30, 200, and 1500 cbar (Taylor & Ashcroft, 1972). Ear yields were 18 368, 12 768, and 12 544 kg/ha, respectively, for the three irrigation treatments. There was very little difference in ear yield between the 200 and 1500 cbar treatments. However, there was a 10 000 kg/ha difference in total yield between the two treatments.

Irrigation scheduling procedures such as frequency of application and amount of water per application can have a significant effect on corn yields in sandy soils of southeastern United States (Rhoads & Stanley, 1973). Water was applied to recharge the top 15 to 20 cm of soil when soil-water suction at the 15-cm depth reached 20, 40, and 60 cbar (Table 3). Highest yield occurred with irrigation at 20 cbar soil-water suction, which required recharge of the top 15-cm every 3 d from 15 d before to 15 d after tassel. There was approximately a 940 kg/ha grain yield reduction for each 20 cbar increase in soil-water suction (SWS) for scheduling irrigation. Yield increase in response to irrigation was directly proportional to amount of water applied. Irrigation water use efficiency (IWUE) in terms of yield increase per centimeter of water applied was > 200 kg/ha-cm for each schedule procedure. Corn irrigated at 60 cbar in these experiments did not show signs of wilting. Furthermore, yield of corn irrigated when SWS at the 15-cm depth

Table 4—Response of two peanut varieties to irrigation in Florida during a relatively dry season (Gorbet & Rhoads, 1975).

Soil-water suction at irrigation	Number of irrigations	Water applied	Variety: Florigiant	Variety: Florunner
			Yield	
cbar		cm	kg/ha	
40	11	30.5	6310	6091
60	7	19.3	5960	6580
NI†	0	0	4390	4940
LSD (0.05)			550	550

† Not irrigated.

Table 5—Effect of soil-water suction (15-cm depth) when irrigated on utilization efficiency of irrigation water for two peanut varieties (Gorbet & Rhoads, 1975).

Soil-water suction at irrigation	Variety: Florigiant	Variety: Florunner
	Yield increase	
cbar	kg/ha-cm	
40	63	37
60	81	85

reached 60 cbar was not significantly different from corn irrigated at 200 or 500 cbar SWS. Results cited by Taylor and Ashcroft (1972) agree with the concept of high-frequency irrigation for maximum corn yield. Water stress levels that limit yield of corn occur before plant wilt is observed.

Water stress imposed on soybeans during the vegetative stage of growth has very little effect on final yield. When soybeans were stressed during flowering, the yield reduction was 9%. Stress at the beginning of pod development reduced yield by 18%, while stress during the middle of the pod development period reduced yield by 20%. The greatest yield reduction occurred when plants were stressed during the bean-filling period. There was a 34% yield decrease due to stress in early bean fill and a 40% decrease in midbean fill (Shaw & Laing, 1966).

Peanut yield was reduced most by water stress occurring 70 to 145 d after planting (Pallas et al., 1979). Yield reduction was 16% with stress 35 to 70 d after planting, 32% when stressed between 70 and 105 d, 37% for stress from 105 to 145 d, 54% for the 35- to 105-d stress period, and 65% for the 70- to 145-d stress period. Nonstressed plants produced 101 kg/ha-cm for 54 cm of water used. Water applied between 35 and 70 d increased yield by 79 kg/ha-cm. Yield increased by 120 kg/ha-cm for water applied between 70 and 105 d. The yield increase for water applied between 105 and 145 d was 170 kg/ha-cm. Water used between 0 and 35 d contributed 47 kg/ha-cm to total pod yield. The above experiment was conducted on small plots with an automatic rain out shelter.

Irrigation experiments were conducted on peanuts in Florida under natural rainfall conditions (Gorbet & Rhoads, 1975). Water was applied in sufficient amounts to recharge the plow-layer when soil-water suction at the 15-cm depth reached 40 and 60 cbar and yields from these two treatments were compared with a nonirrigated treatment (Table 4). Although more

Table 6—Effect of irrigation on peanut pod loss during a relatively wet season in Florida (Gorbet & Rhoads, 1975).

Soil-water suction at irrigation	Number of irrigations	Water applied	Harvested yield	Pod loss	Total yield
cbar		cm	kg/ha		
40	7	17.8	4150	1490	5640
60	5	12.7	3930	1271	5201
Not irrigated	0	0	4440	640	5080

water was required to irrigate at 40 cbar SWS, yield was not significantly different from the 60-cbar treatment. Both varieties exhibited a much lower IWUE for the 40-cbar treatment than for the 60-cbar treatment (Table 5). Observations over the past 10 yr indicate that it is more efficient to irrigate peanuts when SWS reaches 60 cbar at the 15-cm depth or 30 cbar at the 30-cm depth. Pod loss of peanuts was much greater with irrigation than with no irrigation during both a relatively wet season (Table 6) and a dry season. However, the maximum loss values were much less in the dry than in the wetter season. Pod and stem diseases appear to cause more damage under moist conditions than under dry conditions.

Irrigation increased the yield of barley and decreased water utilization efficiency (WUE) expressed as grain yield divided by total water use in Arizona tests (Stanberry & Lowrey, 1965). Yield average for 3 yr was 4749 kg/ha for barley irrigated when SWS at the 20-cm depth reached 15 cbar, compared with 3046 kg/ha irrigated at 8 to 10 bar suction. Average seasonal water use was 70 ha-cm for the wet treatment and 32 ha-cm in the dry treatment. Water utilization efficiency was 68 kg/ha-cm in the wet treatment and 95 kg/ha-cm in the dry treatment.

Severe moisture stress before heading reduced yield of spring wheat by about 30%, but the greatest yield reduction occurred when plants were stressed during and following heading (Robins & Domingo, 1962).

Yield of winter wheat in Nebraska (Black, 1966) increased with irrigation. Grain yield was 529 kg/ha with no irrigation, 1676 kg/ha with 7.4 cm of irrigation water applied, and 2647 kg/ha with 20.6 cm of irrigation water. Water utilization efficiency also increased with irrigation. The WUE was 17 kg/ha-cm for nonirrigated winter wheat, 44 kg/ha-cm for 7.4 cm of water added as irrigation, and 51 kg/ha-cm for 20.6 cm of irrigation water. Total dry matter production of winter wheat responded to irrigation in a manner similar to grain yield.

Dry matter production of alfalfa was greatest when irrigated to maintain the integrated soil-water suction in the root zone below 40 cbar (Bahrani & Taylor, 1961). The reported yield was 5500 kg/ha irrigated at 40 cbar, 3750 kg/ha at 219 cbar, and 3250 kg/ha at 343 cbar. Yield increase per 100 cbar change in soil-water suction was approximately 1000 kg/ha between 40 and 219 cbar and 400 kg/ha between 219 and 343 cbar. In comparison, grain yield of barley increased 1800 kg/ha per 100 cbar change in soil-water suction between 15 and 70 cbar and 100 kg/ha between 70 and 900 cbar (Stanberry & Lowery, 1965). Yield response of most crop plants seems to be most sensitive to change in soil-water suction in the 0- to 100-cbar range.

III. WATER STRESS AND ROOT NODULES ON SOYBEANS

Moist soil appears to be more favorable for bacteria survival and activity than dry soil. Sprent (1976) made the following observations: nodule number and size on soybean roots were reduced by water stress; water stress reduced N fixing potential as measured by C_2H_2 reduction; an increase in

Table 7—Forage yield, nodule production, and grain yield of irrigated vs. nonirrigated soybeans (Rhoads, 1973, unpublished data).

Treatment	Forage yield		Nodule production			Grain yield
	Fresh wt	Dry wt	Fresh wt	Dry wt	Number per plant	
	g/plant					kg/ha
Irrigated	33.8	6.4	1.06	0.31	38	2419
Nonirrigated	25.5	4.8	0.64	0.16	13	2043
LSD	7.0*	1.4**	0.32*	0.12**	18**	NS†

*,** Significant at the 5 and 1% levels, respectively. † Not significant.

water suction from 4.5 to 18 bar caused the rate of C_2H_2 reduction to decrease by 50% in soybean root nodules; and rate of C_2H_2 reduction in plants stressed for 9 d did not recover to the level in nonstressed control plants after rewatering by 18 d after the treatment started.

The data in Table 7 were taken from a soybean experiment planted on 20 June 1973. Irrigations (2.5 cm per application) were applied 1 July, 8 July, 24 August, and 31 August when tensiometers at the 15-cm depth reached a reading of 30 cbar. Nodule counts were made on 23 July from two samples per plot containing 10 plants per sample dug from four reps. The roots were washed before counting. The sampling procedure was repeated on 30 July, but instead of counting nodules, fresh and dry weights of nodules and forage were measured. Forage yield was greater on irrigated plots at the time nodule weights were taken. Difference in fresh weight was significant at the 5% level, and difference in dry weight was significant at the 1% level. About 30% more dry forage was produced on irrigated plots than on nonirrigated plots. Irrigated plants produced the greatest weight of nodules with a significant difference at the 5% level for fresh weight and at the 1% level for dry weight. There were about three times as many nodules on irrigated as on nonirrigated soybeans. Grain yield was about 18% greater for irrigated soybeans, but the difference was not significant at the 5% level. Since water stress affects other plant processes in addition to nodulation, it is difficult to measure the relationship between water stress, nodulation, and yield of soybeans. Yields of soybeans are not reduced very much due to water stress in the vegetative growth stage; therefore, the plant must be able to obtain a considerable amount of the required N from nodules during the flowering and fruiting cycles.

IV. EFFECT OF N ON WATER USE EFFICIENCY

Ear yield of sweet corn showed very little response to water at N fertilizer levels of 0, 56, and 112 kg/ha, but a marked response to water was observed at 224 kg N/ha (Taylor & Ashcroft, 1972). Yields with 112 kg N/ha were 12 992; 14 560, and 12 992 kg/ha, respectively, for low, medium, and high soil-water treatments. However, with 224 kg N/ha, sweet corn ear yields were 12 544, 12 768, and 18 368 kg/ha, respectively, for low,

Table 8—Effect of scheduling fertilization on irrigation water use efficiency of corn on a sandy soil (Rhoads et al., 1978).

Conventional fertilization	Biweekly application
kg/ha-cm†	
73	150
Total yield increase from 25 cm of irrigation	
kg/ha	
1825	3750

† Yield increase per unit of water applied.

medium, and high soil-water treatments. This amounts to more than a 40% increase in WUE for the high soil water treatment by increasing N from 112 to 224 kg/ha. Total yield of sweet corn responded to water treatments at N levels of 112 and 224 kg/ha.

In Florida tests (Table 1), WUE of field corn was increased by 31% at 30 000 plants/ha where N application was increased from 30 to 90 kg/ha. Water use efficiency was increased 34% with 60 000 plants/ha when N application increased from 60 to 180 kg/ha. At 90 000 plants/ha, WUE was increased 41% when N application increased from 90 to 270 kg/ha. These calculations are based on the assumption that water use was the same for N treatments within populations. Considering that irrigation was applied uniformly to the whole experiment and the leaf area index was approximately the same for all N treatments, the assumption seems reasonable.

Scheduling N fertilization is as important to WUE on sandy soils of southeastern United States as the total amount applied (Table 8). The conventional fertilization procedure consisted of applying P and K preplant and three applications of N. One-third of the N was applied at planting time and one-third during each of weeks 6 and 8 after emergence. Biweekly applications were as follows: 5% of the N, P, and K was applied at emergence, 5% during week 2, 10% during week 4, and 20% during each of weeks 6, 8, 10, and 12 for a total of seven applications. Both fertilization schedules received identical irrigations and the same total amount of N, P, and K. Irrigation water use efficiency was increased by > 100% with the biweekly schedule in comparison with the conventional schedule. Although it is difficult to document, there was obviously a significant N loss from the conventional treatment caused by leaching rains during the growing season. The increased yield of the biweekly treatment is attributed to the application of 80% of the N during the rapid nutrient uptake period of plant growth, sometimes referred to as the grand growth period.

Brown (1971) reported that the amount of water needed to grow winter wheat to the point where grain production begins increased with N rate. Yield divided by evapotranspiration (ET) was 73 kg/ha-cm with no N applied, 114 kg/ha-cm with 67 kg N/ha applied, and 115 kg/ha-cm with 268 kg N/ha applied. Total ET increased from 22 cm with no N applied to 32 cm with the application of 268 kg N/ha. Higher water use efficiency of winter wheat observed by Singh et al. (1975) occurred on both a clay loam

Table 9—Effect of N fertilization on water use efficiency of winter wheat in Nebraska in terms of dry matter yield per unit of water applied as irrigation and rainfall (Black, 1966).

Water applied		N applied (kg/ha)			
Rainfall	Irrigation	0	22	45	90
cm		kg/ha-cm			
31	0	43	63	71	71
31	7.4	60	87	110	124
31	20.6	62	91	115	133

and loamy sand soil with 80 kg N/ha. The clay loam soil produced 77 kg grain/ha-cm of water used with no N applied and 122 kg/ha-cm with 80 kg N/ha. Values for the loamy sand soil were 96 and 140 kg/ha-cm, respectively. However, the clay loam soil produced the highest yield because of its higher content of available water.

Water use efficiency of winter wheat in Nebraska was increased in each of three soil-water treatments (Black, 1966). At the lowest level of available water, maximum efficiency occurred with 45 kg N/ha (Table 9). However, the highest N rate (90 kg/ha) produced highest grain yield per centimeter of water used at the medium and high levels of available water.

The effect of N on water use efficiency in sugarbeets (*Beta vulgaris* L.) was similar to that observed in corn and small grain (Taylor & Ashcroft, 1972). Response of sugarbeets to soil-water was 14% greater with 180 kg N/ha than with 90 kg N/ha.

V. EFFECT OF WATER ON CROP RESPONSE TO N

Sweet corn ear yield did not respond to N levels above 112 kg/ha with medium and low levels of soil-water (Taylor & Ashcroft, 1972). However, response to N was linear in the range of 0 to 224 kg N/ha with a high soil-water level. Soil-water treatments are described earlier in this chapter. The ratio of yield increase to applied N was $>$ 56 for 56 and 112 kg N/ha with low and medium soil-water levels, but it dropped to $\leq$ 27 for 224 kg N/ha. At the high level of soil-water, the ratio was 48 for all N levels.

High-frequency irrigation of field corn increased the yield from 7200 kg/ha of grain with no irrigation to 11 900 kg/ha (Rhoads & Stanley, 1973). Applied N for all irrigation treatments was 336 kg/ha. The ratio of grain yield divided by applied N was 21.5 for nonirrigated and 35.5 for corn irrigated 11 times when the soil-water suction at the 15-cm depth reached 20 cbar.

Irrigation of winter wheat in Nebraska increased dry matter yield response to N fertilization at 22, 45, and 90 kg/ha (Table 10). Yield was highest for each N level with the high amount of irrigation. Maximum yield with no irrigation occurred with 45 kg N/ha. Grain yield response was similar except that maximum grain yield with no irrigation occurred at 22 kg N/ha. Winter wheat responded in a similar way in other studies (Rickman et al.,

Table 10—Effect of irrigation on yield increase of winter wheat due to N fertilization in terms of dry matter yield increase divided by amount of N applied (Black, 1966).

	N applied (kg/ha)		
Irrigation	22	45	90
cm	kg dry matter/kg N/ha		
0	28	20	10
7.4	47	43	27
20.6	68	61	41

1975). Dry matter yield with 90 kg N/ha was 6000 kg/ha with 17 cm of irrigation and 8000 kg/ha with 23 cm of irrigation.

Sugarbeets showed very little response to N fertilization when soil was allowed to dry to the 15-bar suction level before irrigating (Taylor & Ashcroft, 1972). Increase in beet yield divided by fertilizer N with low soil-water was 25 for 60 kg N/ha and 17 for 160 kg N/ha. High soil-water (irrigated at 0.3 bar suction) produced a yield increase of 50 kg of beets/ha per kilogram of N with 80 kg N/ha applied and a 63 kg increase/ha per kilogram of N with 160 kg N/ha applied.

According to Viets (1972), nutrient and water absorption are independent processes in roots of crop plants. The quantity of soil-water affects the nutrient concentration in the soil solution and rate of nutrient movement to the root by diffusion and mass flow. Soil-water content affects rate of root elongation; for example, at a soil-water suction of 0.5 bar, roots elongated over 80 mm in 48 h, while at 8 bar suction, elongation rate was about 30 mm in 48 h. Uptake of N was proportional to dry matter accumulation in both irrigated and unirrigated corn in Nebraska. An important point mentioned by Viets (1972) is that water availability fluctuates most in the surface soil where the concentration of most soil and fertilizer nutrients is highest. Stressed tomato plants (*Lycopersicon esculentum* Mill.) contained less total N at a lower concentration than nonstressed control plants (Slatyer, 1969).

REFERENCES

Bahrani, B., and S. A. Taylor. 1961. Influence of soil moisture potential and evaporative demand on the actual evapotranspiration from an alfalfa field. Agron. J. 53:233–237.

Black, C. A. 1966. Crop yields in relation to water supply and soil fertility. p. 177–206. *In* W. H. Pierre et al. (ed.) Plant environment and efficient water use. Am. Soc. of Agron. and Soil Sci. of Am., Madison, Wis.

Brown, P. L. 1971. Water use and soil water depletion by dryland winter wheat as affected by nitrogen fertilization. Agron. J. 63:43–46.

Gorbet, D. W., and F. M. Rhoads. 1975. Response of two peanut cultivars to irrigation and kylar. Agron. J. 67:373–376.

Hardy, G. W. 1959. Nitrogen fertilization of soybeans. Soybean Dig. 19(8):18.

Lipscomb, R. W., W. K. Robertson, and W. H. Chapman. 1965. Fertilization and spacing of peanuts. Soil Crop Sci. Soc. Fla. Proc. 25:329–334.

Pallas, J. E., Jr., J. R. Stansell, and T. J. Koske. 1979. Effects of drought on florunner peanuts. Agron. J. 71:853–858.

Rhoads, F. M., R. S. Mansell, and L. C. Hammond. 1978. Influence of water and fertilizer management on yield and water-input efficiency of corn. Agron. J. 70:305–308.

Rhoads, F. M., and R. L. Stanley, Jr. 1973. Response of three corn hybrids to low levels of soil moisture tension in the plow layer. Agron. J. 65:315–318.

Rhoads, F. M., and R. L. Stanley, Jr. 1979. Effect of population and fertility on nutrient uptake and yield components of irrigated corn. Soil Crop Sci. Soc. Fla. Proc. 38:78–81.

Rickman, R. W., R. E. Ramig, and R. R. Allmaras. 1975. Modeling dry matter accumulation in dryland winter wheat. Agron. J. 67:283–289.

Robins, J. S., and C. E. Domingo. 1953. Some effects of severe moisture deficits at specific growth stages in corn. Agron. J. 45:618–621.

Robins, J. S., and C. E. Domingo. 1962. Moisture and nitrogen effects on irrigated spring wheat. Agron. J. 54:135–138.

Shaw, R. H., and D. R. Laing. 1966. Moisture stress and plant response. p. 73–94. *In* W. H. Pierre et al. (ed.) Plant environment and efficient water use. Am. Soc. of Agron. and Soil Sci. Soc. of Am., Madison, Wis.

Singh, R., Y. Singh, S. S. Prihar, and P. Singh. 1975. Effect of N fertilization on yield and water use efficiency of dry land winter wheat as affected by stored water and rainfall. Agron. J. 77:599–603.

Slatyer, R. D. 1969. Physiological significance of internal water relations to crop yield. p. 53–83. *In* J. D. Eastin et al. (ed.) Physiological aspects of crop yield. Am. Soc. of Agron. and Crop Sci. Soc. of Am., Madison, Wis.

Smith, R. L., C. E. Hutton, and W. K. Robertson. 1968. The effect of nitrogen on the yield of soybeans. Soil Crop Sci. Soc. Fla Proc. 28:18–23.

Sprent, J. I. 1976. Water deficits and nitrogen-fixing root nodules. *In* T. T. Kozlowski (ed.) Water deficits and plant growth. Vol. IV. Academic Press, New York.

Stanberry, C. O., and M. Lowrey. 1965. Barley production under various nitrogen and moisture levels. Agron. J. 57:31–34.

Taylor, S. A., and G. L. Ashcroft. 1972. Physical edaphology. W. H. Freeman & Co. Publ., San Francisco. p. 414–450.

Viets, F. G., Jr. 1965. The plants need for and use of nitrogen. *In* W. V. Bartholomew and F. E. Clark (ed.) Soil nitrogen. Agronomy 10:503–549.

Viets, F. G., Jr. 1972. Water deficits and nutrient availability. *In* T. T. Kozlowski (ed.) Water deficits and plant growth. Academic Press, New York.

Weber, C. R. 1966. Soybean nodules work for you. Iowa Farm Sci. 20:11.

21

P. F. Pratt
University of California
Riverside, California

Nitrogen Use and Nitrate Leaching in Irrigated Agriculture[1]

The need for a supply of available N for crop production on irrigated land is qualitatively the same as on all croplands. But quantitatively, this need is probably (on the average) greater on irrigated land, because water is usually not a limiting factor and average crop yields are greater. The need for available N is the same whether it comes from inorganic fertilizers, organic materials, or fixation by legume-rhizobium combinations. The 1974 agricultural census data show that about 80% of the irrigated land in the United States is used for nonleguminous crops, indicating that most of these crops receive substantial applications of commercial fertilizers and/or animal manures to supply the needed available N.

With the exception of a few areas where irrigation waters are almost salt-free (for example, surface waters that come from snowmelt in mountain watersheds), irrigated lands must be leached periodically to maintain the rooting zone free of excessive soluble salts. In many areas leaching takes place as a result of rains; in some areas the rainfall is so small or so erratic that management must provide sufficient irrigation water to leach the soil profile. The leaching process, combined with relatively large inputs of N, ensures that one of the sinks for N is removal of NO_3^- in the drainage waters that move beyond the root zone.

A critical factor related to the leaching of NO_3^- from irrigated lands is the subsequent use of drainage waters or waters composed significantly of drainage waters. Irrigated lands are typically located in dry regions that

[1]Contribution from the Department of Soil and Environmental Sciences, University of California, Riverside, CA 92521.

Nitrogen in Crop Production.

naturally have low potentials for dilution of irrigation drainage waters. Because of limits in the total water supply, the demand for reuse of water for municipal supplies as well as for irrigation are high. Thus, the problem of NO_3^- leaching to groundwaters is naturally more crucial in irrigated western valleys where high-value crops with high water and high N demands are grown and where municipalities and irrigation districts are both using the underground supplies.

I. AREA AND DISTRIBUTION OF IRRIGATED LAND

The two sources of statistical data for irrigated land in the United States are (i) the *Census of Agriculture for 1974* of the Bureau of the Census of the U.S. Department of Commerce (U.S. Dep. of Commerce, 1974), and (ii) the "Irrigation Survey" of the *Irrigation Journal* (Anon., 1979). Both sources of data are derived from surveys. The "Irrigation Survey" data are collected from cooperative extension personnel from each of the states. The *Census of Agriculture* data come from a survey of farmers. The "Irrigation Survey" contains estimates of total irrigated area that are greater than those of the *Census of Agriculture*. For example, in 1969, the estimates were 19.4 and 15.6 million ha, respectively, for the "Irrigation Survey" and the *Census of Agriculture*. In 1979 (Anon., 1979), the "Irrigation Survey" showed 21.1 million ha of irrigated land, which indicates a 9% increase during the past decade. The *Census of Agriculture* estimates for 1979 are not available for comparison. The latest available data from the *Census of Agriculture* were for 1974. Extension irrigation specialists (Personal communication, J. L. Meyer, Irrigation & Soil Specialist, Coop. Ext., Univ. of California, Riverside) feel that the total area of irrigated land on which irrigation is necessary for crop production, and for which irrigation provides most of the available water, has remained constant, and that the area receiving supplemental irrigation has increased during the past 10 yr. In the next 10 yr, the total irrigated area is expected to increase slightly if there is a continued increase in supplemental irrigation in the central and eastern sections of the United States.

The 1974 *Census of Agriculture* (U.S. Dep. of Commerce, 1974) as summarized by Tucker and Hauck (1978), estimated that (i) irrigation is practiced to some extent in all of the United States; (ii) four states had $<$ 400 ha each, and 33 states had 10 000 ha or more; (iii) the states with the most irrigated land were California, Texas, Nebraska, Idaho, Colorado, Kansas, and the remaining 11 western states; Florida and Oklahoma each had $>$ 200 000 ha; (iv) in the area occupied, the main crops under irrigation were corn (*Zea mays* L.), sorghum (*Sorghum bicolor* Moench.), wheat (*Triticum aestivum* L.), hay, and cotton (*Gossypium hirsutum* L.); and (v) irrigated area as a percentage of the total area in that crop was 58% for potatoes (*Solanum tuberosum* L.), 52% for vegetable crops, 30% for cotton, 18% for sorghum, 15% for hay, 9% for field corn, and 5% for wheat.

A 1966 estimate (National Technical Advisory Committee, 1968) of the amount of irrigated land in the United States was 17.8 million ha, or about 10% of the total cropland. However, about 25% of the value of all crops harvested was produced on the irrigated land. The percentage of irrigated land has increased slightly, and because this increase likely has been largely from supplemental irrigation of high-value crops, the relatively higher dollar value of irrigated crops has probably been maintained or increased.

This chapter used data from California to illustrate the leaching of NO_3^- from irrigated lands. California has about 4 million ha of irrigated land, which is about 20% of the total irrigated U.S. land and between 2 and 3% of the total harvested U.S. cropland. However, California used 5.7% of the N fertilizer marketed in the United States in 1980 (Hargett & Berry, 1980), indicating that the rate of application of N to irrigated land in California was relatively large, compared with the average rate. Also, many irrigated valleys in California have intensive crop production located above groundwaters that are used for municipal supplies for which the NO_3^- concentration is critical. The author realized that much data existed for NO_3^- leaching from irrigated lands in areas other than California, but felt that the extensive data from California adequately demonstrated the factors and relationships involved.

II. LEACHING OF NITRATE

While a crop is growing and absorbing mineral N, several other processes are competing for the available mineral N in the soil. These competing processes are (i) microbial immobilization, (ii) volatilization of NH_3, (iii) denitrification, (iv) leaching of NO_3^-, and (v) miscellaneous losses such as that caused by erosion. Assuming a well-managed area with no NH_3 and erosion losses, the main sinks for available N over a time span of a decade or more are removal in harvested crops, denitrification, and leaching.

The amount of NO_3^- that leaches from a soil depends on the amount of water that moves through the soil and the amount of NO_3^- in the soil when water drains through and out of the soil profile. Thus, the drainage volume and the leachable NO_3^- are the primary factors, but these are related to many other factors (Fig. 1). Soil, climate, and economics determine to a large extent the selection of the crop to be grown. When the crop cultivar has been selected, management of the crop, irrigation, and N fertilizer must ensure that the crop has a competitive advantage over other sinks for N.

A. Freely Drained Sites

Freely drained soils are those for which tile drains are not necessary to keep the root zone well aerated. Drainage water that leaves the root zone percolates through an unsaturated zone to a saturated zone at some depth below. The rate of percolation through the unsaturated zone is proportional to the volume of drainage water and inversely proportional to the volu-

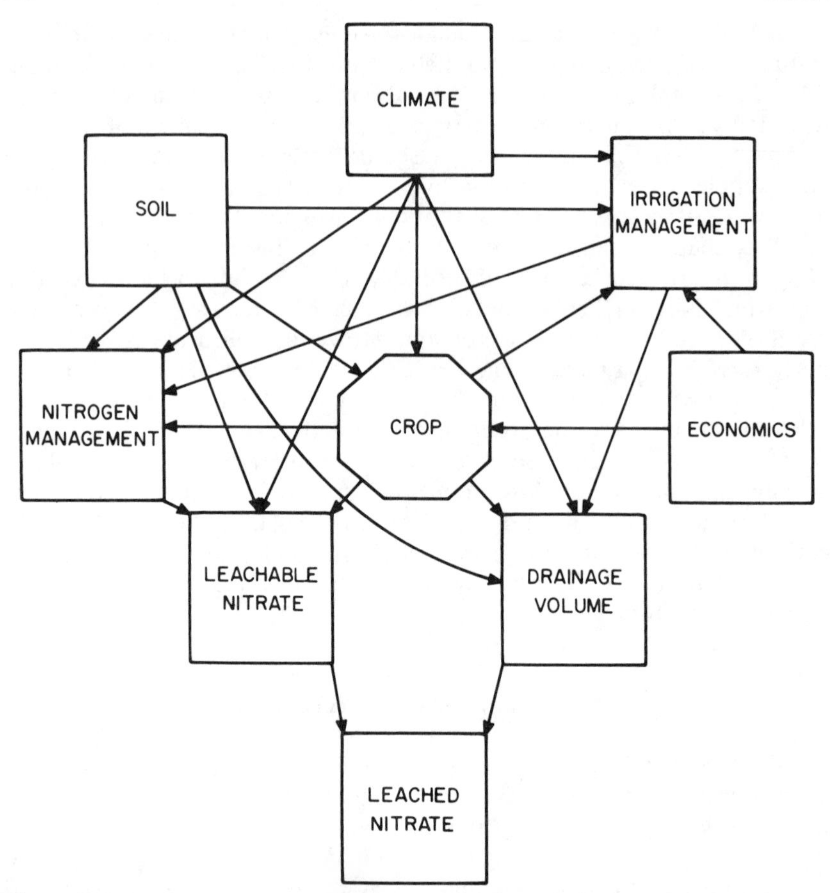

Fig. 1—Relationships among various factors that control drainage volume and leachable NO_3^- in irrigated agriculture. The arrows indicate dominant effects.

metric water content of the soil material through which the water percolates. This rate can range tremendously from high values in areas of high drainage volumes and sandy soil materials of low volumetric water contents to low values in areas of low drainage volumes and clayey soil materials of high volumetric water contents. Estimates from studies in California indicate rates of 0.5 to 1.8 m/yr, corresponding to about 15 to 60 yr for a transit to a saturated zone at a 30-m depth.

Pratt and Adriano (1973) reported data collected from 25 sites in the Santa Ana River Basin, some in commercial fields, and others in experimental plots. The N input ranged from 111 to 1525 kg/ha per year. The leached NO_3^--N ranged from 25 to 912 kg/ha per year, the drainage volume from 18 to 76 cm/yr, and the NO_3^--N concentration in the drainage water ranged from 12 to 123 mg/L. The average NO_3-N leached was 190 kg/ha per year, but when the data were limited to sites with $<$ 400 kg/ha per year inputs of fertilizer N, the average was 99 kg/ha per year. The average NO_3^--N concentration in the drainage water was 52 mg/L, but when sites with N inputs $>$ 400 kg/ha per year were excluded, the average was 36 mg/L.

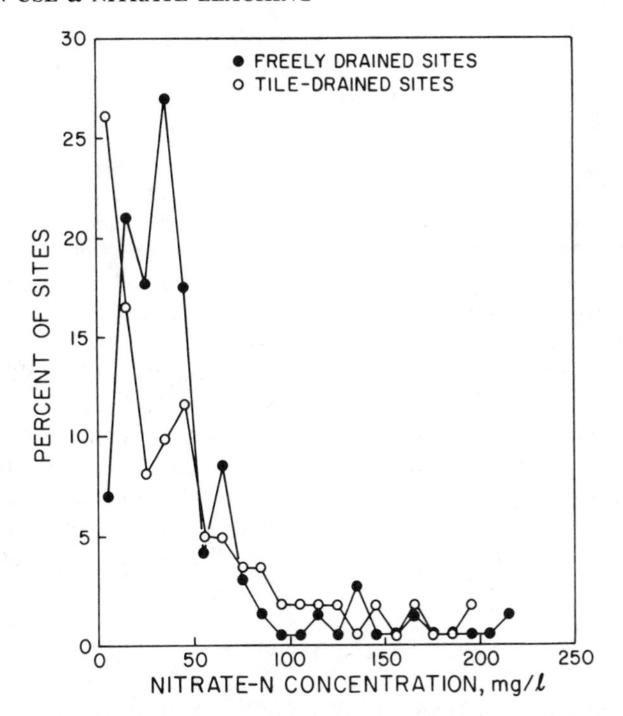

Fig. 2—Frequency distribution of site averages of NO_3^- concentrations in drainage water using class sizes of 10 mg NO_3^--N/L for freely drained and tile-drained sites.

Rible et al. (1979) reported data for 83 freely drained sites located in coastal valleys and in the San Joaquin Valley. The average NO_3^--N concentration in the drainage water was 41 mg/L, with a range of 6 to 220. A frequency distribution of these data (Fig. 2) indicate that typical values were 20 to 40 mg/L and that the average was highly influenced by a few values of much higher concentrations. At sites where no fertilizer N had been used, the average concentration was 19 mg/L, as compared with an average of 41 mg/L for all sites. The average concentration for the inland San Joaquin Valley was 36 mg/L, as compared with an average of 46 mg/L for coastal valleys.

B. Tile-Drained Sites

A report (Letey et al., 1979a) of data from 61 tile-drained sites from six different areas in California showed an average NO_3^--N concentration in tile drainage of 44 mg/L, with a range of averages per site from 1 to 196 mg/L. Data presented in Fig. 2 show that the most frequently encountered concentration was < 10 mg/L, but a total of 28 sites had concentrations between 10 and 90 mg/L. The distributions of NO_3^- concentrations for the free-drained and tile-drained sites were fairly similar at concentrations > 50 mg/L, but differed significantly at lower concentrations.

The results of monitoring tile drainage effluents for about 1 yr in 55 fields and more intensive studies of fewer sites for a longer period showed that (i) NO_3^- concentrations were not well-correlated with fertilizer rate, drainage volume, or soil profile properties and were not an index of the total quantity of NO_3^- leached, (ii) relatively low NO_3^--N concentrations and quantity of NO_3^- leached were measured where alfalfa (*Medicago sativa* L.) was the crop and no N had been applied, and (iii) soil profile characteristics significantly affected the amount of NO_3^- leached.

Letey et al. (1979b) reported significant correlations between fertilizer N applied and drainage volume for both freely drained fields and for fields that were tile-drained. This relationship suggests that farmers learn from experience to apply more N where leaching is greater to compensate for leached NO_3^- and, thus, to maintain maximum crop production.

The data presented indicate that substantial leaching of NO_3^- has occurred, that leaching has been one of the main sinks for the N added to irrigated croplands in California, and that leaching has contributed to the NO_3^- contents of underground water supplies. Because of relatively high rates of application of fertilizer N on intensively cropped lands (double-cropped in many cases) the quantities of NO_3^- leached in California are probably greater than would be expected in other areas of less intense agriculture, using lower N inputs.

C. Correlations

A number of studies have shown that the NO_3^- leached from irrigated land is correlated with mineral N inputs and drainage volume. Letey et al. (1977) reported correlations from a study of the NO_3^- leached from a number of tile-drained, irrigated fields in California. When fields that had received no N and those that had high levels of native NO_3^- were excluded from the correlation analyses, the amount of NO_3^- leached (M) was correlated with N input (N_i) ($r = 0.72$**), drainage volume (D) ($r = 0.83$***), and the product of N_i times D ($r = 0.92$***). When all sites were included, M was correlated with N_i ($r = 0.46$*) and D ($r = 0.84$***). (The symbols *, **, and ***, respectively, indicate statistical significance at probability levels of 0.05, 0.01, and 0.001). Table 1 presents correlation coefficients and regression equations for this tile-drainage study. The relationship between the amount of NO_3^--N leached and drainage volume is presented in Fig. 3.

Letey et al. (1978) reported similar correlations for a number of freely drained, irrigated sites. Nitrate leached was correlated with N_i ($r = 0.68$***), D ($r = 0.77$***), and N_iD ($r = 0.79$***). McNeal and Pratt (1978) reported a correlation ($r = 0.95$***, and $y = 0.97x$) between measured NO_3^- leached and the NO_3^- leached, as calculated from the equation $N_L = 0.20\,(N_iD)^{0.712}$ (where N_L is the NO_3^--N leached and N_i and D are as previously defined), for a large number of freely drained fields (Fig. 4) data, which were provided by Pratt and Adriano (1973), Pratt et al. (1980), and Rible et al. (1979).

Data presented by Nielsen et al. (1979) for NO_3^- leached from the root zone of a Yolo soil (thermic Typic Xerorthents) showed a series of regression equations for the relationship between NO_3^- leached and drainage volume for each of four fertilizer N inputs (Table 2). The highly significant correlation coefficients and the consistent increases in regression coeffici-

Table 1—Regression equations and correlation coefficients for NO_3^--N leached and various factors for sites included in the tile drainage study.†

Equation‡	Correlation coefficient	F
Excluding sites in alfalfa and sites having high native NO_3^-		
$M = -4.52 + 2.66\,D$	0.83***	55.0***
$M = -48.9 + 3.82\,N_i$	0.72**	16.8***
$M = 16.4 + 0.0042\,N_iD$	0.92***	82.9***
$C = 7.92 + 0.36\,N_i$	0.53*	5.9*
$C = 17.6 + 0.119\,D$	0.29 NS§	1.4 NS
All sites		
$M = 7.72 + 2.76\,D$	0.84***	47.8***
$M = 66.1 + 0.40\,N_i$	0.46*	5.2*
$C = 44.9 - 0.182\,D$	0.14 NS	0.4 NS
$C = 29.4 - 0.0007\,N_i$	0.006 NS	0.001 NS

*,**,*** Indicate statistical significance at the 0.05, 0.01, and 0.001 probability levels, respectively.

† From Letey et al., 1978.

‡ M is quantity of NO_3^--N leached in kg/ha per year, C is NO_3^--N concentration in mg/L, D is the drainage volume in cm/yr, and N_i is the fertilizer N input in kg/ha per year.

§ Not significant.

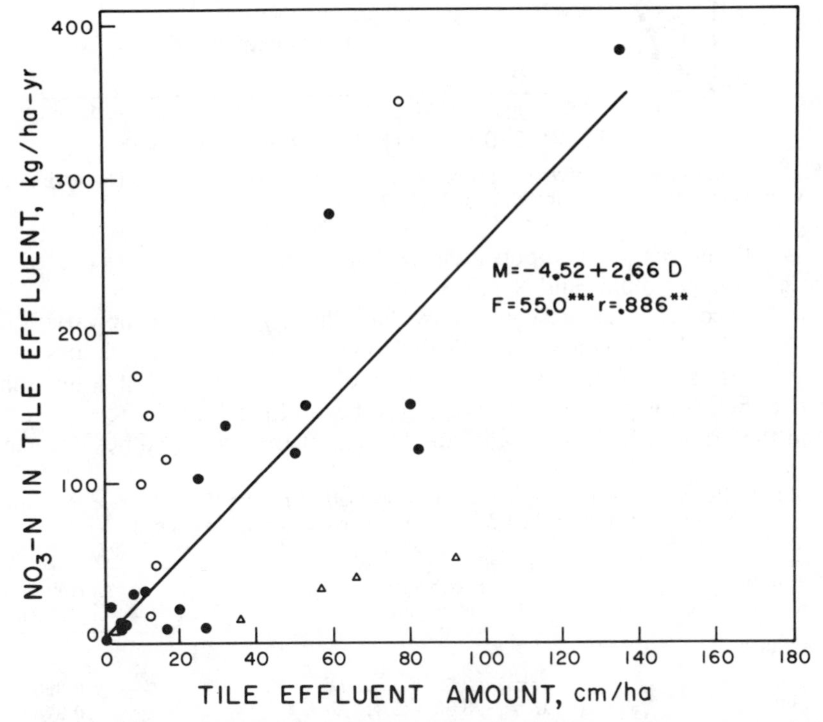

Fig. 3—The amount of NO_3^--N removed with the tile effluent as a function of the amount of tile effluent during the monitoring year. The open circles represent data from sites in the San Joaquin Valley suspected of having high native NO_3^-. The triangles are for sites with alfalfa and no fertilizer application. The linear regression analysis is for the data represented by the solid circles only. (Taken from Letey et al., 1977).

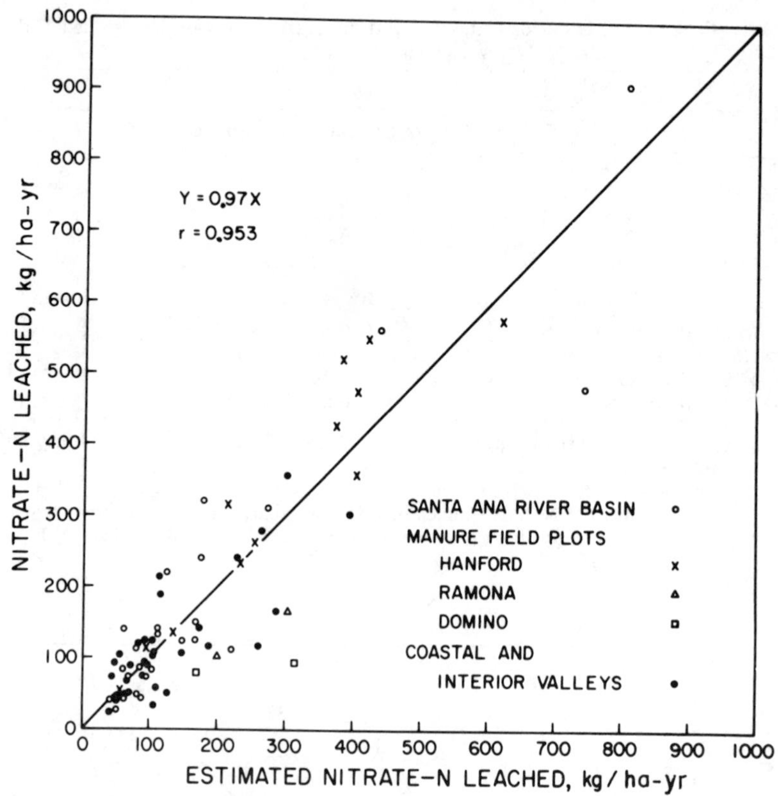

Fig. 4—Relationship between measured and estimated NO_3^- leached. The estimated values are from the equation $N_L = 0.20\ (N_iD)^{0.712}$.

ents with increases in N inputs indicate that the NO_3^- leached was related to both drainage volume and N inputs.

The correlations presented show that a large fraction of the variations in NO_3^- leached was associated with N inputs and drainage volumes. Assuming that the NO_3^- leached is a function of the amount of water that passes below the effective rooting depth (beyond recall of the root system) and the amount of NO_3^- in the root system at the time or times drainage

Table 2—Regression equations and correlation coefficients for the linear relationships between NO_3^--N leached (M) and drainage volume (D) at each of four fertilizer N inputs to a Yolo soil.

Fertilizer N input	Regression equation	Correlation coefficient
kg/ha		
0	$M = -\ 2.52 + 0.895\ D$	0.895
90	$M = -\ 8.83 + 1.31\ D$	0.978
180	$M = -16.1 + 1.89\ D$	0.968
360	$M = -30.0 + 4.15\ D$	0.927

† Calculated from data of Nielsen et al., 1979.

occurs (the leachable NO_3^- of Fig. 1), the interpretation of these correlations is that the main factor controlling leachable NO_3^- was fertilized N input. For a large number of sites representing various soils, crops, climates, and N managements, the significant factor controlling leachable NO_3^- was N inputs. Thus, the overall approach to reductions in NO_3^- leaching in a large basin appears to be to reduce N inputs along with reductions in drainage volume.

The concentration of NO_3^- in drainage waters was poorly correlated with quantities of NO_3^- leached or with N inputs and drainage volume. In fact, high concentrations were sometimes found in cases of low drainage volume and low mass emissions of NO_3^- and low concentrations with high drainage volumes and high mass emissions. The data show that NO_3^- concentrations provide no index to quantities of NO_3^- leached. Therefore, data for NO_3^- concentrations throughout a profile below the root zone or in an effluent from a tile line, without data for drainage volumes, give no indication of how much NO_3^- has been leached. The soil system does not behave like a municipal wastewater treatment system with a constant volume of flow for which a concentration is a direct index to quantity.

III. BASIN STUDIES

A. Upper Santa Ana River Basin

This basin has an area of 144 000 ha overlying extensive groundwater supplies, which are used for agricultural, municipal, and industrial purposes. Because of the increasing concentrations of NO_3^- in waters from wells in a number of locations in the basin, a study was made of sources of NO_3^-, land and water use, and waste disposal. As part of this study, a N balance for the basin gave an estimate of 7.64 million kg N leached from the soil per year, or an average of 53 kg/ha per year. If this amount of N were leached in a drainage volume of 25 cm/yr, the average concentration of NO_3^--N in the drainage water would be 21 mg/L. In comparison, the average NO_3^--N concentration in the unsaturated zone beneath fields in citrus and row crops was 62 mg/L. Rible et al. (1979) studied eight sites in citrus and vegetable crops in southern California and reported an average concentration of 48 mg NO_3^--N/L in the drainage water. The difference between the average for the basin and the measurements in irrigated fields is probably reasonable. During this study, the irrigated cropland in the basin was about 22 000 ha, or about 15% of the land area. The rest of the land area received little or no fertilizer N. The land-use patterns and the limited amounts of direct measurements of NO_3^- leaching indicated that irrigated croplands were one of the main sources of NO_3^- in groundwaters and that measures for more efficient use of N fertilizers, animal wastes, and irrigation water were needed. A summary of this study is available in bulletin no. 861 of the California Agricultural Experiment Station (Ayers & Branson, 1973).

B. Southern San Joaquin Valley

Miller and Smith (1976) calculated a N balance in an area of 1.772 million ha containing 1.039 million ha under irrigation. The removal of N in harvested crops in the year 1971 for field, hay, vegetable, tree, and vine crops was estimated as 54, 73, 43, 19, and 37%, respectfully, of the amount of fertilizer N applied, leaving substantial amounts of N to either be lost by volatilization as NH_3, denitrification, or leaching from the root zone. The average amount of NO_3^- leached was estimated as 44.8 kg/ha per year for the total area, or 76.3 kg/ha per year if all the leached NO_3^- came from the irrigated area. Since the dominant land use was irrigated crop production with low populations of people and animals, it is reasonable to assume that the NO_3^- leached was mainly from irrigated cropland. The 76.3 kg N/ha per year would give an effluent containing 30 mg NO_3^--N/L if the effluent volume were 25 cm/yr. Later measurements by Rible et al. (1979) at 18 irrigated sites in the San Joaquin Valley gave an average NO_3^--N concentration in the drainage water of 38.6 mg/L, although the range in concentration was 9 to 163 mg/L. The lower average concentration in the San Joaquin Valley as compared with the Santa Ana River Basin is probably related to the higher proportion of field and hay crops in the San Joaquin.

C. Santa Maria Valley

Lund et al. (1978) presented data for a N balance for this valley from a study similar to those of the Upper Santa Ana River Basin and the Southern San Joaquin Valley and also presented a balance for a 1-yr study of a given field. The Santa Maria Valley is a coastal valley containing 67 000 ha of which 18 600 ha are intensively farmed with field and vegetable crops. The field crops were mainly beans (*Phaseolus vulgaris* L.) and irrigated pasture, but a very small area was alfalfa or hay. The vegetable crops were mainly broccoli (*Brassica oleracea* var. *botrytis*), cauliflower (*Brassica oleracea* var. *botrytis*), and lettuce (*Lactuca sativa* L.). The average removal by harvested crops of the N input in fertilizers and irrigation water was estimated as 24%. The average leaching was 39%, leaving 37% unaccounted for. The amount of N leached from 18 600 ha of irrigated land was 128 kg N/ha per year, and at a leaching volume of 25 cm/yr the NO_3^--N concentration would be 51 mg/L. In the Simas field (the site of the 1-yr study) 30% of the applied N was removed in harvested crops, 37% was leached below the root zone, and 33% was unaccounted for. In this field, the average measured NO_3^--N concentration was 48, 46, and 29 mg/L, respectively, for three consecutive sampling dates. A more recent report (Lund, 1979) of detailed measurements in seven fields gave an average of 90 mg NO_3^--N/L in the drainage water with a range of averages per field of 40 to 152. The amount of N leached from these seven fields ranged from 90 to 260 kg/ha per year. These seven sites were in a variety of vegetable crops plus some beans and sugarbeets (*Beta vulgaris* L.), which probably is the cause of the higher levels of

Table 3—Mean NO_3^--N in the percolating solution for four sampling times for Blosser field.†

Time	Site‡ 1	2	3	4	5	6	7	8	Field as whole§
	NO_3^--N μg/mL								
1	68.7z	90.2y	92.1z	82.9y	84.5z	69.6z	85.3z	89.8z	82.9a
2	63.7z	82.3y	94.4z	85.3y	94.0z	108.8y	90.9z	97.7z	89.6a
3	100.0z	83.8y	77.2z	73.5y	92.1z	77.9z	90.7z	92.9z	86.0a
4	94.5z	53.9z	76.6z	34.3z	67.3z	59.9z	68.4z	80.6z	67.0b
Mean	81.8ab	77.6ab	85.1ab	69.0b	84.5ab	79.1ab	83.8ab	90.2a	81.4

† Values are means of percolating solution concentrations in the 1.8- to 6.0-m depth.
‡ Values within sites followed by the same letter are not significantly different at the 1% level.
§ Means over time or across sites followed by the same letter are not significantly different at the 1% level.

N leached and NO_3^- concentrations than estimated for all of the irrigated land in the valley. An alternate explanation is that the balance for the valley underestimated the N leached.

Data for one 16-ha field, the Blosser field, which appears to be typical of the irrigated land in the Santa Maria Valley, are presented in Table 3. The soils in this field were classified as Mollisols and Entisols, having surface textures of loamy sand to loam and sandy loam to silty clay subsoils. The initial sampling of the soil was in early 1976, with three subsequent samplings that followed crop harvests. Samples were collected to a depth of 6 m from three holes at each of eight sites at each sampling date. The crops grown during the study period were potatoes, broccoli, and beans. The estimates of N leached were obtained from techniques and procedures presented by Pratt et al. (1978), and the data presented are those of Lund et al. (1978). The field as a whole had relatively constant NO_3^- concentrations with time (Table 3), until the last sampling following a bean crop, when the NO_3^- concentration decreased by about 20 mg N/L.

The Blosser field, which was sprinkler-irrigated and had the typical crops used in the valley, is fairly representative of data for coastal valleys of California. The fertilizer N inputs varied from 225 to 285 kg/ha per crop except on beans, for which none was used.

The summary data from these three basin studies indicate the magnitude of the losses of N by leaching and the savings that could be obtained with greater efficiency of use of N inputs. The total quantity of NO_3^--N leached was estimated as 89 000 t/yr on 1.08 million ha of irrigated land for an average of 82.4 kg N/ha per year. Extrapolation of these estimates to the 3.5 million ha gives an estimate of 311 000 t of N leached from irrigated land in California. This quantity represents about 45% of the estimated N used as fertilizer plus animal manures in 1980, and about 55% of the average amount used during the 1970 to 1978 period, during which these basin studies were conducted. Of course, some of the leached N could have originated from microbial N fixation and from a net reduction in organic matter, but these contributions were probably minor.

IV. MANAGEMENT ALTERNATIVES

From agronomic and water quality points of view, the objective of an irrigated crop management system should be to produce the largest crop output for each unit input of N and water. From the point of view of crop production economics, N and water inputs may represent such small fractions of the total costs of production, that to maximize production per unit of these inputs is not as important as maximizing production per unit of other inputs such as area of land cultivated, labor, or fuel and equipment costs. However, if the costs of removal of NO_3^- from irrigation return flows were to be added to the costs of crop production, the economics might change dramatically in favor of management for maximizing recovery of N by the crop.

Essentially two features are needed for high production with low NO_3^- leaching losses; the first is an excellent production system, and the second is an efficient N management program that requires an efficient water management system. Rauschkolb et al. (1979) have defined the production system as consisting of the *soil, climate, N cycle, crop,* and *irrigation system*; the N management variables as *placement, source, rate,* and *timing* of fertilizer N; and the water variables as *rate, frequency,* and *quantity* of *irrigation water* applied. Of course, the selection of the crop and variety of crop is a management decision, and the irrigation system can be changed. After these decisions are made, they become part of the production system. Also, other crop-management and pest-control decisions are important because they influence crop yields and the ability of the crop to compete with other sinks for available N in the soil.

An ideal system for management of N and water to maximize N utilization by the crop and minimize leaching of NO_3^- has at least two requirements. The first requirement is that N utilization by the crop is so efficient that there are no increases in leachable NO_3^- with increases in available N inputs up to the point of maximum yield on a yield vs. N input curve. The second requirement is that the N needs of the crop for maximum yield or for maximum net economic return can be accurately predicted, so that large increases in leachable NO_3^- resulting from excess N applications can be avoided. The ideal system herein defined serves as a goal and also can be used as a model against which research and practice can be compared.

Broadbent and Carlton (1978, 1979) have reported data from two field experiments where the first requirement for the ideal management system was nearly met. On a Yolo fine sandy loam cropped to corn on which maximum yields were obtained at a N rate of 180 kg/ha, there were only small increases in soil NO_3^- at the end of the cropping season as the N application increased up to the 180 kg/ha rate. On a Hanford sandy loam (thermic Typic Xerorthents) for which maximum corn yield was obtained at 224 kg/ha, there was a very small increase in soil NO_3^- as the N input was increased from 0 to 112 kg/ha, and a larger increase as the N input was increased from 112 to 224 kg/ha. At rates of 224 to 560 kg/ha there was a

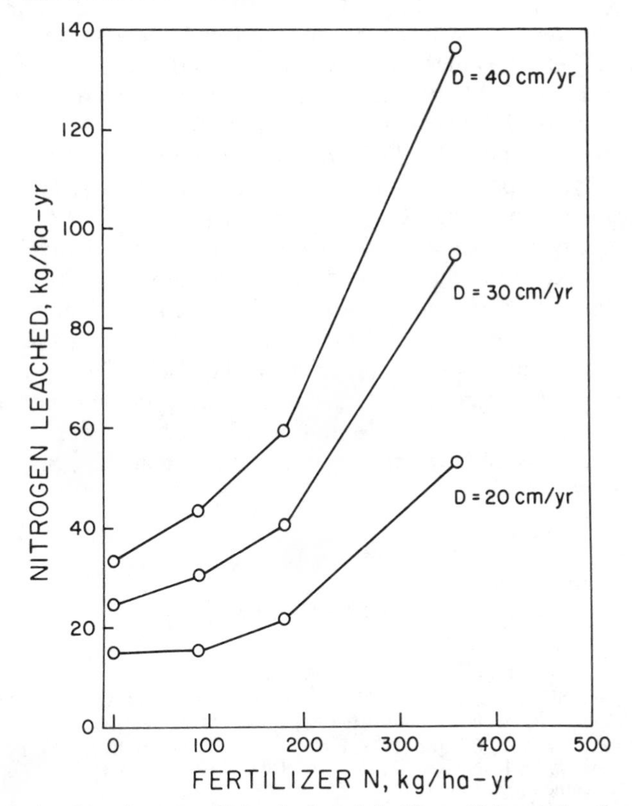

Fig. 5—Relationships between N leached and fertilizer N input for three drainage volumes. The data were calculated from the equations in Table 2.

linear increase in soil NO_3^- with increase in rate of applied N. The data suggest that the management system for the Yolo soil more closely met the first requirement for the ideal system than that for the Hanford soil.

The equations in Table 2, relating direct measurements of NO_3^- leached to drainage volume for the Yolo soil (Nielsen et al., 1979), were used to calculate the amounts of NO_3^--N leached for constant drainage volumes of 20, 30, and 40 cm/yr for each fertilizer input (Fig. 5). When the N fertilizer input increased from 0 to 90 kg/ha, the increase in NO_3^--N leached was 17, 24, and 31%, respectively, for drainage volumes of 20, 30, and 40 cm/yr. As the N input increased from 90 to 180 kg/ha, the increase in NO_3^--N leached averaged 35%, with no consistent relationship to drainage volume. With the next increment in N input, the increase in N leached was 66%, with no consistent relationship to drainage volume. The data for the first increment of N input suggest that the lowest drainage volume more closely approximates the first requirement of the ideal system. However, at each drainage volume the slope of the curves increased with increase in N inputs, indicating the need for a capability to predict quantities of N that do not exceed crop requirements.

Amounts of NO_3^- in the root zone at the end of the season (Broadbent & Carlton, 1978; 1979) increased only slightly with increase in N inputs up to the rate at maximum yield. However, direct measurements of NO_3^- leached for one of the same experiments (Nielsen et al., 1979) showed substantial increases in NO_3^- leached, with increase in N input below the point of maximum yield, particularly at higher drainage volumes. This difference between the two approaches suggests that measurements of NO_3^- in the root zone may not be an accurate index to NO_3^- leaching and that we need to develop inexpensive and reliable methods for direct measurements.

The relationships presented for the lowest leaching volume in Fig. 5 are possible using (i) high-yielding cultivars that respond well to N inputs, (ii) excellent fertilizer management in terms of placement, timing, and source, (iii) excellent water management to keep fertilizer N in the root zone during the growing season, (iv) adequate pest control to ensure excellent growth, and (v) optimum overall crop management to avoid any limiting factors on crop productivity.

REFERENCES

Anon. 1979. Irrigation survey. Irrig. J. 29(6):58A–58H.

Ayers, R. S., and R. L. Branson. 1973. Nitrates in the Upper Santa Ana River Basin in relation to groundwater pollution. California Agric. Exp. Stn. Bull. 861.

Broadbent, F. E., and A. B. Carlton. 1978. Field trials with isotopically labeled nitrogen fertilizer. p. 1–41. *In* D. R. Nielsen and J. G. McDonald (ed.) Nitrogen in the environment. Vol. 1. Academic Press, New York.

Broadbent, F. E., and A. B. Carlton. 1979. Field trials with isotopes—plant and soil data for the Davis and Kearney site. p. 433–465. *In* P. F. Pratt (ed.) Nitrate in effluents from irrigated lands. Rep. to Natl. Sci. Foundation, May 1979. National Technical Information Service, Springfield, Va.

Hargett, N. L., and J. T. Berry. 1980. Fertilizer summary data. Natl. Fertilizer Development Center, Tennessee Valley Authority, Muscle Shoals, Ala.

Letey, J., J. W. Biggar, L. H. Stolzy, and R. S. Ayers. 1978. Effect of water management on nitrate leaching. p. 231–249. *In* P. F. Pratt (ed.) National Conference on Management of Nitrogen in Irrigated Agriculture, May 1978. Dep. of Soil & Environmental Sciences, Univ. of California, Riverside.

Letey, J., J. W. Blair, D. Devitt, L. J. Lund, and P. Nash. 1977. Nitrate-nitrogen in effluent from agricultural tile drains in California. Hilgardia 45:289–319.

Letey, J., J. W. Blair, D. Devitt, L. J. Lund, and P. Nash. 1979a. Nitrate in effluent from specific tile-drained fields. p. 247–296. *In* P. F. Pratt (ed.) Nitrate in effluents from irrigated lands. Rep. to Natl. Sci. Foundation, May 1979. National Technical Information Service, Springfield, Va.

Letey, J., P. F. Pratt, and J. M. Rible. 1979b. Combining water and fertilizer management for high productivity low water degradation. Calif. Agric. 33:8–9.

Lund, L. J. 1979. Nitrogen studies for selected fields in the Santa Maria Valley. A. Nitrate leaching and nitrogen balances. p. 355–415. *In* P. F. Pratt (ed.) Nitrate in effluents from irrigated lands. Rep. to Natl. Sci. Foundation, May 1979. National Technical Information Service, Springfield, Va.

Lund, L. J., J. C. Ryden, R. J. Miller, A. E. Laag, and W. E. Bendixen. 1978. Nitrogen balances for the Santa Maria Valley. p. 395–413. *In* P. F. Pratt (ed.) National Conference on Management of Nitrogen in Irrigated Agriculture, May 1978. Dep. of Soil & Environmental Sciences, Univ. of California, Riverside.

McNeal, B. L., and P. F. Pratt. 1978. Leaching of nitrate from soil. p. 195–230. *In* P. F. Pratt (ed.) National Conference on Management of Nitrogen in Irrigated Agriculture, May 1978. Dep. of Soil & Environmental Sciences, Univ. of California, Riverside.

Miller, R. J., and R. B. Smith. 1976. Nitrogen balance in the southern San Joaquin Valley. J. Environ. Qual. 5:274–278.

National Technical Advisory Committee. 1968. Water quality criteria. Fed. Water Pollut. Control Admin., U.S. Department of the Interior, U.S. Government Printing Office, Washington, D.C.

Nielsen, D. R., C. S. Simmons, and J. W. Biggar. 1979. Flux of nitrate from a spatially variable field soil. p. 487–501. *In* P. F. Pratt (ed.) Nitrate in effluents from irrigated lands. Rep. to Natl. Sci. Foundation, May 1979. National Technical Information Service, Springfield, Va.

Pratt, P. F., and D. C. Adriano. 1973. Nitrate concentrations in the unsaturated zone beneath irrigated fields in southern California. Soil Sci. Soc. Am. Proc. 37:321–322.

Pratt, P. F., L. J. Lund, and J. M. Rible. 1978. An approach to measuring leaching of nitrate from freely drained irrigated fields. p. 223–256. *In* D. R. Nielsen and J. G. MacDonald (ed.) Nitrogen in the environment. Vol. 1. Academic Press, New York.

Pratt, P. F., L. J. Lund, and J. E. Warneke. 1980. Nitrogen losses in relation to soil profile characteristics. p. 33–45. *In* A. Banin and U. Kafkafi (ed.) Agrochemicals in soils. Pergamon Press, New York.

Rauschkolb, R. S., J. R. Hills, A. B. Carlton, and R. J. Miller. 1979. Nitrogen management relative to crop production factors. p. 647–687. *In* P. F. Pratt (ed.) Nitrate in effluents from irrigated lands. Rep. to Natl. Sci. Foundation, May 1979. National Technical Information Service, Springfield, Va.

Rible, J. M., P. F. Pratt, L. J. Lund, and K. M. Holtzclaw. 1979. Nitrates in the saturated zone of freely drained fields. p. 297–320. *In* P. F. Pratt (ed.) Nitrate in effluents from irrigated lands. Rep. to Natl. Sci. Foundation, May 1979. National Technical Information Service, Springfield, Va.

Tucker, T. C., and R. D. Hauck. 1978. Removal of nitrogen by various irrigated crops. p. 135–167. *In* P. F. Pratt (ed.) National Conference on Management of Nitrogen in Irrigated Agriculture, May 1978. Dep. of Soil & Environmental Sciences, Univ. of California, Riverside.

U.S. Department of Commerce. 1974. Census of agriculture for 1974. U.S. Department of Commerce, Bureau of the Census, Washington, D.C.

22

R. A. Olson
University of Nebraska
Lincoln, Nebraska

Nitrogen Use in Dryland Farming Under Semiarid Conditions

The management of soil and crop for most economic production in dryland farming requires substantially different operational procedures than are employed in the more humid regions. Moisture conservation is of the essence for preventing drought effects on the crop grown; all that is possible must be done to store and preserve the limited rainfall received for crop use. This basic requirement adds complication to the management of fertilizer N, which, when improperly used, can accentuate the drought hazard.

Wheat is the predominant crop grown in the semiarid regions of the world, partly because of its acceptance as food grain and partly because of the fit of its growth cycle with climatic, especially moisture, factors. It will be the purpose of this chapter to elaborate results achieved in adapting fertilizer N use into dry farming practice, particularly with the wheat (*Triticum aestivum* L.) crop.

I. DRYLAND FARMING DEFINED

The term *dryland farming* implies agricultural production under conditions of limited moisture availability. A common textbook definition is the tillage of land for crop production where the subsoil is continuously dry. Such a delineation embraces a major portion of the arable land of the world, including the dry topics, the Sahel of Africa, the Middle East, much of the People's Republic of China and USSR, and the western half of North America. Soils so situated will normally have a zone of lime accumulation

Nitrogen in Crop Production.

in the solum, and, where dependent on annual rainfall for moisture storage, will have dry farming methods employed in their effective management.

No specific amount of mean annual rainfall can connote a semiarid dryland farming situation because of differential evapotranspiration accompanying varied conditions of temperature, humidity, rainfall distribution, and wind action. Whereas 550 mm rainfall permits intensive annual crop production in Oregon or northern Minnesota, at least periodic fallowing is required with this amount in the warm and windy high plains of western Kansas and Nebraska. Perhaps a more apt definition for *dryland farming* is: Any farming situation where potential evapotranspiration exceeds the growing season rainfall by a factor of 2. This gives allowance for a lesser water loss from a growing crop than from a free water surface and for a certain amount of moisture storage before planting. Soil depth and the actual amount of storage are key variables. Thus, a soil with 75 cm of solum over bedrock in New York and 90 cm annual rainfall could represent an equal moisture limitation to a 180-cm depth soil in the northern Great Plains with only 60 cm rainfall. Figure 1 expresses the temperature–annual rainfall relationship as might be represented for a cool temperate-to-semitropical climatic regime. The range provided between semiarid and humid region cropping accommodates in some degree for varied wind action and periodicity of precipitation with respect to when the crop grows. Successful moisture storage in the soil and controlling losses obviously becomes the essence of dryland farming in the semiarid regions.

Small grains and grain sorghum (*Sorghum bicolor* Moench.) are the predominant cultivated crops produced under semiarid farming in the United States. Growth during the fall and spring months when temperature is lower and the incidence of rainfall greatest accounts for the former, and a combination of drought resistance and low water requirement for the latter.

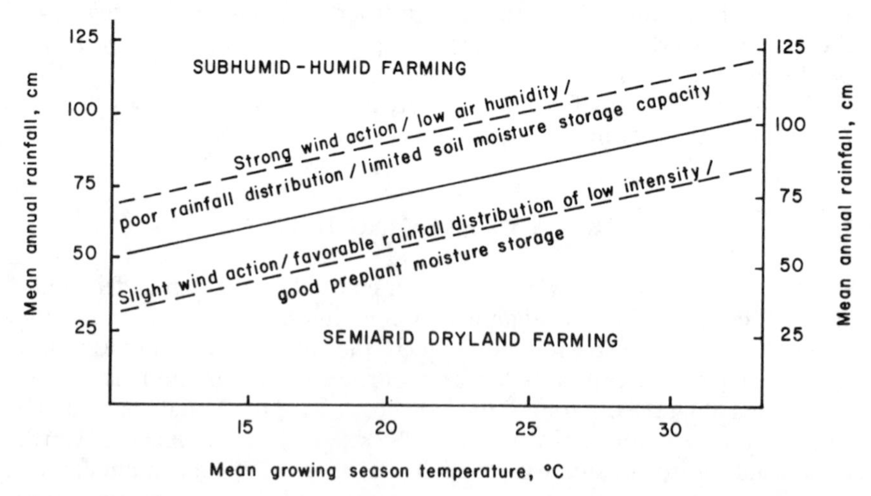

Fig. 1—Classification of semiarid dryland farming compared with humid region farming based on annual rainfall, growing season temperature, and related environmental factors.

Table 1—Nitrogen harvested in good yields of important crops produced in dryland farming regions.

Crop	Yield	N Removed†	
		Harvested portion	Total crop
	kg/ha		
Cotton‡	300–600	25–40	50–80
Corn	2000–3500	30–55	45–80
Barley	1900–2250	36–45	45–65
Oats§	1500–1800	30–35	45–55
Grain sorghum	2200–4000	35–60	50–80
Wheat	1400–3000	30–70	40–90
Nonlegume hay	2200–6000	50–120	50–120
Safflower¶	1000–1500	30–50	40–60
Millet	1500–2500	25–40	35–55

† Based on values in Anon. (1972), Nelson (1968), and Tucker & Hauck (1978).
‡ *Gossypium hirsutum* L.
§ *Avena sativa* L.
¶ *Carthamus tinctorius* L.

Other crops of lesser importance include millet [*Setaria italica* (L.) Beauv.], corn (*Zea mays* L.), sunflower (*Helianthus annuus* L.), rape (*Brassica rapus* L.), and mustard (*Brassica perviridis*).

Not to be overlooked are very large acreages of forage crops, especially grasses, that support extensive livestock operations in the dry regions. But the use of N on rangelands is quite limited to date, since the responses–although reasonably assured—are marginally economic. The N requirement of the grain crops is quite variable depending on the yield potential, primary determinant of which is the supply of moisture available for producing the crop. Table 1 gives amounts of N removed by representative yields of the several crops listed, and therewith, likely long-term annual N replacement requirements.

II. N RESPONSE IN RELATION TO AVAILABLE SOIL MOISTURE

As expressed in Fig. 1, the total amount of precipitation received is not the only criterion of moisture availability in dryland farming. Soil depth for moisture storage as it controls moisture availability for the crop between rains can be as important as the mean annual precipitation. Timeliness of the precipitation received in relation to when the crop grows can be of major significance. The winter-spring rainfall pattern of the Pacific Northwest permits high annual yields of wheat from total amounts that would require alternate-year cropping in the High Plains. As a consequence, large response to applied N is obtained in the former region but limited response in the latter with most representative soils. It is especially the added development and survival of tillers in wheat affording an increased number of heads that accounts for the grain yield increase from applied N (Power & Alessi, 1978).

Table 2—Effect of preplanting supplemental moisture and rate of N fertilization on yield and water use efficiency of winter wheat at North Platte, Neb., 1954–1956.†

N Applied	Preplant irrigation water applied, cm‡			
	0	7	15	21
kg/ha		Grain yield, kg/ha		
0	473	1013	1283	1350
22.5	675	1418	1890	1958
45	608	1688	2160	2363
90	540	1688	2498	2633
		Water use efficiency, kg/ha-cm		
0	16	26	32	30
22.5	22	36	45	41
45	19	44	48	47
90	16	41	54	51

† Ramig & Rhoades, 1963.
‡ There was no available water to the 180-cm depth in the nonirrigated plots; 60 cm of soil moistened to field capacity with 7 cm irrigation; 120 cm moistened with 15 cm added; and 180 cm moistened with 21 cm added.

Deep subsoil moisture storage in dry regions can be the prime determinant of yield and crop response to applied or residual N, as evidenced by experiments in western Nebraska, Colorado, and Montana (Smika et al., 1969). Such reserves in a drought year from a prior year of favorable rainfall have given the false indication that fertilizer N substituted for water when it was observed that the fertilized crop produced fairly well and the unfertilized crop was a failure. A more correct interpretation is that a more active root system that penetrated deeper into the soil because of appropriate fertilization made possible the extraction of existing deep subsoil moisture (Brown, 1971; Kmoch et al., 1957; Pesek et al., 1955). Recognition of the role of these reserves has made projections possible of quantity of fertilizer N to be applied by the farmer with different levels of moisture storage in the soil rooting profile at planting. In the case presented in Table 2 for winter wheat production in the semiarid environment at North Platte, Neb., 22 kg N/ha would be justified in the soil represented with no supplemental moisture storage at planting, 45 kg N with 7 cm storage, and 90 kg N with 15 and 21 cm preplant irrigation. It will be noted that the two heavier N rates were actually depressive to yield with the low moisture regime. Maximum water use efficiency accompanied the 15-cm supplemental water and 90 kg N/ha in this study on continuously cropped soil where seasonal rainfall averaged 30 cm. A general recommendation forthcoming was that 22 kg N/ha could be used for each 60-cm depth of moist soil at the time of wheat planting in September. Such recommendation could obviously be valid only for the soil represented, as other soils might have quite different amounts of residual mineral N and inherent N mineralization potentials. But the data do serve the point that moisture storage is a major factor controlling crop response to N in semiarid farming of western Nebraska as has been reported for spring wheat and barley (*Hordeum vulgare* L.) in North

Dakota (Bauer et al., 1965), for stubble spring wheat in Saskatchewan (Read & Warder, 1974), and for winter wheat in Oregon (Leggett, 1959).

Where rainfall is so limiting as to prevent continuous, annual cropping, the fallowing of soil for one out of every two or three years has become common practice. Therewith, additional moisture storage is accomplished in the subsoil below the evaporation zone, making cropping feasible. The proportion of the rainfall stored during the fallow period is usually quite low, as in the order of 20 to 30%, but when added to cropping-season rainfall allows more economic production than from annual cropping. Recently conceived chemical methods for weed control with minimum tillage (ecofallow) have made possible a reduction of the area for which fallowing had been considered essential in the past. When ecofallow is practiced, efficiency of moisture storage is markedly enhanced up to the 40 to 50% level (Fenster & Peterson, 1979). Holding snow on the soil surface by crop residues, grass strips, and/or artificial barriers during the winter months is especially important to moisture storage efficiency (Greb, 1979). The reduced- and no-tillage systems are rapidly being accepted as the best measure available for soil erosion control and energy conservation, an associated soil N fertility problem notwithstanding (Phillips et al., 1980).

The fallow practice stores not only moisture, but accomplishes a build-up of NO_3^--N in the soil from the mineralization that occurs during the fallow period. As a result, response to applied N is notably less on noneroded, fallowed soils than occurs in nonfallowed soils supplied with a comparable amount of moisture for crop use. With continued decay of the native organic matter, an equilibrium can eventually be expected where even the fallowed soils require substantial amounts of fertilizer N for optimum yields. This state has not been reached to date in much of the High Plains other than on sandy and eroded soils. Where it has, long-term studies indicate that N-fertilized continuous wheat is as effective for maintaining soil N as the production of N-fertilized crested wheatgrass or a grass-legume mixture (Haas et al., 1976).

Moisture presence is absolutely essential to effective crop response to applied N, but at the same time, the crop must experience a sufficient level of N nutrition to utilize efficiently the water available for it. Correlation has been shown to exist between stored soil moisture to the 122-cm depth at planting and yield response of wheat and barley to N fertilization in North Dakota (Young et al., 1967). Consumptive water use studies at 68 locations under dry farming conditions in western Nebraska with winter wheat and grain sorghum demonstrated that both crops used slightly more total water when fertilized for most economic yield, but in the process used the water 12 and 28%, respectively, more efficiently for grain production than the nonfertilized crops (Olson et al., 1964). Thus, Viets (1962), in reviewing prior investigations dealing with moisture-fertility interaction, observed that enhanced water use efficiency usually accompanied fertilizer treatment that increased yield. He concluded that an adequate level of crop nutrition was essential to effective utilization of water resources. This does mean, of course, that residual soil moisture following an adequately fertilized crop is likely to be somewhat less than that remaining without fertilization.

III. N NEED DETERMINANTS

The tillage system employed by the farmer has a considerable impact on the level of N nutrition derived from the soil by the crop. Most studies have revealed greatest N uptake from the soil with plowing, less with a disking of soil, and least with mulch or minimum tillage. The reduction in N availability with minimum tillage has been attributed especially to a lower temperature caused by the shading and insulating effects of surface residues, to leaching loss, and to retarded decomposition and N release from the residues accumulating on or near the soil surface (Phillips et al., 1980). Acidification is a further potential factor in at least some low buffer-capacity soils of the dry regions. As a consequence, mulch tillage has been found to raise the level of organic N in the surface few centimeters of soil in a relatively few years. Recent evidence also suggests that denitrification losses may be accentuated with mulch tillage (Doran, 1980). The combined processes have a depressing effect on N availability as long as mulch tillage is practiced, necessitating more fertilizer N for top grain and protein yields, but a probable enhancement in nitrification capacity when the soil is eventually plowed. On the other hand, greater responses to applied N with minimum tillage have been recorded under drier than normal weather conditions when yield failures occurred with conventional tillage methods. The difference in this case was the greater crop growth and higher yields made possible by enhanced moisture storage in the soil with minimum tillage.

With the advent of extensive fertilizer N use for grain production during the past 25 yr, the predominant factor controlling rate of N required for many soils today has become the amount of residual mineral N existing in the rooting profile when the new crop is planted (Olson et al., 1964; Soper & Huang, 1963; Young et al., 1967). Perhaps the first comprehensive report establishing the significance of residual NO_3^--N in soil on wheat yields comes from the dryland area of eastern Washington (Leggett, 1959). Equations were developed for predicting fertilizer N needed for maximum yield based on stored soil moisture and residual NO_3^--N to a 180-cm depth at planting. Figure 2 indicates that a yield reduction from fertilizer N was measured in dry farming wheat production in Nebraska where residual NO_3^--N exceeded 135 kg/ha, but that yield response continued through 67 kg N applied when residual was < 45 kg/ha. Intermediate residual levels required intermediate N rates for maximum yield expression. Residual mineral N, however, has not proved to be as much of a yield determinant as is storage of soil moisture in some soils in the western portion of the Great Plains (Smika et al., 1969).

Efficiency with which the small grains can utilize residual NO_3^--N from increasing depth in soil remains a somewhat conjectural issue. It is well established that winter wheat has a root system capable of extracting water and nutrients from depths in excess of 180 cm by the time grain filling begins, where such soil depth is favorable to root exploration. Grain sorghum and corn have equal or even greater capabilities in this respect. Research with the latter has shown a strong feeding capacity for residual NO_3^--N at

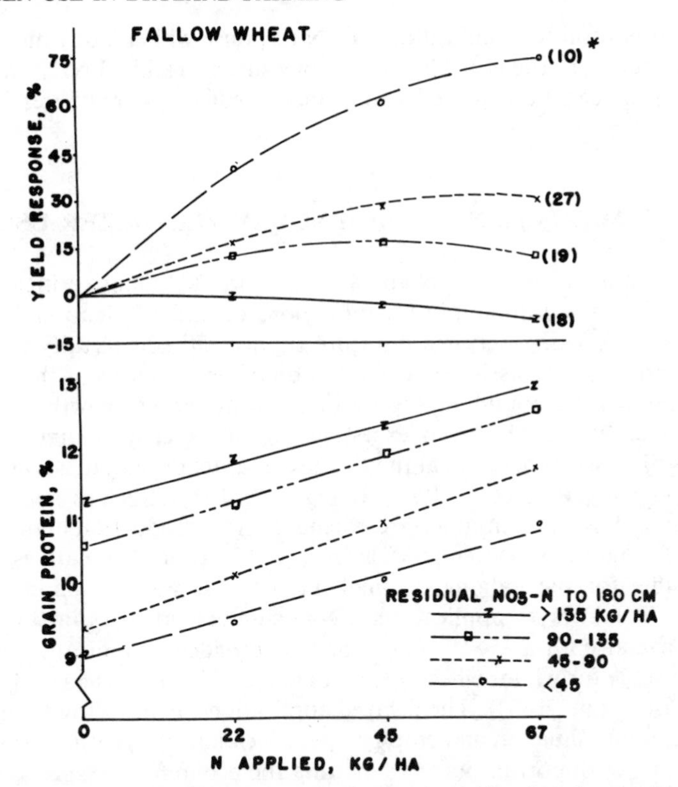

Fig. 2—The impact of residual NO_3^--N in soil on grain yield and grain protein response of fallow-wheat to fertilizer N (Olson et al., 1976). *Number of field experiments involved (a total of 74 experiments in Nebraska during 1962–1968).

the 180-cm depth if moisture is available there and if the plant has not been supplied already with adequate N from shallower depths (Gass et al., 1971) —similarly for grain sorghum through at least 150 cm (Bourg, 1961). It seems likely that wheat would perform similarly, especially under circumstances of limited N accessability in the upper root zone and declining moisture availability with depth in the soil as the season progresses—the norm in most dry farming regions of the United States. Adding credibility to this concept is the known efficient uptake and translocation into the developing grain proteins of N taken up late in the season by deeply penetrating roots under the winter rainfall–dry summer conditions of eastern Washington (Cochran et al., 1978).

Important as the residual NO_3^--N has become with recent intensive fertilizer use in dry farming regions, it is by no means the only criterion of fertilizer N need. In fact, it has been found quite inadequate in some situations. The soil's inherent capacity to nitrify from the organic N pool can be the dominant factor. This applies especially for soils of relatively high organic matter content that are near neutral-to-moderately alkaline in reaction. A measure in this case of hydrolyzable or readily nitrifiable N as described elsewhere in this book becomes of the essence. It seems probable that eventual soil-test calibrations for N response will embody measures of both N sources to the crop. Further, a measure of residual NH_4^+ in at least surface soils can be expected to enhance predictive capabilities in many cases.

IV. MAXIMIZING EFFICIENCY IN FERTILIZER USE

The timing of fertilizer N application is probably more critical for dry farming than for cropping in humid regions or under irrigation. The delicate balance existing between moisture supply and soil fertility as growth control processes necessitates the definition of one in terms of the other, as noted before. An abundant supply of N from earlier growth stages can result in early drought from excessive vegetative growth that dissipates stored soil moisture reserves at the expense of grain development later in the growth season (Olson et al., 1964; Ramig et al., 1975). Such overstimulation is recognized as a potential hazard whenever crop production under limited moisture conditions is attempted, as observed in Table 2. It can, as well, be responsible for the lodging of small grains observed in humid and dry regions alike. Delayed application of N as summer side-dressing after grain sorghum and corn are well established or broadcast application of N at jointing stage for winter wheat alleviates in large degree this hazard (Bourg, 1961; Olson et al., 1964). The delayed application further allows evaluation of the moisture situation and crop prospects well into the growing season and affords better opportunity for estimating the economic prospects of fertilizer application. Thus, long-term comparison of fall vs. spring application of N for winter wheat in the Great Plains have shown quite similar grain yield responses as an average for most locations (Olson et al., 1964; Schlehuber & Tucker, 1967)—similarly for the Pacific Northwest (Hunter et al., 1961). But in the year of limited moisture supply late in the growing season, spring-applied N has afforded higher grain yields because of less water consumption and consequent higher water use efficiency as expressed in Fig. 3. Evidence has also been shown of substantial losses of N from fall N treatments in dryland of Alberta from a series of 20 field experiments over a 4-yr period. Losses measured by ^{15}N labeling were 41, 30, and 16% from KNO_3, $CO(NH_2)_2$, and $(NH_4)_2SO_4$, respectively, and attributed primarily to denitrification. Evidence could not be found for leaching loss by analyzing for $^{15}NO_3^-$ to a 120-cm depth (Mahli & Nyberg, 1978).

A further likely benefit for delayed N application is a higher protein content of the resulting grain as compared with early treatment, probably due to a more direct translocation of absorbed N into grain with less vegetative entrapment and to reduced N loss from the root zone. Occasions may exist, however, where positional unavailability is responsible for poor results when extended dry periods follow delayed application with no percolating moisture to carry the applied N into the active portions of the root zone.

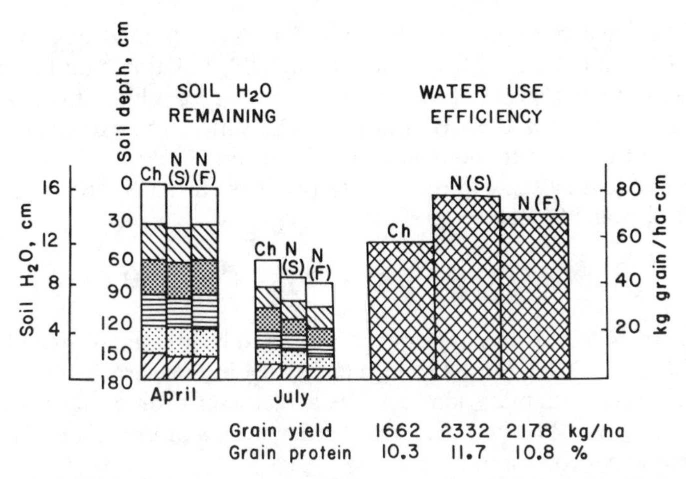

Fig. 3—Soil water depletion, water use efficiency, grain yield, and protein percentage in 16 field experiments with winter wheat in western Nebraska as affected by time of applying 45 kg N/ha as NH_4NO_3. *S* = Spring, *F* = Fall-applied N; *Ch* = Check (Olson et al., 1964).

There are other agronomic factors influencing the optimum time of N application for grain crops in dryland farming. The prior comments have been derived from studies with conventional solid and solution N carriers. Anhydrous ammonia, when placed at depths of approximately 20 cm or more just prior to planting, does not afford the early stimulative action on vegetative growth that occurs with shallower placement of the solids and solutions. It thereby has a delayed action not greatly different from slow-release N carriers or the effect imposed by nitrification inhibitors. These delayed-action materials when applied at or before planting will commonly give comparable results to delayed application of solids and solutions in most cases (Olson et al., 1964). Their appropriate use is detailed elsewhere in this book.

Fertilizer N rates are of necessity notably lower for grain crops in dryland farming than under more humid conditions. With moisture short and N excessive, the result is usually too much vegetative growth for the late-season moisture supply to support with consequent yield reduction. Thus, rates in excess of 60 kg N/ha are seldom required on other than sandy and eroded soils that have been fallowed, and the most economic rate is more likely in the order of 20 to 30 kg/ha. Higher rates are required, of course, in those areas with sufficient moisture to allow cropping without fallow. But appropriate rate is also very much dependent on mineral N stored in the soil at planting as noted in Fig. 2. Rates are likely to increase in time with N depletion effected by crop removal, and perhaps even sooner by the introduction of higher yielding semidwarf varieties into the dry regions.

Greater care is also required in respect to fertilizer N placement in dryland farming than is necessary with higher rainfall situations. An adequate depth of incorporation is essential for NH_3 (anhydrous or aqua), $CO(NH_2)_2$, and $(NH_4)_2SO_4$ for minimizing NH_3 volatilization losses as the soil dries in these regions of usually high soil pH. Loss by this channel is accentuated the higher the pH, the lower the soil cation exchange capacity (CEC), and the higher the temperature (Allison, 1966). Another potential pathway of loss exists with the use of urea as N carrier by the following reaction (Allison, 1963):

$$(NH_2)_2CO + 2HNO_2 = 2N_2 + CO_2 + 3H_2O.$$

The loss of N by this mechanism is expected to be greatest at low pH where the oxidation of NO_2^- to NO_3^- in the nitrification process is inhibited. But low moisture availability may also be a factor in promoting this reaction contributing to the often-noted lower effectiveness of urea than ammonium nitrate or anhydrous ammonia.

Greater care is also necessary in dryland farming in reducing contact between seed and fertilizer for preventing salt and toxicity effects on germination and growth (Olson & Dreier, 1956; Toews & Soper, 1978). Such effects are accentuated by higher moisture tensions in soil that limit movement of the fertilizer salt from the site of placement by leaching and diffusion. Temperature, air humidity, and wind action are related factors, as they influence the rate of water loss from the dryland farming soil.

Not to be overlooked in the maximizing of fertilizer N efficiency is the interaction that occurs between applied N and other fertilizer nutrients. Many reports exist in the literature detailing the supportive role of N on uptake and utilization of applied S, P, and Zn where deficient in soil, affording greater efficiency of both N and the companion element.

V. N USE AND CROP QUALITY

The enhancement in protein content of all grain crops by appropriate N fertilization is well established. The usually observed relationship with severe deficiency of N is a rapid increase in yield from the first increments of applied N with no increase or even slight decrease in protein percentage of the grain. Thereafter, both yield and protein percentage advance with further N increments to the maximum yield level. Following this plateau in yield, additional N eventually turns yield level downward, even though protein percentage continues upward for several increments of N. The end result, as expressed in Fig. 4, is that maximum protein yield requires a significantly higher N rate than is needed for maximum grain yield (Olson et al., 1976; Sander et al., 1985). Maximum grain yield of hard red spring wheat in the more northerly latitudes of wheat production will usually be accompanied by protein percentages in the range of 13 to 14%, whereas hard red winter wheats in the central Great Plains will normally contain 11 to 12% protein at peak yield level, and the soft white wheats of the Pacific Northwest 9%. Climatic conditions from one year to the next markedly condition these values.

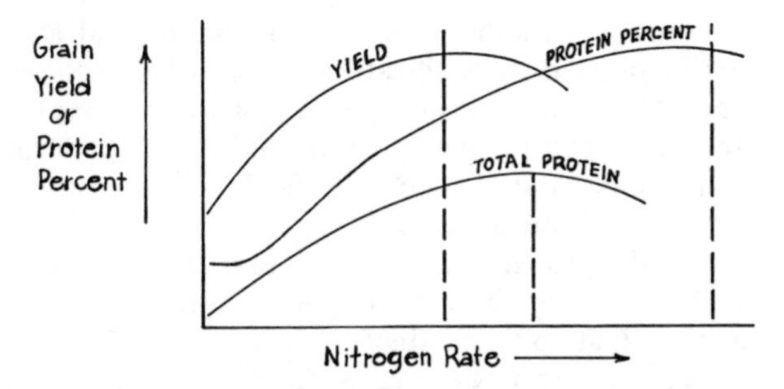

Fig. 4—Relative response curves of grain yield and protein content of grain crops to applied N as commonly expressed in field experiments (Sander et al., 1985).

Protein content of grain as a nutritional factor is not of great importance in the United States with the national dependence on meat and dairy products for this essential component of the diet. The level can be of significance, however, to the livestock feeder who can adjust the amount of high-cost protein supplement downward when feeding a grain of higher protein content. It is also an important consideration for the miller of wheat whose product is to be used in baking. In some years when environmental conditions contribute to a region- or country-wide crop of low protein, substantial premiums are paid for grain of higher protein that can be blended with the bulk of lower quality grain to provide an acceptable product for baking. On the other hand, high protein levels must be avoided for malting-type barleys or for the soft white wheats used for pastry flour. The marketplace, thus, determines the feasibility of a farmer applying more N than is needed for the optimum grain yield.

Protein level is obviously of much greater significance in those developing countries where food grains provide most of the human intake of protein. There, development of improved genotypes for higher protein and associated fertilization practice for protein enhancement takes on a more important role.

With normal growing conditions, the predominant factor controlling grain protein is the quantity of N available to the crop during its growth, as evidenced in Fig. 2 and 3. The time at which the N is readily available also has a significant bearing, with protein percentage increasing progressively as time of N application is delayed. Thus, delayed summer side-dressing of N for corn and grain sorghum, broadcasting N on rice (*Oryza sativa* L.) shortly before primordial initiation, and application of N on winter wheat in the spring have commonly afforded higher protein in the grain than is accomplished with N applications at planting. When applied foliarly as late as the bloom stage for wheat, however, the entire N effect may be expressed in protein and not at all on yield (Finney et al., 1957).

The quantity of residual mineral N accumulation in the rooting profile of soil in dryland farming appears to exert a controlling influence on protein that is not equalized with any fertilizer N rate likely to be used by farmers (Fig. 2). The explanation apparently lies in declining root activity for nutrient absorption from the soil surface downward with seasonal moisture depletion that prevents full utilization of heavier N treatments made to the surface soil. The data cited imply that some quantity of available N must be present deeper in the soil where roots are active during the later stages of grain filling and protein setting for full protein expression (Cochran et al., 1978; Olson et al., 1976).

Prediction of the protein content of a grain crop beforehand for any given year is difficult. There are so many interacting factors involved that can have a bearing right up to the final stages of grain maturation. Factors such as pest attacks, moisture deficiency, excessive heat, lodging, etc. that restrict yield are usually responsible for poor starch filling and elevated protein percentage in grain kernels. When all measurable parameters are put together, however, a reasonably accurate explanation of protein variability can be made. Thus, Smika and Greb (1973) showed that 96% of the variation in wheat protein could be accounted for when the four factors—available water and residual NO_3^--N in the top 1.5 m of soil at planting, maximum air temperature for the 15- to 20-d period before maturity, and precipitation during the 40- to 55-d period before maturity—were combined in multiple correlation.

REFERENCES

Allison, F. E. 1963. Losses of nitrogen from soil by chemical mechanisms involving nitrous acid and nitrates. Soil Sci. 96:404–409.

Allison, F. E. 1966. The fate of nitrogen applied to soil. Adv. Agron. 18:219–258.

Anon. 1972. Better Crops Plant Food 56:1–2.

Bauer, A., R. A. Young, and J. L. Ozbun. 1965. Effects of moisture and fertilizer on yields of spring wheat and barley. Agron. J. 57:354–356.

Bourg, J. L. 1961. Leaf composition and soil properties for evaluating the response of grain sorghum to fertilizer nitrogen and phosphorus. M.S. Thesis. Univ. of Nebraska.

Cochran, V. L., R. L. Warner, and R. I. Papendick. 1978. Effect of N depth and application rate on yield, protein content, and quality of winter wheat. Agron. J. 70:964–968.

Brown, P. L. 1971. Water use and soil water depletion by dryland winter wheat as affected by nitrogen fertilization. Agron. J. 63:43–46.

Doran, J. W. 1980. Soil microbial and biochemical changes associated with reduced tillage. Soil Sci. Soc. Am. J. 44:765–771.

Fenster, C. R., and G. A. Peterson. 1979. Effects of no-tillage fallow as compared to conventional tillage in a wheat-fallow system. Nebraska Agric. Exp. Stn. Research Bull. no. 289.

Finney, K. F., J. W. Meyer, F. W. Smith, and H. C. Fryer. 1957. Effect of foliar spraying of Pawnee wheat with urea solutions on yield, protein content, and protein quality. Agron. J. 49:341–347.

Gass, W. B., G. A. Peterson, R. D. Hauck, and R. A. Olson. 1971. Recovery of residual nitrogen by corn (*Zea mays* L.) from various soil depths as measured by ^{15}N tracer techniques. Soil Sci. Soc. Am. Proc. 35:290–294.

Greb, B. W. 1979. Reducing drouth effects on cropland in West Central Great Plains. USDA Information Bull. no. 420. U.S. Government Printing Office, Washington, D.C.

Haas, H. J., J. F. Power, and G. A. Reichman. 1976. Effect of crops and fertilizer on soil nitrogen, carbon, and water content, and on succeeding wheat yields and quality. Agric. Res. Service, North Central Region Circ. no. 38. USDA, Washington, D.C.

Hunter, A. S., L. A. Alban, C. J. Gerard, W. E. Hall, H. E. Cushman, and R. G. Peterson. 1961. Fertilizer needs of wheat in the Columbia Basin dryland wheat areas of Oregon. Oregon Agric. Exp. Stn. Tech. Bull. no. 57.

Kmoch, H. G., R. E. Ramig, and F. E. Koehler. 1957. Root development of winter wheat as influenced by soil moisture and nitrogen fertilization. Agron. J. 49:20–25.

Leggett, G. E. 1959. Relationships between wheat yield, available moisture, and available nitrogen in eastern Washington dryland areas. Washington Agric. Exp. Stn. Bull. no. 609.

Mahli, S. S., and M. Nyberg. 1978. The fate of fall-applied nitrogen in northern Alberta as determined by the ^{15}N technique. Int. Congr. Soil Sci. 11th 1:370.

Nelson, L. B. (ed.). 1968. Changing patterns in fertilizer use. Soil Sci. Soc. of Am., Madison, Wis.

Olson, R. A., and A. F. Dreier. 1956. Fertilizer placement for small grains in relation to crop stand and nutrient efficiency in Nebraska. Soil Sci. Soc. Am. Proc. 20:19–24.

Olson, R. A., C. A. Thompson, P. H. Grabouski, D. D. Stukenholtz, K. D. Frank, and A. F. Dreier. 1964. Water requirement of grain crops as modified by fertilizer use. Agron. J. 56: 427–432.

Olson, R. A., A. F. Dreier, C. Thompson, K. Frank, and P. H. Grabouski. 1964. Using fertilizer nitrogen effectively on grain crops. Nebraska Agric. Exp. Stn. Bull. no. 479.

Olson, R. A., K. D. Frank, E. J. Deibert, A. F. Dreier, D. H. Sander, and V. A. Johnson. 1976. Impact of residual mineral N in soil on grain protein yields of winter wheat and corn. Agron. J. 68:769–772.

Pesek, J. T., R. P. Nicholson, and C. Spies. 1955. What about fertilizers in dry years? Iowa Farm Sci. 9:3–6.

Phillips, R. E., R. L. Blevins, G. W. Thomas, W. W. Frye, and S. H. Phillips. 1980. No tillage agriculture. Science 208:1108–1113.

Power, J. F., and J. Alessi. 1978. Tiller development and yield of standard and semidwarf spring wheat varieties as affected by nitrogen fertilizer. J. Agric. Sci. 90:97–108.

Ramig, R. E., and H. F. Rhoades. 1963. Interrelationships of soil moisture level at planting time and nitrogen fertilization on winter wheat production. Agron. J. 55:123–127.

Ramig, R. E., P. E. Rasmussen, R. R. Allmaras, and C. M. Smith. 1975. Nitrogen-sulfur relations in soft white winter wheat. I. Yield response to fertilizer and residual sulfur. Agron. J. 67:219–224.

Read, D. W. L., and F. G. Warder. 1974. Influence of soil and climatic factors on fertilizer response of wheat grown on stubble land in southwestern Saskatchewan. Agron. J. 66: 245–248.

Sander, D. H., W. H. Allaway, and R. A. Olson. 1985. Modification of nutritional quality by environment and production practices. *In* R. A. Olson (ed.) Nutritional quality of cereal grains. Agronomy Monograph (In preparation).

Schlehuber, A. M., and B. B. Tucker. 1967. Culture of wheat. p. 117–179. *In* K. S. Quisenberry and L. P. Reitz (ed.) Wheat and wheat improvement. Agronomy 13.

Smika, D. E., A. L. Black, and B. W. Greb. 1969. Soil nitrate, soil water, and grain yields in a wheat-fallow rotation in the Great Plains as influenced by straw mulch. Agron. J. 61:785–787.

Smika, D. E., and B. W. Greb. 1973. Protein content of winter wheat grain as related to soil and climatic factors in the semiarid central Great Plains. Agron. J. 65:433–436.

Soper, R. J., and P. M. Huang. 1963. The effect of nitrate nitrogen in the soil profile on the response of barley to fertilizer nitrogen. Can. J. Soil Sci. 43:350–358.

Toews, W. H., and R. J. Soper. 1978. Effect of nitrogen source, method of placement, and soil type on seedling emergence and barley crop yields. Can. J. Soil Sci. 58:311–320.

Tucker, T. C., and R. D. Hauck. 1978. Removal of nitrogen by various irrigated crops. p. 135–167. *In* P. F. Pratt (ed.) Management of nitrogen in irrigated agriculture. Natl. Sci. Foundation, USEPA, Univ. of California, Riverside.

Viets, F. G., Jr. 1962. Fertilizers and the efficient use of water. Adv. Agron. 14:223–264.

Young, R. A., J. L. Ozbun, A. Bauer, and E. H. Vasey. 1967. Yield response of spring wheat and barley to nitrogen fertilizer in relation to soil and climatic factors. Soil Sci. Soc. Am. Proc. 31:407–410.

23

B. R. Wells
University of Arkansas
Fayetteville, Arkansas

F. T. Turner
Texas A&M University
Agriculture Research and Extension Center
Beaumont, Texas

Nitrogen Use in Flooded Rice Soils

Nitrogen is the mineral element most limiting rice (*Oryza sativa* L.) growth and yield on soils throughout the rice-growing regions of the world. Nitrogen use efficiency by the rice plant under flooded culture is also lower than that measured for arable crops on most soils, because of the additional N loss pathways present under flooded soil conditions. Research to improve the uptake efficiency of fertilizer N has moved toward an understanding of the loss pathways and of manipulating N fertilizer to maximize plant uptake and minimize losses. Results from extensive, applied research in such areas as N source, N rate, N time and placement, and water management vary widely depending on such factors as climate, soil type, cultural practices, etc.

Studies of N transformations and movement in flooded soils emphasize the poor efficiency of N fertilizer. The most comprehensive publication reviewing these transformations is *Nitrogen and Rice,* published by the International Rice Research Institute (IRRI, 1979). This chapter does not provide an exhaustive review of N use in the rice-plant, flooded-soil system, but it provides an assessment of current knowledge and research needs concerning practical aspects of N use by rice plants, especially as related to rice culture in the United States.

Nitrogen in Crop Production.

I. CHARACTERISTICS OF MAJOR N TRANSFORMATION PROCESSES IN THE WATER-SOIL-RICE SYSTEM

To accelerate progress toward improving N efficiency in lowland rice culture, current research is directed toward N transformations within the entire plant-soil-floodwater system. Figure 1 identifies the current concepts of the major N pathways and transformations in lowland rice culture. The floodwater restricts O_2 diffusion into the soil (process 1, Fig. 1), thereby inhibiting aerobic N transformation (primarily nitrification, process 8) and enhancing anaerobic N transformation (primarily denitrification, process 12). Aerobic N transformations are not completely inhibited, because O_2 diffuses into the soil-plant system sufficiently to cause some oxygenation of the floodwater, soil surface, and rice rhizosphere. The O_2 content of the floodwater influences the degree of oxidation and thickness (range from 0 to 2 cm) of the oxidized soil surface layer. The maintenance of the oxidized surface layer appears to depend on O_2 diffusion rate into the floodwater relative to the soil's O_2 consumption (Patrick & Reddy, 1976a). The degree of rhizosphere oxidation (process 17) is less clearly understood, but there is evidence that O_2 is transported through aerenchyma to rice roots that have the potential to release O_2. Such O_2 may promote nitrification in the rhizosphere of rice roots in anaerobic soils.

The occurrence of oxidized and reduced zones with abrupt and diffuse boundaries in flooded soils provides an ideal environment for simultaneous nitrification and denitrification (processes 8 and 12), resulting in substantial potential for loss of fertilizer and soil N through denitrification (Patrick & Reddy, 1976b). Generally poor efficiency of fertilizer N use by rice is attributed to denitrification, yet data confirming this are limited because of the lack of information on the fate of N not absorbed by plants. This lack of supportive data has encouraged researchers to reevaluate all aspects of N loss, particularly NH_3 volatilization (process 7) (Mikkelsen et al., 1978). As in nonflooded soils, NH_3 volatilization losses from flooded soils are dependent on water pH, proton conditions, P_{CO_2}, aquatic CO_3^{2-} systems, water temperature, wind velocity, etc. One condition unique to flooded soils, which has enhanced NH_3 volatilization, is the high pH experienced in floodwater when an incomplete canopy allows algae to utilize sunlight and photosynthesize sufficiently to deplete the floodwater of CO_2 (process 20). The lack of CO_2 can result in floodwater pH's of 9 or more, causing NH_4^+-N located at the soil surface to volatilize as NH_3 (Mikkelsen et al., 1978).

Algal blooms can occur in rice that has not developed sufficient canopy to shade the floodwater. Under controlled conditions, algal blooms have immobilized (process 13) up to 30% of the added N (Craswell & Vlek, 1979). This immobilization and the indication that mineralization of organic N in flooded soils (process 14) is slower and less complete than in aerobic soils (Patrick & Mikkelsen, 1971) are two practical ways in which N transformation in flooded soils appears to be different from nonflooded soils. However, some researchers indicate that organic N mineralization

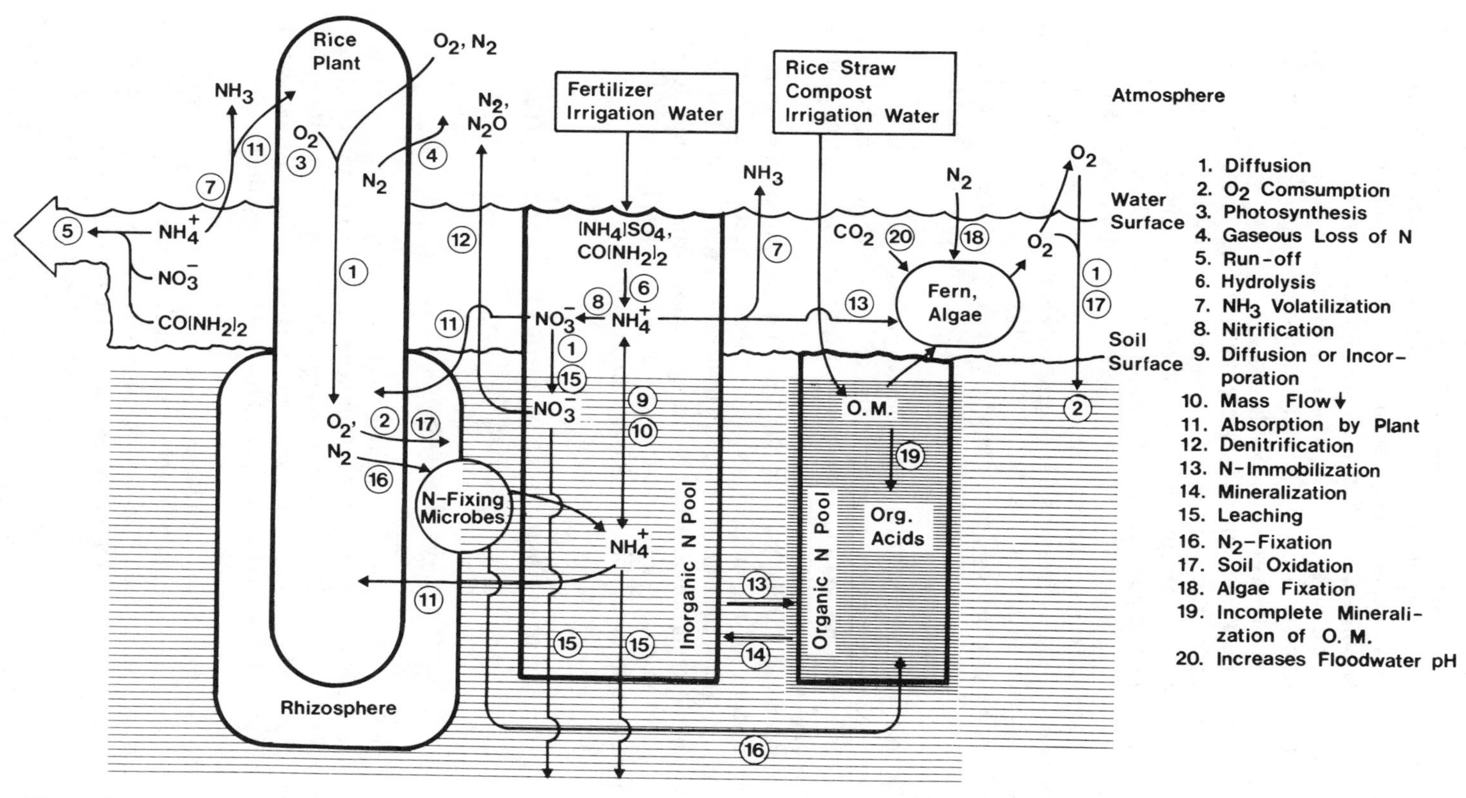

Fig. 1—Diagram of major N transformations and pathways in the flooded-soil–rice-plant system, showing the three main components (rice plant, inorganic N pool, organic N pool) in relation to the 20 identified processes. The shaded area represents the least oxidized zones.

during growth of lowland rice is higher than that for a dryland crop (Broadbent, 1979). The greater N supply from some flooded soils relative to nonflooded soils may be explained by a greater amount of atmospheric N_2 being fixed in flooded soils rather than increased mineralization. The chemical and biological immobilization of N, mineralization, and fixation in flooded soils continue to be subjects of research interest because of their practical influence on N availability (IRRI, 1979).

Long-term field experiments indicate that more N_2 fixation (processes 16 and 18) occurs in flooded rice culture than in aerobically grown crops. Blue-green algae, heterotrophic bacteria, *Azolla* (water fern), and/or rhizosphere N_2 fixation appear to add 15 to 50 kg N/ha to flooded rice crops under certain conditions (IRRI, 1979). The exact amount of N fixed by the various processes remains to be established.

Nitrogen leaching (process 15) is generally not significant in flooded rice soils because such soils usually have very low saturated hydraulic conductivities. Yet, there are a few areas where rice is planted on more permeable soils, and, as a result, $\geq$ 20 kg N/ha can be lost through leaching annually (IRRI, 1979).

Runoff water from flooded rice land (process 5) contains very low NO_3^--N levels because of the ideal conditions for denitrification; however, NH_4^+-N concentration in runoff can be significant within 5 d after topdressing of N (Turner et al., 1980).

Nitrogen can also be fixed as NH_4^+ by clay colloids or tied up in the organic fraction of the soils. These mechanisms are not actually losses; however, they have the short-term effect of depriving the rice plant of N.

The most recently established N loss mechanism from lowland rice culture is that of gaseous N loss from rice leaves (process 4). The magnitude of this loss has not been established (Silva & Stutte, 1979).

II. PRACTICAL ASPECTS OF APPLYING FERTILIZER N TO RICE

A. Source of N

Fertilizer N is normally required to maximize rice yield, especially where rice varieties, climate, cultural practices, and economic conditions make possible economical responses. Currently, research into natural N supply is expanding, but historically, most effort to increase N use efficiency in lowland rice has been directed toward increasing the efficiency of fertilizer N.

Of the commercially available N fertilizers, NH_4^+-N sources have proven more effective than NO_3^--N sources under lowland rice conditions because of the likelihood of denitification of NO_3^--N. Nitrate N can be almost as effective as NH_4^+-N for increasing rice yields when applied in small amounts ($\leq$ 30 kg/ha) at the panicle initiation stage after a root mat has formed at the soil surface and the plants can rapidly absorb N. Research has shown that fertilizer N applied at or shortly after panicle initiation is absorbed by the plant within 4 to 8 d (Wells, 1960).

A summary of the relative effectiveness of various N sources applied in a variety of ways to lowland rice grown in eight areas of the world showed ammonium sulfate (AS) to be slightly-to-considerably more effective than urea (Patrick & Mikkelsen, 1971). The better performance of AS over urea under certain conditions could be due to (i) the greater potential of urea to diffuse horizonally out of research plots when topdressed into floodwater [which would not be of practical importance under field conditions (Turner & Stansel, 1978)], (ii) an increase in floodwater pH following urea topdressing can enhance the potential for NH_3 volatilization (see Fig. 2), (iii) the possibility of the soil being S deficient, and (iv) possibly differences in rates of nitrification and hydrolysis.

Nitrogen solutions have been used successfully, but their use for topdressing has been limited because of foliar burn, often associated with topdress applications. Recently, U.S. manufacturers have reduced the free NH_3 content of the material to minimize burning. Currently, N solutions are gaining favor with Arkansas farmers due to the ability to apply a liquid uniformly using herbicide spray equipment. Aqua ammonia (AQA) and anhydrous ammonia (AA) are used extensively as preplant N sources for water-seeded rice in California. They are injected 5 to 10 cm into the soil just prior to flooding and seeding.

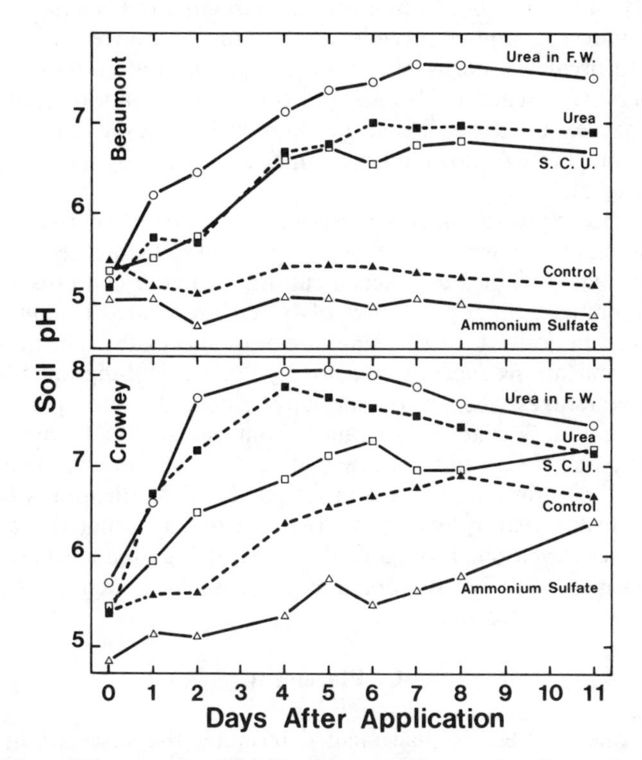

Fig. 2—Effect of N source on the pH of the top 0.5 cm of surface soil of a flooded Crowley vfsl and a Beaumont c soil (fine, montmorillonitic, thermic Entic Pelluderts). Urea in floodwater (F.W.) represents the floodwater pH after urea was applied. Nitrogen was topdressed on a flooded soil at 112 kg N/ha (unpublished data by F. T. Turner).

The increased energy required to topdress N encourages development of N sources or cultural practices to protect preplant N applications from loss. Controlled-release N sources such as sulfur-coated urea (SCU), super granules of urea, and urea-aldehyde products are being evaluated as sources of N (Wells & Shockley, 1975; Patnaik & Rao, 1979). Treating of NH_4^+-N with nitrification inhibitors has been effective in soils incubated in the laboratory and in small-scale research plots in dry-seeded rice on certain soils. Results under commercial field conditions when inhibitors are used at recommended rates has been more variable because of the short lifetime of the inhibitors. The volatile nature of the current nitrification inhibitors allows soil texture and floodwater management to influence the inhibitor's effectiveness. Reduction of nitrification for increased N efficiency appears promising, and research continues toward developing nitrification inhibitors for flooded soils. Nitrification inhibitors probably have limited effectiveness when NH_4^+-N is incorporated into the soil just prior to flooding in a water-seeded, permanently flooded rice culture.

B. Rate and Frequency of N Application

In the past, chemical extraction of soil samples for determining N availability to rice has not been useful in predicting the N supply available to rice, but recent studies provide some optimism concerning use of extractants to predict N needs of rice soils (Dolmat et al., 1980). Generally, rice farmers have relied on N rate and timing information from research and their past experience to arrive at economical N rates for their rice fields. The optimum N rate for lowland rice can be influenced by season and planting date (Fig. 3).

When rice is grown under a water-seeded, permanently flooded culture system, preplant N placed 5 to 10 cm deep in the soil consistently results in higher grain yields. For a dry-seeded culture, or under conditions of intermittant flooding, split topdressings of N fertilizer timed to coincide with peak demand periods of the rice plant are more successful. The number and time of N applications depends on cultivar, climate, and the soil's N supply. Plant growth stages when N is especially critical are during tillering and panicle initiation. Variation in climate from one season to the next may cause the "best" N rate and frequency of application to vary. Figure 3 illustrates how the optimum N rate and frequency of application varied with planting date and year at Beaumont, Tex. Use of accumulated heat units in predicting rice development stage and N topdressing time has been successful in Arkansas and Mississippi and is being evaluated in Texas and Louisiana.

C. Placement

The zone of NH_4^+-N placement determines the susceptibility of the fertilizer N to loss, and thus determines its efficiency of use by rice plants. Lack of O_2 in the reduced zone lessens the possibility of nitrification and

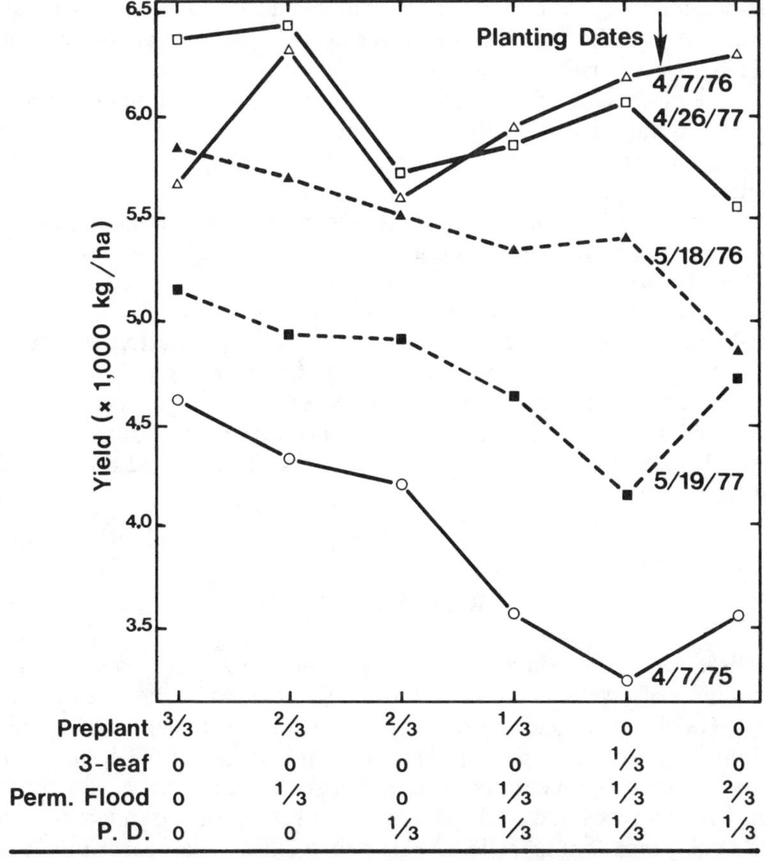

Fig. 3—Effect of planting date, year, time, and rate of AS application on yield of 'Labelle' variety of long-grain rice on a Beaumont clay soil at Beaumont, Tex. The total N rate was 134 kg N/ha; *PD* = growth stage when panicle is 2 mm long (unpublished data by F. T. Turner).

subsequent denitrification in flooded soils and reduces the potential for NH_3 volatilization for NH_4^+-N placed within the zone. The mechanics of placing the N well within the reduced zone, thus preventing NH_4^+ diffusion out of the reduced zone (Reddy et al., 1976) and preventing oxidation of the reduced zone, are more difficult to accomplish under field conditions for clay soils than for silt loam soils.

Anhydrous NH_3 knifed into the soil is used in several rice-growing areas of the United States. Practical problems limiting this approach appear to be ability to seal the NH_3 into the soil and the excessive time between N application and the period when dry-seeded rice plants will tolerate floodwater (in some areas, developing seedlings will not tolerate submergence). Use of the "Cold Flo" method of applying AA shows promise of reducing energy requirements for application. Aqua NH_3 can also be knifed to a 5- to

10-cm depth into the soil, with very little loss and with lower energy requirement than AA. This method is used extensively in California, with excellent results in their water-seeded culture.

Hand placement of supergranules of urea and "mudballs" (prilled urea encased into mudballs) 10 cm deep into flooded rice soil has proven effective in increasing N use efficiency in transplanted rice (Prasad & DeDatta, 1979).

Research work in Louisiana, Arkansas, and Texas has shown that the technique of applying granular urea to dry, cracked soil just prior to flooding allows the irrigation water to transport the urea into the soil to the zone that subsequently becomes reduced. Use of this technique in Texas has increased rice yield by 10 to 12% over plots that received N after, rather than before, flooding. Similar or greater yield differences have occurred in Louisiana and Arkansas tests. Recent research in Arkansas with rice growing on silt loam soils show NH_3 volatilization to be almost nil where the N is broadcasted onto dry soil prior to the initial flood and into the floodwater at midseason.

D. Residual Fertilizer N

There are limited data on which to draw conclusions concerning the contribution of residual fertilizer N to the rice plant's N needs. Reddy and Patrick (1978), in a field experiment on Crowley silt loam, (fine, montmorillonitic, thermic Typic Albaqualt) found 34 to 40 kg N/ha of the residual fertilizer N remaining in the soil after harvest, but < 10% of this residual N was recovered in the rice grain and straw of the following crop. Surprisingly, even though residual N supplied only a small part of the plant needs in the Crowley soil, the major source of N for the plant during the second half of the 120-d growing season was not N fertilizer (Patrick & Reddy, 1976a).

E. Runoff from Rice Fields

Placement of N also affects potential for N loss in runoff from rice fields. Applying AS into the soil prior to flooding eliminated the potential for N pollution of streams receiving runoff from rice fields. Nitrogen leaching loss is usually small because of lack of seepage in most lowland rice soils (Brown et al., 1978). Most of the N runoff and seepage studies have used AS as the N source; thus, data on the potential for loss of urea-N in runoff and seepage from lowland rice are lacking.

Nitrate-N loss in runoff from flooded rice land is usually very low because of the ideal conditions for denitrification. The potential for NH_4^+-N in runoff is significant for a short interval (3–5 d) after topdressing N into continuous flow water management systems or where levees erode (Turner et al., 1980).

F. Effects of N on Rice Plant Development

Development of the rice plant normally is divided into three phases: vegetative, reproductive, and ripening. Work by Tanaka et al. (1964), Matsushima (1976), and others has shown that N demand by the rice plant is greatest during the active tillering portion of the vegetative phase and during the first part of the reproductive phase.

Fertilizer N applied either preplant or prior to tillering of dry-seeded rice increases the number of tillers, length and width of leaves, number of florets, panicles per unit area, and grain yield. Leaf area index increases with increasing N rate and is accompanied by an increase in net photosynthesis (Tanaka et al., 1964) with adequate sunlight up to the point where further increases in N fertilizer cause excessive mutual shading and eventually reduce net photosynthesis and rice yield. Work by Matsushima (1976) in Japan indicates that a rice plant with short, erect leaves, stiff stems, and numerous but relatively small panicles is an ideal plant type because it exhibits minimum mutual shading, thus utilizing more of the existing sunlight. Using solution culture methods, he could alter plant type by the amount of nutrient solution N made available to rice plants at a given developmental stage.

Field research in Arkansas by Wells and Johnston (1970) has also shown that plant height, lodging, and grain yields can be altered by controlling the N supply to the plant by means of split applications. Split applications of N applied prior to tillering and at the early internode elongation stage resulted in maximum grain yields, shorter plants, and less lodging than when all N was applied before or during the tillering stage. The alteration in plant type and improvement in grain yield was greatest for tall, weak-strawed varieties. Therefore, timing N fertilizer application to match development stages can be an important, practical tool in rice production.

Research by Brandon and Mikkelsen (M. Brandon, personal communication) in California using the new semidwarf cultivar M9 in a water-seeded culture further illustrates the differences in N response associated with cultural practices. Maximum grain yields and panicles per square meter were associated with 135 kg N/ha applied preplant.

In Arkansas, N applied prior to tillering did not increase number of panicles unless plant population was low. However, N applied at the panicle initiation stage increased the number of filled grains per panicle and grain yield, regardless of plant population (Wells & Faw, 1978). Generally, N fertilizer applied at midseason increased rice yield by increasing filled grains per panicle and unit area rather than increasing individual seed weight. However, the time of application of N fertilizer can increase or actually decrease average seed weight under certain conditions as shown in Fig. 4. Where N was applied to a soil very deficient in N, average seed weight increased from about 18 to 20 mg. Applying part of the N at the 2-mm panicle stage actually decreased average seed weight, even though it increased grain yields. Apparently, the N applied at the panicle development stage caused more florets to be formed than could be maximumly filled under the condi-

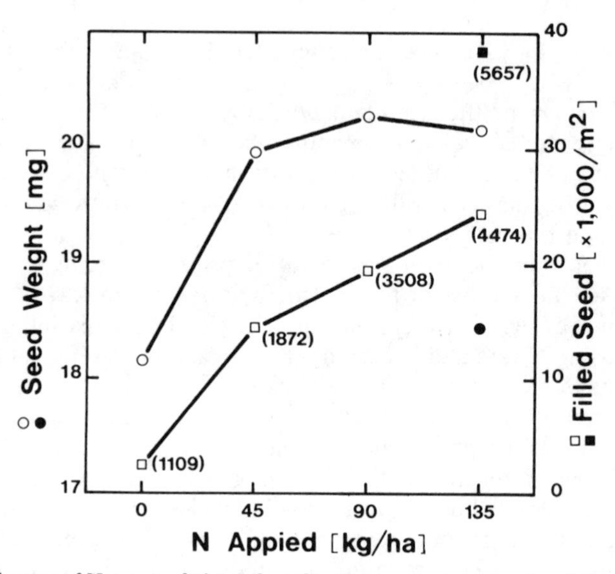

Fig. 4—Influence of N rate and time of application on average seed weight and number of filled grains of long-grain 'Labelle' rice on Beaumont clay at Beaumont, Tex. All the urea N was applied preplant except when the 134 kg N/ha rate was split 2/3 preplant and 1/3 at 2-mm panicle stage (solid data points). Numbers in parenthesis are grain yield in kg/ha (unpublished data by F. T. Turner).

tions for photosynthesis in this experiment. Figure 4 also shows how N management can influence number of filled grains and further illustrates how N fertilizer management can influence practical aspects of plant development.

There is also evidence that plants reduce N fertilizer loss from flooded soils (Reddy & Patrick, 1980). Their research showed reduced losses of applied urea and AS in the presence of plants.

G. Varietal Response to N

As mentioned above, most varieties with short, stiff straw and erect leaves require higher rates of N to produce maximum grain yields, as compared with the tall, easily lodged varieties. One recent example is reported from California, where a rapid shift from tall, easily lodged varieties to semidwarfs is taking place. Mikkelsen (personal communication) reports that semidwarf varieties required 20 to 45 kg/ha more fertilizer N for maximum grain yield than tall plant types. These new varieties have a yield potential of 11 000 to 13 000 kg/ha.

III. REGIONAL DIFFERENCES IN N FERTILIZER MANAGEMENT IN THE UNITED STATES

Typical N management practices for rice production in Arkansas, California, Louisiana, and Texas are discussed to illustrate how N fertilizer management for rice production varies with cultural practices, soils, and climate. Nitrogen fertilizer management for rice-planted areas in Mississippi will not be discussed individually because it includes the techniques practiced n Louisiana and Arkansas.

A. California

In California, rice is water-seeded primarily on alluvial clay soils and a continuous flood is maintained until time to drain for harvest. Nitrogen sources used are urea, AS, AQA, and AA. The dry materials are drilled or broadcasted and then incorporated to a depth of 10 cm prior to flooding for seeding. Aqua and anhydrous NH_3 are injected to a soil depth of 10 cm prior to flooding. To assure rice seedlings of an adequate N supply prior to roots reaching the AA, an application of 20 to 40 kg N/ha as urea or AS may be applied and incorporated to a shallow depth. Wet, spring weather will decrease the use of AA, because dry fertilizer can be applied more rapidly by airplane. However, when weather conditions are favorable, AA is widely used because of its lower application cost.

In California, highest grain yields are produced with preflood fertilization, and topdressing is used primarily as a supplement to adjust N rates to variable seasonal N needs.

Rice tissue analysis is used to predict the need to topdress N during the growing season. When tissue N concentrations in the most recently matured leaf at the beginning of the reproductive growth phase drop below 2.8 to 3.0% (depending on variety), N topdressing is recommended.

B. Arkansas

Most rice in Arkansas is either drill- or broadcast-seeded with the expectation that soil moisture or subsequent rainfall will germinate the seed. Flooding is initiated from 4 to 6 weeks following seeding, although some farmers may delay flooding until near panicle initiation if sufficient rain falls. The lengthy delay between seeding and flooding allows time for almost complete nitrification of preplant, soil-incorporated NH_4^+-N. Therefore, most Arkansas farmers use aerial split applications of N fertilizers to minimize lodging and increase N use efficiency. Urea is the predominate N source because of availability and favorable economics; some AS and 32% N solutions are used.

In recent years, some growers have begun to use the "Cold-Flo" method of applying preplant AA. The most successful method has been to apply the NH_3 in sufficient quantity to satisfy 75 to 80% of the N requirement of the rice variety, then topdress at internode elongation with one urea application.

The typical topdressing system used in Arkansas involves applying approximately 50% of the total N needs to dry soil just prior to permanent flooding when the rice is 10 to 15 cm tall. The remaining 50% of the N is applied shortly after panicle initiation. This N is applied into the floodwater in two applications at a 10 to 14 d interval when > 50 kg/ha are needed.

In Arkansas, N rate is keyed to soils, past cropping history, and variety. The currently grown varieties vary from 70 to 140 kg/ha in their N requirements when grown on previously cropped soils in a rice-soybean rotation.

C. Louisiana

Rice is usually presoaked and water-seeded in southwest Louisiana, where silty clay and silt loam soils predominate. Seed are drilled or broadcast into a prepared seedbed in the alluvial clay soils of the Mississippi Delta in northeast Louisiana. A permanently to semipermanently flooded system is employed in the Southwest, whereas in the Northeast, rice is either rainfed or flushed until the early tillering stage, then permanently flooded after plants are $\geq$ 10 cm in height.

In the Southwest, urea, AS, or N–P–K blends are applied preplant and incorporated 5 to 10 cm deep prior to flooding or topdressed on the soil surface during the time when fields are temporarily drained to encourage rooting and anchorage of the rice seedling. The preplant, soil-incorporated N is normally utilized more efficiently than is surface-applied N.

Urea is used in split, topdress applications on the northeast Louisiana clay soils. The preflood, early season N is applied to a dry soil surface within 7 to 10 d of flooding. Normally, two-thirds to three-fourths of the total N rate is applied prior to flooding and the remainder topdressed at the beginning of the reproductive growth stage.

In general, under Louisiana conditions, the long-grain varieties of rice will require 20 to 40% more N than the medium-grain varieties if maximum grain yields are to be obtained. This difference is often attributed to increased lodging resistance of the long-grain varieties.

D. Texas

Urea, AS, or blends of N–P–K are applied mostly by airplane to obtain 80 to 120 kg N/ha. The actual N rate used depends on rice variety, soil, and climate during the growing season, while the rate of any N topdressing is governed by the color and leafiness of the rice plants. The lower N rates and minimum number of applications are optimum for coarser-textured (fine, sandy loam and silty) soils or when rice is planted during warmer weather or after the optimum planting date. Early or cool-weather plantings yield best when higher N rates are applied in at least two and sometimes three applications. This time of planting and frequency of application interaction is more pronounced on clay soils, which apparently have a higher demand for N and/or cause fertilizer N to be less efficiently used by rice plants.

The three best times to apply N are: (i) preplant to three-leaf stage, (ii) just prior to permanent flood, which is 25 to 30 d after seedling emergence, and (iii) at the 2-mm length panicle stage, which is equivalent to the internode elongation. Most of the N is applied before the permanent flood, with no more than 35% applied at the 2-mm panicle stage.

Long-grain varieties generally perform best when fertilized with about 25 to 35 kg/ha more N than the medium grains, regardless of planting date, soil, or climate. Varieties that produce a good ratoon crop are fertilized with 35 to 55 kg N/ha just before reflooding following harvest of the first rice crop.

IV. FUTURE RESEARCH NEEDS

Even with the worldwide research effort on practically all phases of N in flooded soils, many unanswered questions exist about the gains and losses of N in rice soils. Basic studies of all the N_2 fixation processes in flooded soils, factors influencing N_2 fixation, and availability of fixed N to rice plants are needed. Some soils produce relatively high yields without N fertilizer, apparently by utilizing processes other than N_2 fixation. We need to identify and enhance the N supply of all rice soils in order to decrease our dependence on energy-consuming, synthetically produced N fertilizer. We also need to continue our research into ways of improving the efficiency with which rice plants use fertilizer N. Understanding the influence of the rice rhizosphere on N uptake and availability will help us manage soil and fertilizer N. Fate of fertilizer N not absorbed by plants needs more research emphasis; for example, how much of the volatilized NH_3 is absorbed by the rice plant canopy?

Applied research studies to improve N fertilizer use efficiency are needed in such areas as: (i) new fertilizer formulations, (ii) interrelationships of N_2 fixation processes and N fertilizer, with the goal of developing complementary rather than separate programs, (iii) fertilizer source and new techniques of placement and timing as related to soil, climate, and water management, (iv) the possible role of nitrification inhibitors, (v) rice breeding with emphasis on increased rice produced per unit of N applied rather than grain yield increase with response to high rates of fertilizer N, and (vi) development of diagnostic aids.

Decreasing water resources and increasing pumping costs in the United States indicate the need for devising techniques for optimizing N fertilizer efficiency under intermittently and/or nonflooded rice cultures. Also, especially in the applied research area, researchers should look toward a multidisciplinary approach, including all aspects of rice culture to integrate soil and fertilizer N management into a total program to maximize rice yields with a minimum input of expensive and energy-requiring N fertilizer.

ACKNOWLEDGMENT

The authors wish to recognize and express their gratitude to Dr. Duane S. Mikkelsen, University of California-Davis; Dr. D. Marlin Brandon, Louisiana State University, Rice Exp. Stn., Crowley; and Mr. Bobby Huey, University of Arkansas Cooperative Extension Service for their constructive suggestions to this chapter.

REFERENCES

Broadbent, F. E. 1979. Mineralization of organic nitrogen in paddy soils. p. 105–118. *In* Nitrogen and rice. IRRI, Los Banos, Philippines.

Brown, K. W., F. T. Turner, J. S. Thomas, and M. E. Keener. 1978. Water balance of flooded rice paddies in Texas. Agric. Water Manage. 1:277–291.

Craswell, E. T., and P. L. G. Vlek. 1979. Fate of fertilizer nitrogen applied to wetland rice. p. 175–192. *In* Nitrogen and rice. IRRI, Los Banos, Philippines.

Dolmat, M. T., W. H. Patrick, Jr., and F. J. Peterson. 1980. Relation of available soil nitrogen to rice yield. Soil Sci. 129:229–237.

International Rice Research Institute. 1979. Nitrogen and rice. IRRI, Los Banos, Philippines.

Matsushima, S. 1976. High yielding rice cultivation. Univ. of Tokyo Press, Tokyo.

Mikkelsen, D. S., S. K. DeDatta, and W. N. Obcemea. 1978. Ammonia volatilization losses from flooded rice soils. Soil Sci. Soc. Am. J. 42:725–730.

Patnaik, S., and M. V. Rao. 1979. Sources of nitrogen for rice production. p. 25–44. *In* Nitrogen and rice. IRRI, Los Banos, Philippines.

Patrick, W. H., Jr., and D.S. Mikkelsen. 1971. Plant nutrient behavior in flooded soil. p. 187–215. *In* R. A. Olson et al. (ed.) Fertilizer technology and use. 2nd Ed. Soil Sci. Soc. of Am., Madison, Wis.

Patrick, W. H., Jr., and K. R. Reddy. 1976a. Fate of fertilizer nitrogen in flooded rice soil. Soil Sci. Soc. Am. J. 40:678–681.

Patrick, W. H., Jr., and K. R. Reddy. 1976b. Nitrification-denitrification reactions in flooded soils and water bottoms: dependence on oxygen supply and ammonium diffusion. J. Environ. Qual. 5:469–472.

Prasad, R., and S. K. DeDatta. 1979. Increasing fertilizer nitrogen efficiency in wetland rice. p. 465–484. *In* Nitrogen and rice. IRRI, Los Banos, Philippines.

Reddy, K. R., and W. H. Patrick, Jr. 1978. Residual fertilizer nitrogen in a flooded rice soil. Soil Sci. Soc. Am. J. 42:316–319.

Reddy, K. R., and W. H. Patrick, Jr. 1980. Losses of applied ammonium ^{15}N, Urea ^{15}N and Organic ^{15}N in flooded soils. Soil Sci. 130:326–330.

Reddy, K. R., W. H. Patrick, Jr., and R. E. Phillips. 1976. Ammonium diffusion as a factor in nitrogen loss from flooded soils. Soil Sci. Soc. Am. J. 40:528–533.

Silva, P. R. F., and C. A. Stutte. 1979. Loss of gaseous N from rice leaves with transpiration. Arkansas Farm Res. 28(4):3.

Tanaka, A., S. A. Navaseso, C. V. Garcia, F. T. Pareo, and E. Ramirez. 1964. Growth habit of the rice plant in the tropics and its effect on nitrogen response. Tech. Bull. no. 3. IRRI, Los Banos, Philippines.

Turner, F. T., K. W. Brown, and L. E. Deuel. 1980. Nutrients and associated ion concentrations in irrigation return flow from flooded rice fields. J. Environ. Qual. 9:256–260.

Turner, F. T., and J. W. Stansel. 1978. Horizontal diffusion as a mechanism of N loss from research plots. p. 66–67. *In* Proc. 17th Rice Technical Working Group. Tex. A&M Univ., College Station, Tex.

Wells, B. R. 1960. Nitrogen absorption by rice as determined by use of the stable isotope ^{15}N. M.S. Thesis. Univ. of Arkansas, Fayetteville.

Wells, B. R., and W. F. Faw. 1978. Short-saturated rice response to seeding and nitrogen rates. Agron. J. 70:477–480.

Wells, B. R., and T. H. Johnston. 1970. Differential response of rice varieties to timing of mid-season nitrogen applications. Agron. J. 62:608–612.

Wells, B. R., and P. A. Shockley. 1975. Conventional and controlled release nitrogen sources for rice. Soil Sci. Soc. Am. Proc. 39:549–551.

24

J. H. Sherrard, R. J. Lambert, M. J. Messmer,
F. E. Below, and R. H. Hageman
University of Illinois
Urbana, Illinois

Plant Breeding for Efficient Plant Use of Nitrogen

Increased crop yields have been a function of improved production practices and improved genotypes of cultivars or hybrids adapted to the new production practices. Selection of improved and adapted genotypes has been the major contributor to the overall gain in productivity of crop plants. In the past 10 yr, several studies have been completed with different crops to estimate the gain in productivity resulting from use of improved cultivars. Comparisons of this type have some bias for measuring the contribution of new production practices; however, the observed responses are useful as an estimate of genetic improvement. Experiments to measure genetic improvement in cultivars as a function of their time of release or development have been conducted with corn (*Zea mays* L.), oats (*Avena sativa* L.), peanuts (*Arachis hypogeae* L.), rice (*Oryza sativa* L.), soybeans [*Glycine max* (L.) Merr.], and wheat (*Triticum aestivum* L.).

Increases in crop yields as a function of improved genotypes during the past 40 to 60 yr vary with the individual crop. Duncan et al. (1978) observed a 100% increase in peanut yields in Florida over a 40-yr period (1930–1970). For rice, 60 to 100% of the increase in grain yields in the tropics was the result of producing semidwarf, lodging-resistant cultivars. Estimates of grain yield improvement in corn varieties as a function of time of development and release of the hybrids, vary from 57 to 63%. The rate of grain yield increase during this 40 yr period was 78 kg/ha per year (Russell, 1974). Thirty-five to sixty percent of the grain yield increase for wheat in the central United States during the last 60 yr was due to varietal improvement (Frey, 1970). This improvement was due chiefly to development of semidwarf, lodging-resistant cultivars capable of responding favorably to higher N environments. Soybean seed yields in the central United States increased

Nitrogen in Crop Production.

by 25% during the last 50 yr (Wilcox et al., 1979). In addition, the most recently released varieties were more lodging resistant. Sixty-six oat varieties developed during a 40-yr period (1932–1972) showed an average increase of 9% in grain yield due to genetic improvement when tested in the Iowa variety trials (Langer et al., 1978); however, this increase occurred predominantly in the first 10 yr of that period. The reason for the lack of response to grain yield improvement in oats is probably due to the necessity of producing cultivars with single gene resistance to the major races of stem and leaf rust during this period. Oat breeders were forced to select for qualitative disease resistance and were unable to select for grain yield as such. From the published data, the conclusion is that significant amounts of genetic variability exist in major crop plants, thus permitting continued gains in grain yield.

Selection of cultivars or hybrids with higher grain yield may have been dictated in part by the plant's responsiveness to fertilizer N and higher plant densities. This is especially valid for corn. In a comparison of 19 commercial hybrids that were developed and released for farm production at intervals during the period 1939 through 1971, Duvick (1977) found that when grown at a "moderately high" fertility level, the more recently released hybrids were more productive. Welch (1979) found that for recently developed corn hybrids, grain yields were decreased (46%) when the annual application of fertilizer N (180 kg/ha) to highly fertile central Illinois soil was omitted. These data indicate that fertilizer N has a direct effect on grain production. The newly developed semidwarf rice and wheat varieties that utilize greater amounts of N fertilizer without lodging also produce greater yields of grain.

These varieties could be considered to be more efficient in use of N where the concept of efficiency is defined as the plant's ability to convert soil-applied N into increased grain yield. This concept takes into account the plant's capacity for uptake, assimilation, translocation, and deposition of N within the grain. Although the more recently developed corn genotypes may be more efficient at converting soil-applied N into increased yield at a particular N level (Duvick, 1977), yield has not increased in proportion to the rate of increase in N applied (Fig. 1). The average production of corn per tonne of N applied was 376 Mg (3760 quintals) during the 1950 to 1955 period, but only 39 Mg (390 quintals) for the 1974 to 1979 period. These data (Fig. 1) indicate that N utilization is related to grain production and that the potential exists for increasing yield by identifying genotypes with efficient abilities for N uptake and metabolism. Enhanced uptake and efficient use of N could result in increased corn productivity through maximization of sink size, enhancement of grain development, or by developing and maintaining an effective photosynthetic apparatus that should be conducive to vegetative development and sustained grain filling.

The potential for breeding for efficient use of N (conversion of applied N to yield) in crop plants is dependent on the genetic variability present in the species for the trait(s) that determine efficient N utilization, and the development of procedures to accurately measure parameters that reflect N

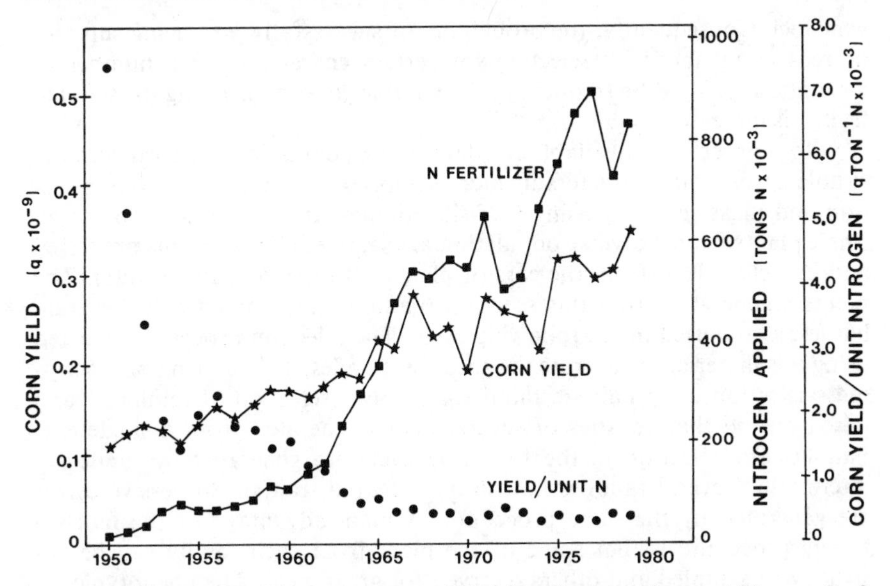

Fig. 1—Trends in N fertilizer usage, statewide yields of corn, and corn yield per unit of N applied over 30 yr in Illinois. Approximately 90% of the fertilizer was applied to corn (update of data originally presented by Hageman, 1979).

use in the plant. Of these components, the identification of heritable physiological or biochemical traits, causally related to grain yield, is the most formidable barrier. Currently, additional knowledge is needed on the metabolic role of N in grain production to permit identification of such traits.

The chapter format is as follows: Section I—a brief outline of standard corn-breeding procedures that can be adapted to the use of physiological and biochemical traits; section II—a review of the interrelationship between N, nitrate reductase, and yield of grain and grain protein; section III—a discussion of preliminary results obtained when biochemical and physiological traits were used in conjunction with phenotypic recurrent divergent selection and mass selection with half-sib families; and section IV—a description of a hypothetical ideotype corn plant with the potential for efficient conversion of fertilizer N to yield.

I. STANDARD CORN-BREEDING PROCEDURES

Corn planted in the United States is largely F_1 single-cross hybrids. The production of such hybrid seed requires the use of two inbred lines. The source or parental material for development of these lines can be germ plasm pools or synthetic varieties that are heterogenous and contain genes favorable for development of high-quality inbred lines. For corn, almost all the traits important in developing superior hybrids are controlled by many gene loci. Consequently, the probability of successfully producing superior inbreds is enhanced by selecting for certain characters for a number of generations so that the frequency of favorable genes controlling the selected trait(s) is increased.

Two procedures available for developing populations of corn with desirable agronomic characters include phenotypic recurrent divergent selection and mass selection using half-sib families. In the former procedure, single plants from a diverse population are sampled. These plants are selfed and kernels bulked from the ears of plants with the desirable trait(s). The plants regenerated from this seed are randomly intermated and the grain harvested and used in the following cycle. The selection process is repeated using this material to generate the successive cycles. The second approach—mass selection using half-sib families—involves selection of families (ears) based on the charcteristics of several plants generated from a single ear. Families for selection in the following cycle are generated by randomly intermating several families having the required trait(s). Successive cycles are generated by the same procedure. A main advantage of the half-sib family procedure is that some of the plant from each family can be destructively sampled and others reserved for grain yield. The phenotypic recurrent divergent selection procedure allows only limited sampling, because each plant must produce seed for regeneration. In addition, the half-sib family procedure can be used to obtain estimates of the genetic variance and genetic correlations among traits. These estimates allow prediction of the response to selection, and the information generated can be used to determine which selection criteria should enhance progress towards increasing grain yield. Selection theory dictates that for a trait to be an aid in selecting for grain yield, the trait must (i) have a higher heritability than grain yield, and (ii) have a high genetic correlation (0.7–0.9) with yield.

A more detailed description of the procedures described is presented by Sprague and Eberhart (1977). The results summarized by these authors show that certain selection procedures have been very effective in improving agronomic traits in corn. These principles should also be effective when applied to physiological traits. In fact, by combining selection for the proper physiological traits with grain yield, a greater rate of selection progress should be made than by selecting for yield alone. It is envisioned that by selecting for some type of improved N utilization, higher grain yields should be obtained in certain environments than by selecting for grain yield alone.

II. INTERACTIONS BETWEEN N, NITRATE REDUCTASE, AND YIELD

In view of the limited knowledge of the role of N in achieving maximum grain yields, it is a major problem to identify a single physiological or biochemical trait, concerned with N metabolism, that can serve as a selection criterion. Although in cereal crops, potential grain yield may be affected by the availability of N at various stages of plant growth, the effects of N are confounded by interactions with other nutrients and metabolic systems, especially C metabolism. Further complications are imposed by metabolic variation among genotypes and environmental effects.

It is known that N is in some way involved in the early establishment of a ground-covering canopy for development and maintenance of an effective photosynthetic system, plant mass, and development of sink capacity. However, the amount of N that should be available to the various plant parts at various stages of development to ensure that N is not the factor limiting grain yield is not clear. This section will review a limited number of the studies concerned with the role of N in grain yield and the initial attempts to identify selection traits.

A. Source–Sink Relationships

Some insight into the relative importance of N availability during vegetative and reproductive growth phases is provided by evaluation of the importance of sink or source in limiting grain yield. Studies conducted by Earley et al. (1967) in Illinois and Prine (1973) in Florida showed that the level of irradiation during flowering had more influence on grain yield than during the grain-filling period. Light energy in the form of reduced ferredoxin is required in the assimilation of NO_2^-, as well as for the generation of NADPH for the fixation of CO_2 by green leaves (Beevers & Hageman, 1980). In another study, Allison and Watson (1966) found the stalk to contain excess capacity for supplying dry matter to the grain during the grain-filling period. These and other reports led Tollenaar (1977) to conclude that sink capacity rather than source capacity commonly limits grain yield of corn grown in regions unaffected by a short growing season (Hume & Campbell, 1972). Source limitations appear to exist in short growing season environments. Tollenaar (1977) suggested that where sink limitation exists, improved grain yield could be achieved by selecting for factors that influence photosynthate supply to the ear during the flowering period. Alternatively, N—rather than carbohydrate supply—could be the factor limiting the development of sink capacity. This latter view is supported by the following considerations. Kernel initiates, pollen, and developing embryo all have relatively high N/C ratios.

Environmental Effects on N Metabolism

Grain production is sharply reduced by adverse environmental conditions that occur during the onset and early phases of reproduction. Several studies indicate that N metabolism is more sensitive to such adverse conditions than carbohydrate metabolism. Nitrate reductase (NR) is more sensitive to decreases in leaf water potential than either photosynthesis or respiration, and NR is more sensitive to elevated temperatures than either phosphoenolpyruvic or ribulose 1,5-diphosphate carboxylase (Hageman, 1979). Nitrogen metabolism in corn was more adversely affected by decreased irradiation than carbohydrate metabolism (Knipmeyer et al., 1962). A study of the rate of change in leaf, stalk, and ear carbohydrates and reduced N contents of the corn hybrid 'Mo17' × 'B73' showed little or no remobilization of carbohydrate from the vegetation throughout ear development (Below et al., 1981). In contrast, there was a rapid depletion of reduced N from the leaves and stalks, which was initiated concurrent with the linear phase of ear growth. The early and extensive loss of reduced N from the leaves is closely associated with, if not the cause of, the diminished rate of photosynthesis commonly noted between anthesis or shortly after until maturity.

An adequate N supply is considered to be essential in the establishment of maximum sink capacity and in maintaining photosynthetic capacity in rice (Murata & Matsushima, 1975). However, Murata and Matsushima suggested that excessive use of N is undesirable since it can result in excessive leaf area expansion, which is inversely correlated with photosynthetic capacity. Thomas and Thorne (1975) found that large applications of fertilizer N to spring wheat at planting resulted in larger plants with increased leaf area, but did not increase yield. In contrast, Hucklesby et al. (1971) increased grain yield and grain protein in three cultivars of winter wheat with increased fertilizer N application on three dates during late spring. Application of supplemental fertilizer N to corn during the reproductive stage resulted in an increase in grain protein production (Deckard et al., 1973), indicating that the rate of supply of reduced N is limiting grain N accumulation. The data of Hucklesby et al. (1971) indicate it is the unavailability of NO_3^- to and in the plant during the grain filling period that could be responsible for the constancy of the grain protein percentages commonly observed.

C. N Metabolism and Grain Yield

In a field study of 10 corn genotypes (diverse in grain yield and protein), several components of N metabolism were related to yield, grain protein, and percent grain protein (Below, 1981). With measurements made at 7 d following anthesis, total leaf N content and dry weights of stalks, leaf sheaths, and upper leaves were positively correlated with grain yield, while percent N in leaf sheaths and stalks was negatively correlated. Leaf-reduced N content and stalk NO_3^- content were positively correlated with amount of grain crude protein, while percent N in all plant parts was positively correlated with percent grain protein.

The initial experiments by Zieserl et al. (1963) indicated that level of leaf nitrate reductase activity (NRA) might serve as a simple selection criterion relating N metabolism and yield of grain and grain protein. This view was supported by the work of Brunetti and Hageman (1976) who found a highly significant correlation between NRA and actual accumulation of N by wheat seedlings. However, in subsequent experiments, correlations between NRA and either grain yield or grain N among diverse genotypes have not been consistent. Dalling (personal communication) found significant, but low, correlations between in vivo NRA of samples from growth chamber seedlings and grain yield and grain N production for 14 Australian wheat cultivars grown under field conditions. Similarly, using six corn hybrids, Deckard et al. (1973) found significant, but low, correlations between seasonal canopy NRA and grain yield and plant and grain reduced N content. The low correlations indicate that measured NRA does not accurately reflect the in situ assimilation of NO_3^- among diverse genotypes and/or that factors other than the level of NRA are involved in grain yield and grain protein. Subsequent work provides some insight as to why NRA may not be a good measure of in situ reduction. Shaner and Boyer (1976) have shown that it is the flux of NO_3^- rather than the NO_3^- content of the leaf that regulates the level of NRA in the corn leaf. Later, Reed and Hageman (1980a, b) noted that a given flux rate of NO_3^- will not result in the same amount of NRA among corn genotypes. From their work with two wheat varieties, Rao et al. (1977) indicated that absorption and translocation of NO_3^-, as well as the level of NRA, affected in situ accumulation of reduced N. They concluded that no single identifiable trait can be used as a criterion in selecting wheat genotypes for efficient N utilization. Among wheat genotypes, Dalling et al. (1975) found that variations in efficiency of remobilization of vegetative N to the grain precluded a close relationship between NRA and accumulation of N in the grain. When they classed the genotypes on basis of N remobilization, correlations between NRA and grain protein were obtained. In corn, ear removal depressed both photosynthesis and NO_3^- reduction, but had little effect on N remobilization, indicating that remobilization was controlled independently of sink capacity (Christensen et al., 1981). The concurrent onset of proteolysis, loss of leaf-reduced N, amino N, and chlorophyll for both the eared and earless plants in this study indicate that factors other than N supply could also affect the development of plant senescence.

With bean plants (*Phaseolus vulgaris* L.), Neyra et al. (1977) observed seasonal patterns of nitrogenase activity to be maximal at flowering and decline rapidly thereafter. In contrast, the observed NRA of the leaf canopy was maximal after flowering. Soil application of fertilizer N at this time caused a marked increase in leaf NRA and a near doubling of bean yields. Thus, it appears that in some instances, there is a causal relationship between NRA and accumulation of N in the grain.

The increase of leaf NRA in vivo in response to added NO_3^- that occurs at most stages throughout the growth cycle of the corn plant, but particularly during the reproductive phase, also implies the leaf has the potential to assimilate more NO_3^- in situ, with a resultant increase in grain protein, if NO_3^- was available to the leaf blades. An increased NO_3^- supply is difficult to achieve, however, because there is a progressive decrease in ability of the plant to accumulate NO_3^- from the soil with plant maturation. The decrease could result from root senescence, a diminishing rate of NO_3^- absorption by the roots, through recycling of NO_3^- from the plant to the soil or a decrease in soil NO_3^- supply or availability. This problem might be overcome by the application of suitable forms of N directly to the plant foliage.

Reed et al. (1980) found that two corn genotypes low in NR terminated NO_3^- absorption earlier than two high NR genotypes. In addition, the low NR genotypes had higher proteolytic activities that increased earlier, and had a higher percentage of grain N and harvest indices than the high NR genotypes. They concluded that leaf proteolytic activity was more closely related to the accumulation of grain N than leaf NRA. In a study of five corn genotypes, Below et al. (1981) found the remobilization of both vegetative dry weight and N contributed to ear development, but the N contributed to a greater degree. Although the rate of N accumulation in the developing ear was not related to the rate of dry matter accumulation, these data do not preclude the possibility that dry matter accumulation is dependent upon simultaneous movement of N to the developing grain. Excessive remobilization could result in an increased percentage of grain N—either directly or indirectly—by decreasing the rate of assimilate supply through simultaneous degradation of the leaf photosynthetic apparatus. Tsai et al. (1978) suggested that protein accumulation might be an important factor in grain development in corn, based on the correlation found between the accumulation of zein and grain yield.

In summary, the successful use of physiological and biochemical criteria in selecting superior cultivars is dependent upon a better understanding of the N metabolism of the plant in relation to yield. In particular, there is a need to know how much reduced N should be available to permit development and maintenance of an effective photosynthetic apparatus, maximum sink initiation, effective pollination, and a high rate and duration of grain filling among genotypes grown under different environmental conditions.

III. SELECTION USING PHYSIOLOGICAL TRAITS RELATED TO N METABOLISM

The following two projects concern specific attempts to utilize physiological and biochemical traits associated with N metabolism as the selection criterion in standard breeding procedures. The breeding procedures have been described.

A. Selection for Nitrate Reductase Activity (NRA)

A divergent phenotypic recurrent selection program was initiated in 1974 at the University of Illinois (Dunand, 1980) in a corn synthetic. Stiff-stalk synthetic was chosen because several superior inbreds had been developed from this population. The sole selection criterion was leaf blade in vivo (plus NO_3^-) NRA measured in upper canopy leaves at 10 and 20 d after anthesis using the method described by Hageman and Hucklesby (1971). Leaf blade NRA was chosen as the selection criterion for the following reasons. An increased level of NRA was taken as an approximation of an increased amount of reduced N made available to the plant (based on data available in 1974) during the critical stage of grain development. The increased amount of newly reduced N should increase the amounts available for partitioning for development of sink capacity, increased grain protein, and maintenance of the photosynthetic apparatus. The divergent low selection would be the converse of these changes. At the two sampling dates, NRA was integrated and modified to reflect the level and rate of change in activity between the two dates. Selection in the high strain was for plants with a high level of NRA at both sampling dates and which maintain activity between dates. Selection in the low strain was for low NRA and a decrease in activity between dates. In each cycle of selection, 200 self-pollinated plants from each strain were assayed for leaf blade NRA. The modified integrated NRA value (MIA) obtained was used to select the highest or lowest 40 plants in each strain.

During 1978 and 1979, cycle 0 through cycle 4 were grown in replicated trials at Urbana, Ill. to determine the response to selection for NRA and the correlated response for certain other traits. The data for four cycles of selection for high and low leaf blade NRA during grain filling are presented in Table 1. Selection for leaf blade NRA in the high strain increased the MIA by 16% from cycle 0 to cycle 4, with most of the progress occurring between cycle 2 and cycle 4. For the low NRA strain, a 45% decrease was observed between cycle 0 and cycle 4. In addition, the low NRA strain had a continuous response to selection from cycle 0 to cycle 4. A precise assessment of the reasons for the difference in response of the two NRA strains to selection is not apparent; however, selection theory dictates that the rate of response to selection is a function of the genetic variance. Since the maximum genetic variance at a locus occurs between gene frequencies of 0.4 and 0.6, a synthetic variety that had a large number of loci with alleles for high NR levels at frequencies > 0.6 would have less genetic variance and respond less

Table 1—Response to selection for four cycles of divergent phenotypic selection for nitrate reductase activity in maize averaged over 2 yr (1978–1979) and three environments at Urbana, Ill.

Cycle	MIA†		Stover dry weight/ plant at maturity		Grain yield	
	High	Low	High	Low	High	Low
			g		kg/ha	
0	31	31	188	188	7330	7330
1	31	25	190	193	7370	8120
2	32	20	211	198	7370	7960
3	34	19	208	189	7260	7590
4	37	17	202	203	7270	8150
LSD 0.05	4	2	12	NS‡	NS	540

† Modified integrated area was computed as follows: Integrated area = (slope + 1) × $[10(NRA_{10} + NRA_{20})/2] \times [(NRA_{20} - NRA_{10})/10 + 1]$, where NRA_{10} and NRA_{20} represents the nitrate reductase activity (μmol NO_2^- g fresh wt^{-1} h^{-1}; in vivo + NO_3^-) at 10 and 20 d postanthesis, respectively; 10 represents the 10-d interval between sampling, and 1 is a constant.

‡ NS = not significant.

to selection. Selection for low NRA in a strain described above would increase the frequency of alleles for low NRA into the 0.4 to 0.6 range, which would increase the genetic variance and response to selection for low NRA. Since the stiff-stalk synthetic strain used in this study had undergone several generations of selection for grain yield, there could have been a correlated response for high NRA, which may have resulted in gene frequencies for high NRA alleles to be above the 0.4 to 0.6 range.

B. Correlated Response to Selection

The correlated response for stover dry weight and grain yield are presented in Table 1. Stover dry weight at maturity showed a significant increase in the high NRA strain and no significant change in the low NRA strain. The largest increase in the high strain occurred between cycle 2 and cycle 4. The low strain showed little change in stover dry weight until cycle 4, when a small increase was noted. The change in stover dry weight of the high NRA strain was not reflected in increased grain yields of the high strain (Table 1). The high NRA strain showed no change in grain yield. In contrast, the low NRA strain had a significant correlated response for grain yield. There was an 11% increase in grain yield in the low NRA strain from cycle 0 to cycle 1 and a small increase observed for cycle 4. The correlated response of the low NRA strain for grain yield was unexpected and difficult to explain; however, under source-limiting conditions, the low NRA strain may have more carbohydrate available for grain filling, because less would be required for N assimilation than in the high NRA strain. Alternatively, a low level of NRA might reflect an increase in remobilization of leaf N and translocation to the grain, which, in turn, might be necessary for deposition of carbohydrate in the grain. Under conditions where the source is nonlimiting, this would increase grain yield. The correct explanation must await a better understanding of the relationship between N metabolism and grain yield.

The percent grain protein content of the NRA strains is presented in Table 2. In four cycles of selection, there was a significant correlated increase of 6% in grain protein content of the high NRA strain when grown at 224 kg/ha of supplemental N in both 1978 and 1979. The low NRA strain showed little change in percent grain protein during four cycles of selection. In contrast, when the two NRA strains were grown in a low-N environment (112 kg/ha) in 1979, the response was unexpected. At this low level of N, no significant grain protein response was observed for either NRA strain for five cycles of selection (Table 2). However, this field had been in continuous corn production for 10 yr previous to 1979, and only 112 kg/ha of N was applied each year. The two NRA strains may not have been able to express grain protein differences because of the low supply of soil N. The low supply of soil N, especially after anthesis, could have been the reason for the lack of a correlated response for grain protein content of the high NRA strain. The data illustrate that selection for high leaf NRA during the grain-filling period in corn resulted in a correlated response for grain protein as

Table 2—Percent grain protein for the high and low nitrate reductase activity strains selected for four cycles in 1978 and five cycles in 1979 at two different N levels.

	224 kg N/ha				112 kg N/ha	
	1978–1979†		1979		1979	
Cycle	High	Low	High	Low	High	Low
	%					
0	9.5	9.5	9.2	9.2	9.2	9.2
1	10.1	9.5	10.1	9.3	9.0	9.0
2	9.9	9.8	9.5	9.4	9.1	8.7
3	9.7	9.7	9.3	9.5	9.1	9.3
4	10.0	9.7	9.6	9.3	9.4	8.9
5			9.7	9.4	9.3	9.2
LSD 0.05	0.4	NS‡	0.6	NS	NS	NS

† Mean value for percent grain protein. ‡ NS = not significant.

predicted. In addition, the increase in grain protein content as a correlated response to selection for NRA did not show the classical decrease in grain yield usually observed in corn when percent grain protein, as such, is used as the sole selection criterion. The low NRA strain did not change in grain protein content with selection, but did change in grain yield. In the low NRA strain, the low level of NRA in the corn leaf blade during the grain-filling period did not reduce grain protein relative to starch. In fact, on a unit-per-area basis, the low NRA strain produced more protein per hectare than the high NRA strain because of the increased grain yield of the low NRA strain. The results illustrate the complexity of the relationship between N metabolism, photosynthesis, and grain N and yield.

C. Mass Selection for N Utilization

The divergent selection program for high and low NRA was started with little knowledge of the parameters involved in N utilization in corn. Additional results accumulated on N utilization indicate that other N parameters may be more efficient in estimating N utilization than NRA. The half-sib mass selection breeding procedure has advantages for determining which N parameters should be used as selection criteria. The method allows for the analysis of a number of characters. A breeding program using the half-sib family procedure was initiated in May 1980 at Urbana, Ill. This method and slight modifications have been effective in improving grain yield of corn populations with increases of 2.4 to 5.3% observed per cycle (Bshadur, 1974; Dudley et al., 1974). The objective of the program is to select genotypes with improved efficiency in N utilization and higher grain yield for several cycles. Selection increases the probability that improved inbreds with the desired expression of the traits can be developed from the population.

In the program initiated at Urbana, Ill., 100 families (ears) were planted from RSSSC (Illinois version of stiff-stalk synthetic), which had been selected for three cycles for high grain and protein yield by the modified-ear-to-row method of breeding. Nitrogen was supplied at the rate of 224 kg/ha. The soil type was a Drummer silty clay loam (fine-silty, mixed, mesic Typic Haplaquolls).

During the first year of the program, whole plants were harvested at 40 d after planting, at anthesis, and at physiological maturity (black layer formation). The harvested plants were subdivided into leaf and stalk, and dried, weighed, and assayed for NO_3^- and reduced N. The leaf below the ear was also removed from other plants at 15 and 40 d following anthesis and, after subdivision into lamina and midrib, similarly analyzed in addition to being assayed for NR. Leaf area per plant and plant height were determined shortly after anthesis. Grain yield and harvest index (stover) were determined at physiological maturity. The grain protein percent was determined by infrared reflectance according to the procedure of Hymowitz et al. (1974).

Data from the half-sib families are presented in Tables 3 and 4 for traits that showed significant differences between family means. Significant amounts of additive genetic variance as measured by the heritability (h^2) value were found for a number of the traits measured (Table 3). The h^2 values for the traits listed are all high enough to indicate that progress from selection could be expected. In addition, the genetic coefficient of variability indicates that most traits have about the same relative amount of genetic variation (4–10%). In general, maturity traits (harvest index for stover, percent grain protein, and grain yield) had lower h^2 values (31–37%) than other N or vegetative traits (42–68%). The range in family mean values (Table 3) also indicates considerable diversity for the traits measured. The data have to be interpreted with some caution because of the lack of an

Table 3—Means, range in means, heritabilities, and genetic coefficient of variability for several physiological and morphological traits of 100 half-sib families of the RSSSC (a version of stiff stalk synthetic is described in section IIIC) corn synthetic grown at Urbana, Ill. in 1980.

Traits	Mean	Range in means	h^2†	GCV‡
			%	%
Reduced N (g/total leaves§) (anthesis)	1.5	1.8–1.1	44 ± 12	5.9
Reduced N (g/plant) (maturity)	3.8	5.3–3.0	31 ± 15	6.2
Nitrate-N (mg/plant) (maturity)	174	284–76	34 ± 15	17.1
Nitrate reductase activity (units/leaf¶) (15 d after anthesis)	42	60–29	46 ± 3	9.1
Total-N# (g/plant) (anthesis)	2.7	3.6–2.0	39 ± 19	5.7
Leaf dry wt (g/plant) (anthesis)	54	64–41	60 ± 11	6.1
Leaf width (cm) (15 d after anthesis)	11.2	13.1–9.8	68 ± 16	4.2
Plant height (cm) (15 d after anthesis)	224	247–205	67 ± 11	2.9
Harvest index (stover)	0.46	0.55–0.27	37 ± 12	6.1
Percent grain protein	11.9	13.4–10.5	38 ± 17	2.7
Grain yield (g/plant)	116	161–73	31 ± 13	8.4

† Heritability in the "narrow-sense."
‡ Genetic coefficient of variability.
§ Total leaves/plant.
¶ Single leaf lamina.
# Reduced-N plus nitrate-N.

estimate of the genotype × environment interaction. The lack of this estimate inflates the h^2 value; however, additional cycles of selection will allow for additional estimates and comparison of the observed selection response to the predicted.

Estimates of the genetic and phenotypic correlations beween certain N parameters and grain yield are presented in Table 4. High, positive, genetic correlations were observed between grain yield and several N parameters (reduced N, nitrate N, NRA) at various stages throughout the life cycle of the plant. Several vegetative traits relating directly or indirectly to N metabolism were also positively correlated with grain yield. The high genetic correlations illustrate the importance of the role of N metabolism in determining grain yield. The phenotypic *r* values (Tables 4) in general were low, indicating the influence of environmental factors on the estimates. Based on the genetic correlation and heritability estimate, reduced N at anthesis was chosen as the N selection criterion. Using the appropriate prediction equations, the gain in grain yield using yield as the sole selection criterion should be 6% in the next cycle. Using the selection criteria of yield and reduced N at anthesis, the predicted gain should be 8%.

A similar approach was taken by Muruli and Paulsen (1981), who investigated the feasibility of improving N use efficiency in corn to increase yields by selecting from half-sib families of the cultivar 'Mex-Mix' grown at low and high levels of N (0 and 200 kg/ha). Synthetic lines developed from the highest yielding families grown at the low level (N-efficient) and high level of N (N-inefficient) were then tested for yield at 0, 50, 100, and 200 kg N/ha. At low levels of N, the N-efficient selection outyielded both the original population and the N-inefficient selection, but was not responsive to high levels of N. The N-inefficient selection yielded most at high levels of N. It was concluded that N use efficiency could be improved through selection. The relationship between yield and several biochemical and physio-

Table 4—Genetic and phenotypic correlations between grain yield per plant and certain other traits for 100 half-sib families of RSSSC corn synthetic grown at Urbana, Ill. in 1980.

	r_g†	r_p‡
Reduced N/leaf§ (anthesis)	+0.70 ± 0.19	+0.20 ± 0.10
Reduced N/plant (maturity)	+0.79 ± 0.28	+0.77 ± 0.04
Nitrate-N/plant (maturity)	+0.85 ± 0.19	+0.23 ± 0.10
Nitrate reductase (activity/leaf¶)		
(15 d after anthesis)	+0.62 ± 0.21	+0.38 ± 0.09
Total N#/plant (anthesis)	+0.88 ± 0.19	+0.24 ± 0.09
Leaf dry wt/plant (anthesis)	+0.52 ± 0.20	+0.16 ± 0.11
Leaf width (15 d after anthesis)	−0.41 ± 0.20	−0.06 ± 0.10
Plant height (15 d after anthesis)	+0.75 ± 0.22	+0.37 ± 0.09
Harvest index (stover)	1.24 ± 0.21	+0.41 ± 0.08
Percent grain protein	−0.41 ± 0.19	−0.14 ± 0.10

† Genetic correlation coefficient and standard error.
‡ Phenotypic correlation coefficient and standard error.
§ Total leaves/plant.
¶ Single leaf lamina.
# Reduced N plus nitrate-N.

logical traits of N metabolism was also examined in the same study, at both silking and maturity. It was found that the quantity of N accumulated between silking and maturity provided the largest difference between the 10 highest yielding lines in each selection cycle and the means of the original population.

Interpretation of the results presented in this section should be viewed with some caution, since only a limited amount of data are presently available on studies using physiological and biochemical criteria in plant breeding. However, preliminary results indicate that it should be possible to enhance progress towards increased grain yield basing selections on these criteria in combination with traditional breeding methods. Based on this premise, several parameters that might be successfully used in a plant-breeding program with the primary objective of increasing grain yield are presented in the following section.

IV. PROPOSED PLANT IDEOTYPE FOR HIGH GRAIN YIELD AND QUALITY

The proposed "ideotype" plant for increased grain yield and N efficiency in corn has a greater capacity for converting N supplied into increased yield, and would envisage a plant where leaf senescence is delayed. It has the following characteristics:

1) High dry matter and high reduced N accumulated by anthesis,
2) High stalk NO_3^- content at anthesis,
3) High rate of N uptake and assimilation during the grain-filling period,
4) High rate of N movement to the developing grain,
5) High photosynthetic rates during grain fill, and
6) Prolonged grain-filling period without a later physiological maturity date.

High harvest index for N could be included as a further desirable character in improving the efficiency of use of N supplied, although it might not be complementary to yield increase. High harvest index for N could be achieved through remobilization of leaf N, but could, in turn, adversely affect the photosynthetic apparatus. A high harvest index for N might be achieved together with high grain yield through remobilization of N later in the grain-filling period. A high level of reduced N at anthesis has been correlated with increased yield and presumably influences the amount of N available during grain filling. However, the level of reduced N at anthesis might also reflect the level of kernel initiation and consequently sink capacity and yield, although this relationship is still unclear.

Several heritable, physiological, and biochemical traits have been described for corn, which, in a suitable breeding program, have the potential for increasing grain yield through more efficient utilization of N. Additional information on the role of N metabolism in the plant in relation to yield should allow further progress in this area. In particular, there is a need to know the metabolic requirements for development and maintenance of an effective photosynthetic apparatus, maximum sink size, and a high rate and duration of grain filling. This objective would be difficult to achieve, however, due to the complexity and interaction of the processes involved. If this information could be obtained, it would not only permit selection of plants more efficient in N utilization, but would also provide knowledge for achieving an optimal N environment for plant growth.

ACKNOWLEDGMENT

This work was supported in part by a grant from the Science & Education Administration of the U.S. Department of Agriculture, and Grant no. 5901-0410-8-0144-0 from the Competitive Grants Office and a grant from Pioneer Hi-Bred International, Inc.

REFERENCES

Allison, J. C. S., and D. J. Watson. 1966. The production and distribution of dry matter in maize after flowering. Ann. Bot. (London) 30:365–381.

Beevers, L., and R. H. Hageman. 1980. Nitrate and nitrite reduction. p. 115–168. *In* B. J. Miflin (ed.) The biochemistry of plants. Vol. 5. Academic Press, New York.

Below, F. E. 1981. Nitrogen metabolism as related to productivity in ten maize (*Zea mays* L.) genotypes diverse for grain yield and grain protein. M.S. Thesis. Univ. of Illinois, Urbana-Champaign.

Below, F. E., L. E. Christensen, A. J. Reed, and R. H. Hageman. 1981. Availability of reduced N and carbohydrates for ear development of maize. Plant Physiol. 68:1186–1190.

Brunetti, N., and R. H. Hageman. 1976. Comparison of in vivo and in vitro assays of nitrate reductase in wheat (*Triticum aestivum* L.) seedlings. Plant Physiol. 58:583–587.

Bshadur, K. 1974. Progress from modified ear-to-row selection in two populations of maize. Ph.D. Dissertation. Univ. of Nebraska, Lincoln.

Christensen, L. E., F. E. Below, and R. H. Hageman. 1981. The effects of ear removal on senescence and metabolism of maize. Plant Physiol. 68:1180–1185.

Dalling, M. J., G. M. Halloran, and J. H. Wilson. 1975. The relation between nitrate reductase activity and grain nitrogen productivity in wheat. Aust. J. Agric. Res. 26:1–10.

Deckard, E. J., R. J. Lambert, and R. H. Hageman. 1973. Nitrate reductase activity in corn leaves as related to yield of grain and grain protein. Crop Sci. 13:343–350.

Dudley, J. W., D. E. Alexander, and R. J. Lambert. 1974. Gene improvement of modified protein maize. p. 120–135. *In* Int. Symp. on Protein Quality in Maize, Mexico. 4–8 Dec. 1972. Dowden, Hutchinson & Ross, Inc., Stroudsburg, Pa.

Dunand, R. T. 1980. Divergent phenotypic recurrent selection for nitrate reductase activity and correlated responses in maize. Ph.D. Dissertation. Univ. of Illinois, Urbana-Champaign.

Duncan, W. G., D. E. McCloud, R. L. McGraw, and K. J. Boote. 1978. Physiological aspects of peanut yield improvement. Crop Sci. 18:1015–1020.

Duvick, D. V. 1977. Genetic rates of gain in hybrid maize yields during the past 40 years. Maydica 22:187–196.

Earley, E. B., W. O. McIlrath, R. D. Seif, and R. H. Hageman. 1967. Effect of shade applied at different stages in corn (*Zea mays* L.) production. Crop Sci. 7:151–156.

Frey, K. J. 1971. Improving crop yields through plant breeding. p. 15–58. *In* J. D. Eastin and R. D. Munson (ed.) Moving off the yield plateau. Am. Soc. of Agron., Madison, Wis.

Hageman, R. H. 1979. Integration of nitrogen assimilation in relation to yield. p. 591–611. *In* E. J. Hewitt and C. V. Cutting (ed.) Nitrogen assimilation of plants. Academic Press, New York.

Hageman, R. H., and D. P. Hucklesby. 1971. Nitrate reductase from higher plants. *In* A. San Pietro (ed.) Methods in Enzymology. 23:491–503. Academic Press, New York.

Hucklesby, D. P., C. M. Brown, S. E. Howell, and R. H. Hageman. 1971. Late spring applications of nitrogen for efficient utilization and enhanced production of grain and grain protein of wheat. Agron. J. 63:274–276.

Hume, D. J., and D. K. Campbell. 1972. Accumulation and translocation of soluble solids in corn stalks. Can. J. Plant Sci. 52:363–368.

Hymowitz, T. J., J. W. Dudley, F. I. Collins, and C. M. Brown. 1974. Estimations of protein and oil concentrations in corn, soybean, and oat seed by near-infrared light reflectance. Crop Sci. 14:713–715.

Knipmeyer, J. W., R. H. Hageman, E. B. Earley, and R. D. Seif. 1962. Effect of light intensity on certain metabolites of the corn plant (*Zea mays* L.). Crop Sci. 2:1–5.

Langer, I., K. J. Frey, and T. B. Bailey. 1978. Production response and stability characteristics of oat cultivars developed in different eras. Crop Sci. 18:938–942.

Murata, Y., and S. Matsushima. 1975. Rice. p. 73–99. *In* L. T. Evans (ed.) Crop physiology. Cambridge Univ. Press, London.

Muruli, B. I., and G. M. Paulsen. 1981. Improvement of nitrogen use efficiency and its relationship to other traits in maize. Maydica 22:63–73.

Neyra, C. A., A. A. Franco, and J. C. Pereira. 1977. Seasonal patterns of nitrate reductase and nitrogenase activities in *Phaseolus vulgaris* (L.). Plant Physiol. 59:S-127.

Prine, G. M. 1973. Critical period for ear development among different ear-types of maize. Soil Crop Sci. Soc. Fla. Proc. 33:27–30.

Rao, K. P., D. W. Rains, C. O. Qualset, and R. C. Huffaker. 1977. Nitrogen nutrition and grain protein in two spring wheat genotypes differing in nitrate reductase activity. Crop Sci. 17:283–286.

Reed, A. J., F. E. Below, and R. H. Hageman. 1980. Grain protein accumulation and the relationship between leaf nitrate reductase and protease activities during grain development in maize (*Zea mays* L.). Plant Physiol. 66:164–170.

Reed, A. J., and R. H. Hageman. 1980a. The relationship between nitrate uptake, flux and reduction and the accumulation of reduced N in maize (*Zea mays* L.): I. Genotypic variation. Plant Physiol. 66:1179–1183.

Reed, A. J., and R. H. Hageman. 1980b. The relationship between nitrate uptake, flux and reduction and the accumulation of reduced N in maize (*Zea mays* L.): II. The effect of nutrient nitrate concentration. Plant Physiol. 66:1184–1189.

Russell, W. A. 1974. Comparative performance of maize hybrids representing different eras of maize breeding. p. 81–101. *In* Proc. of the 29th Annu. Corn and Sorghum Industry Res. Conf., Chicago, Ill. 10–12 Dec. 1974. Am. Seed Trade Assoc., Washington, DC.

Shaner, D. L., and J. S. Boyer. 1976. Nitrate reductase activity in maize (*Zea mays* L.) leaves: I. Regulation by nitrate flux. Plant Physiol. 58:499–504.

Sprague, G. F., and S. A. Eberhart. 1977. Corn breeding. *In* G. F. Sprague (ed.) Corn and corn improvement. Agronomy 18:303–362.

Thomas, S. M., and G. E. Thorne. 1975. Effect of nitrogen fertilizer on photosynthesis and ribulose 1,5-diphosphate carboxylase activity in spring wheat in the field. J. Exp. Bot. 26: 43–54.

Tollenaar, M. 1977. Sink-source relationships during reproductive development in maize. A review. Maydica 22:49–75.

Tsai, C. Y., D. M. Huber, and H. L. Warren. 1978. Relationship of the kernel sink for N to maize productivity. Crop Sci. 18:399–404.

Welch, L. F. 1979. Nitrogen use and behaviour in crop production. Agric. Exp. Stn., College of Agric., Urbana, Ill. Bull. 761. p. 31–33.

Wilcox, J. R., W. T. Schapaugh, Jr., R. L. Bernard, R. L. Cooper, W. R. Fehr, and M. H. Niehans. 1979. Genetic improvement of soybeans in the Midwest. Crop Sci. 19:803–805.

Zieserl, J. F., W. L. Rivenbank, and R. H. Hageman. 1963. Nitrate reductase activity, protein content, and yield of four maize hybrids of varying plant populations. Crop Sci. 3:27–32.

25

Michael A. Cole
University of Illinois
Urbana, Illinois

Legume Seed Inoculation[1]

Recurrent questions from agronomists in the temperate Midwest and in tropical wet or semiarid areas have been of the type: "How often must I re-inoculate if (i) the winter was colder or warmer than average; (ii) the summer was wetter or drier or hotter or colder than average; (iii) the same crop was grown in past years?" These questions and other similar ones can be summarized by the questions: "Do rhizobia exist in soils?" and "How long can they survive?" This chapter is intended to provide some broad guidelines to answer these questions where detailed on-site studies are not yet available.

I. NEED FOR AND EFFICACY OF INOCULATION

Legumes must be inoculated when planted in an area for the first time unless an appropriate, efficient *Rhizobium* that also nodulates a different plant species is already present. Where environmental conditions include extended dry periods, high temperatures, or saline, acidic, or alkaline soils, re-inoculation with each planting may be necessary. In temperate climates similar to the midwestern United States, inoculation is usually required only upon introduction of the legume, as long as the legume is replanted in the field within 3 to 5 yr of any previous planting and if the initial crop was well-nodulated. Johnson and Boone (1976) found no yield increase of soybeans [*Glycine max* (L.) Merr.] in two Illinois soils with either seed-applied or granular inoculant; other workers in midwestern states have reported similar results. Inoculation should also be performed if the previous legume crop was not well-nodulated. Detailed criteria for determining whether or

[1] Contribution of the Department of Agronomy, Illinois Agricultural Experiment Station, Urbana, Ill.

Nitrogen in Crop Production.

not inoculation is needed are given by Meisner and Gross (1980) and Date (1977).

Readily demonstrable efficacy of an inoculum requires two major conditions: (i) that no rhizobia that nodulate and fix N_2 with the legume are already in the soil; and (ii) that available soil N levels are sufficiently low to dramatically limit plant growth in the absence of N_2 fixation. The first condition can be readily recognized by planting uninoculated seed and examining the root systems for nodules at flowering or early podfill (by which time, nodulation will have occurred). For this kind of inspection, root systems must be carefully removed from the soil by digging, because many nodules will be detached from the roots if the plants are merely uprooted by grasping the stem and pulling. When performing a nodulation test, care must be taken to minimize accidental inoculation due to spread of rhizobia by cultivating and planting equipment that was previously used in inoculated areas or by mass soil transfer by wind or water. The second condition is quite difficult to establish, as the simple soil fertility tests ordinarily used measure only a limited portion of the total N available to the plants over the growing season (Bezdicek et al., 1974). However, a yield increase upon application of N fertilizer may be taken as presumptive evidence that soil N cannot fully satisfy plant requirements and that the plants would benefit from inoculation.

The amount of N_2 fixed by inoculant *Rhizobium* will vary because of differences in N_2-fixing capabilities of specific legumes, of specific *Rhizobium* strain–legume cultivar combinations (Caldwell & Vest, 1968; Burton, 1980), and environmental factors. In general, forage legumes fix more N_2 than grain legumes (Erdman, 1967). The amount of N_2 fixed by any *Rhizobium*-legume association will be determined ultimately by the degree to which the environment will permit or restrict plant growth. Detailed examination of important environmental factors, particularly temperature, moisture, and soil factors, has been limited primarily to laboratory and/or greenhouse studies and to situations where obvious problems existed. Hence, there is a substantial need for research under field conditions to provide data of value to the legume grower (Thompson et al., 1974).

Whether or not reinoculation is necessary with each planting of the legume is best determined by experimentation in a specific locale, since there are many variables in soil and climatic conditions that influence survival of rhizobia. The majority of reports indicate that soil rhizobia will survive high temperatures better in dry than in moist soil (Danso & Alexander, 1974; Wilkins, 1967), and in heavy-textured soils better than light-textured soils (Marshall, 1964), although substantial differences in heat tolerance exist between *Rhizobium* species (Chatel & Parker, 1973). Based on observation of the survival of rhizobia in midwestern states, a period of at least several months when the soil is frozen does not kill the population. Key factors affecting *Rhizobium* survival were discussed in depth in Lowendorf's (1980) excellent review, to which the reader is referred.

II. FORMS OF INOCULANT AND CURRENT PRACTICES

The majority of commercially produced inoculants are supplied as *Rhizobium* cells mixed with a carrier. The most satisfactory carrier is peat, but there are substantial differences among peats in their ability to maintain *Rhizobium* viability (Vincent, 1974). Partial sterilization of peat—which reduces the number of nonrhizobial cells—improves *Rhizobium* survival and thereby provides a better quality inoculant (Strijdom & Van Rensburg, 1981). A similar improvement would be expected with other carriers.

When a suitable peat is not available, other materials may be used. It appears that any relatively inert material that is locally available and/or inexpensive may be suitable. The list includes such diverse materials as ground coconut hulls (*Cocos nucifera*), plaster ($CaSO_4 \cdot 1/2\ H_2O$) granules (Fraser, 1975), and filter mud from sugarcane (*Saccharum officinarum* L.) mills (Philpotts, 1976). The majority of these materials are not as satisfactory as a good peat, but are acceptable if a suitable peat is not available. Rigorous testing of any new carrier material is strongly recommended before adopting it for inoculant production.

A deficiency shared by all solid carriers is that the *Rhizobium* cells remain metabolically active. Therefore, the effective shelf life of the product is limited and the inocula must be handled and stored under conditions that minimize detrimental effects on the cells. In practice, this means that temperatures over 40 to 45°C or below freezing must be avoided. Thus, inocula must not only be of high quality when they are shipped by the manufacturer, but also require careful handling during shipment, in storage, and in the hands of the user. Common misconceptions are that the *Rhizobium* cells are very durable and that little care is needed to protect them from temperature extremes and sunlight. Consequently, farmers may leave bags of inoculant in bright sunlight, or inoculate seeds and leave the planter in the sun for several hours. Both of these practices are likely to kill most of the cells in the inoculant (Burton, 1980). While manufacturer's directions stress careful handling of the inocula, the reason for avoiding temperature extremes and exposure to sunlight is not obvious (and therefore often considered unimportant) by shippers and final users.

Solid carrier inoculants are available in two forms, as a powder to be applied to the seed prior to planting, or as granules applied in the row at planting. Preinoculated seed—i.e., seed to which inoculum is added days or weeks before planting—is also available, but is generally unsatisfactory because of poor survival of the *Rhizobium.* Seed-applied inoculant is generally satisfactory if the seeds are planted within a few hours of inoculation and if environmental conditions allow rapid seed germination. Granular inoculants provide more rhizobia and permit longer survival of rhizobia in soil than do seed inoculants. However, granular inoculants are much more expensive than powdered forms, and the user must carefully consider the advisability of using a preparation whose cost may be a significant portion of overall production costs. At current prices, seeds may be inoculated for a few cents per hectare, whereas granular inoculants may cost several dollars per hectare. A detailed cost comparison for various forms and rates of inoculant is not provided, since local cost of inoculant may vary greatly and because recommendations about method and rate of inoculation should be locally made.

Because of the limitations of solid carrier inocula, other inoculant forms such as freeze-dried cells, frozen concentrates, and metabolically inactive cells in a liquid carrier such as mineral oil have been tested. In spite of the greater shelf life of these forms, they generally have not been as satisfactory as solid carrier inoculants for several reasons (Burton, 1967). Freeze-drying, followed by resuspension of cells in liquid, causes death of a high percentage ($> 99\%$) of the *Rhizobium* cells. Unless the initial cell concentration is very high, these inoculants may not supply sufficient cells for good nodulation. Further, the rehydrated cells are damaged and require a recovery period before they can nodulate the plant. If the cells are introduced into a comparatively hostile environment (e.g., hot, dry soil), they may die before they can recover, and inoculant failure will occur. Similarly, extensive cell damage is caused by freezing inoculants, with the additional restriction that the inoculant must remain solidly frozen until used. A few cycles of thawing and refreezing will result in death of cells in the concentrate. Further discussion of the merits and deficiencies of the various inoculant forms may be found in articles by Vincent (1974), Strijdom and Deschodt (1976), and Burton (1967).

III. INOCULUM QUALITY

A high-quality inoculum will contain a sufficiently large number of live *Rhizobium* cells to ensure abundant nodulation under low-stress conditions when the manufacturer's recommendations are followed (Burton, 1967). Methods for evaluating inoculant quality have been described by Speidel and Wollum (1980) and Vincent (1970). Depending on the specific form of inoculant (seed-applied liquid or powder, or liquid or granular inoculant applied in the furrow as the seed is planted), the minimum number of rhizobia required to ensure good nodulation will vary. For seed-applied inoculants, maximum nodulation of clover (*Trifolium pratense*) occurs with 1000 to 10 000 rhizobia per seed (Dart, 1974), while small numbers may be sufficient with soybeans. Commercial inocula, immediately after production, will contain enough rhizobia to provide good nodulation (Burton, 1967), but postproduction handling of the inocula may adversely affect performance (Ayanaba, 1977; Scudder, 1975). Strict adherence to the manufacturer's recommendations for shelf life, storage conditions, and the like are vital. When inoculation is conducted with the expectations of high soil temperatures ($\geq 35°C$), or where immediate seed germination is not expected because of dry soil, the producer's recommended rate of inoculation may still be used, although there is some chance of inoculant failure (Day et al., 1978; Philpotts, 1967; Marshall, 1964; Wilkins, 1967; Brockwell &

Phillips, 1965; Bowen & Kennedy, 1959). The advisability of increasing inoculum levels is a controversial point, since the increased cost—particularly with granular inoculants—is difficult to justify.

If adverse conditions can be anticipated, the best solution at the present time is use of a granular inoculant implanted 5 to 10 cm (2–4 inches) below the soil surface. This suggestion is based on the fact that granular inoculants supply a greater number of rhizobia than other inoculant forms. Greater numbers of cells, combined with the lower temperatures found deeper in the soil profile, will increase the probability of survival of rhizobial cells in adequate numbers to obtain good nodulation (Philpotts, 1967).

Because of the numerous variables involved, the best way to determine how much inoculum should be used is to consult a local researcher who is familiar with local conditions and problems. The articles by Dennis (1977), Keya (1977), and Chowdhury (1977), which document a variety of problems encountered in Africa, indicate a great need for regional studies on the specific difficulties encountered when attempting to establish a new legume in an existing cropping regimen.

IV. *RHIZOBIUM* STRAIN SELECTION

Assessing the adequacy of a *Rhizobium* strain actually requires simultaneous assessment of the strain, the host plant, and the ability of the environment to satisfy plant growth requirements. In practice, unambiguous demonstration of the superiority of one strain over another is difficult, particularly under field conditions. In an effort to minimize some of the potential variability, testing is frequently conducted in the greenhouse or growth chamber with a single legume cultivar. The deficiencies of this approach are discussed below. A detailed consideration of strain evaluation may be found in the booklet by Stowers and Elkan (1980).

Under ordinary circumstances, *Rhizobium* inoculants are generally optimized for the legume that is to be inoculated. All inoculant manufacturers produce inoculants for the major pasture and grain legumes grown in the United States such as alfalfa (*Medicago sativa* L.) and soybeans. Some producers also provide inoculants for uncommon legumes on a special-order basis. Thus, for routine use, one is obliged to rely upon the inoculant company to furnish rhizobial strains that are good N_2 fixers and possess other necessary characteristics. Most strains are satisfactory when used with the appropriate legume and when soil or other environmental conditions are not too inimical to survival. Since most inoculants that are produced in the United States are also consumed in the United States, the strains employed can be expected to perform reasonably well under "average" U.S. conditions, but may not perform as well under more rigorous conditions. Specific problems may arise where soil temperatures exceed 40°C (104°F), in soils with a high sand or soluble salt content, in acidic (Bouton et al., 1981a, b) or alkaline soils, or where a gap of 2 weeks or more occurs between planting and sufficient rainfall to allow seed germination. While it would be desirable to have *Rhizobium* strains available that would survive under these adverse conditions, there is not sufficient economic incentive for the companies to develop and market such strains, nor does there exist sufficient evidence that such special-purpose strains can be developed at the present time. The lack of more durable strains is not a major impediment for most usage in the United States—with the possible exception of peanut (*Arachis hypogaea* L.) inoculants in Florida—but may be a significant barrier to introducing legumes developed in the United States into other countries, where soil and climatic conditions may not be as favorable for inoculation as in the United States.

In numerous studies concerned with leguminous N_2 fixation, objectives such as increasing N_2-fixing capabilities or increasing the competitiveness of the *Rhizobium* strains have been emphasized. Considering the amount of effort devoted to these objectives, there has been relatively little progress, not because of incompetent research, but rather of the inherent difficulties of working with a two-organism relationship that is so dependent on environmental components for success or failure. The work of Brill and coworkers provides a good example of the problems encountered when attempting to relate promising growth chamber or greenhouse results to field conditions. Maier and Brill (1978) described two *R. japonicum* mutants whose acetylene-reducing (N_2-fixing) activity was nearly twice that of the parent strain from which they were derived. Plants inoculated with one isolate (SM35) also showed a 50% increase in N content, compared with plants inoculated with the parent strain. In spite of the large increase in N_2-fixing capability shown by strain SM35 in the growth chamber, it did not fix a significantly greater amount of N_2 when tested under field conditions (Burton, 1980).

A similar example of laboratory-derived results not being verified under field conditions is the work of Shubert and Evans (1976). These two investigators suggested that H_2-utilizing strains of *R. japonicum* were much more efficient users of available energy (i.e., efficient strains fixed more N_2 with a given amount of substrate than did inefficient strains that were unable to utilize H_2. Their calculations suggested that in circumstances where N_2 fixation was energy-limited, efficient strains would provide about 10 to 50% more N for plant growth than inefficient strains. However, these expectations have not been reproducibly fulfilled under field conditions (Lepo et al., 1981).

In both of the cited examples, highly promising results were obtained in the laboratory, but not under field conditions. The results suggest that, if we are to achieve a useful increase in N_2 fixation by judicious choice of the rhizobial symbiont, the selection of strains by growth chamber and greenhouse testing must be regarded as a preliminary step, and ultimate confirmation of the activity of promising strains must be achieved under a broad range of field conditions.

There are strong indications that N_2 fixation is influenced by *Rhizobium* strain × legume cultivar interactions (Caldwell & Vest, 1968; Burton, 1980). A logical extension of this concept would be to match the *Rhizobium* strain in the inoculant to the specific cultivar being planted in order to maximize N gains by the legume. In practice, it has not been possible to achieve this result because of competition between rhizobial cells already in the soil (indigenous strains) and rhizobial cells in the inoculant. In all legumes tested, the plants are preferentially nodulated by indigenous rhizobia, even though inoculant rhizobia may be added at ratios of several thousand inoculant cells for each indigenous cell (Amarger, 1974; Kuykendall & Weber, 1978; Roughley et al., 1976). Thus, strain-cultivar matching is currently feasible only in cases where the legume is introduced for the first time or where soil conditions do not allow rhizobial survival from one planting to the next. A frequently suggested solution to this problem is to find *Rhizobium* strains that compete successfully with indigenous strains. At this time, there has been little success with this approach, in part because we do not know if this competition is due to an "incumbent" effect—i.e., environmental conditions in some manner favor indigenous cells—or to intrinsic differences in competitiveness among *Rhizobium* strains. Although differences in competitive ability among strains simultaneously added to soil have been documented (Caldwell, 1969; Roughley et al., 1976), the results do not suggest a clear choice between the two explanations. Competitiveness also has a large environmental component, as demonstrated by Weber and Miller (1972). An inoculant strain that was a good competitor under one set of conditions was a poor competitor under another set of conditions.

Available data do suggest that there is sufficient genetic variability among *Rhizobium* strains to allow eventual improvement in symbiont performance (Gibson et al., 1976; Day et al., 1978), but it is equally obvious that there are several formidable problems yet to be solved. A major difficulty is that, genetically speaking, the rhizobia have not been amenable to the facile gene exchange that is possible with other microorganisms. The discovery in the past few years that symbiotic competence of *Rhizobium* is at least partially determined by nonchromosomal (plasmid carried) genes (Buchanan-Wollaston et al., 1980; Brewin et al., 1980; Hirsch et al., 1980; Nuti et al., 1979) in conjunction with the recent advances in genetic engineering (Hennecke, 1981; Hooykaas et al., 1981) may facilitate *Rhizobium* strain improvement.

V. SOIL FACTORS AFFECTING EFFICIENT USE OF INOCULUM

In the context of this discussion, *efficient use* means using no more inoculum than is needed to obtain good nodulation initially and reinoculating only when the surviving *Rhizobium* population is too low to ensure good nodulation.

Where soil conditions are not overly hostile to rhizobia, there is little or no benefit from using higher rates of inoculum than the manufacturer recommends (Johnson & Boone, 1976). Burton and Curley (1965) reported that 2×10^5 *Rhizobium* cells/seed were sufficient for good nodulation of soybeans, but Hamedi (1976) found that about 3×10^8 *Rhizobium* cells/seed were required for good nodulation under greenhouse conditions in Egypt, with no nodulation in the field even at this rate. Smith et al. (1981) reported an increase in nodule numbers with 10-fold increases in rhizobial numbers from about 3 $\times$ 10^5 to 3 $\times$ 10^8 cells/cm of row. They recommended that greater inoculum levels be used when introducing soybeans in tropical soils. It is evident from these examples that a recommended rate appropriate for one region may be either insufficient or excessive in a different region, and that experiments adequate for determining the proper inoculum rate must be conducted when any legume is introduced into an area where previous experience with it is not available.

In previous sections, it was shown that substantial differences exist among *Rhizobium* species and strains in their heat-tolerance, competitiveness, and other features. When considering inoculum success in overall terms (i.e., crop yield), one must simultaneously consider the following: the specific legume being planted, its appropriate *Rhizobium,* climatic factors (rainfall, temperature, etc.), and edaphic factors (soil pH, salt content, physical characteristics, etc.). With the number of variables to be considered, it is obvious that no simple generalizations about how much inoculum to use or how often to inoculate are possible. Environmental factors affecting nodulation and N_2 fixation has been a comparatively well-studied area. The articles by Holding and King (1963), Lie (1974), and Edwards (1977) provide a good background for study of this topic. One should be particularly careful about trying to extrapolate recommendations based upon experience with one legume to other legumes. Since the earlier work on legume establishment on acid soils was done with pasture legumes of low acid tolerance, the conventional wisdom (until quite recently) was that introduced legumes could not be successfully grown in acid soils without major modifications of soil pH. While the statement is true for some pasture legumes, it is not true for soybeans or for all cultivated legumes that are native to acid soils. We now know that soybeans can be successfully introduced into regions where it was previously thought they could not be grown without extensive soil amendments. In working with the legumes and their associated rhizobia, the researcher should constantly keep in mind that the legume family is very large and diversified and that there may be more differences than similarities among *Rhizobium* species. Strains within a single *Rhizobium* species may share few common characteristics other than their ability to nodulate the same host plant species.

REFERENCES

Amarger, N. 1974. Competition pour la formation des nodosites sur la feverole entre souches de *Rhizobium leguminosarum* apportees par inoculation et souches indigenes. C. R. Seances Acad. Sci. Ser. D 279:527–530.

Ayanaba, A. 1977. Toward better use of inoculants in the humid tropics. p. 181–187. *In* A. Ayanaba and P. J. Dart (ed.) Biological nitrogen fixation in farming systems in the tropics. John Wiley & Sons, Inc., New York.

Bezdicek, D. F., R. F. Mulford, and B. H. Magee. 1974. Influence of organic nitrogen on soil nitrogen, nodulation, nitrogen fixation, and yield of soybeans. Soil Sci. Soc. Am. Proc. 38:268–273.

Bouton, J. H., M. E. Sumner, and J.E. Giddens. 1981a. Alfalfa, *Medicago sativa* L., in highly weathered, acid soils. II. Yield and acetylene reduction of a plant germplasm and *Rhizobium meliloti* inoculum selected for tolerance to acid soils. Plant Soil 60:205–211.

Bouton, J. H., J. K. Syers, and M. E. Sumner. 1981b. Alfalfa, *Medicago sativa* L., in highly weathered, acid soils. I. Effect of lime and P application on yield and acetylene reduction (N_2-fixation) of young plants. Plant Soil 59:455–463.

Bowen, G. D., and M. M. Kennedy. 1959. Effect of high soil temperatures on *Rhizobium* spp. Queensl. J. Agric. Sci. 16:177–197.

Brewin, N. J., J. E. Beringer, and A. W. B. Johnston. 1980. Plasmid-mediated transfer of host-range specificity between two strains of *Rhizobium leguminosarum*. J. Gen. Microbiol. 120:413–420.

Brockwell, J., and L. J. Phillips. 1965. Survival at high temperatures of *Rhizobium meliloti* in peat inoculant on lucerne seed. Aust. J. Sci. 27:332–333.

Buchanan-Wollaston, A. V., J. E. Beringer, W. J. Brewin, P. R. Hirsch, and A. W. B. Johnston. 1980. Isolation of symbiotically defective mutants in *Rhizobium leguminosarum* by insertion of the transposon Tn5 into a transmissible plasmid. Mol. Gen. Genet. 178:185–190.

Burton, J. C. 1967. *Rhizobium* culture and use. p. 1–33. *In* H. J. Peppler (ed.) Microbial technology. Van Nostrand-Reinhold, New York.

Burton, J. C. 1980. *Rhizobium* inoculation and soybean production. p. 89–100. *In* F. T. Corbin (ed.) World Soybean Research Conference: II. Proceedings. Westview Press, Boulder, Colo.

Burton, J. C., and R. L. Curley. 1965. Comparative efficiency of liquid and peat-base inoculant on field-grown soybeans (*Glycine max*). Agron. J. 57:379–381.

Caldwell, B. E. 1969. Initial competition of root-nodule bacteria on soybeans in a field environment. Agron. J. 61:813–815.

Caldwell, B. E., and G. Vest. 1968. Nodulation interactions between soybean genotypes and serogroups of *Rhizobium japonicum*. Crop Sci. 8:680–682.

Chatel, D. L., and C. A. Parker. 1973. Survival of field-grown rhizobia over the dry summer period in western Australia. Soil Biol. Biochem. 5:415–423.

Chowdhury, M. S. 1977. Response of soybean to *Rhizobium* inoculation at Morogor, Tanzania. p. 245–253. *In* A. Ayanaba and P. J. Dart (ed.) Biological nitrogen fixation in farming systems in the tropics. John Wiley & Sons, Inc., New York.

Danso, S. K. A., and M. Alexander. 1974. Survival of two strains of *Rhizobium* in soil. Soil Sci. Soc. Am. Proc. 38:86–89.

Dart, P. J. 1974. The infection process. p. 381–420. *In* A. Quispel (ed.) The biology of nitrogen fixation. North-Holland Publ. Co., Amsterdam.

Date, R. A. 1977. The development and use of legume inoculants. p. 169–180. *In* A. Ayanaba and P. J. Dart (ed.) Biological nitrogen fixation in farming systems in the tropics. John Wiley & Sons, Inc., New York.

Day, J. M., R. J. Roughley, A. R. J. Eaglesham, M. Dye, and S. P. White. 1978. Effect of high soil temperatures on nodulation of cowpea, *Vigna unguiculata*. Ann. Appl. Bot. 88: 476–481.

Dennis, E. A. 1977. Nodulation and nitrogen fixation in legumes in Ghana. p. 217–232. *In* A. Ayanaba and P. J. Dart (ed.) Biological nitrogen fixation in farming systems in the tropics. John Wiley & Sons, Inc., New York.

Edwards, D. G. 1977. Nutritional factors limiting nitrogen fixed by rhizobia. p. 189–204. *In* A. Ayanaba and P. J. Dart (ed.) Biological nitrogen fixation in farming systems in the tropics. John Wiley & Sons, Inc., New York.

Erdman, L. W. 1967. Legume inoculation: what it is—what it does. USDA Farmers Bull. 2003. U.S. Government Printing Office, Washington, D.C.

Fraser, M. E. 1975. A method for culturing *Rhizobium meliloti* on porous granules to form a pre-inoculant for lucerne seeds. J. Appl. Bacteriol. 39:345–351.

Gibson, A. H., R. A. Date, J. A. Ireland, and J. Brockwell. 1976. A comparison of competitiveness and persistance among five strains of *Rhizobium trifolii*. Soil Biol. Biochem. 8:395–401.

Hamedi, Y. A. 1976. Field and greenhouse experiments on the response of legumes in Egypt to inoculation and fertilizers. p. 289–298. *In* P. S. Nutman (ed.) Symbiotic nitrogen fixation in plants. Cambridge Univ. Press, Cambridge, England.

Hennecke, H. 1981. Recombinant plasmids carrying nitrogen fixation genes from *Rhizobium japonicum*. Nature 291:354–355.

Hirsch, P. R., M. van Montagu, A. W. B. Johnson, N. J. Brewin, and J. Schell. 1980. Physical identification of bacteriocinogenic, nodulation, and other plasmids in strains of *Rhizobium leguminosarum*. J. Gen. Microbiol. 120:403–412.

Holding, A. J., and J. King. 1963. The effectiveness of indigenous population of *Rhizobium trifolii* in relation to soil factors. Plant Soil 18:191–198.

Hooykaas, P. J. J., A. A. N. van Barussel, H. den Dulk-Ras, G. M. S. van Slogteren, and R. A. Schilperoot. 1981. *Sym* plasmid of *Rhizobium trifolii* expessed in different rhizobial species and in *Agrobacterium tumefaciens*. Nature 291:351–353.

Johnson, R. R., and L. V. Boone. 1976. Soybean inoculation: is it necessary? Illinois Research (Fall, 1976) 18(9):3–4. University of Illinois Agric. Exp. Stn.

Keya, S. O. 1977. Nodulation and nitrogen fixation in legumes in East Africa. p. 233–243. *In* A. Ayanaba and P. J. Dart (ed.) Biological nitrogen fixation in farming systems in the tropics. John Wiley & Sons, Inc., New York.

Kuykendall, L. D., and D. F. Weber. 1978. Genetically marked *Rhizobium* identifiable as inoculum strain in nodules of soybean plants grown in fields populated with *Rhizobium japonicum*. Appl. Environ. Microbiol. 36:915–919.

Lepo, J. E., R. E. Hickok, M. A. Cantrell, S. A. Russell, and H. J. Evans. 1981. Revertible hydrogen uptake-deficient mutants of *Rhizobium japonicum*. J. Bacteriol. 146:614–620.

Lie, T. A. 1974. Environmental effects on nodulation and symbiotic nitrogen fixation. p. 555–582. *In* A. Quispel (ed.) The biology of nitrogen fixation. North-Holland Publ. Co., Amsterdam.

Lowendorf, H. S. 1980. Factors affecting survival of *Rhizobium* in soil. Adv. Microb. Ecol. 4:87–124.

Maier, R. J., and W. J. Brill. 1978. Mutant strains of *Rhizobium japonicum* with increased ability to fix nitrogen for soybeans. Science 201:448–450.

Marshall, K. C. 1964. Survival of root-nodule bacteria in dry soils exposed to high temperatures. Aust. J. Agric. Res. 15:273–281.

Meisner, C. A., and H. D. Gross. 1980. Some guidelines for the evaluation of the need for and response to inoculation of tropical legumes. North Carolina Agric. Res. Service Tech. Bull. no. 365.

Nuti, M. P., A. A. Lepidi, R. K. Prakash, R. A. Schilperoort, and F. C. Cannon. 1979. Evidence for nitrogen fixation (*nif*) genes on indigenous *Rhizobium* plasmids. Nature 282:533–535.

Philpotts, H. 1967. Effect of soil temperature on nodulation of cowpeas (*Vigna sinensis*). Aust. J. Exp. Agric. Anim. Husb. 7:372–376.

Philpotts, H. 1976. Filter mud as a carrier for *Rhizobium* inoculants. J. Appl. Bacteriol. 41: 277–281.

Roughley, R. J., W. M. Blowes, and D. F. Herridge. 1976. Nodulation of *Trifolium subterraneum* by introduced rhizobia in competition with naturalized strains. Soil Biol. Biochem. 8:403–407.

Scudder, W. T. 1975. Florida tests favor inoculation. Soybean Dig. 35(9):16.

Shubert, K. R., and H. J. Evans. 1976. Hydrogen evolution: a major factor affecting efficiency of nitrogen fixation in nodulated symbionts. Proc. Natl. Acad. Sci. USA 73: 1207–1211.

Smith, R. S., M. A. Ellis, and R. E. Smith. 1981. Effect of *Rhizobium japonicum* inoculant rates on soybean nodulation in a tropical soil. Agron. J. 73:505–508.

Speidel, K. L., and A. G. Wollum II. 1980. Evaluation of leguminous inoculant quality: a manual. North Carolina Agric. Res. Service Tech. Bull. no. 266.

Stowers, M. D., and G. H. Elkan. 1980. Criteria for selecting infective and efficient strains of *Rhizobium* for use in tropical agriculture. North Carolina Agric. Res. Service Tech. Bull. no. 264.

Strijdom, B. W., and C. C. Deschodt. 1976. Carriers of rhizobia and the effects of prior treatment on the survival of rhizobia. p. 151–168. *In* P. S. Nutman (ed.) Symbiotic nitrogen fixation in plants. Cambridge Univ. Press, Cambridge, England.

Strijdom, B. W., and H. J. van Rensburg. 1981. Effect of steam sterilization and gamma irradiation of peat on quality of *Rhizobium* inoculants. Appl. Environ. Microbiol. 41: 1344–1347.

Thompson, J. A., R. J. Roughley, and D. F. Herridge. 1974. Criteria and methods for comparing the effectiveness of *Rhizobium* strains for pasture legumes under field conditions. Plant Soil 40:511–524.

Vincent, J. M. 1970. A manual for the practical study of root-nodule bacteria. I. B. P. Handb. no. 15. Blackwell Scientific Publishers, Oxford.

Vincent, J. M. 1974. Root-nodule symbioses with *Rhizobium*. p. 265–341. *In* A. Quispel (ed.) The biology of nitrogen fixation. North-Holland Publ. Co., Amsterdam.

Weber, D. F., and V. L. Miller. 1972. Effect of soil temperature on *Rhizobium japonicum* serogroup distribution in soybean nodules. Agron. J. 64:796–798.

Wilkins, J. 1967. The effects of high temperature on certain root-nodule bacteria. Aust. J. Agric. Res. 18:299–304.

Management of Soils and Pests for Nitrogen Utilization

section IV

26

J. J. Meisinger
Agricultural Research Service, USDA
Beltsville, Maryland

Evaluating Plant-Available Nitrogen in Soil-Crop Systems

Nitrogen plays a central role in modern crop production because it is an essential macronutrient, an energy-intensive input, and the major limiting nutrient in most U.S. agricultural soils. Modern soil-crop systems must supply N in quantities needed for high crop productivity (primarily through mineralization, residual NO_3^--N, and fertilizer N) and also ensure high N utilization by optimizing the fertilizer N rate, timing, and application methods. Optimizing the fertilizer rate is the largest single N management factor affecting crop productivity and N use efficiency, provided the N is applied conscientiously. Achieving optimal N rates requires the evaluation and integration of *several* elements including crop production factors (e.g., potential crop yield) and soil N availability factors (e.g., mineralization, residual NO_3^--N) into an overall N evaluation scheme. I shall, therefore, consider the general principles of such a "systems" approach and will then relate these principles to current soil N tests. Readers interested in a detailed discussion of soil N tests per se are referred to other recent reviews (Bremner, 1965b; Dahnke & Vasey, 1973; Keeney, 1982; Stanford, 1981).

I. GENERAL PRINCIPLES

The underlying principle for evaluating N in soil-crop systems is the N balance. A diverse, and seemingly unrelated, group of recommendation systems has emerged as researchers have studied specific systems. However, by applying basic N balance principles, it can be shown that these divergent approaches have common features. Considering basic principles will also (i) establish an overall framework to place various soil N tests in perspective,

Nitrogen in Crop Production.

(ii) outline the situations where various soil N tests will give the greatest information, (iii) enumerate the important factors needed to evaluate plant-available N, and (iv) describe the circumstances where simplifying assumptions are in order.

A. General Mass Balance Principles

A description of plant-available N for any soil-crop system can be given through mass balance principles once the system is defined in space and time, i.e.

$$\text{N inputs} - \text{N outputs} = \text{Change in N storage } (\Delta N_s). \qquad [1]$$

For example, one might consider the N inputs and outputs from 1 Jan. 1980 to 1 Jan. 1981 for a 10-ha tract of unmanured Marshall silt loam growing continuous corn (*Zea mays* L.) (Fig. 1). The next step is to establish the system boundaries. This is the underlying cause of several diverging viewpoints. Specifically, should the system boundaries be drawn at the bottom of the root zone or at the soil surface? This choice dominates the entire subsequent analysis.

1. WHOLE CROP N BALANCE

If the boundary is drawn just below the root zone (line AB of Fig. 1), one adopts a "whole crop N uptake" view point and Eq. [1] becomes

$$N_f + N_{misc} - N_{ch} - N_e - N_l - N_g = \Delta N_{son} + \Delta N_{sin}$$

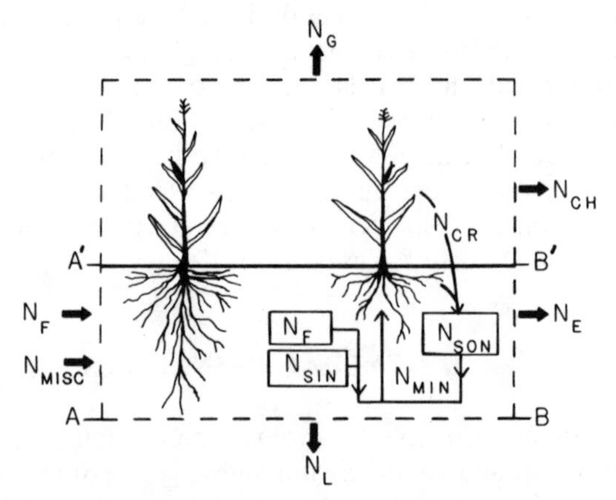

Fig. 1—Schematic representation of a soil-crop system with N inputs of fertilizer (N_f) and miscellaneous sources (N_{misc}); N outputs of leaching (N_l), erosion (N_e), gaseous losses (N_g), and harvested crop products (N_{ch}); and internal N pools of soil organic N (N_{son}), soil inorganic N (N_{sin}), returned crop residues (N_{cr}), and N mineralization (N_{min}).

where

N_f = fertilizer N input,
N_{misc} = rainfall N, NH_3 adsorption, irrigation, etc.,
N_{ch} = N removed in harvested crops,
N_e = N lost by erosion,
N_l = N lost by leaching,
N_g = gaseous N loss (denitrification and NH_3 losses),
ΔN_{son} = the change in soil organic N content; the difference between N immobilized (N_{imm}) in root and stover residues plus microbial immobilization, and N produced from mineralization (N_{min}), i.e., $\Delta N_{son} = N_{imm} - N_{min}$, and
ΔN_{sin} = the change in soil inorganic N content; the difference between the soil mineral N content at the end of the time period (N_{sine}) and that at the beginning (N_{sinb}), i.e., $\Delta N_{sin} = N_{sine} - N_{sinb}$.

Limiting our attention to N processes that are > 20 kg N ha^{-1} yr^{-1} (or by assuming that $N_{misc} \cong N_e$) and solving for N_f gives

$$N_f = N_{ch} + N_l + N_g + \Delta N_{son} + \Delta N_{sin}.$$

Nitrogen leaching (N_l) can be conveniently expressed as a certain percentage (P_l) of crop-available N so that $N_l \cong P_l\,(N_f + N_{min} + N_{sinb})$. Gaseous losses ($N_g$) can also be expressed as a percentage (P_g) of available N so that $N_g \cong P_g\,(N_f + N_{min} + N_{sinb})$. Substituting these relations into the preceeding equation along with the appropriate ΔN_{son} and ΔN_{sin} terms (see above definitions) gives

$$N_f = \frac{N_{ch}}{E} + \frac{N_{imm}}{E} - N_{min} + \frac{N_{sine}}{E} - N_{sinb}. \quad [2]$$

Here, E is the fractional total N_f recovery within the soil-crop system and equals $(1 - P_l - P_g)$. Equation [2] contains both crop and soil factors; N_f requirements are directly proportional to harvested crop N plus the change in the soil organic N status $[(N_{imm}/E) - N_{min}]$ plus the change in the soil inorganic N status $[(N_{sine}/E) - N_{sinb}]$. It is important to note that E refers to the fractional recovery of N_f *within the entire soil-crop system* (i.e., the N_f efficiency for the whole system); typical values would be 65 to 85%. Soil N transformations appear directly in this description; some cross system boundaries, e.g., N_l and N_g, while others remain within the system and appear as internal flows or phase changes, e.g., M_{min} and N_{imm}.

Equation [2] assumed that P_l and P_g were the same for N_f, N_{min}, and N_{sinb}. But it is a straight forward matter to generalize the solution to include manure (N_m) and miscellaneous (N_{misc}) inputs, erosion losses (N_e), and to allow each N input to have its own independent proportional loss to leaching, erosion, and gaseous output. Making these substitutions and solving for N_f gives the following general solution:

$$N_f = \frac{N_{ch}}{E_f} + \frac{N_{imm}}{E_f} + \frac{N_{sine}}{E_f} - \frac{E_{min}\,N_{min}}{E_f} - \frac{E_{sinb}\,N_{sinb}}{E_f} - \frac{E_m\,N_m}{E_f} - \frac{E_{misc}\,N_{misc}}{E_f}, \quad [2a]$$

where the efficiency or "E" terms refer to the fractional total N recovery of a specific N source within the entire soil-crop system (e.g., E_f is the fractional recovery of fertilizer N). Each "E" term equals $(1 - P_l - P_e - P_g)$ for the specific N source, where P_l, P_e, and P_g are the percentage losses of the specific N source to leaching, erosion, and gaseous output, respectively. This general cereal crop solution has been included for completeness, but Eq. [2] adequately describes the whole crop N balance system for our purposes.

2. ABOVEGROUND N BALANCES

If the system boundary is drawn at the soil surface (line A′B′ of Fig. 1) one adopts an "aboveground N uptake" viewpoint, i.e., the system is the aboveground crop with inputs from fertilizer and soil N.

Applying Eq. [1] and solving for N_f gives

$$N_f = \frac{N_{ch} + N_{cr} - e_m N_{min} - e_s N_{sin}}{e_f}, \quad [3]$$

where

N_f = fertilizer N input,
N_{min} = estimated soil N mineralization,
N_{sin} = estimated soil mineral N content,
N_{ch} = N contained in harvested product,
N_{cr} = N contained in aboveground crop residues,
e_f = the fraction of N_f in the aboveground crop,
e_m = the fraction of N_{min} in the aboveground crop, and
e_s = the fraction of N_{sin} in the aboveground crop.

This approach has been used by Parr (1973), Stanford (1973, 1981), and others (Carter et al., 1976; Fox & Piekielek, 1978a). Both soil and crop factors appear in Eq. [3]; N_f needs are directly related to the N contained in the standing crop, less that contributed by the soil through mineralization and residual mineral N. It is important to note that *all soil N transformations are outside* this system and appear indirectly in the efficiency terms (e_f, e_m, and e_s), which represent the final net effect of all the soil N transformations, i.e., leaching, gaseous losses, immobilization, etc. Consequently, such efficiency terms are very complex parameters that are difficult to predict for a specific soil-crop system. General values would be 40 to 60%.

B. General Applications to Soil-Crop Systems

Either of the above approaches (section A1 or A2) can be used to evaluate plant-available N; but they are quite distinct in their application and potential simplification.

1. STEADY-STATE CONDITIONS

Steady-state conditions occur when the change in N storage within the system is zero, or is acceptably small (i.e., ΔN_s of Eq. [1] $\cong$ 0). The aboveground N uptake approach (Eq. [3]) does not lead to a steady-state condition for annual crops because it does not consider N stored within the system from year to year (although perennial crops could store N). The whole crop approach (Eq. [2]) contains a steady-state simplification if ΔN_{son} and ΔN_{sin} equal zero; that is, N mineralized from organic sources is balanced by organic N returned in root and stover residues, etc., [$(N_{imm}/E) \cong N_{min}$] and the soil mineral N content at the end of the time increment equals that at the beginning [$(N_{sine}/E) \cong N_{sinb}$]. The unmanured plots of the Broadbalk continuous wheat experiment at Rothamstad are examples of steady-state systems, while the continuously manured plot at Rothamsted and the Morrow plots of Illinois are examples of systems slowly approaching steady-state conditions (Fig. 2). Under steady-state conditions, Eq. [2] simplifies to $N_f = N_{ch}/E$. Fertilizer N needs are directly related to anticipated N removals in harvested products and no soil N terms appear because the soil is within the system boundaries and is assumed to be at steady state. For example, a given soil-crop-management system under average climatic conditions may have $P_l \cong 15\%$ and $P_g \cong 15\%$ so that $N_f = N_{ch}/0.70$. If corn grain contains 15 kg N/t (0.70 pounds N/bushel), then the preceeding equation be-

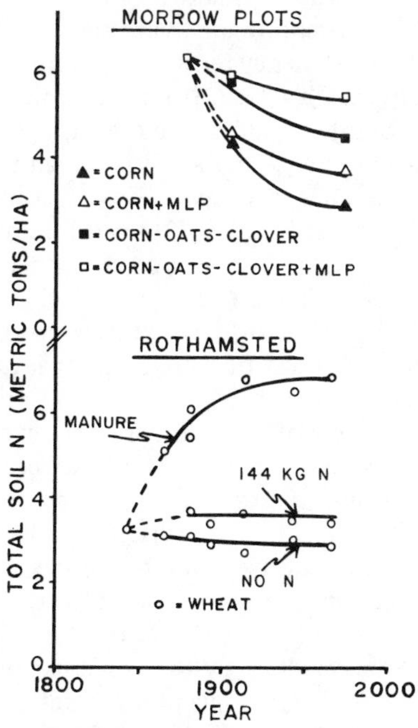

Fig. 2—Topsoil total N content of the Morrow plots (MLP refers to manure + lime + P) and the Broadbalk plots at various times. Dashed lines refer to estimated changes since the beginning of the experiment [data summarized from (Illinois Agric. Exp. Stn. and Dep. of Agronomy, 1982) and (Jenkinson, 1977)].

comes $N_f = 21$ kg N/t ($N_f = 1.0$ pounds N/bushel). For wheat grain (*Triticum aestivum* L.) containing 25 kg N/t (1.3 pounds N/bushel), the equation becomes $N_f = 36$ kg N/t ($N_f = 1.86$ pounds N/bushel).

It should be emphasized that the steady-state case is applicable *if and only if* the change in soil N is very small. Such a situation is often difficult to justify experimentally because a small change in the large soil organic N pool can result in the mineralization or immobilization of significant quantities of N relative to crop N requirements. For example, an annual decline of only 0.002% N in the surface 30 cm of soil could mineralize 90 kg N/ha. Another problem with the steady-state approach is that the time frame must cover several years due to the slow reaction time of soil-crop systems (note time scale of Fig. 2). The farmer's time frame, however, is 1 yr, because outside N inputs usually have to be made annually. A system may be at steady state in the long term (5- to 10-yr time scale), yet experience a dry year in which significant quantities of N are carried over to next year's crop. These sporadic perturbations would affect the annual N_f requirements but the long-term, steady-state approach would ignore them. In view of these arguments, it is important to emphasize that the above steady-state approach is a long-term first approximation, and should be used only after careful justification.

2. NON-STEADY-STATE CONDITIONS

Applying the aboveground N uptake approach (Eq. [3]) to non-steady-state conditions involves estimating the quantities of N in the N_{sin} and N_{min} pools and the crop uptake efficiency terms e_s, e_m, and e_f (see section II).

The whole crop approach (Eq. [2]) may contain the non-steady-state condition in either (or both) the soil organic N component ($\Delta N_{son} \neq 0$) or the mineral N component ($\Delta N_{sin} \neq 0$). In areas subject to large leaching losses between growing seasons, such as shallow soils in humid regions, one might assume that $\Delta N_{sin} \cong 0$ so that Eq. [2] would simplify to $N_f = (N_{ch}/E) + (N_{imm}/E) - N_{min}$. Similarily, a soil-crop system may have $\Delta N_{son} \cong 0$ so that Eq. [2] becomes $N_f = (N_{ch}/E) + (N_{sine}/E) - N_{sinb}$. However, the complete form of Eq. [2] should be utilized whenever doubts exist regarding steady-state conditions. Methods of estimating the various terms of this equation are discussed in section II.

A graphical presentation of the general non-steady-state case for the whole crop approach is given in Fig. 3. The first case (Fig. 3A) depicts a large net mineralization over several years, as occurs when grasslands are brought under cultivation (e.g., Illinois Agric. Exp. Stn. & Dep. of Agronomy, 1982; Jenny, 1941). The reverse situation occurs (Fig. 3B) if organic N inputs exceed mineralization, as in the continuous heavy manuring of a cereal crop system (e.g., Greenland, 1971; Jenkinson, 1977) or the reversion of cultivated soils to reduced-tillage systems. Here, the N_f needs would increase in order to meet crop N removals plus the net immobilization. Oscillating systems are also common in which organic N accumulates for a time and is then depleted (Fig. 3C), e.g., rotation systems involving legumes and cereal crops. The length of the accumulation phase will vary with the dura-

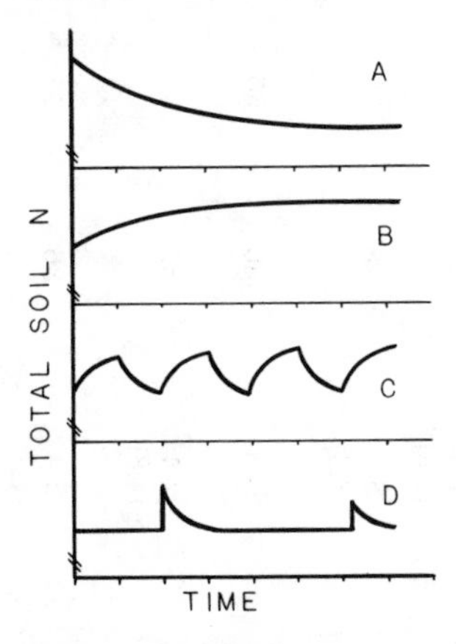

Fig. 3—Generalized changes of soil total N with time for various soil-crop systems. A = net N mineralization; B = net N immobilization; C = oscillating systems; D = intermittent oscillations (see text for examples).

tion of the legume stand and the height will depend on the difference between N inputs from symbiotic fixation and N harvested as forages. Other systems involve periodic N inputs followed by subsequent relaxation to initial conditions (Fig. 3D), e.g., periodic manure additions. Important factors to determine in these latter two systems are the steepness and duration of the N depletion phase.

II. CURRENT METHODS OF EVALUATING PLANT-AVAILABLE N

Equations [2] or [3] are used implicitly in virtually all areas of the United States to evaluate plant-available N (e.g., Engle et al., 1975; Gardner et al., 1980; New York State College of Agriculture and Life Sciences, Cornell University, 1978; Wagner et al., 1977; Weise & Penas, 1979; and Table 1). However, there is considerable variation among areas in the methods used to estimate various terms.

A. Crop N Requirement

The crop N requirement factor is essential to every soil N evaluation system and can be defined as the crop N uptake near maximum yield. Equation [2] requires an estimate of the total N uptake, i.e., grain N (N_{ch}) and stover plus root-soil immobilization (N_{imm}). Equation [3] requires an estimate of aboveground N uptake, i.e., grain N (N_{ch}) and stover N (N_{cr}). These terms are most practically expressed in terms of crop yield and can be estimated from the total plant N distribution. For example, consider that corn contains 25% of its total N in the root-soil complex, 25% in the stover, and 50% in the grain [grain contains about two-thirds of the total aboveground N (Hanway, 1962)]. For corn grain containing 15 kg N/t (0.70

Table 1—General summary of current soil N evaluation systems in various regions of the United States.

Region—states included	Soils data commonly solicited	N factor for crop yield goal		Soil N evaluation		Average N credits		
							Legumes	
		Crop	Avg value	Inorganic	Mineralization	Manure	Soybeans	Alfalfa
			kg N/1000 kg (pounds N/ bushel)			kg N/1000 kg (pounds N/ 2000 pounds)	kg N/ha	
Northeast Conn, NY, Pa	Soil type, drainage	Corn	24.5 (1.16)	None	Based on soil type in one state	2.3 (4.5)	NA†	150
Mid-Atlantic Del, Ky, Md, NJ, Va, WVa	Soil type, texture, drainage	Corn	24.9 (1.18)	None	None	3.0 (6.0)	18	80
Southeast Ga, NC, Miss, SC	Geographic area, soil type	Corn	26.4 (1.25)	None	None	NA	30	55
Midwest Ill, Ind, Iowa, Mich, Minn, Mo	Soil type, texture, drainage, geographic area	Corn	26.2 (1.24)	One state tests for NO_3^--N	Based on total N and soil texture in one state	2.4 (4.8)	32	100
West Colo, Kan, Neb, ND, SD	Soil type, drainage, geographic area	Corn Wheat	29.8 (1.41) 40.5 (2.10)	NO_3^--N	Two states use total N and fixed mineralization rates	2.4 (4.8)	36	90
Southwest Ariz, Okla, Tex	Soil association, topographic area, texture	Wheat	41.4 (2.15)	NO_3^--N	None	NA	NA	90
Northwest Mont, Ore, Wash, Idaho	Soil type, drainage, geographic area	Wheat	46.2 (2.40)	NO_3^--N	A region within one state uses aerobic mineralization	NA	NA	NA

† NA refers to inadequate data, or infrequently used in area.

pounds N/bushel) the whole crop N requirement (N_{ch} + N_{imm}) for Eq. [2] is 30 kg N/t (1.4 pounds N/bushel), while the (N_{ch} + N_{cr}) terms of Eq. [3] is 22 kg N/t (1.05 pounds N/bushel). Using the same plant N distribution for wheat grain containing 25 kg N/t (1.3 pounds N/bushel), the corresponding crop N requirements would be 50 kg N/t (2.6 pounds N/bushel) for (N_{ch} + N_{imm}) and 37 kg N/t (1.95 pounds N/bushel) for (N_{ch} + N_{cr}). Crop N requirements will also vary with cultivar or crop type as evidenced by the different N needs of soft white vs. hard red winter wheat (Engle et al., 1975; Gardner et al., 1980).

After estimating the crop N requirement, there remains the problem of estimating the yield goal. The objective is to predict the limiting yield of a specific soil-crop-climate system so that N rates do not greatly exceed crop assimilation capacity. The yield goal should not be the absolute maximum yield, because such yields require a stress-free environment; nor are they the best yield the farmer has ever produced, because such environmental conditions occur infrequently. The yield goal should be somewhat above the average yield of a specific soil-crop-climate system considering both yield limitations (drought hazard, past erosion, etc.) and yield potentials (crop cultivar, pest management, planting date, etc.). Some states have the farmer set the yield goal, others set yield goals by considering soil resources at the soil association level or the productivity class level, while others assign yield goals for each soil type (e.g., Engle et al., 1975; New York State College of Agriculture and Life Sciences, Cornell University, 1978; Soltanpour et al., 1979; Vitosh et al., 1974; Wagner et al., 1977). If an environmental factor limits yield, it should be used to help predict yield goals. For example, in the western United States, available water limits yield and yield goals can often be estimated from direct soil water measurements (Engle et al., 1975; Kresge & Halvorson, 1979; Ramig & Rhoades, 1962; Young et al., 1967) or from the likely rainfall of an area.

B. Inorganic Soil N

Evaluating available N through soil mineral N contents was considered to be of very limited value as recently as 1965 (Bremner, 1965b; Harmsen & VanSchreven, 1955), but over the past 15 yr, the importance of residual mineral N has been underscored by research in both semiarid and humid regions (Herron et al., 1971; Maples et al., 1977; Muller & Weigert, 1980; Nyborg et al., 1976; Olsen et al., 1970; Soper et al., 1971; Wehrmann & Scharpf, 1979). Soil NO_3^--N determinations are routinely made in areas where winter leaching is small, after summer fallow, and after heavy N inputs from fertilizer or manure. These studies will not be examined in detail because they have been discussed in recent reviews (Dahnke & Vasey, 1973; Keeney, 1982; Stanford, 1981). It is sufficient to state that the soil mineral N content has been conclusively shown to be an important component of available soil N in modern agriculture, particularly in semihumid areas; its usefulness in humid areas has also been demonstrated under certain conditions such as low winter rainfall, deep soils, and large previous N inputs.

1. ESTIMATING THE SOIL MINERAL N CONTENT

Both Eq. [2] and [3] contain soil mineral N terms (N_{sinb} and N_{sine} or N_{sin}) which require (i) an accurate and precise chemical analysis and (ii) the collection of a representative soil sample. The analytical aspects have been adequately discussed (Bremner, 1965a; Dahnke, 1971; North Central Region-13 Soil Testing Committee, 1975). They present no major problems, since NO_3^--N is easily extracted with various salt solutions and is readily analyzed with modern equipment (Best, 1976; Bremner, 1965a; Jackson et al., 1975), although quality control is a continuing concern. In some cases, both NO_3^--N plus NH_4^+-N are routinely analyzed; in others, they are analyzed only if NH_4^+ fertilizers have been recently applied, but in most cases, just NO_3^--N is determined (Gardner et al., 1980; Soltanpour et al., 1979; Wagner et al., 1977; Weise & Penas, 1979). The major problem with mineral N determinations is the collection of a representative soil sample.

Soil NO_3^- variability is multidimensional and involves spatial components in the horizontal and vertical directions, as well as temporal components. These sampling difficulties stem from (i) the solubility and mobility of NO_3^--N, (ii) the nonuniform application of N (e.g., anhydrous NH_3, manure), (iii) the natural spatial heterogeneity of soil organic matter and soil water, and (iv) the spatial and temporal heterogeneity of soil N transformations such as leaching, denitrification, and immobilization. Difficulties also arise from sample preparation and economic limitations of sample collection.

a. Sampling Depth and Spatial Variability.—The desired sampling depth is the crop root zone, but the recommended depth might be a fraction of this if predictions of the entire depth can be made from the shallower sample. Recommended sampling depths vary from relatively shallow samples of 30 cm (Nyborg et al., 1976; Onken & Sunderman, 1972), to deep samples of 120 to 180 cm (Soltanpour et al., 1979; Weise & Penas, 1979). The most commonly used depths vary between 60 and 120 cm (e.g., Engle et al., 1975; Neeteson & Smilde, 1983; North Central Region-13 Soil Testing Committee, 1975; Wagner et al., 1977; Whitney, 1976). This variability exists due to different:

1) Crops—shallow- vs. deep-rooted crops (Herron et al., 1971; Olsen et al., 1970);
2) Soil types—texture, hard pans, etc. (Maples et al., 1977; Olsen et al., 1970; Onken & Sunderman, 1972);
3) Prevailing soil conditions—depth of soil water, etc. (Onken & Sunderman, 1972; Young et al., 1967); and
4) Previous history—rainfall, previous crop, irrigation, etc. (Broadbent & Carlton, 1978; Herron et al., 1968; Smith, 1980; Vander Paauw, 1962).

Deep NO_3^--N samples will be most useful with extensively rooted, full-season crops [winter wheat, sugarbeets (*Beta vulgaris* L.), corn], grown on deep, well-drained soils under irrigation. Shallower samples will likely

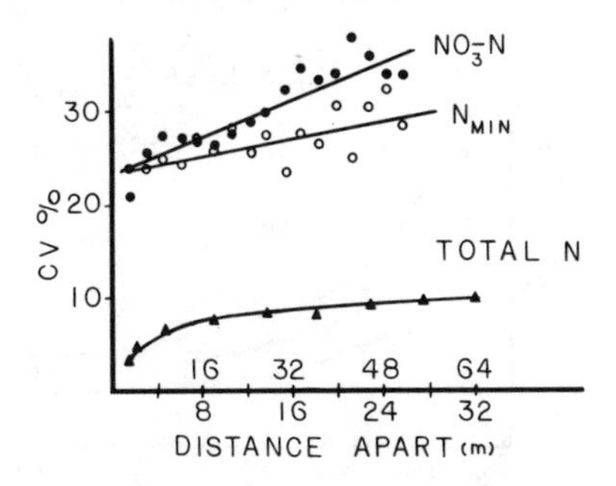

Fig. 4—Relationship of the CV to the distance between soil samples; symbols for NO_3^--N (●) and aerobic mineralization (○) relate to the lower distance scale (up to 32 m). Total N data (▲) relate to the upper distance scale (up to 64 m). Data summarized from Waynick (1918) and Waynick and Sharp (1919).

suffice for shallow-rooted, short-season crops [potato (*Solanum tuberosum* L.), other vegetable crops] grown on soils with restricted root volumes (fragipan, plow pan, etc.). Local experimentation over several years is required to determine the minimum sampling depth for a given soil-crop-climate condition.

Soil scientists are well aware of the fact that soil is a heterogeneous body, yet we have not adequately studied the spatial variability of NO_3^--N, despite early recognition that such variability was large (Prince, 1923; Waksman, 1923; Waynick, 1918). The spatial variability of NO_3^--N is characterized by a slowly increasing coefficient of variation (CV), as the area under study increases and a large small-scale component that is not uniformly dispersed over the field. It is common to find that over 50% of the variability present within 1 ha is already present in a few square meters (Beckett & Webster, 1971; Cameron et al., 1971; Waynick, 1918; and Fig. 4). Due to these characteristics, a field CV will often bear little relation to field size.

The magnitude of the field NO_3^--N CV's is usually large, which relates directly to the problem of determining sample size. In a Colorado study, the 0- to 60-cm NO_3^--N CV's averaged 52% over 12 sites (Reuss et al., 1977, and Dr. P. N. Soltanpour, Colorado State Univ., personal communication). Nitrate N CV's for three central Alberta fields were 40, 72, and 47% (Cameron et al., 1971), while CV's from 0.16-ha areas of 15 different southern Alberta fields had a median of 33% (Bole & Pittman, 1976). A median CV of 57% was reported from 10 fields sampled by Hunt et al. (1979), but they ranged from 41 to 109%. The mean CV value among all these field studies is about 45%. If we assume that individual cores from a given area will be composited and that the NO_3^--N content of the composited sample is approximately normally distributed (i.e., the central limit theorm holds), then one can use the equation of Harris et al. (1948) to estimate the number of samples required for the length of the 80% confidence interval about the mean to be within a given percentage of the mean (see Fig. 5). For example, curve B of Fig. 5 shows that by compositing 22 cores, the length of the 80%

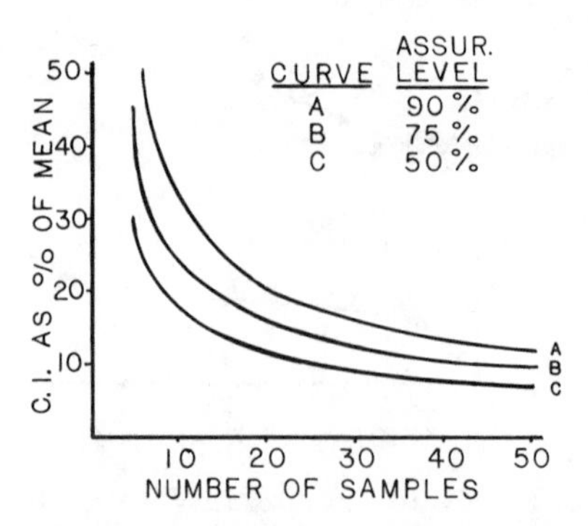

Fig. 5—Graphical relation between the 80% confidence interval (C.I.) of the mean, expressed as a percentage of the mean, and the sample number as calculated from Harris et al. (1948). Curves *A, B,* and *C* refer to assurance levels of 90, 75, and 50%, respectively. See text for further explanation.

confidence interval about the sample mean will be within 15% of the mean and this degree of precision should be realized on about 75% of the areas samples. A 45-core composite would be required for the 80% confidence interval to be within 10% of the mean on 75% of the areas sampled. Large numbers of cores are thus needed to accurately and reliably estimate NO_3^--N means. Common soil sampling instructions call for a 10 to 20 core composite from an area that is selected to reflect similar past management, topography, soil type, etc. (Bole & Pittman, 1976; Cameron et al., 1971; Reuss et al., 1977; Smith, 1980). This sampling intensity would estimate the mean to approximately ± 20% in about 75% of the areas, or to ± 25% of the mean in about 90% of the cases (curve A of Fig. 5). Although the upper limit of sample number is also affected by economic considerations (cost of sampling, equipment costs, time, etc.), it is quite unlikely that a field NO_3^--N mean can be estimated to better than ± 20% of the mean without an extensive sampling program. This spatial variability problem is lessened if the soil NO_3^--N test is used to classify the area into a low, medium, or high category as long as the class interval is large in relation to the sampling variability. Of course one will always have problems assigning classes to areas which fall near a class boundary.

Current NO_3^--N sampling procedures usually composite cores from large tracts (ca., 5 ha), which masks underlying high and low areas within that tract. The spatial pattern of NO_3^--N can be just as important as the mean value of an area, especially for crops that suffer yield or quality reductions at high N rates [e.g., malting barley (*Hordeum vulgare* L.) (Geist et al., 1970), sugarbeets (Carter et al., 1976), and cotton (*Gossypium hirsutum* L.) (Maples et al., 1977)]. Reuss et al. (1977) reported large NO_3^--N variations within irrigated fields and estimated that, in the average field, 7% of the area would have NO_3^--N levels below 67% of the geometric mean, while 9% of the area would exceed 150% of geometric mean.

The above sampling problems are a relatively new aspect of N management. Unfortunately, our current sampling procedures were developed in an era when fields were more uniformly low in N or were adapted from P and K sampling plans. Consequently, there is a need to expand our field sampling research to develop efficient schemes to accurately and precisely estimate the mean NO_3^--N content, together with some estimate of the spatial variation of the area.

b. Temporal Variability and Sample Preparation.—It is generally recognized that NO_3^--N samples should be collected annually and that the time between sampling and crop N uptake should be as short as practical. Thus, samples are usually collected in the spring of the crop year. If sampling is done the preceeding fall, it is also necessary to estimate the subsequent change in mineral N due to the opposing processes of ammonification and nitrification vs. leaching plus denitrification. In dry areas, these overwinter losses are not a major factor and an estimate of overwinter mineralization can be added in directly (Gardner et al., 1980; Nyborg et al., 1976). In more humid areas it is necessary to adjust for overwinter losses through appropriate climatic and soil variables such as rainfall, water percolation, soil depth, drainage, etc. (Muller & Weigert, 1980; Vander Paauw, 1962, 1963). The adjustments for sample time should be determined through local experimentation over several years on representative soil types.

Sample preparation has been investigated on a number of occasions, but different investigators have reported conflicting results with various procedures (Bremner, 1965a; Keogh & Maples, 1980; Selmer-Olsen et al., 1971; Storrier, 1966; Westfall et al., 1978). Suggested procedures usually include freezing or field extraction (Bremner, 1965a; Ross et al., 1979; Selmer-Olsen et al., 1971), but these are often impractical on a commercial scale. Significant increases in NO_3^--N content have been reported when moist soils are held at ambient temperatures for prolonged periods (Ross et al., 1979; Selmer-Olsen et al., 1971). One investigation reported large increases after 1 or 2 d storage at room temperature (Westfall et al., 1978), although others have not observed large changes, except when the sample had been contaminated with NH_4^+ fertilizer (Keogh & Maples, 1980). The generally recommended method is to rapidly air-dry the soil right after collection to promptly arrest biological activity. If samples cannot be dried within several hours after collection, they should be frozen and dried as soon as practical.

2. ESTIMATING THE EFFICIENCY OF MINERAL N UPTAKE

The remaining factor of the mineral N component involves the ratio of the mineral N efficiency to the fertilizer N efficiency, i.e., the e_s/e_f term of Eq. [3] and the E_{sinb}/E_f term of Eq. [2a]. This latter term does not appear directly in Eq. [2], because in deriving this equation, it was assumed that both N_{sinb} and N_f suffered the same P_l and P_g losses so that $E_{sinb}/E_f = 1$. Considerable field research has shown that this efficiency ratio approximately equals one if N losses are small between sampling and crop N uptake. This situation is also implicit in many N recommendation systems that simply subtract N_{sinb} from the crop N requirement (Engle et al., 1975; Gardner et al., 1980; Wagner et al., 1977).

Many circumstances, however, could result in apparent or real differences in these efficiency terms. For example, Haby et al. (1983) concluded that soil NO_3^--N to 1.2 m was less efficient than N_f in increasing winter wheat yields. Crops probably experience a somewhat different N_{sin} pool than is actually estimated before the growing season. These changes in pool size may be due to leaching losses, mineralization additions, or to incomplete sampling of the root zone; but the overall adjustment for such changes may occur in the efficiency terms. The efficiency terms of Eq. [3] are particularly susceptible to these types of changes because all soil N transformations are outside the system. Other conditions may result in true differences in these efficiencies, such as smaller gaseous losses of deep NO_3^--N (low available C), varying root densities or varying soil water contents down the soil profile, and greater NH_3 losses from surface-applied NH_4^+ fertilizers.

3. INDIRECT MINERAL N EVALUATIONS

It is noteworthy that one state (Vitosh et al., 1974) uses an indirect mineral N method that allows a NO_3^--N credit only after adverse years (drought, disease, etc.). The procedure assumes that 50% of the difference between the previous year's N_f and N_{ch} is carried over on medium- and fine-textured soils. Such indirect systems are useful when NO_3^--N carry-over is an occasional event associated with easily documented conditions.

C. Soil N Mineralization

Mineralization has been known to be a significant source of N for many years. In fact, a 1908 study evaluated a biological incubation procedure and concluded that "the production of active N (NO_3^--N plus NH_4^+-N) in soil can probably be developed into a method for the determination of the needs of the soil for nitrogenous fertilizer" (Fraps, 1908). In the intervening 75 yr, hundreds of scientific papers and procedures have investigated methods to estimate mineralization. It is not necessary to examine these studies in detail, because they have been reviewed on several occasions (Bremner, 1965b; Dahnke & Vasey, 1973; Harmsen & VanSchreven, 1955), most recently by Keeney (1982) and Stanford (1981).

1. ESTIMATING SOIL N MINERALIZATION

Procedures to estimate the N_{min} term of Eq. [2] and [3] can be classified into crop vegetative tests, microbial tests, total analysis procedures, chemical extraction procedures, and indirect estimates.

a. Vegetative Procedures.—Vegetative tests include both field and greenhouse procedures that measure total dry matter, yield, N concentration, or preferably total N uptake. Greenhouse procedures have limited application to field conditions, but serve a useful purpose in the preliminary screening of more empirical chemical procedures. The NO_3^--N or total N concentrations of a plant or plant part have often been used as indicators of plant N stress (Pierre et al., 1977; Stanford, 1973; and Chapt. 16, this book), especially in N-sensitive crops like sugarbeets (Carter et al., 1976; Giles et al., 1979) and cotton (Maples et al., 1977). These nonquantitative measures supply useful supplementary data, which should be utilized in a long-term N recommendation system, but they require extensive calibration to overcome sampling, environmental (water stress), crop maturity, and crop variety effects.

Determining the N uptake of a field crop receiving no N fertilizer is the most satisfactory method of estimating the soil N supply in a given soil-crop-climate system, because it integrates the factors of crop growth and soil N dynamics under natural conditions. Nitrogen mineralization and residual mineral N will be the major N sources on zero N_f plots. Mineralization can be estimated by subtracting an estimate of the change in soil mineral N or by conducting the evaluation for at least 2 yr, because most crops exhaust the residual NO_3^--N during the first year (e.g., Broadbent & Carlton, 1978). This procedure has been generally accepted as the standard by which other more empirical methods are evaluated. It is, therefore, an indispensible part of soil N calibration studies. The most frequent modification is to measure grain yield instead of N uptake. Grain yields, however, are strongly influenced by weather during anthesis, while N uptake is a cumulative parameter integrated over the whole growing season. Moreover, the conversion of grain yield to N uptake requires an estimate of the grain percent N and the partitioning of the total plant N into the grain. These latter factors are difficult to estimate and it is thus advisable to measure N uptake directly whenever possible.

The major disadvantage of vegetative tests are their large requirements for time and labor. Since they integrate field conditions over time, they are also quite site specific, and therefore, are not easily transferred to other environmental conditions.

b. Microbial Procedures.—Microbial tests incubate a soil sample under temperature and water conditions conducive to mineralization and measure the total mineral N produced after a given time period. Many variations of this basic procedure exist, which employ various inert bulking materials, incubation times, temperatures, removal or nonremoval of initial mineral N, aeration conditions, etc. Despite the large number of specific procedures, it is generally recognized that biological incubations that measure NO_3^--N plus NH_4^+-N production provide a sound *relative* measure of the soils mineralization ability (Bremner, 1965b; Harmsen & VanSchreven, 1955; Keeney, 1982; Stanford, 1981). This is because the soil organic N is broken down by the same biological agents that are active under field conditions. Correlations between biological procedures and greenhouse results are usually good, but when testing progresses to field conditions, the correlations are usually considerably lower (Bremner, 1965b; Fox & Piekielek, 1985; Gasser & Kalembasa, 1976; Jenkinson, 1968; Keeney, 1982; Stanford, 1981). This is especially true if the field trial covers a range of climates, soil types, or years. Microbial tests are markedly affected by sample pretreatment, especially those involving short-term aerobic incubation (Dahnke & Vasey, 1973; Harmsen & VanSchreven, 1955; Keeney, 1982; Keeney & Bremner, 1966). Mineralization values generally increase with air-drying and storage, but current information does not permit any definite guidelines for proper techniques to prepare and store samples. In order to obtain meaningful comparisons among soils, it is necessary to standardize sample pretreatment and incubation conditions.

Recent investigations have shown that it may be possible to extend laboratory mineralization data to field conditions, at least within a limited climatic area (Smith et al., 1977; Stanford et al., 1977). This approach summarizes long-term aerobic mineralization data through statistical parameters designated as the mineralization potential (N_o) and an associated rate constant. The rate constant is adjusted for temperature based on a doubling of the rate for every 10°C rise in temperature (Stanford et al., 1973). An adjustment for water is made by a direct percentage correction based on the actual water content expressed as a fraction of the optimum water content, i.e., mineralization will be 50% at a water content of 50% of optimum (Stanford & Epstein, 1974). These adjustments offer the prospect of transferring mineralization data across temperature and water regimes. This approach is less empirical, but requires long-term mineralization data and is tied to average climatic conditions for predictive purposes. It is being further evaluated as a research approach, but a simplified version may become practical in future years.

The spatial variability of the microbial procedures has not been adequately documented, but the scant information available indicates that it is quite large and may approach that of NO_3^--N (see Fig. 4). Field sampling is, therefore, an important unresolved concern.

The principle drawbacks to biological procedures are the sample handling and pretreatment effects, and the long incubation times (at least 1-week incubations). The currently recommended biological index is the NH_4^+-N produced after 1 week of anaerobic incubation at 40°C (Keeney, 1982). Biological indexes were used in several statewide N recommendations systems 25 yr ago, but are currently used in only one state where aerobic mineralization is applied to a relatively narrow climatic region of irrigated agriculture (Keeney, 1982; Dr. R. E. McDole, Univ. of Idaho, personal communications).

c. Total Analysis Procedures.—Total analysis procedures usually estimate total N directly by a Kjeldahl analysis, or indirectly through a Walkley-Black organic matter determination. These procedures are not greatly influenced by sample pretreatment, sample preparation, or time of sampling. Topsoil CV's for total N are frequently near 15% for plot areas (ca., 0.01 ha), but usually increase slowly to about 30% for field areas (ca., 20 ha). This variability is common for areas including a single soil mapping unit and may be much larger for areas with more than one mapping unit (Beckett & Webster, 1971; Keogh & Maples, 1967). The CV's for total N are usually considerably less than those of NO_3^--N or microbial procedures (Prince, 1923; Waynick, 1918; Waynick & Sharp, 1919; and Fig. 4). Ten states currently use a total analysis in their N recommendation systems (Brown et al., 1980; Keeney, 1982; Soltanpour et al., 1979; Vitosh et al., 1974; and Table 1). Mineralization is typically estimated by taking a fixed percentage of the total N, amounting to a 1 to 3% annual rate, which may vary with soil texture (Brown et al., 1980). Some states use organic N indirectly as a general indicator of soil fertility, which is used to predict yield goals rather than mineralization. Researchers often consider total N procedures to be a poor indicator of mineralization (Harmsen & VanSchreven, 1955; Stanford, 1981), because they measure a slowly available N component that is not sensitive to changes in the small active N pool. Others consider total N to be directly related to mineralization, especially for soils originating within a limited area that contain low amounts of N (Klemmedson & Jenny, 1966; Robinson, 1968a). However, the current use of total N in 10 states is an indication that it can be used to provide a useful, albeit limited, estimate of mineralization.

d. Chemical Extraction Procedures.—Separating all or a part of the active N pool is the goal of procedures based on chemical extraction. Proposed extracting agents vary greatly in intensity, ranging from strong acids/bases to neutral salts or water. The extracted material has also been analyzed in a number of ways including C content, total N, distillable NH_3-N, and ultraviolet adsorption. The general conclusion from recent research (Jenkinson, 1968; Kenney & Bremner, 1966; Livens, 1959b) and reviews (Bremner, 1965b; Dahnke & Vasey, 1973; Keeney, 1982; Stanford, 1981) is that intensive extracts (e.g., boiling soils with strong acids/bases) remove large amounts of soil N and are usually highly related to total N; while extracts of mild intensity [e.g., boiling water or salt solution (0.01*M* $CaCl_2$)] remove an active soil N component [see Keeney (1982) and Stanford (1981) for recent, detailed reviews]. Current research has, therefore, centered on mild extractants with subsequent determination of total N or distillable NH_3-N (Kenney & Bremner, 1966; Smith & Stanford, 1971). The basic method was proposed 20 yr ago by Livens (1959a, b) and later modified by Keeney and Bremner (1966) and Smith and Stanford (1971). The currently recommended method consists of an overnight (16 h) autoclaving in 0.01*M* $CaCl_2$ and subsequent determination of the NH_4^+-N released (Keeney, 1982). This method is not greatly affected by sample pretreatment, is rapid and ammenable to automated analysis, and is highly correlated to N availability as measured in greenhouse evaluations (Gasser & Kalembasa, 1976; Jenkinson, 1968; Keeney, 1982; Kenney & Bremner, 1966). Field results to date contain both encouraging (Fox & Piekielek, 1978b; Roberts et al., 1972; Robinson, 1968b) and discouraging results (Fox & Piekielek, 1985), but the method has not been conclusively evaluated under a series of field calibration programs. There is virtually no information regarding the spatial variability of chemical extraction procedures. No state or region is routinely employing a chemical extraction procedure to estimate N mineralization. Thus, despite decades of research and development, there has been virtually no acceptance of these procedures into routine soil N evaluation programs.

e. Indirect Procedures.—Indirect measures of mineralization are used in virtually all states and usually involve documenting a local variable that directly affects mineralization or an easily measured indicator variable that is linked to a change in mineralization. Examples of the former are crop histories that give a legume N credit of 50 to 150 kg N/ha for previous alfalfa (*Medicago sativa* L.) crops (depending on the stand, years since alfalfa, etc.) or 15 to 50 kg N/ha for previous soybean [*Glycine max* (L.) Merr.] crops (Table 1). These historical adjustments attempt to adjust for oscillating levels of soil N in rotation systems (see Fig. 3C). Other indirect adjustments account for manure additions, above average residue returns, or other organic inputs that are inadequately accounted for in a direct soil N test (see Fig. 3D and Table 1). An example of an associated variable used to estimate mineralization is the use of rainfall in Washington; here, N_{min} is directly proportioned to average annual rainfall (Engle et al., 1975). The rainfall-mineralization associated is related to greater total N contents with greater rainfall and to greater microbial activity in moist soils under semiarid conditions. Another indirect approach assigns a N_{min} value based on a site category (e.g., soil drainage) and then adjusts for legume history, manure, etc.

Indirect estimates are easy to integrate into N recommendation systems, but they must be supported by a strong field calibration program. They are a first step toward adjusting N_f needs to specific conditions of a given site.

2. ESTIMATING THE UPTAKE EFFICIENCY OF N MINERALIZATION

The efficiency factor for N_{min} (E_{min} or e_m) has not been determined directly. This is because the estimates of N_{min} are empirical and in the field calibration process, the product of E_{min} and N_{min} are calibrated concurrently. This is most evident when field calibrations indicate a different mineralization relation for different soil drainage classes. Although the value of E_{min} is not known, it should be relatively high (i.e., > 0.8) whenever crops are actively growing, since conditions that favor mineralization (warm, moist soils) also favor crop growth. Without an actively growing crop, E_{min} should be markedly lower, as occurs during the fall when soils are warm and moist and annual crops have matured.

D. Fertilizer Efficiency

Fertilizer efficiencies are functions of several interacting N transformation processes such as leaching, denitrification, and NH_3 volatilization, as well as several management variables such as N source, placement, and timing. It has not been possible to predict them from first principles. Therefore, soil scientists have approached this problem empirically by estimating fertilizer efficiencies over a range of soil and climatic conditions. The end result is that one can usually estimate a general value of E_f (ca., 75%) or e_f (ca., 50%), but one cannot accurately predict this parameter for a specific soil-crop system.

One of the main elements influencing fertilizer efficiency is soil water content, because water directly affects denitrification and leaching losses. In drier areas, water dominates fertilizer efficiency by determining the crop N uptake capacity. Therefore, some measure of soil water content or soil drainage should definitely be considered when efficiency terms are forecast. It should also be noted that E_f and e_f are functions of N_f, since fertilizer efficiency decreases as N rates approach maximum yields. Obviously, further research is needed to investigate the factors affecting fertilizer efficiency so that better predictions can be made for specific soil-crop systems.

E. Overview

Nitrogen is a mobile and dynamic nutrient; its fate is governed by several competing physical and biological factors that interact with each other and the environment over time. The terms of Eq. [2] and [3] are functions of uncontrolled variables (rainfall, temperature, etc.), functions of site-specific variables such as soil properties (available water, other nutrients, etc.), and functions of controlled management variables such as crop cultivar, fertilizer source, and application method. Our current inability to forecast long-range weather places an upper limit on the accurate prediction of several of these terms such as N_{ch} and the efficiency terms. However, we are able to predict many of these terms in a probabilistic manner, based on current conditions or long-term averages (e.g., probable yields as a function of stored soil water or anticipated rainfall). Some terms can be estimated from field research experiments over several years (e.g., the efficiencies of N sources, application methods, etc.). Other terms have to be estimated in a preliminary manner based on literature results and experience (e.g., the P_l and P_g components of the E term for new tillage systems). Therefore, our current inability to quantitatively predict all the terms of Eq. [2] or [3] should not detract from their usefulness with the best available data, or from their usefulness as a conceptual framework to evaluate and design field experiments to improve soil N evaluation systems.

Combining the above factors into a working N recommendation system requires an important final step—field calibration. Calibration studies are an art and a science in their own right, but one that has received too little support over the years. Hanway (1973) has emphasized the need for a comprehensive approach. Since N is a mobile and dynamic nutrient, it is the one element where a comprehensive approach is not only suggested—it is required. Nitrogen calibration studies should (i) document the crop response in terms of economic yield and N recovery, (ii) identify and measure the important site variables and the management variables that affect the crop response, (iii) measure or empirically estimate the N components of Eq. [2] or [3] such as N_{sin} and N_{min}, (iv) be conducted over a range of controlled and uncontrolled variables so a representative sampling of the soil-crop-climate system is obtained, (v) determine the magnitude and variability of the N components among sites and their controlling or associated variables, and (vi) combine the above results into a coherent framework for N recommendations. The serious need for current calibration data is shown by the results of a recent poll of state soil fertility extension specialists; the majority indicated that the "weakest link" in their current N recommendation systems was up-to-date calibration studies over a range of conditions.

Calibration studies must identify the important components of Eq. [2] or [3] that result in different N_f needs. If soils vary significantly in N_{min} and N_{sin}, then both factors need to be included in the final N fertilizer expression. Some terms may be relatively small and can be dropped; terms that do not vary greatly among sites might be combined. The latter case occurs frequently when N_{min} does not vary widely among soils (e.g., Stumpe, 1977) and is combined with the crop N requirement by expressing it in terms of yield goal. For example, if the N requirement of corn is 22 kg N/t (1.05 pounds N/bushel, see section IIA) and the soil supplies about 9.5 kg N/t (0.45 pounds N/bushel) with an e_f equal to 0.50, then by combining these terms of Eq. [3] one obtains a crop N factor of 25 kg N/t (1.2 pounds N/bushel), which compares well with the values listed in Table 1. The crop N requirement term is often subtly combined with the N_{min} term during field testing when the "crop N requirement" is calculated from the quantity of N_f needed to produce the actual yield observed at a given site without assessing N_{min} directly.

Calibration studies should also determine the within-field variability for the components of Eq. [2] or [3] in order to devise efficient sampling schemes. These data are essential in determining the cost effectiveness of various N recommendation systems and are an important input when new systems are compared or when old systems are updated.

III. FUTURE TRENDS AND IMPROVED METHODS OF EVALUATING PLANT-AVAILABLE N

The choice of the most cost-effective N recommendation system will depend heavily on economic conditions, particularly on the cost of N. Current systems were developed in an era of inexpensive energy and, consequently, inexpensive N. Present and future needs will no longer reflect these conditions. Cost-effective N recommendation systems for expensive N will require a detailed evaluation of each site's N needs and a more complete evaluation of the soil N status.

A. Extending and Updating Proven Techniques

Extending and updating current techniques is the most straight forward method to improve N evaluation systems. For states already using a soil NO_3^--N test, large advances can be made with improved field sampling strategies that estimate the NO_3^--N content more accurately, as well as the within-field variability (see section IIB and Fig. 5). The NO_3^--N test could also be extended from the western United States to the states adjacent to the Mississippi River and possibly into the humid East. Such an extension would involve the current problems of sample intensity, sample depth, etc., as well as estimating the likelihood of significant mineral N carry-over as influenced by fall NO_3^--N level, precipitation (amount, intensity, time of year, etc.), and soil characteristics (drainage, rooting depth, mineralization, nitrification) (Addiscott, 1982; Muller & Weigert, 1980). Using NO_3^--N tests in humid areas will thus require more detailed site-specific inputs (climate, soil type, etc.).

Other areas that need to be updated are the legume and manure N credits. Although a N mineralization test could potentially adjust for these N inputs, the past performance of such tests indicates that an indirect method, such as site history, will be more readily integrated into current systems [e.g., Pennsylvania's recent updating (Northeastern Regional Comm.-39, Origin, Transformation, and Management of Nitrogen in Soils, Water, and Plants, 1980; Fox & Piekielek, 1984). Revising these credits should improve the accuracy of the first year affect, and more importantly, improve the residual N credit beyond one year, which is especially important for manures. More work will also be needed to determine the N credits for legumes grown as cover crops in reduced-tillage systems and as intercrops in conventional-tillage systems. Improving the N credits from legumes and manures will allow farmers to better utilize these alternate N sources in an integrated N management system (i.e., a system that utilizes expensive fertilizer N to supplement the N contributed from biological fixation and recycling of wastes).

Assessing the N requirements of new varieties is also important. The percent N in corn grain and the translocation or partitioning of N into the grain at maximum yield is greatly influenced by plant genotype (Beauchamp et al., 1976; Russell & Pierre, 1980; Tsai et al., 1984). The N_{ch} term of Eq. [2] and [3] may, therefore, change with varieties, but there is insufficient data to assess the practical significance of this point. New varieties could have a greater whole crop N requirement (a greater $N_{ch} + N_{imm}$ term) or they may simply partition more N into the grain. Updating the crop N requirement should thus consider both the harvested N and the total N required to produce the crop.

B. Improving Techniques and New Approaches

Improvements in current N evaluation systems will most likely occur by putting more of the pieces of the puzzle together, rather than through large single discoveries. For example, several investigators have significantly improved N fertilizer forecasts by using both a mineral N term and a mineralizable N term, i.e., using a more complete form of Eq. [2] or [3] (Carter et al., 1976; Geist et al., 1970; Neeteson & Smilde, 1983; Stanford et al., 1977). Previous discussion has revealed several aspects of Eq. [2] and [3], which need further investigation. Equation [2] needs evaluation of (i) the quantity of N required to support the root-soil complex, (ii) the environment and management factors determining the soil total N content at steady-state, (iii) the proximity of our soil-crop systems to steady-state conditions, (iv) the total N decomposition or accumulation rate for non-steady-state systems, and (v) the total N_f recovery within the entire soil-crop system. Equation [3] needs evaluation in the areas of (i) the best method to estimate crop-available N from mineralization, (ii) the efficiency terms for residual mineral N at various soil depths, and (iii) the prediction of e_f for specific soil-crop systems as influenced by the interacting processes of leaching, denitrification, and immobilization, as well as management factors such as N source, placement, and timing. Evaluating these areas will require experiments on research stations to study controlled management factors and experiments in farmers' fields, which will give a wider sampling of soils and climates. An extensive, large-scale experimental program such as this will require a comprehensive approach (see section IIE) in order to document and interpret the results.

A new soil N test would make the greatest contribution in the area of a quick method to estimate mineralization. Soil scientists have recognized this for over 75 yr (see section IIC), yet have been unable to develop a satisfactory method. We must, therefore, recognize that it is unlikely that such a quick method will be developed in the near future. We should not abandon such efforts, but we should realize that a rapid breakthrough is unlikely. Perhaps it is time to modify the search for a *single* chemical method, implied by the use of Eq. [3], and turn to a combination approach that utilizes indirect estimates, such as legume and manure histories, plus a total analysis and a chemical extraction method as implied by Eq. [2].

Detailed site information will be needed in order to make more accurate N fertilizer recommendations. This information is difficult to obtain, but there is a valuable source of detailed information that should be fully exploited in the future—soil type. Knowledge of soil type gives a deeper insight into the field soil N dynamics that directly affect several important terms of Eq. [2] or [3] (e.g., e_f as affected by drainage, texture, rooting depth, etc.). Unfortunately, there is a paucity of data that evaluates soil type as related to N evaluation systems. The limited data suggest that soil series information is important in N evaluation systems (Keogh & Maples, 1967; Nelson & McCracken, 1962), but one should not expect soil type to be a universal integrating factor. Rather, soil series data should be viewed as a readily available source of information that is relevant to the problem of assessing the potential movement and likely N transformations under field conditions. If Eq. [2] or [3] were evaluated for each soil type, it would allow yield goals, efficiency factors, and other terms to vary from soil to soil. It would also allow the N_{sin} and N_{min} estimates to become more relevant and more interpretable, because they would be placed within a narrower framework of soil N dynamics. By knowing the soil type, Eq. [2] or [3] could be extended down to an individual field level that would produce a very site-specific N recommendation system.

The recent development of the affordable personal computer will directly impact N fertilizer evaluation systems. These devices will give farmers, consultants, and extension personnel a means to monitor, store, and process large volumes of data from specific locations. The personal computer could allow development of field scale N budget-recommendation models for a specific site (e.g., Greenwood, 1982; Kresge & Halvorson, 1979). For example, a management level N budget-recommendation model could be applied to a given field by developing routines for estimating (i) crop N requirement as a function of soil water, plant population, and planting date; (ii) mineralization as a function of soil temperature, water, manure additions, and hot water-extractable N; (iii) residual inorganic N as a function of soil mineral N content, prior N_f, prior crop yield, overwinter rainfall, and rooting depth; and (iv) N use efficiency as a function of N fertilizer source, placement, and timing as well as the likelihood of leaching or denitrification losses during the year for that particular soil type. The major problem facing such a "systems approach" is our ability to collect, summarize, and interpret field data that describe soil N cycle processes over a range of soil-crop-climate conditions. Our soil N evaluation systems must therefore be undergirded by a comprehensive field calibration program that is able to document and interpret site-specific data that affect basic soil N transformations. It will be impossible to develop a site-specific N fertilizer evaluation system without a continuing committment to comprehensive field calibration.

REFERENCES

Addiscott, T. M. 1982. Computer assessment of the N status during winter and early spring of soils growing winter wheat. p. 15-26. *In* Assessment of the N status of soils. Research Workshop, Leuven, Belgium. 12-14 Jan. Katholieke Univ., Leuven, Belgium.

Beauchamp, E. G., L. W. Kannenberg, and R. B. Hunter. 1976. Nitrogen accumulation and translocation in corn genotypes following silking. Agron. J. 68:418-422.

Beckett, P. H. T., and R. Webster. 1971. Soil variability: a review. Soils Fert. 34:1-15.

Best, E. K. 1976. An automated method for determining nitrate-nitrogen in soil extracts. Queensl. J. Agric. Anim. Sci. 33:161-166.

Bole, J. B., and U. J. Pittman. 1976. Sampling southern Alberta soils for N and P soil testing. Can. J. Soil Sci. 56:531-535.

Bremner, J. M. 1965a. Inorganic forms of nitrogen. *In* C. A. Black et al. (ed.) Methods of soil analysis. Part 2. Agronomy 9:1179-1237.

Bremner, J. M. 1965b. Nitrogen availability indexes. *In* C. A. Black et al. (ed.) Methods of soil analysis. Part 2. Agronomy 9:1324-1345.

Broadbent, F. E., and A. B. Carlton. 1978. Field trials with isotopically labeled nitrogen fertilizer. p. 1-41. *In* D. R. Nielsen and J. G. MacDonald (ed.) Nitrogen in the environment. Vol. I. Nitrogen behavior in field soil. Academic Press, New York.

Brown, J. R., R. G. Hanson, and D. D. Buchholz. 1980. Interpretation of Missouri soil test results. Agronomy Dep., Univ. of Missouri Misc. Pub. no. 80-04.

Cameron, D. R., M. Nyborg, J. A. Toogood, and D. H. Laverty. 1971. Accuracy of field sampling for soil tests. Can. J. Soil Sci. 51:165-175.

Carter, J. N., D. T. Westermann, and M. E. Jensen. 1976. Sugarbeet yield and quality as affected by nitrogen level. Agron. J. 68:49-55.

Dahnke, W. 1971. Use of the nitrate specific ion electrode in soil testing. Commun. Soil Sci. Plant Anal. 2:73-84.

Dahnke, W. C., and E. H. Vasey. 1973. Testing soils for nitrogen. p. 97-114. *In* L. M. Walsh and J. D. Beaton (ed.) Soil testing and plant analysis. Soil Sci. Soc. of Am., Madison, Wis.

Engle, C. F., F. E. Koehler, K. J. Morrison, and A. R. Halvorson. 1975. Fertilizer guide for dryland wheat nitrogen needs. Washington State Extension Service Reprint no. FG-34.

Fox, R. H., and W. P. Piekielek. 1978a. A rapid method for estimating the nitrogen-supplying capability of soil. Soil Sci. Soc. Am. J. 42:751–753.

Fox, R. H., and W. P. Piekielek. 1978b. Field testing of several nitrogen availability indexes. Soil Sci. Soc. Am. J. 42:747–750.

Fox, R. H., and W. P. Piekielek. 1984. Relationships among anaerobically mineralized nitrogen, chemical indexes, and nitrogen availability to corn. Soil Sci. Soc. Am. J. 48:1087–1090.

Fraps, G. S. 1908. The production of active nitrogen in soil. Texas Agric. Exp. Stn. Bull. no. 106.

Gardner, H., N. R. Goetzen, and P. E. Rasmussen. 1980. Oregon State University fertilizer guide for winter wheat (non-irrigated-Columbia Plateau). Oregon State Extension Service no. FG54, Corvallis, Oreg.

Gasser, J. K. R., and S. J. Kalembasa. 1976. Soil nitrogen. IX. The effect of leys and organic manures on the available-N in clay and sandy soils. J. Soil Sci. 27:237–249.

Geist, J. M., J. O. Reuss, and D. D. Johnson. 1970. Prediction of nitrogen fertilizer requirements of field crops: II. Application of theoretical models to malting barley. Agron. J. 62:385–389.

Giles, J. R., A. E. Ludwick, and J. O. Reuss. 1979. Prediction of late season nitrate-nitrogen content of sugarbeet petioles. Agron. J. 69:85–88.

Greenland, D. J. 1971. Changes in the nitrogen status and physical condition of soils under pastures, with special reference to the maintenance of the fertility of Australian soils used for growing wheat. Soils Fert. 34:237–251.

Greenwood, D. J. 1982. Models for predicting N fertilizer requirements of vegetable crops. p. 27–31. *In* Assessment of the N status of soils. Research Workshop, Leuven, Belgium. 12–14 Jan. Katholieke Univ., Leuven, Belgium.

Haby, V. A., C. Simons, M. S. Staber, R. Lund, and P. O. Kresge. 1983. Relative efficiency of applied N and soil nitrate for winter wheat production. Agron. J. 75:49–52.

Hanway, J. J. 1962. Corn growth and composition in relation to soil fertility: II. Uptake of N, P, and K and their distribution in different plant parts during the growing season. Agron. J. 54:217–222.

Hanway, J. J. 1973. Experimental methods for correlating and calibrating soil tests. p. 55–66. *In* L. M. Walsh and J. D. Beaton (ed.) Soil testing and plant analysis. Soil Sci. Soc. of Am., Madison, Wis.

Harmsen, G. W., and D. A. VanSchreven. 1955. Mineralization of organic nitrogen in soil. Adv. Agron. 7:299–398.

Harris, M., D. G. Horvitz, and A. M. Mood. 1948. On the determination of sample sizes in designing experiments. J. Am. Stat. Assoc. 43:391–402.

Herron, G. M., A. F. Dreier, A. D. Flowerday, W. L. Colville, and R. A. Olson. 1971. Residual mineral N accumulation in soil and its utilization by irrigated corn (*Zea mays* L.) Agron. J. 63:322–327.

Herron, G. M., G. L. Terman, A. F. Dreier, and R. A. Olson. 1968. Residual nitrate nitrogen in fertilized deep loess-derived soils. Agron. J. 60:477–482.

Hunt, J., Y. Ng Wai, A. Barnes, D. J. Greenland. 1979. A rapid method for estimating nitrate-nitrogen concentration in field soils. J. Sci. Food Agric. 30:343–353.

Illinois Agric. Exp. Stn. and Dep. of Agronomy. 1982. The Morrow plots: a century of learning. Illinois Agric. Exp. Stn. Bull. no. 775.

Jackson, W. A., C. E. Frost, and D. M. Hildreth. 1975. Versatile multirange analytical manifold for automated analysis of nitrate-nitrogen. Soil Sci. Soc. Am. Proc. 39:592–593.

Jenkinson, D. S. 1968. Chemical tests for potentially available nitrogen in soil. J. Sci. Food Agric. 19:160–168.

Jenkinson, D. S. 1977. The nitrogen economy of the Broadbalk experiments. I. Nitrogen balance in the experiments. Rothamsted Exp. Stn. Rep. for 1976, part 2 (1977):103–109.

Jenny, H. 1941. Factors of soil formation. McGraw-Hill Book Co., Inc., New York.

Keeney, D. R. 1982. Nitrogen—availability indices. *In* A. L. Page et al. (ed.) Methods of soil analysis. Part 2. 2nd ed. Agronomy 9:711–733.

Keeney, D. R., and J. M. Bremner. 1966. Comparison and evaluation of laboratory methods of obtaining an index of soil nitrogen availability. Agron. J. 58:498–503.

Keogh, J. L., and R. Maples. 1967. A statistical study of soil sampling of Arkansas alluvial soils. Arkansas Agric. Exp. Stn. Rep. Series 157.

Keogh, J. L., and R. Maples. 1980. Nitrate changes in processing soil samples in eastern Arkansas. Commun. Soil Sci. Plant Anal. 11:557–569.

Klemmedson, J. O., and H. Jenny. 1966. Nitrogen availability in California soils in relation to precipitation and parent material. Soil Sci. 102:215–222.

Kresge, P. O., and A. D. Halvorson. 1979. Flexcrop users manual. Montana State Univ. Coop. Ext. Service Bull. no. 1214.

Livens, J. 1959a. Contribution to a study of mineralizable nitrogen in soils (In French). Agricultura 7:27–44.

Livens, J. 1959b. Studies concerning ammoniacal and organic soil nitrogen soluble in water (In French). Agricultura 7:519–532.

Maples, R., J. G. Keogh, and W. E. Sabbe. 1977. Nitrate monitoring for cotton production in Loring-Calloway silt loam. Arkansas Agric. Exp. Stn. Bull. no. 825.

Muller, S., and I. Weigert. 1980. Studies on the change of soil inorganic nitrogen content in the period from late autumn to beginning of the growing season in dependence on soil class, precipitation, and inorganic nitrogen content (In German). Arch. Acker Pflanzenbau Bodenkd. 24:425–432.

Neeteson, J. J., and K. W. Smilde. 1983. Correlation methods of estimating the optimum nitrogen fertilizer rate for sugar beet as based on soil mineral nitrogen at the end of the winter period. p. 409–421. *In* Nitrogen and sugar beet. Int. Inst. Sugar Beet Res. Symp., Brussels, Belgium. 16–17 Feb. Institut International de Recherches Betteravieres, Brussels, Belgium.

Nelson, L. A., and R. J. McCracken. 1962. Properties of the Norfolk and Portsmouth soils: statistical summarization and influence on corn yields. Soil Sci. Soc. Am. Proc. 26:497–502.

New York State College of Agriculture and Life Sciences, Cornell University. 1978. Cornell field crops handbook. Coop. Ext. Service Pub. 10/78 HH 15M 5516-A, Cornell Univ., Ithaca, N.Y.

North Central Region-13 Soil Testing Committee. 1975. Recommended chemical soil test procedures for the North Central Region. N. Dakota Agric. Exp. Stn. Bull. no. 499.

Northeastern Regional Comm.-39, Origin, Transformation, and Management of Nitrogen in Soils, Waters, and Plants. 1980. Efficient use of nitrogen on cropland of the northeast. Conn. Agric. Exp. Stn. Bull. no. 792.

Nyborg, M., J. Neufeld, and R. A. Bertrand. 1976. Measuring crop available nitrogen. p. 102–117. *In* Proc. Western Canada Nitrogen Symp., Calgary, Alberta, Canada. 20–21 Jan. 1976. Alberta Agric., Edmonton, Alberta, Canada.

Olsen, R. J., R. H. Hensler, O. J. Attoe, S. A. Witzel, and L. A. Peterson. 1970. Fertilizer nitrogen and crop rotation in relation to movement of nitrate nitrogen through soil profiles. Soil Sci. Soc. Am. Proc. 34:448–452.

Onken, A. B., and H. D. Sunderman. 1972. Applied and residual nitrate-nitrogen effects on irrigated grain sorghum yield. Soil Sci. Soc. Am. Proc. 36:94–97.

Parr, J. F. 1973. Chemical and biochemical considerations for maximizing the efficiency of fertilizer nitrogen. J. Environ. Qual. 1:75–84.

Pierre, W. H., L. Dumeneil, V. D. Jolley, J. R. Webb, and W. D. Shrader. 1977. Relationship between corn yield, expressed as a percentage of maximum, and the N percentage in the grain: I. Various N-rate experiments. Agron. J. 69:215–220.

Prince, A. L. 1923. Variability of nitrates and total nitrogen in soils. Soil Sci. 15:395–406.

Ramig, R. E., and H. F. Rhoades. 1962. Interrelationships of soil moisture level at planting time and nitrogen fertilization on winter wheat production. Agron. J. 54:123–127.

Reuss, J. O., P. N. Soltanpour, and A. E. Ludwick. 1977. Sampling distribution of nitrates in irrigated fields. Agron. J. 69:588–592.

Roberts, S., A. W. Richards, M. G. Day, and W. H. Weaver. 1972. Predicting sugar content and petiole nitrate of sugarbeets from soil measurements of nitrate and mineralizable nitrogen. J. Am. Soc. Sugar Beet Technol. 17:126–133.

Robinson, J. B. D. 1968a. Measuring soil nitrogen availability to crops in east Africa. East Afr. Agric. For. J. 33:269–280.

Robinson, J. B. D. 1968b. Chemical index for available soil nitrogen. East Afr. Agric. For. J. 33:299–301.

Ross, D. J., B. A. Bridger, A. Cairns, and P. L. Searle. 1979. Influence of extraction and storage procedures and soil sieving on the mineral nitrogen content of soils from Tussock Grasslands. N.Z. J. Sci. 22:143–152.

Russell, W. A., and W. H. Pierre. 1980. Relationship between maize single crosses and their parent inbred lines for N content in the grain. Agron. J. 72:363–369.

Selmer-Olsen, A. R., A. Oien, B. Aerug, and I. Lyngstad. 1971. Pretreatment and storage of soil samples prior to mineral nitrogen determination. Acta Agric. Scand. 21:57-63.

Smith, C. M. 1980. Scope and possibilities of soil testing for nitrogen. p. 122-133. *In* Soil and plant testing and analysis. Report of expert consultation, Rome, Italy. 13-17 June 1977. FAO Soils Bull. 38/1. Rome, Italy.

Smith, S. J., and G. Stanford. 1971. Evaluation of a chemical index of soil nitrogen availability. Soil Sci. 111:228-232.

Smith, S. J., L. B. Young, and G. E. Miller. 1977. Evaluation of soil nitrogen mineralization potential under modified field conditions. Soil Sci. Soc. Am. J. 41:74-76.

Soltanpour, P. N., A. E. Ludwick, and J. O. Reuss. 1979. Guide to fertilizer recommendations in Colorado—soil analysis and computer process. Colorado State Univ. Coop. Ext. Service, Ft. Collins, Colo.

Soper, R. J., G. J. Racz, and P. I. Fehr. 1971. Nitrate nitrogen in the soil as a means of predicting the fertilizer nitrogen requirements of barley. Can. J. Soil Sci. 51:45-49.

Stanford, G. 1973. Rationale for optimum nitrogen fertilization in corn production. J. Environ. Qual. 2:159-166.

Stanford, G. 1981. Assessment of soil nitrogen availability. *In* F. J. Stevenson (ed.) Nitrogen in agricultural soils. Agronomy 22:651-688. Am. Soc. of Agron., Madison, Wis.

Stanford, G., J. N. Carter, D. T. Westermon, and J. J. Meisinger. 1977. Residual nitrate and mineralizable soil nitrogen in relation to nitrogen uptake by irrigated sugarbeets. Agron. J. 69:303-308.

Stanford, G., and E. Epstein. 1974. Nitrogen mineralization-water relationships in soils. Soil Sci. Soc. Am. Proc. 38:103-106.

Stanford, G., M. H. Frere, and D. H. Schwaninger. 1973. Temperature coefficient of soil nitrogen mineralization. Soil Sci. 115:321-323.

Storrier, R. R. 1966. The pre-treatment and storage of soil samples for nitrogen analysis. J. Aust. Inst. Agric. Sci. 32:106-113.

Stumpe, H. 1977. The influence of proceeding crops and fertilization practices on the soil inorganic and mineralizable nitrogen content and on the efficiency of nitrogen fertilization on winter wheat (In German). Arch. Acker Pflanzenbau Bodenkd. 21:575-586.

Tsai, C. Y., D. M. Huber, D. V. Glover, and H. L. Warren. 1984. Relationship of N deposition to grain yield and N response of three maize hybrids. Crop Sci. 24:277-281.

Vander Paauw, F. 1962. Effect of winter rainfall on the amount of nitrogen available to crops. Plant Soil 16:361-380.

Vander Paauw, F. 1963. Residual effect of nitrogen fertilizers on succeeding crops in a moderate marine climate. Plant Soil 19:324-331.

Vitosh, M. L., R. E. Lucas, and R. J. Black. 1974. Effect of nitrogen fertilizer on corn yield. Michigan State Univ. Coop. Ext. Service Bull. no. E-802.

Wagner, D. F., W. C. Dahnke, and E. H. Vasey. 1977. Fertilizing small grains. N. Dakota Coop. Ext. Service Circular S-F 2.

Waksman, S. A. 1923. Microbiological analysis of soil as an index of soil fertility. VI. Nitrification. Soil Sci. 16:55-67.

Waynick, D. D. 1918. Variability in soils and its significance to past and future soil investigations. I. A statistical study of nitrification in soils. Univ. Calif. Berkeley Publ. Agric. Sci. 3:243-270.

Waynick, D. D., and L. T. Sharp. 1919. Variability in soils and its significance to past and future soil investigations. II. Variation in nitrogen and carbon in field soils and their relation to the accuracy of field trials. Univ. Calif. Berkeley Publ. Agric. Sci. 4:121-139.

Wehrmann, J., and H. C. Scharpf. 1979. The mineral nitrogen of soil as a measure of fertilizer nitrogen requirement (In German). Plant Soil 52:109-126.

Weise, R. A., and E. J. Penas. 1979. Fertilizer suggestions for corn. Univ. of Nebraska Coop. Ext. Service Publ. no. G74-174.

Westfall, D. G., M. A. Henson, and E. P. Evans. 1978. The effect of soil sample handling between collection and drying on nitrate concentration. Commun. Soil Sci. Plant Anal. 9: 169-185.

Whitney, D. A. 1976. Soil test interpretations and fertilizer recommendations. Kansas State Univ. Coop. Ext. Service Circular C-509.

Young, R. A., J. L. Ozbun, A. Bauer, and E. H. Vasey. 1967. Yield response of spring wheat and barley to nitrogen fertilizers in relation to soil and climatic factors. Soil Sci. Soc. Am. Proc. 31:407-410.

27

Fred Adams
*Auburn University
Auburn, Alabama*

John B. Martin
*National Fertilizer Development Center
Tennessee Valley Authority
Muscle Shoals, Alabama*

Liming Effects on Nitrogen Use and Efficiency[1]

Liming acid soils improves N use by crops, primarily by maintaining or restoring a soil pH environment favorable for plant growth and microbial activity. Potential toxic elements such as Al and Mn will be at subtoxic levels, while many essential nutrients will be at or near optimum availability. It is under such optimum growth conditions that crops are able to use N most efficiently.

There is also a strong interdependence between soil pH and microbial N transformations. Mineralization of organic N is slowed by low soil pH, N_2 fixation by some host-rhizobium combinations may vanish at low soil pH, nitrification is a major cause of increased soil acidity in many cropping systems, and N_2 fixation by rhizobia causes increased soil acidity because legumes take up more cations than anions from the soil (Andrew & Johnson, 1976; Nyatsanga & Pierre, 1973). Where the climate favors development of acid soils, maintaining a favorable soil pH for both soil microbes and plants is accomplished through a good liming program.

[1] The authors are indebted to the Fertilizer Abstract Section and Technical Library Staff, NFDC, TVA, Muscle Shoals, Ala., for their assistance in the literature search, which contributed greatly to the preparation of this chapter.

Nitrogen in Crop Production.

I. MINERALIZATION OF ORGANIC N

Low soil pH slows the rate at which organic N is mineralized, probably because only a part of the microbial population is able to adapt to a more acid environment. The soil chemical factor responsible for inhibiting mineralization will probably be some combination of H and Al toxicities and Ca deficiency. The dominant controlling factor will be determined by characteristics of the particular soil. For example, organic soils are less likely to contain toxic levels of Al than are mineral soils.

Although low pH slows the mineralization rate, it does not prevent it. For example, Harmsen and Van Schreven (1955) reported that the mineralization rate of N in acid, peat soils was usually high. They concluded that N mineralization was not especially sensitive to a particular optimum pH. Similar results were reported by Renacle (1977) for *Penicillum* isolated from a forest soil and by Ivarson (1977) for an acid, peat-bog soil.

Mineralization of organic N is probably inhibited more by low pH in mineral soils than in peats and other high organic-matter soils. Added factors of Al and Mn toxicities are likely to be present in mineral soils, but missing in organic soils. For example, Nyborg and Hoyt (1978) incubated 40 acid (pH 4.0–5.6) mineral surface soils with and without lime and found that liming to about pH 6.7 almost doubled the amounts of N mineralized. However, they were unable to delineate the relationship between the amounts of mineralized N and initial soil pH, base saturation, or soluble Fe, Al, or Mn. In three associated field experiments, they also reported that crop uptake of soil N was increased 15 to 42 kg/ha by liming the first year, but only 7 to 10 kg/ha during the third year.

Liming was also reported to greatly increase N mineralization in a tropical peat, resulting in a fourfold increase in the yield of napiergrass (*Pennisetum purpureum*) (Chew et al., 1976). A comparison of the relative effects of gypsum and lime on mineralization showed that the positive response was due to an increase in soil pH and not to an increase in soil Ca.

II. NITRIFICATION

Oxidation of NH_4^+ to NO_2^- by *Nitrosomonas* and NO_2^- to NO_3^- by *Nitrobacter* is performed by bacteria more sensitive to low soil pH than the microorganisms responsible for mineralization. The rate of nitrification decreases below pH 6.0 and generally becomes negligible below pH 4.5, although NO_3^- may be present as low as pH 4.0 in some soils. The capacity for nitrification to occur in some acid soils and not in others can be caused by the presence of more acid-tolerant strains of microbes in those soils, as well as by differences in soil characteristics. The optimum pH for nitrifying bacteria is 6.6 to 8.0 or higher, depending upon the geographical and soil areas from which they originate.

Soil acidity per se affects bacterial population number as well as rate of activity. Microbial numbers vary from zero to > 1 million per gram of soil, the greatest numbers occurring in neutral and alkaline soils (Alexander, 1977).

It has been known since the 1920's that NO_3^- production in soils incubated with NH_4^+ was strongly inhibited by soil acidity and that liming benefited the rate of nitrification even when soils were initially near neutral (Schmidt, 1982). The optimum pH in pure cultures is about 8.0 for *Nitrobacter* and > 7.6 for *Nitrosomonas.* However, nitrifying organisms have been found in soils with acidity levels as high as pH 4.0 to 4.7, even though nitrification in liquid cultures ceases below pH 6.0. The reason for seemingly different responses in pure culture and soils has not been satisfactorily resolved (Schmidt, 1982); however, possibilities include microbial adaptation and localized favorable pH zones.

Recent studies by Williams and Cooper (1976) showed that nitrification was greatly increased when acid, coal-mine spoils (pH 5.0) were limed. In an experiment involving five soil pH levels, Dancer et al. (1973) reported similar nitrification rates at pH 5.3, 6.0, 6.3, and 6.6, but a considerably lower rate at pH 4.7. It is axiomatic that nitrification depends upon the population of nitrifiers, regardless of soil pH. This was demonstrated by Pang et al. (1975), who compared nitrification in two soils, one with a much higher nitrifier population than the other. Increasing soil pH with the low population of nitrifiers increased nitrification, but it still remained considerably below that for the soil with a high native population. Nakos (1975) also found that liming a nonnitrifying Greek forest soil did not stimulate nitrification, because nitrifying microorganisms were absent. In other incubation studies with forest soils, Hovland and Ishac (1975) found that simulated acid-rain treatments (pH 3.0 or 4.0) totally inhibited nitrification in a pine forest soil. In a comparable field study with acid precipitation on coniferous forest ecosystems, Abrahamsen et al. (1977) found that nitrification did not occur at pH 4.1 to 4.4. These data showed that nitrification ceased when soil pH was much below 5.0, regardless of the mechanism by which soil acidity accumulated. In a study involving nitrification and a nitrifying inhibitor, nitrapyrin, Hendrickson and Keeney (1979) reported that effectiveness of nitrapyrin was increased by increasing soil pH. Further verification of this observation is needed.

III. DENITRIFICATION

Nitrate is reduced by soil microbes under anaerobic soil conditions to N_2O and N_2; the type and quantity of N gas produced during reduction depends upon soil conditions and the effects of these conditions upon the microbial population (Russell, 1973). In reviewing the effect of soil acidity on denitrification, Firestone (1982) and Focht and Verstraete (1977) concluded that optimum pH for denitrification was 7.0 to 7.5. The rate of denitrification is much slower in acid soils (< pH 5) than in near-neutral soils (> pH 6), probably because of the inhibitory effect of low pH on bacterial growth. The relative proportion of N_2O and N_2 produced during denitrification is pH dependent; N_2O is readily reduced to N_2 > pH 7.0 but not < pH 6.0. This was also demonstrated by Balasubramanian and Kanehiro (1976) on acid Hawaiian soils.

Low soil pH may be less inhibitory on denitrification where there is a supply of easily decomposable organic material. This was illustrated by the work of Gilliam and Gambrell (1978) on acid, Coastal Plain subsoils where soil pH was as low as 4.5.

The effects of pH on chemically mediated and microbially mediated denitrification are different. Chemical denitrification is defined by the reaction

$$NO_3^- + 6H^+ + 5e^- \rightleftharpoons 1/2\, N_2(g) + 3H_2O.$$

Thus, denitrification increases with decreasing soil pH and decreasing redox potential, as demonstrated by Van Cleemput and Patrick (1974). They found an almost linear inverse relationship between redox potential and denitrification rate. At each redox level they also found that denitrification was more rapid at pH 4.5 than at pH 6.0. At pH 8.0, denitrification occurred with the redox potential at +400 mV, but it was much greater with the redox potential at zero.

IV. N_2 FIXATION

Symbiotic N_2 fixation is a complex legume-rhizobium association influenced by interrelations of soil acidity factors (H, Ca, Al, and Mn) and biological diversities of legume species and rhizobial strains (Munns, 1978). Because of these host-rhizobial biological variations, a general discussion of the effects of liming on N_2 fixation is diverse. Differential tolerance of legume species to soil acidity has been clearly established, and rhizobial strains also vary significantly in growth rates and in their capacity to nodulate a given host effectively at low pH. Most slow-growing rhizobia can grow in more acid media than can the fast-growing group. Nodule development is sensitive to both Ca deficiency and low pH; detrimental effects of Al and Mn probably occur, but are not definitely understood. Generally, growth of legume roots appears less sensitive to soil acidity than does nodulation. Thus, acid-related inhibition of nodulation due to inhibition of root growth is unlikely unless the legume is affected by severe Al toxicity.

Limited evidence indicates that low pH (and probably Ca depletion in the rhizosphere) can reduce rhizobial growth sufficiently to prevent nodule formation. Infection of root cells is the acid-sensitive step in the nodulation process. In a few species, this step has been found to be sensitive to low Ca as well as low pH, but not especially sensitive to Al or Mn. Also, Ca, pH, and probably Al and Mn can influence biological functioning of established nodules.

Generally, results reported by Evans et al. (1980), Graham and Hubbell (1974), Holding and Lowe (1971), Russell (1973), and Vincent (1965) are in agreement with those presented by Munns (1978).

V. EFFECTS OF SOIL ACIDITY ON AMMONIA VOLATILIZATION

The pH dependency of NH_3 losses from soils is governed by the reaction

$$NH_4^+ + OH^- \rightleftharpoons NH_3(g) + H_2O.$$

As the reaction clearly shows, NH_3 volatilization increases with increasing pH. The uncertainty about this reaction is the critical pH that separates insignificant NH_3 losses from significant losses. There is not complete unanimity in the literature on a critical pH, but that is probably related to the fact that the above chemical reaction depends upon temperature, soil-water content, cation exchange capacity (CEC), and salt content of soil solution. In an early review on the topic, Gardner (1965) concluded that NH_3 losses were negligible from soils below pH 7.0 unless localized alkaline spots were present at or near the soil surface. A more recent review by Terman (1979) cited considerable evidence to substantiate the earlier findings. There is little or no loss of NH_3 from acid soils (except for surface-applied urea), but it may reach > 50% of the applied N at pH 8.0. The findings that NH_3 losses were greatest from alkaline soils being dried are not at all surprising. Drying a soil even caused significant NH_3 losses at pH 6.7 (15% of that added). Because of the effect that hydrolyzed urea has on localized soil pH levels, surface-applied urea is particularly subject to high volatilization losses. Irrigation waters that contain NH_3 will also lose much of their NH_3 if the water is very alkaline.

The effects of environmental factors on NH_3 losses from soils can be explained on the basis of how each factor affects the above chemical reaction. Increasing temperatures will increase the escaping tendency of a gas, thereby increasing NH_3 loss. Increasing pH increases OH^- activity, thereby pushing the reaction toward $NH_3(g)$ formation. Drying removes H_2O from the right-hand portion of the equation, thereby increasing $NH_3(g)$ formation. High CEC lowers NH_3 losses from a constant N rate because more NH_4^+ is adsorbed [and not subject to $NH_3(g)$ formation] and less is present in soil solution where it seeks equilibrium with $NH_3(g)$.

The above principles have recently been demonstrated to apply equally well to flooded tropical soils used for rice (*Oryza sativa* L.) production (Mikkelsen et al., 1978; Ventura & Yoshida, 1977).

VI. INFLUENCE OF SOIL ACIDITY ON MINERAL N UPTAKE

In strongly acid soils, where nitrification is inhibited, NH_4^+ may be an important N source for plants. Species adapted to low soil pH are more tolerant of high levels of soluble Al and Mn and also may tolerate higher NH_4^+ levels. In fact, some plants adapted to acid soil conditions may grow better with NH_4^+ than with NO_3^-. Examples of such species are cranberry (*Vaccinium* sp.), sugarcane (*Saccharune officinarum* L.), blueberry (*Vaccinium* sp.), birch (*Betula* sp.), and certain acid-tolerant grasses (Foy & Flemming, 1978).

Andrews (1954), citing data from Alabama, stated that cotton seedlings (*Gossypium hirsutum* L.) took up more NO_3^- than NH_4^+ on very acid soils, but took up relatively more NH_4^+ as pH approached 7.0. Munn and Jackson (1978), using sweet potato cuttings (*Ipomoea batatas* Lam.), also showed that NH_4^+ uptake increased while NO_3^- uptake decreased as solution pH was increased from 4.5 to 6.5. Similarly, rice plants were found to absorb a maximum of NH_4^+ at pH 8.0, but a maximum of NO_3^- at pH 4.0 (Haynes & Goh, 1978).

In a review of N source effects, Viets (1965) pointed out the effect of NH_4^+ and NO_3^- uptake on root-medium pH. When NH_4^+ is absorbed by plant roots, it is replaced in solution by H^+, resulting in a lower root-medium pH. When NO_3^- is taken up by plants, it is replaced by HCO_3^-, resulting in a higher root-medium pH. A practical consequence of differential ion uptake by plants is its effect on soil-mineral solubilities. For example, NH_4^+ absorption will lower the pH and thereby enhance Al solubility and toxicity; NO_3^- uptake will have the opposite effect. Availability of the cationic micronutrients is affected by NH_4^+ and NO_3^- absorption in a similar manner.

There is adequate documentation that NH_4^+ uptake from soils lowers the pH and NO_3^- uptake raises the pH, just as previously demonstrated for culture solutions (Smiley & Cook, 1973; Soon & Miller, 1977).

VII. N FERTILIZERS AND SOIL ACIDITY

Soil acidity is increased by the use of NH_4^+ fertilizers, the rate of acidity formation being dependent upon the integrated effects of soil characteristics, cropping systems, rate of application, method of application, and N source. Conversely, effects of soil acidity upon efficiency of N fertilization depends upon the critical pH value for a given soil and also on the differential tolerances of plant species to soil acidity, especially Al and Mn toxicity. In a review of N and soil acidity relationships, Adams (1984) summarized data obtained in the southern United States. When urea or ammonium nitrate (AN) was surface-applied to sods of bermudagrass (*Cynodon dactylon* L.), bahiagrass (*Paspalum notatum*), or pangolagrass (*Digitaria decumbens*) on Lakeland fine sand for 2 yr (total N rate of 940 kg/ha), the pH in the 0- to 5-cm depth decreased from 6.2 to 5.4 with urea and to 4.9 with AN. At a N rate of 504 kg from AN, the pH declined to 5.2. Thus, more acidity was developed from 504 kg of N as AN than from 940 kg of N as urea. The amount of acidity produced per kilogram of N is affected by the rate of application, the crop, and the form of fertilizer N. The acidity developed from AN on a 'Coastal' bermudagrass sod on a Cecil sandy loam was equivalent to 1.1 kg of $CaCO_3$/kg of N at the 284 kg of N rate. A similar effect was observed when ammonium sulfate (AS) was applied to napiergrass on a Toa clay loam (Abruna et al., 1958). A more efficient use of N at the lower N rate probably caused the differences in developed acidity. Although limits of soil acidity that can develop from each N source and rate can be calculated, actual amounts are highly variable and hardly calculable because of the many factors that influence it.

High rates of NH_4^+ fertilizers also increase subsoil acidity. Surface-applied AS to napiergrass in Puerto Rico for a 3-yr period decreased the pH of the 15- to 30-cm depth from 6.1 to 5.4 and to 4.2 from the 900 and 4000 kg/ha of N rate, respectively (Abruna et al., 1964).

Jolley and Pierre (1977), using two N-rate experiments with corn (*Zea mays* L.) that had been grown continuously for 15 to 17 yr, found that the amount of acidity that developed per unit of applied N increased as the rate of N exceeded that needed for optimum yield. They also found no significant increases in acidity below a depth of 30 cm. Hiltbold and Adams (1960) incubated soil samples containing seven N sources under conditions favorable for nitrification and compared theoretical acidity with measured acidity. The difference between theoretical acidity and observed acidity was accounted for by equivalent N volatilization.

Blue (1974) evaluated the efficiency of AS, AN, calcium nitrate (CN), urea, and ureaform for the production of Pensacola bahiagrass forage on a Leon fine sand at different lime rates for a 10-yr period. Lime treatments had little effect on forage yield and N uptake and, consequently, little effect on N efficiency. Thompson (1977) studied the effects of N and lime on yield and winterkill of Coastal bermudagrass in Arkansas from 1968 to 1977. By the mid-1970's, high rates of AN had markedly increased soil acidity. Lime was then applied in 1974 and 1976 to one-half of each plot, but it had no effect on the 1974–1976 hay yields. Thus, N fertilizer efficiency for production of Coastal bermudagrass was unaffected by liming. Application of lime, however, did improve survival of the bermudagrass sward during the winter of 1976; plant counts in June 1977 showed that the limed plots had 43% more ground cover. D. Mays (1974, unpublished data, Soils & Fertilizer Research Branch, TVA) conducted a 5-yr experiment with three AN rates (0, 336, and 672 kg/ha of N) on an established Coastal bermudagrass sod (original pH 5.2) in which lime rates of 0, 11.2, and 22.4 t/ha were surface-applied, disked, or plowed into the soil. Soil pH was lower throughout the profile where the high N rate was applied, with the difference generally being greatest in the uppermost 15 cm. The increased acidity did not affect yields, however, and consequently did not affect the efficiency of N applied to acid-tolerant bermudagrass.

Simms and Atkinson (1974) reported that high rates of AN for burley tobacco (*Nicotiana tabacum* L.) on a silt loam soil (original pH 6.2) decreased soil pH, exchangeable Ca, available Mo, available P, and increased exchangeable Mn. Nitrification was somewhat slower at the highest N rate, 360 kg/ha. Correlation and regression analysis indicated that early reduction in dry-matter accumulation was due to increased soil acidity; thus, efficiency of N for burley tobacco production was reduced by soil acidity. Awad and Edwards (1977) studied effects of AS (336 kg/ha of annually for 4 yr followed by 672 kg/ha annually for 2 yr) on production of a kikuyugrass (*Pennisetum clandestinum*) pasture in Australia. At the end of the 6-yr study, soil pH had decreased from 5.0 to 4.0, which increased soluble Al and decreased exchangeable Ca, Mg, and K. Liming to pH 5.5 increased yield and reduced the amount of N required for maximum growth from 672 kg/ha on the unlimed soil to 134 kg/ha on the limed soil; thus, liming increased N efficiency. Further liming from pH 5.5 to 6.0 barely increased yields; thus, it only slightly affected N efficiency above pH 5.5.

REFERENCES

Abrahamsen, G., R. Horntvedt, and B. Tveite. 1977. Impacts of acid precipitation on coniferous forest ecosystems. Water Air Soil Pollut. 8:57–73.

Abruna, F., R. W. Pearson, and C. B. Elkins. 1958. Quantitative evaluation of soil reaction and base status changes resulting from field application of residually acid-forming nitrogen fertilizers. Soil Sci. Soc. Am. Proc. 22:539–542.

Abruna, F., J. Vincente-Chandler, and R. W. Pearson. 1964. Effects of liming on yields and composition of heavily fertilized grasses and on soil properties under humid tropical conditions. Soil Sci. Soc. Am. Proc. 28:657–661.

Adams, F. 1984. Crop response to lime in the southern United States. *In* F. Adams (ed.) Soil acidity and liming. 2nd ed. Agronomy 12:211–265.

Alexander, M. 1977. Introduction to soil microbiology. 2nd ed. John Wiley & Sons, Inc., New York.

Andrew, C. S., and A. D. Johnson. 1976. Effect of calcium, pH, and nitrogen on the growth and chemical composition of some tropical and temperate pasture legumes. II. Chemical composition (calcium, nitrogen, potassium, magnesium, sodium, and phosphorus). Aust. J. Agric. Res. 27:625–636.

Andrews, W. B. 1954. Crop responses to sources of nitrogen. p. 1–39. *In* W. B. Andrews (ed.) The response of crops and soils to fertilizer and manures. 2nd ed. Publ. by W. B. Andrews, State College, Miss.

Awad, A. S., and D. G. Edwards. 1977. Reversal of effects of heavy ammonium sulfate on growth and nutrient status of a kikuyu pasture. Plant Soil 48:169–183.

Balasubramania, V., and Y. Kanehiro. 1976. Denitrification potential and pattern of gaseous N loss in tropical Hawaiian soils. Trop. Agric. 53:292–303.

Blue, W. G. 1974. Efficiency of five nitrogen sources for Pensacola bahiagrass on Leon fine sand as affected by lime treatments. Proc. Soil Crop Sci. Soc. Fla. 33:176–180.

Chew, W. Y., C. N. Williams, K. T. Joseph, and K. Ramli. 1976. Studies on the availability to plants of soil nitrogen in Malaysian tropical pligotrophic peat. I. Effect of liming and pH. II. Effect of N, P, K, and micronutrients. Trop. Agric. 53:69–87.

Dancer, W. S., L. A. Peterson, and A. Chesters. 1973. Ammonification and nitrification of N as influenced by soil pH and previous N treatments. Soil Sci. Soc. Am. Proc. 1:67–69.

Evans, L., K. F. Lewin, and F. A. Vella. 1980. Effect of nutrient medium pH on symbiotic nitrogen fixation by *Rhizobium leguminosarum* and *Pisum sativum*. Plant Soil 56:71–80.

Firestone, M. K. 1982. Biological denitrification. *In* F. J. Stevenson (ed.) Nitrogen in agricultural soils. Agronomy 22:289–326.

Focht, D. D., and W. Verstraete. 1977. Biochemical ecology of nitrification and denitrification. p. 135–214. *In* M. Alexander (ed.) Advances in microbial ecology. Vol. 1. Plenum Press, New York.

Foy, C. D., and A. L. Flemming. 1978. The physiology of plant tolerance to excess available aluminum and manganese in acid soils. p. 301–328. *In* G. A. Jung (ed.) Crop tolerance to suboptimal land conditions. ASA Spec. Pub. no. 32. Am. Soc. of Agron., Madison, Wis.

Gardner, W. R. 1965. Movement of nitrogen in soil. *In* W. Bartholomew and F. E. Clark (ed.) Soil nitrogen. Agronomy 10:550–572.

Gilliam, J. W., and R. P. Gambrell. 1978. Temperature and pH as limiting factors in loss of nitrate from saturated Atlantic Coastal Plain soils. J. Environ. Qual. 7:524–532.

Graham, P. H., and D. H. Hubbell. 1974. Soil plant-rhizobium interaction in tropical agriculture. p. 211–227. *In* E. Bornemisza and A. Alvarando (ed.) Soil management in tropical America. Soil Sci. Dep., N.C. State Univ., Raleigh.

Harmsen, G. W., and D. A. van Schreven. 1955. Mineralization of organic nitrogen in soil. Adv. Agron. 7:300–383.

Haynes, R. J., and K. M. Goh. 1978. Ammonium and nitrate nutrition of plants. Biol. Rev. 53:465–510.

Hendrickson, L. L., and D. R. Keeney. 1979. A bioassay to determine the effect of organic matter and pH on the effectiveness of nitrapyrin (N-Serve) as a nitrification inhibitor. Soil Biol. Biochem. 11:51–55.

Hiltbold, A. E., and F. Adams. 1960. Effect of nitrogen volatilization on soil acidity changes due to applied nitrogen. Soil Sci. Soc. Am. Proc. 24:45–47.

Holding, A. J., and J. F. Lowe. 1971. Some effects of acidity and heavy metals on the rhizobium-leguminous plants association. p. 153–156. *In* T. A. Lie and E. G. Mulder (ed.) Biological nitrogen fixation in natural and agricultural habitats. Plant Soil. Spec. Vol. The Hague, Martinus Nighoff.

Hovland, J., and Y. Z. Ishac. 1975. Effect of acid precipitation on forest and fish (In Norwegian). Internal Rep. 14/75. Mikrobiol. Inst. Norges Landerukshogeskole 1432 As-NLH, Norway.

Ivarson, K. C. 1977. Changes in decomposition rate, microbial population, and carbohydrate content of an acid peat bog after liming and reclamation. Can. J. Soil Sci. 57:129–137.

Jolley, V. D., and W. H. Pierre. 1977. Soil acidity from long-term use of nitrogen fertilizer and its relationship to recovery of the nitrogen. Soil Sci. Soc. Am. J. 41:368–378.

Mikkelson, D. S., S. K. de Datta, and W. N. Abcemea. 1978. Ammonia volatilization losses from flooded rice soils. Soil Sci. Soc. Am. J. 42:725–730.

Munn, D. A., and W. A. Jackson. 1978. Nitrate and ammonium uptake by rooted cuttings of sweet potato. Agron. J. 70:312–316.

Munns, D. N. 1978. Legume-rhizobium relations, soil acidity and nodulation. p. 247–263. *In* C. S. Andrew and E. J. Kamprath (ed.) Mineral nutrition of legumes in tropical and subtropical soils. CSIRO, Australia.

Nakos, G. 1975. Absence of nitrifying organism from a Greek forest soil. Soil Biol. Biochem. 7:335–336.

Nyatsanga, T., and W. H. Pierre. 1973. Effect of nitrogen fixation by legumes on soil acidity. Agron. J. 65:936–940.

Nyborg, M., and P. B. Hoyt. 1978. Effects of soil acidity and liming on mineralization of soil nitrogen. Can. J. Soil Sci. 58:331–338.

Pang, P. C., C. M. Cho, and R. A. Hedlin. 1975. Effect of pH and nitrifier population on nitrification of band-applied and homogenously mixed urea nitrogen in soil. Can. J. Soil Sci. 55:15–21.

Renacle, J. 1977. The role of heterotrophic nitrification in acid forest soils—preliminary results. Ecol. Bull. (Stockholm) 25:560–561.

Russell, E. W. 1973. Soil conditions and plant growth. 10th ed. Longman Group, Ltd., London.

Schmidt, E. L. 1982. Nitrification in soil. *In* F. J. Stevenson (ed.) Nitrogen in agricultural soils. Agronomy 22:253–283. Am. Soc. of Agron., Madison, Wis.

Sims, J. L., and W. O. Atkinson. 1974. Soil and plant factors influencing accumulation of dry matter in burley tobacco growing in soil made acid by fertilizer. Agron. J. 66:775–778.

Smiley, R. W., and J. R. Cook. 1973. Relationship between take-all of wheat and rhizosphere pH in soils fertilized with ammonium versus nitrate nitrogen. Phytopathology 63:882–890.

Soon, Y. K., and M. H. Miller. 1977. Changes in the rhizosphere due to ammonium and nitrate fertilization and phosphorus uptake by corn seedlings (*Zea mays* L.). Soil Sci. Soc. Am. J. 41:77–80.

Termen, G. L. 1979. Volatilization losses of nitrogen as ammonia from surface-applied fertilizers, organic amendments, and crop residues. Adv. Agron. 30:189–223.

Thompson, L. 1977. Effect of fertilizers and limestone on yield and winterkill of coastal bermudagrass. Arkansas Farm Res. 26:7.

Van Cleemput, O., and W. H. Patrick, Jr. 1974. Nitrate and nitrite reduction in flooded soils at controlled redox potential and pH. Int. Congr. Soil Sci., Trans. 10th (Madison, Wis.) 9:152–159.

Ventura, W. B., and T. Yoshida. 1977. Ammonia volatilization from flooded tropical soils. Plant Soil 46:521–531.

Viets, F. G., Jr. 1965. The plant's needs for and use of nitrogen. *In* W. Bartholomew and F. E. Clark (ed.) Soil nitrogen. Agronomy 10:503–549.

Vincent, J. M. 1965. Environmental factors in the fixation of nitrogen by the legume. *In* W. Bartholomew and F. E. Clark (ed.) Soil nitrogen. Agronomy 10:384–435.

Williams, P. J., and J. E. Cooper. 1976. Nitrogen mineralization and nitrification in amended colliery spoils. J. Appl. Ecol. 13:533–543.

28

Ellery L. Knake and Marshal D. McGlamery
University of Illinois
Urbana, Illinois

Nitrogen Use and Weed Control

In today's energy-conscious era, an appropriate objective for agriculture is to convert as much of the available energy as possible into a form useful to man. There is a limited amount of energy available on each hectare of land —primarily in the form of moisture, light, and nutrients. Because of the finite amount of energy available on a hectare of land, there is competition between weeds and crops for this energy (Knake & Slife, 1962). Weeds are of little or no use to man; any energy used by them is largely wasted. Each unit of energy captured by weeds is one unit less for crop production (Knake, 1960a). Thus, farmers and others have sought diligently to eliminate the competition between weeds and crops (Fig. 1).

I. WEED MANAGEMENT

A. Weed Thresholds

The term *economic threshold* was introduced as part of integrated pest management (IPM). This concept has been used primarily for insects. How many insects can be tolerated before a threshold is reached that justifies the cost of treatment? There have been attempts to apply an analogous concept to weeds, but it is not as well adapted. A single weed of some species may produce over 100 000 seeds. Thus, many weed scientists suggest that although the single weed may not affect yield very much that season, the potential for problems the next season is considerable. They conclude that even one weed is one too many. Seldom can perfect weed control be achieved, even with today's herbicides. But complete control of weeds can be the goal to strive for.

Nitrogen in Crop Production.

Fig. 1—Weeds compete with crops for available energy in the form of nutrients, moisture, and light.

B. Time Schedule for Weed Control

Research at Urbana, Ill. with giant foxtail (*Setaria faberi* Herrm.) in corn (*Zea mays* L.) and soybeans [*Glycine max* (L.) Merr.] suggests that this weed species has little or no significant effect on final yields if removed before it is about 20 to 30 cm high (Table 1). Especially for soybeans, this study suggests that the effect of early weeds during the vegetative growth stage may not be critical. The major competition appears to occur during the reproductive growth stage when soybeans are flowering and setting pods (Knake & Slife, 1969). Results may vary with different weed species, location, and season.

Many weed scientists and farmers have preferred early control. Weeds are frequently easiest to control in early life stages by means of soil-applied herbicides and early cultivation (Knake & Slife, 1965, 1966). Also, field con-

Table 1—Effect of removing giant foxtail at various times on crop yields.

Height (cm) of foxtail when removed	Corn	Soybeans
	Three-year average	Two-year average
	kg/ha	
Weed-free check	9032	2019
7.6	8969	2040
15.2	8906	1999
22.9	8718	1952
30.5	8593*	1885
Foxtail left to maturity	7903*	833*

* Significantly lower than weed-free check at the 5% probability level.

ditions may prevent timely treatments later. However, the trend toward reduced tillage, the "wait and see" philosophy of IPM, and the introduction of more effective and more selective postemergence treatments might encourage increased consideration of postemergence applications.

C. N for Crop and Weeds

Immobile elements such as P and K can become limiting as roots of different species compete for absorptive sites. For N, somewhat like water, one might assume that sufficient amounts of this mobile element could be applied to satisfy the needs of both crop and weeds. However, research suggests that this may not necessarily be true (Table 2) (Knake, 1960b). Applying high amounts of N or removing weeds increased yields compared with having weeds and only a moderate level of N. However, greatest yields were obtained where weeds were controlled in addition to using high rates of N. We conclude that to achieve optimum yields, weeds must be controlled; it is neither logical, practical, nor economical to attempt compensating for weed competition by adding more fertilizer.

Will addition of fertilizer help a crop to compete better with weeds or will it simply help the weeds grow better? The answer depends partly on the life cycle of the crop and of the weeds.

Winter wheat (*Triticum aestivum* L.) seeded in the fall competes quite well with annual weeds that begin growth in the spring. A good fertility program, including addition of N, can help to assure vigorous growth of wheat and greater advantage over weeds. Weeds such as mustards (*Brassica* sp.) and wild garlic (*Allium vineale* L.) that begin their growth in the fall at about the same time as wheat are more difficult to control because the life cycle coincides with that of the crop. But vigorous crop growth is still helpful.

For a crop such as corn, many annual weeds begin growth about the same time as the crop. Application of N early in the season can enhance the growth of weeds as well as of the crop. Thus, good weed control is important if the crop is to achieve the greatest benefit from the N. If early weeds were left to grow for a few weeks and then controlled with postemergence herbicides, some preference might be given to side-dressing N rather than earlier application. However, since most farmers strive for early weed control, other factors should generally dictate timing for N.

Table 2—Corn grain yields as affected by weeds and N.

Weeds		No weeds	
N applied, kg/ha			
90	270	90	270
kg corn/ha			
5018	6147	6397	7338

II. HERBICIDE-N-SOIL-PLANT RELATIONSHIPS

A. Optimum Placement

When deciding whether or not to mix herbicide with fertilizer, consider where each should be placed for optimum results. Most herbicides should be located in the top 5 cm of the soil (Knake et al., 1967). They may be moved there by rainfall or irrigation, or they may be physically placed there with mechanical incorporation. Herbicides that are subject to loss from the soil surface by volatilization or photodecomposition should be incorporated soon after application to avoid this loss. For some herbicides, mechanical incorporation is optional. For some, it may be detrimental. For most forms of N with which herbicides could be mixed, it is optional whether they are applied to the surface or mixed with the soil. Depth of N placement is generally less critical than for herbicides.

The majority of weed seeds that will germinate and present a potential weed problem during a given season are located in the upper 5 cm of the soil. Having herbicide in the weed seed zone per se is usually not the critical factor, since herbicides do not necessarily affect the weed seed. However, after the weed seed germinates, the herbicide is absorbed by the emerging shoot or possibly by the roots of the weed seedling. The relative importance of the shoot vs. the root for herbicide absorption may vary with species and with herbicides. However, for some of our major weed species and major herbicides, shoot uptake appears to be of primary significance (Parker, 1966; Knake & Wax, 1968). The critical factor is having moisture available for absorption of the herbicide by the emerging weed seedling. Mechanical incorporation can move the herbicide into more moist soil, thereby allowing less dependence on rainfall for good herbicide performance. Since the top 2 or 3 cm of soil can dry out relatively fast, moving the herbicide a little deeper can be advantageous. However, placing the herbicide deeper than 5 cm can simply dilute the herbicide excessively and result in poor weed control. The goal should be to have the herbicide concentrated where there is adequate moisture for absorption of the herbicide by the emerging weed seedlings.

Subsurface layering of herbicides in a concentrated band several centimeters below the soil surface has been tested. This may be feasible for rather volatile herbicides in coarse-textured soils or to inhibit emergence of a perennial such as field bindweed (*Convolvulus arvensis* L.). However, subsurface layering is not generally practical in most soils for controlling annual weeds, because the weed seeds may germinate above the herbicide layer and weed seedlings emerge without adequately absorbing the herbicide.

Because N is mobile and is absorbed by crop roots that grow throughout the surface profile, placement of N is usually less critical than placement of the herbicide. However, it is generally recommended that urea fertilizers not be left on the soil surface where large amounts of crop residue are present.

Fig. 2—Equipment for subsurface placement of fertilizer is not usually appropriate for application of most herbicides.

Application methods that "knife-in" the fertilizer or apply it in subsurface layers would not appear appropriate for most herbicides (Fig. 2). However, some trials with certain relatively volatile herbicides mixed with anhydrous NH_3 suggest that this may be feasible if injection bands are adequately close and soil physical conditions and moisture suitable for sufficient diffusion of the herbicide to provide uniform distribution. If these conditions are not provided, streaked weed control can result.

B. Uniform Distribution

Generally, uniform distribution is more critical for herbicides than for N. Most soil-applied herbicides are applied in the rate range of about 0.56 to 3.36 kg/ha active ingredient. For many herbicides, the margin of crop tolerance is rather narrow. Accurate and uniform application is very important. Equipment must be carefully calibrated. Wettable powders and flowables should be kept suspended with good agitation. Overlaps and skips that result in nonuniform application should be avoided. Equipment should be shut off when turning or stopping. Do not double up on field ends. With some herbicides, 1.5 to 2 times the recommended rate may cause significant crop injury or increase risk of herbicide residue affecting the next year's crop.

C. Effect on Plants

Field experience and research thus far does not seem to suggest any significant detrimental effect on the crop from using fertilizer as a carrier for soil-applied herbicides. Very little, if any, difference has generally been noted in degree of weed control in the field when comparing herbicide applied with water to herbicide applied with fertilizer solution. In greenhouse experiments where close observations have been made using N solution compared with water as a carrier for atrazine (2-chloro-4-ethylamino-6-isopropylamino-*s*-triazine), weed control was at least as good with the N carrier. Weeds that emerged and then died may have succumbed slightly sooner where N was used.

However, all common N fertilizers alter the pH of the soil, particularly where surface or shallow applications are used with reduced tillage methods (Berg, 1980; Parochetti, 1980). Altering soil pH can alter performance of some herbicides (McGlamery & Slife, 1966). For example, residual activity of atrazine and effect on subsequent crops is usually greater on soils with a relatively high pH. Thus, while differences in weed control have not been very obvious, the possibility might exist, partly because of a change in soil pH. It has been suggested that no-tillage corn growers sample the top 5 cm of soil, make a pH test, and maintain a pH of about 6.2 or a little above for good activity of triazine herbicides.

When N fertilizer is applied and enhances growth of weeds as well as of crops, the more lush, active growth may allow some postemergence herbicides to be more effective (Penner, 1974). This can be advantageous for herbicides to which the crop is quite tolerant or if the herbicide can be directed to minimize contact with the crop. However, for herbicides with relatively close crop tolerance, which are applied over the crop, the more lush, active growth of the crop may make it more subject to herbicide injury.

There are two major types of postemergence herbicides: those that kill primarily by contact action to "burn" the plant tissue and those that are absorbed into the plant and translocated. Nitrogen solution has been considered for controlling weeds by contact action and at one time was recommended for use with 2,4-D (2,4-dichlorophenoxyacetic acid) or atrazine to help control weeds in corn.

Nitrogen solution or some other fertilizer solutions may "burn" small, actively growing weeds with very sensitive leaf tissue. But, especially as weeds become larger, control with fertilizer solution alone is not generally adequate. Some herbicide labels caution against early postemergence application of herbicide-fertilizer combinations because of increased risk of injury to small crop plants.

If fertilizer solution "burns" leaf tissue rather rapidly, this might interfere with absorption, translocation, and effectiveness of a translocated herbicide applied simultaneously. Results will depend in part on how rapidly the herbicide enters the plant and how rapidly the leaf tissue is burned by the fertilizer.

When considering substitution of fertilizer solution for water to apply postemergence herbicides, some consideration should be given to effect of concentration of the mixture on herbicide performance. For example, paraquat (1,1′-dimethyl-4,4′-bipyridinium ion) usually performs best with relatively high volumes. Glyphosphate [*N*-(phosphonomethyl) glycine] may perform best with low volume, more concentrated solutions.

If herbicides were to have a detrimental effect on certain soil microorganisms, this might interfere with conversion of the various forms of N. However, there has been little or no significant evidence thus far that the commonly used herbicides are significantly altering the soil microflora in a detrimental way. Herbicides are not generally mixed so thoroughly with the soil and do not generally persist long enough to significantly affect the microflora for long periods. Research on this aspect, however, has been somewhat limited and as new compounds are introduced, this aspect might be given some consideration.

III. FERTILIZER-HERBICIDE MIXTURES

Applying both fertilizer and herbicide at the same time can be advantageous for saving energy, time, and trips over the field (McGlamery, 1970). However, there are several factors to consider (Fig. 3).

A. Timing of Application

The persistence of herbicides varies considerably. Some, such as trifluralin (α,α,α-trifluoro-2,6-dinitro-*N,N*-dipropyl-*p*-toluidine), persist long enough that they can be applied several months ahead of spring planting. Fall applications have even been considered. While it is true that there is very little chemical and biological degradation of herbicides during cold winter months, fall applications have not generally given as good results as spring applications and have not been very common for soil-applied herbicides. Especially for those herbicides with relatively short persistence, it is usually preferable to make applications as close to time of planting as possible. This allows the herbicide to last as long into the growing season as possible.

Fig. 3—Uniform application of herbicides is extremely important. Fertilizer-herbicide combinations should be thoroughly mixed. Field applications should be applied with accuracy and precision to assure uniform distribution.

B. Type of Carrier

The material that the herbicide is mixed with for application simply serves as a carrier to allow uniform distribution. Water is the most common carrier. However, fluid fertilizer may be used (Knake & Mulvaney, 1966). For dry applications, the herbicide may be sprayed on dry fertilizer. Certain herbicides may be formulated initially on such carriers as attapulgite clay, ground corn cobs, or vermiculite. The majority of herbicides for agricultural use are currently formulated as emulsifiable concentrates, solutions, wettable powders, flowable suspensions, or water dispersible granules. These are commonly mixed with a suitable carrier for spraying.

C. Amount of Carrier

The amount of carrier should be sufficient to provide uniform distribution of the herbicide. Uniformity is particularly critical for herbicides with limited crop tolerance. For soil-applied herbicides, about 100 to 300 L of water/ha is fairly common. About 50 L may sometimes be adequate, especially for aerial applications where low volumes are desired. Even for ground applications, using high amounts of water may be inconvenient, time-consuming, and costly.

When fertilizer solution is used as the carrier, the deciding factor on amount will usually be the rate of fertilizer desired. A minimum of about 100 to 200 L/ha is usually desired for uniform distribution of soil-applied herbicides. There is no disadvantage to using whatever higher amount is needed to obtain desired fertilizer rate.

To be effective, moisture is needed so soil-applied herbicides can be absorbed by the emerging weed seedlings (Stickler et al., 1969). Moisture is provided by rainfall or irrigation. The small amount of carrier is of no significance for providing this moisture. To obtain the equivalent of 1.25 cm of rainfall needed for some herbicides would require 127 000 L of water/ha. The purpose of carrier is simply to allow uniform distribution with commonly used equipment.

D. Chemical Compatibility

With the commonly used herbicides and fertilizers, there has been little or no evidence thus far that the fertilizer altered the herbicide chemically. However, such may be within the realm of possibility with certain compounds. Caution should be exercised with certain mixtures. For example, when mixed with clay suspension fluid fertilizers, paraquat has been inactivated by being adsorbed onto the clay.

E. Physical Compatibility

Problems are sometimes encountered with physical incompatibility of certain herbicide formulations and liquid fertilizers. The failure of components to remain uniformly dispersed is called *incompatibility.* Some causes of this are inadequate agitation, insufficient spray volume, lack of a stable emulsifier, or improper mixing procedure (McGlamery, 1970).

Some emulsifiable concentrates may be unstable in salty solutions, such as fluid fertilizers. Most manufacturers specify that the emulsion stability be checked and a compatibility agent be added if needed. These agents are usually added at the rate of about 0.25 L of compatability agent to 100 L of spray volume.

Compatibility problems can sometimes occur when attempting to mix wettable powders and emulsifiable concentrates. The wettable powder may draw off the emulsifier and the result is a paste or putty formed from the wettable powder and the "oil." Mixing procedures can make a difference between a satisfactory mix and a "gunky" mess. Always partially fill the spray tank before adding the pesticide. If a compatibility agent is to be used, add it before the pesticides. Always add formulations that form suspensions and thoroughly disperse them before adding emulsifiable concentrates. Premix wettable powders with water to form a slurry unless using an inductor system. Emulsifiable concentrates should be mixed with equal amounts of water (preemulsified) before adding them to a fluid fertilizer.

Water quality, fertilizer type, and solution temperature also affect compatibility. Hard water may cause more problems than soft water. Chlorinated water may also cause problems. Fertilizer quality and grade affect compatibility. Nitrogen solutions containing free NH_3 may cause problems. However, N solutions usually are more compatible than mixed liquids (McGlamery, 1973). Check compatibility when you change water sources or fertilizer batches.

F. Checking Compatibility

Check compatibility of agricultural chemicals in small containers before mixing tankfuls. Simply use proportional amounts. If you plan to use 1 L of herbicide in 100 L of fertilizer solution for field application, you can use 1 mL/100 mL in the laboratory. If you plan to use 1 kg/100 L in the field, that is the same as 1000 g/100 000 mL. Thus, you can use 1 g/100 mL.

Testing procedure:

1) Calculate spray volume per hectare and volume or weight of pesticide per hectare.
2) Place 400 mL of carrier in each of two containers.
3) Add 1 mL of compatibility agent to one container and mark *A*. Mark the other container *B*.
4) Add the proper amount of each pesticide to each container in the proper sequence (note above on mixing procedure).
5) Close the containers and shake or invert to mix.
6) Observe the mixtures immediately and again later.

Compare container *B* with container *A* to determine the value of adding a compatibility agent. If materials remain suspended or if they are easily resuspended in container *B*, mixing is possible with good agitation without a compatibility agent. If they separate, precipitate, or form "gunk" in container *B*, check container *A* to see if adding a compatibility agent will solve the problem. If so, you may want to repeat the test using varying amounts of compatibility agent to determine the optimum amount needed.

G. Impregnation on Dry Fertilizer

Some herbicides are approved for mixing with dry fertilizers (Cates, 1981). Good herbicide performance will depend on thorough mixing and uniform application.

A minimum dry fertilizer rate of 225 kg/ha is recommended to assure that herbicide is uniformly absorbed during impregnation and uniformly applied. The maximum per hectare rate that can be impregnated uniformly with most herbicides is 450 to 500 kg.

The impregnation process begins with placing at least 1 t of the fertilizer mixture to be used on a grower's field in a clean blender. Next, the herbicide is sprayed into the mixer at appropriate pattern and volume—a 15° nozzle with a capacity of 68 L/min at 1.40 kg/cm^2 (atm) pressure.

To assure thorough coverage, the nozzles should be aimed so that spray hits the fertilizer surface inside the blender over the length of the drum. Once all the herbicide has been sprayed into the blender, mixing should continue for another 2 min to assure proper impregnation.

With a herbicide–fertilizer mix, a powdered drying agent such as Microcel E, diatomaceous earth, or finely powdered clay can help condition the mixture so it flows freely. This is especially useful when applying amounts near the 225-kg minimum. Generally, < 2% by weight of absorptive powder is needed.

Some herbicide manufacturers caution against using straight limestone for impregnation because good absorption is not achieved, but fertilizer blends containing some limestone may be suitable. Some labels also caution against attempting to blend certain herbicides with ammonium nitrate, potassium nitrate, or sodium nitrate alone or in blends with other fertilizers.

To help assure more uniform application, fields can be double-spread by overlapping 50% on each adjacent spreader pass. The feather edge of the first pass serves as the center line for the second.

Those interested in impregnating liquid herbicide on dry fertilizer should request information from the herbicide manufacturer to determine what fertilizers are appropriate for use and to obtain additional instructions.

H. Weed and Feed for Lawns

Some dry materials are available for application to turf areas to provide both fertilizer and weed control (Kageyama, 1981). One of the more common herbicides for this use is 2,4-D. Although more commonly used as a spray, the 2,4-D can be released from the dry material and absorbed by leaves and by roots of weeds when adequate moisture is available from dew or from watering (Fig. 4).

Fig. 4—Weed and feed combinations have been popular for lawns (courtesy O. M. Scott & Sons, Marysville, Ohio).

As with field crops, a good fertility program for lawns to encourage vigorous growth of the desirable plants can help considerably in giving the desirable plants a competitive advantage over weeds.

IV. ENVIRONMENTAL AND REGULATORY CONSIDERATIONS

A. Storage and Distribution

Dealers and applicators using fertilizer and herbicides for agricultural purposes usually store each herbicide product separately in original containers or in bulk tanks. Original containers are usually plastic or metal. The drums are sometimes referred to as mini-bulk. For some herbicides it is important to prevent freezing of the herbicides.

For herbicides in bulk quantities, stainless steel tanks are commonly used. Contamination of herbicides with other herbicides, other pesticides, water, or other impurities should be prevented to assure purity. A very small amount of some herbicides contaminating fertilizer or other pesticides can cause serious problems on certain crops. A good storage system should be used that will prevent errors and assure that the proper herbicide is used for each application (Fig. 5).

When herbicides are mixed with liquid or dry fertilizer, the combination is usually prepared shortly before use, and mixing is frequently done directly in the tanks of the application equipment. Attempting to store some herbicide–fertilizer mixtures may allow loss of especially the more volatile herbicides, create adverse working conditions at the facility, and contribute to cross contamination.

Fig. 5—Herbicides to be mixed with fertilizer are frequently stored in bulk tanks. Tanks should be clean and resistant to corrosion and the effect of herbicide solvents. Extreme care should be taken to avoid contamination.

Fig. 6—A unique use for weeds such as waterhyacinth is to remove nutrients such as N from water for purification purposes.

Dealers and applicators should keep informed of current state and federal regulations governing the labeling, storage, sale, and movement of herbicides, especially when dealing with bulk quantities and mixtures. Regulations on container disposal must also be considered.

B. Weeds for Removing N from Water

A rather unique use for weeds has been for removal of excessive nutrients such as N and P from wastewater effluent (Holm & Yeo, 1981; Sutton, 1981). Water treatment engineers report successful removal of nutrients from water by using waterhyacinth [*Eichornia crassipes* (Mart.) Solms], which has suspended fibrous roots and a phenomenal growth rate (Fig. 6). The weed can be harvested and dried and then used for fertilizer or the biomass can serve as an energy source, since it can be digested anaerobically to produce methane gas.

REFERENCES

Berg, G. L. 1980. How to consult on no-till. Ag Consultant and Fieldman 36(6):26–27.

Cates, D. 1981. Mixing herbicides with dry bulk fertilizer. Weeds Today 12(4):12–13.

Holm, L., and R. Yeo. 1981. The biology, control and utilization of aquatic weeds, Part IV. Weeds Today 12(2):3–4.

Kageyama, M. E. 1981. Controlling weeds in turf. Weeds Today 12(3):8–10.

Knake, E. L. 1960a. Giant foxtail—a robber in your fields. Illinois Research 2(3):3–4.

Knake, E. L. 1960b. Weed tax? Patrons Guide 12(3):5–7.

Knake, E. L., A. P. Appleby, and W. R. Furtick. 1967. Soil incorporation and site of uptake of preemergence herbicides. Weeds 15:228–232.

Knake, E. L., and D. L. Mulvaney. 1966. Nitrogen solution vs. water as a carrier for atrazine. North Central Weed Control Conference Research Report 23:5–6. North Central Weed Control Conf., Champaign, Ill.

Knake, E. L., ad F. W. Slife. 1962. Competition of *Setaria faberii* with corn and soybeans. Weeds 10:26–29.

Knake, E. L., and F. W. Slife. 1965. Giant foxtail seeded at various times in corn and soybeans. Weeds 13:331–333.

Knake, E. L., and F. W. Slife. 1966. Control those weeds early! Illinois Research 8(2):8–9.

Knake, E. L., and F. W. Slife. 1969. Effect of time of removal of giant foxtail from corn and soybeans. Weed Sci. 17:281–283.

Knake, E. L., and L. M. Wax. 1968. The importance of the shoot of giant foxtail for uptake of preemergence herbicides. Weeds 16:393–395.

McGlamery, M. D. 1970. Tank-mixed agricultural chemical combinations. Crops Soils 22(7): 10–12.

McGlamery, M. D. 1973. Feed and weed—the perfect marriage. Farm Chem. 136:2–3.

McGlamery, M. D., and F. W. Slife. 1966. The adsorption and desorption of atrazine as affected by pH, temperature, and concentration. Weeds 14:237–239.

Parker, C. 1966. The importance of shoot entry in the action of herbicides applied to the soil. Weeds 14:117–121.

Parochetti, J. V. 1980. Tillage rotation. Crops Soils 3(1):8–9.

Penner, D. 1974. The effectiveness of pesticide interactions. Weeds Today 5(3):13–14.

Stickler, R. L., E. L. Knake, and T. D. Hinesly. 1969. Soil moisture and effectiveness of preemergence herbicides. Weed Sci. 17:257–259.

Sutton, D. L. 1981. Who needs aquatic weeds. Weeds Today 12(3):6–7.

29

J. Mark Scriber
University of Wisconsin
Madison, Wisconsin

Nitrogen Nutrition of Plants and Insect Invasion

In a publication on integrated pest management, Bottrell (1980, p. vi) wrote that "Despite advances in modern chemical control and the dramatic increase (about 10-fold) in chemical pesticides used on U.S. cropland during the past 30 years, annual crop losses. . .caused by insects may have nearly doubled." The most significant factor in this trend is our general production style in agriculture and silviculture. These systems represent major ecological disturbances that impact not only upon the soil and plant insect populations but also upon adjacent and distant ecosystems (Whittaker, 1975; Brink et al., 1977; Nielson & MacDonald, 1978; Bormann & Likens, 1979).

The "Green Revolution" has increased food production in some countries in recent years with new plant varieties; however, this increase is generally possible only with continued inputs of water (irrigation), fertilizers, and pesticides (Pimentel, 1976; cf. Lockeretz et al., 1981). In view of the increasing costs of petroleum products to produce these inputs and the fact that our crop losses to insects have increased in spite of a well-developed insecticide technology, it becomes especially important to develop a broad-scope approach to pest management. We must employ a variety of management tools, including insecticides, with an understanding of the ecological processes involved at all levels of biological organization (e.g., organism, population, community, ecosystem).

Because fertilizers are a major part of our modern agricultural and silvicultural systems, it is logical to question the extent to which increased crop damage due to insects in the last 30 yr might be the result of heavy application of N in various fertilizers or green manure crops. It seems that fertilizers rarely affect insect population dynamics directly; however, their indirect effects through alterations in plant resistance, tolerance, or phenological escape (pseudoresistance) are often very significant.

Nitrogen in Crop Production.

The large differences in N content of insect tissue and plant tissue may be the major reason why less than one-third of the insect orders and higher taxa of terrestrial arthropods have achieved the ability to feed on seed plants (Southwood, 1972). Animals (including insects and mites) consist mainly of protein (ranging from 7 to 14% N), whereas plants or plant parts rarely reach concentrations of 7% N, and are generally much lower (mean = 2.1% N, n = 894 analyses of nearly 400 species of woody plants; Russell, 1947) (see also Mattson, 1980). For the insect faced with the need to grow and reproduce, success depends upon its ability to efficiently and/or rapidly ingest, digest, and convert plant N. The various organism, environmental, and plant factors affecting these processes have recently been reviewed (Scriber & Slansky, 1981). The particular importance of N in insect/plant interactions has been excellently reviewed by McNeill and Southwood (1978) and Mattson (1980). The more specific relationships between N fertilizer regimes and plant resistance to pests has been most recently reviewed by Singh (1970), Leath and Radcliffe (1974), Jones (1976), and Tingey and Singh (1980).

While it is generally assumed that increased plant N increases insect damage or populations, there is a considerable amount of evidence that is inconclusive or indicates the reverse. A survey of literature over the last 100 yr (Scriber, 1984) shows a minimum of 115 different studies in which insect damage, growth, fecundity, or populations increased with increased plant N. Similar increases with increased N are observed for at least 20 studies with mites. On the other hand, a minimum of 44 studies with insects and a few studies with mites indicate a decrease in herbivore populations or damage to plants with high N concentrations (or insect increases with low N).

Fertilization of plants will affect the host selection, survival, growth rates, and reproduction of insect herbivores and may also alter plant microclimate (e.g., larger internodes or more leaf surface), relationships to other plants (i.e., weeds; see Chapt. 28, this book) or plant pathogens (Chapt. 30, this book). These changes alter the value of the plant (or crop) as a home for the herbivore and its natural enemies (Altieri & Whitcomb, 1979). The extent of the effects of such fertilizer-induced physiological and morphological alterations to plants will also depend upon the particular "feeding guild" to which an insect might belong (e.g., phloem or xylem "sappers"; wood, root, shoot, or stalk "borers"; leaf or needle "miners"; root, needle, or leaf "chewers"; fruit and seed feeders; etc.; see Slansky & Scriber, 1982, 1984). We must also remember the fertilization effects upon the desired "invasion" of plants by pollinators.

My objective is to show how insect invasion and population growth rates are related to plant N. My approach is based upon data from bioassays from a variety of crop species of different plant growth forms (i.e., trees, shrubs, ferns, grasses, forbs, etc.) using 30 different species of leaf-chewing Lepidoptera and Hymenoptera. This information provides a conceptual basis or model for interpreting fertilizer-induced nutritional and/or allelochemic effects upon the optimal physiological growth responses to be expected from leaf-chewing insects in general. Behavioral, ecological, and evolutionary considerations must then be superimposed upon this insect/plant interaction response surface (Price et al., 1980; Scriber & Slansky, 1982; Slansky, 1982b). For example, previous feeding experience of an individual armyworm (*Spodoptera eridania*) can affect the N utilization efficiency and reproductive success of the insect in significant ways (Scriber, 1981, 1982).

While agricultural consultants are sometimes unaware of the theoretical bases for effective application, ecologists often draw the wrong conclusions and make erroneous predictions because their assumptions are frequently formed from studies of natural ecosystems (Levins & Wilson, 1980). The hybrid discipline of "agro-ecology," which is emerging from integration of theory and application, provides the potential for an extremely fertile research interface, especially if we begin to accept the concept that natural systems as well as agro-ecosystems are dynamic and perhaps never at an ecological or evolutionary equilibrium.

I. PLANT N AND INSECT GROWTH

A. Nutritional Aspects of Plant Quality

The only data concerning approximate digestibility (AD) response in insects as a function of plant fertilization regimes are for maize plants (*Zea mays* L.) (Fig. 1b and Table 1) and collards (Fig. 2). Fertilization of these crops resulted in both higher leaf water and N contents and a significant increase in digestibility for the insects concerned (Manuwoto, 1984 and unpublished; Slansky & Feeny, 1977).

Leaf water as a general index of plant growth form (i.e., tree vs. forb) is surprisingly valid (Fig. 1a) despite seasonal (McHargue & Roy, 1932; Scriber & Slansky, 1981) and daily variations (Scriber, 1977). Leaf N is generally less useful in this regard, possibly because it varies by as much as 20% during a daily cycle (Pate, 1980). However, the growth rates of forb-feeding insects are generally more closely correlated with N than is the case of tree leaf feeders (Scriber & Feeny, 1979). The relative growth rate (RGR = milligrams gain per day per milligram of mean biomass during that particular instar) of the Umbelliferae-specialized herbivore *Papilio polyxenes* on 11 different food-plant species closely paralleled total organic leaf N (Fig. 3), largely because of the strong correlation of approximate digestibility (AD) with leaf N. The variation in secondary plant metabolites in these species had little effect on the correlation. The poor correspondence of plant N content with calendar date relates to the different phenologies and habitats (meadow, marsh, woodland, etc.) of the plant species. An extremely strong relationship between digestibility and N concentration of ingested food is also observed for phloem-sucking Hemiptera (McNeill & Southwood, 1978) and many other guilds of insects. A literature search (Slansky & Scriber, 1982) reveals the mean AD of 32 species of insect

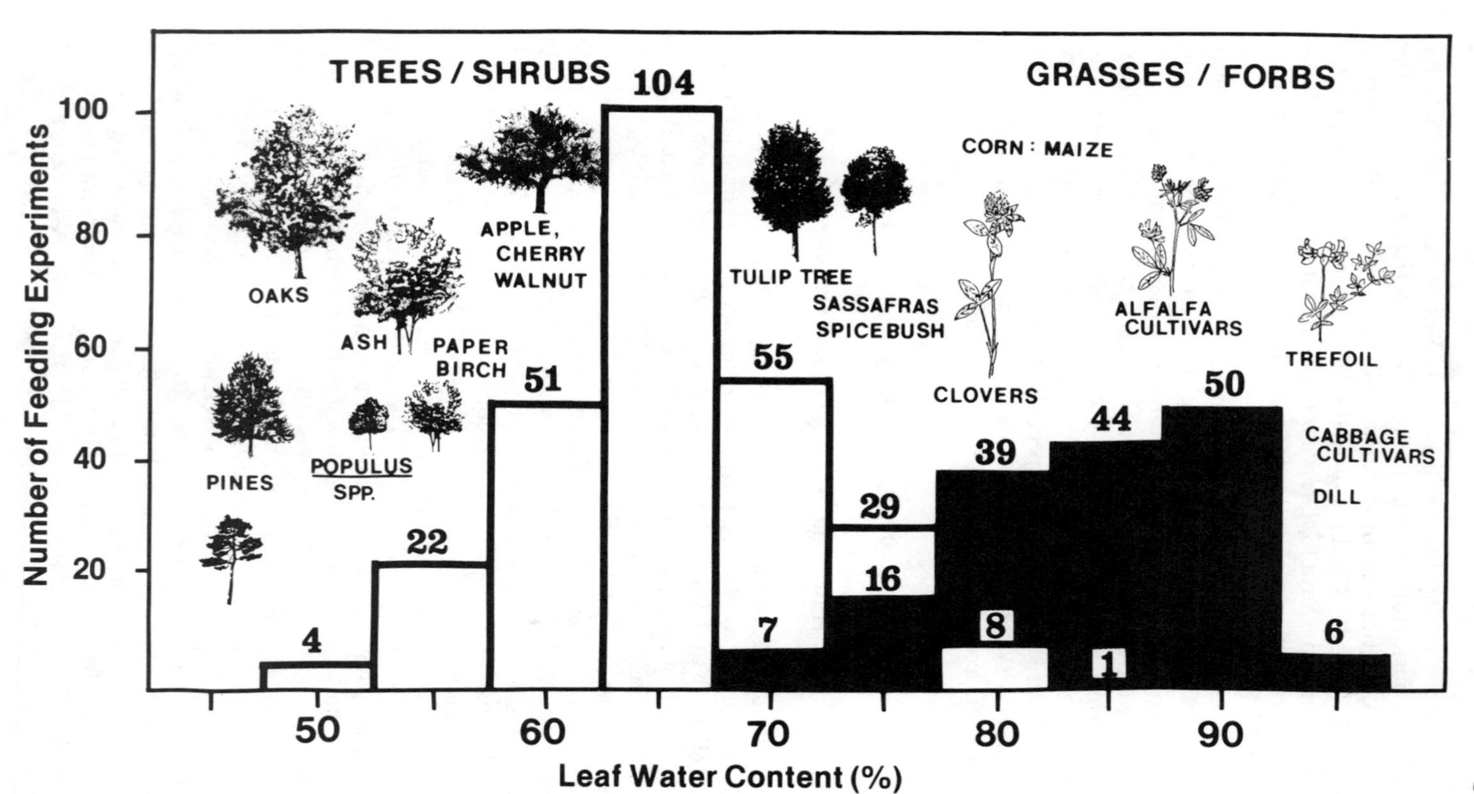

Fig. 1a—The number of penultimate instar insect feeding experiments ($n = 436$, with each represented by numbered points in Fig. 1b) for various plant species along sections of the axis of leaf water content of Fig. 1b. Where overlap exists, the shaded portion represents experiments involving insects fed forbs, and the nonshaded portion represents experiments using leaves of trees or shrubs.

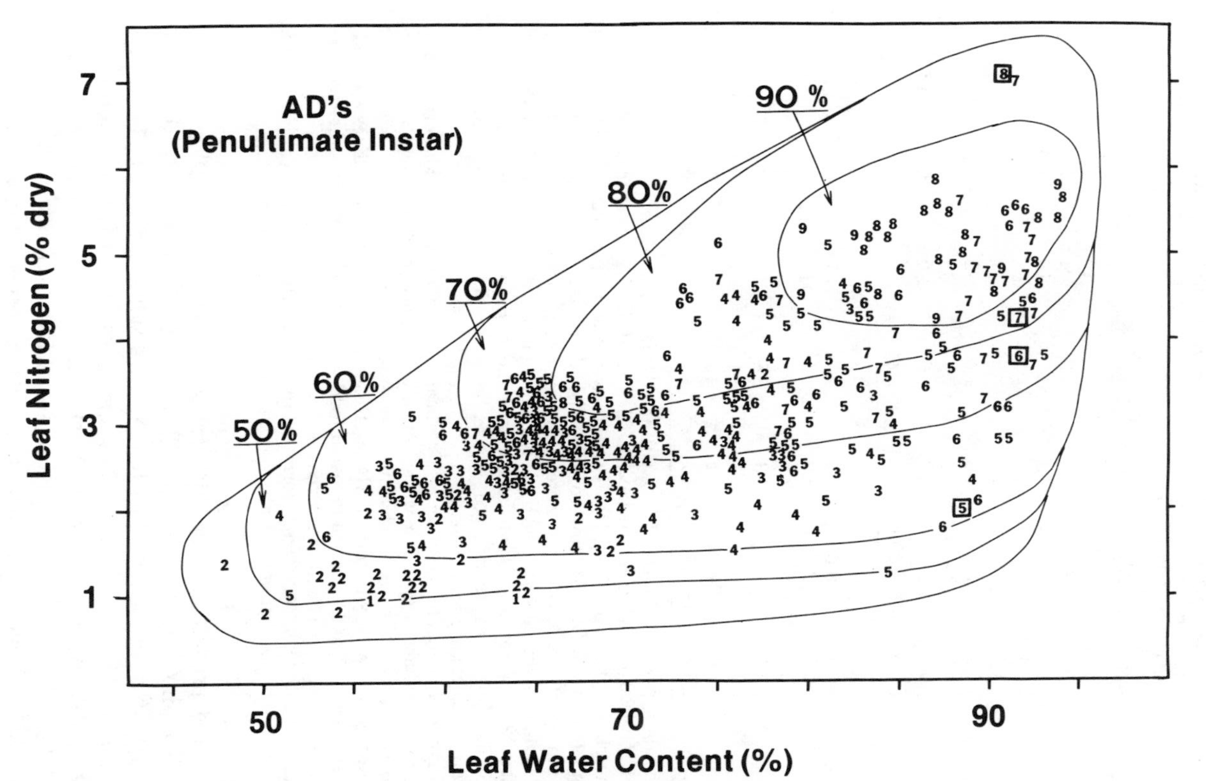

Fig. 1b—The approximate digestibilities (AD's) of penultimate instar insect larvae from 30 species as a function of plant quality as defined by leaf water and total organic N. Lines exclose all digestibilities equal to or greater than those indicated for the zone, but do not exclude lower values. Each number represents the mean AD of replicate larvae for a particular host plant species under controlled environment conditions (0 = 0–4%, 1 = 5–14%, 2 = 15–24%, etc. to 9 = 85–94%; see text; Fogal, 1974; Scriber & Slansky, 1981; and Scriber, 1984). Performance values for *S. eridania* in relation to differential N fertilization of 'B49' maize are indicated by boxes (see also Table 1).

Table 1—Armyworm growth performance in relation to various maize fertility regimes (S. Manuwoto, unpublished results).

Plant† and fertility regime‡	Leaf quality§		Indices¶ of insect# growth performance				
	N	H_2O	AD	ECD	ECI	RCR	RGR
	% dry	% fresh	%			mg/day per mg	
			DIMBOA-less mutant Maize				
A	2.7 ±0.3	88 ±0.7	61 ±3	26 ±2	15 ±1	1.9 ±0.1	0.29 ±0.01
B	4.7 ±0.1	92 ±0.2	71 ±1	27 ±2	19 ±1	1.9 ±0.1	0.35 ±0.01
C	4.9 ±0.1	92 ±0.3	68 ±3	32 ±4	21 ±2	1.9 ±0.2	0.38 ±0.01
D	7.0 ±0.1	91 ±1.0	68 ±2	54 ±6	35 ±0	1.9 ±0.2	0.62 ±0.0
			'B49' Maize (high DIMBOA)				
A	2.0 ±0.4	89 ±0.5	52 ±3	13 ±1	7 ±1	2.3 ±0.1	0.15 ±0.01
B	3.7 ±0.1	92 ±0.2	59 ±2	29 ±3	17 ±1	1.8 ±0.1	0.30 ±0.01
C	4.1 ±0.1	92 ±0.4	73 ±1	18 ±1	13 ±1	2.8 ±0.1	0.36 ±0.01
D	7.0 ±0.2	91 ±0.5	81 ±1	12 ±1	10 ±1	4.8 ±0.2	0.48 ±0.03

† Maize seeds were obtained from W. D. Guthrie (USDA–SEA Corn Insects Lab, Iowa and V. E. Gracen (Cornell Univ., Ithaca, N.Y.). The DIMBOA concentrations at 15-cm plant height are 0.5 and 5.0 mg DIMBOA/g dry tissue, respectively, for DL mutant and 'B49' (treatment D; S. Manuwoto, unpublished results).

‡ Plants in treatments *A*, *B*, and *C* were grown in sterile sand with variable N concentrations and a modification of the basic nutrient solution described by Chevalier and Schrader (1977). Treatment *D* consisted of greenhouse-grown maize in a standard soil mixture (compost, soil, sand, peat moss, and vermiculite) supplemented with Osmocote® (a slow-release fertilizer).

§ Total organic N was determined ($n = 3$ replications) via a micro-Kjeldahl technique (McKenzie & Wallace, 1954). For water content, samples of leaves were taken each day and averaged for the whole larval instar ($n = 18$).

¶ AD = approximate digestibility; ECD = efficiency of conversion of digested food; ECI = efficiency of conversion of ingested food = AD × ECD; RCR = relative consumption rate; RGR = relative growth rate = RCR × AD × ECD.

Penultimate instar southern armyworms were used in these bioassays (Noctuidae: Lepidoptera). The number of replicates per treatment ranged from 8 to 12.

parasites and predators to be 80% (range 37–98%, n = 119 studies); 9 species of grain/seed chewers with 72% (range 46–96, n = 48 studies)—compared with lower mean values for 38 forb-feeding species of 53% (range 16–97%, n = 716 treatments); and 99 species of insects feeding on trees of 39% (range 2–94%, n = 633 treatments).

While digestibilities of southern armyworm larvae (*Spodoptera eridania*), on fertilized 'B49' maize were near upper values expected based on plant N and H_2O (Fig. 1b), the digestibilities of the cabbage butterfly (*Pieris rapae*) were much lower than might otherwise have been expected (Fig. 2). Possibly the glucosinolates in leaf tissues were interfering with assimilation of biomass and/or N of various Cruciferae (Slansky & Feeny, 1977), especially

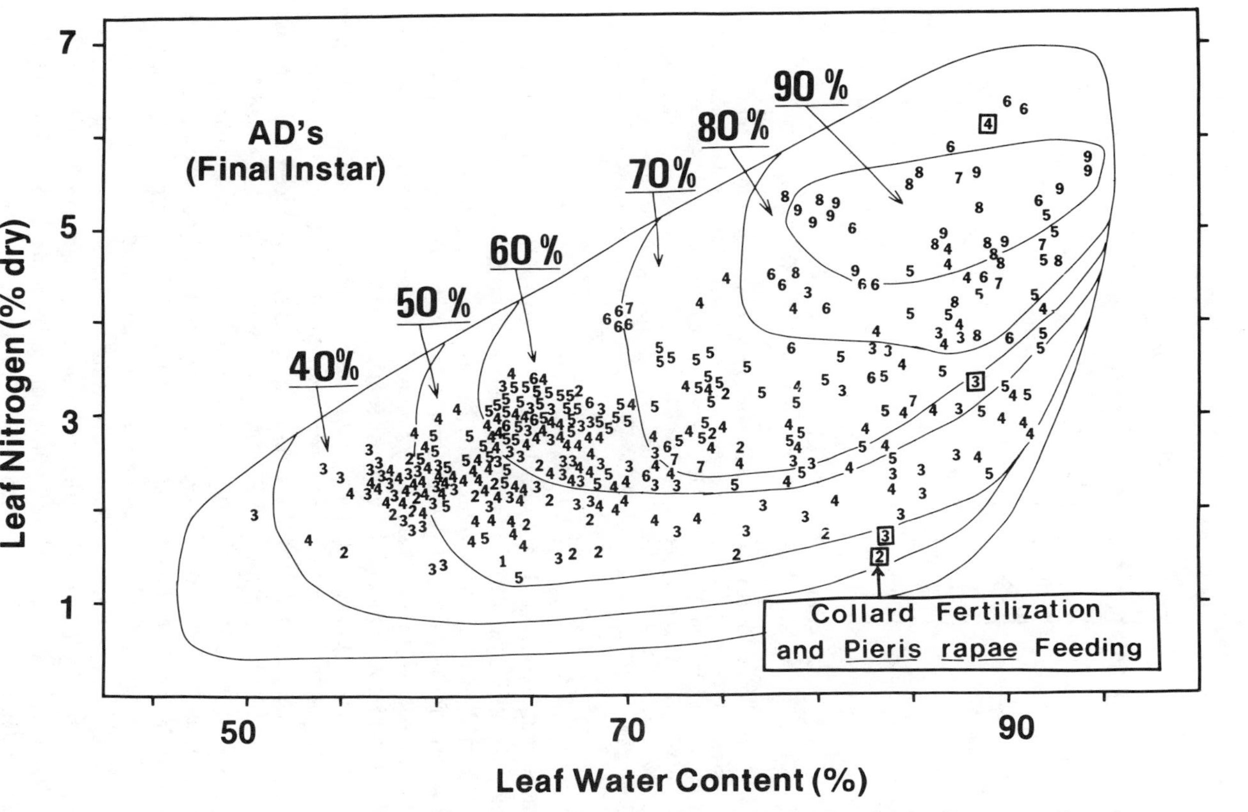

Fig. 2—Digestibilities of final instar larvae as a function of plant quality (see description in Fig. 1b). Performance values for *Pieris rapae* in response to differential N fertilization of collards (*Brassica oleraceae*) (Slansky & Feeny, 1977) are represented by four boxes.

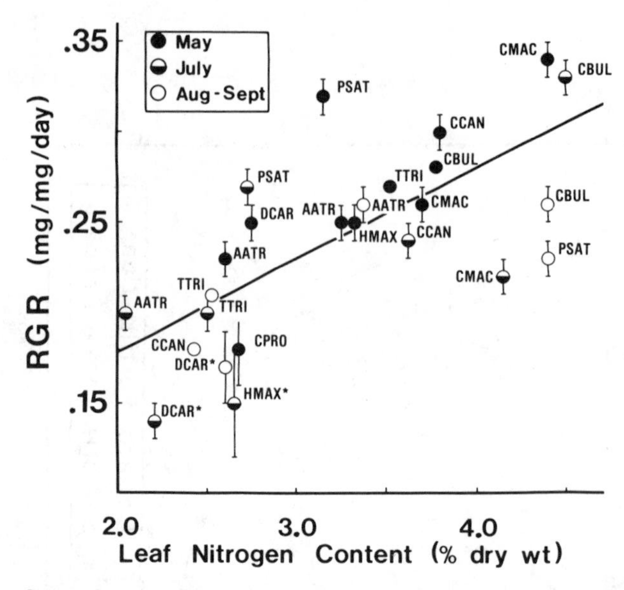

Fig. 3—Larval growth rate RGR (±SE) of *Papilio polyxenes* in relation to total organic N content of leaves of various species of Umbelliferae at various dates (data from Finke, 1977) (Dayton, Ohio). AATR = *Angelica atropurpurea*; DCAR = *Daucus carota*; CCAN = *Cryptotaenia canadensis*; HMAC = *Heracleum maximum*; PSAT = *Pastinaca sativa*; CMAC = *Conium maculatum*; CBUL = *Cicuta bulbifera*; CPRO = *Chaerophyllum procumbens*; and TTRI = *Thaspium trifoliatum* ($n = 25$; $r = 0.685$).

because these compounds are known to be gut irritants in mammals and to reduce digestibility in certain Lepidoptera (Erickson & Feeny, 1974).

Crop digestibility for many phytophagous insects may generally increase with fertilization. However, certain allelochemics such as DIMBOA [2,4-dihydroxy-7-methoxy(2H)-benzoxazin-3(4H)-one] in maize may impose a large enough metabolic expense (perhaps reflecting costs of detoxifying or otherwise processing such plant tissues) that the overall efficiency (ECI) may actually decline. For southern armyworms fed 'B49' maize tissue, this decline in ECI was the result of a severely reduced efficiency of conversion of digested food (ECD) in spite of very high AD's (see Manuwoto & Scriber, 1982; and Table 1). Another interesting mechanism employed by armyworms on deleterious plant species or cultivars (i.e., with toxins and/or digestibility reducers) is an increased consumption rate (see 'B49' Table 1). With this "power vs. efficiency" trade-off mechanism in their adaptive repertoire, it is not surprising that little or no resistance to armyworms has yet been identified for most crops (see also Scriber, 1979). Thus, in certain crops, invasion and damage by some insects may remain largely independent of fertilization and plant-breeding programs. This also suggests that in addition to the limits upon growth performance of insects imposed by quantity and/or quality of leaf N and/or water, fertilization may alter secondary metabolites (especially N-containing allelochemics; Bernays, 1983), and thus, cause differential acceptability or suitability for various potential pests.

B. Allelochemic Aspects of Plant Quality

If toxins are generally concentrated in young or growing tissues of plants (Rosenthal & Janzen, 1979), if concentration of total organic N is positively correlated with metabolic activity of plants (Mattson, 1980), and if both decline with maturity or leaf age, it is logical to assume that N fertilization reverses or slows the processes correlated with aging and increases the levels of toxic allelochemics. However, new vegetative growth competes effectively with plant allocation of C for other modes of protection (e.g., digestibility-reducing allelochemics), structure (cellulose, lignin), and reproduction (see Mattson, 1980).

Examples exist for all plant growth forms in which the hydrocyanic acid (HCN) potential or concentration of cyanogenic glycosides decline with age: black cherry trees (*Prunus serotina*) (Smeathers et al., 1973); the chapparal shrub (*Heteromeles arbutifolia*) (Dement & Mooney, 1974); the fern (*Pteridium aquilinum*) (Lawton, 1976; Cooper-Driver et al., 1977); the forb (*Lotus corniculatus*) (Scriber, 1978); and the grass (*Sorghum bicolor*) (Woodhead & Bernays, 1978). While HCN is certainly important in plant defense against various potential enemies (Jones, 1972; Conn, 1979), its role in primary plant processes (see Seigler & Price, 1976) and its antiherbivore role relative to other cooccurring allelochemics is still generally uncertain (Conn, 1980).

Studies with sudangrass (*Sorghum* sp.) indicate that N fertilization increases the concentration of HCN and NO_3^- in plant tissues, whereas they are negatively associated with P fertilization (Harms & Tucker, 1973 and references therein). These and other N-based allelochemics are affected by light intensity and other environmentl variables, as well (Fluck, 1963; Harms & Tucker, 1973), and may increase more when fertilized by NH_4^+ or urea than by NO_3^- sources (Marten et al., 1974).

Variable effects of N fertilization upon other classes of allelochemics are reported by Fluck (1963). Nitrogen deficiency, as one form of plant stress, may result in greater concentrations of starch in C_4 plants (Gallaher & Brown, 1977) and increases in chlorogenic acids in sunflowers (*Helianthus annuus* L.) and other plants (de Moral, 1972). The importance of these and other changes in plant defensive chemicals (including the digestibility-reducing quantitative defenses such as tannins, lignin, silica, waxes, resins, etc., which result from plant stresses) will likely become one of the highest priorities of research involving insect/plant interactions in relation to herbivore population dynamics during the next 10 yr (Rhoades, 1979; Haukioja, 1980; Mattson, 1980; Schultz & Baldwin, 1982).

The behavioral and physiological adaptability of insects provides ecological options that are not always available to fungus, bacteria, and virus pathogens. While fertilizers can modify antibiosis and nonpreference (antixenosis), it is usually not certain whether these effects are due to alterations in allelochemics, nutrients, or water content. Furthermore, it is uncertain if desirable effects could be implemented in field practice where several pests of varying degrees of feeding specialization may present conflicting requirements.

The primary role that N fertilization plays in insect pest management is to help plants compensate for lost or damaged parts (i.e., tolerance), rapidly pass critical developmental stages, and escape attack (Jones, 1976). In general, the form of N fertilizer is less responsible for plant susceptibility to insects than to plant pathogens (see Chapt. 30, this book); however, heavy manuring may favor mites or certain insects such as the seedcorn maggot (*Hylemya* sp.), which are attracted to soil organic material of high N content (Dicke, 1977). In addition, plants of the Chenopodiaceae, Compositae, Cruciferae, and Solanaceae families tend to store N as NO_3^-, sometimes in amounts up to 5% of tissue dry weight (Pate, 1980). Levels of NO_3^- even lower than this can be harmful to both mammalian grazers (Raymond, 1969; Noller & Rhykerd, 1974) and insect herbivores (Manglitz et al., 1976). On the other hand, trees tend to transport organic N (specific proteins and/or amino acids) instead of inorganic N (Terman et al., 1976; Mattson, 1980; Pate, 1980). The implications of this phenomenon in explaining fertilizer-induced differences of insect responses on forbs vs. trees are as yet unknown.

These considerations again emphasize that total organic leaf N is a useful index of maximum insect growth performance, especially when combined with water content. However, the specific form (NO_3^-, NH_4^+, protein, amino acids, glucosinolates, cyanogenic glycosides, alkaloids, etc.), the balance, and the interactions of nutrients and allelochemics will be especially critical to insect invasion of a particular plant tissue. For example, while balsam wooly aphid (*Mindarus abietinus* Koch) populations are promoted on fir trees fertilized with urea, the opposite effects are observed when fertilized with ammonium nitrate (AN) (Carrow & Betts, 1973).

Tannins reduce digestibility and availability of leaf N for herbivores (van Sumere et al., 1975) and may thus serve as formidable defenses against many Lepidoptera (Feeny, 1970; cf. Zucker, 1983). These compounds are, however, less effective against certain Orthoptera and Coleoptera. Growth of Coleoptera on 11 different species of *Eucalyptus* trees was better correlated with small differences in N (over a range of 0.5–1.9%) than with large differences in tannins or phenolics (Fox & Macauley, 1977).

C. Plant N and Beneficial Insects

Nectar is far more than just a sugar-water solution and may contain a variety of chemicals such as amino acids, enzymes, lipids, organic acids, and alkaloids, among others (Maurizio, 1975). Differential concentration of these various compounds and diel changes in nectar quality are significantly influenced by fertilization, and will significantly affect insect visitation (Beutler, 1953; Corbet et al., 1979). A delicate balance between pollinators and flower quality is critical to the coevolutionary ecology and practical management of the species involved (Heinrich, 1979; Gojmerac, 1980). While "sugar value" may be most critical for honey bees, other pollinating insects such as carrion and dung flies are attracted to amino acids and odors of decay (Baker & Baker, 1980).

Nitrogen fertilization can affect the color, number, and size of flowers, but generally has a deleterious effect upon nectar secretion and the "pollination index" (Maurizio, 1975; Shasha'a et al., 1976). The reduced nectar secretion at high N levels may be related to increased vegetative growth, whereas low levels of N fertilization would favor limited growth, accumulation of a surplus of carbohydrates, and copius nectare secretion (Shuel, 1957).

For examples of short-term plant benefits from insect herbivory exist. However, Owen (1980) has suggested that the deposition of sugary honeydew exudates of aphids may increase the rate of N fixation beneath the plant by providing an energy source for free-living bacteria. The extent of insect "cropping" (via eating plant growing tissue) and the role of arthropod feces in N cycling and vegetational responses of various ecosystems can be significant (Mattson & Addy, 1975), and the long-term advantages of such activities warrant additional research.

D. Insect Population Dynamics and the Synoptic Model

Theories of host plant selection by insects have usually placed a major emphasis upon either the role of plant nutritional quality or allelochemics. However, fertilizer-induced differences in leaf water may also determine or influence oviposition choices for certain Lepidoptera (Wolfson, 1980). Again, it is the dynamic interaction of various chemicals that determines the behavioral, physiological, ecological, and evolutionary interactions between insects and plants (Thompson, 1982; Ahmad, 1983; Bell & Carde, 1984). Interpopulation and intrapopulation differences in host plant utilization abilities have significant implications for plant resistance to insects (Scriber, 1983; Diehl & Bush, 1984).

In attempting to understand the role of N as it affects the population dynamics of insects, we must interpret its potential to increase or enhance (i) individual growth rate, (ii) survival, (iii) reproduction, or (iv) some combination of responses of herbivores (e.g., McClure, 1980) in relation to their various plant communities (Atsatt & O'Dowd, 1976) and different life styles (Southwood, 1977). At a gross level of comparison, we are beginning to see what might be considered major classes or potential categories of plant/herbivore interaction, which are related in large part to plant successional status and plant growth forms (Feeny, 1976; Rhoades & Cates, 1976; Southwood, 1977; Scriber, 1984). Mattson (1980) provides an excellent discussion of the role of N in determining the ecological "strategies" of plants. Whitham (1981, 1983) reviews within-plant variation in chemistry as a fascinating insect pest defensive strategy.

By using the conceptual (synoptic) model of Southwood (1977), we can obtain an important ecological perspective of many key factors that affect population dynamics of pest species simultaneously across a spectrum of crops from annuals to perennial trees. For example, we can observe how small nutritional differences (such as those resulting from N fertilization) can enhance the rate of increase of insect herbivores to such an extent that they escape (outgrow) the numerical response capabilities of their natural enemies (Huffaker et al., 1969; Hurd & Wolf, 1974; White, 1978). This change from the endemic levels normally maintained by natural enemies, to epidemic levels may occur along all points of the habitat-axis in the synoptic model. But, insect parasites and/or predators are least effective at either extreme, and biological control is successful only in limited types of situations. Enrichment of a crop system with N may destabilize the insect population dynamics to such a significant level that some insect species may be driven to extinction (Rosenzweig, 1971; Reibesell, 1974; Myers & Post, 1981; Prestidge, 1982).

When combined with key considerations of the apparency theory of plant defense (Gilbert, 1979), the synoptic model of insect population dynamics explains why breeding plants for insect resistance works for only certain types of pests and why cultural practices (including various irrigation, crop rotation, or fertilizer regimes) may reduce colonization and damage for certain insects but not others. Understanding the needs for insecticides or miticides, especially at the extremes of Southwood's (1977) habitat stability axis, and the variable potential for other control options that might work for insects typical of these different cropping situations is central to effective designs of ecologically comprehensive, environmentally sound, and economically feasible pest management systems for all crops.

II. PRACTICAL APPROACHES

A crop that is fertilized with N changes quality as a resource for insect herbivores as well as for livestock or human consumers. The magnitude and direction of the effects of fertilization upon insect populations vary significantly, depending upon many factors. However, except for tank mix compatibility of fertilizer-insecticide combinations, there are essentially no existing insect management strategies based directly upon the type (NH_3, AN, urea, etc.), form (solid or fluid), or timing of crop fertilization. Except for preliminary assessment of the potential direct deleterious effects of NH_3 or urea upon certain soil insects [e.g., rootworms (*Diabrotica* sp.) at South Dakota, and clover root weevils (*Sitona* sp) at Kentucky], serious attempts have not been made in this regard. The direct effects upon herbivorous insects of foliar applications of fertilizer are also basically unexplored at this time. The indirect effects of N and/or N–P–K fertilizers upon the plant or the plant community seem to be much more important considerations regarding insect responses. Carefully designed experiments that address specific physiological plant responses to different types or methods of fertilization with respect to different guilds of pests at specific stages of plant growth under diverse yet controlled conditions will contribute significantly to our understanding of herbivore or pollinator population dynamics. However, such an autecological approach is of limited usefulness unless we also simultaneously investigate the community (i.e., synecological) aspects of insect population management under natural conditions. In these considerations we must include fertilizer-induced changes in crop or community microclimate (e.g., with weed growth) and corresponding changes in numbers or composition of the insect natural enemies (i.e., parasites and/or predators).

A large part of the confusion in the literature regarding differential effects of plant fertilization upon insect invasion, survival, and growth is in direct proportion to our ignorance of the effects of fertilization upon physiology of the plant itself. A critical area that needs more research concerns the insect responses with respect to biochemical dynamics of allelochemics as well as key nutrients and water (cf. Huber & Watson, 1974; Beevers, 1976; Rosenthal & Janzen, 1979; Scriber & Slansky, 1981). Such interactions between insects and plant chemistry will vary with the degree of stress encountered from moisture deficits (Viets, 1972; White, 1974; Mattson, 1980); soil N (e.g., Bartholomew & Clark, 1965; McKey et al., 1978); weed competition (Chapt. 28, this book); and previous damage by nematodes and/or plant pathogens (Ritter, 1976; Lewis, 1979; McIntyre, 1980; Chapt. 30, this book) and/or by other insects (Rhoades, 1979; Ryan, 1979).

In addition to these complex interactions, we must also consider the physiological and ecological effects of fungicides, herbicides, and insecticides and other environmental variables upon the susceptibility of the crop to pathogen, weed, or insect invasion (Altman & Campbell, 1977; Oka & Pimentel, 1976; Pimentel, 1977; Tingey & Singh, 1980). Furthermore, the biochemical interactions between soil microflora and various agricultural chemicals are important in understanding the environmental fate (i.e., degradation) or synergisms of particular natural and/or synthetic insecticides (Lichtenstein et al., 1974; Ferris & Lichtenstein, 1980; Alexander, 1981). This aspect of environmental chemistry becomes especially interesting in view of recent evidence that insect resistance to pesticides may be induced by natural plant chemicals (Brattsten, 1979; Yu et al., 1979).

Other soil-management programs (e.g., contour strips, no-till, limited tillage, monoculture with rotation, etc.) have direct and significant effects upon insect invasion of various crops. These practices differentially influence not only the soil fertility and soil loss by erosion, but they also contribute to different insect pest complexes with which we must subsequently deal. This aspect of crop pest management is not new (see Fleming, 1957; Gregory & Musick, 1976), nor is it confined to insects (see Chapt. 28 and 30, this book). There is insufficient space here to discuss all of the various insect problems that are enhanced by soil conservation practices or poor weed control. It is certainly one area in which increased research effort by entomologists will be required.

Northern corn rootworm adults find no-tillage systems much more attractice than conventional-tillage systems for egg laying; however, there is no increase in root damage because larval/egg survival is reduced (Musick & Collins, 1971). This study illustrates certain of the difficulties involved in determining cause-effect relationships in one particular insect-plant interaction and is one of the few of its kind. While it so happens that recommended soil insecticides can be used to achieve satisfactory control of northern corn rootworms in both no-tillage and conventional-tillage systems of corn production, this is not always the case for other insects and other crops.

As an example, I would like only briefly to describe a completely new pest problem (hop vine borer) that has developed in corn only within the last few years, but is likely to persist with increasing trends toward no-tillage, reduced-tillage, and contour-tillage practices of continuous corn. This insect problem originates each spring in grassy weeds of approximately 55 counties in southwestern Wisconsin and the three adjacent states. Larvae of the hop vine borer (*Hydraecia immanis* Guenee), (from overwintering eggs that were deposited on grasses the previous fall) kill seedling corn plants by feeding at the growing point just below ground level. The hop vine borer and its more polyphagous eastern relative, the potato stem borer [*H. micacea* (Esper)], were essentially unknown as corn pests previous to the mid-1970's. The last few years has seen both species rise to economic pest status on corn (Muka, 1976; Scriber, 1980; Godfrey, 1981; Deedat & Ellis, 1983). Insecticides appear to be ineffective against the hop vine borer in Wisconsin (Giebink et al., 1981); however, grassy weed control will largely restrict the problem to the field margins and grassy contour strips. Weedy fields or fields with reduced tillage systems may be infested throughout, as is often the case with several other lepidopterous insect pests such as armyworms, some cutworms, and other stalk borers.

The hop vine borer is a caterpillar of a noctuid moth that has apparently made a successful transition from hop plants (*Humulus lupulus* L.) to corn as its primary host (Giebink, 1983). It uses grass in both cases as the early spring food of the first larval instar. Known for over 100 yr primarily as a pest of hops (see Hawley, 1918), the hop vine borer now appears to be slowly extending its range on corn from fairly local areas in southwestern Wisconsin, which seem to represent residual pockets of hops from the last efforts of a once-thriving hop industry. The increased tendencies toward continuous corn production since the 1930's (with the concurrent decline in hop production) has provided the necessary resource for permitting the successful switch of host plants. With few natural enemies, stem-feeding habits that kill corn and make insecticide control nearly impossible, and fueled by an increased trend of conservation tillage, the distribution of the hop vine borer may soon expand southward into the heart of the Corn Belt where it could become a major pest problem.

III. CONCLUSIONS

While insect feeding, growth, and crop damage potential does not seem to be directly related to N fertilization application, the indirect effects upon the plant physiology, the crop microclimate, and the associated plant community can drastically affect changes in insect numbers, the potential for invasion, and associated economic losses. Conservation tillage and reduced tillage systems are increasingly important in preserving our soil resources;

however, we must be prepared to deal with a new complex of potential insect pests whose ecological prominence is favored by such crop management trends. With a holistic, agro-ecosystem management approach demanding serious efforts, especially in the tropics (Janzen, 1973), but also in all other ecosystems (Loucks, 1977), no-tillage agriculture will undoubtedly continue to increase, perhaps to the extent that it will represent the majority of U.S. crop area by the year 2000 (Phillips et al., 1980; Phillips & Phillips, 1984).

The primary role that N fertilization plays in insect pest management is to help plants compensate for damaged or lost parts or to rapidly pass critical development stages and perhaps escape attack. It must be pointed out, however, that N fertilization differentially affects the nutritional and allelochemic makeup for various species of plants, and that what is deleterious for one insect may predispose the plant to attack by others. Designing research that might elucidate the general governing principles and also clarify currently existing contradictory and fragmentary results regarding plant fertilization and insect invasion must carefully define the target hypotheses at various levels of biological organization—from the biochemical and behavioral to nutrient cycles and energy flow in the community and ecosystem. The effort required will be great, but the benefits will certainly be worth it.

ACKNOWLEDGMENT

I would like to thank S. Manuwoto and M. D. Finke for providing unpublished data, and M. Evans, J. Hainze, W. Mattson, F. Slansky, and W. Tingey for helpful comments. Maize germ plasm was obtained from V. E. Gracen and W. D. Guthrie. Research reported here was supported by the following: NSF Grant DEB 7921749; Hatch Project 5134; the NC-105 North Central States regional research project; an Indonesia-Wisconsin MUCIA (AID) Project (S. Manuwoto); EPA Grant L800148 (RF 3884-4); and is a contribution of the Graduate School and the College of Agricultural and Life Sciences of the University of Wisconsin, Madison.

REFERENCES

Ahmad, S. (ed.). 1983. Herbivorous insects: host-seeking behavior and mechanisms. Academic Press, New York.

Alexander, M. 1981. Biodegradation of chemicals of environmental concern. Science 211:132–138.

Altieri, M. A., and W. H. Whitcomb. 1979. The potential use of weeds in the manipulation of beneficial insects. Hortic. Sci. 14:12–18.

Altman, J., and C. L. Campbell. 1977. Effects of herbicides on plant diseases. Annu. Rev. Phytopathol. 15:361–385.

Atsatt, P. R., and D. J. O'Dowd. 1976. Plant defense guilds. Science 193:24–29.

Baker, H. G., and I. Baker. 1980. Studies on nectar-constitution and pollinator-plant coevolution. p. 100–140. *In* L. E. Gilbert and P. H. Raven (ed.) Coevolution of animals and plants. Revised Ed. Univ. of Texas Press, Austin.

Bartholomew, W. V., and F. E. Clark (ed.). 1965. Soil nitrogen. Agronomy 10.

Beevers, L. 1976. Nitrogen metabolism in plants. Edward Arnold, London.

Bell, W. J., and R. T. Carde. 1984. Chemical ecology of insects. Chapman & Hall, London.

Bernays, E. A. 1983. Nitrogen in defense against insects. p. 321–344. *In* J. A. Lee et al. (ed.) Nitrogen as an ecological factor. Blackwell Scientific Publ., Oxford.

Beutler, R. 1953. Nectar. Bee World 34:106–116, 128–136, 156–162.

Bormann, F. H., and G. E. Likens. 1979. Pattern and process in a forested ecosystem. Springer-Verlag, New York.

Bottrell, D. R. 1980. Integrated pest management. Council on Environmental Quality, U.S. Government Printing Office, Washington, D.C.

Brattsten, L. 1979. Biochemical defense mechanisms in herbivores against plant allelochemicals. p. 199–270. *In* G. A. Rosenthal and D. H. Janzen (ed.) Herbivores: their interaction with secondary plant metabolites. Academic Press, New York.

Brink, R. A., J. W. Densmore, and G. A. Hill. 1977. Soil deterioration and growing demand for world food. Science 197:625–630.

Carrow, J. R., and R. E. Betts. 1973. Effects of different foliar-applied nitrogen fertilizers on balsam wooly aphid. Can. J. For. Res. 3:122–139.

Chevalier, P., and L. E. Schrader. 1977. Genotypic differences in nitrate absorption and partitioning of N among plant parts in maize. Crop Sci. 17:897–901.

Conn, E. E. 1979. Cyanide and cyanogenic glycosides. p. 387–412. *In* G. A. Rosenthal and D. H. Janzen (ed.) Herbivores: their interaction with secondary plant metabolites. Academic Press, New York.

Conn, E. E. 1980. Cyanogenic compounds. Annu. Rev. Plant Physiol. 31:433–451.

Cooper-Driver, G., S. Finch, T. Swain, and E. Bernays. 1977. Seasonal variation in secondary plant compounds in relation to the palatability of *Pteridium aquilinum.* Biochem. Syst. Ecol. 5:177–183.

Corbet, S. A., D. M. Unwin, and O. E. Prys-Hones. 1979. Humidity, nectar and insect visits to flowers with special reference to *Crataegus, Tilia,* and *Echium.* Ecol. Entomol. 4:9–22.

Deedat, Y. D., and C. R. Ellis. 1983. Damage caused by potato stem borer (Lepidoptera: Noctuidae) to field corn. J. Econ. Entomol. 76:1055–1060.

del Moral, R. 1972. On the variability of chlorogenic acid concentration. Oecologia (Berlin) 9: 289–300.

Dement, W. A., and H. A. Mooney. 1974. Seasonal variation in the production of tannins and cyanogenic glucosides in the chaparral shrub, *Heteromeles arbutifolia.* Oecologia (Berlin) 15:69–76.

Dicke, F. F. 1977. The most important corn insects. *In* G. F. Sprague (ed.) Corn and corn improvement. Agronomy 18:501–590.

Diehl, S. R., and G. L. Bush. 1984. An evolutionary and applied perspective of insect biotypes. Annu. Rev. Entomol. 29:471–504.

Erickson, J. M., and P. P. Feeny. 1974. Sinigrin: a chemical barrier to the black swallowtail butterfly, *Papilio polyxenes.* Ecology 55:103–111.

Feeny, P. P. 1970. Seasonal changes in oak leaf tannins and nutrients as a cause of spring feeding by winter moth caterpillars. Ecology 51:565–581.

Feeny, P. P. 1976. Plant apparency and chemical defense. Recent Adv. Phytochem. 10:1–40.

Ferris, I. G., and E. P. Lichtenstein. 1980. Interactions between agricultural chemicals and soil microflora and their effects on the degradation of [^{14}C] parathion in a cranberry soil. Agric. Food Chem. 28:1011–1019.

Finke, M. D. 1977. Factors controlling the seasonal food-plant utilization by larvae of the specialized herbivore, *Pepilio polyxenes* (Lepidoptera). M.S. Thesis. Wright State Univ., Dayton, Ohio.

Fleming, W. E. 1957. Soil management and insect control. p. 326–333. *In* Yearbook of agriculture. U.S. Government Printing Office, Washington, D.C.

Fluck, H. 1963. Intrinsic and extrinsic factors affecting the production of secondary plant products. p. 163–185. *In* T. Swain (ed.) Chemical plant taxonomy. Academic Press, New York.

Fogal, W. H. 1974. Nutritive value of pine foliage for some diprionid sawflies. Proc. Entomol. Soc. Ont. 105:101–118.

Fox, L. R., and B. J. Macauley. 1977. Insect grazing on *Eucalyptus* in response to variation in leaf tannins and nitrogen. Oecologia (Berlin) 29:145–162.

Gallaher, R. N., and R. H. Brown. 1977. Starch storage in C_4 vs. C_3 grass leaf cells as related to nitrogen deficiency. Crop Sci. 17:85–88.

Giebink, B. L. 1983. Biology, phenology, and control of the hop-vine borer (*Hydraecia immanis* Guenee) in Wisconsin corn. M.S. Thesis. Univ. of Wisconsin, Madison.

Giebink, B., M. Scriber, and J. Wedberg. 1981. Update on hop vine borer. Proc. 1981 Fert., Aglime, and Pest Management Conf., Univ. of Wisconsin-Madison. 20:147–150. College of Agric. and Life Sci., Univ. of Wisconsin Ext., Madison, Wis.

Gilbert, L. E. 1979. Development of theory in the analysis of insect-plant interactions. p. 117–154. *In* D. J. Horn et al. (ed.) Analysis of ecological systems. Ohio State Univ. Press, Columbus.

Godfrey, G. L. 1981. Identification and descriptions of the ultimate instar larvae of *Hydraecia immanis* (hop vine borer) and *H. micacea* (potato stem borer) (Lepidoptera: Noctuidae). Ill. Natl. History Survey Biol. Notes no. 114. Illinois Natl. History Survey, Champaign, Ill.

Gojmerac, W. 1980. Bees, beekeeping, honey and pollination. AVI Publ., Westport, Conn.

Gregory, W. W., and G. J. Musick. 1976. Insect management in reduced tillage systems. Bull. Entomol. Soc. Am. 22:302–304.

Harms, C. L., and B. B. Tucker. 1973. Influence of nitrogen fertilization and other factors on yield, prussic acid, nitrate, and total nitrogen concentrations of sudangrass cultivars. Agron. J. 65:21–26.

Haukioja, E. 1980. On the role of plant defenses in the fluctuation of herbivore populations. Oikos 35:202–213.

Hawley, I. M. 1918. Insects injurious to the hop in New York, with special reference to the hop grub and the hop redbug. p. 141–224. Cornell Univ. Exp. Stn. Memoir no. 15.

Heinrich, B. 1979. Bumblebee economics. Harvard Univ. Press, Cambridge, Mass.

Huber, D. M., and R. D. Watson. 1974. Nitrogen form and plant disease. Annu. Rev. Phytopathol. 12:139–165.

Huffaker, C. B., M. van de Vrie, and J. A. McMurty. 1969. The ecology of tetranychid mites and their natural control. Annu. Rev. Entomol. 14:125–174.

Hurd, L. E., and L. L. Wolf. 1974. Stability in relation to nutrient enrichment in arthropod consumers of old field successional ecosystems. Ecol. Monogr. 44:465–482.

Janzen, D. H. 1973. Tropical agroecosystems. Science 182:1212–1219.

Jones, D. A. 1972. Cyanogenic glycosides and their function. p. 103–124. *In* J. B. Harborne (ed.) Phytochemical ecology. Academic Press, London.

Jones, F. G. W. 1976. Pests, resistance and fertilizers. p. 233–258. *In* Fertilizer use and plant health. Proc. 12th Int. Potash Inst., Worblaufen, Bern, Switzerland. Der Bund, Bern, Switzerland.

Lawton, J. H. 1976. The structure of the arthropod community on bracken. Bot. J. Linn. Soc. 73:187–216.

Leath, K. T., and R. K. Radcliffe. 1974. The effect of fertilization on disease and insect resistance. p. 481–503. *In* D. A. Mays (ed.) Forage fertilization. Am. Soc. of Agron., Madison, Wis.

Levins, R., and M. Wilson. 1980. Ecological theory and pest management. Annu. Rev. Entomol. 25:287–308.

Lewis, A. 1979. Feeding preference for diseased and wilted sunflower in the grasshopper, *Melanoplus differentialis*. Entomol. Exp. Appl. 26:202–207.

Lichtenstein, E. P., T. T. Liang, K. R. Schultz, H. K. Schnoes, and G. T. Carter. 1974. Insecticidal and synergistic components isolated from dill plants. Agric. Food Chem. 22: 658–664.

Lockeretz, W., G. Shearer, and D. H. Kohl. 1981. Organic farming in the corn belt. Science 211:540–547.

Loucks, O. L. 1977. Emergence of research on agro-ecosystems. Annu. Rev. Ecol. System. 8: 173–192.

Manglitz, G. R., H. J. Gorz, F. A. Haskins, W. R. Akeson, and G. L. Beland. 1976. Interactions between insects and chemical components of sweetclover. J. Environ. Qual. 5:347–352.

Manuwoto, S. 1984. Feeding and growth of Lepidoptera species and subspecies as influenced by natural and altered nutrient and allelochemic concentrations in their diets. Ph.D. Thesis. Univ. of Wisconsin, Madison.

Manuwoto, S., and J. M. Scriber. 1982. Consumption and utilization of three maize genotypes by the southern armyworm, *Spodoptera eridania.* J. Econ. Entomol. 75:163–167.

Marten, G. C., A. B. Simons, and J. R. Frelich. 1974. Alkaloids in reed canarygrass as influenced by nutrient supply. Agron. J. 66:363–368.

Mattson, W. J. 1980. Herbivory in relation to plant nitrogen content. Annu. Rev. Ecol. Syst. 11:119–161.

Mattson, W. J., and N. D. Addy. 1975. Phytophagous insects as regulators of forest primary productivity. Science 190:515–522.

Maurizio, A. 1975. How bees make honey. p. 77–105. *In* E. Crane (ed.) Honey: a comprehensive survey. Heineman, London.

McClure, M. S. 1980. Foliar nitrogen: a basis for host suitability for elongate hemlock scale, *Fiorinia externa* (Homoptera: Diaspididae). Ecology 61:72–79.

McHargue, J. S., and W. R. Roy. 1932. Mineral and nitrogen content of the leaves of some forest trees at different times in the growing season. Bot. Gaz. (Chicago) 94:381–393.

McIntyre, J. L. 1980. Defenses triggered by previous invaders: nematodes and insects. Plant Dis. 5:333–343.

McKenzie, H. H., and H. S. Wallace. 1954. The Kjeldahl determination of nitrogen: a critical study of digestion conditions—temperature, catalyst, and oxidizing agent. Aust. J. Chem. 7:55–70.

McKey, D., P. G. Waterman, C. N. Mbi, J. C. Gartlan, and T. T. Struhsaker. 1978. Phenolic content of vegetation in two African rain forests: ecological implications. Science 202:61–64.

McNeill, S., and T. R. E. Southwood. 1978. The role of nitrogen in the development of insect/plant relationships. p. 77–79. *In* J. Harborne (ed.) Biochemical aspects of plant and animal coevolution. Academic Press, London.

Muka, A. 1976. A new corn pest is south of the border. Hoard's Dairyman 121:688.

Musick, G. J., and D. L. Collins. 1971. Northern corn rootworm affected by tillage. Ohio Rep. 56(6):88–91.

Myers, J. H., and B. J. Post. 1981. Plant nitrogen and fluctuations of insect populations: a test with the Cinnabar moth-tansy ragwort system. Oecologia 48:151–156.

Nielson, D. R., and J. G. MacDonald. 1978. Nitrogen in the environment. Vol. 1 and 2. Academic Press, New York.

Noller, C. H., and C. L. Rhykerd. 1974. Relationship of nitrogen fertilization and chemical composition of forage to animal health and performance. p. 363–394. *In* D. A. Mays (ed.) Forage fertilization. Am. Soc. of Agron., Madison, Wis.

Oka, I. N., and D. Pimentel. 1976. Herbicide (2,4-D) increases insect and pathogen pests on corn. Science 193:239–240.

Owen, D. F. 1980. How plants may benefit from the animals that eat them. Oikos 35:230–235.

Pate, J. S. 1980. Transport and partitioning of nitrogenous solutes. Annu. Rev. Plant Physiol. 31:313–340.

Phillips, R. E., R. L. Blevins, G. W. Thomas, W. W. Frye, and S. H. Phillips. 1980. No-tillage agriculture. Science 208:1108–1113.

Phillips, R. E., and S. H. Phillips (ed.). 1984. No tillage agriculture: principals and practices. Van Nostrand-Reinhold, New York.

Pimentel, D. 1976. World food crisis: energy and pests. Bull. Entomol. Soc. Am. 22:20–26.

Pimentel, D. 1977. The ecological basis of insect pest, pathogen and weed problems. p. 3–31. *In* J. M. Cherrett and G. R. Sagan (ed.) Origins of pest, parasite, disease and weed problems. Blackwell Scientific Publ., Oxford.

Prestidge, R. A. 1982. The influence of nitrogenous fertilizer on the grassland Auchenorrhyncha (Homoptera). J. Appl. Ecol. 19:735–749.

Price, P. W., C. E. Bauton, P. Gross, B. A. McPheron, J. N. Thompson, and A. E. Weis. 1980. Interactions among three trophic levels: influence of plants on interactions between insect herbivores and natural enemies. Annu. Rev. Ecol. Syst. 11:41–65.

Raymond, W. F. 1969. The nutritive value of forage crops. Adv. Agron. 21:1–108.

Rhoades, D. F. 1979. Evolution of plant chemical defense against herbivores. p. 1–55. *In* G. A. Rosenthal and D. H. Janzen (ed.) Herbivores: their interaction with secondary plant metabolites. Academic Press, New York.

Rhoades, D. F., and R. G. Cates. 1976. Toward a general theory of plant antiherbivore chemistry. Recent Adv. Phytochem. 10:168–213.

Riebesell, J. F. 1974. Paradox of enrichment in competitive systems. Ecology 55:183–187.

Ritter, M. P. 1976. The interaction between nutrients and host resistance to nematodes with reference to Mediterranean crops. p. 291–299. *In* Fertilizer use and plant health. Proc. 12th Int. Potash Inst., Worblaufen, Bern, Switzerland. Der Bund, Bern, Switzerland.

Rosenthal, G. A., D. H. Janzen. 1979. Herbivores: their interaction with secondary plant metabolites. Academic Press, New York.

Rosenzweig, M. L. 1971. Paradox of enrichment: destabilization of exploitation ecosystems in ecological time. Science 171:385–387.

Russell, F. C. 1947. The chemical composition and digestibility of fodder shrubs and trees. Jt. Publ. Commonw. Agric. Bur. Pastures Field Crops For. Anim. Nutr. 10:185–231.

Ryan, C. 1979. Proteinase inhibitors. p. 599–618. *In* G. A. Rosenthal and D. H. Janzen (ed.) Herbivores: their interaction with secondary plant metabolites. Academic Press, New York.

Schultz, J. C., and I. T. Baldwin. 1982. Oak leaf quality declines in response to defoliation by gypsy moth larvae. Science 217:149–151.

Scriber, J. M. 1977. Limiting effects of low leaf-water content on nitrogen utilization, energy budget, and larvae growth of *Hyalophora cecropia* (Lepidoptera: Saturniidae). Oecologia (Berlin) 28:269–287.

Scriber, J. M. 1978. Cyanogenic glycosides in *Lotus corniculatus*: their effect on growth, energy budget, and nitrogen utilization of the southern armyworm, *Spodoptera eridania.* Oecologia (Berlin) 34:143–155.

Scriber, J. M. 1979. Post-ingestive utilization of plant biomass and nitrogen by Lepidoptera: legume feeding by the southern armyworm. J. N.Y. Entomol. Soc. 87:141–153.

Scriber, J. M. 1980. The potato stem borer and the hop vine borer: new problems in corn? Proc. Wis. Fert., Aglime, and Pest Management Conf. 19:28–31. College of Agric. and Life Sci., Univ. of Wisconsin Ext., Madison, Wis.

Scriber, J. M. 1981. Sequential diets, metabolic costs, and growth of *Spodoptera eridania* (Lepidoptera: Noctuidae) feeding upon dill, lima bean, and cabbage. Oecologia 51:175–180.

Scriber, J. M. 1982. The behavior and nutritional physiology of southern armyworm larvae as a function of plant species consumed in earlier instars. Entomol. Exp. Appl. 31:359–369.

Scriber, J. M. 1983. The evolution feeding specialization, physiological efficiency, and host races. p. 373–412. *In* R. F. Denno and M. S. McClure (ed.) Variable plants and herbivores in natural and managed systems. Academic Press, New York.

Scriber, J. M. 1984. Plant-herbivore relationships: host plant acceptability. p. 159–202. *In* W. Bell and R. Carde (ed.) The chemical ecology of insects. Chapman & Hall, London.

Scriber, J. M., and P. P. Feeny. 1979. Growth of herbivorous caterpillars in relation to feeding specialization and to growth form of their food plants. Ecology 60:829–850.

Scriber, J. M., and F. Slansky, Jr. 1981. The nutritional ecology of immature insects. Annu. Rev. Entomol. 26:183–211.

Shasha'a, N. S., W. F. Campbell, and W. N. Nye. 1976. Effects of fertilizer and moisture on seed yield of onion. HortScience 11:425–426.

Shuel, R. W. 1957. Some aspects of nectar secretion and nitrogen, phosphorus, and potassium nutrition. Can. J. Plant Sci. 37:220–236.

Seigler, D. S., and P. W. Price. 1976. Secondary compounds in plants: primary functions. Am. Nat. 110:101–104.

Singh, P. 1970. Host-plant nutrition and composition: effects on agricultural pests. Can. Dep. Agric. Res. Inst. Inf. Bull. no. 6. Canada Dep. of Agric., Ottawa, Ontario, Canada.

Slansky, F. 1982a. Insect nutrition: an adaptationists' perspective. Fla. Entomol. 65:45–71.

Slansky, F. 1982b. Toward a nutritional ecology of insects. p. 253–259. *In* Proc. 5th Int. Symp. Insect-Plant Relationships. Pudoc, Wageningen.

Slansky, F., Jr., and P. P. Feeny. 1977. Stabilization of the rate of nitrogen accumulation by larvae of the cabbage butterfly on wild and cultivated food plants. Ecol. Monogr. 47: 209–228.

Slansky, F., and J. M. Scriber. 1982. Selected bibliography and summary of quantitative food utilization by immature insects. Bull. Entomol. Soc. Am. 28:43–55.

Slansky, F., and J. M. Scriber. 1984. Food consumption and utilization. *In* G. A. Kerkut and L. I. Gilbert (ed.) Vol. 4, Comprehensive insect physiology, Biochemistry and pharmacology. Permagon Press, Oxford (in press).

Smeathers, D. M., E. Gray, and J. H. James. 1973. Hydrocyanic acid potential of black cherry leaves as influenced by aging and drying. Agron. J. 65:775–777.

Southwood, T. R. E. 1972. The insect/plant relationship—an evolutionary perspective. p. 3–30. *In* H. F. van Emden (ed.) Insect/plant relationships. R. Entomol. Soc. London, Symp. no. 6. Blackwell Scientific Publ., Oxford.

Southwood, T. R. E. 1977. The relevance of population dynamic theory to pest status. p. 35–54. *In* J. M. Cherrett and G. R. Sagan (ed.) Origins of pest, parasite, disease, and weed problems. Blackwell Scientific Publ., Oxford.

Terman, G. L., J. C. Noggle, and C. M. Hunt. 1976. Nitrate-N and total N concentration relationships in several plant species. Agronomy J. 68:556–560.

Thompson, J. N. 1982. Interaction and coevolution. John Wiley & Sons, Inc., New York.

Tingey, W. M., and S. R. Singh. 1980. Environmental factors influencing the magnitude and expression of resistance. p. 89–113. *In* F. G. Maxwell and P. R. Jennings (ed.) Breeding plants resistant to insects. John Wiely & Sons, Inc., New York.

van Sumere, C. F., J. Albrecht, A. Dedonder, H. DePooter, and I. Pe. 1975. Plant proteins and phenolics. p. 211–264. *In* J. B. Harborne and C. F. van Sumere (ed.) The chemistry and biochemistry of plant proteins. Proc. Phytochem. Soc. Symp., Ghent, Belgium. Academic Press, London.

Viets, F. G., Jr. 1972. Water deficits and nutrient availability. p. 217–239. *In* T. T. Kozlowski (ed.) Water deficits and plant growth. Academic Press, New York.

White, T. C. R. 1974. A hypothesis to explain outbreaks of looper caterpillars, with special reference to populations of *Selidosema suavis* in a population of *Pinus radiata* in New Zealand. Oecologia (Berlin) 16:279–301.

White, T. C. R. 1978. The importance of a relative shortage food in animal ecology. Oecologia (Berlin) 33:71–86.

Whitham, T. G. 1981. Individual trees as heterogeneous environments: adaptation to herbivory or epigenetic noise? p. 9–27. *In* R. F. Denno and H. Dingle (ed.) Insect life history patterns: habitat and geographic variation. Springer-Verlag, New York.

Whitham, T. G. 1983. Host manipulation of parasites: within-plant variation as a defense against rapidly evolving pests. p. 15–41. *In* R. F. Denno and M. S. McClure (ed.) Variable plants and herbivores in natural and managed systems. Academic Press, New York.

Whittaker, R. H. 1975. Communities and ecosystems. 2nd ed. The MacMillan Co., New York.

Wolfson, J. L. 1980. Oviposition response of *Pieris rapae* to environmentally induced variation in *Brassica nigra.* Entomol. Exp. Appl. 27:223–232.

Woodhead, S., and E. A. Bernays. 1978. The chemical basis of resistance of *Sorghum bicolor* to attack by *Locusta migratoria.* Entomol. Exp. Appl. 24:123–144.

Yu, S. J., R. E. Berry, and L. C. Terriere. 1979. Host plant stimulation of detoxifying enzymes in a phytophagous insect. Pest. Biochem. Physiol. 12:280–284.

Zucker, W. V. 1983. Tannins: does structure determine function? An ecological perspective. Am. Nat. 121:335–365.

30

Thor Kommedahl
University of Minnesota
St. Paul, Minnesota

Interaction of Nitrogen Use and Plant Disease Control

Nitrogen, the fourth most important element in plants, is equally important to a pathogen, whether it is growth in a soil or in a host. Nitrogen is assimilated by plants both as NH_4^+ and NO_3^-. Huber and Watson (1974) list 14 crops that prefer NO_3^--N and 11 crops that prefer NH_4^+-N This host preference for N form also has an effect on pathogens. Sometimes the form of N has a greater effect on host resistance or disease severity than does the amount of N.

It is difficult to generalize about the effects of N on plant disease. Huber (1981), for example, tabulated 156 diseases in which 103 were increased and 53 were decreased in severity by N application, regardless of the form of N. Moreover, many other factors affect the role of N, such as time and rate of N application, stability, ratio of NH_4^+ to NO_3^-, and the residual N in soil. Also, cropping history, crop residues in soil, soil microorganisms present, disease complexes, tillage and irrigation practices, and amendments with other elements in either organic or inorganic form can affect the role of N in plant health and disease. Frequently, the N that lessens disease does so more by its effect on the host than its effect on the pathogen. Thus, the effect of N on plant health or disease has to be determined for each crop and disease combination within a given environment.

I. EFFECT OF N ON DISEASE SUSCEPTIBILITY

Nitrogen favors disease resistance by promoting plant health, thereby enabling the physiologic, morphologic, or functional resistances inherent in the host to actively ward off infection. If plant growth is promoted, roots and shoots can sometimes outgrow the pathogens, or plants can escape the disease entirely. Nitrogen, by delaying maturity, can delay senescence of

Nitrogen in Crop Production.

pith cells in corn stalks (*Zea mays* L.) and thereby postpone stalk rot. By increasing rates of NH_3, less stalk rot of corn can occur when infected naturally or artificially with *Diplodia zeae* or *Colletotrichum graminicola* (Warren et al., 1975; White et al., 1978). Nitrapyrin [2-chloro-6-(trichloromethyl)-pyridine] can be mixed with NH_3 just prior to application to reduce stalk rot incidence. The reduced stalk rot from adding N, or from using lower rates of N plus nitrapyrin, can be attributed to the continuous supply of N made available to the plant throughout the growing season. These results are in disagreement with earlier work in which stalk rot incidence was reported to increase with increasing amounts of N. However, earlier work frequently involved sources of N other than NH_3.

Nitrogen fertilizer can reduce the incidence of take-all in barley (*Hordeum vulgare* L.) and wheat (*Triticum aestivum* L.), probably because it enables plants to produce crown roots faster than the take-all fungus can destroy them. Even though the added N increases the intrinsic susceptibility of individual roots to infection, it also promotes disease escape, and the result is an increase in yield of the entire plant (Garrett, 1948).

Soil moisture and nutrients must be considered. The application of ammonium sulfate (AS) can reduce take-all of barley and wheat in moist soil (Glynne, 1951), but is not as effective in relatively dry soil (Butler, 1961). This disease reduces yields of poorly nourished plants very much and well-nourished plants very little (Glynne, 1953), so fertilizer is recommended to lessen the effects of disease and increase the yields of grain.

Pathogens in the host can affect the sucrose–amino acid cycle. Sugar that normally moves to roots through the phloem combines (in part) with inorganic N and is transported upward through the xylem as amino acids and amides. Each transport system is affected by a change in the other. For example, parasitic seed plants accumulate sugar and intercept the amount that would normally go to roots of the host. With less root sugar, there is less organic N produced and transported. Because the xylem of both host and parasitic seed plant are connected, there is a continuous flow of amino acids and amides to the parasite, thereby reducing the supply of N available to the host. The resulting slight N deficiency causes a further reduction in sugar movement, separate from the sugar that collects at infection sites, and this sugar restriction by the parasite results in less N and even less sugar—the effect on the host is chlorosis, less growth, and yield loss.

While sucrose and N effects are related internally in the plant, the effect is somewhat different at the root-soil interface with soil-borne pathogens. Nitrogen application favors establishment of pathogenicity and parasitic development of root rot fungi such as *Fusarium solani* on bean (*Phaseolus vulgaris* L.), whereas glucose application weakens pathogenesis.

Although N promotes vigorous growth essential for the production of amino acids and new protoplasm, it can, if available in excess, encourage succulence, increase plant density, delay maturity, and cause cell walls to be thin and vulnerable to penetration by fungi. This increase in succulence makes plants more nutritious for the pathogen and increases consumption of amino acids by the pathogen in competition with the host. Moreover, with prolonged growth, leaves are exposed to pathogens over a longer season. Leaf lesions may be larger than normal, and plants, especially cereals, may lodge more readily. Although there is little difference in succulence when plants are fertilized with different forms of N, one form may enhance resistance and another susceptibility to disease. These effects are especially pronounced with obligate parasites such as the rusts and the mildews. With wheat stem rust (*Puccinia graminis* f. sp. *tritici*), excess N favors more rust infections but does not necessarily alter the infection type. The increase in susceptibility of wheat to powdery mildew after N fertilization is associated with changes in growth rates: the more N, the greater the leaf area and the greater the total infection. However, N-deficient plants continuously resist infection.

Excessive applications of N are not always detrimental to plants. For example, excess N would not be detrimental to wheat in a pathogen-free soil, but it is a hazard with soil infested with the foot rot pathogen (*Fusarium roseum* 'Culmorum') (Papendick & Cook, 1974).

The addition of N to sweet corn seedlings increases their susceptibility to bacterial wilt, but not because of increased succulence. This bacterium lives almost exclusively in the tracheal tubes and depends on inorganic N of the tracheal sap for its parasitic existence. The severity of invasion is directly correlated with the total amount of N supplied to the seedling and not with the growth of the host resulting from such applications (McNew & Spencer, 1939). At a low concentration of N in tracheal sap, there is competition between host and bacterium, but at high concentrations there is more N than the plant can use; this favors bacterial growth and increases disease severity.

II. CONTROLLING PLANT DISEASES WITH N FERTILIZERS

Sometimes, for N to be effective in favoring disease resistance, it must be used in certain forms, be combined with certain other elements, or be used with certain agricultural practices. Examples of these relations follow.

A. NH_3, NO_3^-, and Other Forms

The topic of N form and plant disease has been reviewed thoroughly and recently by Huber and Watson (1974) and by Huber (1981). In the latter paper, application of NO_3^- was reported to increase severity in 26 diseases and decrease severity in 32 diseases; application of NH_4^+ was found to increase severity in 27 diseases and decrease severity in 26 diseases. These diseases include those caused by fungi, bacteria, nematodes, and viruses that produce symptoms of seedling blights, root and cortical rots, vascular disorders, galls, leaf spots, and yellows.

Nitrate-N and residues stimulate nitrification and moderate severity of many diseases such as black scurf (*Rhizoctonia solani*) of potato (*Solanum tuberosum* L.), root rots of bean and peas (*Pisum sativum* L.), and foot rot of wheat, but these diseases are aggravated by the NH_4^+ forms of N. On the other hand, soil-borne diseases such as potato scab (*Streptomyces scabies*) and early dying of potato are moderated by the NH_4^+ forms but exacerbated by the NO_3^- forms. Many other examples are given by Huber (1981).

For a given amount of N, deviation towards too much NH_4^+ causes more disease than deviation towards too much NO_3^-. Severity of take-all of wheat can be reduced by selecting amounts and ratio of NH_4^+-N to NO_3^--N that is most favorable for wheat growth (Hornby & Goring, 1972). However, it is possible to reduce severity, even with application of NH_4^+-N in relation to NO_3^--N (Smiley, 1978a). To control Verticillium wilt (*Verticillium albo-atrum*) of cotton (*Gossypium hirsutum* L.), it is best to apply equivalent amounts of NH_4^+- and NO_3^--N than to apply NH_4^+, NO_3^-, or urea singly (Ranney, 1962), which also emphasizes the importance of the NH_4^+/NO_3^- ratio.

The delicate balance between resistance and susceptibility can be upset by nutrition as has been shown for stem rust of 'Thatcher' wheat (Daly, 1949). Here, a mesothetic (resistant) reaction can be produced in a susceptible cultivar by fertilizing plants with NH_4^+, whereas a susceptible reaction is more likely to occur from NO_3^- applications. 'Mindum' wheat, however, is resistant regardless of N source. Thus, application of N as NH_4^+ is more important in reducing disease in a rust-susceptible cultivar than in a resistant one.

Nitrogen is important in reducing parasitic and saprophytic activities in sandy loam of parasitic fungi such as *Rhizoctonia solani* on snap bean; here the soluble ammonium nitrate (AN) is more effective than the slowly soluble Aramite®, a urea-formaldehyde product (discontinued by Uniroyal, Bethany, Conn.) (Davey & Papavizas, 1960). With *R. solani* on potato, various N salts and cornmeal not only reduce disease incidence, but they enable the pathogen to persist in soil (Sanford, 1947). The value of N in reducing disease incidence or severity, however, outweighs the effect of persistent survival.

Anhydrous ammonia (AA) can be injected into soil, not only to provide plants with a readily available form of N but also to kill or inhibit soil-borne pathogens. The injection of AA into the soil can reduce populations to zero within the injection zone of *Fusarium* species that cause foot rot of wheat or root rot of pea (Smiley et al., 1970). However, this treatment does not affect appreciably the numbers of *Fusarium* propagules in the tillage layer (Smiley et al., 1972). Injection is better than band application for Fusarium root rot of winter wheat, probably because injection gives better distribution of NH_3 within the tillage layer. Too little NO_2^- accumulates with band treatment to kill *Fusarium,* except in the tillage layer itself. Even though NH_3 can destroy *Fusarium* species that cause root rots of both pea and wheat in the NH_3-retention zone, it is not possible or feasible to control these root diseases by the use of NH_3.

B. Ca and N Applications

Nitrate-N is sometimes more effective in reducing disease severity when accompanied by application of Ca. In fact, N may increase susceptibility of plants to disease, but reduce susceptibility if lime is present, also. For example, the growing of chrysanthemum (*Chrysanthemum* × *morifolium* Ramat.) under high N nutrition for 3 weeks before inoculation with *Fusarium oxysporum* favored wilt more than growing plants under low N nutrition (Littrell, 1966). Liming alone can reduce wilt severity, and NO_3^-, together with $Ca(OH)_2$, decreases wilt additively. The decrease in soil pH also contributes to the decrease in wilt severity (Wolz & Englehard, 1973).

Complete control of Fusarium wilt can be obtained by growing potted chrysanthemums in a high-lime, all NO_3^--N regime and drenching with benomyl, 50% wettable, and BASF 3201-F,[1] 50% wettable (Englehard & Wolz, 1973). The lime, all NO_3^--N fertilization, and the chemotherepeutant are additive in their effect. Each factor contributes some increment of control, but to be effective all have to be applied at the same time. Also, when lime is adequate, either NH_4^+ or NO_3^- can be used to reduce severity of Fusarium root and foot rot of tomato (*Lycopersicon esculentum* Mill.) (Jarvis & Thorpe, 1980).

Granular calcium nitrate (CN) applied as a side-dressing (113 kg N/ha) as soon as transplants are established can either reduce or delay appearance of disease caused by *Sclerotium rolfsii* on tomato (Sitterly, 1962). If a cultural procedure is followed in which soil is applied to stem bases of plants as they mature, then it is not necessary to incorporate CN into soil, but it is important to place the granules next to the tomato stems. With tomato, both Ca and N are important—Ca for periderm formation to protect plants against the pathogen and N to promote plant health and growth. If soils are sandy and low in organic matter, bacterial wilt can be a problem, and the added N can arrest wilt besides increasing salt balance and osmotic values that affect the water-holding capacity. Moreover, organisms antagonistic to pathogens are likely to be more abundant where N is applied. Nitrogen applications are not as likely to control wilt in heavy soils; also, too much N can favor Verticillium wilt of tomato. So if N is used to control disease, soil conditions and cultural practices used in a given area have to be considered.

C. P and N Applications

The effect of N is sometimes linked with that of P. The number of local lesions on leaves may be large because of applications of both N and P, but generally only when there is also greater growth. This effect is indirect because of the increase in plant size and applies to such diseases as viruses of tomato and tobacco (*Nicotiana tabacum* L.) and over critical ranges. On the

[1] [methyl 1-(methylthioethylcarbamoyl)-2-benzimidazole carbamate], BASF Wyandotte Corp., Parsippany, N.J.

other hand, the added N makes plants less susceptible while still increasing plant growth. Nutritional conditions that favor plant growth are those that increase susceptibility, but neither N nor P specifically favors susceptibility. With adequate available P, the addition of N can make plants grow better but be more susceptible to disease, whereas without P, the addition of N can reduce growth and susceptibility. Considering the fertility usually present in most field crops, it seems likely that varying the nutrition will mainly affect the chance that plants will become infected because of the change in plant size. However, the P/N ratio may have effects additional to those on plant size.

The P/N ratio is one of the important ratios associated with disease incidence. For example, wide P/N ratios retard and narrow ratios promote Pythium root rot of wheat (Vanterpool, 1935). A narrow P/N ratio early in the season allows root rot to develop and delay maturity, whereas a wide ratio maintains the crop in an apparently healthy state and hastens maturity. Heavy applications of N and P fertilizer can control Phoma root rot of chrysanthemum in both compost or soil and in greenhouse or field (Menzies & Colhoun, 1976a). The mechanism of control can be attributed to the increased root growth made possible by the availability of nutrients (Menzies & Colhoun, 1976b).

D. Cultural Factors

1. IRRIGATION

One way to curtail disease is to add N fertilizer to irrigation water. The incidence of sugarbeets (*Beta vulgaris* L.) infected with *Sclerotium rolfsii* can be reduced significantly by dissolving AA or AS in irrigation water. This effect can be explained by the work of Henis and Chet (1967, 1968), who found that sclerotia of this fungus failed to germinate when $\geq$ 0.2% NH_3 was added to sandy loam. The NH_3 also increased the pH of the soil to a value toxic to the fungus for > 24 h, as well as to favor microorganisms, especially bacteria and actinomyces, at the sclerotium-soil interface. This larger population of organisms is antagonistic to the pathogen and can lower the germination percentage of sclerotia to bring about less disease. Thus, N controls germination of the survival structures (sclerotia), fosters plant growth, and lessens disease.

Adding N to irrigation water can also reduce incidence of white heads in winter wheat caused by the take-all fungus (Huber et al., 1968). Ammonium sulfate reduces take-all by restricting the amount of infected tissue. The application of 45 kg N/ha as AS can increase yields by reducing the disease index and halving the yield loss (Syme, 1966). Thus, losses from take-all in irrigated wheat can consistently be reduced by fertilizer application.

2. SPRING, FALL, OR SPLIT APPLICATIONS

Spring applications of N affect take-all differently from fall applications (Huber, 1972). It is possible to increase yields by 50% over non-fertilized plots in spring by application of AS at the rate of 83 kg N/ha.

However, fall applications are not recommended, because they may give half the yield of a spring application.

Nitrogen applied in the fall may be lost during the winter. To prevent this loss, a nitrification inhibitor such as nitrapyrin can be added to maintain the NH_4^+ form throughout winter. This principle can be used to control stalk rot of corn by combining nitrapyrin with fall-applied AA. The lower stalk rot incidence from this treatment may have resulted from altered N metabolism in plants due to uptake of substantial amounts of NH_3, and the plants may grow more vigorously in nitrapyrin-treated than in nontreated plots. This principle does not always work and sometimes there are no differences between fall- and spring-applied N, even though more of the N is likely to be in the NO_3^- form in the fall than in the spring.

Instead of comparing fall and spring applications, one can compare split applications within the same season. Where increased severity of stem rot of rice (*Oryza sativa* L.) followed increases in N rates as preplant treatments, stem rot can become less severe and yields can be greater by making properly timed split applications of N. For example, 28 kg N/ha before seeding and 84 kg N/ha after seeding delayed disease development and increased yield compared with plants receiving topdressings at earlier stages of growth (Keim & Webster, 1974).

If high rates of N (45 g N/m^2) are applied to some bluegrass (*Poa pratensis*) cultivars during summer, Fusarium blight (*F. roseum* and *F. tricinctum* in combination) becomes more severe, but if lower rates are used in early fall, Sclerotinia dollar spot (caused by *Sclerotinia homoeocarpa*) occurs (Turgeon & Meyer, 1974). To control turfgrass diseases, a program should be developed to apply N lightly in the spring and more in August and September to reduce the incidence of both Fusarium blight and dollar spot.

3. RESISTANT CULTIVAR EFFECT

Nitrogen is likely to increase susceptibility of cultivars inherently susceptible to disease but have much less or no effect on cultivars resistant to disease. For example, high N applications ($>$ 100 kg/ha) can increase disease severity and reduce yield in a rice cultivar susceptible to bacterial blight, but have little effect on blight-resistant cultivars (Reddy et al., 1979). This means that an application rate should be chosen that gives the highest yield with the least amount of disease whenever a disease-susceptible cultivar is grown (for the susceptible rice cultivar, that rate was 76 kg/ha). Despite the fact that most lines of bluegrass tested by Gibbs et al. (1973) suffered the greatest severity of disease from Helminthosporium leaf spot under high N applications and the least severity under low N rates, host resistance proved a greater factor than N in influencing disease severity.

Sometimes N can be recommended only if disease-resistant cultivars are grown, e.g., in California, N fertilization of sugarbeets resulted in greater incidence of rot from *Erwinia carotovora* in susceptible than in resistant cultivars (Thomson et al., 1981). Healthy beets respond to N fertilization by increasing yield, but diseased beets do not. Thus, fertilization is recommended only when disease-resistant beet cultivars are planted.

4. N EFFECTS ON ANTAGONISTS AND PATHOGENS IN SOIL

Nitrogen may affect plants and plant disease by its effect on soil fungi. Fertilization with N affects individual soil fungi and favors their multiplication, while P and K have much less of an effect. Root-infecting fungi are confronted with a multitude of antagonists competing for sites at the root surface in response to the added N fertilizer. Sometimes reduced fertilization over a period of time can reduce disease such as Verticillium wilt when fertilizer is applied in limited amounts. Apparently, pathogens cannot thrive in infertile soil, so they increasingly colonize the root surface. Again, N applications affect different diseases differently.

Anhydrous ammonia, applied by conventional applicators with knife-type injectors, produces large concentrations of NH_3 in localized areas and reduces populations of fungi and nematodes in soil; concentrations of NH_4^+ from 150 to 750 mg/kg (ppm) are effective (Eno et al., 1955). A concentration of 600 mg/kg NH_4^+-N in soil, which occurs regularly in the retention zone, can reduce nematodes to $< 1\%$ and fungi to $< 5\%$ of the population prior to injection, and the greatest reduction in both fungi and nematodes occurs when NH_4^+-N exceeds 365 mg/kg.

The form of N may not alter the countable population of microorganisms on or near the root surface, but it can affect the number of bacteria (including streptomyces) that inhibit pathogen growth. Hyphae of pathogens grow more slowly on wheat roots in soil fertilized with NH_4^+ than with NO_3^- (Smiley, 1978a, b). The germination of sclerotia of *Sclerotium rolfsii* can be reduced significantly by amending soil with NH_3, urea, AN, chitin, peptone, CN, and NH_4 acetate, (Henis & Chet, 1968). In soil amended with urea, peptone, NH_4 acetate, and ammonium carbonate, the number of fungi associated with sclerotia is especially high, and this favors destruction of sclerotia. The amendments act to increase the antagonistic activity of microorganisms at the sclerotium-soil interface to reduce sclerotial germinability.

5. SIGNIFICANCE OF THE RESIDUE EFFECT ON SOIL N TRANSFORMATION

Crop residues can modify disease severity through their effects on nitrification, which in turn determines the form of N available in soil. Residues such as those from alfalfa (*Medicago sativa* L.) and corn stimulate the biological oxidation of NH_4^+ to NO_3^- and reduce the severity of bean root rot caused by species of *Fusarium* and *Rhizoctonia,* whereas amendments that decrease nitrification, such as barley straw or glucose, result in severe root rot. The form of N, perhaps more than the C/N ratio, may be the determining factor that modifies the severity of the soil-borne disease. The C/N ratio alone may not influence the control of root rot.

Huber and Watson (1970) think that organic amendments and crop rotation probably influence the severity of soil-borne disease by (i) increasing the biological buffering capacity of soil, (ii) reducing the number of pathogens during anaerobic decomposition of organic matter, (iii) affecting nitrification, which influences the form of N that predominates in soil, and (iv) denying the pathogen a host during the interim of unsuitable species. The specific form of N available to plant and soil microflora, in turn, influences the specific microbial associations and host physiology.

In the field, turned-under crop residues need to contain 11 to 12 kg N/t of residue to prevent competition between growing plants and microorganisms that assimilate N. Ammonium-N is the preferred source for assimilation by microorganisms that assimilate N. Ammonium-N is the preferred source for assimilation by microorganisms in contact with straw, especially under neutral-to-alkaline conditions. Under acid conditions, there is little or no difference between NH_4^+ and NO_3^- forms (Frederick & Broadbent, 1966). Nitrogen can be applied to increase the effectiveness of corn and oat (*Avena sativa* L.), but not soybean [*Glycine max* L. (Merr.)], residues and sawdust in reducing Rhizoctonia disease of snap bean (Davey & Papavizas, 1960).

The greatest biological immobilization of N in soil occurs when large quantities of readily decomposable crop residues of wide C/N ratios such as straw or cornstalks are added. Under such conditions, heterotrophic microorganisms grow so rapidly that frequently all available mineral soil N may be used by them. If the C/N ratio of crop residues is greater than about 25 or 30, an external source of N is needed to get the greatest amount of decomposition and biological immobilization. If the C/N ratio of residues is $\leq$ 20, no external N is needed; in fact, N as NH_3 may be released (Allison, 1966).

It is possible to reduce the inoculum potential of a pathogen such as *Verticillium albo-atrum* on tomato by amending soil with materials rich in N, e.g., AS, blood meal, cottonseed meal, and fish meal, when applied in amounts giving approximately equivalent N values (0.1–0.13% dry wt soil) (Wilhelm, 1951). Organic soil amendments affect the number of microbes in soil and on root surfaces of beans and lessen the severity of attack on bean by *Rhizoctonia solani.* Incorporating green corn and oat residues into soil stimulate streptomyces antagonistic to *R. solani* and provide the host with a relatively long period of protection. Although mature grain straw offers protection to beans soon after its incorporation into soil, it loses its protective capability as the straw decays (Davey & Papavizas, 1959; Papavizas & Davey, 1960). Mature legume residues are not effective in protection because of the considerable lag time between incorporation of residue and suppression of *R. solani.*

The population of bean rhizosphere organisms can be encouraged to multiply and be antagonistic to several bean pathogens (*Fusarium solani* f. sp. *phaseoli, Rhizoctonia solani, Thielaviopsis basicola,* and *Verticillium albo-atrum*) by amending soil with ground, mature oat straw (C/N ratio = 83) or oat straw enriched with AN (C/N ratio = 30 or 50) (Papavizas, 1963). It is not advisable to apply high N (200 mg/kg N as AN) either alone or with oat straw (C/N ratio = 10), because antagonists in the rhizosphere are reduced substantially and disease is enhanced. This is likely to be more important in acid soil (pH < 5) than in slightly acid, neutral, or mildly alkaline soil, and can be more pronounced with NO_3^- than with NH_4^+. It is better to add ammonium chloride, casamino acid, or 2-chloro-6-(trichloromethyl) pyridine (with or without ammonium chloride) to soil to reduce survival of root pathogens such as *R. solani* than to apply sodium nitrate or AN, which can increase survival, if soil is loamy sand (Papavizas, 1969).

III. SUMMARY

Nitrogen affects disease probably more by its effect on plant growth than by its effect on plant pathogens, especially if the N is in the NO_3^- form. With AA, there can be a direct killing or inhibition of root-infecting fungi in soil. Nitrogen is most effective in disease control when it is part of a management practice that takes into account items such as crop history, host resistance to disease, disease complexes, crop residues, tillage, physical soil factors, spring or fall application, ratio of NO_3^--N to NH_4^+-N, as well as ratios of N to P, K, Ca, C, and other elements needed for plant growth. Some diseases can be controlled by timely applications of N, but others are made worse. Nitrogen that in one crop may increase succulence and delay maturity, thereby extending the time that plants are vulnerable to disease, may in another crop prolong the time that the resistance of the host remains effective, even until harvest. The role of N present in or added to soil as amendments to crop residues is a complicated ecological relationship and the effect obtained frequently depends upon many factors including the form of N. Residual N and the C/N ratios have to be determined for each crop and disease combination to enable an evaluation to be made of the benefits of N in controlling disease and in promoting plant health. Despite this variability in response to N, plant health is generally improved by N at the expense of plant disease, especially with appropriate management practices.

REFERENCES

Allison, F. E. 1966. The fate of nitrogen applied to soils. Adv. Agron. 18:219–258.

Butler, F. C. 1961. Root and foot rot diseases of wheat. New South Wales Dep. Agric. (Aust.) Sci. Bull. no. 77.

Daly, J. M. 1949. The influence of nitrogen source on the development of stem rust of wheat. Phytopathology 39:386–394.

Davey, C. B., and G. C. Papavizas. 1959. Effect of organic soil amendments on the Rhizoctonia disease of snap beans. Agron. J. 51:493–496.

Davey, C. B., and G. C. Papavizas. 1960. Effect of dry mature plant materials and nitrogen on *Rhizoctonia solani* in soil. Phytopathology 50:522–525.

Engelhard, A. W., and S. S. Wolz. 1973. Fusarium wilt of chrysanthemum: complete control of symptoms with an integrated fungicide-lime-nitrate regime. Phytopathology 63:1256–1259.

Eno, C. F., W. G. Blue, and J. M. Good, Jr. 1955. The effect of anhydrous ammonia on nematodes, fungi, bacteria, and nitrification in some Florida soils. Soil Sci. Soc. Am. Proc. 19:55–58.

Frederick, L. R., and F. E. Broadbent. 1966. Biological interactions. p. 198–212. *In* M. H. McVicker et al. (ed.) Agricultural anhydrous ammonia technology and use. Agric. Ammonia Inst., Memphis, Tenn., and Am. Soc. of Agron. and Soil Sci. Soc. of Am., Madison, Wis.

Garrett, S. D. 1948. Soil conditions and the take-all disease of wheat. IX. Interaction between host plant nutrition, disease escape, and disease resistance. Ann. Appl. Biol. 35:14–17.

Gibbs, A. F., R. D. Wilcoxson, and H. L. Thomas. 1973. The effect of nitrogen fertilization and mowing on Helminthosporium leaf spot and sugar content of bluegrass leaves. Plant Dis. Rep. 57:544–548.

Glynne, M. D. 1951. Effects of cultural treatments on wheat and on the incidence of eyespot, lodging, take-all and weeds. Field experiments 1945–8. Ann. Appl. Biol. 38:665–668.

Glynne, M. D. 1953. Wheat yield and soil-borne diseases. Ann. Appl. Biol. 40:221–224.

Henis, Y., and I. Chet. 1967. Mode of action of ammonia on *Sclerotium rolfsii.* Phytopathology 57:425–427.

Henis, Y., and I. Chet. 1968. The effect of nitrogenous amendments on the germinability of sclerotia of *Sclerotium rolfsii* and on their accompanying microflora. Phytopathology 58:209–211.

Hornby, D., and C. A. I. Goring. 1972. Effects of ammonium and nitrate nutrition on take-all disease of wheat in pots. Ann. Appl. Biol. 70:225–231.

Huber, D. M. 1972. Spring versus fall nitrogen fertilization and take-all of spring wheat. Phytopathology 62:434–436.

Huber, D. M. 1981. The use of fertilizers and organic amendments in the control of plant disease. p. 357–494. *In* D. Pimentel (ed.) Handbook of pest management. Vol. 1. CRC Press, Inc., Boca Raton, Fla.

Huber, D. M., C. G. Painter, H. C. McKay, and D. L. Peterson. 1968. Effect of nitrogen fertilization on take-all of winter wheat. Phytopathology 58:1470–1472.

Huber, D. M., and R. D. Watson. 1970. Effect of organic amendment on soil-borne plant pathogens. Phytopathology 60:22–26.

Huber, D. M., and R. D. Watson. 1974. Nitrogen form and plant disease. Annu. Rev. Phytopathol. 12:139–165.

Jarvis, W. R., and H. J. Thorpe. 1980. Effects of nitrate and ammonium nitrogen on severity of Fusarium foot and root rot and on yield of greenhouse tomatoes. Plant Dis. 64:309–310.

Keim, R., and R. K. Webster. 1974. Nitrogen fertilization and severity of stem rot of rice. Phytopathology 64:178–183.

Littrell, R. H. 1966. Effects of nitrogen nutrition on susceptibility of chrysanthemum to an apparently new biotype of Fusarium oxysporum. Plant Dis. Rep. 50:882–884.

Menzies, S. A., and J. Colhoun. 1976a. Control of Phoma root rot of chrysanthemums by the use of fertilizers. Trans. Br. Mycol. Soc. 67:455–462.

Menzies, S. A., and J. Colhoun. 1976b. Mechanisms of control of Phoma root rot of chrysanthemums by the use of fertilizers. Trans. Br. Mycol. Soc. 67:463–467.

McNew, G. L., and E. L. Spencer. 1939. Effect of nitrogen supply of sweet corn on the wilt bacterium. Phytopathology 29:1051–1067.

Papavizas, G. C. 1963. Microbial antagonism in bean rhizosphere as affected by oat straw and supplemental nitrogen. Phytopathology 53:1430–1435.

Papavizas, G. C. 1969. Survival of root-infecting fungi in soil XI. Survival of *Rhizoctonia solani* as affected by inoculum concentration and various soil amendments. Phytopathol. Z. 64:101–111.

Papavizas, G. C., and C. B. Davey. 1960. Rhizoctonia disease of bean as affected by decomposing green plant materials and associated microfloras. Phytopathology 50:516–522.

Papendick, R. I., and R. J. Cook. 1974. Plant water stress and development of Fusarium foot rot of wheat subjected to different cultural practices. Phytopathology 64:358–363.

Ranney, C. D. 1962. Effects of nitrogen source and rate on the development of Verticillium wilt of cotton. Phytopathology 52:38–41.

Reddy, A. P. K., J. C. Katyal, D. I. Rouse, and D. R. MacKenzie. 1979. Relationship between nitrogen fertilization, bacterial leaf blight severity, and yield of rice, Phytopathology 69: 970–973.

Sanford, G. B. 1947. Effects of various soil supplements on the virulence and persistence of *Rhizoctonia solani.* Sci. Agric. 27:533–544.

Sitterly, W. R. 1962. Calcium nitrate for field control of tomato southern blight in South Carolina. Plant Dis. Rep. 46:492–494.

Smiley, R. W. 1978a. Antagonists of *Gaeumannomyces graminis* from the rhizoplane of wheat in soils fertilized with ammonium- or nitrate-nitrogen. Soil Biol. Biochem. 10:169–174.

Smiley, R. W. 1978b. Colonization of wheat roots by *Gaeumannomyces graminis* inhibited by specific soils, microorganisms and ammonium-nitrogen. Soil Biol. Biochem. 10:175–179.

Smiley, R. W., R. J. Cook, and R. I. Papendick. 1970. Anhydrous ammonia as a soil fungicide against *Fusarium* and fungicidal activity in the ammonia retention zone. Phytopathology 60:1227–1232.

Smiley, R. W., R. J. Cook, and R. I. Papendick. 1972. Fusarium root rot of wheat and peas as influenced by soil applications of anhydrous ammonia and ammonia-potassium azide solutions. Phytopathology 62:86–91.

Syme, J. R. 1966. Fertilizer and varietal effects on take-all in irrigated wheat. Aust. J. Exp. Agric. Anim. Husb. 6:246–249.

Thomson, S. V., F. J. Hills, E. D. Whitney, and M. N. Schroth. 1981. Sugar and root yield of sugar beets as affected by bacterial vascular necrosis and rot, nitrogen fertilization, and plant spacing. Phytopathology 71:605–608.

Turgeon, A. J., and W. A. Meyer. 1974. Effects of mowing height and fertilizer level on disease incidence in five Kentucky bluegrasses. Plant Dis. Rep. 58:514–516.

Vanterpool, T. C. 1935. Studies on browning root rot of cereals III. Phosphorus-nitrogen relations of infested fields IV. Effects of fertilizer amendments. V. Preliminary plant analyses. Can. J. Res., Sect. C 13:220–250.

Warren, H. L., D. M. Huber, D. W. Nelson, and O. W. Mann. 1975. Stalk rot incidence and yield of corn as affected by inhibiting nitrification of fall-applied ammonium. Agron. J. 67:655–662.

White, D. G., R. G. Hoeft, and J. T. Touchton. 1978. Effect of nitrogen and nitrapyrin on stalk rot, stalk diameter, and yield of corn. Phytopathology 68:811–814.

Wilhelm, S. 1951. Effect of various soil amendments on the inoculum potential of the Verticillium wilt fungus, Phytopathology 41:684–690.

Wolz, S. S., and A. W. Engelhard. 1973. Fusarium wilt of chrysanthemum: effect of nitrogen source and lime on disease development. Phytopathology 63:155–157.

Management of Fertilizer Nitrogen for Crop Production

section V

31

Frank P. Achorn and Michael F. Broder
National Fertilizer Development Center
Tennessee Valley Authority
Muscle Shoals, Alabama

Mechanics of Applying Nitrogen Fertilizer

Of the three major plant foods—N, P, and K—N is required in the largest quantities. Countries that have increased fertility of their soils to a relatively high level will continue to increase consumption of N while use of P and K will increase less rapidly (Achorn, 1981). In the United States, 75% of the energy consumed in manufacturing all fertilizers is used to produce N products (Achorn, 1980). The rising cost of energy mandates efficient use of N through uniform and accurate application. Energy is also conserved by combining fertilizer with herbicides to eliminate a trip across the field. The means for reducing energy consumption in application are dictated by the form of N used.

There are four basic forms of N: pressure liquid, nonpressure solutions, suspensions, and granular. The latter three forms include mixed fertilizers containing N; the application method is the same as N alone.

I. RECENT DEVELOPMENTS IN APPLICATION EQUIPMENT

A. High-Pressure Liquids

1. ANHYDROUS NH_3

The most common form of directly applied N is anhydrous NH_3 (AA). It comprises about 50% of all directly applied N. Popularity of AA results from its low cost, which is largely the result of excellent facilities for transporting and storing liquid NH_3.

Direct application of NH_3 was begun in the early 1940's, using a horse-drawn applicator with a small laboratory gas cylinder. As the agronomic ef-

Nitrogen in Crop Production.

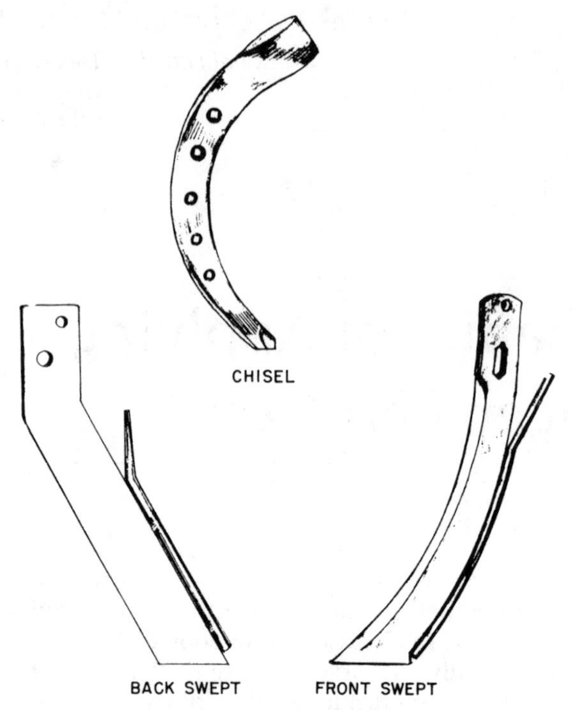

Fig. 1—Application knives.

fectiveness of NH_3 was proven, more sophisticated equipment was developed. Because it is applied as a volatile liquid, NH_3 must be injected 12 to 24 cm below the surface of the soil. This usually is accomplished by application knives such as those in Fig. 1. In sandy, loose soil, NH_3 often is applied by an NH_3 chisel, also shown in Fig. 1. Anhydrous NH_3 usually is metered by a variable-orifice meter or by a piston pump. Rate of application using the orifice meter is determined by applicator speed, swath width, and orifice opening. Piston pumps usually are actuated by a chain driven by a sprocket attached to a wheel of the applicator. Application rate is varied by changing length of stroke of the piston, the rate being independent of applicator speed.

Ammonia applicators range in size from small, 5-row rigs to large rigs with swath widths up to 20 m that are pulled by crawler tractors. Crawler tractors are being replaced by rubber-tired tractors because maintenance costs are much less.

The estimated cost of applying NH_3 is $15/ha. This cost, however, varies widely throughout the country. Most companies report that the cost of applying NH_3 is usually about double that of applying nonpressure solutions.

Speeds of NH_3 application vary from 4 ha/h in the Midwest to 16 ha/h in the western United States. Application speeds are increasing, and a major reason is use of the large, rubber-tired tractors rated at over 300 kW to pull applicators. The large tractors have reduced energy consumption per hectare because of the wider swaths and more efficient engines.

The energy required to apply different types of fertilizers has been calculated by Achorn and Kimbrough (1978). Application of AA requires about 60% more energy than does application of nonpressure fluids and 40% more energy than application of solid fertilizers.

	1000 kcal/t	
	Product	Nutrient
Anhydrous NH_3	346	422
Solids	116	296
Nonpressure fluids	81	270

To help decrease energy consumption, some companies have introduced equipment for applying AA during other tilling operations, thereby avoiding one pass across the field. This concept has been used for many years in Kansas where AA is added during tillage following the wheat (*Triticum aestivum* L.) harvest. In Kansas and other sections of the Midwest, a triangular-shaped plow is passed about 7 to 10 cm beneath the soil. This helps destroy weed growth, tills the soil, and leaves a protective stubble mulch on the surface to prevent wind and water erosion. Figure 2 is a photograph of this tillage implement. Anhydrous NH_3 is released beneath the soil as the soil passes over the implement. Experience with this equipment has shown that 60 to 90 kg NH_3/ha can be applied without noticeable application losses. One dealer reported no noticeable losses with application rates as high as 225 kg/ha when the implement was used on irrigated land. About 40 ha/d can be tilled with one of these implements.

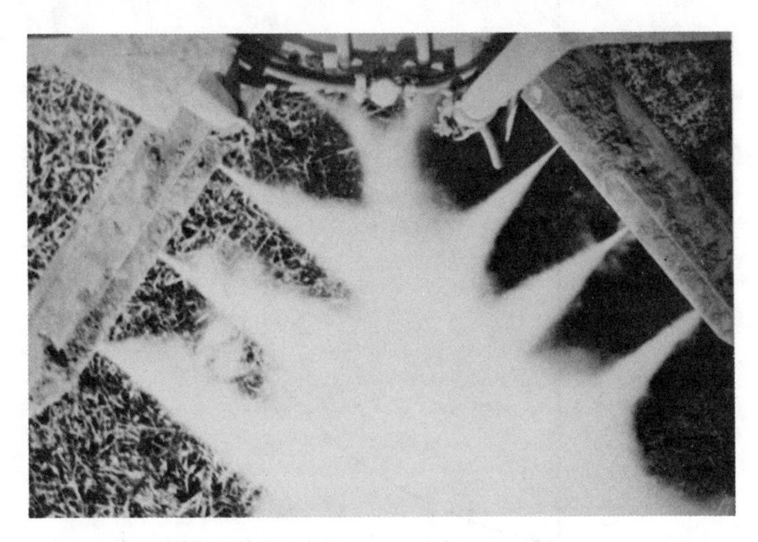

Fig. 2—Ammonia discharge from triangular tillage implement.

2. COLD NH_3

Some applicators keep NH_3 in the liquid state for a prolonged period so that the liquid can be applied and covered with soil during conventional tilling operations, such as disking. Figure 3 is a curve depicting the effect of liquid NH_3 temperature on vapor pressure. The vapor pressure of NH_3 is equal to atmospheric pressure at −34°C. At this temperature, liquid NH_3 does not have to be injected deep into the soil to avoid NH_3 loss.

Cold NH_3 equipment is shown in Fig. 4. This equipment has one evaporator for several application knives. In the evaporator, NH_3 is allowed to expand, which cools the liquid from −32 to −34°C. Cold liquid from the evaporator is added to each applicator knife and gaseous NH_3 is added to alternate knives. Some operators report that with this equipment it is possible to apply NH_3 during tillage operations and thus avoid one pass across the field. One operator reported that with this equipment and a tractor speed of 8.0 km/h it was possible to apply 140 to 180 kg N/ha. Others report that NH_3 can be injected to a shallow depth (10–15 cm), resulting in some energy saving as compared with the conventional applicator, which uses deep injection (20–30 cm). Others report high NH_3 losses when this equipment is used to inject NH_3 to the side of a cultivating disk. This equipment is still being tested to prove its usefulness and value.

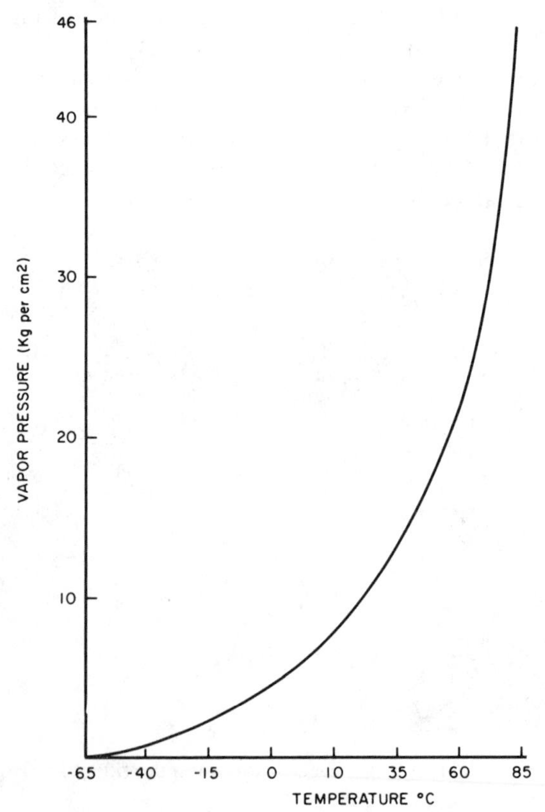

Fig. 3—Effect of temperature on vapor pressure of anhydrous NH_3.

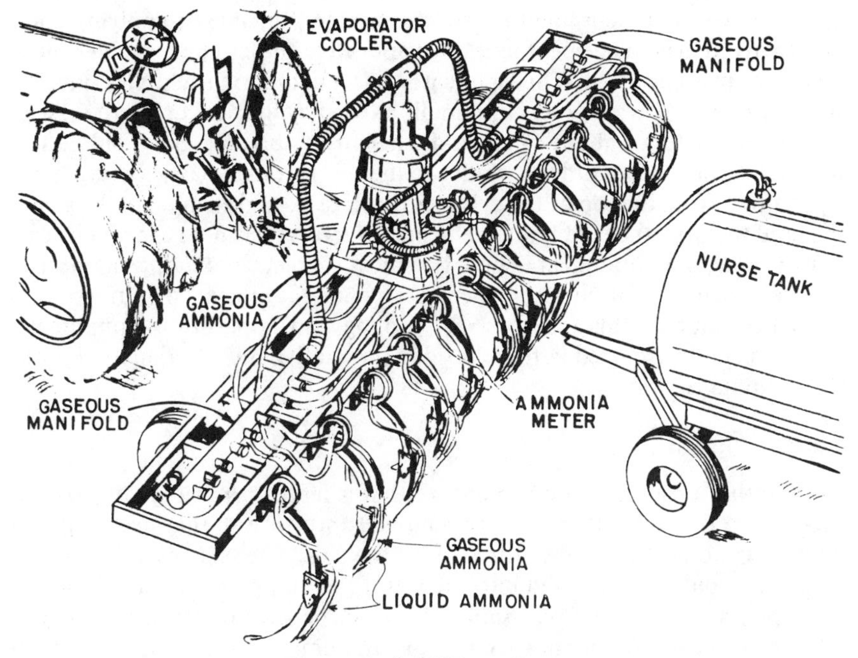

Fig. 4—Cooled NH_3 applicator.

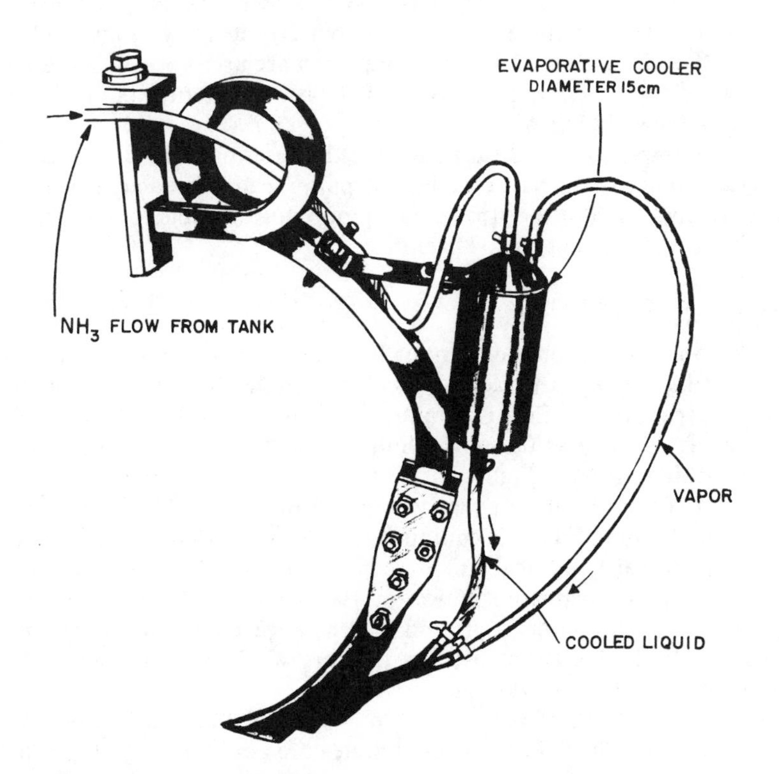

Fig. 5—Small NH_3 evaporator-cooled for each NH_3 tine.

Another equipment manufacturer using the principle of applying cold NH_3 to avoid deep placement uses an evaporative cooler much smaller than the cooler just discussed. This applicator has a cooler on each NH_3 tine, as shown in Fig. 5. These coolers are small, baffled cylindrical pressure vessels having a diameter of about 15 cm and a height of about 25 cm. Cold liquid flows from the bottom of the cylinder to the tine through plastic tubing, and vapor flows from the top of the cylinder through a separate tube to the same tine used to apply the cold liquid. The manufacturer has not reported its experience with this equipment. There is some disagreement among engineers, physical chemists, applicator operators, and equipment manufacturers about the value of this type of NH_3 cooling and application equipment. Therefore, tests are needed to determine its relative merits compared with conventional equipment.

3. HIGH-PRESSURE NH_3

Another procedure used to ensure that a larger percentage of NH_3 is delivered to the soil is that of increasing NH_3 pressure so that nearly all of the NH_3 is liquid before it is forced into the soil. The NH_3 can be kept in liquid form until it is injected into the soil if the pump pressure exceeds the vapor pressure in the delivery tubes. This can be accomplished by using a positive displacement pump, such as a piston pump, and restricting the hole size at the outlet end of the injection tine or injection tube. This outlet hole is drilled at an angle so that the high-pressure stream of NH_3 is forced down into the soil. This equipment, of course, requires delivery tubing and hose capable of withstanding higher pressures than are used for conventional or the new cold NH_3 equipment. A typical piston metering pump for this purpose is shown in Fig. 6.

This pump also has a heat exchanger that uses partially expanded NH_3 from the pump to cool NH_3 entering the pump. Pump capacity is varied by changing the length of the stroke. The pump must be able to deliver liquid NH_3 at pressures as high as 10 kg/cm^2.

4. DUAL APPLICATION

The observed agronomic benefit from placement of fertilizer near the roots of plants has led to development of equipment that also conserves energy. A trip over the field is eliminated by dual application because other plant nutrients, such as those in a liquid mixture, are knifed into the soil directly behind the NH_3. Figure 7 is a sketch of a research applicator used in Kansas to study the agronomic benefit of injecting P_2O_5 fertilizer. The applicator shown in Fig. 7. is equipped for injecting granular fertilizer with NH_3. The vertical cylindrical tank in the figure is for a liquid source, such as ammonium polyphosphate solution of 10–14.8–0 (10–34–0) grade. When applying 10–14.8–0, either a ground driven squeeze pump or piston pump injects the 10–14.8–0 through tubes that are welded "piggyback" to the NH_3 injection tubes.

In spite of the increased energy required to inject fertilizers, farmers in wheat-producing areas in the United States are adopting dual application. An active program for improving application equipment for subsurface

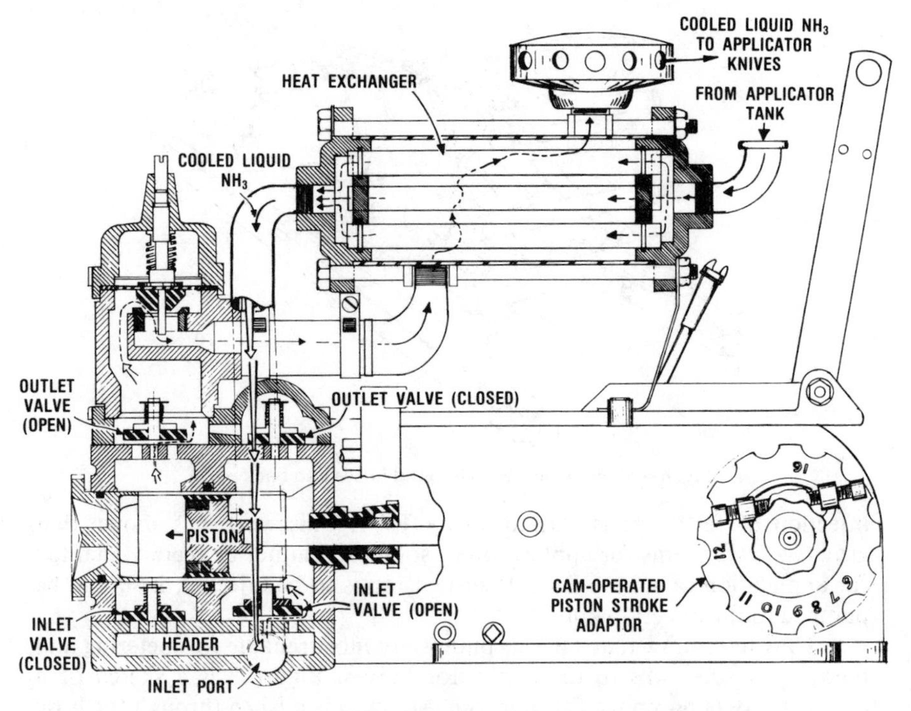

Fig. 6—Anhydrous NH_3 metering pump with heat exchanger.

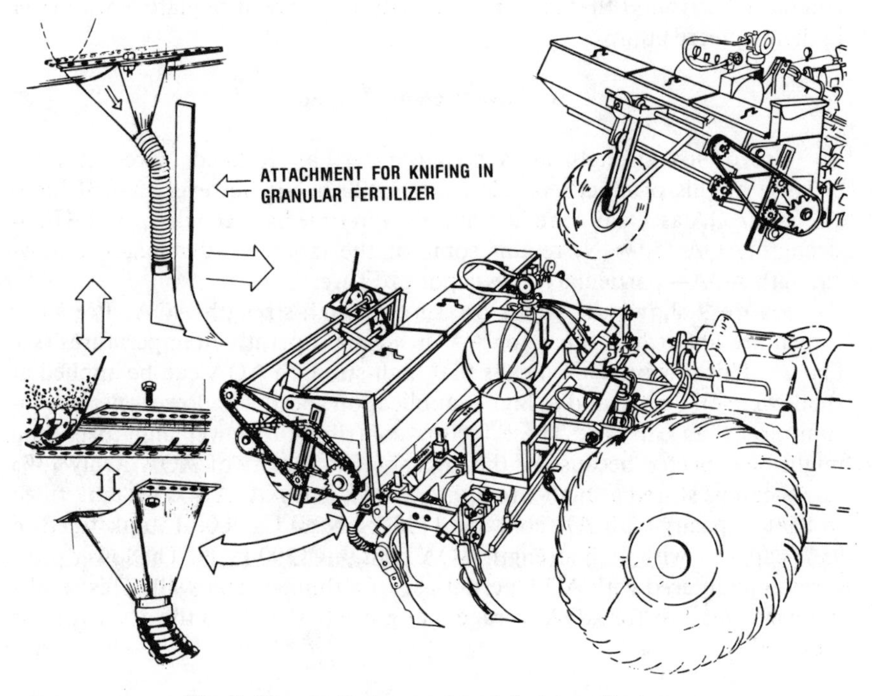

Fig. 7—Kansas State experimental plot applicator.

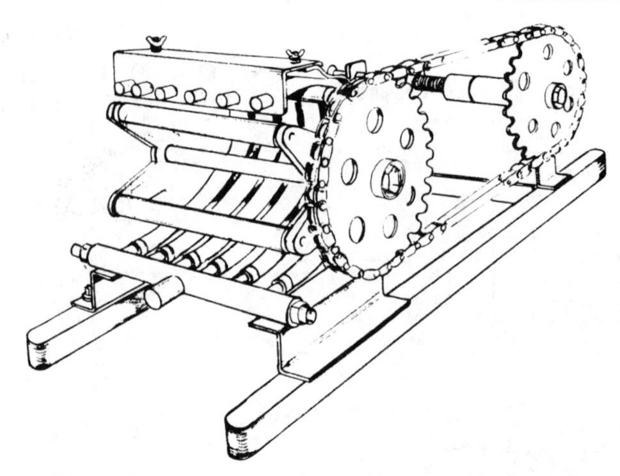

Fig. 8—Squeeze pump with round hoses and back plate.

injection of fertilizers is carried on by TVA. This work has shown that gravity-flow systems for application of solutions cannot be readily adapted for application of suspensions. If suspension is to be injected, it should be pumped to the injection knife.

TVA tests show that squeeze pumps are most reliable for metering and pumping suspensions to the applicator knives. Figure 8 is a sketch of a typical squeeze pump for this purpose. Material is pushed through the hose by rollers as they press over the hose. The spring-loaded back-up plate provides a surface onto which the round hose can be pressed. One equipment company has found that flatter hoses without a pressure plate work better in their squeeze pumps.

B. Low-Pressure Liquids

Use of low-strength aqua NH_3 (AQA) has declined in recent years because of this product's low N content (20% N); however, use of high-strength AQA as a way to reduce application time has been discussed. High-strength AQA (50% N) retains some of the inherent advantages of low-strength AQA—particularly low vapor pressure.

Figure 9 shows the vapor pressure of high-strength AQA. Pressures shown are about half those for AA at normal operating temperatures (see Fig. 3). This low pressure means that high-strength AQA can be applied at shallower soil depths and at higher application speed with lower energy consumption than can AA. These factors contribute to lower unit costs than might be expected because of the high specific gravity of AQA. Only 13% less N can be stored as high-strength AQA than as AA. A 4000-L tank filled to 85% capacity with AA contains 1728 kg N and a 4000-L tank filled to 95% capacity with high-strength AQA contains 1500 kg N. The lower pressures encountered with AQA permit use of a thinner tank wall. This results in lower unit cost for AQA storage that more than offsets the extra volume required.

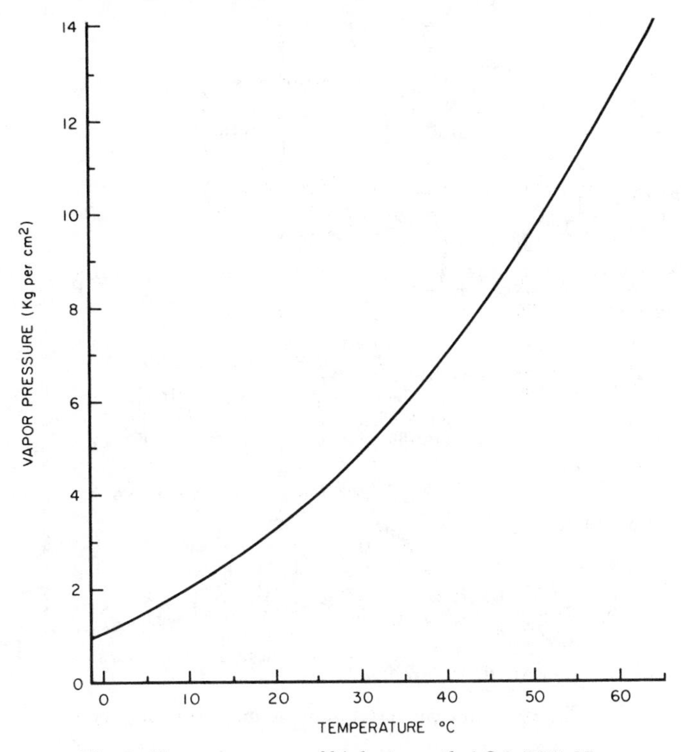

Fig. 9—Vapor pressure of high-strength AQA (50% N).

Another advantage of producing and applying AQA is that the chemical heat of dilution of the NH_3 in water can be used to convert ammonium nitrate and urea to an N solution that can be used in irrigation units. Figure 10 shows a plant for using this chemical heat to produce a nonpressure solution containing 18% N. This solution is added through an irrigation system throughout the growing season for efficient use of N.

C. Nonpressure Solution

1. N SOLUTIONS AND CLEAR LIQUIDS

Most solutions are used for direct application; however, some are used to produce fluid mixtures. Most are applied by broadcasting using various types of nozzles. Figure 11 shows common spray nozzles. The nozzle on the right is a fan-type that emits fine droplets. Some applicator operators like fan-type nozzles to apply mixtures of N solution with herbicides. They report uniform response from the N and uniform kill of weeds. Other operators report that with moderate wind, there is too much fertilizer drift with these nozzles; they prefer to apply solutions through flooding nozzles such as shown on the left side of Fig. 11. Flooding nozzles emit drops of about the same size as average rain drops; there is little or no drift. They

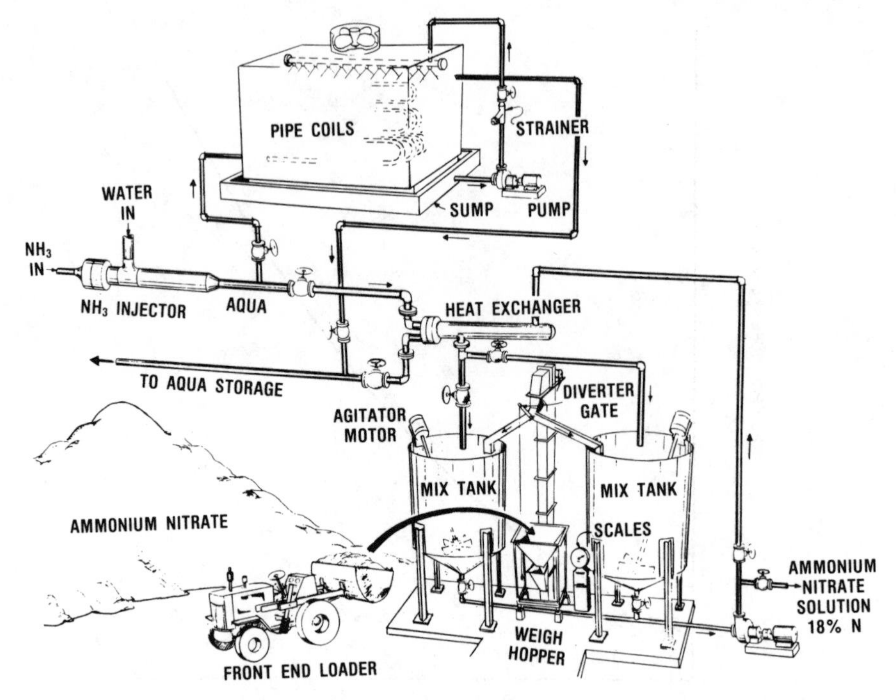

Fig. 10—Aqua converter ammonium nitrate dissolver.

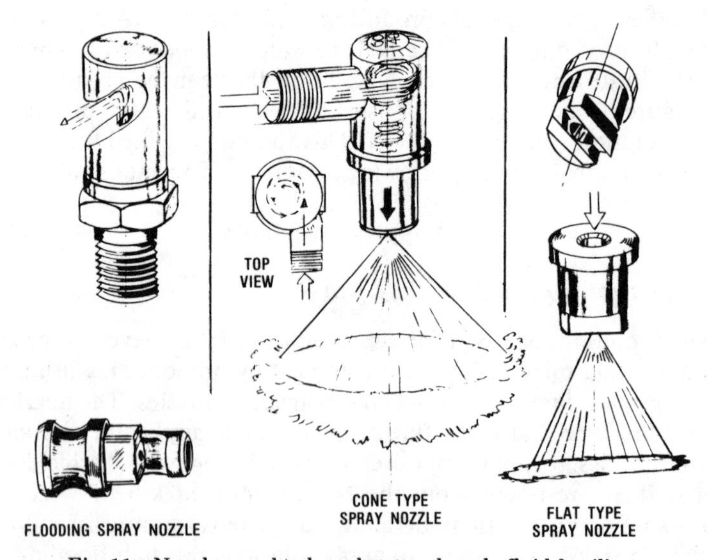

Fig. 11—Nozzles used to broadcast and apply fluid fertilizers.

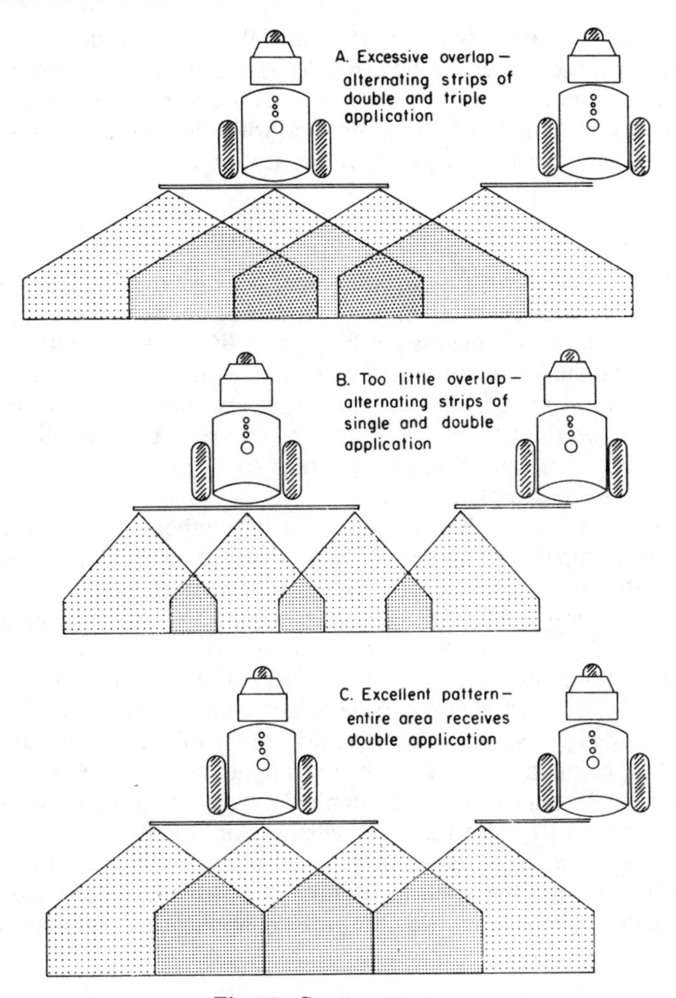
A. Excessive overlap –
alternating strips of
double and triple
application
B. Too little overlap –
alternating strips of
single and double
application
C. Excellent pattern –
entire area receives
double application

Fig. 12—Spray patterns.

also give uniform application across the swath if proper overlap of adjacent nozzle sprays is maintained. Figure 12 shows how to obtain uniform application by proper adjustment of overlap.

A nozzle designed to reduce drift and produce larger droplets than flood nozzles is rapidly being adopted. This nozzle produces a hollow cone pattern and is shown in the center of Fig. 11. The large droplets are produced in a swirl chamber, which also decreases fluid velocity. The orifices in these hollow cone nozzles are, therefore, slightly larger than orifices in flood nozzles of similar capacity.

Since nozzle pressure is generally used to control application rate, nozzle flow rates for particular pressures must be available. This information is determined by a nozzle calibration. Equipment fabricators usually calibrate nozzles with water; a more effective calibration is obtained with the N solution to be applied (Broder et al., 1980). Also, nozzles wear, resulting in changes in flow characteristics of the nozzles. For this reason, it is recommended that nozzles be calibrated with the fluid to be applied. The following formula is used to calculate application rate:

$$R = 600\, FN/SW,$$

where R = application rate in kilograms per hectare; F = single nozzle output in kilograms per minute; N = number of nozzles on applicator; S = applicator speed in kilometers per hour; and W = swath width in meters. This formula can be adapted to English units. Usually the nozzle is calibrated at several pressures. These pressures and flow rates and the above formula can be used to plot a system of curves for the applicator from which the operator can choose a particular applicator speed for a desirable nozzle pressure and application rate.

In most liquid applicators, solution is pumped to the nozzles by a centrifugal pump driven by a power takeoff of the applicator engine. Figure 13 shows the flow diagram of solution in the truck. Usually, a 15.24-m boom with several equally spaced nozzles if used instead of the single nozzle shown in the figure. Note that pressure at the nozzle is usually controlled by the amount of material bypassed to the applicator tank. The regulating valve and the pressure gauge are usually located in the truck cab. Sometimes a spring-loaded pressure regulator is used instead of the control valve.

A few companies use air-pressure applicators. Pressure required to spray solution from the nozzle is supplied by a small air compressor driven by a power takeoff from the truck engine. Air is added to the applicator tank through an air sparger, a pipe with holes oriented to spray air tangentially around the applicator tank. This helps keep herbicides mixed into the solution as it is applied. Air pressure applicators are often equipped

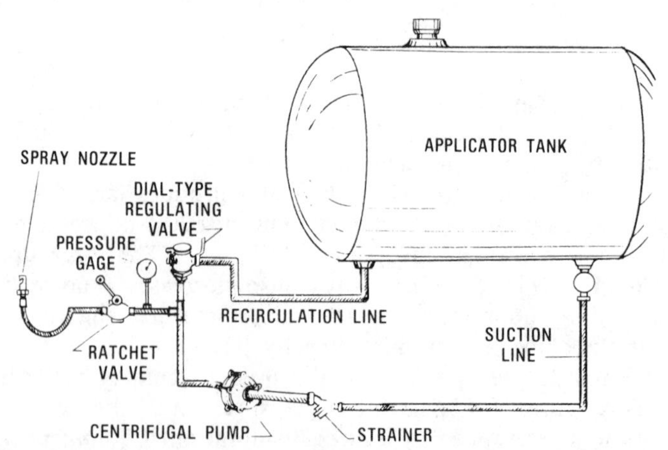

Fig. 13—Flow diagram for N solution applicator that uses pump bypass to control pressure at nozzle.

with a single nozzle that has some advantages over an applicator equipped with a boom. The advantages are (i) fewer problems are encountered with boom carriage and breakage, (ii) a light truck can be used, and (iii) solution can be applied in restricted spaces. Application rate is controlled by pressure at the nozzle, height of nozzle, ground speed of the applicator, and size of nozzle. Pressure in the tank is controlled by the amount of air vented through the vent line. This truck works especially well on small farms. It usually is a small, inexpensive truck with a light, small tank. A similar unit is designed to fit into the bed of a pickup truck. This applicator with a single nozzle has some disadvantages. The swath width is more affected by pressure changes than is the swath width for booms with several nozzles. The swath width is also narrow, usually about 10 m; with booms, swath width is as great as 27 m.

2. INJECTION INTO IRRIGATION WATER

Nonpressure N solutions are added through various irrigation systems such as sprinkler, gated pipe, and drip. Solution is usually injected into the system by a positive displacement pump such as a piston or a diaphragm pump. It usually is added at a rate of one part solution per 1000 parts of irrigation water. Nitrogen is added several times through the growing season so it can be used efficiently—a practice sometimes designated as "spoon feeding." Probably the main advantage of adding N solution through the irrigation system is that it is applied at the time the crop needs it. Corn (*Zea mays* L.), for example, uses most of its N for rapid growth through the "milk" stage. Applying part of the N in irrigation water during this time assures the irrigator that adequate N will be available for crop use during critical stages of growth (Dennis, 1971).

D. Suspensions

Most suspension fertilizers are broadcast with high flotation equipment. Manufacturers have modified conventional solution broadcast equipment, making applicators suitable for both suspensions and solutions. Suspensions require larger plumbing, more pumping capacity, and more recirculation. Main lines are 7.6 cm in diameter with 2.5- or 3.8-cm lines supplying each nozzle. Pumps are rated as high as 1300 L/min at 3.0 kg/cm^2 fluid pressure. Fertilizer flow path is the same as that shown in Fig. 13. Some applicators disperse recycled fertilizer in the tank by use of a sparger.

Nozzles suitable for suspensions are shown on the left and center of Fig. 11—flood nozzles and hollow cone nozzles. Flat fan nozzles have orifices that are too small to allow crystals in suspensions to pass.

A few dealers have successfully applied suspensions with a single nozzle applicator as shown in Fig. 13. These applicators are successful in areas where fields are small and terrain is hilly. Where land is suitable for large-scale farming, booms are preferred because they facilitate uniform application.

The science of suspension broadcasting is still developing. For years nozzle manufacturers have recommended flow correction factors to account for changes in solution density. Suspensions, however, vary considerably in such properties as viscosity, gel strength, solid fraction, and size and shape of suspended solids. Variations in these physical properties may lead to changes in flow properties and errors in suspension application. Research is needed to determine the effect that suspension properties have on nozzle flow.

E. Granular N and Mixed Fertilizers

Broadcast application of granular fertilizer containing N has not increased as rapidly as has fluid application. One reason for this is that uniform application is more difficult to achieve with most dry broadcast equipment. Most dry broadcast equipment uses one or two spinning disks to broadcast fertilizer in 12- to 15-m swaths. Two factors are largely responsible for nonuniform application from spinner applicators, segregation of blends, and poor maintenance of equipment. *Segregation* is defined as the unmixing or separation of different components in a blend during handling, transport, or application.

Size variation is the main cause of segregation. For example, if coarse KCl with many fines is blended with granular urea, the fines sift to the bottom of holding bins and applicator hoppers. When this same blend is poured into a pile, the larger urea will flow to the outside of the pile, while the fine KCl will remain in the center. This phenomenon, known as *coning,* can only be remedied by making several piles in storage or by installing partitions in bins and hoppers. Segregation is compounded by the spinner applicator, because fines will not travel as far from the spinners as will granular material. Segregation can also be minimized by proper sizing of blended components. According to TVA studies, shape and density variations contribute little to segregation. Therefore, uniform sizing of components will minimize segregation.

A spread pattern calibration is an essential part of applicator maintenance (Fig. 14). Adjustments are often required to obtain a desirable pattern. An area of concern is the point of delivery of fertilizer onto the spinner or

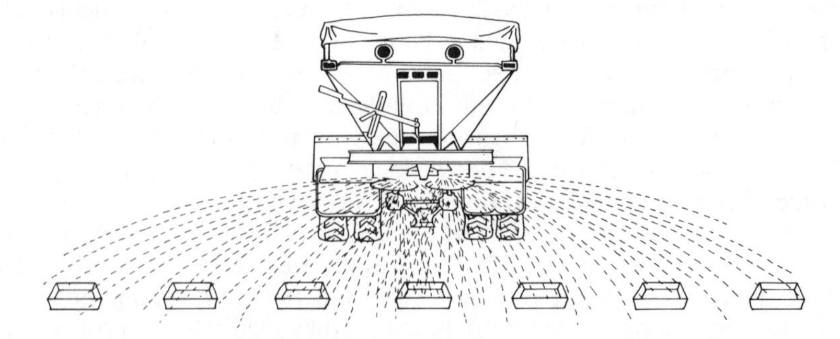

Fig. 14—Collecting samples across swath to determine patterns.

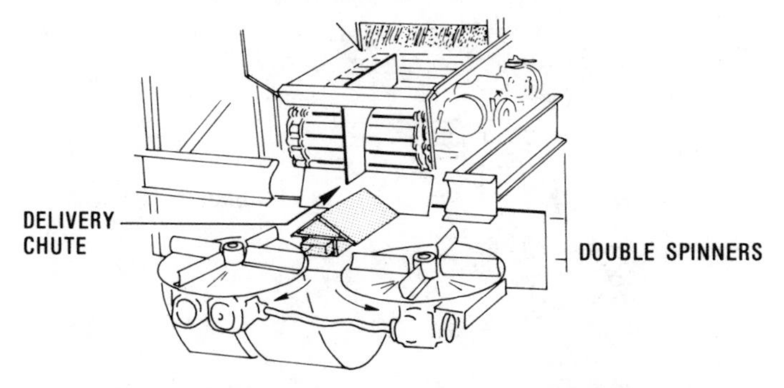

Fig. 15—Delivery chute on double spinner applicator.

spinners (Fig. 15). The most desirable patterns are shown in Fig. 16. A slight change in the location where fertilizer drops onto the spinners can considerably alter the distribution pattern. Location of the spinner blades on the disk also has an effect on the distribution pattern (Fig. 17). Caking of fertilizer on the spinner blades or delivery chute can alter distribution. Spinner speed or a change in particle size of the material applied will also alter distribution of fertilizer behind the applicator. Despite all the causes of

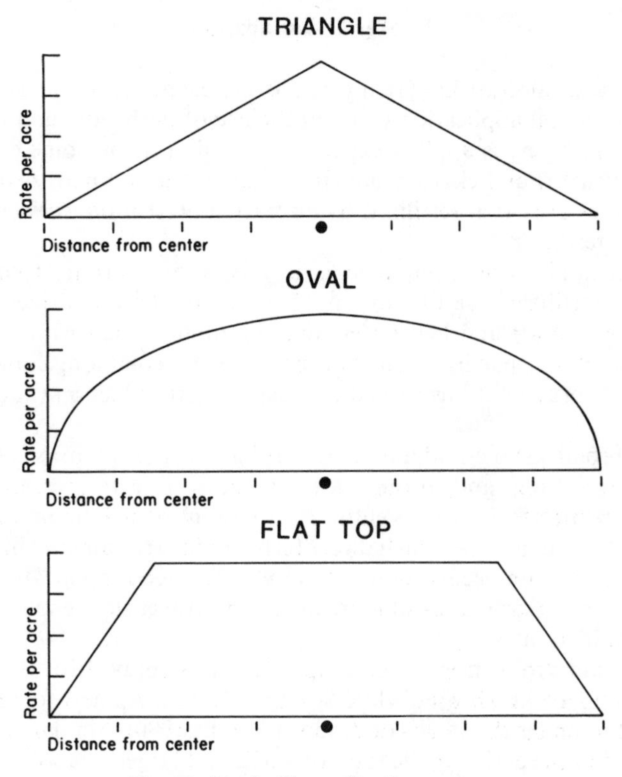

Fig. 16—Desirable application patterns.

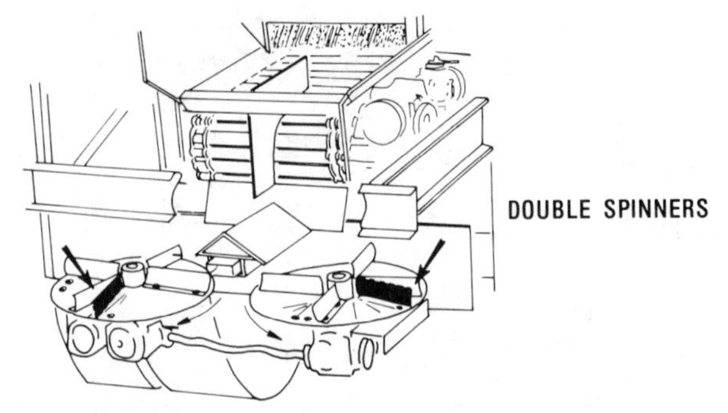

Fig. 17—Spinner blade adjustment on a double-spinner spreader.

nonuniform application, spinner applicators work well when they are well maintained and adjusted.

A few dry applicators use a boom. In one, fertilizer is moved by an auger across adjustable openings in a boom. Another uses high-velocity air to transport granules to as many as 20 nozzles. The nozzles have a deflector to distribute fertilizer. Both applicators give uniform application, but high maintenance and cost have impeded their adoption in the United States.

F. Aerial Application

The use of aircraft in agriculture has increased greatly in the past few years. Most aerial application involves the use of herbicides and pesticides. Fertilizer, however, is applied extensively by aircraft on some crops. Rice (*Oryza sativa* L.) and citrus-producing areas in the southern region of the United States are often fertilized by aircraft. Forest producers use aircraft to apply N fertilizer.

Aerial application equipment is largely derived from ground equipment. Dry fertilizers are usually broadcast by electrically driven spinners. Fixed wing aircraft and helicopters are commonly fitted with booms and nozzles at close spacing. Flat fan nozzles are common; however, the problem of spray drift has turned interest to drift reduction (hollow cone) nozzles.

One broadcast method unique to airplanes is the venturi system. Air is forced into a scoop and through a restricted section to increase velocity. Fertilizer is injected into the venturi at the point of maximum air velocity and forced through ducts which direct fertilizer into the air stream.

Helicopters are becoming more popular for aerial application because they can reach areas inaccessible by plane and travel slow enough to apply fertilizer at high rates.

The main problem with aerial application is spray drift. A fine spray drifts 15 m in a 5 km/h wind while falling only 3 m. An aerosol droplet can travel 34 km under the same conditions (Anon., 1980). Similar results have been found in the aerial application of small and larger size solid materials. For example, much less drift occurs when large granules of urea (0.48–1.27

cm) are used in preference to standard size urea (0.1–0.33 cm). To reduce spray drift, one aerial sprayer manufacturer has placed extra nozzles on the right side of the fuselage to help compensate for prop wash, which moves spray from right to left. Other methods involve wing modifications on airplanes to modify their wake and nozzles that control droplet size.

REFERENCES

Achorn, F. P. 1980. The effects of rising energy costs on the fertilizer industry. p. 22–46. *In* Florida lime and fertilizer conference proceedings, May 15. Vol. 10. Soil Sci. Dep., Inst. Food Agric. Sci., Univ. of Florida, Gainesville.

Achorn, F. P. 1981. Accurate application of fluid nitrogen products. p. 429–456. *In* A. I. More (ed.) Fertilizer nitrogen—facing the challenge of more expensive energy. Proc. British Sulphur Corporation's Fourth Int. Conference on Fertilizer Technology, London. 9–21 Jan. 1981. Part I. Papers. British Sulphur Corp., Ltd., London.

Achorn, F. P., and H. L. Kimbrough. 1978. Energy use and how to conserve it in the fertilizer industry. p. 32–41. *In* Situation 78, TVA Fertilizer Conference, St. Louis, Mo. 15–16 Aug. 1978. Bull. Y-131. Natl. Fert. Dev. Center, TVA, Muscle Shoals, Ala.

Anon. 1980. From wind tunnel to wing tip. Farm Chem. 143(11):26, 29, 32, 34, 37.

Broder, M. F., H. L. Kimbrough, and F. P. Achorn. 1980. Perfecting the art of spray patterns/calibration. Solutions 24(4):72, 74, 76–78, 80–81.

Dennis, E. J. 1971. Fertigation—a production agronomist's point of view. Fert. Solutions 15(3):33–34, 36–38.

32

Bryant R. Gardner and Robert L. Roth
University of Arizona
Yuma, Arizona

Applying Nitrogen in Irrigation Waters

Irrigation water is often the most convenient and inexpensive means of applying N fertilizers. Usually, fertilizer applications in irrigation waters are not as uniform as those made with conventional ground-driven fertilizer spreaders. Efficient applications of water-applied fertilizer can be obtained from properly engineered systems if the operator has competent counsel from soil and irrigation scientists as well as fertilizer dealers. Nitrogen losses to deep percolation for soils with high leaching potentials can be minimized by properly designed irrigation systems. Savings in time, labor, fuel, and equipment costs can offset any fertilizer losses.

The uniformity of distribution from N fertilizers applied in irrigation waters will be the same as the uniformity of water distribution. Direct N fertilizer losses and poor N fertilizer distribution can be prevented by skilled irrigation management and properly designed irrigation systems.

Application of anhydrous NH_3 (AA) or other fertilizer materials such as nonpressure urea ammonium nitrate (UAN) solutions containing free NH_3 to irrigation waters high in Ca, Mg, and HCO_3^- may result in precipitation of calcite. This reaction, which will cause scaling and plugging problems in equipment, can be prevented or corrected by the addition of acid.

I. FLOOD IRRIGATION

A. Irrigation Methods

The three most common methods of applying surface irrigation water are border, basin, and furrow. The border method consists of dividing the field into a series of strips separated by low border dikes. This system is usually used on land that is level in one direction and uniform in grade in

Nitrogen in Crop Production.

the other direction. Water is turned in at the highest end (ditch end) and flows in a sheet over the soil. The borders usually are constructed with no cross slope so that water flows uniformly between the border ridges.

The basin method consists of a level or nearly level area completely surrounded by a dike. Basins may vary in size from those around individual trees in an orchard up to several hectares. The size of the individual basin must be adjusted to the amount of water that can be delivered to the basin and soil permeability. The desired amount of water is applied rapidly and ponded in the area surrounded by the dike until absorbed by the soil. High water application efficiencies with uniform water distribution are possible with properly constructed and managed basins.

Small channels between parallel ridges are used to convey the water over the soil surface in furrow irrigation. Water infiltrates and moves both laterally and vertically downward into the soil. Different shapes of furrows may be used to achieve desired water distribution for different crops and for land with different soil permeabilities. Furrows can be used with border, basin, and a wide variety of conditions, including contours for irrigation of steep slopes with a minimum of erosion. A detailed description of surface irrigation systems is given by Bishop et al. (1967).

B. Injection Equipment

Equipment used for applying dry N fertilizer materials in irrigation water is usually unwieldy and inefficient. Some of the more sophisticated equipment consists of a structure over or along side the ditch with a water- or power-driven mechanisms to meter the dry fertilizer from a hopper into the water. The simplest method probably consists of a farm worker pouring or dribbling dry fertilizer into the irrigation water from a sack or bucket.

The metering of liquid N fertilizer into open irrigation water is usually accomplished with a constant head-metering device attached to a fertilizer nurse tank (Fig. 1A). One of the more common devices used is a float box. The nurse tank is connected to a hard rubber box that contains a valve controlled by a float similar to the ordinary toilet tank valve (Fig. 1B). The desired outflow of fertilizer into the irrigation water can be maintained with this constant head device. Another common method of metering liquid N fertilizer into open irrigation water consists of a clamp on the discharge hose (Fig. 1C) to control the amount of liquid fertilizer flowing from the fertilizer tank. The clamp is adjusted until the desired flow is obtained. This flow will change as the level in the fertilizer tank drops and has to be readjusted periodically during the time of fertilizer application. Fertilizer application accuracies are lower with this method as compared with the use of a constant head-metering device.

More elaborate equipment is necessary for injecting AA into open irrigation systems. The equipment consists of a pressure tank, a liquid control valve, a pressure regulator, a check valve, an orifice, and usually a spreader device at the point of injection. Charts are available from

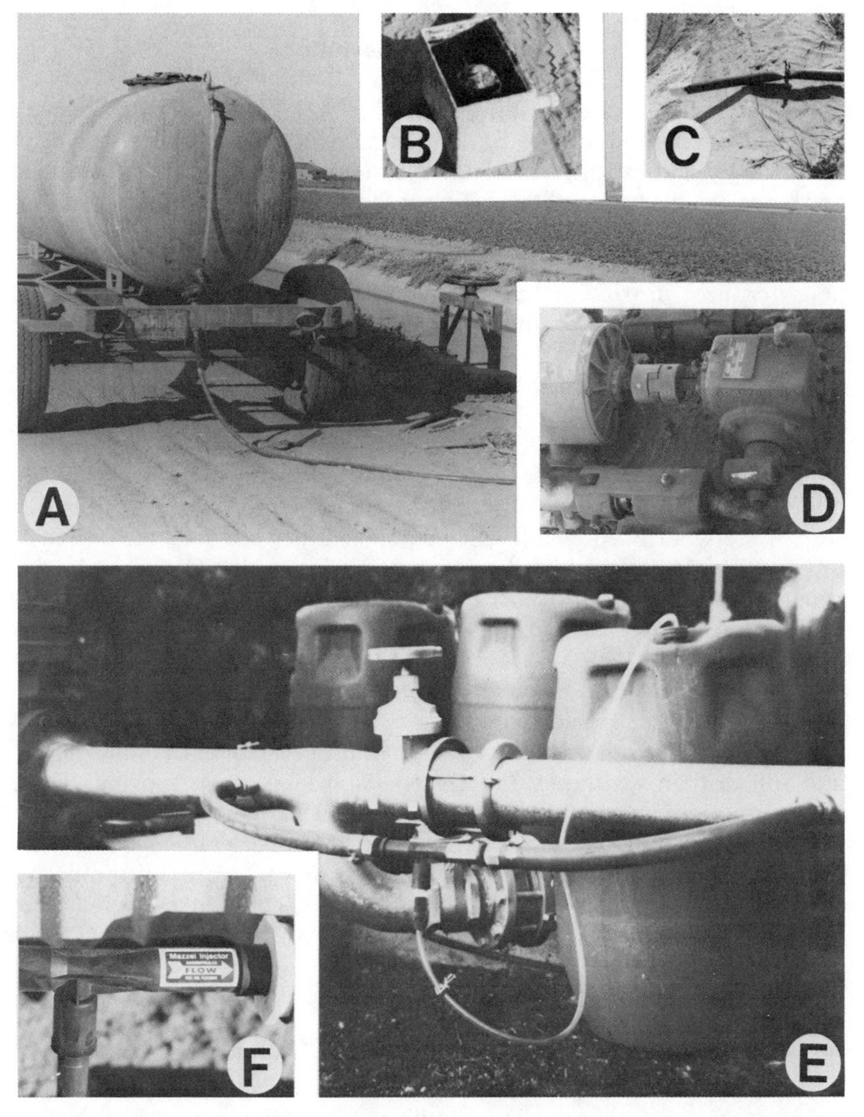

Fig. 1—Equipment and devices used to inject fertilizer into irrigation water: (*A*) fertilizer nurse tank; (*B*) float box; (*C*) hose clamp; (*D*) injector pump; (*E*) venturi installation on irrigation well; and (*F*) venturi. Parts *E* and *F* courtesy of the Mazzei Injector Corp.

manufacturers that relate temperature and/or pressure with orifice size and rate of delivery. These charts should be consulted when selecting an orifice.

Regardless of the type of fertilizer applied, the fertilizer should be applied at a point several meters upstream from the field outlets and preferably at a weir box, check drop, or some point where water turbulence will allow for maximum mixing of the fertilizer.

C. Practical Considerations

Applying N fertilizers in irrigation water is often the most convenient and cheapest method, and becomes even more attractice when energy is in short supply and expensive. Irrigation systems must be properly designed and managed in order to obtain efficient fertilizer applications and distribution. There must be no loss of tail water from surface irrigation if direct losses of fertilizers are to be prevented. Poor water distribution will result in poor fertilizer distribution and conversely good water distribution will result in good fertilizer distribution.

Humbert (1954) used tracer techniques to study water distribution of various surface irrigation systems. As a result of their studies, irrigation systems were redesigned so that water would be evenly distributed, enhancing water-applied fertilizer efficiencies. Leavitt (1957) reported that when irrigation water was properly controlled in furrows, water-applied NH_3 concentrations varied within 5% between the head and end of the furrows. Laser-controlled land-leveling equipment has provided the degree of precision needed to level borders or large basins so that water can be distributed more uniformly. Erie and Dedrick (1979) reported water distribution efficiencies of 92 to 96% with proper irrigation management of level basins.

The chemistry and behavior of the various sources and forms of N fertilizers should be considered when evaluating N distribution and application efficiencies. Both urea and NO_3^--N tend to remain in the soil solution and will drift in whichever direction the soil moisture is moving. These forms are more susceptible to leaching loss if excessive water is applied. Broadbent et al. (1958) showed that both urea and NO_3^--N moved with the wetting front in sandy soils, but that urea was more uniformly distributed through lighter-textured soils (Fig. 2). Their work also indicated that applied urea was almost completely hydrolyzed to NH_3-N within a 24-h period.

Ammonia or NH_4^+ salt applied in water readily converts to exchangeable NH_4^+. Broadbent et al. (1958) showed that this NH_4^+ was captured by the exchange sites in the surface 2 to 3 cm of the soil (Fig. 2). Such NH_4^+ is vulnerable to volatilization losses. Moisture movement from the soil, as by evaporation, will increase NH_3 volatility. A mechanism by which NH_3 is lost from surface applications of NH_4^+ compounds on calcareous soils has been proposed by Fenn and Kissell (1973). They suggest that when an NH_4^+ salt dissolves in a calcareous soil, ammonium carbonate and a Ca^{2+} salt of varying solubility will form. Ammonium carbonate subsequently decomposes, losing CO_2 at a faster rate than NH_3. This causes formation of NH_4OH and an increase in pH and greater NH_3 losses. In later work, results of Feagley and Hossner (1978) indicated that the reaction proceeded through an ammonium bicarbonate intermediate and not ammonium carbonate. Urea may be the best NH_4^+-N source to apply in irrigation water because it penetrates further into the soil with water than NH_4^+.

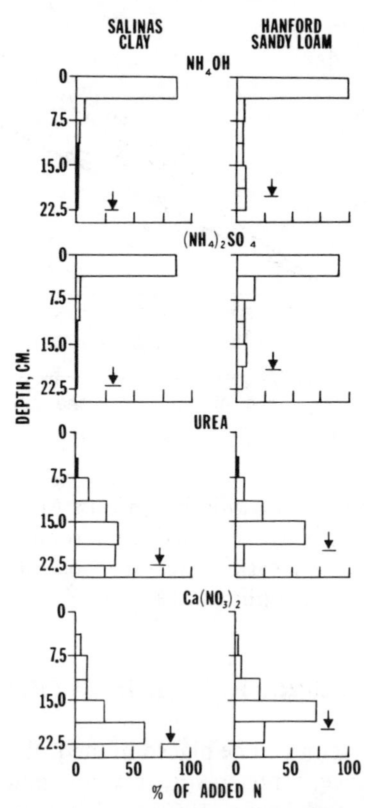

Fig. 2—Concentration profiles of NH_3-N from NH_4OH and ammonium sulfate of NO_3^--N from calcium nitrate and urea in columns of two soils. Arrows indicate depth of water penetration. The NH_4^+ salts were applied with solutions containing 150 mg/L N. Urea and calcium nitrate were placed on top of columns and then leached downward with water. (Redrawn from Broadbent et al., 1958).

Ammonia also can be lost directly from irrigation waters. Miyamoto et al. (1975a) reported NH_3 losses from water of 45 to 73% (Fig. 3). Higher concentrations of NH_3 and higher pH's increased NH_3 volatilization. Higher temperatures, greater water turbulence, and length of exposure all increased the potential for losses of NH_3. Losses of NH_3 from irrigation waters can be decreased by lowering the pH of the water by addition of acids (Miyamoto et al., 1975a).

Application of AA or other fertilizer materials containing free NH_3 to irrigation waters high in Ca, Mg, and HCO_3^- may result in precipitation of calcite. These precipitates may cause scaling and plugging problems in irrigation systems where gated pipe, siphon tubes, etc. are utilized. These problems can be prevented or corrected by the addition of acid to the irrigation water.

Fertilizer applications of N can be made in any irrigation during the growing season. Thus, N can be applied when crop damage caused by

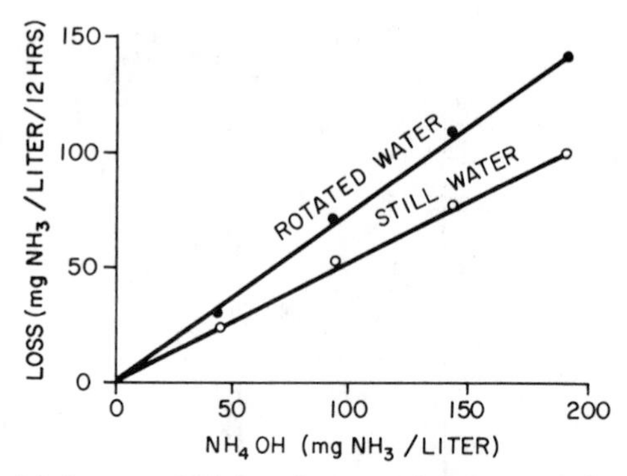

Fig. 3—The relationship between NH_3 loss during 12 h from ponded waters and the concentration of NH_4OH at the moment of exposure to the atmosphere (Miyamoto et al., 1975a).

mechanical applications of N would be serious. Multiple applications of N can be made to highly permeable soils to overcome problems of leaching losses. High N application efficiencies are possible on any soil type or texture as long as irrigation application efficiencies are high.

II. SPRINKLER IRRIGATION

Most sprinkler systems can be placed in the following categories: solid-set, periodic lateral move, center-pivot, traveling gun, and traveling lateral systems. Christiansen and Davis (1967) and Addink et al. (1980) give a general description of the various types of sprinkler irrigation systems. The efficiencies of water-applied N fertilizers are dependent upon the irrigation water application efficiencies, regardless of the type of sprinkler irrigation systems used.

In the solid-set and periodic lateral move, N fertilizer is injected into the system at a rate and duration that results in the desired application. Application times will vary, depending upon application rate, injection equipment, fertilizer type, and size of the irrigation system. The system is usually operated after the N fertilizer has been injected to rinse the fertilizer from the pipes and crops. An estimate of rinsing time would be the time it takes water to flow from the sprinkler nearest the injection point to the sprinkler farthest from the point of injection. Distribution of those N fertilizers that move with the water in the soil can be controlled to some degree by the time of injection, the injection period, and by the amount of water applied after injection.

Nitrogen fertilizer must be injected continuously with center-pivot, traveling gun, and traveling lateral systems from the start of fertilization until the system has moved across the field. Some fertilizer will remain on the crop after irrigation with these systems. Because of the extreme dilution of the fertilizer, little or no damage to plants by N fertilizer applied with this method has been reported.

A. Injection Equipment

There is a variety of injection equipment available to introduce N fertilizer into sprinkler systems. The injector chosen should be of adequate size to provide the quantity of fertilizer that must be applied to satisfy the application requirements. High-pressure pumps, venturi, and Pitot tube injectors normally place N fertilizer in the pressure side of the sprinkler system.

One of the simplest methods of applying N fertilizer through a sprinkler system is to introduce the solution at the suction side of a centrifugal pump. A hose or pipe delivers the fertilizer from the supply tank to the suction pipe. A shutoff valve or metering orifice is used to control the rate of delivery. A pipe from the pump discharge can be used for water to dissolve dry material, dilute concentrated solution, or for rinsing. A variation of this method can be adapted to deep-well turbines. There is a N pollution hazard associated with this method of introducing fertilizer into the suction side of a pump.

High-pressure pumps (Fig. 1D) can be used to inject N fertilizer under pressure into sprinkler lines. A separate pump is used and must be able to develop a pressure greater than that in the sprinkler line. The pump needs to be constructed of materials that are not corrosive to the N fertilizer used. The supply tank and pump should be rinsed after injection to prevent corrosion.

Another common method used to apply N fertilizers through sprinkler systems is the *venturi principle.* A tapered constriction increases the velocity of the water, creating a drop in pressure. The difference in pressure between a connection above and one within the constriction is used to inject the fertilizer solution. Commercial venturi injectors are available that will create a lower pressure than the supply tank to inject liquid chemicals. This type injector can be obtained at selected injection rates to fit most any irrigation system (Fig. 1E and 1F).

The principle of the Pitot tube also creates a pressure differential. Water is circulated through a pressurized supply tank containing the fertilizer. The flow of water is established by placing an open-ended pipe into the stream of water and another is faced in the opposite direction. These pipes are connected to the supply tank with valves to regulate the flow of water into the tank and fertilizer solution into the sprinkler pipe. With this principle, the fertilizer in the supply tank is diluted with time and the rates of N injections will decrease unless regulated.

Chemical injectors are available that will allow the use of dry N fertilizers. This type injector is different from the liquid injectors because a mixing or agitation device must be used to dissolve the N fertilizers. After the dry N fertilizers are dissolved, it is metered into the sprinkler irrigation system. This type injector allows the operator to use dry chemicals; however, only those chemicals that are easily dissolved should be used. Surface conditioners such as kaolin and diatomaceous earth used on dry N sources may interfere in the operation of some irrigation systems. However, in most instances these materials are too small to be a problem. Some operators place an in-line filter after the fertilizer is dissolved, but before the fertilizer is injected. This will filter out any aggregate too coarse to pass through the system. The plugging hazard should be determined for each system and material before dry N fertilizers are used.

Pair et al. (1975) presented a detailed outline for the selection of injection equipment. Among the factors they suggest to be considered are: water flow rate, water pressure, fertilizer flow rate, source of power, and duration of operation.

Injection rates of N fertilizer can be determined by several methods. The area to which fertilizer will be applied must be determined. This can be calculated by using the dimensions of the system and its mode of operation. Most sprinkler irrigation handbooks contain tables that aid in determining both the area covered and the quantity of fertilizer to be injected per set.

B. Practical Considerations

Irrigation waters used in some regions often contain high concentrations of Ca^{2+}, Mg^{2+}, and HCO_3^-. These ions form calcite and magnesite deposits when waters are exposed to the atmosphere or to elevated temperatures. Applications of N fertilizers containing free NH_3 increases this precipitation. These precipitates often cause incrustation and plugging of irrigation systems. These problems can be prevented or corrected by acid treatment. Since the precipitation reactions involving calcite and magnesite are similar, only the calcite reaction will be treated.

Dutt et al. (1972) reported the reactions involved in calcite precipitation as follows:

$$H^+ + HCO_3^- \rightleftharpoons H_2CO_3 \quad [1]$$

$$HCO_3^- \rightleftharpoons CO_3^{2-} + H^+ \quad [2]$$

$$Ca^{2+} + CO_3^{2-} \rightleftharpoons CaCO_3(\downarrow) \quad [3]$$

with the overall reaction as

$$Ca^{2+} + 2HCO_3^- \rightleftharpoons CaCO_3(\downarrow) + H_2CO_3. \quad [4]$$

If the system is open to gas phases, then

$$H_2CO_3 \rightleftharpoons H_2O + CO_2(\uparrow). \quad [5]$$

Calcite precipitation increases with increasing concentrations of Ca^{2+} and with decreasing CO_2 partial pressure (Eq. [4] and [5]). Miyamoto et al. (1975b) and Tanji and Doneen (1966) have reported numerical procedures for estimating the extent of calcite formation from routine water analysis data.

Acid applied to alkaline irrigation waters dissociates almost completely. Hydrogen ions react with HCO_3^- and CO_3^{2-} and convert them to H_2CO_3 as indicated in Eq. [1] and [2]. This results in a reduction of HCO_3^- and CO_3^{2-} and in an increase in H_2CO_3, which limits calcite precipitation. The amount of acid needed to prevent calcite precipitation will depend upon the irrigation water quality, CO_2 partial pressure, and temperature. The necessary acid rates can be estimated if routine water analysis data are available (Miyamoto et al., 1975b).

When N fertilizers that contain free NH_3 are applied to alkaline irrigation waters, the pH of the water rises and calcite precipitation is increased drastically. Miyamoto and Ryan (1976) reported the reactions as follows:

$$NH_4OH \rightleftharpoons NH_4^+ + OH^- \quad [6]$$

$$OH^- + HCO_3^- \rightleftharpoons CO_3^{2-} + H_2O \quad [7]$$

$$Ca^{2+} + CO_3^{2-} \rightleftharpoons CaCO_3(\downarrow). \quad [8]$$

The overall reaction is

$$Ca^{2+} + HCO_3^- + NH_4OH \rightleftharpoons CaCO_3(\downarrow) + H_2O + NH_4^+. \quad [9]$$

Equations [4] and [9] indicate that 1 mol of HCO_3^- is required to precipitate 1 mol of Ca^{2+} from ammoniated water, whereas 2 mol of HCO_3^- are needed for nonammoniated water. Consequently, the extent of $CaCO_3$ precipitated is much greater in ammoniated water.

Hydroxyl ions produced by the reaction given by Eq. [6] are neutralized by acid applied to ammoniated water. The reduction in OH^- concentrations results in less conversion of HCO_3^- to CO_3^{2-} by Eq. [7] and less $CaCO_3$ precipitation. The amount of acid needed to prevent $CaCO_3$ precipitation from ammoniated water is approximately equal to the amount required to neutralize the NH_3 plus the amount needed to prevent precipitation from the nonammoniated water.

Carbonates will be dissolved if acid rates exceed the amount needed to prevent precipitation:

$$2H^+ + CaCO_3 \rightleftharpoons Ca^{2+} + H_2CO_3. \quad [10]$$

When the $CaCO_3$ is all dissolved or if the system does not contain $CaCO_3$, the pH of the water will be lowered, increasing the corrosion hazard of the irrigation system.

In order for dry N fertilizer materials to be injected into sprinkler systems, they should be predissolved and should not contain any coating material that would plug the filter or the system. Coating information is available at most local fertilizer dealers. Liquid N solutions are very convenient for use with pumps and gravity flow supply tanks. Forms of N that are subject to leaching losses can be applied to any desired depth without

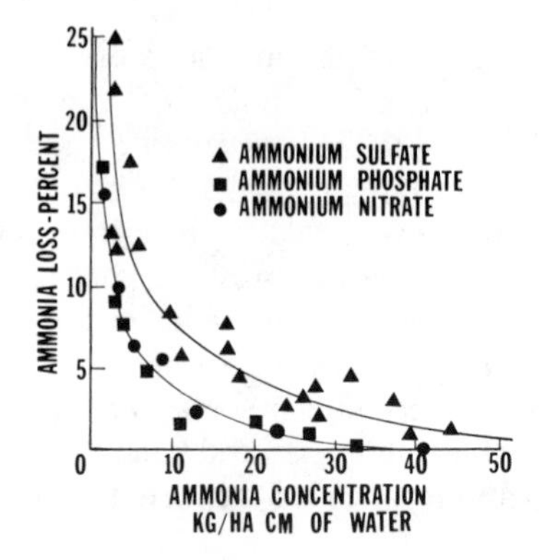

Fig. 4—Losses of NH_3 from fertilizer salt in relation to NH_3 concentration in the irrigation water (Henderson et al., 1955).

danger of leaching with careful irrigation management. Henderson et al. (1955) reported NH_3 losses from NH_4^+-N salts applied through sprinklers were as high as 10% (Fig. 4). This did not include volatilization losses of NH_3 from the soil surface during drying. They reported the losses of NH_3 from AA to be as high as 60% when applied through sprinklers (Fig. 5). Anhydrous or aqua NH_3 (AQA) cannot be applied in sprinkler water without large losses from volatilization and probably shouldn't be used with sprinklers. Urea will enter the soil with the water and is the least susceptible to volatilization losses of the NH_4^+-forming materials.

The uniformity of sprinkler irrigation systems normally is evaluated on the basis of distribution at the soil surface. The crop, however, is dependent on available water and N in the root zone. Water and N will be redistributed within the soil profile due to differences in soil water potential caused by application nonuniformity or plant uptake. Hart (1972) used finite-difference solutions of the two-dimensional diffusion equation to analyze the subsurface redistribution of water. He concluded that there is a substantial redistribution within 1 m and the final soil water distribution is not significantly affected by the water nonuniformity applied on the soil surface within that distance.

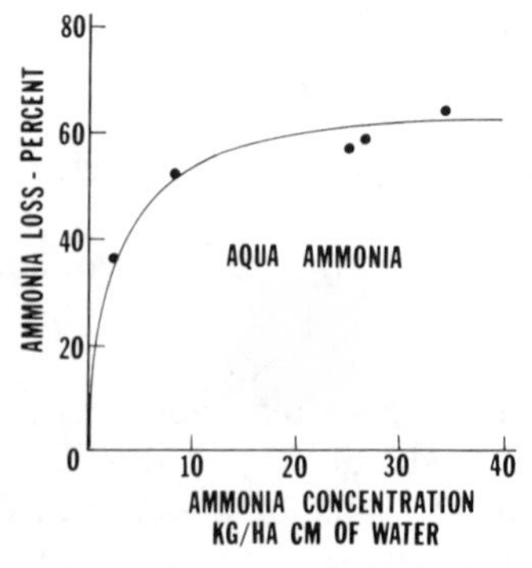

Fig. 5—Losses of AQA in relation to concentration of NH_3 in the irrigation water (Henderson et al., 1955).

Sidedressed N may be more effective for plants that have limited root distribution. Sprinkler-applied N may be positionally unavailable for crops with small root systems that are widely spaced. Thus, the effective non-uniformity of water and N is a function of the root development of individual plants as well as the soil water potential.

Applications of acid-forming N fertilizers to the surface of unbuffered sands may result in serious declines in pH. The buffer capacity of sands should be determined before this method of fertilizer application is used as a routine practice.

Sprinkler-applied N fertilizers result in savings of labor and equipment, and greater flexibility of timing of N applications. Fertilizer can be applied through the sprinkler system when it is difficult to apply by other means.

III. OTHER PRESSURIZED IRRIGATION SYSTEMS

Other pressurized irrigation systems may be categorized as trickle or drip, spray, and bubbler. These systems are considered low-pressure, with ranges of 30 to 200 kPa, while sprinkler pressures are normally 350 to 700 kPa. With trickle or drip irrigation, water under pressure is delivered through pipes to drippers or emitters, or tubing with equally spaced emitters. Water is delivered through these emitters at a very slow rate of 2 to 8 L/h near the plant. The soil is near saturation at the emitters with gradual decrease in moisture content in the soil in all directions away from the source. The flow rate from the source and the soil's water conducting properties regulate the soil moisture at any given location. The amount of water delivered to a plant can be controlled by number and spacing of emitters and the rate of flow from each emitter. These systems may be designed for use on row crops as well as trees and orchards.

Spray irrigation systems as defined here are used primarily for trees and orchards. The systems are similar to a trickle system in that water is delivered through pipes to each plant. The water is supplied through spray heads that deliver the water around each plant. The outlet size is much larger and water is delivered at a rate of 0.5 to 2.0 L/min. The application rate is usually somewhat less than the infiltration rate of the soil so that there is no surface runoff.

Bubbler systems are similar to the spray systems, except that water is delivered to each plant through a device at a rate that exceeds water infiltration. The area of water application must be diked to prevent surface runoff. Bubblers deliver water at a rate of 3 to 5 L/min.

A. Injection Equipment

A differential pressure system (venturi or Pitot) or a positive pressure pump can be used to inject N fertilizer into these pressure irrigation systems. Both systems, including variations of the pressure differential method, are described by Goldberg et al. (1976). In some systems, the flow rates are too low to operate a venturi, and positive pressure pumps must be used. The positive displacement, piston or diaphragm, pumps are the most common. The same type of equipment used to inject dry N fertilizer into sprinkler systems can be used with these systems, also. The main consideration in choosing injection equipment are (i) injection rate; (ii) system capacity and flow rate; (iii) storage tank capacity; and (iv) power source. Injection rates can be calculated from a simple equation reported by Isobe (1974):

$$I = 10^{-6}\, CQA, \qquad [11]$$

where I is the injection rate in kg N/min; C is the desired final concentration of N in milligrams/liter in the irrigation water; Q is the flow rate of water in the system in liters/minute per hectare; and A is the area in hectares. From this, the injection time, T (in hours), required to apply a rate (R) of N (in kilograms/hectare), may be calculated by

$$T = RA/60I. \qquad [12]$$

B. Practical Considerations

Fertilizer N application costs, both equipment and labor, are minimal with these pressurized irrigation systems. These methods of irrigation offer flexibility in N fertilization that is unique to these systems; particularly, they offer more options in timing of N applications because of the high frequency of irrigations. More data are needed to determine the effect of small, frequent applications of N on plant growth, yield, and crop quality. There is a shortage of information on the distribution and losses of N fertilizer applied through these systems. Roth et al. (1974) reported the distribution of NO_3^--N in the root zone of citrus trees irrigated by trickle, bubbler, and basin methods. These trees were grown on sandy soils with a great potential for N leaching losses. Higher levels of NO_3^--N were maintained in the root zone of trees irrigated by the pressurized systems.

Clogging of orifices by particles or organic slime is one of the major problems associated with these pressurized systems. Bucks and Nakayama (1980) suggest any fertilizer or chemical added to a drip system meet the following criteria: (i) not corrode or clog any component in the system; (ii) be safe for field use; (iii) be water soluble; (iv) not react adversely with irrigation water; and (v) not reduce crop yields. Aqua NH_3 is an attractive source of N for use in these systems because it is usually one of the cheapest sources and is in solution form ready for use. This material requires careful management, particularly with waters containing Ca, Mg, or other substances that might form precipitates when NH_3 is injected into the water. Isobe (1974) found that the NH_3 tolerance level was reduced from 30 000 mg/L N to 180 mg/L N when the Ca^{2+} and Mg^{2+} content of irrigation water was increased from 11 to 260 mg/L. Most of the NH_4^+ salts are fairly water-soluble and cause few clogging problems. The one exception is ammonium phosphates. The phosphates in these salts tend to precipitate as Ca and Mg phosphates if Ca and Mg are present in the irrigation water. Urea and NO_3^- sources are highly soluble and rarely result in precipitation problems. Nitrogen could increase microbial growth if applied continuously through the system and thus cause increased clogging problems. This system should be rinsed after each application.

IV. SUMMARY

The application of N fertilizers in irrigation waters is a well-established agronomic practice. Nearly all types of agricultural crops have part or all of their fertilizer requirements applied through irrigation waters. There are several distinct advantages and disadvantages in this practice. These should be weighed and evaluated against other methods of N fertilizer application before water-applied fertilizer is used as a routine practice. Some of the advantages of water application of N fertilizer are: (i) cost savings of energy, labor, and equipment; (ii) greater choice in timing of N applications; and (iii) greater control of N distribution with pressurized irrigation systems, particularly on sandy soils. The disadvantages are: (i) fertilizer distribution is determined by water distribution; (ii) N fertilizer can be lost in runoff water; (iii) N may be lost from NH_4^+-containing fertilizers applied to calcareous soils by volatilization from the soil surface; (iv) AA or AQA applied in surface or sprinkler irrigation water are subject to large volatilization losses; (v) precise application rates are not possible with some of the water application equipment in current use; and (vi) application of NH_3 or other fertilizer materials containing free NH_3 to irrigation waters high in Ca, Mg, and HCO_3^- may result in precipitation of calcite. This reaction may cause scaling and plugging problems in equipment but can be prevented or corrected by the addition of acid.

In spite of these disadvantages and the potential losses of NH_3 from irrigation water and soil surfaces, there is a widespread use of NH_3 and NH_4^+ salts in water application. Simultaneous application of acid with NH_3 reduces the volatile losses. The use of NH_3 and NH_4^+ salts will continue as long as growers find it economically advantageous. It is surprising that so little quantitative data exist on the efficiency of this fertilizer method. A majority of the data found in the literature concerning N fertilizer distribution and losses applied through irrigation water are from laboratory studies. While these studies have laid an excellent foundation for predicting N application efficiencies, there is a great need for field and on-farm evaluations of water-applied fertilizer efficiencies. There is a great potential for increasing N efficiency in crop production in water-applied N fertilizer.

REFERENCES

Addink, J. W., J. Keller, C. H. Pair, R. E. Sneed, and J. W. Wolfe. 1980. Design and operation of sprinkler systems. p. 621–660. *In* M. E. Jensen (ed.) Design and operation of farm irrigation systems. Am. Soc. of Agric. Eng., St. Joseph, Mich.

Bishop, A. A., M. E. Jensen, and W. A. Hall. 1967. Surface irrigation systems. *In* R. M. Hagan et al. (ed.) Irrigation of agricultural lands. Agronomy 11:865–884.

Broadbent, F. E., G. N. Hill, and K. B. Tyler. 1958. Transformations and movement of urea in soil. Soil Sci. Soc. Am. Proc. 22:303–307.

Bucks, D. A., and F. S. Nakayama. 1980. Injection of fertilizer and other chemicals for drip irrigation. p. 166–180. *In* The Irrig. Assoc. 1980 Annu. Tech. Conf. Proc., Houston, Tex. The Irrigation Assoc., Silver Springs, Md.

Christiansen, J. E., and J. R. Davis. 1967. Sprinkler irrigation systems. *In* R. M. Hagan et al. (ed.) Irrigation of agricultural lands. Agronomy 11:885–904.

Dutt, G. R., M. J. Shaffer, and W. J. Moore. 1972. Computer simulation model of dynamic bio-physico-chemical processes in soil. Ariz. Agric. Exp. Stn. Tech. Bull. 196.

Erie, L. J., and A. R. Dedrick. 1979. Level-basin irrigation: a method for conserving water and labor. USDA-SEA Farmers Bull. no. 2261. U.S. Government Printing Office, Washington, D.C.

Feagley, S. E., and L. R. Hossner. 1978. Ammonia volatilization reaction mechanism between ammonium sulfate and carbonate systems. Soil Sci. Soc. Am. J. 42:364–367.

Fenn, L. B., and D. E. Kissel. 1973. Ammonia volatilization for surface application of ammonium compounds on calcareous soils: I. General theory. Soil Sci. Soc. Am. Proc. 37: 855–859.

Goldberg, D., B. Gornat, and D. Rimon. 1976. Drip irrigation: principles, design and agricultural practices. Drip Irrig. Scientific Publ. Kfar Shmaryahu, Israel. p. 174–180.

Hart, W. E. 1972. Subsurface distributions of nonuniformity applied surface waters. Am. Soc. Agric. Eng. Trans. 15:656–661, 666.

Henderson, D. W., W. D. Bianchi, and L. D. Doneen. 1955. Ammonia loss from sprinkler jets. Agric. Eng. 36:398–399.

Humbert, R. P. 1954. Water distribution studies in the Hawaiian sugar industry. Hawaiian Planter's Rec. 54:211–225.

Isobe, M. 1974. Investigations in sugarcane fertilization by drip irrigation in Hawaii. p. 405–410. *In* Second Int. Drip Irrig. Cong. Proc., San Diego, Calif. Library of Congress Catalog no. 74-15261.

Leavitt, H. 1957. The application of liquid gaseous fertilizer. p. 85–95. *In* Natl. Joint Comm. of Fert. Appl. Proc. Natl. Plant Food Inst., Washington, D.C.

Miyamoto, S., J. Ryan, and J. L. Stroehlein. 1975a. Sulfuric acid for the treatment of ammoniated irrigation water: I. Reducing ammonia volatilization. Soil Sci. Soc. Am. Proc. 39:544–548.

Miyamoto, S., R. J. Prather, and J. L. Stroehlein. 1975b. Sulfuric acid for controlling calcite precipitation. Soil Sci. 120:264–271.

Miyamoto, S., and J. Ryan. 1976. Sulfuric acid for treatment of ammoniated irrigation water: II. Reducing calcium precipitation and sodium hazard. Soil Sci. Soc. Am. J. 40:305–309.

Pair, C. J., W. W. Hinz, C. Reid, and K. R. Frost. 1975. Sprinkler irrigation. 4th ed. The Irrig. Assoc., Silver Springs, Md.

Roth, R. L., D. R. Rodney, and B. R. Gardner. 1974. Comparison of irrigation methods, rootstocks, and fertilizer elements on Valencia orange trees. p. 103–108. *In* Second Int. Drip Irrig. Cong. Proc. San Diego, Calif. Library of Congress Catalog no. 74-15261.

Tanji, K. K., and L. D. Doneen. 1966. A computer technique for prediction of $CaCO_3$ precipitation in HCO_3^- salt solution. Soil Sci. Soc. Am. Proc. 30:53–55.

33

Roland D. Hauck
National Fertilizer Development Center
Tennessee Valley Authority
Muscle Shoals, Alabama

Significance of Nitrogen Fertilizer Microsite Reactions in Soil

Fertilizer particles, or globules of concentrated fertilizer solutions or suspensions, when added to soil, form discrete microenvironments at the site of fertilizer application. These microenvironments (microsites, fertilizer-soil reaction zones), when formed by water-soluble fertilizers, are characterized by extremely high concentrations of chemical (or chemicals) and its (their) dissociation and reaction products. The pH within the microsite often is markedly different from the average pH of soil outside of the microsite boundary. The microsite initially may be biologically sterile in its inner, most concentrated zones. Its volume may be determined by the size, chemical nature, and physical form and properties of the fertilizer particle or globule, soil properties, and soil environmental factors. Water diffusing toward the fertilizer creates the microsite; mass flow of solution from the center of the microsite dilutes solution concentrations and extends the microsite boundary. Extensive mass flow (e.g., after heavy rainfall) disperses the microsite until it no longer exists as a discrete entity in the soil. Fertilizers of low water solubility form fertilizer-soil reaction zones that are not markedly different from the surrounding soil; their reaction zones are very small in size, have low reactant concentrations, and are otherwise not comparable to the microsites formed with water-soluble materials. The microsite reactions of soluble and relatively soluble phosphate fertilizers have been discussed in detail (e.g., see Khasawneh et al., 1974). Nitrogen fertilizer–soil microsite reactions have not received widespread attention.

Nitrogen in Crop Production.

I. CONCEPTS

A. Fertilizer N vs. N Fertilizer

Fertilizer N is defined here as a *nutrient* input to soil that affects the activities of microorganisms and crop plants and has other effects in the macroenvironment. In this regard, fertilizer N is part of a dilute soil solution that changes in composition throughout the growing season. As a component of a dilute solution, fertilizer N also interacts chemically and physically with nonliving soil constituents and biochemically with the living biomass.

Nitrogen fertilizers are viewed here as discrete *chemical* entities that progressively lose their identity as their nutrient-supplying components interact with the soil system. As chemical entities, N fertilizers undergo chemical reactions within fertilizer-soil microsites almost immediately after fertilizer application. Biologically-induced N transformations also may occur within, and especially at the peripheries, of the microsites. These microsite reactions can profoundly affect the amount of nutrient supplied and, therefore, the crop production value of the fertilizer N during the growing season.

B. Rate vs. Concentration

When studying N fertilizer reactions and transformations in soil microsites, N concentration can be viewed in either of two ways. In a given volume of soil, fertilizer N concentration can be increased by (i) increasing the number of particles or globules in that volume, this being equated to increasing the N application rate, and (ii) bringing the same number of particles or globules closer together (the application rate remains the same but the spacing between adjacent particles or globules decreases). At the same rate of applied N, particles in a band application interact with a much smaller volume of soil than the same number of particles in a broadcast application. The chemical and biochemical reactions occurring within the microsites of a broadcast application may differ both qualitatively and quantitatively from those occurring in a fertilizer band. Similarly, the volume of soil contacting drops of N solution [e.g., urea–ammonium nitrate (UAN), 32% N] applied by the "dribble-on" method is smaller than that contacting droplets applied from a spray boom. In this case, N concentration at the same rate of N application is affected both by globule distribution and size. Thus, another way of increasing the concentration of fertilizer N in a given volume of soil is to supply the same amount of N in fewer particles or globules, i.e., by increasing their size. This viewpoint of concentration is a modification of the second situation above, whereby particles or globules are brought into such close proximity that they merge or coalesce; even further increase in concentration results when the enlarged particles or globules are brought closer together.

II. PHYSICAL, CHEMICAL, AND BIOLOGICAL CHARACTERISTICS OF N FERTILIZERS

A. Solubility and pH Groupings

Nitrogen fertilizers can be divided into two groups based on their rate of solution in water. Those that are slowly soluble in the soil solution and that do not rapidly mineralize can be considered to be slow-release N fertilizers. Their rate of solution is a complex function of their inherent solubility, physical size and shape, and rate of removal of solubilized constituent from the particle surface through degradation and mass flow (Hauck, 1972). The soil reactions of slow-release N fertilizers will not be discussed here; this class of N fertilizers is reviewed by Allen (Chapt. 13, this book).

Most of the fertilizer N applied to agricultural soils is readily water soluble. Soluble N fertilizers can be divided into three groups based on the pH of their aqueous solutions (Table 1). Upon addition to moist soil, fertilizer salts that are products of a weak base relative to the acid component will dissolve and hydrolyze to form an acid microsite solution. Examples of acid-hydrolyzing materials are ammonium sulfate (AS), ammonium chloride (AC), ammonium nitrate, and monoammonium phosphate (MAP). Calcium nitrate, potassium nitrate, and sodium nitrate, being salts of strong acids and bases, are fully ionized and have a neutral pH. Alkaline-hydrolyzing fertilizers are salts or form salts of relatively strong bases as compared with their acid component. For example, urea hydrolyzes enzymatically to form ammonium carbonate, which, being a salt of a very weak acid, carbonic acid (H_2CO_3), compared with the base, ammonium hydroxide (NH_4OH), hydrolyzes chemically to form a basic solution. Other alkaline-hydrolyzing N fertilizers include anhydrous NH_3 (AA), diammonium phosphate (DAP), and urea-based materials such as urea–am-

Table 1—Solution pH of compounds commonly used as N fertilizers (R. D. Hauck, unpublished work).

	pH of Solution†	
Compound	1.0*M*	0.1*M*
Ammonium sulfate	5.41	5.42
Ammonium chloride	4.69	5.11
Ammonium nitrate	4.76	5.12
Monoammonium phosphate	4.03	4.25
Urea phosphate‡	1.64	1.86
Potassium nitrate	~7	~7
Calcium nitrate	~7	~7
Ammonia	>9	>9
Urea§	>9	>9
Diammonium phosphate	8.14	7.94

† The pH varies with compound purity.
‡ Urea phosphate, pH before urea hydrolysis.
§ After hydrolysis.

monium phosphate and urea–ammonium polyphosphate. However, urea phosphate and urea nitrate have a strong acid component and hydrolyze to form an acid microsite.

B. Effects on N Transformations in Soil

1. NH_3 VOLATILIZATION

Temperature, NH_4^+ concentration, and pH markedly affect the partial pressure of NH_3 in equilibrium with an aqueous solution (for calculations and discussion, see Freney et al., 1983). At a given temperature, the nature of the microsite of NH_4^+-containing or NH_4^+-producing fertilizers determines the influence of the two other important factors that control NH_3 volatilization.

A single urea particle forms a microsite in soil with an average horizontal diameter of about 25 to 30 mm. After dissolution and hydrolysis, the pH at the center of the microsite is about 7.5 when added to an acid soil (pH 5.5–6.0) with a cation exchange capacity (CEC) of about 10 to 15 cmol/kg. Soil pH, buffer capacity, and other factors influence to varying degrees the microsite pH from single particles. A single urea particle of average size (e.g., about 2 mm diam) placed on the surface of moist, nonalkaline soil will lose little N as NH_3. Although the NH_4^+-N concentration at the center of the microsite (volume about 65 mm^3) may exceed 100 mmol/L, the pH (about 7.5) associated with this concentration is not sufficiently high to liberate practically significant amounts of NH_3. The percentages of NH_3 in an NH_4^+ solution at pH 7, 8, and 9 are about 1, 10, and 50; NH_3 liberation from solution increases exponentially at about pH 8. When adjacent urea particles are brought in close proximity, their respective reaction zones overlap, creating a microsite with a higher concentration of reactants. For example, in the above soil on which the single particle was placed, three urea particles (each 2 mm in diam) placed 3 mm apart will form a microsite with a horizontal diameter of about 45 mm. The solution in the central volume of this microsite (about 65 mm^3) will have a pH of about 8.5 and an NH_4^+-N concentration of about 400 mmol/L. Considerable NH_3 can be liberated from this microsite.

The volume of the microsite of single fertilizer particles (i.e., with microsites separate from adjacent fertilizer particle microsites) is determined by the size and nature of the particle, its placement on or within the soil, and soil water content, buffer capacity, and other physical and chemical properties. When placed within the soil, the particle usually forms a pear-shaped microsite narrowing with depth and with its longer axis vertical. The volume of groups of particles with overlapping reaction zones is determined by the above factors plus the spacing among particles. The dimensions, pH values, and NH_4^+-N concentrations given above represent actual measurements but are used for illustrative purposes only. Generally, the microsite chemistry of single fertilizer particles is modified by soil properties. For groups of particles, the chemistry of the fertilizer dominates the initial reactions of the microsite, with soil factors modifying this chemistry mostly at the edges of the microsite.

Table 2—Microsite pH of three urea granules, as affected by granule spacing† (R. D. Hauck, unpublished work).

Spacing (mm)	Days after application	pH
25	3	6.9
	5	7.3
	7	7.8
15	3	7.3
	5	7.7
	7	8.3
8	3	8.4
	5	8.5
	7	8.6

† Urea granules, 2.00–2.38 mm, 5.7 ± 0.05 mg of N per granule; surface placed on Hartsells fsl (Typic Hapludults), pH 5.6, CEC 6.6 cmol/kg; soil water content, 33 kPa; spacing = distance between centers of adjacent particles; 10-mm diam core sample taken to 10-mm depth and containing about 300 mg soil; pH determined at 2.5:1 water/soil ratio.

Table 2 illustrates the effect of particle spacing on pH in the urea microsite. The pH values were obtained on soil cores taken at three time intervals from the geometric centers of the microsites formed by three urea granules placed either 8, 15, or 25 mm apart (further details are given in the table footnote). In this example, microsite pH increased with decrease in distance between adjacent granules.

Because particle spacing affects microsite pH where the individual microsites of adjacent particles overlap, particle spacing can also affect NH_3 evolution from urea fertilizer microsites. Hundreds of laboratory measurements have been made of NH_3 evolved from various distributions of urea granules surface applied to 26 different soils. Results (to be published elsewhere) suggest that soils can be placed into one of four categories regarding how urea granule distribution affects NH_3 evolution: (i) no significant effect; (ii) NH_3 evolution is inversely related to distance between granules; (iii) NH_3 evolution is directly related to distance between granules; and (iv) maximum NH_3 evolution occurs at a particular spacing intermediate between the shortest distance between adjacent granules tested (3 mm) and the greatest distance (15 mm). Most soils studied fell into the fourth category. The reasons for this are not fully understood, but appear to involve rate of urea hydrolysis and the influence of soil factors, largely soil buffering capacity and organic matter content.

2. UREASE ACTIVITY

Urea hydrolysis is affected by urea distribution in several ways: decrease in distance between adjacent particles (i) increases urea concentration, which, at high concentration, decreases hydrolysis rate; (ii) decreases the amount of soil surface in contact with urea, thereby reducing the amount of urease acting upon urea; and (iii) increases microsite pH, which retards urease activity. Table 3 gives evidence in support of the preceding statements. It also shows that urea added in solution hydrolyzed considerably faster than equivalent amounts of urea added in solid form. Differences can also be

Table 3—Urea hydrolysis in three soils, as affected by granule spacing and solution vs. solid form† (R. D. Hauck, unpublished work).

Spacing (mm)	Physical form	% Urea hydrolyzed‡		
		Decatur	Davidson	Paden
6	Granule	17.4	27.4	10.8
9	Granule	20.1	31.6	11.4
12	Granule	20.7	35.4	11.6
†	Solution	38.0	41.7	22.0

† Three urea granules (2.00–2.38 mm, 5.7 ± 0.05 mg of N per granule or urea solution (5.7 mL of 107 m*M*), surface-applied to 200 g of either Decatur sil (pH 5.7, CEC 8.7 cmol/kg), Davidson sil (pH 5.9, CEC 8.9 cmol/kg), or Paden sil (pH 4.7, CEC 8.9 cmol/kg), soil water, 33 kPa. Granules were placed in center soil surface at either of three spacings between adjacent granules. Solution was added to circular area (13 mm diam) at center of soil surface. Urea hydrolysis values were obtained after incubating soil systems for 5 h at 36°C.

‡ SD ± 0.4. Decatur: Rhodic Paleudults. Davidson: Rhodic Paleudults. Paden: Glossic Fragiudults.

measured in hydrolysis rates of, or in the percentage of total urea hydrolyzed in, particles differing in size or drops of urea solution differing in size or urea concentration.

3. NITRIFICATION

The biological oxidation of NH_4^+ to NO_2^- can be expressed by the equation

$$NH_4^+ + 1.5\,O_2 \rightarrow NO_2^- + 2\,H^+ + H_2O. \quad [1]$$

Because the reaction produces two H^+ for each NH_4^+ oxidized, neutralization of the acidity formed favors NH_4^+ oxidation. There is evidence also (Suzuki et al., 1974) that NH_3 transfered through a permeable membrane system is the substrate for oxidation within *Nitrosomonas* cells; therefore, alkaline microsites that liberate NH_3 from NH_4^+ should increase nitrification rate, provided that NH_3 concentrations remain below toxic levels.

The optimum pH for nitrification generally is reported to be in the range 7.9 to 9.0, although both NO_2^- and NO_3^- formation have been observed at pH 13.0. High OH^- concentrations per se do not markedly inhibit nitrifying activity, but high OH^- concentrations coupled with high NH_4^+ concentrations result in the liberation of toxic amounts of NH_3. *Nitrosomonas* are reported to be less sensitive to NH_3 toxicity than *Nitrobacter* (Aleem et al., 1957); *Nitrobacter* activity in soil apparently is restricted to a narrow pH range with an optimum calculated at 7.71 (Boon & Laudelout, 1962). Therefore, the pH of the fertilizer-soil microsite can affect not only the overall oxidation of NH_4^+ to NO_3^- but may also affect the relative rates of NO_2^- production by *Nitrosomonas* and removal by *Nitrobacter.* Nitrite oxidation by *Nitrobacter* is inhibited noncompetitively by NO_2^- and NO_3^-; self-inhibition of NO_2^- oxidation has been observed (Lees & Simpson, 1957) at NO_2^- concentrations as low as 4 mmol/L.

Table 4—Nitrification of NH_4^+ and NH_4^+-producing N fertilizer particles in two soils† (R. D. Hauck, unpublished work).

		% Nitrified			
		Soil, pH 5.6		Soil, pH 8.2	
		mg of N/kg of soil		mg of N/kg of soil	
N Source	pH‡	300	400	300	400
Ammonium sulfate	5.11	13	9	98	61
Ammonium chloride	4.69	6	4	88	43
Monoammonium phosphate	4.05	30	5	87	25
Diammonium phosphate	8.14	58	35	95	73
Urea	>9	75	33	99	82
Urea–ammonium phosphate§	7.60	67	20	87	63

† 2.00–2.38 mm mesh particles in 30 g of Hartsells fsl (pH 5.6) or Webster sicl (Typic Haplaquolls) (pH 8.2), water content, 33 kPa, incubated at 30°C for 21 d.
‡ The pH of 1.0*M* solution after hydrolysis.
§ Co-granule of urea, DAP, and MAP.

The above considerations suggest that in acid soils, alkaline-hydrolyzing NH_4^+ or NH_4^+-producing fertilizers nitrify faster than acid-hydrolyzing ones, as demonstrated in Table 4. For example, in acid soils, urea invariably nitrifies faster than AS; DAP nitrifies faster than MAP. However, the differences in rate of nitrification among NH_4^+ sources decrease markedly in alkaline soils, and, in fact, AS may sometimes nitrify faster than urea. For any soil, there exists a combination of NH_4^+ concentration and pH that is optimal for nitrification. Determining factors include the chemical characteristics of the NH_4^+ source and its physical form (solution or solid), size of particle or globule, N application rate, and geometry (distance between and overlapping of microsites), as well as soil factors (for examples, see Hauck & Stephenson, 1965). Nitrification in the fertilizer-soil microsite generally was found to proceed most rapidly in the pH range of 7.5 to 8.0. Nitrite accumulation began at microsite pH > 8.

Rapid *Nitrosomonas* activity within a microsite at pH 8.1 to 8.3 may lead to NO_2^- accumulation. In laboratory studies, rapid buildup of NO_2^- may be observed within 3 or 4 d after soil incubation, but the rate of NO_2^- formation and conversion to NO_3^- depends on factors already discussed and on the type of experimental system used. In the field, NO_2^- accumulation has been observed near the edges of fertilizer bands containing alkaline-hydrolyzing materials, in poorly buffered soils, and in calcareous soils. The accumulation and presistence of high concentrations of NO_2^- in soil can result in phytotoxicity and/or N loss. The mechanisms leading to N loss from accumulated NO_2^- have recently been reviewed by Nelson (1982); they involve NO_2^- or nitrous acid (HNO_2) instability or reaction with NH_4^+, amide or amino groups, and other constituents of soil organic matter, such as phenolic hydroxyls. These reactions require an acidic medium. Therefore, for N loss from NO_2^- reactions (chemodenitrification) to occur, the reaction conditions must be such that NO_2^- accumulating in an alkaline portion of a fertilizer microsite can diffuse into an acid microenvironment before it is oxidized to NO_3^-. Although pH gradients within a fertilizer-soil microsite have been measured and related to H^+ production during nitrification, no acid portion of a predominantly alkaline microsite has been observed. Nevertheless, some experimental, as well as presumptive evidence discussed by Hauck and Stephenson (1965) and corroborated in their more recent work suggests that N loss from the microsite can occur via chemodenitrification.

As is evident from Table 4, nitrification rate is affected by amount of fertilizer N applied to a given volume of soil. In the acid soils, both the amount of NO_3^- produced and the percent of added N that was nitrified were less at the higher rate of N application. In the calcareous, highly active, soil, the percent nitrified was less for both acid-hydrolyzing and alkaline-hydrolyzing materials added at the higher N rate, but the amount of NO_3^- formed was less at the higher N rate only for the acid-hydrolyzing materials.

As already discussed, increasing the N application rate of a particular N source decreases the spacing between adjacent particles, thereby increasing microsite concentrations. However, for any given N rate, N sources relatively low in N content (e.g., MAP) require the addition of more particles than those having higher N contents (e.g., urea). Therefore, overlapping microsites occur at lower N rates for N sources low in N content. The NH_4^+ concentration within a microsite is determined by the amount of NH_4^+ supplied by a particular N fertilizer and the proximity of the microsite to adjacent microsites.

Precipitation reactions within the microsite that affect NH_4^+ concentrations also can affect nitrification. In soils with free calcium carbonate, relatively insoluble calcium sulfate may be precipitated within the microsite of AS particles, thereby producing a pH continuum within the microsite that differs from that obtained in a noncalcareous soil of similar pH and other characteristics. The precipitation of calcium sulfate in the microsite, resulting in the formation of ammonium carbonate, also explains the liberation of NH_3 from AS, but not from AC particles surface-applied to calcareous soils. Precipitation reactions are suggested as the reason why it is difficult to predict the nitrification pattern in a particular soil of MAP or mixtures containing this salt. For example, in a calcareous soil, ammoniated superphosphate, which forms in the pH range of 5.5 to 8.0, can precipitate, thereby decreasing the NH_4^+ concentration within the microsite.

It is difficult to determine the separate contributions of the factors that affect nitrification in the fertilizer-soil microsite. These factors include salt concentration per se, pH, precipitation reactions that change NH_4^+ concentrations, and, possibly, anion toxicity. One can conclude that any combination of factors that contributes to the persistence of NH_3, low pH, or extremely high salt concentrations within the microsite may markedly affect the course of nitrification.

4. OTHER N TRANSFORMATIONS

Very little information is available on the effect of microsite reactions on fertilizer N immobilization, release of immobilized N, NH_4^+ retention by clays, or denitrification. One would expect some measureable differences in the effects of different N fertilizer particles on soil mineralization-immobilization reactions. Visible darkening of soil around groups of urea particles or near NH_3 injection sites demonstrates the solubilization of soil organic matter by alkaline-hydrolyzing materials. Whether the solubilized organic matter mineralizes faster and releases significantly more plant-available N is a question requiring further study.

In this regard, size and placement of particles might be factors meriting study. One would expect that, at a given N rate, more N from small particles would be immobilized than from large particles, because small particles, being in greater abundance, would form more microsites colonized in total by a larger micropopulation. However, because NH_4^+ rather than NO_3^- is the preferential N form for most heterotrophic soil microorganisms, one would also expect that the amount of N immobilized for any particle size would depend on nitrification rate. Therefore, particle size effects on immobilization might be different for NH_4^+ or NH_4^+-producing fertilizers that differ markedly in their rate of nitrification, and, for a particular fertilizer, the amount immobilized would be different under aerobic soil conditions than under anaerobic conditions, such as those used in flooded rice (*Oryza sativa* L.) production.

Similar speculations could be made regarding NH_4^+ retention by clays in the fertilizer-soil microsite, but, to my knowledge, no studies directly addressing the possibility of enhanced NH_4^+ retention within the microsite have been reported. Ammonium retention has been found to increase slightly with increase in pH (little fixation occurring at pH 5.5) and may be increased in the presence of strongly absorbed anions, such as PO_4^{2-} (for references, see Nommik & Vahtras, 1982). Microsites of different N fertilizers in soil, therefore, differ in at least three factors that affect NH_4^+ retention by clays, i.e., NH_4^+ concentration, pH, and anion composition and concentration.

There is little basis for speculating that biological denitrification occurs in the fertilizer-soil microsite, because the microsite probably would be diluted and dispersed in soil sufficiently water-saturated to sustain denitrification. However, microsite reactions that affect the amount and rate of production of NO_3^-, the substrate for denitrification, indirectly affect this process. The possibility of chemodenitrification occurring within the microsite has already been discussed. Nitrous oxide (N_2O) evolution during nitrification also might be affected by the nature of the microsite. Most reports of this phenomenon indicate N losses to be $< 0.2\%$ of the N applied, with one exception. Emissions of N_2O from three Iowa soils amended with AA ranged between 4.0 and 6.8% of the N applied (Bremner et al., 1981). Because of its effect on nitrification, the kind of N fertilizer and its microsite reactions may directly affect N_2O evolution during nitrification and indirectly during denitrification.

III. APPLICATIONS OF THEORY TO PRACTICE

A. Rationale

Studies of N fertilizer–soil microsite reactions are valuable for the following reasons:

1) They increase our understanding of how small changes in experimental technique can affect the process under study, both qualitatively and quantitatively. For example, the placement of equal numbers of urea particles of equal size and composition in different geometrical distributions has been shown to affect both the rate of NO_3^- production and whether or not NO_2^- will accumulate.

2) They lead to the most valid understanding of fertilizer reactions. The N transformations of N fertilizer particles or of concentrated N solutions, such as 32% UAN solution, are both qualitatively and quantitatively different from those of an equivalent source and amount of N added as a dilute solution.

3) The results of microsite studies increase our understanding of the fertilizer-related literature. They help explain why markedly different results are obtained by investigators conducting apparently similar experiments. Often the method of fertilizer application resulted in differences in microsite reactions, resulting in experiments that were, in fact, not similar.

4) The knowledge of fertilizer N behavior in soils obtained from microsite studies permits one to better predict the behavior of a particular N fertilizer in a particular crop production system.

B. Fertilizer Technology

Several technological approaches to increasing the efficiency of N use are discussed in Chapter 36, this book. One of these approaches is to alter the chemistry and/or biochemistry of the fertilizer-soil microsite by adding substances to the fertilizer that change the microsite pH, cause precipitates to form, or retard biological processes. A rational approach to formulations that produce the desired result requires some understanding of the reactions occurring within the microsite. A particular formulation added to a particular soil may be effective when applied in a band but not when broadcast, or it may be toxic in a band but not in a broadcast application. For example, dicyandiamide (DCD) often is effective as a nitrification inhibitor in formulations containing as little as 2.5% DCD–N (weight of DCD–N per weight of total N) if band applied, but a 10% formulation is needed when the DCD-amended fertilizer particles are broadcasted. The factor controlling nitrification is a fertilizer-inhibitor solution containing about

10 mg of DCD/L. This concentration can be maintained with less DCD in a band application, which contacts less soil than a broadcast application. Other examples could be given.

Formulations that alter the reactions occurring in the fertilizer-soil microsite should be studied in relation to their chemical properties (e.g., acid- or alkaline-hydrolyzing), physical properties (e.g., low or high vapor pressure), physical form (solid or liquid), size of the particle or globule, and placement (e.g., surface-applied vs. incorporated; banded vs. broadcast; with seed vs. apart from seed).

C. Management Practices

1. REDUCING NH_3 LOSS

Knowledge of fertilizer-soil microsite reactions may be used to anticipate problems associated with a particular N fertilizer used in a particular manner and to suggest ways to minimize these problems. Clearly, the reactions occurring within the urea-soil microsite suggest that NH_3 loss from urea applied to the surface of either acid or alkaline soils is highly probable under any set of conditions that preserves microsite integrity. Heavy rain or irrigation soon after urea application would disperse the microsites, wash dilute urea solution into the soil, and virtually eliminate NH_3 loss from acid soils and minimize loss from alkaline soils. Less NH_3 probably would be liberated from urea added to wet soil that remains wet for several days than from wet soil under drying conditions. There is a high probability of NH_3 loss when night air is warm and moist, favoring urea hydrolysis, followed by gradual drying of the soil surface during the day. Alternate moistening and drying of microsites results in the periodic concentration of microsite constituents followed by removal of accumulated NH_4^+ as NH_3.

Review of the literature concerned with NH_3 evolution from surface-applied urea suggests that the percentage of applied urea–N that is lost decreases with N application rate. This observation is understandable because with lower N application rates fewer urea particles cover the same soil surface area, thereby decreasing the opportunity for microsite overlapping.

Only laboratory and greenhouse data are available (unpublished) to demonstrate that the regulation of urea particle spacings might be a practical way of reducing NH_3 loss from surface-applied urea. Distances between adjacent particles can be increased by widening the fertilizer band, splitting the N application rate, or using larger, denser particles (urea prills, which are less dense than granules and contain less N per equal-size particle, hydrolyze and nitrify faster than granules). For example, Nommik (1973) found that large particles of urea lost less NH_3 than small ones when urea was applied to the surface of forest soils. However, the practical significance of changing the geometry of urea microsites cannot be assessed until more information from field studies becomes available.

2. NITRIFICATION-DENITRIFICATION

Review of the literature on N sources for rice production indicates that, at equal rates of N application, rice grain yields usually are lower for urea than for AS when either is applied to direct-seeded rice grown in acid soils. These N sources usually are equally efficient in transplanted rice culture. For direct-seeded rice, fertilizer is applied at time of seeding. Usually, 3 or more weeks later, after seedlings are well-established, floodwater is applied. Nitrification occurs during the preflood period, followed by denitrification after flooding. This nitrification-denitrification sequence involving fertilizer N does not occur (or occurs only to a negligible extent) in a transplant system, in which fertilizer is applied and incorporated into soil immediately before flooding. Because urea nitrifies faster than AS in acid soils (section IIB3), more NO_3^- is produced from urea during the preflood period, more urea-N is lost via denitrification, and this N loss is reflected in lower grain yields. Laboratory experiments have confirmed that, in acid soils, alkaline-hydrolyzing N fertilizers are less efficient for direct-seed rice production than acid-hydrolyzing ones, and that yield reduction is directly related to NO_3^- production during the preflood period (assuming negligible NH_3 loss from alkaline flood water). However, in highly alkaline soils, where AS may nitrify faster than urea, rice yields from urea may exceed those from AS, although neither source will be used efficiently under conditions where marked N loss occurs as a result of nitrification-denitrification. Control of nitrification clearly is a solution for increasing the efficiency of N use for direct-seeded rice, as indicated in chapters 23 and 36, this book.

IV. CONCLUSION

The transformations of fertilizer N that occur soon after its addition to soil affect the crop production value of the fertilizer. These N transformations, when occurring within fertilizer-soil microsites, may occur at different rates or may be qualitatively unlike the N transformations of fertilizer N that is part of the dilute soil solution, i.e., after the microsites have been dispersed. Indeed, the N composition of the dilute soil solution after fertilizer addition is determined by the nature of the microsite reactions and the extent to which they go to completion. Soil solution composition, in turn, determines the rate, extent, and course of transformation of its constituent N forms.

The nitrification process clearly demonstrates the interrelationships among N transformation processes. The rate and extent of nitrification can affect the movement of N in soil (NO_3^- is more mobile than NH_4^+), N loss (leaching, denitrification, chemodenitrification, nitrifier N_2O formation), immobilization (soil heterotrophic microorganisms preferentially assimilate NH_4^+ vis-a-vis NO_3^-), and crop plant nutrition (crop plants differ in their response to NH_4^+ and NO_3^-). Because nitrification does not proceed at the same rate and to the same extent in the microsites of different N fertilizers, or even in different microsites of the same N fertilizer, one would expect differences in the way the N constituents of different microsites interact in the soil-plant system after the microsite has been dispersed.

Whether the microsite reactions of N fertilizers will profoundly affect the value of the fertilizer as a supplier of plant-available N will depend on whether the N transformations within the microsite are markedly different from N transformations occurring in the dilute soil solution, i.e., whether microsite reactions lead either to removal of plant-available N or increase plant use of N. Examples have been given where the differences are sufficiently great to affect N use. For some processes, such as immobilization and NH_3 retention by clays, the possible effects of fertilizer microsite reactions have received little study.

As concluded in another article (Hauck, 1981), the chemistry and biochemistry of soil N transformations is known in detail, but not the specific effects of particular N fertilizers on all major transformation processes. To more fully understand these effects as a function of the physical and chemical form of fertilizers and their manner of application, one must consider the fertilizer both as a supplier of N and as a chemical that reacts with the soil system.

REFERENCES

Aleem, M. I. H., M. S. Engel, and M. Alexander. 1957. The inhibition of nitrification by ammonia. Bacteriol. Proc. 57:9.

Boon, B., and H. Laudelout. 1962. Kinetics of nitrite oxidation by *Nitrobacter winogradskyi.* Biochem. J. 85:440–447.

Bremner, J. M., G. A. Breitenbeck, and A. M. Blackmer. 1981. Effect of anhydrous ammonia fertilization on emission of nitrous oxide from soils. J. Environ. Qual. 10:77–80.

Freney, J. R., J. R. Simpson, and O. T. Denmead. 1983. Volatilization of ammonia. p. 1–32. *In* J. R. Freney and J. R. Simpson (ed.) Gaseous loss of nitrogen from soil-plant systems. Developments in plant and soil science. Vol. 9. Martinus Nijhoff, The Hague.

Hauck, R. D. 1972. Synthetic slow-release fertilizers and fertilizer amendments. p. 633–690. *In* C. A. I. Goring and J. W. Hamaker (ed.) Organic chemicals in the soil environment. Marcel Dekker, Inc., New York.

Hauck, R. D. 1981. Nitrogen fertilizer effects on nitrogen cycle processes. *In* F. E. Clark and T. Rosswall (ed.) Terrestrial nitrogen cycles. Ecol. Bull. (Stockholm) 33:551–564.

Hauck, R. D., and H. F. Stephenson. 1965. Nitrification of nitrogen fertilizers. Effect of nitrogen source, size, and pH of the granule, and concentration. J. Agric. Food Chem. 13: 486–492.

Khasawneh, F. E., E. C. Sample, and I. Hashimoto. 1974. Reactions of ammonium ortho- and polyphosphate fertilizers in soil: I. Mobility of phosphorus. Soil Sci. Soc. Am. Proc. 38: 446–451.

Lees, H., and J. R. Simpson. 1957. Biochemistry of the nitrifying organisms. V. Nitrite oxidation by *Nitrobacter.* Biochem. J. 65:297–305.

Nelson, D. W. 1982. Gaseous losses of nitrogen other than through denitrification. *In* F. J. Stevenson (ed.) Nitrogen in agricultural soils. Agronomy 22:327–363.

Nommik, H. 1973. Effect of pellet size on the ammonia loss from urea applied to forest soil. Plant Soil 39:309–318.

Nommik, H., and K. Vahtras. 1982. Retention and fixation of ammonium and ammonia in soils. *In* F. J. Stevenson (ed.) Nitrogen in agricultural soils. Agronomy 22:123–171.

Suzuki, I., U. Dular, and S. C. Kwok. 1974. Ammonia or ammonium ion as substrate for oxidation by *Nitrosomonas europaea* cells and extracts. J. Bacteriol. 120:556–558.

34

Gyles W. Randall
Southern Experiment Station
University of Minnesota
Waseca, Minnesota

Efficiency of Fertilizer Nitrogen Use as Related to Application Methods[1]

Nitrogen fertilizer will undoubtedly become more expensive in the future because of escalating costs of natural gas, due to difficulty in obtaining sufficient amounts of gas and deregulation of the natural gas industry. Recent information suggests that the cost of NH_3 will more than double during the period 1980 to 1985 (Douglas, 1981). We can expect these higher costs to be passed on to the farmer. This extra cost of N will occupy a greater portion of each grower's variable crop production costs and will not be offset unless commodity prices increase commensurately.

It is unlikely that the prices growers receive for their commodities will increase as rapidly as the potential increase in fertilizer costs. Thus, it becomes more important than ever for the grower to maximize the efficiency of his fertilizer N to maximize his profit. Sometimes this efficiency can be improved simply by using better application methods or techniques that fit current crop production systems more precisely than some of our older, more traditional methods. The purpose of this chapter is to discuss and present data showing: (i) various N application methods, (ii) how N efficiency may be improved as we change application methods, and (iii) how N efficiency, as affected by application method, is highly dependent on the crop, climatic conditions, and the soil in which that crop is grown.

For the purposes of this chapter, efficiency will be defined very simply: either obtaining (i) greater crop yield and N uptake with equal or lesser

[1] A contribution of the Minnesota Agricultural Experiment Station, University of Minnesota. Miscellaneous Journal Series no. 1869.

amounts of N, or (ii) equal crop yield and N uptake with lesser amounts of fertilizer N. In either case the grower can expect more production per unit of N.

Application method may have a tremendous influence on the efficiency of N when considering the potential for N losses or immobilization. These losses can occur through: (i) leaching beyond crop rooting depth, (ii) denitrification, and (iii) NH_3 volatilization and also can include that N which is incorporated into the soil organic matter, commonly called *immobilized N*. The application method may also have a large impact on the positional availability as well as the plant uptake of N. This discussion of various application methods will consider potential N losses, immobilized N, plant uptake of N, and crop yields.

I. APPLICATION OF N TO THE SOIL

A. Broadcast vs. Band

Traditionally, two main types of fertilizer N applications have been considered—broadcast and band methods. In the past few decades, dry forms of N, e.g., ammonium nitrate (AN) and urea, have been broadcast while anhydrous NH_3 (AA) has been banded. These forms have accounted for the majority of the N application market. We can expect broadcast application to become somewhat more complicated with the spraying of liquid solutions on the soil. Questions asked by growers then arise: Should the N that is broadcast-applied to the soil surface be incorporated or can it be left on the soil surface? Should the N, regardless of source, be knifed into the soil to improve efficiency? Can pesticides be combined with N and applied simultaneously? Will reduced tillage methods affect fertilizer efficiency and, hence, application methods? Can the fertilizer N be applied in conjunction with other operations, particularly tillage? If irrigation is a possibility for a grower, should N be applied to the soil and irrigated into the root zone, or should part or all of the N be applied through the irrigation system in a spoon-fed approach throughout that crop's growing season?

Anhydrous NH_3 must be subsurface banded. But the effect of the AA on the nitrification rate of the NH_4^+ in that band may be completely different than when solution forms of urea–ammonium nitrate (UAN) containing 28 to 32% N are banded. Research evidence has indicated that the banding of AA delays the nitrification rate in the band by temporarily suppressing the nitrifying microorganisms. Within a few weeks, the zone is repopulated by these bacteria and the NH_4^+ is nitrified to NO_3^- (Nommik & Nilsson, 1963). When UAN solutions are band-applied in this manner, the temporary suppressing effect on the microorganisms does not occur. Therefore, the NH_4^+ in the UAN should be nitrified much more quickly. This difference in nitrification rate between the band applications of N sources may lead to significant differences in N efficiency. In addition, the application of nitri-

Table 1—Forage yield, N uptake, and calculated NH_3 losses from various N sources (Hargrove et al., 1977).

N				
Source	Rate	Yield	N Uptake	NH_3 Loss
	kg N/ha	kg Dry matter/ha	kg N/ha	%
None	0	3069	42	--
CN	140	8052	148	--
AN	140	7900	146	4
AS	140	5785	102	45

fication inhibitors, e.g., nitrapyrin (N-Serve®) and terrazole (Dwell®), in this band should be considered. Application of these inhibitors with N will also delay nitrification rate and consequently improve the potential for increased N efficiency under certain conditions.

Ammonia volatilization losses can seriously affect the efficiency of surface-applied N (Fenn & Kissel, 1973, 1976; Fenn & Escarzaga, 1976; Fenn & Miyamoto, 1981; Fenn et al., 1981). Hargrove et al. (1977) reported very little loss of NH_3 from calcium nitrate (CN), measurable but small losses from AN, and up to 45% of the NH_3 lost from ammonium sulfate (AS) (Table 1). They also found severe yield depression and reduced N uptake of coastal bermudagrass (*Cynodon dactylon* L.) with the AS. Nitrogen efficiency was severely depressed under these soil conditions with the surface application of AS.

Substantial NH_3 losses from surface-applied urea have also been reported in other Texas laboratory studies (Fenn & Miyamoto, 1981). Surface placement of urea and ammonium carbonate to a moist Harkey sicl (pH 7.8–8.0) resulted in maximum NH_3 losses of 66 and 63% of the total N applied. Ammonia losses from urea were minimal when injected into the soil 2.5 cm or more, whereas NH_3 losses from AS were still high when placed 7.5 cm deep. They concluded that $CaSO_4$ was produced at roughly the chemical equivalent of AS added. Consequently, excess Ca and Mg existed and competed for the exchange sites with NH_4^+, which resulted in increased mobility of the NH_4^+.

Direct measurements of NH_3 volatilized from urea applied to a field of coastal bermudagrass on Houston Black clay (fine, montmorillonitic, thermic Udic Pellusterts) were made by Hargrove and Kissel (1979). They found field losses of NH_3 were small (0–9% of the applied N) compared with those obtained in the laboratory (13–31% of the applied N) and concluded that N losses by NH_3 volatilization under field conditions may not be as severe as previously thought. Results from a 3-yr Minnesota study showed that the incorporation of surface-applied urea or UAN did not improve corn (*Zea mays* L.) yield except in 1 yr when the incorporation of UAN was beneficial (Randall & Meredith, 1982).

Studies conducted in Kansas by Murphy et al. (1978) indicated much greater efficiency when N was knifed into the soil compared with surface applications (Table 2). Wheat (*Triticum aestivum* L.) yields were increased

Table 2—Wheat yield and protein percentage as influenced by N–P placement (Murphy et al., 1978).

Treatment†	Yield 1975	Yield 1976	Protein 1976
	t/ha		%
Control	1.48	1.01	10.9
Knifed N	2.69	2.29	12.3
Surface N	2.42	1.68	11.0
Knifed N–P	3.03	2.29	12.2
Surface N–P	2.69	1.75	11.3

† 84 kg/ha rate of N applied as NH_3 knifed-in and UAN on the surface.

significantly over the control by the addition of N. In addition, the yield from knifed-in N was always greater than that from surface-applied N, regardless of whether it was combined with P in a dual-placement procedure or not. Wheat grain protein percentage also improved significantly with knifed-in application compared with the surface-applied N.

Additional research conducted in Kansas by Lamond and Moyer (1983) indicates significantly greater tall fescue (*Festuca arundinacea*) yields over a 2-yr period with the knifing-in of UAN (28% N) compared with surface broadcasted applications (Table 3). These results indicate the possibility of N loss through volatilization or immobilization of UAN surface-applied to cool season grass pastures.

Results from the injection of N into the soil do not always agree, however, in a 9-yr study, Nebraska researchers found improved utilization of AN left on the soil surface as compared with UAN (32% N), which was injected 10 cm below the soil surface (R. A. Olson, Univ. of Nebraska, personal communication). Their data indicate that the efficiency of the injection method was not influenced by the time of application (either preplant or sidedress) and was not greatly affected by the rate of N application. Percent utilization of the N by the irrigated corn ranged from 41 to 46% with the injected UAN solution, compared with a range from 51 to 71% with the surface application of AN in this irrigated long-term trial.

Under nonirrigated conditions in Minnesota, marked improvement in corn yields occurred with the sidedress application of N at the eight-leaf

Table 3—Tall fescue yield as influenced by method of N application (Lamond & Moyer, 1983).

Method	N rate†	1980–1981 Avg yield
	kg/ha	kg/ha
Broadcast	14	3070
Broadcast	112	4080
Broadcast	168	4640
Knifed	14	4170
Knifed	112	5280
Knifed	168	5530

† As UAN (28% N); P and K (20 + 37 kg/ha, respectively) were applied to all plots.

stage compared with spring and fall broadcasted applications (G. W. Randall, Univ. of Minnesota, unpublished data). These results were obtained on a tile-drained Webster clay loam soil (fine-loamy, mixed, mesic Typic Haplaquolls) where denitrification is a potential problem. The sidedress application also resulted in higher grain/stover ratios. As the grower evaluates the sidedress or topdress time of application (postemergence), he must consider a method of application that will not harm the growing crop. To do so may necessitate the banding of either AA or the use of solids or solution materials that are dribbled or direct-sprayed on the soil surface as compared with broadcast spraying over the top of the crop canopy.

B. Deep Placement

When discussing application methods and their role toward the improvement of N efficiency, it is important to consider both the soil and crop. The most efficient method of N application may also be dependent upon the climatic and environmental factors under which a crop is grown. Dry land areas of the Great Plains, the northwestern states, and the prairie provinces of western Canada may need N deeper in the profile where the roots can obtain it for late-season grain development. Studies by Cochran et al. (1978) at Washington State indicate a substantial improvement in hard red winter wheat yields and grain protein with deep placement of N as compared with shallow placement. They suggest that in low-precipitation areas where the shallow soil layers become dry during the later stage of crop growth, deep-placed N is more available to the roots extracting water from below the dry layers and that water use efficiency may be improved. This same phenomenon may hold for other nonirrigated crops throughout the Great Plains of the United States and Canada, especially those crops that depend upon N uptake throughout their development up until physiological maturity (Smika & Grabouski, 1976).

C. Spray Applications

Fertilizer N can be either sprayed on the soil surface or on the plant surface with the intention of utilizing that N either by direct absorption into the plant or by the N falling off the plant onto the soil and then being absorbed through the root system. This method of application is becoming more popular with custom applicators as they meet the farmer's demand for quick application of fertilizer over a large land area. The broadcast application of UAN satisfies this demand. In many cases, UAN is used as a carrier for pesticides and that combination is applied to the soil and/or small growing crop at a preemergence or early postemergence stage. A common term used for the application method that combines herbicides with solution N is called the *weed and feed* program. This application method allows the grower to apply his N and herbicide at the same time in one operation. The benefits of this system are reduced time and labor needs as well as improved weed kill with some fertilizer-herbicide combinations. The improved weed kill is primarily due to the additive effect of the N solution, which acts both as a surfactant and a desiccant and to the uniform distribution of the herbicide.

Results from Minnesota studies (Randall, 1980) indicate that UAN (28% N) can be applied over the top of corn through the four-leaf stage without seriously damaging the plants or reducing yields if N rates do not exceed 134 kg/ha. Yields were also not affected when 2.1 kg atrazine [2-chloro-4-(ethylamino)-6-(isopropylamino)-*s*-triazine]/hectare was applied with UAN at the four-leaf stage as long as the N application rate did not exceed 68 kg/ha. Weed control with this sytem was excellent. Application of UAN over the top of the crop canopy at the eight-leaf stage at rates > 34 kg N/ha resulted in severe yield depressions. Delaying the N application until the 12-leaf stage and then side-dressing the UAN directly onto the soil surface resulted in yields that were identical to the control where all of the N was applied to the soil and incorporated at the time of planting.

D. Lime–UAN Suspensions

Lime suspension technology has been developed in recent years to maximize use of fluid fertilizer applicators and to satisfy customer need for lime. Use of UAN as the suspension carrier rather than water has gained popularity due to time and cost advantages of the simultaneous application of N and lime. Research conducted by Winter et al. (1981) in Kansas showed substantial losses of NH_3 may occur while lime–UAN suspensions are being prepared only if the lime contains Ca oxide. They also concluded that NH_3 losses can occur in the absence of Ca oxide if the lime–UAN is allowed to remain on the soil surface. Thus, incorporation of the suspension by the grower is necessary for improved N efficiency with this method of N application.

E. Effect of Reduced Tillage

Changes in tillage are occurring rapidly throughout the United States. The amount and degree of tillage is being reduced, which leaves more of the residue on the soil surface to reduce erosion. Surface residue accumulations will increase the potential for greater immobilization of broadcasted fertilizer N, will create conditions conducive for NH_3 volatilization of surface-applied ammoniacal N sources, and will also provide a condition for greater potential denitrification losses. In addition, as farmers consider no tillage, this becomes a deterrent towards the incorporation or injection of N and may necessitate the surface application of N. Results shown in Table 4 from Indiana demonstrate the importance of applying the N below the soil surface area in a no-tillage system. Corn yields from the injection of either AA or UAN (28% N) were superior to the surface application of either urea or UAN. These yield differences indicate that N efficiency was much improved with the injection application method. Reasons for the apparent inefficiency of the surface-applied N could have been due to volatilization of NH_3 from the soil surface, greater immobilization during

Table 4—Corn grain yield as affected by placement method and N source in a no-tillage system in seven experiments in Indiana (Mengel, 1982).

Placement method	N Source	1978–1980 Yield avg.
		t/ha
Injected	AA	8.71
Injected	UAN (28%)	8.49
Surface	UAN (28%)	7.40
Surface	Urea	7.72

breakdown of accumulated crop residues, and/or greater denitrification losses.

An alternative to no tillage is the ridge-plant system, which involves a slight amount of tillage by the planter as it places the seed into a preformed ridge area. This system results in an accumulation of residue on the soil surface (approximately 20–30% surface coverage) and presents many of the same problems that no tillage does, yet provides a warmer seed zone in the spring in the northern Corn Belt. Nitrogen losses could be expected to be significant due to volatilization and immobilization with surface appliction of N to this system; therefore, the injection of AA may improve N efficiency. Three sources of N (UAN, urea, and AA) were applied at three times of application (preplant, emergence, and eight-leaf) to a ridge-plant study conducted in Minnesota (Randall & Langer, 1982). Grain yields from 1981 shown in Table 5 indicate losses of N and poor N efficiency with UAN and AA applied at the emergence stage. The UAN applied to the soil surface that was 23% covered by residue may have been either volatilized, immobilized, or denitrified. On the other hand, AA escaped at the time of injection from the soil due to sealing problems caused by the large amount of residue that was incorporated between the ridges by the planter. This residue was not well decomposed at this stage; therefore, the slit caused by the application knife did not seal well and NH_3 vapors escaped from the soil. Application of N immediately prior to planting and incorporation of

Table 5—Effect of N source and time/method of application on corn yields in 1981 (Randall & Langer, 1982).

	Application time/method		
N Source	Preplant (broadcast)†	Emergence (broadcast)‡	Eight-leaf (sidedress)§
	t/ha		
UAN (28%)	10.42	9.60	10.61
Urea	10.55	10.48	10.42
AA¶	10.30	9.92	10.55

† UAN and urea were applied preplant and incorporated with the planter.
‡ UAN and urea were broadcast applied and not incorporated to soil with 23% surface residue cover.
§ UAN and urea side-dressed in a band near the row and covered lightly.
¶ Knifed-in between 75-cm spaced rows.

the surface-applied materials by the planter resulted in high yields and no difference among the three N sources. Sidedress application at the eight-leaf stage resulted in highest yields and greatest N efficiency.

With less tillage, it is likely that more N will be knifed-in or injected below the surface residue layer to improve N efficiency. This may necessitate a number of modifications in fertilizer application equipment such as specialized pieces of equipment for specific crops and soil areas (i.e., equipment that injects N and as many as two or three other forms of fertilizer through the applicator knife in one application pass).

F. Increasing Efficiency of Fall-Applied N

Fall application of N fertilizers to medium- and fine-textured soils is often less effective than spring application; hence, lower N efficiency. This inefficiency is generally attributed to denitrification, although leaching losses have been responsible under some soil-climatic conditions.

Within the last 10 yr, nitrification inhibitors have become available for application with ammoniacal forms of N. These inhibitors slow the nitrification of NH_4^+ to NO_3^-; thus, the intent is to keep as much of the fall-applied N as possible in the immobile NH_4^+ form until the next spring. Meisinger et al. (1980) thoroughly discuss this and other relevant aspects of nitrification inhibitors.

Recently, application techniques have been described, which lead to more efficient use of fall-applied N. Laboratory studies by Malhi and Nyborg (1979) showed that increasing the pellet size of urea from 0.01 to 0.21 g reduced the hydrolysis rate by 12%, but when one part thiourea was pelleted with two parts urea, the hydrolysis was reduced by 50%. In field studies, they reported only 22 to 39% of the thiourea plus urea (1:2) hydrolyzed after 2 weeks when applied as larger pellets. Field experiments conducted in Alberta and Saskatchewan by Nyborg and Malhi (1979) indicated that fall-applied urea was more effective in increasing barley (*Hordeum vulgare*) yield when urea was placed in widely spaced big pellets (2.1 g/pellet)

Table 6—Effect of method and time of application of urea on the yield and N uptake of barley grain (Nyborg & Malhi, 1979).

Location	Treatment†	Time of application	Grain yield	Apparent fertilizer N in the grain‡
			t/ha	%
Saskatchewan	Control	--	1.33	--
	Urea, mix	23 Oct.	2.42	18
	Urea, nest	23 Oct.	3.62	40
	Urea, mix	8 May	2.99	29
Alberta	Control	--	0.78	--
	Urea, mix	27 Oct.	1.47	9
	Urea, nest	27 Oct.	2.44	23
	Urea, mix	2 June	3.75	43

† 84 kg N/ha.
‡ Total N content of control subtracted from the N content of the urea treatments.

or nests (Table 6). These nests were centered on areas measuring 60 by 60 cm, and the method is sometimes referred to as *chutki placement*. They concluded that this method appears to be a practical application method and that it may eliminate or reduce the amount of nitrification inhibitors necessary to improve the efficiency of fall-applied urea where losses of mineral N are a problem.

II. APPLICATION OF N TO THE FOLIAGE

A. Plant Uptake

Much attention has been given to the foliar application of fertilizer solutions containing N along with P, K, and S (N–P–K–S solutions) to various crops in the last few years. One of the most dramatic responses to this method of application was obtained by Garcia and Hanway (1976), where soybean [*Glycine max* (L.) Merr.] seed yields were increased by approximately 1.5 t/ha from the application of 96 kg N/ha between the R5 and R7 stages (Table 7). If it is assumed that the 1.5-t/ha yield increase contained approximately 6% N and that this N was due solely to that fertilizer N applied to the foliage, then an efficiency of 90% can be calculated for this method of application. This efficiency level is much higher than we normally expect and has not been obtained in many other studies.

Mixed results to the foliar application of fertilizer solutions containing N have been obtained in subsequent studies (Parker & Boswell, 1980; Syverud et al., 1980; Vasilas et al., 1980; Harder et al., 1982). Leaf burn due to the foliar fertilizer salts was identified by all of these investigators. Vasilas et al. (1980) reported increased soybean yields with foliar application of N–P–K–S solutions when other limiting factors, particularly soil moisture, were minimized and when measures were taken to prevent leaf burn. Modest soybean yield increases from N additions were obtained by Syverud et al. (1980). Leaf damage due to foliar application of N–P–K–S was sufficiently severe in studies by Parker and Boswell (1980) to reduce soybean yield in most cases. Harder et al. (1982) found no effect of foliar fertilization on corn yields in the first year, while a 6.4% yield reduction was noted in the second year. Therefore, additional basic research should be conducted to determine which factors affect the uptake of foliar-applied nutrients and result in improved N efficiency and greater yields.

Table 7—Effect of foliar application of N–P–K–S on soybean yield (Garcia & Hanway, 1976).

Variety	Yield not sprayed	Yield increase from spraying†
	t/ha	
'Corsoy'	3.54	1.57
'Amsoy'	3.85	1.49

† 96 kg N + 9.6 kg P + 28.8 kg K + 4.8 kg S/ha.

B. Desiccation

The use of N for desiccation of a crop to speed its maturity has received the attention of growers and researchers. In addition, carry-over of the N would then be available for the succeeding crop if only the grain was removed. Broadcast application of UAN (32% N) solution to grain sorghum [*Sorghum bicolor* (L.) Moench] has been investigated by Donnelly et al. (1977) in Kansas. Their results indicated that UAN did not accelerate grain drying significantly unless applied before physiological maturity, in which case yields were reduced. Moreover, low recovery of the applied N in the treated crop at harvest and low carry-over to the succeeding crop suggested volatilization losses following application were high and N efficiency low.

III. APPLICATION WITH IRRIGATION WATER

Increased use of irrigation to meet the water demand of crops has occurred in recent years. In addition to meeting water needs of the plant, N can be injected into the system to supply N during the season or at critical stages of crop development. In many soils where supplemental water needs are greatest, N management is also critical because of the potential for leaching losses and, consequently, low N efficiency. Three primary methods of irrigation (center pivot, furrow and trickle/drip systems) may affect the efficiency of the N applied through them differently.

The application of N through the irrigation water has some decided advantages. First, the time and rate of N application can be regulated carefully so that the proper amount of N can be applied at that plant stage when needs are greatest. Hence, N efficiency may be improved. Also, this method allows the injection of a number of N sources, e.g., aqua NH_3 (AQA), AA, AS, ammonium phosphates, urea, AN, and CN or mixtures of these compounds (Univ. of Calif., 1981a). A factor that sometimes causes problems, regardless of the method of irrigation, is the injection of NH_3 or AQA into water containing appreciable amounts of Ca and Mg (Univ. of Calif., 1981b). The increase in pH of the water will cause precipitation of insoluble Ca and Mg salts and may lead to clogging of drip irrigation systems in the presence of substantial concentrations of HCO_3^-. A more detailed discussion of this phenomenon and the application of N with irrigation water in general is presented by Gardner and Roth in Chapter 32, this book.

Volatilization can be a significant problem when NH_3 is injected into furrow irrigation water. Studies conducted by Denmead et al. (1982) indicate that the potential for NH_3 loss is very high. This loss is a function of the windspeed and crop height. With short crops they noted a 7% loss of NH_3 per hour; the loss with tall crops only about 1% per hour. Volatilization losses resulted in very uneven applications of N. The N content of the irrigation water decreased by 84% over a distance of 400 m along the

Table 8—Marketable ear weight of sweet corn as influenced by three irrigation systems (Phene & Beale, 1976).

Irrigation method	Marketable ear weight
	t/ha
Trickle	9.5
Furrow	8.5
Sprinkler	8.3

furrow in the short crop, and by 59% in the tall crop. The mineral N content in the top 10 cm of soil 1 week after application to the short crop was significantly less at the end of the 400-m run compared with the head of the furrows. Practical remedies suggested by these authors included irrigating at night when volatilization rates were one-half those observed in the day.

Application of N in a trickle irrigation system also presents some unique problems. Volatilization losses of N as NH_3 are thought to be relatively small with trickle irrigation but can occur in the vicinity of the emitter when AQA or AA is injected into the water. Localized zones of acidity have been shown to develop in those areas where the N-containing water is continuously being emitted. Saturated or near saturated soil conditions in the vicinity of each emitter might lead to denitrification of NO_3^--N containing sources of fertilizer N (Univ. of Calif., 1981a). A significant management factor is the nonuniformity of the discharge rates of the emitters, which leads to leaching of NO_3^--N. Emitters that release more water than required by evapotranspiration will cause nitrates to be leached out of the rooting profile of some shallow-rooted crops.

Research conducted in South Carolina by Phene and Beale (1976) showed higher marketable sweet corn ear weights with the trickle system when compared with the furrow and sprinkler methods of irrigation (Table 8). They also detected less NO_3^--N leaching loss from the root zone with the trickle irrigation system. The application of AA through a subsurface, trickle irrigation system has been investigated by Mitchell (1981) in Delaware. He noted that AA applied through the subsurface irrigation system increased N use efficiency when compared with broadcast AN. However, soil pH was depressed within a 10- to 12-cm radius of the emitter lines.

IV. FUTURE RESEARCH AREAS

There are a number of aspects relating to application methods that will have bearing on improved N efficiency in the future. Some of those that are most pressing include:

1) The interactions of N with reduced tillage. How can N efficiency with various N sources be improved? What depths of application are appropriate? What application time results in greatest efficiency under reduced tillage conditions? Comprehensive models that consider crop, soil, and climatic conditions need to be developed to predict N losses in various types of tillage systems.

2) Ammonia volatilization has received increased attention in the last few years, but more specific information is needed on the soil, climatic, environmental, and plant conditions that may affect NH_3 volatilization losses under field conditions.

3) Additional research in the area of irrigation-N management is needed, especially in those soils where the potential of N leaching below the root zone is extremely high.

4) New and improved methods of application must be developed where the fertilizer and agrichemical industry can combine pesticides with N to not only improve the performance of the pesticide, but also to apply N in a manner that is most efficient to the plant.

5) Foliar application has been tried by a number of researchers in the last 10 yr, some with a great degree of success, many others with very little or negative results. Basic research in the area of plant absorption, translocation, and assimilation of foliar-applied materials is necessary.

6) As crop production systems change, developments and improvements must be made in application equipment. The precision of application, whether it be the uniformity among knives of an NH_3 toolbar or the distribution of dry N sources applied with mechanical spin-type spreaders, has to be improved. The capability to increase or decrease application rates on fertilizer application equipment as a farmer or a custom applicator moves through the field from areas with low production potential to those that have a higher potential productivity must be developed. In other words, can electronic sensors or similar devices be adapted to application equipment to measure important soil properties (organic matter, moisture, etc.) so that application rates can be adjusted more precisely to meet the crop's need throughout the field? These are some of the future areas in application methodology that need investigation so that N efficiency can be improved.

REFERENCES

Cochran, V. L., R. L. Warner, and R. I. Papendick. 1978. Effect of N depth and application rate on yield, protein content, and quality of winter wheat. Agron. J. 70:964–968.

Denmead, O. T., J. R. Freney, and J. R. Simpson. 1982. Dynamics of ammonia volatilization during furrow irrigation of maize. Soil Sci. Soc. Am. J. 46:149–155.

Donnelly, K. J., R. L. Vanderlip, and L. S. Murphy. 1977. Desiccation of grain sorghum by foliar application of nitrogen solution. Agron. J. 69:33–36.

Douglas, J. 1981. Fertilizer costs—1985: can farmers afford them? Fert. Prog. 12:14–34.

Fenn, L. B., and R. Escarzaga. 1976. Ammonia volatilization from surface applications of ammonium compounds on calcareous soils: V. Soil water content and method of nitrogen application. Soil Sci. Soc. Am. Proc. 40:537–541.

Fenn, L. B., and D. E. Kissel. 1973. Ammonia volatilization from surface applications of ammonium compounds on calcareous soils: I. General theory. Soil Sci. Soc. Am. Proc. 37: 855–859.

Fenn, L. B., and D. E. Kissel. 1976. The influence of cation exchange capacity and depth of incorporation on ammonia volatilization from ammonium compounds applied to calcareous soils. Soil Sci. Soc. Am. Proc. 40:394–398.

Fenn, L. B., J. E. Matocha, and E. Wu. 1981. Ammonia losses from surface-applied urea and ammonium fertilizers as influenced by rates of soluble calcium. Soil Sci. Soc. Am. J. 45: 883–886.

Fenn, L. B., and S. Miyamoto. 1981. Ammonia loss and associated reactions of urea in calcareous soils. Soil Sci. Soc. Am. J. 45:537–540.

Garcia, R. L., and J. J. Hanway. 1976. Foliar fertilization of soybeans during the seed-filling period. Agron. J. 68:653–657.

Harder, H. J., R. E. Carlson, and R. H. Shaw. 1982. Corn grain yield and nutrient response to foliar fertilizer applied during grain fill. Agron. J. 74:106–110.

Hargrove, W. L., and D. E. Kissel. 1979. Ammonia volatilization from surface applications of urea in the field and laboratory. Soil Sci. Soc. Am. J. 43:359–363.

Hargrove, W. L., D. E. Kissel, and L. B. Fenn. 1977. Field measurements of ammonia volatilization from surface applications of ammonium salts to a calcareous soil. Agron. J. 69:473–476.

Lamond, R. E., and J. L. Moyer. 1983. Effects of knifed vs. broadcast fertilizer placement on yield and nutrient uptake by tall fescue. Soil Sci. Soc. Am. J. 47:145–149.

Malhi, S. S., and M. Nyborg. 1979. Rate of hydrolysis of urea as influenced by thiourea and pellet size. Plant Soil 51:177–186.

Meisinger, J. J., G. W. Randall, and M. L. Vitosh (ed.). 1980. Nitrification inhibitors—potentials and limitations. Am. Soc. of Agron. Spec. Pub. 38. Am. Soc. of Agron., Madison, Wis.

Mengel, D. B., D. W. Nelson, and D. M. Huber. 1982. Placement of nitrogen fertilizers for no-till and conventional till corn. Agron. J. 74:515–518.

Mitchell, W. H. 1981. Subsurface irrigation and fertilization of field corn. Agron. J. 73:913–916.

Murphy, L. S., D. F. Leikam, R. E. Lamond, and P. J. Gallagher. 1978. Applying N and P at the same time into the same soil shows promise for winter wheat. Better Crops 62:16–23.

Nommik, H., and K. O. Nilsson. 1963. Nitrification and movement of anhydrous ammonia in soil. Acta Agric. Scand. 13:205–219.

Nyborg, M., and S. S. Malhi. 1979. Increasing the efficiency of fall-applied urea fertilizer by placing in big pellets or in nests. Plant Soil 52:461–465.

Parker, M. B., and F. C. Boswell. 1980. Foliar injury, nutrient intake, and yield of soybeans as influenced by foliar fertilization. Agron. J. 72:110–113.

Phene, C. J., and O. W. Beale. 1976. High-frequency irrigation for water nutrient management in humid regions. Soil Sci. Soc. Am. J. 40:430–436.

Randall, G. W. 1980. Postemergence application of UAN (28% N) to corn. Agron. Abstr. p. 174.

Randall, G. W., and D. K. Langer. 1982. Nitrogen efficiency as affected by ridge-planting. p. 136–139. *In* A Report on field research in soils. Univ. of Minnesota Agric. Exp. Stn. Misc. Pub. 2 (revised).

Randall, G. W., and H. L. Meredith. 1982. Incorporation of surface-applied N to improve corn production and N efficiency. Agron. Abstr. p. 277.

Smika, D. E., and P. H. Grabouski. 1976. Anhydrous ammonia applications during fallow for winter wheat production. Agron. J. 68:919–922.

Syverud, T. D., L. M. Walsh, E. S. Opplinger, and K. A. Kelling. 1980. Foliar fertilization of soybeans (*Glycine max* L.) Commun. Soil Sci. Plant Anal. 11:637–651.

University of California. 1981a. Applying nutrients and other chemicals to trickle-irrigated crops. Univ. of California Div. of Agric. Sci. Bull. 1893. Univ. of California, Davis.

University of California. 1981b. Drip irrigation management. Univ. of California Div. of Agric. Sci. Leaflet 21259. Univ. of California, Davis.

Vasilas, B. L., T. O. Legg, and D. C. Wolf. 1980. Foliar fertilization of soybeans: absorption and translocation of ^{15}N-labeled urea. Agron. J. 72:271–275.

Winter, K. T., D. A. Whitney, D. E. Kissel, and R. B. Ferguson. 1981. Ammonia volatilization from lime urea ammonium nitrate suspensions before and after soil application. Soil Sci. Soc. Am. J. 45:1224–1228.

35

K. L. Wells
University of Kentucky
Lexington, Kentucky

Nitrogen Management in the No-Till System

No-till crop production has expanded from nearly nothing 15 yr ago to a currently important technique in agricultural production. This is not surprising due to the inherent advantages of the technique in erosion control and soil water conservation. Rapidly rising fuel costs during the past 5 yr have also provided another distinct advantage, since considerably less fuel is required for no-tillage techniques. Phillips et al. (1980) have reviewed no-tillage agriculture and provide a thorough discussion of the topic. As pointed out in their article, the USDA estimated that 2.23 million ha of cropland were no-tilled in the United States during 1974, and predicted it to reach 65 million ha by the year 2000.

Although the no-tillage technique is used for production of several important crops, major application has been to corn (*Zea mays* L.), soybean [*Glycine max* (L.) Merr.], and grain sorghum [*Sorghum bicolor* (L.) Moench] production in the United States. It is worth noting that much of the technical information we now have was developed coincidentally with commercial expansion of the practice, most of which has occurred within the past 15 yr. While most previous measurements have been made largely on yield effect, we are just now to the point that long-term effects of the practice can be evaluated with respect to soil properties. Knowledge of these effects is important in adapting the technique to various soil and climatic conditions.

I. EFFECT OF NO-TILL PRODUCTION ON SOIL PROPERTIES

The no-tillage system was defined by Phillips et al. (1980) as: ". . .one in which the crop is planted either entirely without tillage or with just sufficient tillage to allow placement and coverage of the seed with soil to allow

Nitrogen in Crop Production.

it to germinate and emerge. Usually no further cultivation is done before harvesting. Weeds and other competing vegetation are controlled by chemical herbicides. Soil amendments, such as lime and fertilizer are applied to the soil surface." Review of several papers (Triplett & Van Doren, 1969; Moschler et al., 1973; Bandel et al., 1975; Blevins et al., 1977, 1978, 1980; Doran, 1980; Phillips et al., 1980) shows the following effects of no-tillage on soil properties:

1) *Soil Moisture Content*—Because of less surface evaporation and better infiltration of rainfall, there is usually 15 to 25% more available soil water during the growing season with no-till as compared with conventional tillage.

2) *Soil Temperature*—Due to the insulation effect from the surface mulch of killed sod, killed small grain, or previous crop residues, soil temperature changes more slowly under no-till conditions. This results in cooler soil temperatures during the growing season and much less diurnal fluctuation.

3) *Organic Matter*—In continuous no-till systems, organic residues from the previous crop collect on the soil surface with subsequent redistribution of soil organic matter as compared with conventional tillage. Greater root growth activity in the top 5 cm of soil resulting from no-till also adds to increased organic matter content in the uppermost part of the soil profile. As the result, organic matter content becomes concentrated at the soil surface with continuous no-tillage, whereas it is mixed somewhat uniformly to plow depth in conventional tillage.

4) *Microbial Activity*—As would be expected from more moisture and accumulation of a concentrated layer of organic residues at the soil surface, there is greater surface microbial activity in no-tilled soils as compared with conventionally tilled soils. Greater numbers of both aerobic and anaerobic bacteria have been measured under no-till as compared with conventional tillage. Even though large numbers of aerobes are present, the relatively large presence of anaerobes, together with a high concentration of organic materials and a higher soil moisture content, results in a greater potential for less oxidation in no-tilled soils as compared with conventionally tilled soils.

5) *Residual Soil N*—Residual soil N content increases in continuous no-till production despite the increased potential for both leaching and denitrification of NO_3^--N. It appears, based on the few studies reported, that this increased residual soil N results from the increased concentration of organic residues near the soil surface, and that it is in organic form. Indications are that mineralization of this labile pool of organic N is enough slower in no-till that an increased residual N content develops, even though it is only slowly available to crops in any 1 yr. Because of this, NO_3^--N soil tests are not likely to indicate the presence of this labile pool of organic N.

6) *Bulk Density*—Most studies reported have shown little difference in long-term effect of no-tillage on bulk density. Even though annual seedbed tillage in conventional systems lowers the plow layer bulk density relative to no-till, subsequent recompaction of the plow layer through the growing

season and over winter results in bulk densities of near the same value as those measured in no-till.

7) *Soil pH*—Studies indicate that after 3 to 4 yr of continuous no-till production of corn, a very acid layer of 1- to 5-cm thickness develops at the surface of the mineral soil. This accelerated acidification related to no-till has been explained as resulting from decomposition of the concentrated layer of organic residues on the surface with subsequent intensive leaching by the resultant organic acids in a thin layer at the top of the soil profile. As with conventional-tillage systems, this acidification process is increased by use of acid-forming fertilizer N.

It should be emphasized here that the effects listed above are largely based on continuous, long-term use of no-till techniques. From the practical standpoint of developing crop production systems to utilize no-till techniques, particularly those involving rotations, it is debatable as to whether producers will use no-till continuously in the same field. Most studies reporting on such long-term effects have been 5 to 10 yr duration, due to the relatively recent advent of no-till production systems. While these studies provide the best current knowledge of no-till effects on soil properties, it should be kept in mind that 5 to 10 yr is probably not long enough to evaluate long-term effects, and at best likely represents only directions of change that take place. Care should be taken in interpolating these effects to individual production systems where the field is not continuously no-tilled for long periods of time.

Along these lines, it is also worth noting that continuous no-till production in some fashion places vegetation in a more influential role as related to on-going factors of soil formation. With these thoughts in mind, a discussion of N management in the no-till system can be more meaningful. Since most experiments reported involved corn as the major crop, the following discussion relates specifically to N management for corn production using no-till techniques.

II. MOVEMENT OF N IN NO-TILL

Besides the obvious erosion control benefits from use of no-till techniques, early research reports showed about 15 to 25% more available soil moisture during the growing season than for conventional tillage. This led to an early concern about N efficiency by no-till corn, particularly since yields for suboptimal levels of fertilizer N had been shown to be lower than those from conventional tillage.

Studies were reported from Kentucky, which were conducted to measure NO_3^--N movement in no-tilled soil as compared with conventional tillage. In these studies, Thomas et al. (1973) grew corn by no-till and conventional techniques on a Maury silt loam soil (Typic Paleudalfs) and studied NO_3^--N movement to a soil depth of 90 cm. They reported that a large part of NO_3^--N was lost to depths below 90 cm during June from the no-till corn growing in a killed bluegrass (*Poa pratensis* L.) sod. This was at-

Table 1—Average annual grain yield over 10 yr continuous corn production on a Maury soil (Blevins et al., 1980).

N Rate	Grain yield	
	No-till	Conventional tillage
kg/ha	kg/ha	
0	4767	5958
84	7715	8028
168	8028	7840
336	8342	8216

tributed to lower surface evaporation from the no-till corn and deep penetration of water and NO_3^- through large pores in the wetter, no-tilled soil. In a follow-up study, McMahon and Thomas (1976) reported on studies of NO_3^- and Cl^- movement through the 90-cm depth of Maury silt loam and Lowell silt loam (Typic Hapludalfs) soils. They reported that subsequent NO_3^--N losses by leaching beyond the 90-cm depth, particularly in view of lower mineralization of residual soil N in no-till, caused the lower no-till corn yields observed at low fertilizer N rates. They indicated that denitrification loss of N from the Maury soil was of little importance. Tyler and Thomas (1977) reported on NO_3^- content of soil leachate collected from field lysimeters at a depth of 106 cm below corn grown either no-till or by conventional tillage on a Maury silt loam soil, which had been in a bluegrass sod for at least 50 yr. This study showed greater NO_3^- leaching under no-till, particularly during the month of June. Their data from the 1974 growing season showed that about 18 kg N/ha was collected in the leachate at the 106-cm depth during June under no-till while only 6 kg N/ha leached from the conventionally tilled plots during the same time.

Results from these studies, most of which were conducted on the same plots on the Maury silt loam soil that had been in bluegrass sod for at least 50 yr prior to their initiation in 1970, led the Kentucky researchers to speculate that more fertilizer N was required for optimum no-till corn production than for conventionally tilled corn. This was assumedly to compensate for the high risk of leaching loss of NO_3^--N from the Maury soil, particularly during the early growing season, and for the lower rates of mineralization of residual soil N.

Long-term yield results from the Kentucky study on the well-drained Maury soil (Blevins et al., 1980) are shown in Table 1. These results show little difference in response to fertilizer N in comparing no-till to conventional tillage, except that, due to lower no-till check yields, the rate of yield increase associated with use of the first 84 kg of fertilizer N was greater for no-till. Assumedly, this reflected greater mineralization of residual soil N in conventional tillage.

Reporting on a 1-yr Maryland experiment conducted at three sites, Bandel et al. (1975) showed little difference between no-till and conventional tillage in accumulation of mineral N in the 0- to 45-cm depth during the growing season, except for one site where rainfall was limiting.

III. EFFICIENCY OF FERTILIZER N

Triplett and Van Doren (1969) reported a 6-yr study conducted on a Canfield silt loam soil (Aquic Fragiudalfs) in Ohio, showing higher no-till corn yields than conventionally grown corn at either 67, 134, or 268 kg fertilizer N/ha per year. Moschler et al. (1972) also reported higher yields from several years' no-till corn in Virginia grown on Lodi silt loam (Typic Hapludults), Davidson clay loam (Rhodic Paleudults), and Cecil clay loam (Typic Hapludults) soils at the same level of fertilizer N at each site. While the Virginia experiments did not enable a classical N response evaluation, the authors suggested that their results indicated a more efficient utilization of fertilizer with no-till production as compared with conventionally grown corn.

Other studies were reported in 1975 from Virginia and Maryland on fertilizer N use efficiency by no-till corn as compared with conventionally tilled corn. The Virginia study reported by Moschler and Martens (1975) was conducted for a 3-yr period on an initially acid, infertile Jefferson soil (Typic Hapludults) on which no-tillage was begun into a rye (*Secale cereale* L.) cover crop. Rye was seeded each fall to provide winter cover. It was used as a mulch for no-tilling corn the following spring and was turned under for the conventionally tilled corn. All fertilizers used in this test were broadcasted and disked into the soil during final seedbed preparation for conventional corn, and topdressed onto the surface for no-till corn. They concluded that at the highest rate of fertilizer N, and at all rates of P and K, no-

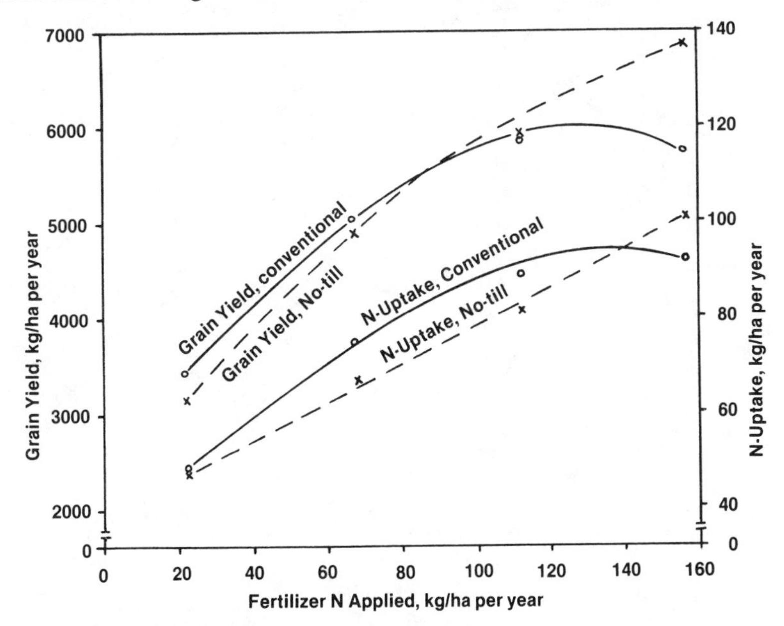

Fig. 1—Effect of conventional tillage and no-till production techniques on grain yields and N-uptake by corn (Moschler & Martens, 1975).

till increased the efficiency of applied fertilizers. Figure 1 shows a graph of their N response data. The grain yield response curve shown in Fig. 1 is typical of that which has been obtained from experiments in several studies. The unique characteristic shown is that, although yielding less at suboptimum levels, no-till yields are higher at the higher tested N rates. Lower grain yields and N uptake observed with no-till at suboptimal rates of fertilizer N application probably result from either greater immobilization of fertilizer N, losses of N from denitrification or leaching, lower mineralization of soil organic N, or some combination of these factors.

Grain yields for conventional tillage peak at a lower rate of fertilizer N than no-till, which would suggest less fertilizer N is required for conventional till to reach maximum production attainable. However, grain yield from no-till at the high rate of fertilizer N exceeds that of conventional till by a greater margin than is observed with either zero N or lower rates of fertilizer N. A likely explanation is that soil moisture becomes more yield-limiting in conventional tillage than in no-till, making it possible for grain yield of no-till to reach higher levels with additional fertilizer N. Future work comparing conventional and no-till corn production with varying fertilizer N rates with or without irrigation is needed in order to more accurately understand these observations. Currently, the higher level of fertilizer N necessary to reach optimum grain yield levels in no-till can be justified, since

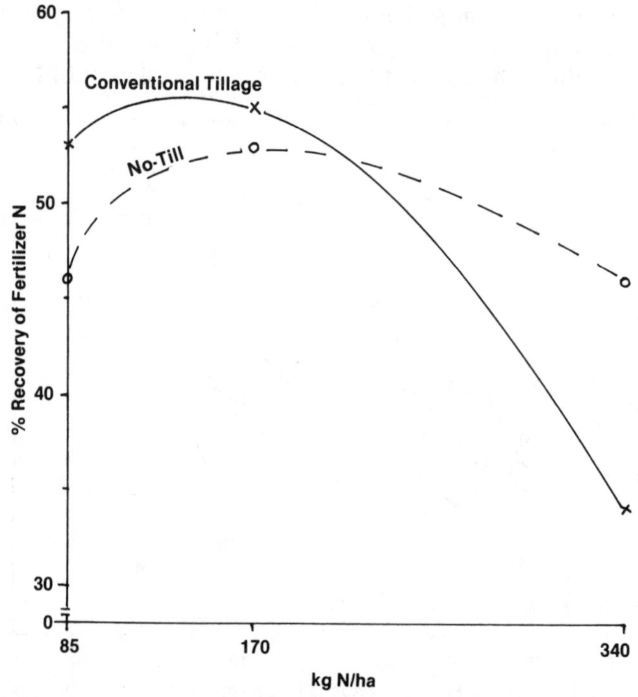

Fig. 2—Recovery of fertilizer N by no-till and conventional grown corn (Legg et al., 1979).

Table 2—Dry matter production and N accumulation of corn silage grown on two soils either with no-till or conventional techniques, 4-yr average (Murdock & Wells, 1978).

	Huntington silt loam		Pope silt loam	
Parameter	No-till	Conv. till	No-till	Conv. till
Dry matter production, kg/ha	15 386	16 033	16 621	13 406
N accumulation, kg/ha	183	190	199	162
kg Dry matter/ha per kg N/ha accumulated	84	84	83	83
Ratio of N accumulated to N applied†	0.86	0.90	0.94	0.76

† Fertilizer N applied at 212 kg/ha per year.

grain yield continues to increase with no-till. Additionally, fertilizer N recovery data reported by Legg et al. (1979) show a similar trend for no-till, i.e., lower recovery with suboptimal fertilizer N but higher recovery at the highest fertilizer N rate tested. Figure 2 shows a graph of their data.

The Maryland studies reported by Bandel et al. (1975) showed similar results. With suboptimum levels of applied fertilizer N, deficiency symptoms of N were more noticeable on no-till, although there was little difference in the amount of fertilizer N required for maximum yields.

Wells et al. (1979) reported results from a 4-yr study of continuous corn silage production on a highly productive Pope soil (Fluventic Dystrochrepts) in eastern Kentucky. Although dry matter yields and the ratio of N accumulated to N applied were higher with no-till, the ratios of dry matter accumulated per unit of N accumulated were the same. Results from a similar study conducted at the same time on a Huntington soil (Fluventic Hapludolls) in western Kentucky (Wells et al., 1983) showed little difference between yields and N accumulation from no-till and conventionally grown silage. Data from both soils are summarized in Table 2, with both sites showing a nearly constant ratio of accumulation of dry matter to total N. At both sites, crop N removal was equivalent to a high proportion of the fertilizer N added.

Legg et al. (1979) reported on results from use of labeled N in a fertilizer experiment with no-till corn conducted for 4 yr in West Virginia. The 4-yr average of percent recovery of fertilizer N is shown in Fig. 2. They indicated that although total plant uptake of residual soil N tended to decrease with increased rates of applied fertilizer N, there was no general relationship in N uptake with tillage system used. In 2 yr of the experiment, rainfall was limiting and uptake and recovery of fertilizer N was substantially greater from no-till. In the 2 normal yr of rainfall, this was apparent for no-till only at the highest fertilizer N rate. Table 3 shows the fertilizer recoveries they measured, and Table 4 shows total aboveground dry matter yield of the mature crop.

Much unpublished work has been conducted in Kentucky (K. L. Wells and M. J. Bitzer, Univ. of Kentucky) to test split applications of fertilizer N on no-till corn as a means of minimizing early season N losses, which are presumed to be important, particularly on soils with slow percolation of

Table 3—Fertilizer N recovery by corn as measured from use of labeled-N fertilizer, 4-yr average (Legg et al., 1979).

	Recovery of fertilizer N applied			
	By midseason		In mature crop	
N Rate	No-till	Conventional	No-till	Conventional
kg/ha	%			
85	42	35	46	53
170	32	27	53	55
340	26	16	46	34

Table 4—Effect of fertilizer N rates and tillage system on dry matter accumulation by corn, 4-yr average (Legg et al., 1979).

	Aboveground dry matter accumulation			
	By midseason		In mature crop	
N Rate	No-till	Conventional	No-till	Conventional
kg/ha	t/ha			
85	4.26	3.06	11.65	10.45
170	4.42	3.43	14.08	12.08
340	4.72	3.21	14.47	12.17

water through the root zone and under influence of the no-till soil surface microenvironment that favors denitrification as compared with conventional tillage. Some of these data are shown in Table 5. For all soils shown in Table 5, N source was ammonium nitrate (AN) applied either all at time of planting or all 5 to 7 weeks after emergence when corn was about 45.7 cm (18 inches) tall. The delayed application of granular N was broadcasted directly over the corn. This practice results in some foliar burn, but it has been shown by unpublished University of Kentucky data to have no detrimental effect on yield when compared to applying the delayed N

Table 5—Effect of delayed application of fertilizer N on grain yields of no-till corn grown in soils with high denitrification potential during the early growing season (unpublished data, K. L. Wells & M. J. Bitzer, Univ. of Kentucky).

	Corn grain yields				
		84 kg N/ha		168 kg N/ha	
Soil type†	Zero N	At planting	Delayed	At planting	Delayed
	kg/ha				
Lowell silt loam	6523	8530	10 223	12 356	11 352
Hampshire silt loam	2 383	4 641	10 223	6 836	10 600
Melvin silt loam	4 328	3 889	6 084	5 896	7 589
Tilsit silt loam	5 582	6 648	7 526	8 279	7 903

† Lowell silt loam: Typic Hapludalfs. Hampshire silt loam: Ultic Hapludalfs. Melvin silt loam: Typic Fluvaquents. Tilsit silt loam: Typic Fragiudults.

directly onto the soil between the rows. As shown by the data, the delayed application on these soils, which tend to waterlog during the spring and early summer, results in considerable fertilizer N efficiency, particularly for the first 84 kg/ha applied. In fact, a delayed application of 84 kg N/ha produced more grain than 168 kg N/ha applied at planting on two soils, and produced almost as much on the other three soils. This increased efficiency due to delaying application of fertilizer N is probably due to less risk of denitrification loss when applied to these soils, which are usually drier than at planting time.

Frye et al. (1980) reported on Kentucky studies conducted to test use of the nitrification inhibitor, nitrapyrin, with fertilizer N as another means of lowering early season N losses. Nitrapyrin was sprayed onto granular urea and AN, which was immediately topdressed on no-till corn at planting time. Results showed increased corn yields of near 25% for use of nitrapyrin at suboptimum N levels when all fertilizer N was applied at planting. The soils used had a potential for early season denitrification losses due to slow water percolation rates through the rooting zone. The positive responses obtained were attributed to lowered early season nitrification rates than those presumed to have occurred on control plots, with the subsequent effect of less fertilizer N being present as NO_3^- and thereby subject to denitrification losses.

IV. FERTILIZER N SOURCES

Reports on fertilizer N responses by no-till corn as influenced by N source have shown variable results. Kentucky studies reported by Wells et al. (1976) and Wells (1979) indicated that although the results obtained were variable from use of AN, urea, or N solution, there was little basis to discriminate among them agronomically for use on no-till corn. However, McKibben (1975) reported on some Illinois studies, indicating severe loss of N from urea as compared with AN on no-till corn. Bandel et al. (1980) reported a study that perhaps puts the fertilizer N source question in proper perspective. They concluded from a 3-yr study in Maryland, comparing AN, urea, and N solutions, that loss of NH_3 from surface-applied urea can seldom be predicted with reasonable certainty under field conditions. In their studies on acid soils, AN was superior to granular or prilled urea or N solution in three of four studies where a source evaluation was possible. They recommended development of machinery to apply urea or N solutions beneath the soil surface on no-till corn. Better response to subsurface application of N solution was also noted by Griffith (1974) in an Indiana study. More recent studies in Indiana have also resulted in more efficient N response from placement of the fertilizer N below the surface of no-till corn (personal communication with Dave Mengel, Purdue Univ.).

Use of urea-N as a surface application for no-till corn should probably be viewed as a practice with a potential but unpredictable risk for N loss of some unpredictable magnitude.

V. USE OF WINTER ANNUAL CROPS TO IMPROVE SEASONAL MINERALIZATION OF ORGANIC N

Substantial field research has been conducted in recent years to test the use of winter annual legumes as winter cover to provide a mulch and to provide N for a subsequently planted corn crop. Mitchell and Teel (1977) reported on a no-till experiment conducted in Delaware on an Evesboro loamy sand soil (Typic Quartzipsamments). They tested eight cover crops, three fertilizer N levels, and two moisture regimes. Fertilizer N was broadcasted over the surface of the cover crops at 0, 56, or 112 kg N/ha just prior to chemically killing them. Hairy vetch (*Vicia villosa* Roth.) and crimson clover (*Trifolium incarnatum* L.) mixtures produced as much grain yield without fertilizer N as did 112 kg N/ha alone. Their study indicated that about one-third of the N content of the mulch cover was released in a single season, with about 90% of that coming from top growth of the mulch crop. Heavy growth of rye was associated with irregular and low population of corn in their studies. Recent work in North Carolina (personal communication with E. G. Krenzer, North Carolina State Univ.) has shown similar detrimental effects from use of rye as mulch cover for no-tillage of corn. Currently, the reasons for such results and the need for concern about them have not been ascertained. Other unpublished work on use of winter annual legumes in no-till corn production have shown similar results to the Delaware study. Kentucky results (personal communication with W. W. Frye, Univ. of Kentucky) showed that a hairy vetch cover crop provided an equivalent of 90 to 100 kg N/ha to the no-till corn crop. Virginia work (personal communication with George Hawkins, Virginia Polytechnic Inst.) has shown Austrian winter pea (*Lathyrus hirsutus* L.) to be an effective annual legume for use in no-till corn production.

In order for the winter annual legume to be effective, it must be well enough adapted to any specific locality that enough dry matter production is possible to make it effective in the total amount of atmospheric N accumulated for subsequent use by corn. Also, an inherent risk from use of any winter cover crop is the degree to which its growth will affect soil moisture content prior to killing it for use as a mulch. North Carolina

Table 6—Effect of fertilizer N rate on the amount of N contained in winter rye when plowed under (Giddens et al., 1965).

Fertilizer N applied to rye	N content of rye when plowed under†	N Content of rye as percent of fertilizer N applied‡
kg/ha	kg/ha	
0	9.7	--
56	47.4	67
112	75.0	58
168	104.4	56

† Total N in tops and roots.
‡ After subtracting check plot N from that of fertilizer N treatments.

studies (personal communication with J. V. Baird, North Carolina State Univ.) have shown under their conditions that soil moisture content was lowered by growth of the cover crop enough that difficulty resulted in no-till planting the corn. On the other hand, this very disadvantage would possibly work as an advantage on soils normally at near moisture saturation during the spring. Growth of the mulch cover would probably help dry the surface of such soils sufficiently to help them warm up sooner and enable earlier planting.

VI. EFFECT OF NO-TILLAGE ON SOIL N CONTENT

While not conducted as no-tillage, Giddens et al. (1965) reported a Georgia study that has bearing on this subject. Their experiment tested the efficiency of fertilizer N (AN) for corn production either by applying all the N to winter rye grown as a cover crop (50% the N at fall seeding of the rye and 50% in February) or all the N directly to the subsequent corn crop (50% at corn planting and 50% when corn was about 25 cm tall). They were testing the hypothesis that a green manure crop would aid in conserving fertilizer N under soil and climatic conditions where it is difficult to maintain soil N levels adequate for good corn production. They concluded after conducting their study for 6 yr on a Cecil sandy loam soil that under Georgia conditions, fertilizer N applied directly to corn was much more effective than applying it to the rye cover crop and plowing it down. However, rye fertilized with 168 kg N/ha did increase soil organic matter and soil N content more than the same amount of N applied directly to corn.

Although rye conserved fertilizer N, they concluded that immobilization of fertilizer N by rye residues was a problem and that N from rye residues (when plowed down) was not mineralized fast enough to meet N requirements for best corn growth. Table 6 is a summary of their data, showing the amount of N contained in rye and plowed under prior to planting corn. They also noted that use of fertilizer N alone on corn increased content of soil N and organic matter, but not as greatly as with the rye cover crop. Barber (1979) reported similar results from long-term studies in Indiana, where all corn residues after grain harvest were plowed under each year. Soil organic matter content was increased from return of the residues and also by the amount of fertilizer N that had been applied to the corn (Table 7).

Table 7—Long-term (12 yr) effect of fertilizer N rate on grain yield of continuously grown corn and organic matter content of soil (Barber, 1979).

N Rate	Avg grain yields	Soil organic matter content after 12 yr
kg/ha per year	kg/ha per year	%
0	2450	2.70
67	5520	2.85
200	8840	2.90

Although neither of the two studies just cited involved no-till corn production, they do emphasize the fact that returning crop residues does increase soil organic matter content, and additionally, that the increase is greater as more fertilizer N is used. This increase represents buildup of a labile pool of organic N in soil, of which the subsequent use by crops will largely be influenced by the amount of such organic N that is mineralized during the growing season. As reported with Kentucky data (Blevins et al., 1977), this labile pool of organic matter is redistributed during continuous no-till production to the top 5 cm of soil, whereas in conventional tillage it is mixed somewhat uniformly throughout the plow depth. The importance of this difference as it affects fertilizer N efficiency by no-till, is that in no-till, organic residues collect near the soil surface where rate of microbial activity at the soil-residue interface increases due to presence of an energy source and moisture (Doran, 1980; M. S. Smith, Univ. of Kentucky, personal communication). Doran (1980) reported that populations of nitrifying organisms increased up to 20-fold and that populations of denitrifiers increased up to 44-fold in the surface layer of soil under no-till corn as compared with conventional tillage.

The Kentucky data of Blevins et al. (1977) show that both soil organic matter content and organic N content is greater with no-till as compared with conventional tillage, and that both increased as fertilizer N was increased. This, together with the Maryland data of Bandel et al. (1975), which showed a greater potentially mineralizable N content in soils that had been no-tilled than in those conventionally tilled, strongly suggests that a larger pool of labile soil N develops from no-tillage of corn as compared with conventional tillage. For this reason, the poorer yield response to fertilizer N commonly observed at suboptimum N levels with no-till corn is possibly not due as much to losses of soil N from the rooting zone as it is to a shift in N content of soil N components, resulting in more total N being immobilized with no-till. This also raises questions concerning losses of fertilizer N from soils no-tilled with corn. Although the early Kentucky data (Thomas et al., 1973; McMahon & Thomas, 1976; Tyler & Thomas, 1977) showed sizeable early season leaching of NO_3^--N to below the rooting depth of the well-drained Maury soil with no-till, cumulative evidence at this time points to immobilization of fertilizer N used at moderate-to-low rates as being a more likely general explanation for lower corn yields measured at suboptimum N rates. Two points are worth noting here:

1) Because of the increased bacterial activity at the mulch-soil contact, there is potentially a greater likelihood for more immobilization, denitrification, or leaching of surface-applied fertilizer N for no-till corn.

2) Even though microbial activity is increased at the surface under no-tillage, the total amount of organic N mineralized during the growing season should be less with no-till. This is due to less surface area of organic residues exposed to bacterial action when the residue exists as an undisturbed surface mulch as compared with plowing down the residues and mixing them into the plow layer. This assumedly accounts for the lower check yields of no-till corn and probably contributes to the lower no-till corn yields observed at suboptimum levels of fertilizer N.

Even though cumulative research findings reported to date indicate the greater buildup of total N with continuous no-till production of corn, more research is needed in order to better understand how this influences grain production and fertilizer N efficiency. In fact, studies have already been initiated by Kladivko and Keeney (1981) in Wisconsin to test the long-term effect of surface accumulation of organic residues on the equilibrium between immobilization and mineralization of organic soil N. They are testing the hypothesis that after an initial buildup of organic soil N resulting from conservation tillage, a new steady-state equilibrium between immobilization and mineralization will be reached at which no further buildup of organic soil N will occur.

VII. MANAGEMENT PRACTICES FOR USE OF FERTILIZER N IN PRODUCTION OF NO-TILL CORN

While most of the experimental data reported in the literature represents continuous corn production, it is questionable that no-till production of corn will be practiced for long periods of time with actual farm-cropping systems. It is more likely that much of the no-till corn will be grown in some type of rotation with soybeans, forage legumes, or forage grasses. From a practical standpoint, use of no-till production systems in this manner should provide for better soil-borne insect and disease control, better broad-spectrum weed control, and the advantage to growing corn following a legume or a grass sod, while still enabling intensive use of erodible land for grain production, depending on the rotation used.

Perhaps first attention for N management should be given to increasing residual soil N uptake in no-tillage due to the lowered mineralization of soil N inherent with the practice. From the N standpoint, one would expect the likelihood for higher residual soil N to be present in a killed forage legume sod as compared with a killed grass sod. This should compensate to some degree for the lower mineralization rate of soil N associated with no-till. Wells et al. (1979) reported on a 4-yr rotation study in which no-till corn was grown during the first 2 yr in a killed grass sod and in a killed red clover (*T. pratense* L.) sod during the last 2 yr of the rotation. Their results from zero-N treatments showed the average yield of no-till corn grown in the grass sod to be 5896 kg grain/ha as compared with an average of 8342 kg grain/ha for the corn no-tilled into the killed red clover sod. While the nature of the experiment precluded comparing these check treatments during each of the 4 yr, they do suggest the likelihood of more residual soil N being used by corn from the killed red clover sod than from the killed grass sod. On this basis, somewhat less fertilizer N should be required for no-till corn production when corn can be grown in a rotation following a forage legume such as red clover. As previously indicated in discussion of the Delaware study on use of winter annual legumes reported by Mitchell and Teel (1977), utilization of atmospheric N fixed by legumes can be of considerable economic value in no-till production of corn.

Table 8—Fertilizer N recommendations for corn production as influenced by previous crop, tillage system, and soil drainage (Lime & Fertilizer Recommendations, 1981).

	N Recommendation				
	Conventional tillage			No-tillage	
Previous crop (index of residual N)	Well-drained soils	Moderately well-drained soils	Poorly drained soils	Well-drained soils	Moderately well-drained soils
	kg N/ha				
Corn, sorghum, soybean, small grain	112–140	168–196†	196–224†	140–168	196–224†
Grass, grass-legume sod (< 4 yr)	84–112	140–168†	168–196†	112–140	168–196†
Grass, grass-legume sod (≥ 5 yr)	56–84	112–140†	140–168†	84–112	140–168†

† Rates of N fertilization can be decreased 40 kg/ha if as much as two-thirds of the N is applied 4 to 6 weeks after planting.

Another concern in no-till corn production is the inherent potential that exists for denitrification losses of surface-applied fertilizer N. Current indications are that, though likely to be more severe on soils with slow water movement through the rooting zone in late spring–early summer, denitrification also takes place to some extent even in well-drained soils during this period. Based on studies reported, this pathway of fertilizer N loss is probably more generally important than the potential for leaching losses associated with no-till (Doran, 1980). Currently, two practices have been shown to reduce these losses, and report of another suggests promise. By far the most widely used method is to delay application of fertilizer N on no-till corn until 5 to 7 weeks after planting, or to split the N application by applying about one-third at planting and two-thirds 5 to 7 weeks later. With this practice, much or all the fertilizer N is surface-applied later in the season after risk of denitrification decreases due to the usually occurring decrease of soil moisture content. Current fertilizer N recommendations for corn production reflect this situation. The Kentucky recommendations shown in Table 8 show increased fertilizer N rates for no-till over conventional till, increased rates as residual soil N decreases, and increased rates as soil drainage decreases. They also suggest a slight decrease if N is split or delayed. The practical risk from delayed application is incurred if unpredictable wet soil conditions prevent use of ground-driven equipment until corn growth is so large that damage from use of ground-driven equipment results.

Kentucky data (Frye et al., 1980) shows that the nitrification inhibitor, nitrapyrin, sprayed onto solid urea or AN just prior to surface application at planting of no-till corn can be useful to counteract this risk. This practice can result in fertilizer N efficiency improvement of about the same magnitude as that possible from the delayed or split application. While more recent studies reported by Bandel et al. (1980) and Kladivko and Keeney (1981) suggest subsurface application of fertilizer N as another means of improving fertilizer N use efficiency by no-till corn, this practice needs further study in order to more conclusively understand the practice.

VIII. SUMMARY

Study of no-till corn experiments reported to the current time indicates that more efficient use of fertilizer N results from no-till corn than from conventionally grown corn. It should be noted that most data published to date come from the upper south and eastern seaboard areas. This better efficiency is assumedly the result of less moisture stress in no-till corn, which provides the potential for higher yields from dryland corn production at higher levels of fertilizer N use than is generally possible with conventional tillage. Even though use of no-till production techniques increases risks for immobilization, denitrification, and leaching of fertilizer N, results have generally shown that corn yields are equivalent to and often better for no-till than conventional tillage at rates of fertilizer N usage likely to be recommended for commercial production. Although such risks are greater on soils likely to be near moisture saturation during the early growing season, delayed or split applications of fertilizer N, or use of the nitrification inhibitor, nitrapyrin, have been shown to be effective in overcoming them for surface N application on such soils. There is some indication that such risks can also be lowered by subsurface application of fertilizer N.

Studies directed at increasing residual soil N content as a means of compensating for the lower mineralization of soil organic N inherent with no-till, have shown good results. This has involved use of winter annual legumes as cover crops and planting no-till corn into killed legume sods.

REFERENCES

Bandel, V. A., S. Dzienia, and G. Stanford. 1980. Comparison of N fertilizers for no-till corn. Agron. J. 72:337–341.

Bandel, V. A., S. Dzienia, G. Stanford, and J. O. Legg. 1975. N behavior under no-till vs. conventional corn culture. I. first year results using unlabeled N fertilizer. Agron. J. 67: 782–786.

Barber, S. A. 1979. Corn residue management and soil organic matter. Agron. J. 71:625–627.

Blevins, R. L., W. W. Frye, and M. J. Bitzer. 1980. Conservation of energy in no-tillage systems by management of nitrogen. p. 14–20. *In* R. N. Gallaher (ed.) Proc. Third Annual No-Tillage Systems Conf., Agronomy Dep., Institute of Food and Agricultural Sciences, Univ. of Florida, Gainesville.

Blevins, R. L., L. W. Murdock, and G. W. Thomas. 1978. Effect of lime application on no-tillage and conventionally tilled corn. Agron. J. 70:322–326.

Blevins, R. L., G. W. Thomas, and P. L. Cornelius. 1977. Influence of no-tillage and nitrogen fertilization on certain soil properties after 5 years of continuous corn. Agron. J. 69: 383–386.

Doran, J. W. 1980. Microbial changes associated with residue management with reduced tillage. Soil Sci. Soc. Am. J. 44:518–524.

Frye, W. W., R. L. Blevins, L. W. Murdock, K. L. Wells, and J. H. Ellis. 1980. Surface application of urea and ammonium nitrate treated with N-Serve 24 nitrogen stabilizer for no-tillage corn. Down to Earth 36(3):26–28.

Giddens, J., S. Arsjad, and T. H. Rogers. 1965. Effect of nitrogen and green manures on corn yield and properties of a Cecil soil. Agron. J. 57:466–469.

Griffith, D. R. 1974. Fertilization and no-plow tillage. *In* Proc. Indiana Plant Food and Agric. Chem. Conf., Purdue Univ., West Lafayette, Ind. 17–18 Dec. Purdue Univ., West Lafayette, Ind.

Kladivko, E., and D. Keeney. 1981. Nitrogen transformations under conservation tillage systems. p. 41–49. *In* Proc. 1981 Fertilizer, Agline, and Pest Management Conf. Univ. of Wisconsin College of Agric., Madison, Wis. 19–21 Jan. Univ. of Wisconsin, Madison.

Legg, J. O., G. Stanford, and O. L. Bennett. 1979. Utilization of labeled-N fertilizer by silage corn under conventional and no-till culture. Agron. J. 71:1009–1015.

Lime and Fertilizer Recommendations. 1981. Publ. AGR-1. Univ. of Kentucky, College of Agric., Lexington, Ky.

McKibben, G. E. 1975. Nitrogen for 0-till corn. p. 87–89. *In* H. A. Cate (ed.) Update, 75, a research report of the Dixon Springs Agricultural Center. Univ. of Illinois, College of Agric.

McMahon, M. A., and G. W. Thomas. 1976. Anion leaching in two Kentucky soils under conventional tillage and killed sod mulch. Agron. J. 68:437–442.

Mitchell, W. H., and M. R. Teel. 1977. Winter-annual cover crops for no-tillage corn production. Agron. J. 69:569–573.

Moschler, W. W., and D. C. Martens. 1975. Nitrogen, phosphorus, and potassium requirements in no-tillage and conventionally tilled corn. Soil Sci. Soc. Am. Proc. 39: 886–891.

Moschler, W. W., D. C. Martens, C. I. Rich, and G. M. Shear. 1973. Comparative lime effects on continuous no-tillage and conventionally tilled corn. Agron. J. 65:781–783.

Moschler, W. W., G. M. Shear, D. C. Martens, G. D. Jones, and R. R. Wilmouth. 1972. Comparative yield and fertilizer efficiency of no-tillage and conventionally tilled corn. Agron. J. 64:229–231.

Murdock, L. W., and K. L. Wells. 1978. Yields, nutrient removal, and nutrient concentrations of double-cropped corn and small grain silage. Agron. J. 70:573–576.

Phillips, R. E., R. L. Blevins, G. W. Thomas, W. W. Frye, and S. H. Phillips. 1980. No-tillage agriculture. Science 208(4448):1108–1113.

Thomas, G. W., R. L. Blevins, R. E. Phillips, and M. A. McMahon. 1973. Effects of killed sod mulch on nitrate movement and corn yield. Agron. J. 65:736–739.

Triplett, G. B., Jr., and D. M. Van Doren, Jr. 1969. Nitrogen, phosphorus, and potassium fertilization of non-tilled maize. Agron. J. 61:637–639.

Tyler, D. D., and G. W. Thomas. 1977. Lysimeter measurements of nitrate and chloride losses from sod under conventional and no-tillage corn. J. Environ. Qual. 6:63–66.

Wells, K. 1979. Why no-till? Solutions. May–June. p. 48–62.

Wells, K. L., G. Armstrong, M. Morrison, and M. J. Bitzer. 1979. Increasing livestock feed productivity from agricultural land in eastern Kentucky. Bull. 715, Univ. of Kentucky, Agric. Exp. Stn.

Wells, K. L., L. Murdock, and H. Miller. 1976. Comparisons of nitrogen fertilizer sources under Kentucky soil and climatic conditions. Agron. Notes. Vol. 9, no. 6. Dep. of Agron., Univ. of Kentucky, Lexington.

Wells, K. L., L. W. Murdock, and W. W. Frye. 1983. Intensive cropping effects on physical and chemical conditions of two soils in Kentucky. Commun. Soil Sci. Plant Anal. 14: 297–307.

36

Roland D. Hauck
Tennessee Valley Authority
Muscle Shoals, Alabama

Technological Approaches to Improving the Efficiency of Nitrogen Fertilizer Use by Crop Plants

The overall objective of fertilizer N research is to maximize the efficiency of plant use of the applied N. Any increase in this efficiency will increase the agronomic and economic value of the fertilizer as a means of increasing crop production, conserve energy and the raw materials needed to make N fertilizers, and minimize possible adverse effects on the environment that may result from inefficient N use (Bremner & Hauck, 1974).

The most common definitions of N fertilizer efficiency are based on plant uptake of N, expressed either as the amount of fertilizer N in the entire plant or in the harvestable crop components, or as the percentage recovery of applied N. Other definitions consider plant use of N, profits, and effects of N use on the environment (as discussed in chapters 5, 17, 18, and 45, this book; and by Hauck, 1978 and Tucker & Hauck, 1978).

Regardless of how it is defined and measured, N use efficiency can be increased by improving plant use of N and by reducing loss of plant-available N (such loss is either through N removal from soil by ways other than plant uptake, movement away from the plant root system, or conversion to relatively unavailable forms in soil). Several approaches have been or are being taken to increase the efficiency of N use by crop plants. They include use of (i) slow-release N fertilizers; (ii) chemicals that inhibit biological N transformations in soils; (iii) amendments to N fertilizers that alter their

Nitrogen in Crop Production.

physical and/or chemical properties; and (iv) improved crop and soil management practices. These approaches are directed mainly toward reducing N losses or maintaining an adequate supply of plant-available N in the plant root zone.

I. CONTROL OF NITRIFICATION

A. Approaches

Major avenues of N loss from the soil-crop system, for which considerable information is available, are by leaching and surface runoff, biological denitrification, and NH_3 volatilization from soil. Other processes, the importance of which is yet to be established, involve the evolution of N_2O during nitrification, chemical decomposition of NO_2^- or its reaction with soil constituents (chemodenitrification), and NH_3 evolution from plant leaf canopies and from floodwater.

Except for N loss via NH_3 volatilization, nitrification is a direct or an indirect cause of N loss by way of those avenues. Nitrification converts a relatively immobile form of N (NH_4^+) to a relatively mobile form (NO_3^-). It produces oxidized forms of N (NO_2^- and NO_3^-) that can be converted in soil to gaseous N (i.e., NO_2^- and NO_3^- are converted to N_2, N_2O, and NO via biodenitrification and NO_2^- is converted to N_2, N_2O, NO, and NO_2 via chemodenitrification). Also, by removing NH_4^+ (the N form used preferentially by soil heterotrophic microorganisms), nitrification probably reduces immobilization of plant-available N. Therefore, control of nitrification should be an effective way to increase the efficiency of N use by reducing loss of plant-available N.

For fertilizers that contain or produce NH_4^+, nitrification can be controlled by (i) slowing the rate at which the fertilizer dissolves in soil, (ii) slowing the rate at which the fertilizer releases N to the soil solution, (iii) adding the fertilizer at a time and in such a manner that allow the plant to more effectively compete with nitrifying and other microorganisms for the applied N, or (iv) adding with the fertilizer a chemical that kills the nitrifying microorganisms or interferes with their metabolism. Slow-release N fertilizers control nitrification by slowing the rate at which substrate NH_4^+ is made available for nitrifier activity. The agronomic use of such fertilizers is restricted to special cropping situations as discussed by Allen (Chapt. 13, this book). Such fertilizers have limited use. The most feasible approaches to controlling nitrification for widespread adoption are improved N management practices, including time of fertilizer application, and use of a nitrification inhibitor. The latter approach will be discussed here.

B. Nitrification Inhibitors

Several intensive searches have been conducted during the past 25 yr to find nonphytotoxic chemicals that can selectively inhibit the biological oxidation of NH_4^+ to NO_2^- in soils. A large number of chemicals have been tested. They include pyridines, pyrimidines, mercapto compounds, succinamides, thiazoles, triazoles, triazines, cyanamide derivatives, and various thio compounds. Agricultural chemicals (fungicides, herbicides, and insecticides) and chemicals that release CS_2 in soils (e.g., thiocarbonates) also inhibit nitrification to a greater or lesser degree.

Seven chemicals are produced commercially worldwide for use as nitrification inhibitors with N fertilizers. Four are produced and used mainly in Japan: 2-amino-4-chloro-6-methyl pyridine (AM, Mitsui Toatsu Co.), 2 mercapto-benzothiazole (MBT, Onodo Chemical Industries), sulfathiazole (ST, Mitsui Toatsu Co.), and thiourea (TU, Nitto Ryuso). Dicyandiamide (Dd, DCD, Dicyan) is produced and marketed in Japan (Showa Denko) and more recently is being produced in West Germany (SKW Trostberg AG). Two chemicals are licensed for use as nitrification inhibitors in the United States: 2-chloro-6-(trichloromethyl)pyridine (nitrapyrin, N-Serve® Nitrogen Stabilizer, Dow Chemical Co., Midland, Mich.), and 5-ethoxy-3-trichloromethyl-1,2,4-thiadiazole (Terrazole®, Dwell®, formerly Olin Corp., now owned by Uniroyal Chemical Co., Bethany, Conn.). Nitrapyrin entered the commercial market in 1976. Terrazole®, already on the market as a fungicide, was registered for use as a nitrification inhibitor in 1982. However, whether or not Terrazole® will continue to be marketed is uncertain, because it is among the agricultural products whose production and marketing rights have recently been sold. Marketing test sales in the United States of dicyandiamide (DCD) produced by SKW began in 1984.

Nitrapyrin is used mainly with anhydrous NH_3 (AA), in which it is soluble. When injected with NH_3, nitrapyrin is present at high concentration within the injection zone. Application with N solutions also has been advocated, but surface application must be followed by immediate incorporation to minimize loss of the volatile nitrapyrin. Moreover, the permissible maximum concentration of nitrapyrin may be too low to effectively retard nitrification when it is broadcasted with fertilizer.

Because of its high vapor pressure, nitrapyrin cannot be granulated with solid N fertilizers without loss of the inhibitor during processing, storage, and handling. This limitation is one reason for renewed interest in use of DCD or thiourea as nitrification inhibitors, because they can be incorporated into solid N sources, as well as into suspensions containing N, and into liquid manures. Dicyandiamide appears to be the most promising of the two materials. For broadcast applications, DCD effectively inhibits nitrification of NH_4^+ in fertilizers containing about 7.5 to 10% DCD (weight of DCD–N to total N in the mixture). For banded fertilizer, less DCD (usually $< 5\%$) in the mixture has proven to be effective. Initial sales of DCD as a nitrification inhibitor in the United States will be for use mainly with solid N fertilizer for direct-seeded rice (*Oryza sativa* L.) (see section III).

Reviews of the hundreds of reports on the effects of chemicals on nitrification in various soil-plant systems include those by Hauck and Koshino (1971), Prasad et al. (1971), Hauck (1972), Meisinger et al. (1980), and

Hauck (1985). An overview of this topic is given by Hoeft (Chapt. 37, this book).

II. CONTROL OF AMMONIA VOLATILIZATION

A. Approaches

Solid urea and N solutions that contain urea account for about 25% of the N used by U.S. farmers. Urea is the dominant form of N fertilizer in world agriculture. For example, it accounts for about 80% of Asian consumption. Favorable economies and conveniences in the manufacture, storage, transport, and distribution of urea and urea-based products account for the continuing increase in urea production and use. Agronomically, urea is an effective N source when used correctly. However, problems associated with its use as a fertilizer include damage to germinating seeds and to seedlings, NO_2^- toxicity, and loss of N as NH_3. These problems result mainly from the rapid hydrolysis of urea to ammonium carbonate through soil urease activity, resulting in the formation of high concentrations of NH_4^+ at high pH and consequent liberation of NH_3 (see Chapt. 33, this book). To avoid or minimize these adverse effects, farmers have been advised not to surface-apply urea without incorporation into soil soon after application and not to band-place urea with or near seed. Fertilizer and crop management systems now being adopted by farmers make such restrictions on urea use impractical or inconvenient. In reduced-tillage agriculture, urea is most often surface-applied without incorporation, creating risk of N loss as NH_3. Being able to safely surface-apply urea under such management conditions would be desirable. Also desirable would be a urea-based fertilizer that could be drilled directly with wheat (*Triticum aestivum* L.) and other cereal grain seed.

Several approaches have been taken to slow the rate of hydrolysis of urea to ammonium carbonate, thereby permitting urea to be surface-applied with minimal risk of losing N as NH_3 or to be placed with seed without creating phytotoxic conditions. They include: (i) coating urea to slow its rate of dissolution, (ii) using a chemical that inhibits soil urease activity; (iii) changing the physical and/or chemical characteristics of urea, thereby changing the chemistry of the urea reaction zone (microsite) in soil; and (iv) improving urea management techniques.

B. Coatings

Coating urea will slow its rate of dissolution, therefore its hydrolysis, but this approach is not a practical solution for most of the urea used in crop production. However, coated urea can be surface-applied with minimum risk of N loss as NH_3, and where economically practical, is a viable approach.

C. Urease Inhibitors

Thousands of chemicals have been tested for their potential as inhibitors of soil urease activity for use with urea fertilizers. Those inhibiting urease activity to some extent include compounds that interfere with the metabolism of urease-producing soil microorganisms (e.g., pyridine-3-sulfuric acid, *o*-chloro-*p*-aminobenzoic acid, γ-benzenehexachloride); urea derivatives that compete with urea for active sites on the urease enzyme complex (e.g., allantoin, methylurea, thiourea, and xanthine); urea complexes with amides; coordination complexes of urea; dithiocarbonates; quinones that probably react with sulfhydryl (SH) groups at active sites on the urease complex (e.g., *p*-benzoquinone, 2,5-dimethyl benzoquinone); polyhydric phenols that can be oxidized to the corresponding quinones (e.g., catechol, hydroquinone, quinhydrone); mercaptans; B- or F-containing chemicals; cyanides; Cu and Ni chelating agents; and salts of metals having atomic weights > 50.

None of these compounds meet the requirements of an effective urease inhibitor for use with urea fertilizer, as discussed by Voss (Chapt. 38, this book). For example, derivatives of acetohydroxamic acid, initially thought promising on the basis of laboratory studies, were subsequently found only nominally effective in inhibiting urea hydrolysis in several soils (for references, see the reviews by Bremner & Mulvaney, 1978; Mulvaney & Bremner, 1981). Phenylmercuric acetate, commonly used to inhibit urease during the extraction of urea from soil, can be effective at concentrations as low as 5 mg of chemical per liter of extractant, but its large-scale use in fertilizer manufacture is environmentally unacceptable because of its Hg component. The benzoquinones effectively retard urea hydrolysis in coarse- and medium-textured soils, but those found most effective, such as 2,5-dimethylbenzoquinone, are carcinogens unsuitable for use with fertilizers.

Phenylphosphorodiamidate (PPDA) is the most potent known inhibitor of soil urease activity. It was first identified as a urease inhibitor during a screening of about 12 000 chemicals for that purpose by East German scientists. Current work at the National Fertilizer Development Center (TVA) and the International Fertilizer Development Center has shown PPDA to be effective in minimizing NH_3 evolution from urea surface-applied to some, but not all, soils in laboratory experiments. Greenhouse studies suggest its usefulness in reducing NH_3 loss from urea added to floodwater in rice culture, but the results of field trials have been negative. Loss of PPDA effectiveness results probably from its rapid degradation in soils; it is chemically stable only within a narrow pH range near neutrality, and it appears to be readily degraded by soil microorganisms. Despite these limitations and its present high cost, PPDA and its homologues (e.g., phosphorotriamide) are receiving considerable attention as urease inhibitors. [See TVA (1981) for a bibliography on potential PPDA use in ruminant feeds and with urea fertilizer.]

There is little question that the development of an effective, low-cost, and safe-to-use urease inhibitor would have immediate practical value in U.S. agriculture and abroad. Unfortunately, no substance patented or under study for use with urea fertilizer can be recommended for commercial production at this time.

D. Altering the N Fertilizer Microsite

1. ACIDULATION

Particles of solid N fertilizers or globules of fluid N fertilizers may form fertilizer-soil microsites with high salt concentrations and pH extremes (Chapt. 33, this book). For urea and other alkaline-hydrolyzing N fertilizers, formation of unionized NH_3 can be reduced by mixing with these fertilizers substances that alter the chemistry of the fertilizer-soil microsite. Adding substances that lower microsite pH is the approach most often considered. As compared with urea alone, less NH_3 was lost from surface-applied urea amended with superphosphate [mainly Ca $(H_2PO_4)_2 \cdot H_2O$], urea added to soil together with phosphoric acid, and urea mixed with acid salts or Ca salts that precipitate carbonate within the urea-soil microsite (for references, see Hauck, 1983). Freney et al. (1983) reviewed the chemistry of these reactions as they affect NH_3 evolution from the microsite.

Addition of acidulating materials to urea to suppress the buildup of high NH_3 concentrations within the microsite is the basis for the urea–urea phosphate formulations being developed by TVA. Urea phosphate (UP) is an adduct-type compound formed from equimolar amounts of urea and orthophosphoric acid. It can be cogranulated with different amounts of urea to produce a family of fertilizers. When UP dissolves on the soil surface, the resulting solution is strongly acid (near pH 1.5). Formulations of urea–UP with N/P ratios < 7.1 show promise of substantially reducing NH_3 loss as compared with unamended urea when surface-applied to soil. In preliminary field trials, these formulations, when banded with wheat seed, resulted in somewhat less damage to seedlings than unamended urea. However, the reduction in phytotoxicity was not sufficient to warrant their application with seed.

The adduct compound urea nitrate (UN) has chemical properties similar to those of UP. It is being considered for development as a fertilizer. Commercial urea–UN products would contain some phosphate ($< 5\%$) to lower their explosion potential, but they would be suppliers primarily of N.

Mixing urea with other acid-hydrolizing N fertilizers also reduces NH_3 liberation from the mixed fertilizer microsite. However, to be effective, the proportion of acid-hydrolyzing fertilizer in the mixture must be high. Many formulations have been tested in TVA laboratories. Typically, about equal parts by weight of acid-hydrolyzing salt, e.g., ammonium sulfate (AS), and urea must be mixed to produce an amended urea particle from which substantially less NH_3 is liberated than from unamended urea, when tested under conditions conducive to NH_3 loss. Some acid-hydrolyzing salts when mixed with urea may increase NH_3 evolution from the mixture, especially when added to calcareous soils; e.g., AS or monoammonium phosphate (MAP) form relatively insoluble precipitates with Ca^{2+}, which removes the acid component from the mixture and liberates NH_3. Mees and Tomlinson (1964) found that on acid soils, adding MAP to urea reduced NH_3 evolution for 3 to 7 d (the time period depending on the proportion of acid salt to urea) with a corresponding decrease in damage to germinating wheat seeds. In calcareous soils, NH_3 evolution and seed damage were increased.

Use of acid-hydrolyzing salts with urea continues to be of interest, but not solely to reduce NH_3 loss from the surface-applied mixed particle. For example, urea–ammonium nitrate (UAN) solutions are manufactured not because NH_3 liberation from surface-applied UAN is less than from solid urea, but mainly because of their other advantages, such as convenience in handling and application, and versatility (Chapt. 12, this book).

2. PRECIPITATION

Mees and Tomlinson (1964) noted that salts that increase Ca^{2+} concentration in soil solution tend to decrease NH_3 evolution from surface-applied urea. They referred to this observation as the *precipitant effect* and speculated that it might be as large as the *acid additive effect.* The precipitant effect is the basis for recent work on the use of calcium chloride ($CaCl_2$) and other divalent alkaline earth salts with urea to minimize the adverse effects of rapid urea hydrolysis. High salt concentrations per se will slightly inhibit urease activity but not to practically significant amounts. However, a mixture of urea and $CaCl_2$, when added to soil, will undergo a sequence of reactions within the fertilizer-soil microsite that results in the lowering of microsite pH, thereby reducing the concentration of unionized NH_3. The main components of the microsite are urea, NH_4^+, ammonium carbonate, Ca^{2+}, Cl^-, ammonium chloride, and calcium carbonate ($CaCO_3$). The $CaCO_3$ is present mainly as a precipitate; the amount of undissociated ammonium chloride and ammonium carbonate present is dependent upon microsite conditions. Within the microsite, urea hydrolyzes biochemically to form ammonium carbonate, which, in turn, hydrolyzes chemically to form an alkaline solution. Concurrently, $CaCO_3$ precipitates, enabling ammonium chloride to hydrolyze to form an acid solution. The acid formed, HCl, is fully ionized in contrast to the base, NH_4OH, formed during ammonium carbonate hydrolysis. The overall effect of these reactions is a net production of H^+, therefore, acidulation of the microsite. The amount of H^+ formed from ammonium chloride hydrolysis and the amount of $CaCO_3$ precipitated is determined by the amount of $CaCl_2$ present in the initial urea mixture plus the Ca^{2+} released into solution from the soil exchange surface.

The foregoing sequence of reactions represents some, but not all, of the events occurring within the microsite of a urea-salt mixture. Of the various salts tested at TVA and elsewhere (e.g., see Fenn et al., 1982), $CaCl_2$ appears to be the most effective in reducing NH_3 evolution from urea. However, the amount of $CaCl_2$ needed to substantially reduce NH_3 evolution is large (e.g., in TVA studies, a Ca/urea mole ratio of 1:4 or narrower was typically needed to reduce NH_3 evolution by 50%). The cost of the amended urea (including the increased costs of storing and transporting a product containing less N) may not be commensurate with the benefits obtained (e.g., prevent crop yield reduction by preventing N loss). Nevertheless, the precipitant effect approach to improving urea use efficiency has been sufficiently successful to merit further study. One application may be to incorporate $CaCl_2$ in suspension fertilizers that contain urea, thereby obviating the problem inherent in the production of a suitable solid product (urea–$CaCl_2$ mixtures have undesirable physical properties). In this regard, adding $CaCl_2$ to the UAN suspensions currently being tested may prove beneficial. The resultant product would be based on both the acid additive effect [from ammonium nitrate (AN)] and the precipitant effect (from $CaCl_2$) for reducing NH_3 loss from urea.

III. CHOOSING APPROPRIATE TECHNOLOGIES

Farmers can choose among several techniques to increase the efficiency of N use in particular cropping situations. Technological advances under development may increase their options. Sometimes two or more approaches can be combined to solve a particular N loss problem.

Where N is to be surface-applied without incorporation, or where immediate incorporation (within 1 or 2 d) is not convenient or practicable, the choices include: (i) using a N fertilizer other than a urea-based one; (ii) adding a urea-based fertilizer immediately before a heavy rain or irrigation, thereby dispersing the urea-soil microsite and minimizing the risk of losing N as NH_3; (iii) using new products such as urea–UP when they become available; (iv) splitting the urea application, thereby decreasing the opportunity for producing microsites of high NH_3 concentrations (Chapt. 33, this book); (v) adding urea under weather conditions (e.g., as in late autumn) that do not favor rapid urea hydrolysis and NH_3 eovlution; and (vi) accepting the risk of losing N as NH_3. Farmers can choose among these options on the basis of materials that are available, cost, convenience, availability of equipment and facilities needed to adopt a given technique, and assessment of risk of losing profit (cost/benefit evaluation).

For direct-seeded, flooded rice, NO_3^--N formed during the preflood period when soils are aerobic is lost via denitrification after flooding. Greenhouse and field experiments have shown that much of this loss can be prevented either through use of a slow-release N fertilizer, such as sulfur-coated urea (SCU), or a nitrification inhibitor. Control of nitrification and, therefore, denitrification loss can also be achieved by shortening the period between N fertilizer application and flooding, or by adding N after flooding. The direct-seeded, flooded rice system provides another example of several approaches available to the farmer for increasing the efficiency of N use. The choice of using SCU or a nitrification inhibitor clearly can be based upon product availability and cost.

IV. PROGNOSIS

Control of nitrification to reduce N loss has received widespread attention for > 20 yr. The need to control rapid urea hydrolysis and the consequent liberation of NH_3 has become important only as farmers began to use increasing amounts of urea and UAN solutions, especially in surface applications that are not incorporated into soil. As a result, the search for urease inhibitors and their evaluation has been less extensive and of shorter duration than the efforts to develop nitrification inhibitors.

These two types of chemicals have markedly different requirements. A nitrification inhibitor ideally should selectively inhibit the activity of *Nitroso* group microorganisms that oxidize NH_4^+ to NO_2^-. Because this oxidation involves sensitive cytochrome electron transport systems of vital importance to the organisms, many types of chemicals can interfere with nitrifier activity by affecting any one of several known enzyme systems. However, to be of value as a nitrification inhibitor for use with NH_4^+ or NH_4^+-producing fertilizers, the chemical must prevent oxidation of the associated NH_4^+ for several weeks. On the other hand, a urease inhibitor for use with urea fertilizer will be of value if it retards urea hydrolysis for 1 or 2 weeks, i.e., until heavy rains disperse the urea-soil microsites and wash urea into the soil, or until the farmer has time to incorporate the urea into soil or apply irrigation water. However, unlike inhibiting nitrification in soil, inhibiting soil urease activity is not readily accomplished. Progress in identifying an acceptable chemical has been limited by: (i) the massive urease activity found in most soils, especially that activity which appears to be protected against chemical attack by a urease-soil organic matter complex; and (ii) our lack of understanding of urease molecular structure, especially the characteristics of soil-bound urease. Although these limitations have impeded an entirely logical quest for a urease inhibitor, they have not limited studies that make use of available information. These studies, in turn, have increased the logical basis for finding and developing a substance that can be used effectively with urea, either as a ruminant feed or as a fertilizer. The prognosis is good that urease inhibitors for commercial use in agriculture will become available and that cropping situations will continue to be identified where nitrification inhibitors and urease inhibitors, or both, will be of value.

Alternatives to use of fertilizer amendments to increase the efficiency of N use also will continue to be pursued. There is need for agricultural engineers and agronomists to study the feasibility of using equipment that can place N fertilizers into soil in a reduced tillage system. Probably the two most effective approaches to increasing the efficiency by which crop plants use agricultural N are: (i) to apply the knowledge already available to an increasing number of farms, and (ii) to strive for an economically acceptable yield and not a maximum possible yield that may leave excessive amounts of inorganic N in the soil after harvest. Technological advances already available and others under study will increase the number of tools that enable farmers to make better use of applied N. Even in our relatively efficient systems of agriculture, improving N fertilizers and the ways that they are used is an achievable goal.

REFERENCES

Bremner, J. M., and R. D. Hauck. 1974. Perspectives in soil and fertilizer nitrogen research. Int. Congr. Soil Sci., Trans. 10th (Moscow) 1974 9:13–27.

Bremner, J. M., and R. L. Mulvaney. 1978. Urease activity in soils. p. 149–196. *In* R. G. Burns (ed.) Soil enzymes. Academic Press, New York.

Fenn, L. B., J. E. Matocha, and E. Wu. 1982. Substitution of ammonium and potassium for added calcium in reduction of ammonia loss from surface-applied urea. Soil Sci. Soc. Am. J. 46:771–776.

Freney, J. R., J. R. Simpson, and O. T. Denmead. 1983. Volatilization of ammonia. p. 1–32. *In* J. R. Freney and J. R. Simpson (ed.) Gaseous loss of nitrogen from soil-plant systems. Vol. 9, Developments in plant and soil science. Martinus Nijhoff, The Hague.

Hauck, R. D. 1972. Synthetic slow-release fertilizers and fertilizer amendments. p. 633–690. *In* C. A. I. Goring and J. W. Hamaker (ed.) Organic chemicals in the soil environment. Vol. 2. Marcel Dekker, Inc., New York.

Hauck, R. D. 1978. Using nitrogen efficiently—theoretical aspects. p. 42–47. *In* Situation 78; TVA fertilizer conference. National Fertilizer Development Center, Muscle Shoals, Ala.

Hauck, R. D. 1983. Agronomic and technological approaches to minimizing gaseous nitrogen losses from croplands. p. 285–312. *In* J. R. Freney and J. R. Simpson (ed.) Gaseous loss of nitrogen from soil-plant systems. Vol. 9, Developments in plant and soil science. Martinus Nijhoff, The Hague.

Hauck, R. D. 1985. Slow-release and bioinhibitor-amended fetilizers. *In* O. P. Engelstad et al. (ed.) Fertilizer technology and use. 3rd ed. Soil Sci. Soc. of Am., Madison, Wis. (in preparation).

Hauck, R. D., and M. Koshino. 1971. Slow-release and amended fertilizers. p. 455–595. *In* R. A. Olson et al. (ed.) Fertilizer technology and use. 2nd ed. Soil Sci. Soc. of Am., Madison, Wis.

Mees, G. C., and T. E. Tomlinson. 1964. Urea as a fertilizer. Ammonia evolution and brairding of wheat. J. Agric. Sci. 62:199–205.

Meisinger, J. J., G. W. Randall, and M. L. Vitosh (ed.). 1980. Nitrification inhibitors—potentials and limitations. Am. Soc. of Agron. Spec. Pub. no. 38. Am. Soc. of Agron., Madison, Wis.

Mulvaney, R. L., and J. M. Bremner. 1981. Use of urease and nitrification inhibitors for control of urea transformations in soils. p. 153–196. *In* E. A. Paul and J. N. Ladd (ed.) Soil biochemistry. Vol. 5. Marcel Dekker, Inc., New York.

Prasad, R., G. B. Rajale, and B. A. Lakhdive. 1971. Nitrification retarders and slow-release nitrogen fertilziers. Adv. Agron. 23:337–383.

Tennessee Valley Authority. 1981. Phenylphosphorodiamidate (1942–1981). TVA Bibliography no. 1675. Technical Library, TVA, Muscle Shoals, Ala.

Tucker, T. C., and R. D. Hauck. 1978. Removal of nitrogen by various irrigated crops. p. 135–165. *In* P. F. Pratt (ed.) Proc. Natl. Conf. Manage. of N in Irrigated Agric. U.S. Environmental Protection Agency, National Science Foundation, and Univ. of California, Sacramento, Calif.

37

R. G. Hoeft
University of Illinois
Urbana, Illinois

Current Status of Nitrification Inhibitor Use in U.S. Agriculture[1]

Of all the nutrients required by plants, N is the one that has the greatest potential for loss from the soil plant system. While there are many ways in which this nutrient may be lost, it is generally agreed that denitrification and leaching are the major loss mechanisms. These two processes are quite different, but the end result is the same, and environmental conditions and management practices that affect these processes are similar.

Both denitrification and leaching occur extensively in wet soils. Therefore, these processes are of importance in those areas where climatic and soil conditions will result in excess precipitation relative to evapotranspiration. Generalized maps have been developed that show those areas of the United States that have high N loss potential (Nelson & Uhland, 1955). In general, this map shows that the potential for loss should be high in the eastern United States, and very low in the western United States. However, within any of these areas, there will be soils that do not fit the general pattern.

In recent years, soil scientists have rated specific soil types for N loss potential (Miller, 1981; Alexander, 1983). Soil properties upon which these ratings are based include natural soil drainage class, soil permeability (both surface and subsurface), slope class and configuration, and subsoil texture. In addition, ratings for specific soils have been adjusted for specific climatic regions.

Denitrification is generally defined as the biochemical reduction of nitrate (NO_3^-) or nitrite (NO_2^-) to gaseous forms of N. The process is

[1] Contribution from the Department of Agronomy, Illinois Agricultural Experiment Station, University of Illinois, Urbana.

Nitrogen in Crop Production.

usually associated with waterlogged, i.e., poorly aerated or anaerobic soils. However, there is some evidence that the process can also occur to a lesser extent in aerobic soils (Freney et al., 1978; Mulvaney, 1983). While soil aeration is a major factor determining the potential for denitrification, other factors including soil pH, soil temperature, decomposable organic matter, and, of course, NO_3^- and NO_2^- supplies also have an impact on the potential for denitrification. The effects of these factors have been reviewed by Kurtz (1980).

Factors affecting the potential for N loss by leaching have been studied and described by several investigators. Reviews of this subject have been prepared by Kurtz (1980) and by Pratt (Chapt. 21, this book). Like denitrification, soil moisture relationships are a primary factor in determining the extent of N leaching.

Since both denitrification and leaching losses are of major concern when N is present as NO_3^- and when soils are excessively wet, N management programs should be designed to minimize the amount of NO_3^--N present during those periods of the year when the probability of excess soil water is high. Factors to consider in designing a N-management program include timing of N application, source of N, soil temperatures, soil type, soil moisture, and time of N need for the particular crop.

In recent years, products have been identified that will reduce the rate of conversion of NH_4^+ to NO_3^-. These products, commonly referred to as nitrification inhibitors, are chemicals that inhibit the activity of *Nitrosomonas* bacteria. Since this bacteria is responsible for the conversion of NH_4^+ to NO_2^-, inhibition of its activity will result in the maintenance of a greater proportion of the applied N as NH_4^+ for a longer period of time. As a result, when nitrification inhibitors are properly used, they should reduce the potential for N loss from denitrification or leaching.

Nitrification inhibitors were first introduced into the commercial market in 1976. At that time, the major emphasis was aimed at the market for fall-applied N on corn (*Zea mays* L.). Since that time, both research and farmer experience have shown that the products could be effectively used on other crops, namely wheat (*Triticum aestivum* L.) and rice (*Oryza sativa* L.), and that they could be effectively used with spring and to some extent side-dress N applications for corn.

Several products have been shown to function as nitrification inhibitors, but as of 1982 only two products were labeled for use in the United States. They were nitrapyrin (N-Serve®) and etradiazol (Dwell®).

Their introduction into the market place raised several questions:

1) Do the compounds reduce the rate of nitrification?
2) How long will they be effective in soil?
3) What N fertilizers can be used with the inhibitors?
4) Will the inhibitors maintain N in the NH_4^+ form too long?
5) Will the use of the inhibitors produce economical increases in grain yield?

The answers to these questions have been obtained in part from research conducted at various locations. This research has identified many factors that will affect the efficacy of inhibitors. Producers and their advisors will need to consider each of these factors for each field when they attempt to ascertain whether inhibitors will likely be beneficial.

I. DO NITRIFICATION INHIBITORS REDUCE THE RATE OF NITRIFICATION?

Research trials conducted in the Midwest, Northwest, and Southwest areas of the United States have consistently shown that nitrification inhibitors do reduce the rate of nitrification of NH_4^+-N (Cochran et al., 1973; Hendrickson et al., 1978; Touchton et al., 1979; Turner et al., 1962). The actual amount of inhibition and/or the length of time for which the products effectively inhibited nitrification varied from region to region and even among soils within a given region. Results obtained in the southeastern United States have been less consistent. In central Georgia, nitrapyrin was effective for over 156 d when applied in late fall (Boswell & Anderson, 1974). However, results obtained in southern Georgia indicated that inhibitors had a short-term effect on nitrification (Touchton & Boswell, 1980). These results from the Southeast appear to indicate that the effectiveness of nitrification inhibitors is much shorter in the more southern areas of the Southeast as compared with the northern areas of the Southeast.

II. HOW LONG WILL NITRIFICATION INHIBITORS BE EFFECTIVE?

Research has identified several factors that influence the length of time for which nitrification inhibitors will effectively maintain NH_4^+-N in soil. These factors include soil temperature, organic matter, soil pH, rate of diffusion, volatilization, and sorption.

Of the various factors that have been identified as having an effect on the efficacy of nitrification inhibitors, temperature appears to have the greatest effect. Increasing temperature from $< 10°C$ to the mid-20°C range markedly increased the rate of nitrapyrin hydrolysis in aqueous solution (Table 1). Similarly, Touchton et al. (1979) and Herlihy and Quirke (1975) have shown that in soil, the half-life of nitrapyrin was markedly reduced when temperatures were increased from $\leq 10°C$ to $> 20°C$. Touchton et al. (1979) also measured the effect of temperature on the half-life of NH_4^+ with and without nitrapyrin. Their results indicated that the NH_4^+ half-life was extended well beyond the nitrapyrin half-life at all temperatures; without an inhibitor, the NH_4^+ half-life was generally less than that of nitrapyrin (Table 1).

Under field conditions, Schmitt (1981) has shown that the recovery of NH_4^+-N applied with inhibitors on 18 November was significantly greater than the recovery of NH_4^+-N applied without an inhibitor on 21 April at all sampling dates through early June. This was true even though the accumulated degree days were 400 to 500 more for the fall than the spring application. Inclusion of inhibitors with N application in mid-October and early

Table 1—Effect of temperature and organic matter on half-life of nitrapyrin and NH_4^+.

Reference	System	Texture	Organic matter	Temperature	Nitrapyrin half-life	NH_4^+ Half-life With inhibitor	NH_4^+ Half-life Without inhibitor
			%	°C	days		
Hendrickson & Keeney (1979)	Solution		--	4	>1000		
			--	25	10		
			--	30	8		
Hendrickson & Keeney (1979)	Soil	Loamy sand	1	25	12		
		Muck	70	25	10		
		Charcoal	100	25	11		
Herlihy & Quirke (1975)	Soil	Loamy sand	2.8	10	77		
				20	15		
		Sandy loam	3.6	10	43		
				20	9		
		Loam	4.6	10	43		
				20	16		
Touchton et al. (1979)	Soil	Silty clay loam	5	4	92	>154	154
				13	70	>154	31
				21	25	77	17
		Silt loam	2	4	22	>154	44
				13	<7	44	11
				21	<7	17	10

November increased the NH_4^+-N recovery as compared with treatments that had not received an inhibitor; by late April, the amount recovered from the inhibitor-treated plots was far less than that recovered from a spring application of N without an inhibitor. Based on results from these field and laboratory studies, many states suggest that fall N applications be delayed until soil temperatures are < 10°C.

The effect of organic matter level on nitrapyrin degradation has not been consistent. Redemann et al. (1964) found increased degradation with increased organic matter on one soil and decreased degradation with increased organic matter on another soil. Briggs (1975) and Touchton et al. (1979) found greater nitrapyrin degradation with low organic matter soils than with higher organic matter soils.

The pH of aqueous solutions has been shown to have no effect on nitrapyrin hydrolysis (Keeney, 1980). However, in a soil system nitrapyrin degradation was more rapid at a pH of 6.4 to 6.8 than at a pH < 6.

III. WHAT N FERTILIZERS CAN BE USED WITH THE INHIBITORS?

Since nitrification inhibitors influence only the reaction of NH_4^+, they can be used effectively only with those compounds that contain NH_4^+ or convert to the NH_4^+ form. This includes anhydrous ammonia (AA), urea-ammonium nitrate solutions (UAN), urea, ammonium sulfate (AS), and ammonium nitrate (AN). The more NO_3^- a material contains, the less effective inhibitors will be.

The method of applying the inhibitor with the various N materials is of more concern than the type of material. Since the compounds are designed to inhibit *Nitrosomonas* organisms, which are present throughout the entire surface soil profile, band applications of the materials should be more effective at a given rate than broadcast applications. The band application of the fertilizer and inhibitor does not contact as much of the soil surface or as many organisms as would be true with broadcast applications.

Both of the nitrification inhibitors currently available require immediate incorporation in order to avoid a loss through volatilization. Briggs (1975) found substantial loss from a surface-broadcast application within 4 h on a cool day. In some instances, the farmer may not be able to incorporate these materials as rapidly as a custom applicator could apply the material. In such instances, it would be more logical to spread the fertilizer without an inhibitor and then apply the inhibitor during seedbed preparation.

IV. WILL NITRIFICATION INHIBITORS MAINTAIN N IN THE NH_4^+ FORM TOO LONG?

The possible but unlikely answer to this question is "yes." In 8 yr of studies in Illinois, only one significant yield decrease was associated with the use of the inhibitor. That occurred in 1976 when nitrapyrin was applied in the fall with NH_3 at 67 kg N/ha. However, when N was applied at $\geq$ 134 kg N/ha, the nitrapyrin did not have a negative effect on yields.

The reason for the decreased yield in 1976 is not known, but one possibility might be *positional unavailability*. During the extended dry period that occurred during the early part of the growing season, the nitrapyrin may have held the N in the NH_4^+ form and thus prevented it from moving down into the soil. Because of the dry surface soil, the roots were absorbing water from deeper depths, and this water did not contain an adequate amount of N for maximum growth.

Since plants can utilize both the NH_4^+ and NO_3^- forms of N, a reduction in the nitrification rate should not reduce N availability as long as the N (either NH_4^+ or NO_3^-) is in the zone of maximum root activity.

V. WILL THE USE OF AN INHIBITOR PRODUCE ECONOMICAL INCREASES IN GRAIN YIELDS?

The potential for benefit from nitrification inhibitors depends on a number of factors. These include soil type, climatic conditions, cultural practices, crop to be grown, and N management. Experiments conducted at many locations throughout the United States clearly show the importance of these various factors. Results shown in Tables 2, 3, and 4 are not all inclusive; rather, they are representative of results that have been documented since inhibitors were introduced into the market place.

Soils that are either excessively drained or poorly drained have a greater potential for N loss than those that are well or somewhat poorly drained. The increase resulting from the use of inhibitors on very sandy

Table 2—Effect of nitrification inhibitors on yield of corn.

Reference	Time of N application	No. of experiments	No. of yield increases to inhibitor	Percent yield# increase from inhibitor
Nelson†	Fall	24	17	12.5
	Spring	51	29	5.8
	Spring (no-till)	12	9	10.0
Hoeft‡	Fall	12	5	5.0
	Spring	14	2	−1.0
Kapusta§	Fall (NH_3)	7	7	4.6
	Spring (NH_3)	9	7	3.4
	Spring (no-till)	2	2	8.5
	Fall (N-solutions)	5	4	3.3
	Spring (N-solutions)	5	2	−1.2
Malzer (1979)	Spring	1	1	21.8
Walsh¶	Fall	2	1	4.7
	Spring	2	0	1.5
Frye et al. (1980)	Spring (no-till)	8	7	14.3
Hale et al. (1982)	Spring	19	11	6.3

† Nelson et al., 1982.
‡ R. G. Hoeft. Factors affecting the efficiency of fall and spring applied nitrogen. Unpublished research reports 1975–1981. Dep. of Agron., Univ. of Illinois.
§ G. Kapusta, and E. C. Varsa. Unpublished progress reports 1972–1981. Plant & Soil Sci. Res. Stn., Southern Illinois Univ.
¶ L. M. Walsh and W. E. Elder. 1978. Effect of nitrogen fertilizer on corn as influenced by use of nitrapyrin. Unpublished research report. Dep. of Soil Sci., Univ. of Wisconsin.
# Average percent increase across all N rates at all locations.

soils in Minnesota was greater than increases observed on finer-textured soils in other areas of the Midwest (Table 2). Similarly, inhibitors were more beneficial on soils in Indiana than in Illinois. Generally, the soils included in the Indiana studies were more poorly drained than those in the Illinois studies.

In most years, reduction in tillage results in moisture conservation. In some years, this increased moisture retention may increase the potential for N loss. In addition, Doran (1980) has shown that reduction in tillage increases the number of nitrifying and denitrifying organisms and thus, it increases the potential for denitrification. Results from Indiana, Kentucky, and Illinois have shown a greater yield increase from inhibitors under no-till conditions than under conventional tillage (Table 2), which is likely the result of the two factors mentioned above.

The nearer N is applied to the time of crop need, the lower the potential for N loss. Therefore, the nearer N is applied to time of crop need, the lower the potential for benefit from nitrification inhibitors. This has been confirmed by results obtained in Indiana, Illinois, and Wisconsin (Table 2). Even though the actual response to inhibitor has been greater when fall-applied than when spring-applied, there are instances where the absolute yield was greater for spring N application without an inhibitor as compared with a fall N application with an inhibitor (Table 3). Therefore, inhibitors do not eliminate a N loss potential on all soils.

The economics of using inhibitors as compared with using additional N

Table 3—Effect of time and rate of N application and nitrapyrin on corn yield. (Adapted from Hoeft, 1980.)

N	Nitrapyrin	Time of application	
		Fall	Spring
kg/ha	kg/ha	—— yield, kg/ha ——	
0	--	4 140	
112	0.0	6 272	9 032
	0.56	7 777	8 405
168	0.0	7 777	10 098
	0.56	9 659	9 972
224	0.0	8 906	10 850
	0.56	9 910	10 788

to offset yield loss will depend on the actual amount of loss that occurs. Using the example in Table 3 and assuming that the cost of the inhibitor would have purchased an additional 45 kg N/ha, it would have been more economical to use the inhibitor at the 168 kg N/ha rate than to have applied 213 kg N/ha; the yield for the inhibitor treatment would have been 975 kg/ha greater with the inhibitor treatment than for the treatment that included extra N. In years when N loss was less severe, there would likely be little difference in economics between use of an inhibitor and use of additional N. The preceding analysis assumes that producers are using the optimum N rate for years in which losses are not severe. If producers use a rate greater than the optimum for their particular conditions, then it is not likely that they will experience a profitable return from the use of inhibitors when considered over several years.

As pointed out earlier, method of application has been shown to influence the efficacy of nitrification inhibitors. Where comparisons have been made, the inhibitors tend to be more effective when used as a band application with NH_3 than when broadcast applied—even when incorporated soon after application with solution N (Table 2).

Work at both Indiana and Southern Illinois University has shown that a fall application of N that includes inhibitors results in equal or higher wheat yield than a spring N application (Table 4). In most cases when spring N applications are made at the proper time, the yield from this treatment will be equal to the fall-applied N with an inhibitor. However, if wet soils delay N application beyond the optimum time in the spring, a fall N application with an inhibitor may yield better. As would be expected, the benefit of inhibitors is generally greater when the N is fall-applied than when spring-applied.

A major advantage of the use of inhibitors for wheat is that farmers may preplant their N at a time when soils are generally at an ideal moisture level for field operations. Otherwise, the farmer must work with the wet soils that frequently occur at the time N needs to be applied in the spring. In some years, excessively wet soils may necessitate the use of the more expensive aerial application technique. When N is topdressed in the spring, producers are limited to dry or solution materials. However, if they preplant their N they have the added option of using AA or of incorporating urea-containing materials.

Table 4—Effect of time of N application and nitrification inhibitors on wheat yield.

Reference	N Source	Nitrification inhibitor	Time of application	No. of experiments	No. of yield increases from inhibitor	Percent yield increase from non-N-treated plot†
Varsa et al. (1981)	Urea	No	Fall	3		48
	Urea	Yes	Fall	3	1	62
	Urea	No	Spring	3		51
	Urea-AN	No	Fall	2		54
	Urea-AN	Yes	Fall	2	1	74
	Urea-AN	No	Spring	2		65
	NH_3	No	Fall	1		55
	NH_3	Yes	Fall	1	1	99
Huber et al. (1980)	Urea	No	Fall	2		54
	Urea	Yes	Fall	2	1	65
	Urea	No	Spring	1		94
	Urea	Yes	Spring	1	1	128
	Urea/NH_3	No	Fall	2		43
	Urea/NH_3	Yes	Fall	2	1	55
	Urea/NH_3	No	Spring	2		48
	AS	No	Fall	1		73
	AS	Yes	Fall	1	1	150
	AS	No	Spring	1		138
	AS	Yes	Spring	1	1	162
	Urea/AS/calcium nitrate (CN)	No	Fall	2		83
	Urea/AS/CN	Yes	Fall	2	2	118
	Urea/AS/CN	No	Spring	1		156
	Urea/AS/CN	Yes	Spring	1	0	159

† Average of all N rates and all locations.

Wells (1977) has attributed a 7 to 11% yield increase to inhibitors when applied for paddy rice. His results have shown that a preplant N application with inhibitors was better than the commonly used split N applications.

McCormick et al. (1981) have reported corn yield response from the use of nitrification inhibitors with liquid swine manure. Response to the inhibitor was generally larger when the manure was fall-applied than when it was spring-applied, but significant response to the inhibitor was also obtained when it was applied in manure in the spring. Since most livestock producers must empty their manure storage pits at least twice a year, it is likely that an economical benefit would be obtained from inclusion of a nitrification inhibitor with manure that is fall-applied on soils that have a medium-to-high N loss potential.

VI. SUMMARY

Nitrification inhibitors should be viewed as a N management tool. The benefit to be derived depends on the soil type, time and rate of N application, and weather conditions between the time the N is applied and absorbed by the crops.

The greatest potential for benefits are with soils that frequently remain saturated with water during the early part of the growing season—primarily the poorly or imperfectly drained soils. However, moderately well-drained soils would also benefit if flooded frequently for periods of $\geq$ 3 d in the spring. Coarse-textured soils (sands) are likely to benefit more than the finer-textured soils, since the use of nitrification inhibitors will reduce the high potential for leaching that exists with such soils.

The longer the time between N application and crop need, the greater the probability of higher yields as a result of using a nitrification inhibitor. Employing nitrification inhibitors could significantly improve the efficiency of fall-applied N on the loams, silts, and clays when wet soil conditions exist in the spring. Since the effectiveness of inhibitors is temperature-dependent, it is suggested that fall applications of N with inhibitors be delayed until temperatures are $< 15°C$ in the fall.

Spring preplant applications of inhibitors may be beneficial on nearly all soil types with which N loss occurs frequently. This is especially true for sandy soils and for soils in the southern United States. The probability of inhibitors resulting in beneficial effects is greater when subsoils are recharged with water than when subsoils are dry at the start of spring.

The greatest potential for benefit will occur when nitrification inhibitors are used at or below the optimum N rates. Where N rates greater than those required for optimum yields are being used, yield benefits are less likely, even when there is excessive moisture.

For some producers, the ability to apply their N fertilizer at a time when labor demands are low with reasonable assurance that their N loss potential will be low may be very beneficial to their total crop management program. The benefits derived may include the need for less hired labor and/or a lessened demand for machinery. In addition, fall applications may result in less soil compaction than might be experienced in some springs.

REFERENCES

Alexander, J. D. 1983. Nitrogen loss potential ratings for Illinois soils and methods of assessing these ratings. Illinois Fert. Conf. Proc. p. 24–33. Univ. of Illinois, Urbana, Ill.

Boswell, F. C., and O. E. Anderson. 1974. Nitrification inhibition studies of soil in field buried polyethylene bags. Soil Sci. Soc. Am. Proc. 38:851–852.

Briggs, G. G. 1975. The behavior of the nitrification inhibitor "N-Serve" in broadcast and incorporated applications to soil. J. Sci. Food Agric. 26:1083–1092.

Cochran, V. L., R. L. Papendick, and W. M. Woody. 1973. Effectiveness of two nitrification inhibitors for anhydrous ammonia under irrigated and dryland conditions. Agron. J. 65: 649–653.

Doran, J. W. 1980. Soil microbial and biochemical changes associated with reduced tillage. Soil Sci. Soc. Am. J. 44:765–771.

Freney, J. R., O. T. Dunmead, and J. Simpson. 1978. Evolution of nitrous oxide from soils at low moisture contents. Abstracts for commission papers. Int. Congr. Soil Sci. 11th (Edmonton, Canada) 1:200–301.

Frye, W. W., R. L. Blevins, L. W. Murdock, K. L. Wells, and J. H. Ellis. 1980. Surface application of urea and ammonium nitrate treated with N-Serve 24 nitrogen stabilizer for no-tillage corn. Down to Earth 36(3):26–28.

Hale, D. B., R. J. Sprenkel, J. L. Stafford, W. H. Gurley, and R. C. Hubbard, Jr. 1982. Performance of N-Serve nitrogen stabilizer in replicated corn plots in the southeastern U.S.—1981. Down to Earth 38(2):1-5.

Hendrickson, L. L., and D. R. Keeney. 1979. Effect of some physical and chemical factors on the rate of hydrolysis of nitrapyrin (N-Serve). Soil Biol. Biochem. 11:47-50.

Hendrickson, L. L., L. M. Walsh, and D. R. Keeney. 1978. Effectiveness of nitrapyrin in controlling nitrification of fall and spring applied anhydrous ammonia. Agron. J. 70:704-708.

Herlihy, M., and W. Quirke. 1975. The persistence of 2-chloro-6-(trichloromethyl)-pyridine in soil. Commun. Soil Sci. Plant Anal. 6:513-520.

Hoeft, R. G. 1980. Nitrification inhibitors. p. 21-26. Illinois Fert. Conf. Proc. Univ. of Illinois, Urbana, Ill.

Huber, D. M., D. W. Nelson, H. L. Warren, C. Y. Tsai, and G. E. Shaner. 1980. Response of winter wheat to inhibiting nitrification of fall applied nitrogen with N-Serve nitrogen stabilizer. Down to Earth 37(1):1-5.

Keeney, D. R. 1980. Factors affecting the persistence and bioactivity of nitrification inhibitors. *In* J. J. Messinger et al. (ed.) Nitrification inhibitors—potentials and limitations. ASA Spec. Pub. 38. Am. Soc. of Agron., Madison, Wis.

Kurtz, L. T. 1980. Potential for nitrogen loss. *In* J. J. Meisinger et al. (ed.) Nitrification inhibitors—potentials and limitations. ASA Spec. Pub. 38. Am. Soc. of Agron., Madison, Wis.

Malzer, G. L. 1979. Progress with nitrification inhibitors. Down to Earth 35(3):1-3.

McCormick, R. A., D. W. Nelson, D. M. Huber, and A. L. Sutton. 1981. Improving the fertilizer value of liquid swine manure with a nitrification inhibitor. Purdue Agric. Exp. Stn. bull. 342.

Miller, G. A. 1981. A qualitative procedure for rating Iowa soil series for nitrogen loss potential. Agron. Abstr. p. 18.

Mulvaney, R. L. 1983. Evolution of dinitrogen and nitrous oxide from soils treated with ^{15}N-labeled fertilizer. Ph.D. Thesis. Univ. of Illinois. Microfilms, Ann Arbor, Mich.

Nelson, D. W., D. W. Huber, and D. B. Mengel. 1982. Current outlook for nitrification inhibitors. Indiana Plant Food and Agric. Chem. Conf. Proc., Purdue Univ. Purdue Univ., W. Lafayette, Ind.

Nelson, L. B., and R. E. Uhland. 1955. Factors that influence loss of fall-applied fertilizers and their probable importance in different sections of the United States. Soil Sci. Soc. Am. Proc. 19:492-496.

Redemann, C. T., R. W. Meickle, and J. G. Widofsky. 1964. The loss of 2-chloro-6-(trichloromethyl)-pyridine from soil. J. Agric. Food Chem. 12:207-209.

Schmitt, M. A. 1983. Effect of temperature, soil type, and nitrification inhibitors on the rate of ammonium-N disappearance in the soil. M.S. Thesis. University of Illinois.

Touchton, J. T., and F. C. Boswell. 1980. Performance of nitrification inhibitors in the southeast. *In* J. J. Meisinger et al. (ed.) Nitrification inhibitors—potentials and limitations. ASA Spec. Pub. 38. Am. Soc. of Agron., Madison, Wis.

Touchton, J. T., R. G. Hoeft, L. F. Welch, and W. L. Argyilan. 1979. Loss of nitrapyrin from soils as affected by pH and temperature. Agron. J. 71:865-869.

Turner, G. O., L. E. Warren, and F. G. Andriessen. 1962. Effect of 2-chloro-6-(trichloromethyl)-pyridine on the nitrification of ammonium fertilizers in field soils. Soil Sci. 93: 270-272.

Varsa, E. C., G. L. Licu, and G. Kapusta. 1981. The effect of nitrification inhibitor on wheat yields and soil nitrogen retention. Down to Earth 37(3):1-5.

Wells, B. R. 1977. Nitrapyrin [2-chloro-6-(trichloromethyl)-pyridine] as a nitrification inhibitor for paddy rice. Down to Earth 32(4):28-32.

38

Regis D. Voss
Iowa State University
Ames, Iowa

Potential for Use of Urease Inhibitors

Urea provides an estimated 40% of the world's current use of 60 million tonnes of fertilizer N. Urea is the dominant N fertilizer in world agriculture and provides an estimated 75, 25, and 10% of the N consumption in Asia, Europe, and the United States, respectively (see Chapt. 2, this book).

The natural process of rapid hydrolysis of urea to ammonium carbonate [$(NH_4)_2CO_3$] in moist soils through soil urease activity with the concomitant rise in pH and liberation of NH_3 can result in seed and seedling injury, NO_2^- accumulation in toxic concentrations, and gaseous loss of urea N as NH_3 (Hauck, 1983). These problems could possibly be eliminated or mitigated by the emerging technology of urease inhibitors.

The potential use of urease inhibitors in production agriculture bears scrutiny. An understanding of urease activity, factors affecting activity of urease in soils, the effect of urea hydrolysis, and the characteristics of urease inhibitors is paramount to making agronomic decisions on the potential use of this technology. Presented here is a brief and very applied discussion to bring agronomic considerations of urease inhibitors into focus. Detailed discussions and extensive literature citations on urease activity in soils, hydrolysis of urea, and the subsequent effect are available (Kiss et al., 1975; Bremner & Mulvaney, 1978; Terman, 1979; Mulvaney & Bremner, 1981; Hauck, 1983).

I. UREA USE IN THE UNITED STATES

Urea is becoming an increasingly important source of N in the United States. During the 1970–1980 period, N use increased > 50%, but use of urea as a straight source of N nearly tripled (Table 1). Nonpressure N solutions containing urea doubled in consumption. Anhydrous NH_3 (AA) con-

Nitrogen in Crop Production.

Table 1—Total N and selected N-source use in the United States for 1970 and 1980 and the percentage increase in use during the period (Hargett & Berry, 1981).

	1970	1980	Percent increase
Total N, tonnes	6 765 317	10 339 353	53
Urea-N, tonnes	222 601	864 468	288
Percent of total N	3.3	8.4	
Solution N (30%)†, tonnes	882 391	1 800 048	104
Percent of total N	13.0	17.4	
NH_3-N, tonnes	2 844 058	4 499 947	58
Percent of total N	38.1	39.4	

† Nitrogen solutions are marketed with different N concentrations, depending on area of the United States; 30% N is used as an average value.

sumption, which is the dominant straight N source in the United States, increased slightly more than total N use.

In the 1970–1980 period, the three straight N sources—urea, nonpressure N solutions, and AA—increased in their share of the United States N market primarily at the expense of fertilizer compounds containing N and to some extent minor straight N sources. Although urea increased its share of the N market, AA still maintained its position in the market with solutions containing urea showing a marked gain.

Although urea use increased in all regions of the United States, the largest tonnage and percentage increased occurred in the East and West North Central and South Central states in the 1970–1980 period (Tables 2 and 3). In 1980, the five states consuming the most urea were Iowa followed by Arkansas, Minnesota, Ohio, and Illinois.

Table 2—Total N and urea-source N use in the eastern regions of the United States for 1970 and 1980 and the percentage increase in use during the period (Hargett & Berry, 1981).

		1970		1980		
Region	N Source†	Tonnes	Percent of total	Tonnes	Percent of total	Percent increase
New England	Total	37 063		38 364		4
	Urea	1 790	4.8	4 898	12.8	174
	Solution	1 232	3.3	1 474	3.8	20
Middle Atlantic	Total	214 574		285 261		33
	Urea	18 349	8.6	24 717	8.7	35
	Solution	22 561	10.5	59 784	21.0	165
South Atlantic	Total	731 087		856 505		17
	Urea	4 459	0.6	5 481	0.6	23
	Solution	225 707	30.9	314 063	36.7	39
East North Central	Total	1 287 056		2 210 938		72
	Urea	31 835	2.5	201 899	9.1	534
	Solution	209 317	16.3	494 305	22.4	136
East South Central	Total	500 623		530 727		6
	Urea	7 554	1.5	54 966	10.4	628
	Solution	51 362	10.3	57 828	10.9	13

† Nitrogen content of N solutions was computed at 30%.

II. UREA HYDROLYSIS

The enzyme urease catalyzes the reaction of urea with water to form ammonium carbonate:

$$NH_2CONH_2 + 2H_2O \xrightarrow{\text{urease}} (NH_4)_2CO_3.$$

The effect of this hydrolysis is an accumulation of NH_4^+ in the application zone and a concomitant increase in pH, which can lead to loss of NH_3 and/or an increase in NO_2^- concentration.

Ammonia loss is affected by several factors, including cultural operations, soil chemical properties, and the soil surface environment. The non-incorporation of surface-applied urea provides no soil cover to retain NH_3 from the hydrolysis product, ammonium carbonate, which decomposes readily as follows:

$$(NH_4)_2CO_3 \xrightarrow{H_2O} 2NH_3\uparrow + CO_2\uparrow.$$

Surface crop residue, which contains urease, enhances the hydrolysis of urea, and thus, losses as NH_3. Mixing urea into the soil or rainfall moving urea into the soil will greatly reduce or eliminate these losses.

The dominant soil chemical properties affecting loss of the NH_3 are buffer capacity (cation exchange capacity) and soil pH. As the buffer capacity increases, the loss of NH_3 decreases because of the greater retension of NH_4^+. As soil pH increases, ammonium carbonate is less stable due to reaction products in the soil, and NH_3 loss is increased.

Table 3—Total N and urea-source N use in the western regions of the United States for 1970 and 1980 and the percentage increase in use during the period (Hargett & Berry, 1981).

		1970		1980		
Region	N Source†	Tonnes	Percent of total	Tonnes	Percent of total	Percent increase
West North Central	Total	2 120 141		3 656 992		72
	Urea	24 974	1.2	297 996	8.1	1 098
	Solution	229 411	10.8	471 797	12.9	106
West South Central	Total	961 296		1 222 425		27
	Urea	57 392	6.0	138 204	11.3	141
	Solution	61 625	6.4	114 318	9.4	86
Mountain	Total	339 829		548 825		62
	Urea	21 372	6.3	48 947	8.9	129
	Solution	29 820	8.8	86 082	15.7	189
Pacific	Total	585 840		940 023		60
	Urea	48 636	8.3	83 656	8.9	72
	Solution	51 030	8.7	194 810	20.7	282

† Nitrogen content of N solutions was computed at 30%.

The soil surface environment, as characterized by surface temperature, surface moisture, and air movement above the surface, affects NH_3 loss from surface-applied urea. As soil temperature increases from 10 to 70°C, urea hydrolysis increases. An increasing soil temperature and air movement in conjunction with a moist soil surface leads to more water evaporation causing more NH_3 loss.

Estimates of N loss as NH_3 from surface-applied urea range from < 1 to $> 50\%$ of the N applied. Terman (1979) cited several references that reported losses of NH_3 from urea surface-applied to a variety of crops. He noted (i) the variability in the results from year to year among N sources, application times, crops, and cultural practices, and (ii) much research needs to be done to forecast a given NH_3 loss resulting from urea fertilizer application under field conditions.

Nitrite accumulation in excess can cause inhibition of seed germination and seedling injury. Cultural practices and soil chemical properties are the dominant factors affecting NO_2^- accumulation. Banding of urea, which after hydrolysis results in a zone of high soil pH that inhibits conversion of NO_2^- to NO_3^- by nitrobacter bacteria, can result in a high NO_2^- concentration. Banding of urea too close or with crop seed can result in poor germination. The soil chemical properties of high soil pH and low buffer capacity contribute to NO_2^- accumulation sufficient to be toxic.

These are the ramifications of urea hydrolysis, but do not provide insight to urease and factors affecting its activity.

III. SOURCES OF UREASE

Plants, bacteria, and soil are sources of urease. There are indications that urease from each of these three sources is different.

Each soil seems to have a distinct capacity for urease or, perhaps more correctly, a capacity to protect a certain level of urease. The soil urease level in each soil does seem to be stable. The hydrolysis of urea is a function of soil urease present before the fertilizer urea is applied to the soil.

IV. FACTORS AFFECTING UREASE ACTIVITY

Soil physical and chemical properties, soil environment, and certain cultural practices affect urease activity as measured by amount of urea hydrolyzed from a given addition to soil. Urease activity in soil increases as either organic C, cation exchange capacity, or the clay content of soil increases. Conversely, as the sand content increases, urease activity decreases. There are conflicting reports on the effects of soil pH on urease activity.

Soil temperature does affect urease activity. As soil temperature increases from 10 to 70°C, urease activity increases. Further temperature increase from 70 to 80°C decreases urease activity. There is slight urease activity at below freezing temperatures of -20 to -10°C, but this activity would seem to have little practical significance. Water and oxygen content of the soil seem to have no affect on urease activity.

Some common cultural practices affect urease activity. Increasing the urea concentration in the soil (higher fertilizer applications) increases the rate of hydrolysis, because the substrate is increased; urease is not increased. If large organic additions from manure or crop residue are made to the soil, urease activity is increased, but then declines to its original level in the soil. Common fertilizer salts have no affect on urease activity. Limestone ($CaCO_3$) seems to have no affect, but lime (CaO) additions to soil have increased urease activity. The measured effect of limestone or lime on soil urease activity may be due to the time span in which the analysis were made. Urease activity has not been affected by the nitrification inhibitors and pesticides that have been evaluated.

V. UREASE INHIBITORS

Although urease in soils and factors affecting its activity seem to be complex and not thoroughly and clearly understood, soil urease activity can be inhibited by several compounds, including inorganic salts of Hg, Ag, and Cu; dihydric phenols; and quinones such as hydroquinone, *p*-bensoquinone, and certain substituted *p*-bensoquinones (for references, see Chapt. 36, this book). Very recently, phenylphosphorodiamide and phosphorotriamide have been shown to have promise as inhibitors of soil urease activity. No chemical is in commercial production as a urease inhibitor for use with urea fertilizers, but several are under study.

Several factors affecting urease activity have been identified, but the effort has not been put forth on identifying factors and their effects on urease inhibition. As might be expected, some factors that increase urease activity, also decrease inhibition of urease activity. Increasing organic C and cation exchange capacity of the soil and soil urease activity itself decrease inhibition of soil urease activity.

VI. DESIRABLE CHARACTERISTICS OF INHIBITOR

As basic research progresses on identifying and evaluating urease inhibitors and the factors affecting their efficiency, agronomic input would seem to be desirable to help determine the potential and proper use of this new technology in soil-plant systems.

From an agronomic and applied crop production point-of-view, there are many characteristics of an inhibitor that would be desirable. First and foremost, it should be inexpensive and effective. Any inhibitor should be safe to use by the applicator, nonphytotoxic to plants, a noncontributor to toxic materials in soil, and innocuous to other beneficial enzymes in the soil-plant system. A urease inhibitor should be easily applied and retained on solid urea and compatible in urea solutions. It should be stable in storage and not subject to decomposition by air, light, or water prior to application to the soil. After application to the soil, it should require no additional cultural operations.

VII. AGRONOMIC CONSIDERATIONS

An important agronomic consideration is whether a urease inhibitor should be water soluble. Water solubility of the inhibitor would permit it to stay with urea as water moves urea into and through the soil. Excessive movement could result in urea being leached below the crop root zone. Water solubility cannot be separated from the consideration of persistence of an inhibitor in a soil-plant system or the sufficient time of inhibition of urease activity. Assuming that urease from plant roots can hydrolyze urea at a rate similar to soil urease, would inhibition of this plant urease be desirable? If this would occur, then the roots would need to absorb urea in sufficient quantities to meet crop N requirements. Also, plants' ability to absorb and utilize urea relative to NH_4^+ and NO_3^- would need to be known.

If the urease inhibitor would be water insoluble, it would not move with urea. Protection of urea at the soil surface would be obtained, but this may not be sufficient in sandy soils or some calcareous soils. Hydrolysis of urea could occur below the soil surface and would eliminate the questions of inhibition of plant root urease and plants' ability to absorb urea.

VIII. AGRONOMIC EVALUATION

If the foregoing questions are to be answered and the agronomic decisions are to be made, then evaluations of urease inhibitors should be done concurrently with basic research. Field evaluations should provide information on verification of basic research and concepts and provide information for agronomic decisions. Field research would determine if urease inhibitors do in fact reduce NH_3 loss or NO_2^- accumulation in a production environment for the soil-plant system, and, if so, whether this increases plant recovery of N. Primary requirements for field evaluations are (i) that a crop response to fertilizer N be obtained and (ii) soil environmental conditions are conducive to NO_3 loss and/or for NO_2^- accumulation.

IX. POTENTIAL

There is a potential for urease inhibitors if they are cost effective, are agronomically beneficial, and improve N management in a production environment for the soil-plant system (i.e., we must be economically, agronomically, and environmentally better off for using them than not).

REFERENCES

Bremner, J. M., and R. L. Mulvaney. 1978. Urease activity in soils. p. 149–196. *In* R. G. Burns (ed.) Soil enzymes. Academic Press, London.

Hargett, N. L., and J. T. Berry. 1981. 1980 Fertilizer summary data. Bull. Y-165. National Fertilizer Development Center, TVA, Muscle Shoals, Ala.

Hauck, R. D. 1983. Agronomic and technological approaches to minimizing gaseous nitrogen losses from croplands. p. 285–312. *In* J. R. Freney and J. R. Simpson (ed.) Gaseous loss of nitrogen from plant-soil systems. Martinus Nijhoff, The Hague, Netherlands.

Kiss, S., M. Dragan-Bularda, and D. Radulescu. 1975. Biological significance of enzymes accumulated in soil. Adv. Agron. 27:25–87.

Mulvaney, R. L., and J. M. Bremner. 1981. Control of urea transformations in soils. p. 153–195. *In* E. A. Paul and J. N. Ladd (ed.) Soil biochemistry. vol. 5. Marcel Dekker, Inc., New York.

Terman, G. L. 1979. Volatilization losses of nitrogen as ammonia from surface applied fertilizers, organic amendments, and crop residues. Adv. Agron. 31:189–223.

39

Robert C. Gray and Gary W. Akin
National Fertilizer Development Center
Tennessee Valley Authority
Muscle Shoals, Alabama

Foliar Fertilization

Foliar fertilization has existed in natural environments since the beginning of plant life. The most obvious examples are related to atmospheric disturbances. Lightning caused prairie and range fires, which poured oxides of N and S into the winds to be deposited on plants kilometers away. However, intense agricultural practices of today require man to add nutrients to overcome the unreliability of nature. For several centuries, man successfully applied the major nutrients for crop production while relying upon the earth to provide those used in lesser quantities. However, by the early 1800's, man recognized that in some locations, in some environments, and with some plants, more than just major nutrient fertilization was required. Elements thought to be relatively unimportant made substantial changes in plant growth, yield, and color when sprayed onto the foliage. The water-soluble salts of Fe were perhaps the first to be used in this manner (Wittwer et al., 1963).

Foliar application of fertilizers has been studied since the 1800's. The greatest use has been with high-value crops such as fruits and vegetables. Limited application has been practiced in the production of cereal, leguminous, and fibrous crops. However, attempts of modern scientists to stave off the dire predictions of Malthus has caused them to turn to foliar fertilization as a means of avoiding the undesirable physiological traits of crops.

I. NUTRIENT UPTAKE BY LEAVES

The absorption mechanism of the plant leaf determines the effectiveness of foliar fertilization. Effectiveness depends upon nutrient penetration of the cuticle and epidermal cells and upon transport across the epidermal cells. Some penetration may occur through stomata but more likely by dif-

Nitrogen in Crop Production.

fusion through the cuticle (Noggle & Fritz, 1976). Discontinuties of the leaf surface may facilitate absorption, while surface vagaries (such as rodlets) and waxes prevent wetting of the cuticle. Adding surfactants to the foliar nutrient solution aids in wetting the cuticle.

Once penetration has occurred, nutrient absorption by leaves is probably not greatly different from absorption of the same nutrients by roots. The major difference is the environment in which each of these plant parts exist. Roots are bathed in a soil solution that is usually quite dilute. The concentration of the nutrients is regulated to some extent by the surrounding soil. In foliar applications, the aqueous phase evaporates quickly after application and absorption must take place from concentrated solution. The internal moisture status of the leaf is an important factor in nutrient uptake. When leaves are turgid, the cuticle is more permeable and penetration and absorption is facilitated. As the leaf loses turgidity, the waxy platelets in the cuticle shrink together, which tends to seal the leaf and retard absorption.

Urea is the most rapidly absorbed fertilizer compound when the leaf is not under moisture stress. Other nutrient sources are more slowly absorbed and substantial nutrient loss may occur by washing, particularly within the first 24 h of application. Although rain is recognized as the major precipitation factor involved in washing, the effects of dew, fog, mist, or even snow cannot be discounted.

While many environmental factors affect the potential losses due to washing, type of plant, specific nutrient(s) to be applied, and the objective of foliar fertilization are prime considerations in the process. The objectives of foliar fertilization may include an attempt to change chemical composition, relieve physiological stress, increase yields, improve fruit quality, increase efficiency of fertilizers, or provide an easier or more feasible application technique.

II. FOREST FERTILIZATION

This area of fertilization alone justified serious consideration of foliar fertilization. Fertilizing forests by hand is time-consuming. Aerial application by either airplane or helicopter greatly reduces labor requirements. However, as the need to harvest "renewable resources" continues to grow, efficiency of fertilizer use also must be considered.

Miller and Young (1976) reported that foliar-applied N solution caused a growth response comparable to that obtained using urea granules. They discovered that burning of Douglas-fir (*Pseudotsuga menziesii*) and Western hemlock (*Tsuga heterophylla*) foliage during the summer occurred at rates of $\geq$ 108 kg N/ha. However, more than a 30% burn was required before benefits were offset by damage. Although foliar application of N solution costs more than urea granules, costs can be reduced by combining fertilizer and herbicide applications. For nutrient deficiencies in young seedlings, the absorption rate through the foliage probably is greater than the combination of movement in the soil and root absorption. This frequently justifies foliar over soil fertilization when corrective measures are needed.

III. CHANGING CHEMICAL COMPOSITION

Many studies have been conducted on wheat (*Triticum aestivum* L.) in an effort to improve the efficiency of N use relative to grain yield. Mistra and Nadagoudar (1977) studied the application of N through soil and foliar spray at different growing stages. They found that 80 kg N/ha applied as four foliar sprays at seeding, tillering, jointing, and flowering gave higher yields than soil application alone. This was due to the decreased exposure to leaching and denitrification that multiple applications afford. Penny et al. (1978) reported increased wheat yields when liquid urea–ammonium nitrate (UAN) was foliar-applied in late spring. The N content of the grain was also increased. Pushman and Bingham (1976) observed a considerable response by grain protein when 45 kg N/ha as liquid urea were applied at anthesis.

Although increased protein levels may be obtained using foliar applications of N, the extra protein must have economic value to justify the cost and logistics of foliar applications. Farmers feeding wheat produced on their farms readily stand to benefit. Farmers selling wheat with higher protein levels are not in a better marketing position than neighboring growers selling wheat with conventional protein levels.

Spraying urea solution on rice (*Oryza sativa* L.) at the heading stage produced a rice with 11% protein (Nishizawa et al., 1977). This was 44% higher than rice without the urea spray. Quality of protein in both the high-protein and the regular rice was similar. Other experiments have found that foliar application of N at later stages of growth is superior to applications of granular N as measured by yield of grain. Also, late applications of foliar N with micronutrients delay senescence and increase yields.

IV. RELIEVING PHYSIOLOGICAL STRESS

This area has the greatest potential for foliar fertilization. When a plant goes from a vegetative stage to the reproductive stage, photosynthate produced by the leaves is translocated to the developing seed and the amount going to the roots is greatly reduced. Root extension and growth of new root hairs ceases. The root still functions, but it operates at reduced efficiency and the uptake of nutrients is reduced. In legumes, nodules no longer receive energy; thus, they die and slough off. Root uptake of N becomes inadequate to supply both vegetative and reproductive growth, and N necessary for seed production must be translocated from the leaves. The pool of protein and N in the vegetative parts is depleted and early senescence of leaves occurs. The plant "self-destructs."

Species of plants vary in the efficiency with which they produce seed from photosynthate and in the amount of N required per unit of photosynthate. Sinclair and de Witt (1975) analyzed efficiencies and N requirements for 24 food crops. They found that cereal grains and oil crops usually could get sufficient N from the soil to complete growth because the N-required/photosynthate-produced ratio was low. Legumes have a higher ratio and may be more subject to early senescence. However, because of varying soil conditions, all species may be subject to early senescence as a result of N translocation from the leaves. An application of N as foliar spray during the reproductive stage of growth may delay senescence and create a potential for increased yield.

The present rationale for foliar fertilization with primary nutrients is the hypothesis that leaf senescence can be delayed and yield increased if nutrients are foliar-applied at seed development. Allied Chemical Company conducted and supported research in this area as early as 1974. It was not until Garcia and Hanway (1976) released their 1975 results of foliar fertilization on soybean [*Glycine max* (L.) Merr.] yields that widespread research efforts began. Garcia and Hanway reported that the optimum ratio of N–P–K–S was 10–1–3–0.5. This ratio was similar to the ratio in the seeds. This ratio gave the greatest increase in soybean yield in 1975. They prepared this ratio from TVA potassium polyphosphate (KPP) solution having a N–P–K ratio of 0–11–21 (0–25–25). They also reported that omitting any one of the elements in the ratio diminished the increase in yield and that two to four applications were needed during the seed-filling period. More frequent applications were not as effective, nor were applications at the beginning and end of the seed-filling period. Garcia and Hanway also reported leaf burning when excessive rates of application were used in any one spraying.

The Iowa results prompted widespread interest and in 1976 TVA provided research and demonstration material to 84 individuals in 28 states for laboratory, greenhouse, and field trials (Gray, 1977). The results from 214 comparisons between soybeans that were not foliar fertilized and those that received from one to five applications showed little yield response to foliar fertilization and, instead, showed an average yield decrease of 5.2%. According to Gray, most of the soybeans showed some degree of leaf burn from the foliar spray. He attributed this partially to the unusually low air humidity in August 1976. Meanwhile, Allied Chemical Corporation continued extensive field trials with their trademarked Folian®. Their results were sufficiently positive that they decided to make Folian® commercially available.

Parker and Boswell (1980) more recently reported similar negative yield responses to foliar fertilization on soybeans. They also reported that ammonium polyphosphate (APP) caused greater foliage injury than KPP or urea. The results of using foliar fertilization during reproductive growth have been both encouraging and disappointing. The increases in yield show that potential exists; however, the failure of other experiments indicates that much more needs to be known about the process before general use should be made.

V. OTHER USES

Cotton (*Gossypium hirsutum* L.) has responded well to foliar applications of N in many experiments. Probably the greatest response occurs where soil N is inadequate, either from less than optimum application at planting or from climatic conditions causing the soil N to be unavailable to the plant. When extra N is needed, foliar applications of N, preferably as urea solution, increase yield by increasing number of bolls per plant, boll weight, and number of seeds per boll. Balance between N and P is important in boll development. The University of Arkansas conducts a program of weekly petiole analysis for growers during the boll-forming stage and advises the grower on his nutrient balance and possible need for more N by aerial application. The program works well in Arkansas but has not been as successful in other areas.

Little effect can be expected from foliar applications of N to corn (*Zea mays* L.) or grain sorghum [*Sorghum bicolor* (L.) Moench] if adequate N fertilizer is applied at planting. A small increase in grain protein is sometimes found in corn that has been foliar sprayed with urea, but this has little economic value. A 32% UAN solution was applied to grain sorghum in Kansas to dessicate foliage before harvest (Donnelly et al., 1977). Grain drying was accelerated only when the N solution was applied before physiological maturity, but this reduced grain yield. Leaf loss reduced harvestable forage, and there was no increase in protein content of either grain or forage.

Among other crops, hop (*Humulus lupulus* L.) yields seem to consistently benefit from foliar N applications. The number of cones per vine is increased with no decrease in quality. Apple (*Malus domestica*) yields are increased by applications of urea solution through better retention on the tree and increased size of fruit. Quality as measured by color is also improved. In Hawaii, a large percentage of the necessary N for pineapple (*Ananas sativus*) is applied as a urea solution. Legume crops seldom show increased yields from foliar application of N alone.

VI. CONCLUSIONS

Foliar application of N is more efficient than soil application in terms of percentage uptake of applied N. But the necessity of multiple applications of very dilute solutions to get enough N to meet the requirements of the plant makes it uneconomical as a primary method of fertilization.

Care must be taken to prevent foliage burn when N solutions are applied to some crops. Low humidity and high temperatures increase chances of burning the leaves, as does a too concentrated solution. Selection of the N source is also important. Urea is less likely to cause foliage burn than other N sources, because it has a lower salt index and is more rapidly absorbed into the leaf.

Foliar application of N is best used as a supplement to soil applications. It can correct an observed N deficiency, improve appearance of plants where appearance is a market factor, and increase protein content of grain when this is desirable. If N can be applied in combination with necessary applications of pesticides, the cost is low and has proven beneficial to fruits, pineapples, and some vegetable crops.

Until more is known about timing and methods of application, allowable concentrations of N in solutions, limits imposed by climatic conditions, and responses of crops, the use of N in foliar application will probably not increase.

REFERENCES

Donnelly, K. J., R. L. Vanderlip, and L. S. Murphy. 1977. Dessication of grain sorghum by foliar applications of nitrogen solution. Agron. J. 69:33–36.

Garcia, L. R., and J. J. Hanway. 1976. Foliar fertilization of soybeans during the seed-filling period. Agron. J. 68:653–657.

Gray, R. C. 1977. Results of foliar fertilizer application studies. p. 54–58. *In* Situation 77. bull. Y-115. National Fertilizer Development Center, Muscle Shoals, Ala.

Miller, R. E., and D. C. Young. 1976. Forest fertilization: foliar applications of nitrogen solution proves efficient. Fert. Solutions 20(2):36, 40, 42, 44, 46, 48, 59, 60.

Mistra, N. M., and B. S. Nadagoudar. 1977. Soil versus foliar application of nitrogen at different growth stages of wheat under rainfed conditions. Mysore J. Agric. Sci. 11:181–184.

Nishizawa, N., I. Kitahara, T. Noguchi, S. Hareyama, and K. Honjyo. 1977. Protein quality of high protein rice obtained by spraying urea on leaves before harvest. Agric. Biol. Chem. 41:477–485.

Noggle, G. R., and G. J. Fritz. 1976. Introductory plant physiology. p. 264–265. Prentice-Hall, Inc., Englewood Cliffs, N.J.

Parker, M. B., and F. C. Boswell. 1980. Foliage injury, nutrient intake, and yield of soybeans as influenced by foliar fertilization. Agron. J. 72:110–113.

Penny, A., F. V. Widdowson, and J. F. Jenkyn. 1978. Spring top-dressings of 'Nitro-Chalk' and late sprays of a liquid N fertilizer and a broad spectrum fungicide for consecutive crops of winter wheat at Saxmundham, Suffolk. J. Agric. Sci. 90:509–516.

Pushman, F. M., and J. Bingham. 1976. Effects of a granular nitrogen fertilizer and a foliar spray of urea on the yield and bread-making quality of ten winter wheats. J. Agric. Sci. 87:281–292.

Sinclair, T. R., and C. T. de Witt. 1975. Photosynthate and nitrogen requirements for seed production by various crops. Science 189:(4202):565–567.

Wittwer, S. H., M. J. Bukovac, and H. B. Tukey. 1963. Advances in foliar feeding of plant nutrients. p. 429–455. *In* M. H. McVickar et al. (ed.) Fertilizer technology and usage. Soil Sci. Soc. of Am., Madison, Wis.

40

J. F. Power and J. W. Doran
Agricultural Research Service, USDA
University of Nebraska
Lincoln, Nebraska

Nitrogen Use in Organic Farming[1]

Organic farming is a form of agriculture that has been practiced throughout the history of mankind. This system of farming depends primarily upon indigenous soil organic matter, crop residues, legumes, and animal manures as sources of N for crop production (USDA, 1980b). Until widespread use of chemical fertilizers became common within the last few decades, most agricultural production enterprises could, to a large degree, be classified as organic-farming systems.

In this chapter, we will review the sources of N used in organic farming systems, discuss the relative availability of these sources, how availability is affected by management practices, and attempt to summarize knowledge of the effects of organic farming on soil biology and N transformations. On many of these subjects, little or no scientific information was available to the authors. Because of lack of specific data, we are drawing on data from related experiments (such as tillage methods) that have some application to organic farming.

I. HISTORICAL CONSIDERATIONS

The requirements for N in crop production can be met by various means. In prehistoric agriculture, and even today in some cultures, N for crop production was provided almost entirely by mineralization of indigenous soil organic N. Systems of shifting cultivation were employed wherein cropping of a soil was discontinued after several decades, at which time the nutrient-supplying power of the soil was reduced. However, as the availability of

[1] Contribution from the Agricultural Research Service, USDA.

Nitrogen in Crop Production.

Table 1—Crop yields, legume seed production (with and without alfalfa), and fertilizer N usage in the United States, 1959 through 1979 (USDA, 1980a).†

Year	Average crop yield (corn, wheat, soybean, sorghum)	Legume seed		N fertilizer applied
		With alfalfa‡	Without alfalfa	
	kg/ha	kg × 10^6		t × 10^6
1959	2274	180	123	2.4
1960	2531	172	110	2.7
1961	2621	151	94	3.0
1962	2710	145	91	3.5
1963	2822	170	98	3.5
1964	2598	160	97	3.9
1965	2912	136	81	4.1
1966	2934	128	75	4.7
1967	2979	104	53	5.4
1968	3024	100	48	6.1
1969	3270	103	56	6.3
1970	3002	116	54	6.7
1971	3494	99	45	7.2
1972	3606	78	30	7.1
1973	3360	77	30	7.4
1974	2733	81	30	8.2
1975	3230	74	31	7.7
1976	3248	68	32	9.3
1977	3360	71	27	9.5
1978	3606	65	23	8.9
1979	3853	67	22	9.5

† Corn, *Zea mays* L.; wheat, *Triticum aestivum* L.; soybean, *Glycine max* (L.) Merr.; sorghum, *Sorghum bicolor* (L.) Moench; alfalfa, *Medicago sativa* L.; red clover, *Trifolium pratense* L.; sweet clover, *Melilotus* sp.; lespedeza, *Lespedeza* sp.; crimson clover, *Trifolium incarnatum* L.; hairy vetch, *Vicia villosa* Roth; white clover, *Trifolium repens* L.; ladino clover, *Trifolium repens* L.

‡ Includes alfalfa, red clover, sweet clover, lespedeza, crimson clover, hairy vetch, white clover, and ladino clover.

noncultivated land declined, N needs for crop production were met by supplementing existing soil N levels with legumes, animal manures, or commercial fertilizers.

During the last 60 yr, there has been a steady increase in usage of synthetic N fertilizers in American agriculture, with a corresponding decrease in dependence upon legumes as a source of N. This change is illustrated by data in Table 1 showing both the tonnage of fertilizer N used annually for the past 20 yr and the production of legume seed. Since alfalfa is widely used in many livestock enterprises primarily as a source of livestock feed rather than as a soil-improving crop, alfalfa production probably was not as much affected by fertilizer N usage as were other legumes. Thus, data are provided in Table 1 for seed production of other legumes—primarily sweet, red, ladino, and other clovers and the vetches. The data clearly indicate that, during this period, there has been a widespread substitution in American agriculture of fertilizer N for N derived from biological fixation by legumes.

In these past several decades, the philosophy has developed that, by eliminating N as a growth-limiting factor through fertilizer application, we can increase the area devoted to cash grain crop production, thereby enabling a farmer to specialize in monocultures or simple cash-crop rotations. Such specialization has allowed each farmer to cultivate more land, use larger equipment, maintain or increase monoculture yields, and reduce manpower requirements per unit of production. The philosophy indicated above has received widespread approval and usage in American agriculture and, in general, has proven to be economically feasible these last several decades.

This system of crop production is not without problems, however. In addition to greater potential for buildup of weeds, diseases, and insect populations, many farmers report an increase in power requirements for tillage and a less favorable soil physical condition, compared with farming systems that included legumes. Although this complaint is common, to date documentation of the factors responsible has not been achieved using the common methods for measurement of specific soil physical properties.

In addition to these problems, considerable evidence has accumulated to imply that improper usage of fertilizers can contribute significantly to surface and groundwater pollution (Asmussen et al., 1979; Viets, 1975). In the majority of field experiments using N isotopes, usually 10 to 30% of the fertilizer applied cannot be accounted for in the soil-plant system at the end of the growing season (Hauck & Bremner, 1976). The N not accounted for is presumably lost from the system, either in soluble form by leaching or runoff, or has escaped into the atmosphere in one of several gaseous forms.

Since almost all fertilizer N is manufactured from natural gas, the recent dramatic increases in cost of natural gas and the anticipated future increases require us to reevaluate our present philosophy, which depends so heavily upon use of fertilizer N for crop production. Also, because of the greater potential for environmental pollution, because of the adverse effects of fertilizers on some soil properties, and for ideological reasons, a significant segment of the population is asking for such a reappraisal of crop production practices. Many such people, both for ideological and economical reasons, have been advocating some degree of conversion to organic-farming systems wherein little or no fertilizer N is utilized. In the following pages of this chapter, several aspects of the N cycle in organic farming will be discussed. No attempt will be made to summarize magnitudes or trends for organic farming because of lack of statistical information and divergence in definition of terms.

II. SOURCES OF N IN ORGANIC FARMING

Information on the amount of natural fertilizers used in organic farming is limited. The results of a survey conducted by *New Farm* magazine indicate that the major sources of organic N used are farmyard manure and crop residues (Table 2). Average rates of application for both sources range from 3 to 11 t/ha. A major purpose for adding residues to soil is to build up

Table 2—Predominant residues used in organic farming and average rates of application.†

Source	Organic farmers using	Application rate
	%	t/ha (tons/acre)
Farmyard manure	78	3–11 (1.3–4.9)
Sludge/sewage	4	2–11 (0.9–4.9)
Processing wastes	3	1–12 (0.45–5.4)
Crop residues	59	3–10 (1.3–4.5)

† Statistics compiled from a survey of readers of *New Farm* magazine in the United States (299 respondents).

the soil organic matter level, which then serves as a source of slowly available N for crop plants. Therefore, the major N sources for crop production in organic farming can be placed in three categories—animal manures, plant sources, and indigenous soil organic N.

A. Animal Manures

Animal manures and other animal waste products are widely used as a source of N for farming in the United States (Table 3). Animal manure production is estimated at 156 million tonnes annually, and an estimated 90% of this is returned to the soil (USDA, 1978). Since there are approximately 200 million ha of cultivated land in the United States, even with 100% utilization there would be only about 0.75 t of animal manure available per hectare, providing only about 34 kg N/ha. Thus, it is evident that there is insufficient animal manure available in the United States to meet the national N requirement for total crop production. With the advent of the multithousand-head feedlots within the past few decades, animal manures are not uniformly available in all farming areas. For those producers located close to a supply of animal wastes, however, sufficient waste products

Table 3—Annual production, N content, and N usage of organic wastes in the United States (USDA, 1978).

Organic waste	Annual production	Total N content	N currently used
		$t \times 10^3$	
Animal manure	156 000	6 875	4 980
Crop residues	385 000	4 287	3 056
Sewage sludge	3 900	167	39
Food processing	2 860	29	5
Industrial organics	7 336	no data	no data
Logging and milling	31 888	negligible	negligible
Municipal refuse	129 500	725	0
Total	716 484	12 083	8 080
Commercial fertilizer (1978)			9 052

can be applied to meet crop production needs, but in other areas, other sources of N would be required.

The availability of N in animal manures and wastes varies widely, depending on a number of factors. These include type and age of livestock, presence or absence of bedding material, method of manure handling and storage, degree of soil incorporation, soil temperature and water content, and possibly other factors. Because of variation in composition, N availability in fresh manures may range from 1% to well over 50%. In composts or well-decomposed corral manures, variability is usually less, and availability values of 10 to 25% have frequently been reported (USDA, 1978, 1979).

Numerous research projects have been conducted to study the effects of animal manures on soil properties and crop production. If applied in sufficient quantity over a period of several years, manures usually result in increased organic C and N content, increased soil porosity, reduced bulk density, increased mineralizable N, and enhanced microbial activity (Baldock & Musgrave, 1980; Halvorson & Hartman, 1975; Martyniuk & Wagner, 1978). Over 100 yr of manure application to soils at the Rothamsted Experiment Station has resulted in restoring soil organic-matter levels to near that of perennial grass sod (Warren, 1956). Animal manures are commonly used on eroded soils or on areas where topsoil has been removed, as in leveling for gravity irrigation (Carlson et al., 1961).

B. Plant Sources

A very important source of plant N for crop production is biologically fixed N. Most research has concluded that almost all biologically fixed N in American agriculture is derived from N fixation by *Rhizobium* sp. growing in symbiotic relationship with legume crops. The quantity of N fixed annually varies widely, depending on crop, management, water availability and temperature, and a number of other factors (Heichel & Barnes, 1984). Although estimates of N fixation by an alfalfa crop may be as great as 600 kg N/ha, most textbook estimates of average annual fixation are about 150 to 200 kg N/ha (Table 4). With about 12 million ha of alfalfa and an esti-

Table 4—Rates of biological N fixation for various legumes (adapted from Evans & Barber, 1977).

Legume crop†	N_2 fixed
	kg/ha per year (pounds/acre per year)
Soybean	57–94 (51–84)
Cowpeas	84 (75)
Clover	104–160 (93–143)
Alfalfa	128–600 (114–535)
Lupins	150–169 (134–151)

† Soybean, *Glycine max* (L.) Merr.; cowpeas, *Vigna sinensis* L.; clover, *Trifolium* sp.; alfalfa, *Medicago sativa*; L.; lupins, *Lupinus* sp.

mated 5 million ha of other forage legumes and with 150 to 200 kg N fixed per ha, average annual N fixation in the United States would be about 2.6 to 3.4 million tonnes. In addition, estimating N fixation by soybeans and other cultivated legume seed crops at 60 kg N/ha and with an average of 33 million ha, average annual N fixation by these crops would be about 2 million tonnes. Thus, N fixation by legume crops could equal approximately 5 million tonnes of N annually.

In organic farming systems, perennial legumes commonly occupy 25 to 50% of the area cultivated (USDA, 1980b) and provide a major part of the N required for the crops produced (Voss & Shrader, 1984). In some midwestern organic farming systems, the cropping sequence used is often similar to a corn–soybeans–corn–oats–alfalfa–alfalfa–alfalfa rotation. If 150 kg N/ha is fixed each year by alfalfa and 60 kg N/ha by soybeans, in the 7 yr, N addition by fixation would be 510 kg N/ha. This would provide an average of about 130 kg N/ha for each of the 4 yr of the rotation during which seed crops are produced. Little information has been published concerning the availability of the organic N added to a soil by growth of perennial legumes. About 15% of the ^{15}N incorporated in medic clover residues in three South Australian soils was taken up by the following wheat crop, and about 75% of the labeled N was found in soil organic matter (Ladd, 1981). In New York, Baldock and Musgrave (1980) concluded that alfalfa in a grain-legume rotation fixed 135 kg N/ha per year and that manures provided 5 kg N/t (wet wt).

An important source of organic N is the crop residues remaining after harvest. Crop residues are estimated to be 385 million tonnes annually in the United States. Total N content of crop residues is estimated at slightly over 4 million tonnes, about 75% of which (3 million tonnes) is normally returned to the soil each year. About 75% of these residues come from three crops: corn, wheat, and soybeans. There are sufficient crop residues to provide an average of about 20 kg N/cultivated ha. Again, the quantity of N returned in aboveground residues varies widely, from near zero (silage, fodder crops, etc.) to about 150 kg N/ha in residues remaining after harvest of an excellent corn crop. Information on the availability of N returned in crop residues for future plant growth is variable because of differences in crop, soil, and climatic factors. However, representative data for corn and wheat residues suggest that 5 to 20% of the N in these residues is mineralized and taken up by the next crop (Broadbent & Nakashima, 1974; Paul & McGill, 1977).

A number of research results have been published showing that returning crop residues to the soil aids in maintaining soil organic matter contents and improved soil physical conditions (Barnhardt et al., 1978; Black, 1973).

Because of the shallow tillage commonly used in organic farming, crop residues are left on or near the surface and thereby improve the soil N status by building up the soil organic N pools and by reducing N loss by erosion. Presence of crop residues on the soil surface also reduces evaporation rates, increases water storage and subsequent plant growth, and reduces wind and water erosion (Barnhardt et al., 1978; Black, 1973; Cochran et al., 1980).

C. Indigenous Soil Organic N

In a stabilized environment, the organic matter content of soil is fairly constant. Equilibrium is reached when C losses, chiefly as CO_2, are balanced by the gain of C through additions of plant residue or other organic residues. The ultimate source of replacement is the photosynthetic fixation of atmospheric CO_2 by crop plants. Much of the N released during organic matter oxidation is conserved in the cells of soil microorganisms or taken up by crop plants. To maintain or increase soil organic matter levels, the N removed as harvested crop products or lost through leaching or volatilization must be replaced using external sources of N. In organic farming, this is accomplished by adding organic residues or through fixation of atmospheric N_2 by legumes.

A major goal of organic farming practices is increasing soil organic matter levels. The increased organic matter improves the physical, chemical, and biological environment of soil and serves as an internal source of N for plant growth. Organic N levels between soils range from 2000 to 10 000 kg/ha. Under ideal conditions, the yearly mineralization potential of temperate-region soils is usually between 1 and 3% of the total organic N content; however, the actual amount of N mineralized during the growing season depends on microbial activity, which is influenced by the availability of water and oxygen, soil temperature, and the accessibility of organic matter. Thus, within a climatic region, the release of N from soil organic matter is directly related to agricultural management practices that influence soil aeration, residue placement, and the amount of water retained in soil.

III. CONSERVATION AND AVAILABILITY OF N

Through selection of management practices (cropping sequences, tillage practices, fertilizer practices, irrigation, etc.), the farmer controls the nature of the soil microenvironment at soil microsites (Fig. 1). Microenvironmental conditions may affect plant root development and activity directly, but they also determine the nature and size of the microbial populations that develop. The unique characteristics of the microbial community influence the composition and concentration of substrates available for utilization from the soil by growing plants. The residues left after harvest of the plant are returned to the soil, which, combined with the soil and crop management practices used on the next crop, set the stage for an annual continuation of this cycle. From this concept, it is evident that the management decisions applied during crop production determine to a large extent the quantity and timing of nutrients—especially N—that are available to the growing crop.

As mentioned earlier, organic farming systems have several essential features. First, they depend upon legumes and manures for the addition of N to the N cycle. Second, they commonly employ shallow tillage, leaving

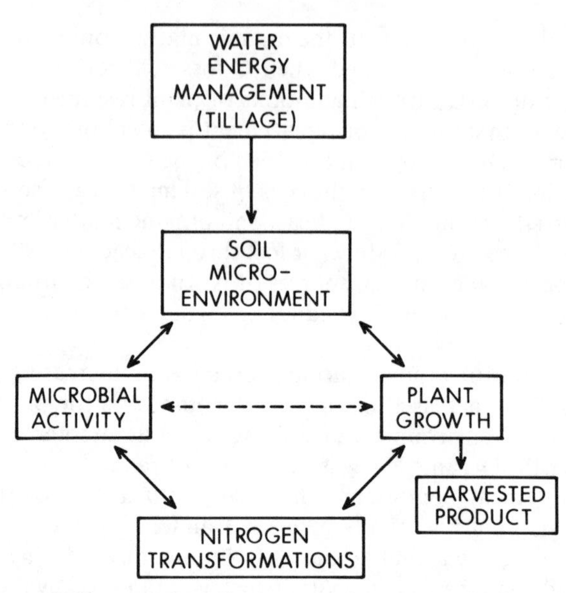

Fig. 1—Influence of tillage on the soil microenvironment as related to microbial activities, N transformation, and plant growth.

crop residues near the surface. The use of shallow tillage, along with surface additions of manures or composts, green manures, and crop residues, creates a relatively large storehouse of nutrients in the organic debris near the soil surface. This accumulation of organic residues near the soil surface reduces evaporation losses from the soil, thereby increasing surface soil water content and reducing soil temperatures. Thus, compared with more conventional methods, the organic farming techniques used create a completely different microenvironment in the surface soil, altering soil water status, soil temperature, and placement of soluble N and C sources needed for microbial activity.

Agricultural management practices that conserve crop residues and organic matter also increase the numbers of microorganisms in surface soil. Doran (1978), in a nationwide study, found that by maintaining crop residues on the soil surface with no tillage (compared with plowing), the numbers of all groups of soil microorganisms in the upper 7.5 cm are greatly increased. It appears that the presence of surface residues reduces evaporation rates, thereby increasing the water content of the surface soil. Also, with no tillage (compared with plowing with extensive secondary tillage), the oxidative loss of organic matter is lessened, resulting in higher steady-state organic matter levels in the surface of no-tillage soil. As a result of the increased organic matter levels, soil water contents, and microbial populations, a greater amount of N in surface soils of no tillage is in a potentially mineralizable form (Table 5). The magnitude of these differences

Table 5—Effect of tillage practices on changes in water and organic matter contents, levels of microbial biomass, and potentially mineralizable N (after Doran, 1978, 1981).

Soil depth	Relative difference†			
	Water content	Organic matter	Microbial biomass	Potentially mineralizable N
cm	%			
0 - 7.5	+28	+27	+32	+37
7.5–15.0	+ 1	− 4	− 5	− 2

† Calculated by [(No tillage/Plow) − 1] × 100.

in microbial biomass, water content, and organic matter levels depends to a large extent upon the degree to which surface residues and soil are disturbed by tillage (Table 6).

As a consequence of these effects, one would expect that organic farming techniques of reducing tillage and adding plant and animal residues would build up soil organic matter levels and decrease the quantity of readily available N; however, the quantity of potentially mineralizable N should increase. Higher microbial biomass levels would increase the highly labile pool of soil N, a pool that is subject to mineralization as the soil microenvironment changes during the year. The increased quantity of microbial by-products (cell walls, etc.) and partially humified materials added to soil would also increase the pool of very slowly labile N.

Table 6—Effect of degree of soil disturbance on the levels of microbial biomass, water content, and organic matter of surface soil (0–7.5 cm) (J. W. Doran, unpublished data).

Management practice	Degree of disturbance	Microbial biomass	Water content (volumetric)	Organic matter
		kg C/ha	%	
Sod control	None	1070	17.7	4.49
No-tillage	Minimum	884	14.3	3.80
Subtillage	Moderate	828	12.2	3.28
Plow	Maximum	657	10.6	2.42

Table 7—Efficiency of recovery of soil and applied N for several crops after 50 yr of management (after Smith, 1942).

Fertility management practice	Percent N recovery†
Inorganic fertilizer	46–58
Manure (13.4 t/ha per year)	52–87
Crop rotations (3, 4, and 6 yr)	84–103 (57–76)‡
6-yr Rotation + manure and inorganic fertilizer	66 (57)
Continuous timothy§	101

† (N removed in crops + soil N after 50 yr ÷ soil N at start + N added) × 100.
‡ Values in parentheses represent N recovery using an estimate of 112 kg N/ha per year fixed by clover crop in rotation.
§ Timothy, *Phleum pratense* L.

Over short periods of time during the cropping season, the availability of N with organic farming practices is probably lower than with conventional systems using inorganic fertilizers. This results predominantly from the increased capacity for storage of N in moderately and slowly available pools, as discussed earlier. However, since N is more slowly released to crop plants with organic farming, the efficiency of N utilization should be greater. A 50-yr study was conducted at Sanborn Field in Missouri involving continuous cropping, rotations, manure, and inorganic fertilizers (Smith, 1942). Results of this study indicated that efficiencies of N recovery were greater when manure or crop rotations were used to supply N to crop plants than where inorganic fertilizer was used (Table 7). Smith also found that a combination of manure and inorganic fertilizer resulted in higher yields and better N utilization than either souce applied alone or in combination. The reduction of manure application rate from 13.4 to 6.7 t/ha per year on continuous wheat increased the recovery of N by 15% but did not influence yield appreciably.

IV. ORGANIC FARMING AND SOIL PRODUCTIVITY

Reliable data on soil productivity and biology from experiments making valid comparisons between organic and conventional farming methods are limited; however, representative information from some studies follows.

A. Crop Yields

Several field experiments have been conducted to compare crop production for organic farming vs. conventional farming (Lockeretz et al., 1984; Voss & Shrader, 1984). Almost universally, those converting to organic farming report that about 3 to 5 yr are required for production problems (weeds, insects, tillage requirements, etc.) to stabilize under organic farming, so yields during this conversion period are often less than those later achieved. Also, the local soil and climate have a major effect upon relative crop yields. For example, the deep-rooted legumes, such as alfalfa and sweet clover, that provide N input to organic farming deplete soil water reserves to several meters in depth. In semiarid climates, this depletion of soil water overrides any beneficial effects of enhanced soil N levels resulting from legume culture. Subsequent crop growth, consequently, is reduced because of the greater water deficit (Haas et al., 1976). In more humid regions or under irrigation, this limiting effect of water is less evident.

With these factors in mind, it does not seem appropriate to attempt any comprehensive review of the literature on comparative crop yields. However, data in Table 8, taken from the USDA organic farming report (USDA, 1980b), illustrate the type of results commonly reported in midwestern United States. Data in this table suggest that corn yields are usually about 10% greater for conventional farming than for organic farming, while oats and hay yields are often slightly higher for organic farming. Ef-

Table 8—Yield of several crops produced on selected farms in several midwestern states by organic and conventional farming methods (USDA, 1980b).

Year	Crop†	Organic farms	County average	Conventional farms	County average
		kg/ha			
1974	Corn	4 641	4 696	4 445	4 570
	Soybean	2 150	1 678	1 879	1 678
	Wheat	1 879	2 079	1 946	1 946
	Oat	2 111	1 968	2 111	2 111
	Hay	11 200	--	7 616	--
1975	Corn	4 641	5 633	5 889	5 510
	Soybean	2 280	2 012	2 549	2 079
	Wheat	1 744	2 549	2 750	2 416
	Oat	2 005	--	2 039	--
	Hay	10 080	--	8 960	--
1974–1976	Corn	4 809	5 058	5 176	4 871
	Soybean	2 020	1 811	2 148	1 811
	Wheat	1 953	2 280	2 549	2 280
	Oat	2 177	1 968	2 216	2 076
1977	Corn	4 878	5 284	--	--
	Soybean	2 273	2 253	--	--
	Oat	2 369	2 143	--	--
1978	Corn	6 172	7 387	--	--
	Soybean	2 381	2 563	--	--
	Oat	2 430	2 283	--	--

† Corn, *Zea mays* L.; soybean, *Glycine max* (L.) Merr.; wheat, *Triticum aestivum* L.; oat, *Avena sativa* L.

fects of farming method on wheat and soybean yields are more variable. Unpublished data from eastern Nebraska (W. W. Sahs, Univ. of Nebraska, Lincoln, personal communication) show similar trends as those illustrated by data in Table 8.

The discussion in the preceding section of this chapter has indicated that, compared with conventional farming, organic farming techniques tend to increase total organic matter and mineralizable N. With organic farming, seldom would an appreciable pool of inorganic N exist in the soil. Consequently, crops may be under slight N stress during periods of rapid N uptake. It has been suggested that greatest crop production may occur when moderate rates of N fertilization are incorporated into farming systems that include legumes and manures (Baldock & Musgrave, 1980; Brinton, 1979; Spratt & McIver, 1979). Such a practice may ensure adequate N at all times without unduly increasing production costs or potential for environmental degradation.

B. Soil Biology

The addition of crop residues, animal manures, sewage sludges, and other organic residues to soil often results in increased organic matter levels in soil and a general improvement of the physical, chemical, and biological state of soil. Martyniuk and Wagner (1978) found microbial populations in

Table 9—Influence of fertilizer sources on surface soil organic C content and biological activities for a wheat/clover-grass/potato/beet rotation (adapted from Brinton, 1979).†

Fertility treatment	Total N applied	Soil organic C	Relative activity‡		
			Earth-worms	Soil respiration	Soil de-hydrogenase
	kg N/ha per year	%			
Control	0	2.3	1.0	1.0	1.0
Raw manure	93	2.6	5.0	1.3	1.8
½ Manure + ½ NPK	61	2.1	2.4	1.1	1.4
NPK (low)	28	2.5	1.7	1.0	1.1
NPK (high)	111	2.7	0.7	1.0	1.2

† Potato, *Solanum tuberosum* L.; beet, *Beta vulgaris* L.
‡ Activity relative to control (no treatment). N added as calcium nitrate or ammonium phosphate, P as superphosphate, K as potassium sulfate.

soil management plots at Sanborn Field, Mo. were closely related to cultivation practices, changes in soil moisture content, and temperature. Microbial populations were lowest in untreated soil, intermediate in soils treated with chemical fertilizer, and highest in soils receiving annual applications of manure. Population differences for various cropping and soil-management practices were primarily affected by the quantities of plant residue produced and the amounts of soil organic matter accumulated.

Increases in soil biological activity with organic-management techniques can also result from factors unrelated to soil organic matter content. The influence of organic vs. inorganic fertilizers on biological activity was evaluated in an 18-yr study for a wheat/clover-grass/potato/beet rotation at Jarna, Sweden (Brinton, 1979). Inorganic fertilizers were added to the soil at rates comparable to the available N supplied by manure treatments, and tillage was identical for all treatments. As shown in Table 9, organic C levels for soils treated with manure or inorganic fertilizer were similar (2.5 to 2.7%), but higher than that of the control soil (2.3%). Earthworm (*Lumbricus* sp.) activity was greatly enhanced by the addition of manure and moderately stimulated by the addition of inorganic fertilizer at the low rate that supplied amounts of available N comparable to that for manure. Soil respiration and dehydrogenase activities were also stimulated in soil receiving manure. The use of inorganic fertilizers had little or no effect on soil respiration rates or dehydrogenase activities as compared with those of the control soil. The increases in soil biological activity between fertility treatments in this study were apparently related to factors other than soil organic matter content.

V. CONCLUSIONS AND FUTURE RESEARCH NEEDS

Because of diversity in techniques, climate, soil, management practices, cropping systems, and other factors, comparisons of the N cycle for organic vs. conventional farming techniques are often difficult to make. A few generalized conclusions, however, are as follows:

1) Organic farming techniques tend to conserve N in the soil-plant system, often resulting in a buildup in soil organic N.
2) Organic farming practices often result in a greater number of soil organisms and enhanced levels of potentially mineralizable soil N.
3) Because of several factors, the net rate of mineralization of N in organically farmed soils is often slower than for conventional farming, possibly resulting in mild N stress during periods of rapid N uptake.
4) Presence of organic residues (and possibly other factors) in organic farming systems aids in reducing N losses from the soil-plant ecosystem.
5) The effects of organic farming on the N cycle are most pronounced in the surface soil.

Research information on the effects of organic farming on the various N transformations that occur in diverse soils, climates, parent materials, and cropping systems is extremely limited. Well-controlled and well-conducted experiments are severely needed to understand some of the basic effects of organic farming on soil physical, chemical, and biological properties. Reliable information on the availability of N in animal manures, green manures, biologically fixed N, composts, organic wastes, and other organic N sources is presently limited. Also, we need to know how the addition of these materials affects availability of soil and fertilizer N.

Such studies as indicated above need to be sufficiently comprehensive to provide a much broader spectrum of information than just plant growth or N cycling. Data are needed on the effects of organic management practices on microbial populations and activity, soil physical conditions, soil water and temperature regimes, plant physiological and phenological response, and other related parameters.

REFERENCES

Asmussen, L. E., J. M. Sheridan, and C. V. Booram, Jr. 1979. Nutrient movement in streamflow from agricultural watersheds in the Georgia Coastal Plain. Trans. ASAE 22:809–815.

Baldock, J. O., and R. B. Musgrave. 1980. Manure and mineral fertilizer effects in continuous and rotational crop sequences in central New York. Agron. J. 72:511–518.

Barnhardt, S. L., W. D. Shrader, and J. R. Webb. 1978. Comparison of soil properties under continuous corn grain and silage cropping systems. Agron. J. 70:835–837.

Black, A. L. 1973. Soil properties associated with crop residue management in a wheat-fallow rotation. Soil Sci. Soc. Am. Proc. 37:943–946.

Brinton, W. F., Jr. 1979. Effects of organic and inorganic fertilizers on soils and crops. Results of a long-term field experiment in Sweden. (Translated from Swedish and German of the original work by B. D. Pettersson and E. V. Wistinghaussen.) Misc. Pub. no. 1. Woods End Agric. Inst., Temple, Me.

Broadbent, F. E., and T. Nakashima. 1974. Mineralization of carbon and nitrogen in a soil amended with carbon-13 and nitrogen-15 labeled plant material. Soil Sci. Soc. Am. Proc. 38:313–315.

Carlson, C. W., D. L. Grunes, J. Alessi, and G. A. Reichman. 1961. Corn growth on Gardena surface and subsoil as affected by application of fertilizer and manure. Soil Sci. Soc. Am. Proc. 25:44–47.

Cochran, V. L., L. F. Elliott, and R. I. Papendick. 1980. Carbon and nitrogen movement from surface-applied wheat (*Triticum aestivum*) straw. Soil Sci. Soc. Am. J. 44:978–982.

Doran, J. W. 1978. Soil microbial and biochemical changes associated with reduced tillage. Soil Sci. Soc. Am. J. 44:765-771.

Doran, J. W. 1981. Soil biomass as regulated by tillage practices. *In* Abstracts, Annu. Meeting of the Am. Soc. of Microbiol., 1-6 Mar. 1981. Dallas, Tex. Am. Soc. of Microbiol., Washington, D.C. p. 185.

Evans, H. J., and L. E. Barber. 1977. Biological nitrogen fixation for food and fiber production. Science 197:332-339.

Haas, J. J., J. F. Power, and G. A. Reichman. 1976. Effect of crops and fertilizer on soil nitrogen, carbon, and water content, and on succeeding wheat yields and quality. ARS-NC-38. ARS-USDA, Peoria, Ill.

Halvorson, A. D., and G. P. Hartman. 1975. Long-term nitrogen rates and sources influence sugarbeet yield and quality. Agron. J. 67:389-393.

Hauck, R. D., and J. M. Bremner. 1976. Use of tracers for soil and fertilizer nitrogen research. Adv. Agron. 28:219-266.

Heichel, G. H., and D. K. Barnes. 1984. Opportunities for meeting crop nitrogen needs from symbiotic nitrogen fixation. p. 49-59. *In* D. F. Bezdicek (ed.) Organic farming. Spec. Pub. no. 46. Am. Soc. of Agron., Madison, Wis.

Ladd, J. N. 1981. The use of ^{15}N in following organic matter turnover, with specific reference to rotation systems. Plant Soil 58:401-411.

Lockeretz, W., G. Shearer, D. H. Kohl, and R. W. Klepper. 1984. Comparison of organic and conventional farming in the Corn Belt. p. 37-48. *In* D. F. Bezdicek (ed.) Organic farming. Spec. Pub. no. 46. Am. Soc. of Agron., Madison, Wis.

Martyniuk, S., and G. M. Wagner. 1978. Quantitative and qualitative examinations of soil microflora associated with different management systems. Soil Sci. 125:343-350.

Paul, E. A., and W. B. McGill. 1977. Turnover of microbial biomass, plant residues, and soil humic constituents under field conditions. p. 149-157. *In* Soil organic matter studies. Vol. I. Int. Atomic Energy Agency, Vienna.

Smith, G. E. 1942. Sanborn Field: Fifty years of field experiments with crop rotations, manure, and fertilizers. Missouri Agric. Exp. Stn. Bull. 458.

Spratt, E. D., and R. N. McIver. 1979. The effect of continual use of phosphate fertilizer and barnyard manure on the yield of wheat and the fertility status of a clay chernozem soil. Can. J. Soil Sci. 59:451-454.

U.S. Department of Agriculture. 1978. Improving soils with organic wastes. USDA Rep. no. PL 95-113. USDA, Washington, D.C.

U.S. Department of Agriculture. 1979. Animal waste utilization on cropland and pastureland. USDA Utilization Res. Rep. no. 6. USDA, Washington, D.C.

U.S. Department of Agriculture. 1980a. Agricultural statistics. USDA, Washington, D.C.

U.S. Department of Agriculture. 1980b. Report and recommendations on organic farming. USDA, Washington, D.C.

Viets, F. G. 1975. The environmental impact of fertilizers. p. 423-453. *In* CRC critical reviews in environmental control. Chemical Rubber Co., Cleveland, Ohio.

Voss, R. D., and W. D. Shrader. 1984. Rotation effects and legume sources of nitrogen for corn. p. 61-68. *In* D. F. Bezdicek (ed.) Organic farming. Spec. Pub. no. 46. Am. Soc. of Agron., Madison, Wis.

Warren, R. G. 1956. N, P, and K residues from fertilizers and farmyard manure in long-term experiments at Rothamsted. Proc. Fert. Soc. 37:1-10.

Nitrogen Management and Quality of Crop and Environment

section VI

41

E. L. Deckard
North Dakota State University
Fargo, North Dakota

C. Y. Tsai
Purdue University
West Lafayette, Indiana

T. C. Tucker
University of Arizona
Tucson, Arizona

Effect of Nitrogen Nutrition on Quality of Agronomic Crops

Fertilizer N has been, and still is, used increasingly to supplement soil N for producing the needed quantity of food, feed, and fiber for an increasing world population. Unfortunately, this use of supplemental N is not always associated with improved quality. Therefore, decisions to sacrifice quantity or quality are sometimes made. Quantity is easily measured and understood, but quality is an ambiguous term, meaning different things to different segments of the population. In addition, the effect of N nutrition on quality and the relationship between quantity and quality are strongly influenced by other factors such as genetics and the environment. Therefore, it is difficult to discuss briefly the effect of N nutrition on crop quality where many cultivars over a wide geographical area are involved. Nevertheless, there are common ties, and the intent of this chapter is to address these common areas regarding the effect of N nutrition–induced yield increases on the quality of agronomic crops.

I. CEREALS

High-quality protein from the cereal crops has been, and will continue to be, a major food shortage as the cereals provide a major portion of the

Nitrogen in Crop Production.

calories and protein in human diets. It has long been recognized that N fertilization increases grain protein percentage, especially when soil N levels are increased above that required for maximum grain yield. The effect of N on grain protein percentage also is dependent upon the crop cultivar and other environmental variables (especially available water). Smika and Greb (1973) showed that soil water and soil NO_3^--N at seeding predicted grain protein percentage of winter wheat (*Triticum aestivum* L.) in the Great Plains of the United States with considerable accuracy. Soil NO_3^--N content was positively associated with grain protein percentage, whereas the opposite relationship was noted between grain protein percentage and available soil water. Therefore, in addition to the genetic potential for high grain protein, adequate soil moisture and fertility are necessary for high yields of high-protein grain.

Cereal endosperm proteins may be fractionated into four fractions according to their solubilities in different solvents. The four fractions are albumin, globulin, prolamin, and glutelin. The amino acid composition and chemistry of these protein fractions have been reviewed for several cereals (Mosse, 1966; Rhodes & Jenkins, 1976; Wall, 1979; Wall & Paulis, 1978). Briefly summarized, the proportions of the four protein fractions are dependent upon the specific cereal, cultivar, and production environment. For example, the prolamin fraction may account for 50 to 60% of the total protein as in corn (*Zea mays* L.) and sorghum [*Sorghum bicolor* (L.) Moench], or only 10 to 15% as in oats (*Avena sativa* L.). Wheat and barley (*Hordeum vulgare* L.) prolamins account for 35 to 45% of the total protein. Prolamins in sorghum, maize, wheat, barley, and oats are designated as kafirin, zein, gliadin, hordeim, and avenin, respectively. The glutelin fraction in wheat also is called glutenin.

Each of the four protein fractions has a different amino acid composition. Prolamins have low levels of several essential amino acids, especially lysine, and high levels of others such as glutamine and proline. Albumins and globulins have considerably less proline and amide-N and more cysteine, glycine, and basic amino acids. The amino acid composition of glutelins generally is intermediate between the prolamins and alubumin-globulins. These are trends in composition, since each fraction contains many different proteins, each having a characteristic amino acid composition.

The four protein fractions are not synthesized simultaneously during kernel development. The albumins and globulins probably represent the enzymes found in the soluble cytoplasm and, therefore, are synthesized early in kernel development. The proportion of these proteins declines as the kernel matures. Glutelins are synthesized at a more uniform rate throughout development, and at least some of these proteins probably represent structural, membrane, or bound proteins. A portion of the glutelins also may be storage proteins. The prolamins, although synthesized throughout development, generally increase as a proportion of total grain protein during the late stages of kernel development. The prolamins are thought to be the storage proteins deposited in protein bodies.

A. Wheat, Barley, and Oats

1. PROTEINS AND NUTRITIONAL QUALITY

The specific proteins that make up the four protein fractions are genetically determined. However, the proportions of these specific proteins or the proportions of the four fractions may be affected by environmental factors and soil fertility, especially by the level of available N. When N fertilizer increases the grain protein levels in oats, the globulin fraction increases more than the remaining fractions (Peterson, 1976). This might be predicted, since the major (75–80%) protein fraction in oats is globulin. However, for wheat and barley, the major fraction increasing as a result of N fertilization is generally prolamin. Frequently, this predominant increase in the prolamin fraction results in a decrease in the proportion of albumins and globulins and only a small decrease in the proportion of the glutelin fraction. These changes in the protein fractions are undoubtedly not general changes in the many proteins making up that fraction. For example, a differential response of the barley hordein components (Koie et al., 1976) and of the wheat glutenin components (Dubetz et al., 1979) to N supply has been suggested.

The amino acid balance of oats is not substantially altered by N fertilization, since the globulins are the major protein fraction and have an amino acid balance similar to that of the total protein (Peterson, 1976). However, the major protein fraction increased by N fertilization of wheat and barley is the prolamin fraction, and this fraction contains low levels of essential amino acids (especially lysine). Therefore, increasing the protein concentration by N fertilization frequently decreases the general nutritional quality of the protein. That is, the lysine concentration (percent of protein) declines with increasing protein percentage, even though the concentration of lysine expressed as a percentage of dry weight increases. The negative relationship between the percentage of lysine in the protein and the percentage of protein in the kernel is curvilinear (at least among wheat varieties), since the reduction in lysine percentage is proportionally smaller as the grain protein percentage increases (Johnson et al., 1975). This curvilinear response would suggest that the prolamin fraction is not present in proportionally greater amounts at high protein percentages when protein level is a function of genotype.

Not all experiments have shown a negative relationship between lysine percentage and protein percentage when N fertilizer is varied (Hucklesby et al., 1971). These experiments have involved relatively high protein percentages, so the lack of a negative relationship probably can be explained by the curvilinear response. These experiments also would suggest that the prolamin fraction is not present in proportionally greater amounts at high protein percentages. Therefore, most results indicate that changes in amino acid composition of wheat proteins as a result of increasing soil N supply are probably of minor nutritional significance except when compared with low kernel protein percentages. The effect of increasing seed protein on lysine percentage also may be influenced by timing of fertilizer application or other management and environmental factors.

2. USE OR INDUSTRIAL QUALITY

As indicated, the protein concentration and the specific proteins present in the kernel affect the nutritional quality. The proteins also affect the industrial quality for wheat and barley. Although most wheat is used for human food, different foods are made from different wheat classes (classes of U.S. Grain Standards Act). However, most wheat is milled into flour and made into a form of bread. The quality of these wheats has been discussed by Finney and Yamazaki (1967) and can refer to seed, milling, or baking quality factors. Therefore, no single factor measures general quality, but the kernel proteins are especially important in determining bread-making quality. Total protein percentage often is used to estimate bread-making quality, but the components of the protein are also important. Gliadins, glutelins, and residue proteins (both amount and proportion) are thought to be the key proteins for bread making. Gliadins have been associated with loaf volume and glutenins with mixing requirement (Finney, 1971). The relationship between the various protein fractions and bread making recently has been reviewed (Wall, 1979).

The fact that cultivars of similar protein percentage can produce flours with different bread-making potential shows that total protein alone does not determine bread-making quality. Also, durum wheat is high in protein but generally does not make a high-quality loaf of bread.

Pushman and Bingham (1976) reported that increased grain protein percentage resulting from N applied post-heading frequently is not associated with better baking quality. Certain studies have suggested that this late-applied N (soil or foliage) can reduce baking quality (Finney et al., 1957; McNeal et al., 1963). These workers suggested that the reduced baking quality is associated with an increased proportion of the albumin and globulin proteins.

There is no evidence that N fertilization prior to heading can change the proportions of the protein fractions (i.e., excessive gliadin) to the extent that baking quality is altered. Although N fertilizer apparently increases the gliadin fraction preferentially, the ratio between the gliadin and glutenin fraction evidently remains in balance. Baking quality, within a cultivar, generally is thought to increase in direct proportion to protein percentage. The major difference between high-quality hard red spring and hard red winter wheats with respect to bread making is that the spring wheats have a higher protein percentage. Bread making quality can be increased by utilizing cultural practices such as proper and efficient use of N fertilizer to increase the protein level of a high-quality cultivar. However, the economic return for the increased quality must, of course, be sufficient to justify the extra production cost.

Breeding higher protein cultivars is a second method that is just as effective (for similar protein increases), if production potential and the many other quality factors can be combined. However, as previously noted, these higher protein cultivars will require relatively high levels of plant-available soil N for the high grain protein potential to be realized. Durum wheat is a specialty wheat, the majority of which is milled into semolina (granular flour) and made into pasta products. Different factors are important for quality in durum than in bread wheats, but again, the kernel proteins are important in determining pasta-making quality (Grzybowski & Donnelly, 1979). High protein percentages are associated with low cooking losses (loss of solids during cooking), good cooked firmness, and good cooking stability. In addition to protein percentage, the types of protein also are important (Grzybowski & Donnelly, 1979). A high gliadin content is associated with poor cooked firmness and high cooking losses. There are no reports indicating that the high kernel protein resulting from high levels of soil N results in lowered cooking quality.

Although achieving a high protein percentage is the general problem for most cereals, malting barley must be grown under cultural practices that provide a low kernel protein percentage. Therefore, soil N must be maintained at a level providing near-maximum grain yields without exceeding acceptable protein percentages. Peterson and Foster (1973) have reviewed the production practices and quality aspects of malting barley in great detail. Although many factors are used to evaluate malting quality, protein percentage is very important. The maximum acceptable protein percentage of midwestern six-rowed malting barley is 13.5% on a dry-weight basis, but a lower percentage is preferred. Although protein is necessary for malt production, excessive amounts have undesirable effects on the malting and brewing processes. The undesirable effects include an increased steeping time, uneven germination, increased malting losses, and increased soluble protein in the brewing product.

3. QUANTITY VS. QUALITY

The increased grain yields resulting from the application of N fertilizer to N-deficient soils and the increased protein percentages resulting from excessive N fertilizer applications have long been recognized. The kernel protein percentage increases most rapidly in response to N supply after the yield response levels off. However, the response of yield and protein to N supply is strongly influenced by environmental conditions, especially the quantity and timing of water available to the crop. Under conditions of low water availability, the positive yield response to added N levels off at low soil N levels, and the increased protein response is shown more readily. Therefore, the relationship between yield and protein percentage across soil N levels can vary from negative to positive, depending on environmental conditions. Grain yield and protein percentage are relatively unrelated at (i) soil N levels that increase yield but have little effect on protein and (ii) at soil N levels in excess of those showing a yield response, but which still show a protein response. Positive relationships occur when both yield and protein percentage are increasing. Negative relationships are less common and generally result from increasing low soil N levels at high available water levels or when high soil N levels result in lodging of the crop.

The relationship between grain yield and protein percentage among cultivars at a fixed environment frequently is negative, although the relationship between grain yield and protein yield is seldom negative and often highly positive. Therefore, protein production can be increased by increasing grain yields, but protein percentage is the character that is a major quality factor in nutrition, bread making, malting, and brewing. Protein percentage is dependent upon two components: the weight of protein and the remaining kernel weight. Therefore, protein percentage can vary without altering the amount of protein, which tends to make this character a difficult variable to research. For example, protein percentage could increase, even though protein per kernel or per unit of land area decreases.

Nevertheless, attempts have been made to explain the apparent negative relationship between grain yield and grain protein. One possible explanation is the report that more photosynthate is required to produce protein than carbohydrate (Penning de Vries et al., 1974). Therefore, given equal amounts of photosynthate, a high protein–low carbohydrate cultivar was predicted to produce less grain yield than a low protein–high carbohydrate cultivar (Bhatia & Robson, 1976). However, the calculations for photosynthate cost to produce these two components are based on dark metabolism, whereas much of the energy for N metabolism is produced by the light reactions. Also, the use of photosynthate in N metabolism could result in increased fixation of CO_2. Therefore, determining the energy cost of producing seed protein vs. seed carbohydrate is difficult. Similar calculations have shown that producing high-quality protein (increased content of essential amino acids) requires more photosynthate than producing low-quality protein (Mitra et al., 1979). These calculations would appear to be more valid.

Another possible explanation for the negative relationship between grain yield and grain protein percentage is provided by the presently used breeding methods for improving crop cultivars. These methods have resulted in the inadvertent improvement of harvest index without the improvement of biomass (Kramer, 1978). More than two-thirds of the grain protein, but only about one-half of the grain dry weight, is in the plant at anthesis. Therefore, Kramer (1978) has proposed that altering the plant environmentally or genetically to increase or decrease harvest index is expected to have the opposite effect on grain protein percentage. For example, N fertilizer generally decreases harvest index and increases protein percentage. This hypothesis could be negated by differences in the protein percentage of the vegetation or differences in the reduced N harvest index. The assumption that these differences do not exist is not always valid (Deckard et al., 1977; Deckard & Busch, 1978). Therefore, no obvious biological reason appears to support a mandatory negative relationship between grain yield and grain protein percentage.

Recent reports suggesting that grain yield and grain protein percentage are potentially compatible were reviewed recently by Frey (1977). Briefly, these include wheat research (Johnson et al., 1975) that shows increased protein percentage without a corresponding yield loss, and oat research

(Frey, 1977) that shows increased yield without a corresponding decrease in protein percentage. Therefore, proper cultural practices and cultivar improvement programs aimed at improving both yield and protein percentage should allow increased quantity of high-quality crops.

B. Corn and Sorghum

To date, the most successful approach to improving the relative nutritional value of cereal proteins, other than by direct supplementation, has been through the use of high-lysine mutants that increase endosperm lysine and tryptophan levels by reducing prolamin content. Breeding for improved nutritional quality of cereal grains has been reviewed recently by Axtell (1982). These high-lysine mutants in maize (Glover, 1976; Lambert et al., 1969; Nass & Crane, 1970; Sreeramulu & Bauman, 1970), sorghum (Axtell, 1982), and barley (Doll & Koie, 1975) have reduced seed size and grain yield. Studies with maize indicate that zein is the major factor influencing kernel weight, grain yield, and protein nutritional quality (Tsai et al., 1978, 1980).

Zein and glutelin combined contain > 90% of the total N in the endosperm of corn and 60% of the total N in the plant at maturity. Tsai et al. (1978, 1980) suggest that these two protein fractions serve as N sinks in the kernel for the deposition of nitrogenous compounds. A reduction of N deposition into the kernel N sink should result in the accumulation of free amino acids and the generation of a more negative osmotic potential. This more negative potential would favor water but reduce solute movement into the kernel and a reduced grain yield would result. Unlike glutelin, however, zein is the major storage protein and its synthesis can be manipulated readily by genetic means and N fertilization. Thus, increased N in the kernel due to N fertilization is confined primarily to zein (Frey, 1951; Keeney, 1970; Randig & Broadbent, 1979; Sauberlich et al., 1953; Schneider et al., 1952; Tsai et al., 1980); nonzein protein increases only slightly (Fig. 1). Increasing grain yield and dry weight during kernel development in response to fertilization are positively correlated with zein content (Tsai et al., 1978). Early termination of zein synthesis in the *o2* mutant results in a reduced movement of photosynthetic assimilates into the kernel (Tsai et al., 1980). These dynamic relationships among zein synthesis, N fertilizer, and dry matter accumulation indicate the importance of this system as a yield-determinant in maize. There are indications that it may be an important system for other cereal grains as well.

By using the much-reduced zein content of the *o2* mutant and its isogenic normal counterpart in a B14 × B37 hybrid as a model system, C. Y. Tsai et al. (Fig. 1, unpublished data) clearly indicate that zein is a major factor affecting grain yield and protein nutritional quality in response to N fertilization. At low levels of N fertilizer, normal kernels appeared similar to *o2*, produced only a small amount of zein (mg/kernel), and yielded (grain) comparably to *o2*. However, as the rate of N fertilizer increased, zein synthesis in the normal hybrid continued to increase in contrast to the

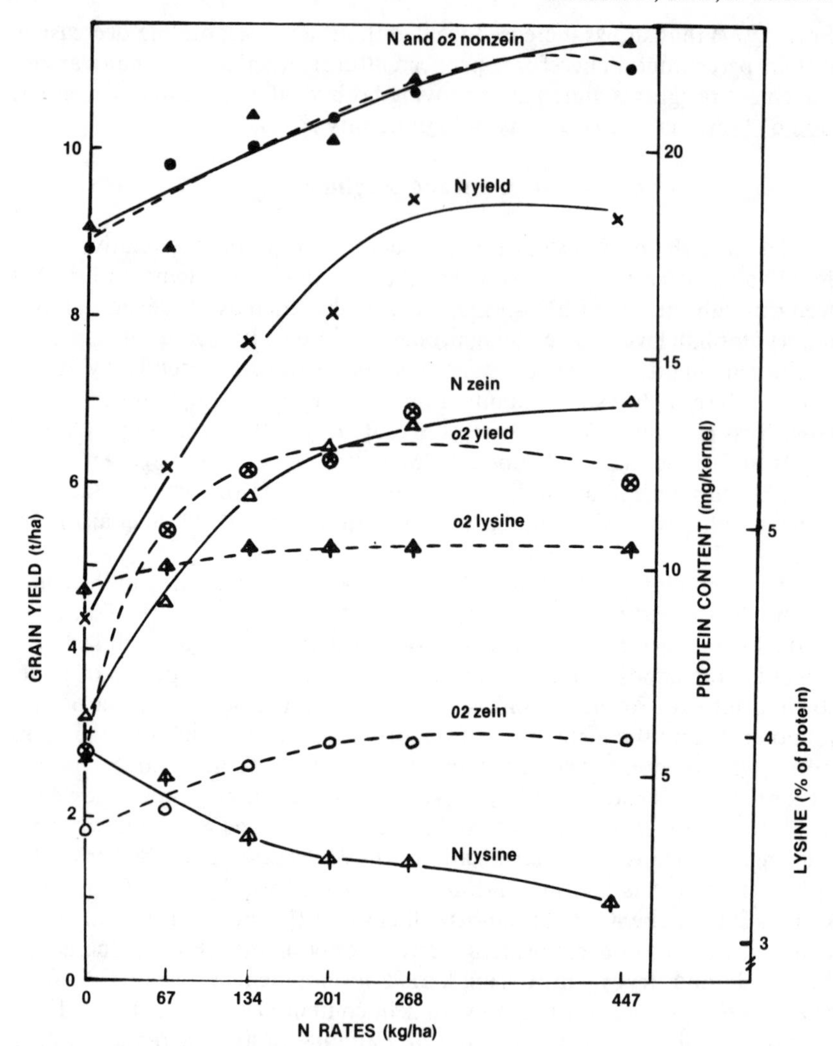

Fig. 1—Lysine concentration, zein and nonzein protein content, and grain yield of B14 × B37 (N, solid line) and B14*o2* × B37*o2* (*o2*, broken line) hybrids grown under different rates of N fertilizer (C. Y. Tsai et al., unpublished data).

o2 mutant, and differences in grain yield between these two genotypes became greater. Nonzein protein content (mg/kernel) increased at a similar rate in both genotypes (Fig. 1).

Kernel protein percentage generally increases as the rate of N fertilizer increases. Maize hybrids grown without N fertilizer may contain only 6% protein, but increase to about 12% when grown under high-N conditions (Pierre et al., 1977; D. M. Huber et al., Purdue Univ., unpublished data), and a much wider range in protein percentage has been shown through genetic selection (Dudley et al., 1974). The increased protein induced by N fertilization in normal hybrids is primarily zein, while the increased protein of the *o2* mutant due to N fertilization appears to be proportionally distributed to all four protein fractions (Tsai et al., 1980).

The maintenance of an adequate supply of N throughout the growing season is essential if crop yield and grain protein percentage are to be increased. In this connection, the addition of a nitrification inhibitor to preplant-applied N (Warren et al., 180) or applying N close to the time of plant uptake (Stevenson & Baldwin, 1969) frequently results in higher grain yield and protein percentage.

The synthesis of large amounts of zein under high rates of N fertilizer may increase yield but reduce protein nutritional quality, because zein is deficient in lysine and tryptophan. Indeed, these two amino acids, as a percent of protein, decrease with increasing rate of N fertilizer applied (Keeney, 1970; MacGregor et al., 1961; Randig & Broadbent, 1979; Sauberlich et al., 1953; Tsai et al., 1978). The *o2* mutant, on the other hand, fails to deposit additional N as zein, so the lysine concentration, on a percent protein basis, remains relatively constant regardless of N rates (Fig. 1). Although the *o2* mutant contains more lysine and tryptophan than its normal counterpart on a percent basis, the yield of these amino acids on an area basis may be comparable (D. V. Glover, Purdue Univ., personal communication; C. Y. Tsai et al., unpublished data).

Like maize, sorghum also contains large quantities of prolamin (kafirin) in the grain. In general, the grain yield and protein response of sorghum and maize to N fertilizer is similar. Lysine concentration in the grain is negatively correlated with total protein concentration (Deosthale et al., 1970; Virupaksha & Sastry, 1968). Thus, the relative concentration of lysine and tryptophan decreases in the normal but remains constant in the high lysine cultivars, as the N fertilizer–induced grain protein percentage increases (J. D. Axtell, Purdue Univ., personal communication). Both maize and sorghum are photosynthetically C_4-type plants possessing a more efficient CO_2 fixation system than C_3-type plants. This system provides the additional C skeletons and energy necessary for NH_3 assimilation, which, in conjunction with a large N sink in the kernel, leads to a greater potential producitivty with N fertilization than attainable with C_3-type plants. However, because of their large N sink in the kernel, the protein nutritional quality of grain sorghum and maize is somewhat poor as compared with that of wheat, barley, oats, and rice (*Oryza sativa* L.).

II. COTTON

Cotton (*Gossypium hirsutum* L.) quality is usually expressed in terms of fiber and boll properties. Cotton fiber quality evaluation is generally made from lint samples removed from bales and is expressed in terms of the fiber properties length, strength, and fineness.

The effect of N nutrition and other agronomic factors is evaluated by collecting mature boll samples prior to harvest. The boll and fiber characteristics of these samples are determined in the laboratory. These values do not reflect market quality changes that sometimes occur due to unfavorable weather before or after harvest or the detrimental effects of poor handling, ginning, or storage. It is well known that different cultivars reflect the genetic range and limits imposed on boll and fiber properties. Nonnutritional environmental factors also cause variations in boll and fiber properties of a given cultivar and may differ in extent of effect with various cultivars. Water stress and temperature are known to cause major variations in boll and fiber properties.

For the effect of N nutrition on cotton boll and fiber properties to be evaluated, it is necessary that a range of N nutrition from acute deficiency to levels more than adequate be considered. Only a few studies reported in the literature have fulfilled this requirement.

A. Boll Properties

Although under conditions of N shortage boll properties are significantly affected by N application (Table 1), yield as a function of N application is primarily affected by number of bolls per unit of land area (Gardner & Tucker, 1967). Severe N shortage dramatically reduces the number of flowers and bolls, but reduces boll size by only 10 to 15%. Generally, boll size is more characteristic of cultivar and environmental conditions than N nutrition. Nitrogen applications that cause a yield response may also cause an increase in mean weight per boll. The increase in boll size is a result of an increase in number of seeds per boll, seed index (mean seed weight), and lint index (mean lint weight per seed). Of these factors, a change in the seed index is the most consistent single measurement associated with N effect on boll size. When lint index is significantly affected by N application, the magnitude of the effect is small.

Table 1—Boll and fiber properties as influenced by N fertilization, a summary from Hamilton et al., 1956; MacKenzie & Van Schaik, 1963 (three experiments); and Gardner, 1963 (four experiments).†

Source	Yield increase	Boll size	Seeds/ boll	Percent lint	Seed index	Lint index	Fiber length‡		Strength	Fineness
							UHM	M		
Hamilton et al.	+	+	+	0	+	+	0	0	0	+
MacKenzie	+	+	+	0	0	+	0	0	0	+
&	+	+	+	−	+	+	0	0	+	0
Van Schaik	+	+	+	0	+	+	0	0	0	0
Gardner	+	+	0	0	+	0	+	+	0	+
	0	0	0	0	+	0	0	0	0	0
	+	+	0	0	+	0	0	0	0	0
	+	+	+	0	+	0	0	0	0	+

† + = Significant increase; 0 = no significant change; and − = significant decrease.
‡ UHM = upper half mean; M = mean.

B. Fiber Properties

Fiber length appears to be increased by N application only when N deficiency is severe; fiber strength is even less subject to influence of N nutrition of cotton. Under conditions ranging from severe N deficiency to an excess, increasing rates of N application result in only slight increases in fiber strength. Similarly, N applications have resulted in significant changes in fiber fineness, but the changes are very small. Apparently, N supply has little practical effect on fiber fineness. Most of the changes in fineness were associated with late N application, which tends to delay fiber maturity resulting in immature fibers with less secondary filling and an increase in fiber fineness.

Within the range of N nutrition commonly found under average to good management, the fiber properties of cotton are unlikely to be influenced by N fertilization. The cotton plant can adapt to a rather wide variation in N supply by adjusting the number and size of bolls with little, if any, change in fiber properties.

III. SUGARBEETS

Sugarbeet (*Beta vulgaris* L.) quality is defined by sugar content and purity. Sugar content is the weight percentage of sucrose in the fresh beet root and purity is the ratio of sucrose content to total soluble solids, also expressed as a percentage. Higher values for sucrose content and purity are desirable for the processor as the potential recovery of crystallized sugar per unit weight of beets is enhanced.

As the sugar content decreases, the nonsucrose soluble solids increase and the efficiency of refining sugar declines, which becomes an important consideration for the processor. Each unit weight of nonsucrose soluble solids in the juice can prevent 1.5 to 1.8 units of sucrose from crystallizing and result in loss to molasses (Alexander, 1971). The price paid to the grower by the processor is adjusted to compensate for this loss when low sugar percentages occur in the beets.

Genetic and environmental factors provide limits within which quality is markedly influenced by N supply. Under field conditions, a limited N supply restricts both root and top growth allowing sucrose to accumulate with relatively little accumulation of soluble nitrogenous metabolites and nonnitrogenous organic acids. The products of metabolic activity increase as N supply is increased. Less sucrose accumulates because it is diverted to more cells in growth of both root and top tissue. Thus, purity is lowered by less sucrose accumulation and more nonsucrose soluble solids. Root growth is generally increased by N in a curvilinear pattern; over the same range of N levels, top growth increase is linear. Much of the increased top growth is at the expense of stored sugar.

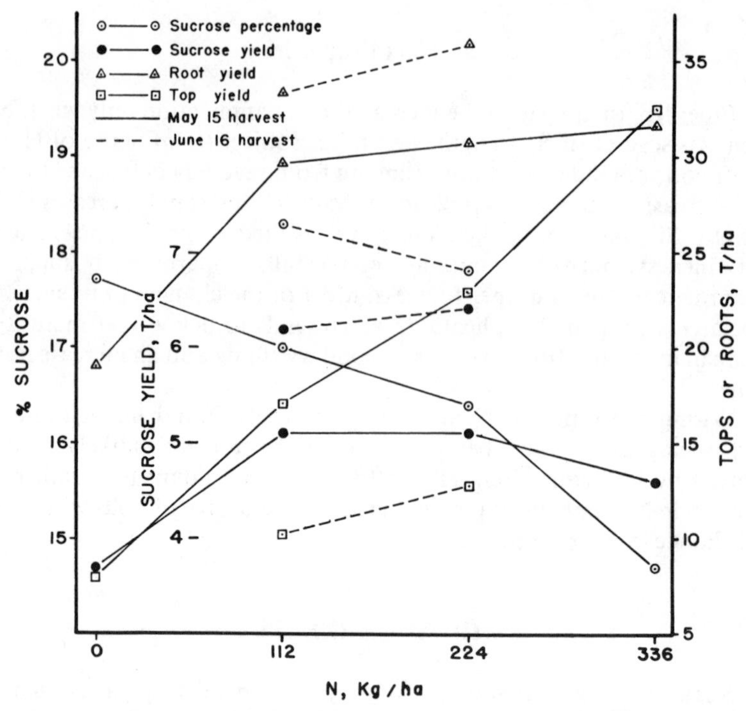

Fig. 2—Sucrose, sugarbeet top, and root yields and sucrose percentage as affected by N application rate (Nelson et al., 1979).

Figure 2 illustrates a relationship between N supply and root, top, and sucrose yield and sugar content. The optimum rate of N application at this site was near the 112 kg/ha treatment with sugar yield near its maximum and sugar percentage reduced only slightly from its maximum. The optimum rate of N application varies, of course, with the N supply of the soil. Delay of harvest can result in more root growth and a higher sugar content increasing utilization of higher rates of N application. Fresh weight of tops decline correspondingly as roots enlarge and accumulate more sugar. The application of N to a soil with an adequate N supply can be increasingly detrimental with each increment, for example, as indicated by the data of Carter et al. (1976).

Nitrogen management is the key to establishing a balance between yield and quality. Other factors being equal, inadequate supplies of N result in a high sucrose content and purity, but reduced beet and sugar yields. Conversely, excessively high N supply results in reduced sucrose concentration and purity. Sugar yield is maximized when both sugar percentage and beet and top yields are slightly less than maximum. This balance between yield and composition will be discussed for sugarbeet and other root crops in great detail in Chapter 43, this book.

Although N nutrition has a minor effect on the quality of some crop products (e.g., cotton fiber), it has a major effect on the nutritional or industrial quality of most agronomic crops. In this regard, systems of N management may be directed toward different objectives, e.g., toward high protein yield in cereal grains, but not in tobacco leaf (*Nicotiana tabacum* L.). In addition, the preferred management system considers many other factors relating to soil type and geographical area. Of these factors, N supply and its management are keys for meeting the increased need for high-quality crops.

REFERENCES

Alexander, J. T. 1971. Factors affecting quality. p. 371–381. *In* R. T. Johnson et al. (ed.) Advances in sugarbeet production: principles and practices. Iowa State Univ. Press, Ames.

Axtell, J. D. 1982. Breeding for improved nutritional quality. *In* K. J. Frey (ed.) Plant Breeding II. Iowa State Univ. Press, Ames.

Bhatia, C. R., and R. Robson. 1976. Bioenergetic considerations in cereal breeding for protein improvement. Science 194:1418–1421.

Carter, J. N., D. T. Westermann, and M. E. Jensen. 1976. Sugarbeet yield and quality as affected by nitrogen level. Agron. J. 68:49–55.

Deckard, E. L., and R. H. Busch. 1978. Nitrate reductase assays as a prediction test for crosses and lines in spring wheat. Crop Sci. 18:289–293.

Deckard, E. L., K. A. Lucken, L. R. Joppa, and J. J. Hammond. 1977. Nitrate reductase activity, nitrogen distribution, grain yield, and grain protein of tall and semidwarf near-isogenic lines of *Triticum aestivum* and *T. turgidum.* Crop Sci. 17:293–296.

Deosthale, Y. G., V. S. Mohan, and K. V. Rao. 1970. Varietal differences in protein, lysine and leucine content of grain sorghum. J. Agric. Food Chem. 18:644–646.

Doll, H., and B. Koie. 1975. Evaluation of high lysine barley mutants. p. 55–59. *In* Breeding for seed protein improvement using nuclear techniques. Int. Atomic Energy Agency, Vienna.

Dubetz, S., E. E. Gardiner, D. Flynn, and A. Ian DeLaroche. 1979. Effect of nitrogen fertilizer on nitrogen fractions and amino acid composition of spring wheat. Can. J. Plant Sci. 59:299–305.

Dudley, J. W., R. J. Lambert, and D. E. Alexander. 1974. Seventy generations of selection for oil and protein concentration in the maize kernel. p. 181–212. *In* J. W. Dudley (ed.) Seventy generations of selection for oil and protein in maize. Crop Sci. Soc. of Am., Madison, Wis.

Finney, K. F. 1971. Fractionating and reconstituting techniques to relate functional (bread making) to biochemical properties of wheat flour components. Cereal Sci. Today 16: 342–356.

Finney, K. F., J. W. Meyer, F. W. Smith, and H. C. Fryer. 1957. Effects of foliar spraying of Pawnee wheat with urea solutions on yield, protein content and protein quality. Agron. J. 49:341–347.

Finney, K. F., and W. T. Yamazaki. 1967. Quality of hard, soft, and durum wheats. *In* K. S. Quisenberry and L. P. Reitz (ed.) Wheat and wheat improvement. Agronomy 13:471–503.

Frey, K. J. 1951. The interrelationships of proteins and amino acids in corn. Cereal Chem. 28:123–132.

Frey, K. J. 1977. Protein of oats. Z. Pflanzenzeucht. 78:185–215.

Gardner, B. R. 1963. A study of factors influencing the nitrogen fertilization of acala cotton. Ph.D. Diss. Univ. of Arizona, Tucson.

Gardner, B. R., and T. C. Tucker. 1967. Nitrogen effects on cotton: I. Vegetative and fruiting characteristics. Soil Sci. Soc. Am. Proc. 31:780–785.

Glover, D. V. 1976. Improvement of protein quality in maize. p. 64–97. *In* H. L. Wilchke (ed.) Improving the nutrient quality of cereals. II. Report of second workshop on breeding and fortification. U.S. Agency for Int. Development, Washington, D.C.

Grzybowski, R. A., and B. J. Donnelly. 1979. Cooking properties of spaghetti: factors affecting cooking quality. J. Agric. Food Chem. 27:380–384.

Hamilton, J., C. O. Stanberry, and W. M. Wooten. 1956. Cotton growth and production as affected by moisture, nitrogen, and plant spacing on the Yuma Mesa. Soil Sci. Soc. Am. Proc. 20:246–252.

Hucklesby, D. P., C. M. Brown, S. E. Howell, and R. H. Hageman. 1971. Late spring applications of nitrogen for efficient utilization and enhanced production of grain and grain protein of wheat. Agron. J. 63:274–276.

Johnson, V. A., P. J. Mattern, and K. P. Vogel. 1975. Cultural, genetic and other factors affecting quality of wheat. p. 127–139. *In* A. Spicer (ed.) Bread: social, nutritional and agricultural aspects of wheat and bread. Applied Science Publ., Ltd., London.

Keeney, D. R. 1970. Protein and amino acid composition of maize grain as influenced by variety and fertility. J. Sci. Food Agric. 21:182–184.

Koie, B., J. Ingversen, A. J. Andersen, H. Doll, and B. O. Eggum. 1976. Composition and nutritional quality of barley protein. p. 55–61. *In* Evaluation of seed protein alterations by mutation breeding. Int. Atomic Energy Agency, Vienna.

Kramer, T. 1978. Environmental and genetic variation for protein content in winter wheat (*Triticum aestivum* L.). Euphytica 28:209–218.

Lambert, R. J., D. E. Alexander, and J. W. Dudley. 1969. Relative performance of normal and modified protein (opaque-2) maize hybrid. Crop Sci. 9:242–243.

MacGregor, J. M., L. T. Taskevitch, and W. P. Martin. 1961. Effect of nitrogen fertilizer and soil type on the amino acid content of corn grain. Agron. J. 53:211–214.

MacKenzie, A. J., and P. H. van Schaik. 1963. Effect of nitrogen on yield, boll, and fiber properties of four varieties of irrigated cotton. Agron. J. 55:345–347.

McNeal, F. H., C. A. Watson, and H. A. Kittams. 1963. Effects of dates and rates of nitrogen fertilization on the quality and field performance of five hard red spring wheat varieties. Agron. J. 55:470–472.

Mitra, R. K., C. R. Bhatia, and R. Robson. 1979. Bioenergetic cost of altering the amino acid composition of cereal grains. Cereal Chem. 56:249–252.

Mosse, J. 1966. Alcohol soluble proteins of cereal grains. Fed. Proc. Fed. Am. Soc. Exp. Biol. 25:1663–1669.

Naas, H. G., and P. L. Crane. 1970. Effect of endosperm genes on dry matter accumulation and moisture loss in corn (*Zea mays* L.). Crop Sci. 10:276–280.

Nelson, J. M., J. L. Abbott, and T. C. Tucker. 1979. Growth and nitrogen and phosphorus uptake of sugarbeets in central Arizona. Arizona Coop. Ext. Series and Agric. Exp. Stn. Series P-45:25–29.

Penning de Vries, F. W. T., A. H. M. Brunsting, and H. H. Van Laar. 1974. Products, requirements, and efficiency of biosynthesis: a quantitative approach. J. Theor. Biol. 45: 339–377.

Peterson, D. M. 1976. Protein concentration, concentration of protein fractions, and amino acid balance in oats. Crop Sci. 16:663–666.

Peterson, G. A., and A. E. Foster. 1973. Malting barley in the United States. Adv. Agron. 25:328–378.

Pierre, W. F., L. Dumenil, V. D. Jolley, J. R. Webb, and W. D. Schrader. 1977. Relationship between corn yield expressed as a percentage of maximum and the N percentage in the grain. I. Various N-rate experiments. Agron. J. 69:215–220.

Pushman, F. M., and J. Bingham. 1976. The effects of a granular nitrogen fertilizer and a foliar spray of urea on the yield and bread making quality of ten winter wheats. J. Agric. Sci. 87:281–292.

Randig, V. V., and F. E. Broadbent. 1979. Proteins and amino acids in grain of maize grown with various levels of applied N. Agron. J. 71:509–512.

Rhodes, A. P., and G. Jenkins. 1976. Improving the protein quality of cereals, grain legumes, and oil seed by breeding. p. 207–226. *In* G. Norton (ed.) Plant proteins. Butterworths, London.

Sauberlich, H. E., W. Y. Chang, and W. D. Salmon. 1953. The amino acid and protein content of corn as related to variety and nitrogen fertilization. J. Nutr. 51:241–250.

Schneider, E. O., E. B. Earley, and E. E. DeTurk. 1952. Nitrogen fractions of the component parts of the corn kernel as affected by selection and soil nitrogen. Agron. J. 44:161–169.

Smika, D. E., and B. W. Greb. 1973. Protein content of winter wheat grain as related to soil and climatic factors in the semiarid Central Great Plains. Agron. J. 65:433–436.

Sreeramulu, C., and L. F. Bauman. 1970. Yield components and protein quality of opaque-2 and normal diallels of maize. Crop Sci. 10:262–265.

Stevenson, C. K., and C. S. Baldwin. 1969. Effect of time and method of nitrogen application and source of nitrogen on the yield and nitrogen content of corn (*Zea mays* L.). Agron. J. 61:381–384.

Tsai, C. Y., D. M. Huber, and H. L. Warren. 1978. Relationship of the kernel sink for N to maize productivity. Crop Sci. 17:399–404.

Tsai, C. Y., D. M. Huber, and H. L. Warren. 1980. A proposed role of zein and glutelin as N sinks in maize. Plant Physiol. 66:330–333.

Virupaksha, T. K., and L. V. S. Sastry. 1968. Studies on the protein content and amino acid composition of some varieties of grain sorghum. J. Agric. Food Chem. 16:198–203.

Wall, J. S. 1979. The role of wheat proteins in determining baking quality. p. 275–311. *In* D. L. Laidman and R. G. Wyn Jones (ed.) Recent advances in the biochemistry of cereals. Academic Press, New York.

Wall, J. S., and J. W. Paulis. 1978. Corn and sorghum proteins. Adv. Cereal Sci. Technol. 2:135–219.

Warren, H. L., D. M. Huber, C. Y. Tsai, and D. W. Nelson. 1980. Effect of nitrapyrin and N fertilizer on yield and mineral composition of corn. Agron. J. 72:729–732.

42

S. J. Locascio, W. J. Wiltbank, D. D. Gull,
and D. N. Maynard
University of Florida
Gainesville, Florida

Fruit and Vegetable Quality as Affected by Nitrogen Nutrition

Fruits and vegetables are consumed in large quantities in human diets and are important sources of carbohydrates, proteins, vitamins, and minerals and provide desirable fiber to the diet. Based on average consumption and the recommended dietary allowance, fruits and vegetables can contribute 92% of the required vitamin C, 50% of the carotene or provitamin A, 19% of the Fe, 18% of the carbohydrate, but only 9% of the calories required.

To be most acceptable to consumers, the product grown by the farmer must be of high external and internal quality. The major external characteristic of quality is product appearance, including size, shape, color, and freedom from blemishes. Other quality criteria include texture, flavor, and composition. All of these quality components are influenced by N nutrition. For most fruits and vegetables there is a close relationship between crop yield and quality. Generally, environmental conditions that lead to production of the highest yields are associated with fruits and vegetables of the highest quality.

Numerous studies have assessed the influence of soil fertility on the quality of fruits and vegetables. Such studies are not without compounding variables such as cultivar, soil, location, season, maturity at harvest, and postharvest handling procedures. Conventional research in mineral nutrition has emphasized growth rates and yield. In reviews of the effects of plant nutrition on quality of vegetables, Lorenz (1964) and Minotti (1975) emphasized that the level of nutrition must be considered; they reported different responses when observations were made at deficient, optimum, or excessive nutrient levels.

Plant dry matter normally contains 2 to 4% N and most species are in the deficient zone when concentrations fall below 1.5% N (Maynard, 1979). Nitrogen is an integral part of numerous plant compounds including amino

Nitrogen in Crop Production.

acids, protein, nucleic acids, and chlorophyll. Increased N concentration in a plant generally results in an increase in these compounds, whereas under conditions of N stress, some plants accumulate carbohydrates.

Fruits and vegetables are intensively grown crops. Highest yields, the goal of most producers, are generally achieved with adapted cultivars, grown on the best land, in the absence of environmental or nutritional stress, and in a low or pest-free situation. Use of labor has decreased for many crops in recent years because of advances in mechanization, but for most fruits and vegetables, labor requirements for production and harvesting are large in comparison with those for agronomic crops. With a once-over mechanical harvest, the effects of N on uniformity of maturity become very important.

Fruits and vegetables respond to fertilization under most situations, and almost always respond to N fertilization. An exception to this is when crops are grown on organic soils where at high soil temperatures, N release is generally adequate for high yields. Because of the rapid visual response in plant appearance and in yield potential, and because fertilizers represent only a small proportion of production and harvesting costs, fruit and vegetable growers tend to fertilize at rates equal to or above those recommended. Fertilizer costs for the major intensively grown Florida vegetables ranged from 3 to 16% of the total production and harvesting costs in the 1978–1979 season. Overall, fertilizer accounted for only 7% of costs (Brooke, 1980).

Yields of tree fruits have increased dramatically in the last 50 to 75 yr, due mostly to improved nutrition and irrigation practices. Improved N nutrition has been a major factor in increasing tree vigor, fruit bud formation, and fruit set, and has resulted in increased yields. These increased yields are often secured at the expense of fruit quality. Therefore, recent research has been directed towards evaluating the effect of increased N application levels on fruit quality.

The effect of N nutrition on fruit quality is complicated by the physiological complexity of the fruit tree. A mature tree is a long-lived perennial that concurrently produces new vegetative growth, forms new fruit buds, blooms, sets fruit, and develops fruit. Since the fruit buds on many fruiting species are initiated in the year prior to bloom and much of the spring growth of the tree is dependent on stored materials from the preceeding growing season, the number of fruit buds, bloom, fruit set and early season vegetative vigor is strongly influenced by the nutritional status and vigor of the tree in the year(s) preceeding the current season's crop. During the crop year, a delicate balance between vegetative growth and reproductive growth must be maintained and differing seasonal nutritional needs must be satisfied. Thus, N application in spring may have a different effect on fruit quality than summer or fall application (Faust, 1979). Since there is competition for photosynthate and other growth substrates between fruits and new shoot growth and competition among fruits, crop load affects fruit size and quality. Fruit thinning is commonly practiced to reduce fruit number and alter leaf/fruit ratio and ultimately fruit size. Environmental factors such as temperature, light, moisture, and cultural practices, primarily pruning and irrigation, also affect the vegetative/reproductive balance of the tree and influence fruit quality.

A further complication arises when attempts are made to define fruit quality, which is a combination of highly subjective evaluations made by the consumer, merchandizer, processor, handler, and/or producer. Large differences exist among factors that comprise quality of a fresh fruit, marketed essentially as harvested from the tree, and a processed fruit, which might be sold entire or modified by peel removal, juicing, or other means. Thus, quality must be considered from the standpoint of a fruit's external, internal, and/or processed product quality.

I. FRUIT

A. Effects of N on Appearance

Nitrogen nutrition influences the appearance of fruits, primarily as it affects size, color, and characteristics of the peel. For the fresh market, mid- to large-sized fruits are preferred. In deciduous fruits, e.g., apple (*Malus domestica* Bork), peach [*Prunus persica* (L.) Batsch], cherry (*Prunus* spp.), and pear (*Pyrus communis* L.), increased N application has generally increased fruit size (Ballenger et al., 1966; Bramlage et al., 1979; Westwood : Wann, 1966; Kenworthy : Silsby, 1974; Zieger, 1978), but not invariably. Increasing N application rate decreases fruit size in citrus (Embleton & Jones, 1975; Embleton et al., 1978; Koo & Reese, 1977; Reese & Koo, 1975). In California, where citrus fruits are primarily grown for the fresh market, decreased fruit size from N levels above optimum for maximum growth can seriously lower market prices, particularly for 'Valencia' orange [*Citrus sinengis* (L.) Osbeck] (Embleton & Jones, 1975; Embleton et al., 1978). Where citrus is primarily processed, as in Florida, increased N application reduces fruit size, but increases total number and yield of fruit and soluble solids per hectare. This encourages higher rates of N application since returns are based on soluble solids production (Koo & Reese, 1977; Reese & Koo, 1975).

In tree fruits, high exterior color, earliness-to-market, and uniformity of maturity improve fresh market grade and price. Raising leaf N above the range for optimum yield lowers fruit color, delays maturity, increases time-to-color-break, and results in irregular ripening (Ballenger et al., 1966; Claypool, 1975; Embleton & Jones, 1975; Embleton et al., 1978; Koo & Reese, 1977; Reese & Koo, 1975). High N rates decrease peel color in apples (Boynton & Oberly, 1966; Bramlage et al., 1979; Zieger, 1978) and peaches (Ballenger et al., 1966). In deciduous fruits, good color is one of the most important factors in fresh fruit quality; often, the only negative factor affecting the use of N fertilizer is its adverse effect on fruit color (Claypool, 1975; Shear & Faust, 1975). Some of these adverse affects can be overcome by timing N applications to reduce N levels during fruit maturation (Boynton & Oberly, 1966; Faust, 1979; Shear, 1979). In citrus, high N application rates increase the intensity of the orange color in the peel, but also increase traces of green color in the fruit and the incidence of regreening in cultivars such as 'Valencia'. However, environment appears to be the major factor influencing exterior color development (Embleton & Jones, 1975; Embleton et al., 1978; Kee & Reese, 1977; Reese & Koo, 1975). High N application rates have resulted in a slight increase in juice color (Koo & Reese, 1977).

High N causes increased peel coarseness and thickness in citrus (Koo & Reese, 1977). The incidence of creasing, a physiological peel disorder, decreases with an increase in N level where N is below the range for maximum production in California (Embleton et al., 1978); in Florida, creasing increases with N from the deficient to the above-optimum level (Koo & Reese, 1977). High N influences field and harvest blemishes in citrus, increasing the severity of scab (*Elsinoe fawcetti* Bitanc. and Jenkins) on susceptible cultivars but decreasing russet, wind-scar, and plugged fruit (Koo & Reese, 1977). Citrus growers apply N for maximum fruit production in areas where fruit is grown for processing (Koo & Reese, 1977), whereas a lower yield is sometimes accepted for an increase in external quality of fruit grown for fresh market (Embleton & Jones, 1975; Embleton et al., 1978).

B. Effects of N on Flavor and Composition

Preferable flavor of deciduous fruits is closely associated with soluble solids in the fruit juice and flesh. When N is deficient, N applications have increased soluble solids. However, at higher N rates, soluble solids may diminish (Ballenger et al., 1966; Claypool, 1975), but not consistently (Zieger, 1978). In oranges and grapefruit (*Citrus paradisi* Macf.), the ratio of soluble solids to acid is the major indicator of internal quality. Soluble solids are little influenced by N but high N increases acidity, thus decreasing the solids/acid ratio and flavor early in the fruit maturation period (Embleton & Jones, 1975; Embleton et al., 1978; Reese & Koo, 1975). In lemons [*Citrus limon* (L.) Burm. f.], internal quality and flavor are based on acid content and N does not affect quality (Embleton et al., 1978).

Compositional characters affecting fruit quality include peel color, flesh or juice color, juiciness, soluble solids, organic acids, peel oil in citrus, vitamin C, and the content of various glycosides, esters, and alkaloids, which give fruits their special taste and flavor. Increased N increases the peel oil content of oranges and reduces the ascorbic acid (vitamin C) content of orange (Embleton et al., 1978) and peach (Ballenger et al., 1966). Peel oil is one of the more important by-products of citrus processing. Increased N has been shown to decrease (Embleton et al., 1978) or increase (Reese & Koo, 1975) juice content in citrus.

C. Effects of N on Fruit Storage

Storability and quality of fresh fruit and processed products can be influenced by N. Increased N application rates on pome fruits (apples, pears) result in increased fruit size and softer fruit with higher respiration rates.

High N rates interfere with Ca nutrition and cause increased storage losses to physiological disorders like internal breakdown, internal browning, cork spot, bitter pit and scald, although this effect may be weak, indirect, and/or inconsistent (Boynton & Oberly, 1966; Bramlage et al., 1979; Shear & Faust, 1975). High N rate increases the variability of fruit maturity on the tree. When harvest of deciduous fruit is delayed until a majority of the fruits are in a marketable maturity range, many fruits may become sufficiently advanced in maturity that they bruise easily (Claypool, 1975) and have less resistance to stresses in the market system, either as fresh fruit or in a processed form. Irregular ripening within individual fruits due to high N level has also been the cause of serious canning problems in apricot (*Prunus armeniaca* L.) and the peaches (Claypool, 1975). In citrus, N levels affect physiological breakdown of the peel in storage. High N increases rind staining of navel oranges (Embleton & Jones, 1975; Embleton et al., 1978) and decreases the severity of stem-end rot (Koo & Reese, 1977).

II. VEGETABLES

A. Effects of N on Appearance

The effects of N nutrition on vegetative growth rates are well known and they have important effects on quality of most vegetables. When crops are grown below the level of sufficient N, a reduction in appearance, size of product, and internal quality generally accompanies reduced yield. Nitrogen deficiencies result in an immediate reduction in the vegetative growth rate and a reduction in the total amount of chlorophyll in the plant. This loss of green color appears first on the older leaves of the plant and progresses to younger leaves with time. In the early stages of the deficiency, leaves become light green, they yellow, and later may become necrotic. With crops such as lettuce (*Lactuca sativa* L.), spinach (*Spinacia oleracea* L.) (Maynard, 1970), cabbage [*Brassica oleracea* L. (Capitata Group)], and celery (*Apium graveolens* L.), where the economic part of the plant is vegetative, the reduction in crop quality as manifested by appearance is as drastic as the reduction in yield. In addition, N deficiency also results in an increase in percent dry matter and a reduction in succulence in such crops as broccoli [*Brassica oleracea* L. (Italica Group)], asparagus (*Asparagus officinalis* L.), and cauliflower [*Brassica oleracea* L. (Botrytis Group)], and they become stringy and less palatable.

The effects of N deficiency on quality of vegetables that produce edible roots, tubers, and fruits are indirect. With a reduction in N level, the size of the plant canopy of many crops including tomato (*Lycopersicon esculentum* Mill.) and pepper (*Capsicum annuum* L.) is reduced. Smaller leaves and plant size result in reduced plant carbohydrate assimilation. Fruits such as tomato, pepper, and watermelon (*Citrullus lunatus* Thumb.) produced on such plants are fewer in number, smaller, develop poor color, and are more subject to sunburn than fruit on plants with sufficient N. Pew and Gardner (1972) reported that N-deficient muskmelon plants (*Cucumis melo* L.) produced fruit with low netting, poor general appearance, and had more cull melons. The severity of the effects on yield and quality was directly related to the degree of deficiency.

For most vegetables, the range of sufficiency for N is quite broad; in this range, the effects on yield and external quality factors are small. A decrease in N stress in most plants is accompanied by an increase in green color, leaf size, and plant size (Jones, 1966). Because of the increase in plant size, plants remain in the juvenile stage longer and maturity in many crops is often delayed. If these vegetables are harvested by size rather than physiological maturity, they may be harvested in an immature stage. These immature vegetables would probably be poorer in color and internal quality components than those harvested in a more mature state (Winsor, 1979), and therefore store more poorly.

Excessive levels of N, but below those that reduce yield, have little negative effect on the appearance of most leafy crops. Effects on plant composition are important. Excessive N influences fruiting, root, and tuber producing vegetables by seriously reducing yield. With nutritional conditions that result in lush vegetative growth, fruit development, and carbohydrate accumulation in roots and tubers is restricted and the quality of the product is generally low.

High or excessive N fertilization of leafy vegetables occurs more commonly than with those vegetables that produce edible fruit, roots, or tubers. With the leafy crops, such practices result in premature splitting of cabbage and other heading crops, and the production of excessively succulent vegetables that may be subject to postharvest decay (Maynard, 1979). Aside from visual and postharvest breakdown effects of excessive N, certain physiological and compositional changes that affect quality may occur. When NH_4^+ constitutes the principal N form available to vegetable crops, symptoms of excess are manifested as stem lesions, leaf chlorosis or necrosis, and in severe cases, death of the plant. Ammonium toxicity and resulting quality detriments occur when NH_4^+ fertilizers are applied to acidic soils and when conditions are not suitable for rapid nitrification (Maynard & Barker, 1969).

B. Effects of N on Composition and Flavor

Vegetables contribute a significant amount of nutrition to the human diet. Although these nutritional constituents are not readily apparent (as they do not contribute to texture or flavor), consumers recognize their importance as essential ingredients. The most frequent deficiencies in diets of Americans are Fe, vitamins A and C, and riboflavin. Nitrogen affects amino acid composition and concentration and the resulting protein. Plants grown in N deficient soil are delayed in growth, produce yellow leaves, and have poor yield and lower protein content. Liberal N applications to corn (*Zea mays* L.) increase crude protein by 2.5%, but with a resulting decrease in lysine (Harris, 1975); therefore, the nutritional value of the high-protein corn is definitely lower. Generally, an increase in protein content of corn is associated with a decrease in the biological value, as determined by rat-feeding studies. However, corn cultivars have been developed that produce increased protein content of high nutritional quality (Harris, 1975).

Maximum yields of canning beets (*Beta vulgaris* L.) in New York State occur with application of 89 kg N/ha preplant. Highest quality, as measured by the sugar/glutamine ratio, also results from this N rate. At higher rates, sugar concentration decreases and glutamine concentration increases. Poor quality, as manifested by objectionable flavor of the processed product, occurs when pyrrolidone carboxylic acid, a degradation product of glutamine, is produced during processing. Maximum quality and yield of canning beets, therefore, occur at the same N rate (Peck et al., 1974).

Potato (*Solanum tuberosum* L.) amino acid content is directly correlated with N fertility, whereas specific gravity, starch, and dry matter are inversely correlated (Minotti, 1975; Vittum, 1963). Fertility programs are generally designed to maximize yields; however, potatoes that have high specific gravity and dry matter content are more desirable for processing. Excess N rates result in the production of certain amino acids that are associated with dark chips and also tend to stimulate production of succulent foliage and delay maturity. Measurable differences in composition, as related to differential rates of applied N, however, are very small when the effect of maturity is eliminated.

Although N has a profound effect on growth of tomato plants and fruit development, the composition of the fruits is little affected by differential N rates (Lorenz, 1964).

In leafy vegetables such as spinach, increased N resulted in more protein but with a reduced methionine content (Arthey, 1975). As with other vegetables, the addition of N at the deficient level increases protein but reduces one or more essential amino acids. At a level where N does not limit growth, the effect on protein content would be expected to be minimal. Succulence is generally more important than composition and flavor in leafy vegetables. Succulence can only be obtained under optimum growing conditions; therefore, situations of N deficiency should be avoided.

Nitrates accumulate in plants when uptake exceeds reduction to NH_4^+. Consequently, conditions favoring high soil NO_3^- concentration or those interfering with reduction may cause NO_3^- concentration to be elevated in the plant. Adverse effects on human health may occur when NO_3^--rich vegetables are consumed by certain individuals, particularly infants, whose systems may favor reduction to NO_2^-. If NO_2^- is absorbed in the blood, Fe^{2+} of hemoglobin may be oxidized to the Fe^{3+} form, producing methemoglobin, which cannot transport O_2. Another possible adverse effect of consumption of vegetables and other foods rich in NO_3^- is that if particular conditions exist, nitrosamines may be formed. Such compounds have been shown to be carcinogenic, teratogenic, and mutagenic in tests with laboratory animals. Factors influencing and possible consequences of NO_3^- accumulation in vegetables have been reviewed (Maynard et al., 1976; Maynard & Barker, 1979).

Vitamin C content in vegetables is more closely associated with exposure to sunlight than variation in soil fertility. Fruits that are shaded by a vigorously growing plant would be lower in vitamin C content than those exposed to the sun (Allaway, 1975). The direct effect of N application on vitamin C in vegetables is variable. With increasing N applications, leaf-type vegetables such as lettuce and cabbage develop more vitamin C; however, fruit-types such as tomatoes, peppers, and muskmelons have less because of the shading effect (Somers & Besson, 1948).

Vitamin A and C contents in vegetables are primarily under genetic control. Where plants grow normally, without loss of green color due to N deficiency, no change in vitamin content results from application of additional N, except as noted above (Allaway, 1975; Somers & Besson, 1948).

Vegetable flavor is a composite of taste, aroma, and texture. Most vegetables have a characteristic mild flavor with the exception of those of the genus *Allium* and the Cruciferae family. Prime contributors to vegetable flavors are plant acids, sugar, and flavor volatiles (Mnotti, 1975; Peck et al., 1974).

When soils are deficient in N, the flavor of vegetables can be improved by N additions. Optimum flavor of vegetables is attained when nutrients are sufficient for optimum growth and are balanced with respect to N, P, and K. Excessive N application results in a decrease of soluble solids and a general loss of flavor (Arthey, 1975; Peck et al., 1974). *Brassica* species also react to excessive N and develop strong and undesirable flavor and odor. Additional studies are needed to establish the role of mineral nutrition on flavor volatiles in vegetables.

From the foregoing examples, it is apparent that vegetable growers should avoid the practice of N fertilization above that which is recommended. Furthermore, researchers should continue fertilization research to determine if recommendations for N may be reduced without a reduction in yield or quality.

C. Effects of N on Vegetable Storage

Most of the research with N effects on vegetable production have been conducted to obtain highest yields with little regard to N effects on storability of the product. Most vegetables are consumed or processed in a relatively short time after harvest. Some vegetables such as potato, sweet potato [*Ipomoea batatas* (L.) Lam.], onions (*Allium cepa* L.), and cabbage may be stored for several months, whereas most others are stored for 2 to 3 weeks or less. Most vegetables are highly perishable and must be stored under rather exact conditions of temperature and humidity. Since most of these products respire at relatively high levels, quality is generally reduced by increased storage time. Many factors contribute to this gradual loss of quality. Storage life of vegetables that contain deficient or excessive N would probably be reduced since excessive N concentrations increase crop succulence, delay maturity, and result in an earlier breakdown of the stored vegetable. Within the optimum range of N, little difference in postharvest breakdown occurs in most vegetables (Lorenz, 1964). However, published information is limited on the effects of N nutrition on the storability of vegetables.

III. SUMMARY

In tree fruits, effects of N nutrition on fruit size are variable, but high N tends to increase size of deciduous fruits but decrease size of citrus fruits. Above-optimum N lowers color development, delays maturity, increases time-to-color break, and causes uneven fruit ripening. These effects are variable and highly dependent upon species, cultivar, environmental conditions, and management practices. Optimum N results in production of fruit with highest soluble solids. High N can increase the acid content in citrus. Above-optimum N increases the incidence of some postharvest and storage problems. Increasing levels of N have also been shown to increase orange juice color, orange peel oil, and lower vitamin C content of several fruits.

In vegetables, N nutrition that results in maximum yield generally produces vegetables of the highest quality. At deficient N levels, vegetables are poor in color, reduced in size, irregular in maturity, and less succulent. With excessive N, but below that which reduces yield, visual effects on leafy vegetables are minimal. Where the marketable portion of the vegetable is a fruit, tuber, or root, high levels of N indirectly effect yield and quality.

The effects of N fertilization on vegetable composition and flavor are variable and inconsistent. Composition and flavor are affected when N is added to N-deficient plants. When N is sufficient for optimum growth, there is little or no effect on composition and flavor. There is a tendency for commercial growers to use sufficient or excess N so minimal effects on composition could be expected. Some of the inconsistent results reported in the literature are due to variables in maturity and nutrient level being studied (deficient, optimum, excess). Fertilizer practices have the greatest effect on N content in leafy vegetables, smaller effects in bulb and tuber crops, and little effect in fruits such as melons, peppers, and tomatoes.

REFERENCES

Allaway, W. H. 1975. The effect of soils and the fertilizers on human and animal nutrition. p. 1–52. USDA Agric. Info. Bull. 378. U.S. Government Printing Office, Washington, D.C.

Arthey, V. D. 1975. Quality of horticultural products. John Wiley & Sons, Inc., New York.

Ballenger, W. E., H. K. Bell, and N. F. Childers. 1966. Peach nutrition. p. 276–390. *In* N. F. Childers (ed.) Nutrition of fruit crops. Horticultural Publications, Rutgers Univ., New Brunswick, N.J.

Boynton, D., and G. H. Oberly. 1966. Apple nutrition. p. 1–50. *In* N. F. Childers (ed.) Nutrition of fruit crops. Horticultural Publications, Rutgers Univ., New Brunswick, N.J.

Bramlage, W. J., M. Drake, and W. J. Lord. 1979. The influence of mineral nutrition on the quality and storage performance of pome fruits grown in North America. Acta Hortic. 92:29–40.

Brooke, D. L. 1980. Costs and returns from vegetable crops in Florida, season 1978-79 with comparisons. Economic Information Rep. 127. Food and Resource Economics Dep., Univ. of Florida, Gainesville. p. 1-24.

Claypool, L. L. 1975. Plant nutrition and deciduous fruit crop quality. HortScience 10:45-47.

Embleton, T. W., and W. W. Jones. 1975. Plant nutrition and citrus crop quality and yield. HortScience 10:48-50.

Embleton, T. W., W. W. Jones, C. Pallares, and R. G. Platt. 1978. Effects of fertilization of citrus on fruit quality and ground water nitrate-pollution potential. Proc. Int. Soc. Citric. 2:280-285.

Faust, M. 1979. Modern concepts in fruit nutrition. Acta Hortic. 92:11-16.

Harris, R. S. 1975. Effects of agricultural practices on foods of plant origin. p. 5-35. *In* R. S. Harris and E. Karmas (ed.) Nutritional evaluation of food processing. 2nd ed. AVI Publ. Co., Westport, Conn.

Jones, W. W. 1966. Nitrogen. p. 310-323. *In* H. D. Chapman (ed.) Diagnostic criteria for plants and soils. Univ. of California, Div. of Agric. Sci., Berkeley, Calif.

Kenworthy, A. L., and L. Silsby. 1974. Red tart cherry quality: as related to location, season, fungicides and nitrogen. Mich. Agric. Exp. Stn. Res. Rep. 250.

Koo, R. C. J., and R. L. Reese. 1977. Influence of nitrogen, potassium and irrigation on citrus fruit quality. Proc. Int. Soc. Citric. 1:34-38.

Lorenz, O. A. 1964. Effect of mineral nutrition on quality of vegetables. p. 23-35. *In* G. A. Taylor and N. F. Childers (ed.) Factors affecting raw product quality in vegetables. New Jersey Agric. Exp. Station, Rutgers. New Brunswick, N.J.

Maynard, D. N. 1970. The effects of nutrient stress on the growth and composition of spinach. J. Am. Soc. Hortic. Sci. 95:598-600.

Maynard, D. N. 1979. Nutritional disorders of vegetable crops: a review. J. Plant Nutr. 1:1-23.

Maynard, D. N., and A. V. Barker. 1969. Studies on the tolerance of plants to ammonium nutrition. J. Am. Soc. Hortic. Sci. 94:235-239.

Maynard, D. N., and A. V. Barker. 1979. Regulation of nitrate accumulation in vegetables. Acta Hortic. 93:153-162.

Maynard, D. N., A. V. Barker, P. L. Minotti, and N. H. Peck. 1976. Nitrate accumulation in vegetables. Adv. Agron. 28:71-118.

Minotti, P. L. 1975. Plant nutrition and vegetable crop quality. HortScience 10:54-56.

Peck, N. H., D. J. Cantliffe, R. S. Shallenberger, and J. B. Bourke. 1974. Table beets (*Beta vulgaris* L.) and nitrogen. N.Y. State Agric. Exp. Stn., Geneva. Search 4(6):1-25.

Pew, W. D., and B. R. Gardner. 1972. Nitrogen effects on cantaloupes. Soil Sci. Plant Anal. 3:467-476.

Reese, R. L., and R. C. J. Koo. 1975. Effects of N and K fertilization on internal and external fruit quality of three major Florida orange cultivars. J. Am. Soc. Hortic. Sci. 100:425-428.

Shear, C. B. 1979. Interactions of nutrition and environment on mineral composition of fruits. Acta Hortic. 92:41-50.

Shear, C. B., and M. Faust. 1975. Preharvest nutrition and postharvest physiology of apples. p. 35-42. *In* N. F. Haard and D. L. Sakunkhe (ed.) Postharvest biology and handling of fruits and vegetables. AVI Publ. Co., Westport, Conn.

Somers, G. F., and K. C. Besson. 1948. The influence of climate and fertilizer practices upon the vitamin and mineral content of vegetables. Adv. Food Res. 1:291-324.

Vittum, M. T. 1963. Effect of fertilizers on the quality of vegetables. Agron. J. 55:425-429.

Westwood, M. N., and F. B. Wann. 1966. Cherry nutrition. p. 158-178. *In* N. F. Childers (ed.) Nutrition of fruit crops. Horticultural Publications, Rutgers Univ., New Brunswick, N.J.

Winsor, G. W. 1979. Some factors affecting the quality and composition of tomatoes. Acta Hortic. 93:335-346.

Zieger, D. C. 1978. Nitrogen fertilizing and pruning of apple trees as they affect yield, fruit quality and tree growth in North Carolina. North Carolina Agric. Exp. Stn. Tech. Bull. 254.

43

Walter A. Hill
Tuskegee Institute
Tuskegee Institute, Alabama

Effect of Nitrogen Nutrition on Quality of Three Important Root/Tuber Crops

Potato (*Solanum tuberosum* L.), sugarbeet (*Beta vulgaris* L.), sweet potato [*Ipomoea batatas* (L.) Lam.], and cassava (*Manihot esculenta* Crantz) are the major root/tuber crops produced in the world, and in 1980 had respective yields of 225 718, 247 225, 107 254, and 122 134 t × 10^3 (FAO, 1980; USDA, 1980). In 1980, production yields of the former three crops in the United States were 13 653, 24 115, and 497 t × 10^3, respectively.

Potato tubers are produced for the fresh market and seed, and for processing into chips, shoestring, dehydrated, frozen, and canned products, and for starch, flour, and livestock feed. Potato tubers are an important source of carbohydrate, ascorbic acid, protein, K, Fe, Mg, and several B vitamins. Sugarbeet roots are grown almost exclusively for processing into raw sugar and sugar products. Sweet potato roots are predominantly used for fresh marketing and seed, and are processed into canned or flaked (dehydrated) products, pies, and cakes. Other products from sweet potato roots include starch, flour, and ethanol. Sweet potato roots are an excellent source of carbohydrate and beta carotene (orange flesh varieties) and are a good source of vitamin C, protein, K, Ca, and Fe.

The foliage of sugarbeet and sweet potato can be used as livestock feed, but the high alkaloid content of potato foliage limits its use as a feed. Sweet potato leaves are rich in protein and minerals and are used as a green vegetable in many nations. They are presently being evaluated for utilization in the United States.

A unique component of root/tuber crop production is the determination of the optimum balance between yield and quality of roots/tubers and foliage depending on the final product use. Of the many cultural practices

Nitrogen in Crop Production.

involved, N fertilization is one of the factors that most influences the yield and quality of the final product.

I. POTATO

A. N Rate

1. TUBER YIELD

Experiments to determine the fertilizer N rate required for maximum and economically optimum potato yield have resulted in a range of recommended N rates. Most studies indicate that 80 to 150 kg N/ha is optimum for greatest total tuber yield (Laughlin, 1971; Smith, 1977; Tahtinen, 1978). Rates ranging from 269 to 448 kg N/ha (Painter & Augustin, 1976; Painter et al., 1977) produce greatest total yields in some locations.

Greatest yield of no. 1 or grade A (113–454 g) potatoes are often obtained with applications of 100 to 134 kg N/ha (Munro et al., 1977; Smith, 1977), but in some locations 269 kg N/ha (Painter et al., 1977) is required. Yield reduction of no. 1 potatoes can occur at high N rates (269–896 kg N/ha) as a result of an increased percentage of large, malformed, undersized, or rotton tubers (Augustin et al., 1977; Painter & Augustin, 1976; Painter et al., 1977; Rowberry & Johnston, 1980; Smith, 1977).

2. SPECIFIC GRAVITY, DRY MATTER, AND PERCENT SOLIDS

Mealiness or consistency of cooked potatoes is a very important quality factor for processed and fresh market potatoes (Smith, 1977). A mealy texture is associated with a high percent solids and starch content. For a given variety there is a high correlation between dry matter content of potatoes, as measured by their specific gravity, and the mealiness of the cooked product. Most manufacturers of processed potatoes prefer high specific gravity tubers because less fresh product is required per unit of processed product.

Most studies have shown that increasing the fertilizer N level lowers (Smith, 1977; Painter et al., 1977) or does not influence (Rowberry & Johnston, 1980) specific gravity of tubers. Dry matter content and percent total solids of tubers most often decrease with N rates greater than 100 kg N/ha (Laughlin, 1971; Smith, 1977). Nitrogen application reduces mealiness and increases the tendency for cooked potatoes to slough and fall apart (Smith, 1977; Tahtinen, 1978).

3. COLOR AND FLAVOR

Discoloration of potatoes detracts from their appearance and is considered an undesirable quality. Three distinct types of discoloration are (i) enzymatic discoloration of peeled or raw cut potatoes; (ii) after-cooking darkening; and (iii) nonenzymatic browning at high temperatures of chips, french fries, or dehydrated potatoes (Smith, 1977).

Color of cooked tubers is not influenced by N fertilization rates

(Rowberry & Johnston, 1980; Tahtinen, 1978). Differences in color of french fries, chips, or crisps due to N fertilization rate is insignificant or of no commercial importance (Painter et al., 1977), but lightened chip color due to high N rates has been reported (Smith, 1977). Skin color is lightened by increasing N rates above 269 kg N/ha (Painter et al., 1977). Enzymatic darkening of potatoes increases significantly as the level of N increases up to 250 kg N/ha. The enzymatic darkening has been positively correlated with phenolic content and negatively correlated with lipid content of tubers (Mondy et al., 1979).

Flavor, cooking quality (color and flavor), and eating quality are largely unaffected by application of N rates for optimum yield, but higher N rates can weaken the flavor of tubers (Tahtinen, 1978).

4. PROTEIN AND AMINO ACIDS

The total N content in tubers increases directly with increasing N application rate, with levels ranging from 1.1 to 2.1% on a dry-weight basis for 0 to 720 kg N/ha (Carter & Bosma, 1974; Painter & Augustin, 1976). At economically optimum N rates for tuber production, total N ranges from 1.2 to 1.5%.

Crude protein in tubers averages 10.4% of total solids and increases as fertilizer N rates increase (Shukla & Singh, 1976; Smith, 1977). The crude protein of potato tubers comprises a number of N fractions, the most important of which are the protein fractions and the nonprotein N fractions, the amides, and free amino acids (Eppendorfer et al., 1979). Under normal growing conditions, the two fractions constitute about 50 to 40% of the crude protein, respectively. The amino acid pattern of individual proteins is genetically determined and hence cannot be influenced by fertilizers or other growth factors; the concentration of free amino acids and particularly of amides may be strongly affected. For most potato varieties there is apparently a negative relationship between total N content in tubers and the concentration in the crude protein of many amino acids. Increasing the fertilizer N rate in pot studies increases the N content in tubers and decreases the concentrations of most amino acids, whereas asparagine and glutamine concentrations increase (Eppendorfer et al., 1979). Increases in asparagine and glutamine are probably a result of increases in their amides.

Under field conditions, increasing fertilizer N rate from 40 to 376 kg/ha almost doubled the total free amino acids (alcohol-soluble N) (Hoff et al., 1971). The increases for individual amino acids ranged from none (tyrosine) to 2.7-fold (glutamic acid + glutamine). Aspartic acid and glutamic acid and their corresponding amides constituted the bulk of the free amino acid pool and increased from 55 to 65% with increasing N fertilization. The amino acids could be grouped into four categories with respect to their response to increasing N levels: those giving almost no response (tyrosine and phenylalanine), those giving full response at low N levels (serine and pyruvate families), those giving full response at a higher but intermediate level (aspartate family), and those that responded to the highest level of fertilization (glutamic acid + glutamine).

Compared with Food and Agriculture Organization (FAO) amino acid reference patterns, potatoes of low N content contained a sufficient amount of all essential amino acids except leucine (Eppendorfer et al., 1979). At the intermediate and especially at the higher N content, the concentrations of essential amino acids in the crude protein were lower than recommended by the FAO. The most limiting amino acids were leucine, isoleucine, and the S-containing amino acids.

5. NITRATE-N

High NO_3^- levels in foods may constitute a health hazard because of the precursor role of this compound in forming nitrates (Augustin et al., 1977). Nitrites may cause methemoglobinemia or form carcinogenic nitrosamines. It has been estimated that white potatoes supply 14% of per-capita ingestion of nitrates in the United States.

Nitrate content in tubers increases directly with increasing N application rates and up to 1200 mg/kg (dry-wt basis) has been reported (Augustin et al., 1977; Carter & Bosma, 1974). Nitrate-N levels range from 82 to 859 mg/kg (dry-wt basis) (Augustin et al., 1977) or from 25 to 131 mg/kg on a wet-weight basis (Carter & Bosma, 1974), depending on N rate, N sources, time of application, irrigation, and/or days in storage after harvest. At fertilizer N rates for optimum production, NO_3^--N levels in tubers ranged from 113 to 445 mg/kg (dry-wt basis); these levels are not expected to contribute substantially to the methemoglobinemia health hazard (Augustin et al., 1977; Carter & Bosma, 1974). When excess N is applied, NO_3^--N levels in tubers exceeds tolerable levels for baby food (67 mg/kg on a fresh-wt basis) in 2 out of 3 yr and may be a potential health hazard to infants.

6. OTHER QUALITY PARAMETERS

Starch content decreases with increasing fertilizer N rate (Shukla & Singh, 1976; Smith, 1977; Tahtinen, 1978). As fertilizer N rate increases from 75–100 to 225–250 kg N/ha, ascorbic acid content increases (Mondy et al., 1979) or decreases (Shukla & Singh, 1976), depending on cultivars and growing conditions. Nitrogen rate has no apparent effect on tuber content of Na, Cu, Fe, Zn, Ba, Al, or Mn (Laughlin, 1971).

7. FOLIAGE

Foliage N content directly increases and percent dry matter decreases as applied N rate increases up to 224 kg N/ha (Laughlin, 1971). Increasing fertilizer N rates up to 179 kg N/ha tends to cause increases in Mn, but decreases Ca, Mg, Al, Ba, and Sr in foliage. Increasing the fertilizer N rate apparently does not influence Na, B, Cu, Zn, Mo, and Cr. Inconsistent responses to N were found for Fe content in foliage.

B. N Sources

When urea, diammonium phosphate, and ammonium nitrate (AN) were compared, specific gravity, texture, color, and flavor of cooked potatoes showed minor differences in some years, but did not follow any regular pattern (Rowberry & Johnston, 1980). Differences in specific gravity of tubers have not been found for ammonium sulfate (AS), AN, or sodium nitrate (Smith, 1977). Ammonium sulfate can result in greater ascorbic acid and starch content in tubers than urea for potatoes grown under dry-season conditions (Shukla & Singh, 1976). Comparisons of the effect of AS and urea as N sources on cooking quality of potatoes have not provided consistent results.

C. Irrigation and Time and Method of N Application

Irrigation increases total yield, grade, and quality of tubers in some locations (Wild & Asfary, 1980), but has no influence in others (Painter & Augustin, 1976). Apparent recovery of fertilizer N was 73 to 77% with irrigation and 58 to 71% without irrigation (Wild & Asfary, 1980), because of increased plant growth and N uptake with the former. Inadequate irrigation and resultant moisture stress, combined with excessive N, result in NO_3^--N levels in tubers that may result in methemoglobinemia (Augustin et al., 1977). Nitrate content in tubers is reduced by increasing the amount of irrigation water and using slow-release forms of N. This reduction results from movement of soil N below the root zone by greater water applications and lower levels of available N from the slow-release fertilizers. Application of N after tuber set and sprinkler-applied N during the growing season significantly increases NO_3^--N levels in tubers when there is inadequate irrigation to move soil nitrates below the root zone.

Increasing from a single application to two or three split applications of fertilizer N results in greater protein content and reduced starch, ascorbic acid, and dry matter contents of tubers grown in sandy loam soils with low organic matter levels. Increased N availability from split applications apparently enhances N uptake and protein synthesis, and depresses carbohydrate and ascorbic acid levels (Shukla & Singh, 1976). Application of 134 to 393 kg N/ha by broadcasting with soil incorporation results in higher yields of U.S. no. 1 grade tubers than does the same rates applied by banding (Painter & Augustin, 1976). Banding results in excessive NO_3^--N levels in petioles, which apparently leads to an increase in cracks and culls. When fertilizer N at 0 to 224 kg N/ha is mixed with soil, both U.S. no. 1 and dry matter yields increase with each N increment through 179 kg/ha. Stand reduction and decreased vigor and foliage dry matter occur when row-placed N exceeds 134 to 179 kg N/ha (Laughlin, 1971). With row placement of the fertilizer, U.S. no. 1 yield and dry matter are highest at 45 to 80 kg N/ha. A number of studies indicate that for ≥ 120 kg N/ha, broadcast application is generally regarded as superior to band placement, but when ≤ 60 kg N/ha are supplied, row application is usually recommended for optimum U.S. no. 1 yield.

II. SUGARBEET

A. N Rate

1. ROOT YIELD

Numerous experiments have been carried out to determine the N rate most efficient for optimum sugarbeet yield and quality. Nitrogen rates usually applied in the United States range from 56 to 180 kg N/ha, depending on the geographical location and NO_3^- content of the soil, but rates of up to 364 kg N/ha are applied in some locations (Hills & Ulrich, 1971).

2. SUGAR CONTENT AND YIELD

Substantial evidence has shown that percent sugar content is inversely related to fertilizer N rate and ranges from 13% sucrose at 448 kg N/ha to 20% at 0 kg N/ha (Carter & Traveller, 1981; Cole et al., 1976; Halvorson & Hartman, 1980; James et al., 1977). Sugar (sucrose) yield has been found to be economically optimum at N rates of 70 to 120 kg N/ha (Carter & Traveller, 1981; Halvorson & Hartman, 1980); these rates may be substantially lower than rates required for maximum root yield (Halvorson & Hartman, 1980). In some studies, decreases in percent sugar resulting from high rates of applied N were compensated for by an increased root yield, and sugar yield was not changed; in most cases, root yield does not compensate for the decline in sucrose content resulting from high N supply (Halvorson & Hartman, 1980; Hills & Ulrich, 1971). When soil test N index is < 30, applied N is required for optimum sugar yield in roots, but N application should be avoided when soil test N index is > 40 (James et al., 1977).

Crown tissue contributes from 12 to 20% of the total recoverable sucrose/ha when N is adequate for highest sucrose production in roots (Halvorson & Hartman, 1980). Sucrose content in crowns decreases from 15.3 to 7.9% as applied N increases from 0 to 448 kg N/ha (Cole et al., 1976).

3. QUALITY

Beet quality is defined by sucrose content on a percent fresh-weight basis and by purity. Purity is measured as a percentage of sucrose relative to the total solids and indicates the content of soluble nonsucrose components (Alexander, 1971; Smith & Martin, 1977). Each gram of nonsucrose substance in extracted juice prevents 1.5 to 1.8 g of sucrose from crystallizing and reduces processed sugar yield. About 70% of the nonsugars in extracted juice are K and Na salts, amino acids, and betaine.

Total and amino N, betaine, NO_3^-, Na, K, P, B, and Zn are nonsucrose constituents that increase with increasing N application rate resulting in decreased solubility and crystallization rate of sucrose (Cole et al., 1976; Smith & Martin, 1977). Ash, amino N of beet pulp, alkalinity, pH, Ca, salt content, and color of thick juice decrease with excess applied N.

In general, juice content decreases with increasing N rate, but for some cultivars and growing conditions, N rates for optimum root yields increase juice content compared with treatments without N. Juice purity is inversely related to fertilizer N rate (Carter & Traveller, 1981; Cole et al., 1976; Halvorson & Hartman, 180), with the coefficient of juice purity (purity index) ranging from 90 to 83.7% for rates from 50 to 200 kg N/ha (Parashar, 1976).

4. FOLIAGE

Increasing fertilizer N rate directly increases weight (dry matter) of sugarbeet tops (Carter & Traveller, 1981; Greenwood et al., 1980; Halvorson & Hartman, 1980). High N rates increase the ratio of leaves to storage root dry weight, even when the total dry matter is unaffected (Greenwood et al., 1980). Highest sucrose yields have been obtained when dry matter root/top ratios were $\geq$ 2 (Halvorson & Hartman, 1975).

B. N Sources

Experiments to compare fertilizer N source effect on yield and sucrose content of roots have not resulted in differences for calcium nitrate (CN), AN, AS, urea, and ammonium hydroxide. If anhydrous ammonia (AA) is used, it must be placed at least 10 cm outside the seed row and 15 cm from the top of the planting beds so as to avoid toxicity damage during germination or seedling growth (Hills & Ulrich, 1971).

C. Time and Method of N Application

Most field experience indicates that a single application at planting or thinning produces results comparable to splitting the total amount required into two or three applications (Hills & Ulrich, 1971). Delaying N application past midseason (mid-July) increases top dry matter, but decreases dry matter production, rate of accumulation of stored sucrose, and sucrose content of roots (Carter & Traveller, 1981; Hills & Ulrich, 1971; Roberts et al., 1981). Applying optimum N at sowing and additional N at singling decreases sugar content and increases K + Na contents (Last & Draycott, 1977).

Applying N in sprinkler irrigation produces lower sucrose content and slightly lower yields when compared with N applied earlier in a topdressed treatment (Roberts et al., 1981). In general, any practice or condition that provides beets with N late in the season will lower quality. Nitrogen rates and application times are considered optimum when plants are adequately supplied for vigorous top and root growth, but at 4 to 6 weeks prior to harvest, plants should become sufficiently deficient in N to retard utilization of sucrose for top growth and allow sugar to accumulate in the storage roots.

Harvest date may significantly influence sucrose yield and N requirement for optimum sucrose yield. The N requiement for optimum sugar yield was found to be 50 to 90 kg N/ha greater for late than early harvested roots. The higher N requirement was due to a more rapid rate of root growth for plants receiving the higher rate (Alexander, 1971; Hills & Ulrich, 1971).

Row placement of N with sowing decreases root yield and increases noxious N content of sugarbeet when compared with broadcasting (Last & Draycott, 1977). Surface application of N with light incorporation produced the same root yields and sucrose content as placement 10 to 15 cm below the furrows (James et al., 1977).

D. Cultivar × N

Evaluation of yield, sucrose, and purity for eight cultivars at five N levels, and five cultivars at eight N levels showed that sugarbeet cultivars responded similarly to N level (Halvorson & Hartman, 1980). Significant differences were found among three cultivars for juice purity, sucrose content, and eight other quality parameters (ash, total N, amino N, NO_3^-, bentaine, Na, K, and Cl) (Smith & Martin, 1977). Plant density and cultivar had more influence on juice purity and the eight quality parameters than soil N (two levels); there was a low incidence of N level, cultivar, or plant population interaction. These and other studies previously summarized (James et al., 1978) suggest that the interaction between soil N fertility and sugarbeet genotype do not influence yield or quality of sugarbeet and that N recommendations need not be varied for cultivar. However, significant interactions were found between 20 genotypes grown at four N levels for root yield, sucrose percent, and gross sugar and impurity index (James et al., 1978). The interactions indicate that genetic variation exists and new cultivars may be developed that give higher sugar production at low N levels or are not affected adversely in sucrose percent or quality at a high N level. The latter results emphasize the need for careful selection of genotypes that produce well under environmental stress in future sugarbeet yield and quality trials.

E. Irrigation × N

With N levels ranging from 0 to 224 kg N/ha, there was little difference between the effect of furrow and sprinkler irrigation on N uptake or sucrose yield of sugarbeet when water supply was adequate (Carter et al., 1975). Variety × N × water interactions have been shown to influence the impurity index and emphasize the need for work to link N × water interactions in soil (denitrification, leaching of N with excess water, etc.) with selection of appropriate sugarbeet genotype, i.e., nonresponsive to N or water or both (James et al., 1978).

F. Plant Density × N

Highest sugar yields are obtained by balancing the supply of N with the plant population for a particular locale. With adequate but not excessive N, studies with plant densities ranging from 22 000 to 134 000 plants/ha resulted in a rapid rate of yield increase up to 50 000 plants/ha, followed by a slower rate that stabilized at > 80 000 plants/ha (Draycott & Durrant, 1974). At the highest density, the amount of applied N needed for maximum sugar yield was least.

Impurities in sugarbeet concentrate generally decrease with increasing plant density from 19 800 to 118 900 plants/ha (Smith & Martin, 1977). Plant populations of 80 000 plants/ha with addition of 80 to 150 kg N/ha, depending on soil NO_3^- content, are needed for maximum sugar content and juice purity (Draycott & Durrant, 1974; Hofer et al., 1979). The adverse effects of high N rates on quality are especially noticeable at low plant densities of 40 000 plants/ha (Hofer et al., 1979).

III. SWEET POTATO

A. N Rate

1. ROOT YIELD

Depending on cultivar, cultural practices, soil type, climatic condition, and product use, recommended N rates for sweet potato range from 0 to 146 kg N/ha (Edmund & Ammerman, 1971; Hill, 1982, 1984; Miller & Covington, 1982; Onwueme, 1978; Wilson et al., 1980). Few studies have been reported that indicate more than 50 to 100 kg N/ha are required to obtain optimum sweet potato root yields. Where recommended rates are ≥ 100 kg N/ha, sweet potatoes are usually grown for commercial production on sandy soils where rapid leaching of nitrates can occur. In studies where root yield was not responsive to fertilizer N application or required low rates of N for optimum yield and quality, N was provided from legumes or manure plowed in prior to planting. Other possible sources of the N include residual N from previous fertilizer N applications or mineralized N from soil organic matter. Recent studies suggest that N_2-fixing bacteria may be associated with sweet potato roots (Hill & Bacon, 1984; Hill et al., 1983).

2. ROOT QUALITY

Important quality parameters for fresh-market and processed sweet potatoes include mouth feel, size and shape, baking quality, color, firmness, flavor, nutrition, carbohydrate content, pithiness, weight loss during storage, and others (Technical Committee S-101, 1980). Favorable texture or mouth feel is based on dryness or moistness, and different ethnic groups or cultures have different preferences. In the United States, most consumers prefer sweet potatoes that have a size of 6.3 to 10.0 cm in diameter and 10 to 22.5 cm in length (U.S. no. 1's). A deep orange color indicates a high beta carotene (vitamin A precursor) content and is preferred for orange flesh varieties, while a low beta carotene content is preferred for white flesh varieties. Firmness is important for consistently producing acceptable canned sweet potatoes. Cooking quality is a subjective parameter composed of flavor (sweetness, moistness, and unknown compounds), texture, color, and fiber content. Carbohydrate content determines energy value and ethanol production potential. Pithiness results from internal breakdown of tissue and is related to the extent of intercellular space. Nutritional status is measured by carbohydrate, vitamin, mineral, and protein content.

Increasing fertilizer N rate between 0 and 100 kg/ha linearly increases firmness and decreases percent fibers of fresh and canned roots (Constantin, 1974, 1984). Yield of no. 1 roots, percent dry matter, and flesh color (carotenoid pigments) increase linearly as the N rate increases from 0 to 67 kg N/ha, but percent dry matter decreases as N applied increases from 101 to 202 kg/ha (Constantin, 1984; Hammett & Miller, 1982). Though starch content of roots can be increased by 0.5 to 1.7% by increasing the N rate from 0 to 90 kg N/ha (Bartolini, 1982), carbohydrates, reducing sugars, sucrose, total sugars, ascorbic acid level, root intercellular space, and cortex thickness are generally not influenced by N rate (Constantin et al., 1974; Hammett, 1981; Hammett & Miller, 1982).

Concentration of N, total N, and percent protein in foliage and roots increase with increasing fertilizer N level (Constantin et al., 1974; Hammett, 1981; Miller & Covington, 1982). When the recommended N rate is applied to sweet potato, protein content in roots ranges from 1.7 to 11.8% for different cultivars and breeding lines (Purcell et al., 1972). Sulfur amino acids and tryptophan are limiting when compared with FAO reference protein. Other essential amino acids are in excess. Total protein is 90% amino acids and nonprotein N is 10% (Purcell et al., 1972). The nonprotein N is composed of asparagine, aspartic acid, glutamic acid, serine, and threonine (Purcell & Walter, 1980). Nonprotein N is a part of a metabolically active N pool, can be made available for synthesis of amino acids and transamination reactions, and may be linked to subsequent sprouting.

B. Irrigation and Time and Method of N Application

A single application or two or three split applications are commonly recommended for application of N to sweet potato (Hill, 1984; Edmund & Ammerman, 1971; Onwueme, 1978; Wilson et al., 1980). Although some studies indicate that timing and placement do not influence root yields, other studies have shown that yield, percent protein, marketable/nonmarketable ratio of roots, other quality parameters, and fertilizer N use efficiency are influenced by timing and placement. Late season, supplemental N applied broadcast at 25 and 50 kg N/ha (Hammett, 1981) or 20 kg N/ha (Miller &

Covington, 1982) reduces root yields. Foliar N at 15 or 60 kg N/ha applied 40 d after transplanting (preplant N was 30 kg N/ha) also caused yield reductions (Miller & Covington, 1982). Delaying the time for application of the recommended N rate causes vine elongation and reduces root growth (Morita, 1970), but application of the recommended N rate in five split applications via drip irrigation produced greater marketable and the same total yields as two split applications of sidedressed N (Hill & Bacon, 1984). The latter result was attributable to optimum N availability from the five split applications during maximum foliage growth and storage root bulking.

A low rate of broadcast (30 kg/ha) plus foliar-applied (15 kg/ha) N results in higher dry matter content of sweet potato roots than conventional N application at the same and higher rates (Miller & Covington, 1982). Late-season supplemental additions of N at 22.4 or 44.8 kg/ha do not influence dry matter of roots when 81 and 47 kg N/ha have been applied as preplant and side-dressed (Hammett, 1981).

Late-applied supplemental N increases Ca and Na, but does not influence P, K, Fe, Mg, N, carotene, and ascorbic acid contents of roots. Root cracking, firmness, and weight loss after 112 d of storage is not influenced by form or rate of late-applied N. The intercellular space of roots is decreased by late addition of N and is lower for late-applied AN than for CN (Hammett, 1981).

Irrigation applied at appropriate times can produce significant increases in sweet potato root yields (Constantin et al., 1974; Li, 1970). Irrespective of N rate from 0 to 100 kg/ha, excessive irrigation reduces yield, percent dry matter, flesh color, and percent protein in roots (Constantin et al., 1974). Drip irrigation also decreases percent dry matter and the percent protein in roots when compared with nonirrigation (Hill & Bacon, 1984).

Starter solutions with 3 to 10% N produce marked increases in marketable (no. 1 and/or no. 2) yields in some locations and moderate or no increases in others (Edmund & Ammerman, 1971).

C. N Source

Sodium nitrate, ammonium nitrate, and urea produced greater yields than AS for 4 yr and AA produced satisfactory results in 3 out of 4 yr (Edmund & Ammerman, 1971). After basal application of AN, side-dressing with AS resulted in lower root yields than side-dressing with AN or urea, but increased the number of roots (Morita, 1970). Acidification of already acid soils by AS apparently causes reduced root size. No differences in root yields were obtained for urea or the slow-release N fertilizers, sulfur-coated urea, or isobutylidene diurea (Hill, 1982).

D. Cultivar × N

It has been suggested that response of sweet potato root yield to fertilizer N may be dependent on type of variety used (Haynes, 1970). Short-vine–bush types, long-vine types with internodes exceeding 10 cm, and intermediate length types have been described; 100 sweet potato varieties have been classified into N-responsive, N-indifferent, and N-depressive types according to the effects of fertilizer N application on final root yield (Wilson, 1977). It has been suggested that plants with small foliage capable of high yields with close spacing and intensive fertilizer use be used in intensive horticultural practice, and that subsistence farmers use plants with long vines capable of adequate yield in the absence of fertilizer applications (Haynes, 1970). Although differences in cultivar yield response to N have occurred, few studies have attempted to correlate cultivar type with yield and quality response to N level. This oversight may explain the apparent lack of response to fertilizer N reported in some studies.

E. K × N

Considerable evidence has been obtained to indicate that yield and quality of sweet potato roots are influenced by the interaction of N and K (Fujise & Tsuno, 1967; Li, 1970; Miller & Covington, 1982; Speights et al., 1967). Storage root dry weight is directly related to K_2O/N ratio in roots and foliage; high K_2O/N ratios cause an increase in tuber water content, acceleration of respiration rate, increased relative growth rate, acceleration of the translocation of photosynthates from leaves to roots, and acceleration of photosynthetic activity of leaves (Fujise & Tsuno, 1967). An imbalance of N/K ratio results in a decrease in secondary xylem and cambial tissue and insufficient production of meristematic tissue (Speights et al., 1967). High N and K rates decrease the L/D (length/diamter) ration and dry matter content of roots (Edmund & Ammerman, 1971; Hammett & Miller, 1982). Some evidence suggests that a N/K imbalance in sweet potato tissue may result in nutrient deficiencies. Good storage quality of sweet potato roots has been obtained by applying N and K in the ratio of 1:1.4 (Hammett & Miller, 1982).

F. Foliage Yield and Quality

Yield and quality of sweet potato foliage are important where used as greens for human consumption or feed for animals. Premature shoot development, lateral branching of shoots, and foliage production of sweet potato are directly related to fertilizer N rate with vine yields increasing with each increment up to 100 to 120 kg N/ha (Li, 1970; Villareal et al., 1979; Wilson, 1977). Weight and number of sweet potato leaf tips grown for human consumption were higher for applications of 120 than 0 or 60 kg N/ha, but oxalate and carbohydrate contents in leaves were reduced by N application (Villareal et al., 1979). Five split applications of the recommended N rate reduced foliage weight and protein content compared with two split applications (Hill & Bacon, 1984) because of lower N availability and uptake prior to storage root bulking.

ACKNOWLEDGMENT

The author expresses special appreciation to Pansy Bacon-Hill for critically reviewing and making invaluable contributions toward finalizing this manuscript.

REFERENCES

Alexander, J. T. 1971. Factors affecting quality. p. 371–381. *In* R. T. Johnson et al. (ed.) Advances in sugarbeet production: principals and practices. Iowa State Univ. Press, Ames, Iowa.

Augustin, J., R. E. McDole, and C. G. Painter. 1977. Influence of fertilizer, irrigation, and storage treatments on nitrate-N content of potato tubers. Am. Potato J. 54:125–136.

Bartolini, P. U. 1982. Timing and frequency of topping sweet potato at varying levels of nitrogen. p. 209–214. *In* R. L. Villareal and T. D. Griggs (ed.) Sweet potato—Proc. 1st Int. Symp., Shanhua, Tainan, Taiwan, China. 23–27 Mar. 1981. Asian Vegetable Research Center, Taiwan.

Carter, J. N., and S. M. Bosma. 1974. Effect of fertilizer and irrigation on nitrate-nitrogen and total nitrogen in potato tubers. Agron. J. 66:263–266.

Carter, J. N., C. H. Pair, and D. T. Westermann. 1975. Effect of irrigation method and late season nitrate-nitrogen concentration on sucrose production of sugar beets. J. Am. Soc. Sugar Beet Technol. 18:332–342.

Carter, J. N., and D. J. Traveller. 1981. Effect of time and amount of nitrogen uptake on sugar beet growth and yield. Agron. J. 73:665–671.

Cole, D. F., A. D. Halvorson, G. P. Hartman, J. E. Etchevers, and J. T. Morgan. 1976. Effect of nitrogen and phosphorus on percentage of crown tissue and quality of sugar beets. N.D. Farm Res. 33(5):26–28.

Constantin, R. J., T. P. Hernandez, and L. G. Jones. 1974. Effects of irrigation and nitrogen fertilization on quality of sweet potatoes. J. Am. Soc. Hortic. Sci. 99:308–310.

Constantin, R. J., L. G. Jones, H. L. Hammett, T. R. Hernandez, and C. G. Kahlich. 1984. The response of three sweet potato cultivars to varying levels of nitrogen. J. Am. Soc. Hortic. Sci. 109(5):610–614.

Draycott, A. P., and M. J. Durrant. 1974. The effect of cultural practices on the relationship between plant density and sugar yield. J. Int. Inst. Sugar Beet Res. 6:176–185.

Edmund, J. B., and G. R. Ammerman. 1971. Sweet potatoes: production, processing, marketing. AVI Publ. Co., Westport, Conn. p. 106–122.

Eppendorfer, W. H., B. O. Eggum, and S. W. Bille. 1979. Nutritive value of potato crude protein as influenced by manuring and amino acid composition. J. Sci. Food Agric. 30:361–368.

Food and Agriculture Organization. 1980. FAO production year book. Vol. 34. FAO, Rome, Italy.

Fujise, K., and Y. Tsuno. 1967. Effect of potassium on the dry matter production of sweet potato. Proc. Int. Symp. Trop. Root Crops 1(2):20–33.

Greenwood, D. J., T. J. Cleaver, M. K. Turner, J. Hunt, K. B. Niendorf, and S. M. H. Loquens. 1980. Comparison of the effects of nitrogen fertilizer on the yield, nitrogen content and quality of 21 different vegetable and agricultural crops. J. Agric. Sci. 95: 472–485.

Halvorson, A. D., and G. P. Hartman. 1975. Long-term nitrogen rates and sources influence sugar beet yield and quality. Agron. J. 67:387–393.

Halvorson, A. D., and G. P. Hartman. 1980. Response of several sugar beet cultivars to N fertilization: yield and crown tissue production. Agron. J. 72:665–669.

Hammett, L. K. 1981. Effects of late-season nitrogen and foliar calcium applications on sweet potatoes. HortScience 16:336–337.

Hammett, L. K., and C. H. Miller. 1982. Influence of mineral nutrition and storage on quality factors of 'Jewel' sweet potatoes. J. Am. Soc. Hortic. Sci. 107:972–975.

Haynes, P. 1970. Some general and regional problems of sweet potato (*Ipomea batatas* (L.) (Lam.) growing. Trop. Root Tuber Crops Tomorrow 1:10–12.

Hill, W. A. 1982. Nitrogen fertility and uses of sweet potato—past, present and future. p. 89–112. *In* Proc. L. T. Kurtz Colloq. on Sci. in a Changing Agric., Univ. of Illinois. Univ. of Illinois, Urbana.

Hill, W. A. 1984. Sweet potatoes. *In* D. L. Plucknett and H. B. Sprague (ed.) Detecting mineral nutrient deficiencies in tropical and temperate crops. Westview Press, Boulder, Colo. (In press).

Hill, W. A., and P. Bacon. 1984. Fertilizer N use efficiency and associative N_2 fixation of sweet potato. Proc. 6th Int. Symp. on Trop. Root Crops, International Potato Center, Lima, Peru. 20–25 Feb. 1983 (In press).

Hill, W. A., P. Bacon-Hill, S. M. Crossman, and C. Stevens. 1983. Characterization of N_2-fixing bacteria associated with sweet potato roots. Can. J. Microbiol. 29:860–862.

Hills, F. T., and A. Ulrich. 1971. Nitrogen nutrition. p. 111–115. *In* R. T. Johnson et al. (ed.) Advances in sugarbeet production: principals and practices. Iowa State Univ. Press, Ames, Iowa.

Hofer, H., P. Vullioud, and U. Walther. 1979. The influence of nitrogen fertilization on yield and quality of sugar beet at different stand densities. Mitt. Schweiz. Landwirtsch. 27: 113–128.

Hoff, J. E., C. M. Jones, G. E. Wilcox, and M. D. Castro. 1971. The effect of nitrogen fertilization on the composition of the free amino acid pool of potato tubers. Am. Potato J. 48:390–394.

James, D. W., D. L. Doney, J. C. Theurer, and R. L. Hurst. 1978. Sugar beet genotype, N, and soil moisture availability interactions in components of beet yield and quality. Agron. J. 70:525–531.

James, D. W., F. J. Francom, R. F. Wells, and D. W. Sisson. 1977. Control of soil fertility for high sugar beet yield and quality. Utah Agric. Exp. Stn. Bull. no. 496.

Last, P. J., and A. P. Draycott. 1977. Methods of fertilizer application for sugar beet. Exp. Husb. 32:56–64.

Laughlin, W. M. 1971. Production and chemical composition of potatoes related to placement and rate of nitrogen. Am. Potato J. 48:1–15.

Li, L. L. 1970. Study of the effect of nitrogen, phosphorus, and potassium on sweet potato yield by response surface. Trop. Root Tuber Crops Tomorrow 1:13–15.

Miller, C. H., and H. M. Covington. 1982. Mineral nutrition studies with sweet potatoes during a three year period. North Carolina Agric. Res. Service Tech. Bull. no. 273.

Mondy, N. I., R. L. Koch, and S. Chandra. 1979. Influence of nitrogen fertilization on potato discoloration in relation to chemical composition. 2. Phenols and ascorbic acid. J. Agric. Food Chem. 27:418–420.

Morita, T. 1970. Effects of the application time of nitrogenous fertilizer on the top growth and the development of the root system during the tuber formation period in sweet potato. J. Jpn. Soc. Hortic. Sci. 39(4):41–47.

Munro, D. C., R. P. White, and J. B. Sanderson. 1977. Effects of applied nitrogen on yields, tuber sizes and specific gravities in two potato cultivars. Can. J. Plant Sci. 57:803–810.

Onwueme, I. C. 1978. The tropical tuber crops. John Wiley & Sons, Inc., New York.

Painter, C. G., and J. Augustin. 1976. The effect of soil moisture and nitrogen on yield and quality of the Russet Burbank potato. Am. Potato J. 53:275–284.

Painter, C. G., R. E. Ohms, and A. Walz. 1977. The effect of planting date, seed spacing, nitrogen rate and harvest date on yield and quality of potatoes in Southwestern Idaho. Univ. of Idaho Agric. Exp. Stn. Bull. no. 571.

Parashar, K. S. 1976. Effect of nitrogen levels on the quality of sugar beet juice. Indian J. Agron. 21:333–335.

Purcell, A. C., H. E. Swaisgood, and D. T. Pope. 1972. Protein and amino acid content of sweet potato cultivars. J. Am. Soc. Hortic. Sci. 97:30–33.

Purcell, A. C., and W. M. Walter, Jr. 1980. Changes in composition of the non-protein-nitrogen fraction of 'Jewel' sweet potatoes (*Ipomoea batatas* Lam.) during storage. J. Agric. Food Chem. 28:842–844.

Roberts, S., W. H. Weaver, and A. W. Richards. 1981. Sugar beets response to incremental application of nitrogen with high frequency sprinkler irrigation. Soil Sci. Soc. Am. J. 45: 448–449.

Rowberry, R. G., and G. R. Johnston. 1980. Alternative sources of nitrogen and phosphorus in potato fertilizer. Am. Potato J. 57:543–552.

Shukla, D. N., and S. J. Singh. 1976. Effect of sources, rates and time of nitrogen application on nutritive and culinary value of potato. Fert. Technol. 13:103–106.

Smith, O. 1977. Potatoes: production, storing, processing. AVI Publ. Co., Westport, Conn.

Smith, G. A., and S. S. Martin. 1977. Effects of plant density and nitrogen fertility on purity components of sugar beet. Crop Sci. 17:469–472.

Speights, D. E., E. E. Burns, D. R. Paterson, and W. H. Thames. 1967. Some vascular variations in the sweet potato root influenced by mineral nutrition. Proc. Am. Soc. Hortic. Sci. 91:478–485.

Tahtinen, H. 1978. Nitrogen and potassium fertilization of potato. J. Sci. Agric. Soc. Finl. 80: 67–77.

Technical Committee S-101. 1980. Sweet potato quality. Southern Coop. Series Bull. 249, USDA-SEA-AR, Athens, Ga.

U.S. Department of Agriculture. 1980. Agricultural statistics. U.S. Government Printing Office, Washington, D.C.

Villareal, R. L., S. K. Lin, L. S. Chang, and S. H. Lai. 1979. Use of sweet potato (*Ipomoea batatas*) leaf tips as vegetables. II. Evaluation of yield and nutritive quality. Exp. Agric. 15:113–116.

Wild, A., and A. F. Asfary. 1980. Uptake of nitrogen in relation to growth of the potato crop. J. Sci. Food Agric. Abstr. 31:841.

Wilson, L. A. 1977. Root crops. p. 187–236. *In* P. Alvim and T. T. Kozlowsk (ed.) Ecophysiology of tropical crops. Academic Press, New York.

Wilson, L. G., C. W. Averre, J. V. Baird, E. A. Estes, K. A. Sorenson, E. O. Beasley, and W. A. Skroch. 1980. Growing and marketing quality sweet potatoes. North Carolina Agric. Ext. Service Bull. no. AG-09.

44

Darrell W. Nelson
Purdue University
West Lafayette, Indiana

Effect of Nitrogen Excess on Quality of Food and Fiber[1]

There has been periodic scientific interest in the accumulation of nonprotein N by plant species. The interest has usually focused on the health aspects related to the presence of nitrate (NO_3^-), nitrite (NO_2^-), or prussic acid (HCN) in human food or livestock feed. However, in recent years, concern has been expressed about the presence of nitrosamines or nitrosamine precursors (secondary amines and NO_2^-) in food. Scientific interest has been spurred by instances of NO_3^- or HCN poisoning of livestock or concerns related to N cycling in the environment. Although a number of reviews (Maynard et al., 1976; Natl. Res. Council, 1972, 1978; Wright & Davidson, 1964) have been completed on various aspects of NO_3^-, NO_2^-, and nitrosamines in the environment, no recent publication has been concerned with the levels of these compounds in plants. The purpose of this paper is to summarize recent information on the accumulation of "undesirable" N compounds in plants and to assess the significance of these accumulations.

I. N METABOLISM IN PLANTS[2]

Crops are normally fertilized with ammoniacal forms of N [e.g., anhydrous ammonia, urea, diammonium phosphate). However, added NH_4^+ is readily converted to NO_3^- through the process of nitrification and, thus, NO_3^- is the form of N normally available in soil for assimilation by plants. Nitrate is present in the soil solution and moves to the root surface by mass flow as plants transpire water. Nitrate is taken into the root by an active

[1] A contribution of the Indiana Agric. Exp. Stn., Purdue University, W. Lafayette, IN 47907. Journal Paper no. 8725.

[2] Scientific names of all plant species mentioned are given in the Appendix.

Nitrogen in Crop Production.

transport mechanism and most of it is transported to the leaves through the plant vascular system. Once in the leaves, NO_3^- is reduced to NH_3 and incorporated into amino acids. Reduction of 1 mol of NO_3^- requires 16.2 kcal of energy, which is derived from metabolism of carbohydrates produced in photosynthesis (Hageman, 1980). Reduction of NO_3^- in leaves is controlled by the level of nitrate reductase (NR) present in the cytoplasm. Nitrate reductase has a high turnover rate and the activity at any time is dependent upon a variety of environmental factors influencing the plant. The level of NR activity in plants is controlled through regulation of enzyme synthesis (Beevers & Hageman, 1969). Thus, adverse environmental conditions that decrease photosynthesis apparently also slow down synthesis of NR so that NO_2^- or NH_3 (both toxic to plants) or other intermediates do not accumulate in the cell.

If environmental conditions such as drought or decreased light intensity decrease photosynthesis, NO_3^- tends to accumulate in plant tissues. Nitrate accumulation may also occur when rates of applied N are much greater than that which the crop can effectively convert to protein. The excess NO_3^- in plants is often stored in stalks, leaf sheaths, and leaf petioles. Plant species vary greatly in amounts of NO_3^- accumulated because of differences in NO_3^- uptake efficiency, NR activity, and photosynthetic capacity.

II. HAZARDS OF NITRATE, NITRITE, PRUSSIC ACID, AND NITROSAMINES IN FOOD

Ingestion of foods containing NO_3^- or drinking NO_3^--contaminated water may lead to health problems in humans and animals. Nitrate, per se, is relatively nontoxic to mammals being readily absorbed and excreted. However, in infants and ruminant livestock NO_3^- can be, under certain circumstances, reduced to NO_2^- by bacteria in the gastrointestinal tract. Nitrite that reaches the bloodstream reacts with hemoglobin to produce methemoglobin, thereby reducing the oxygen-transport capacity of the blood system. The clinical symptoms of methemoglobinemia (blue-baby disease) become detectable when 10% of the hemoglobin has been converted to methemoglobin. Death occurs in infants when their blood contains 50 to 75% methemoglobin.

Methemoglobinemia is much more prevalent in infants under 3 months of age than older children or adults because (i) the stomach pH in infants is higher than in adults, thereby encouraging bacterial growth in the stomach and upper intestine; (ii) red blood cells of infants are deficient in enzymes that convert methemoglobin to hemoglobin; and (iii) fetal hemoglobin may be more readily oxidized to methemoglobin than adult hemoglobin. Treatment of methemoglobinemia in infants usually consists of a reduction in NO_3^- intake.

There are many reports of NO_3^- poisoning (methemoglobinemia) in livestock, particularly ruminants (Cawley et al., 1977; Edwards & McCoy,

1980; Jones & Jones, 1977; Wright & Davidson, 1964). The rumen digestion process provides opportunity for NO_2^- formation from NO_3^-. The LD_{50} for ruminants varies from 160 to 224 mg NO_3^--N/kg of body weight when NO_3^- is included in the roughage portion of the ration. The LD_{50} for cattle (*Bos* spp.) is 45 to 50 mg NO_2^--N/kg of body weight (Wright & Davidson, 1964). Roughage has been considered to be potentially toxic to ruminants if the NO_3^--N content exceeded 0.133% (Edwards & McCoy, 1980), 0.21% (Pratt et al., 1976), 0.34 to 0.45% (Wright & Davidson, 1964), or 0.44% (Rhykerd & Noller, 1977). Monogastric animals are not readily poisoned by NO_3^-, but are highly susceptible to NO_2^- poisoning. The horse (*Equus caballus*) appears to be more susceptible to NO_3^- poisoning than other monogastrics, because the enlarged caecum and colon allows some bacterial reduction of NO_3^- to NO_2^-. Sheep (*Ovis aries*) and cattle fed poor quality diets are more susceptible to NO_3^- poisoning than those fed adequate diets. Animals stressed by heat or cold are more susceptible than unstressed cattle and sheep (Wright & Davidson, 1964). Cattle and sheep adapt, within limits, to high levels of NO_3^- in feed by reduction of both NO_3^- and NO_2^- in the rumen and by an increase in red blood cells.

Oher types of sublethal toxicity associated with NO_3^- ingestion include (i) vasodilation of the circulatory system, (ii) lowered growth and milk production, (iii) abortions, and (iv) impaired vitamin A and iodine (I) nutrition (Wright & Davidson, 1964; Natl. Res. Council, 1972, 1978). It has been known for many years that NO_2^- ingestion results in a drop in blood pressure; however, there is little evidence suggesting that NO_3^- in food causes a similar effect. Several studies have shown that rate of gain and daily milk production decrease as the NO_3^- level in the ration exceeds a "critical" level; however, much of the decrease in production may result from reduced feed intake observed when roughages high in NO_3^- are presented to ruminants (Wright & Davidson, 1964). Feeding roughages containing high levels of NO_3^- has been reported to induce abortions in cattle and sheep. Apparently, levels of NO_3^- in feed high enough to cause abortions in ruminants will also lead to methemoglobinemia (Wright & Davidson, 1964). High NO_3^- levels in feed and water have also been reported to induce vitamin A deficiencies in cattle, sheep, and swine (*Sus scrofa domesticus*) by destroying carotene or interfering with utilization of the vitamin. However, in most cases, vitamin A deficiencies have only been noted in survivors of acute NO_3^- toxicity. Nitrate can also function as an antithyroid substance, thereby necessitating high I levels in the ration. Observations suggest that NO_3^--induced thyroid pathology is most frequent in cold environments. Wright and Davidson (1964) concluded that a ruminant would most probably die from acute toxicity before the NO_3^- or NO_2^- levels in the body became high enough to interfere with I or vitamin A deficiency.

Sudangrass and other types of sorghum sometimes contain a glucoside, dhurrin, that contains bound forms of cyanide (CN^-). Dhurrin is slowly hydrolyzed in healthy plants to HCN by the β-glucosidase, emulsin. If plants are damaged by freezing, drought, chewing, or trampling, large amounts of HCN may be released into plant tissues. Prussic acid is a potent respiratory poison (0.5 g is the toxic dose for cattle), and consumption of forage containing > 600 mg HCN/kg may lead to death of ruminants. Cattle and sheep are more susceptible to HCN poisoning than swine or horses, since ruminants will consume larger quantities of sorghum or sudangrass forage than nonruminants.

Nitrosamines have received increasing attention in recent years because of the extreme carcinogenicity of most compounds in the group (Natl. Res. Council, 1978). Nitrsamines can be formed under "natural" conditions of pH and temperature by the reaction of HNO_2 with secondary amines:

$$HNO_2 + \begin{matrix} R^1 \\ R \end{matrix} \!\!> NH_2 \rightarrow \begin{matrix} R^1 \\ R \end{matrix} \!\!> N-N=O + H_2O. \quad [1]$$

Nitrosamines have been found in meat products cured in sodium nitrite and sodium nitrate, smoked fish, fresh fish, cheese, fermented cabbage, mushrooms, beer, whiskey, and agricultural chemicals (Natl. Res. Council, 1978). Considerable speculation has also occurred concerning in vivo formation of nitrosamines in the human stomach following ingestion of foods or water containing precursors (Natl. Res. Council, 1978).

Ensiling of forages containing high concentrations of NO_3^- can also lead to the formation of toxic gases during the first week of fermentation. Nitric oxide and NO_2 are produced from NO_3^- during the ensiling process and, being heavier than air, tend to accumulate in the silo or flow out the drain pipes or silo chutes (Rhykerd & Noller, 1977; Wright & Davidson, 1964). Humans have been poisoned by entering unvented silos, and animals penned near the base of silos have been severely affected by the gases.

III. NITRATE LEVELS IN CROPS

The levels of NO_3^- in plant tissues vary greatly depending upon many factors such as species, cultivars, age of plant, part of plant, N fertilization rate and time of application, and weather conditions. Table 1 gives representative data on the NO_3^--N concentration in a variety of agronomic crops, vegetables, and other species when grown in the field under "normal" conditions. The values reported in Table 1 tend to bracket the ranges in NO_3^--N content of plant species reported in previous reviews of the subject (Maynard et al., 1976; Natl. Res. Council, 1972; Wright & Davidson, 1964).

A. Factors Affecting NO_3^- Levels in Plants

1. PLANT SPECIES AND CULTIVAR

Data in Table 1 indicate the large differences in the ability of plant species to accumulate NO_3^-. Vegetables (particularly leafy vegetables) and forage crops (including hay made from small grains) are plants noted for NO_3^- accumulation. Foods composed of grain, fruits, and flower parts are

Table 1—Nitrate-N levels in vegetables and forages (tissues sampled are those normally consumed by humans or livestock).

Common name	NO_3^--N conc	Ref.	NO_3^--N conc	Ref.	NO_3^--N conc	Ref.
	% of dry wt		% of dry wt		% of dry wt	
			Vegetables			
Beans	0.04 -0.25	Brown & Smith, 1967	0.047–0.162	Lorenz, 1978	--	
Beets	0.09 -0.84	Brown & Smith, 1967	0.188–0.682	Lorenz, 1978	0.15 -0.43	Cantliffe & Goodwin, 1974
Broccoli	0.01 -0.09	Brown & Smith, 1967	0.018–0.232	Lorenz, 1978	--	
Brussels sprouts	0.01 -0.06	Brown & Smith, 1967	--		--	
Cabbage	0.01 -0.09	Brown & Smith, 1967	0.014–0.207	Lorenz, 1978	--	
Carrots	0.00 -0.30	Brown & Smith, 1967	0.058–0.183	Lorenz, 1978	--	
Cauliflower	0.00 -0.31	Brown & Smith, 1967	0.002–0.203	Lorenz, 1978	--	
Celery	0.11 -1.12	Brown & Smith, 1967	0.003–0.203	Lorenz, 1978	--	
Lettuce	0.02 -1.06	Brown & Smith, 1967	0.026–0.453	Lorenz, 1978	0.95	Moore, 1971
Corn (sweet)	0.00 -0.01	Brown & Smith, 1967	--		--	
Cucumber	0.00 -0.16	Brown & Smith, 1967	0.007–0.144	Lorenz, 1978	--	
Endive	0.06 -0.67	Brown & Smith, 1967	--		--	
Parsnips	0.00 -0.04	Brown & Smith, 1967	--		--	
Peas	0.00 -0.02	Brown & Smith, 1967	0.019–0.033	Lorenz, 1978	--	
Radish	0.41 -1.54	Brown & Smith, 1967	0.046–1.280	Lorenz, 1978	1.31 -1.46	Cantliffe & Phatak, 1974
Spinach	0.07 -0.66	Brown & Smith, 1967	0.030–0.490	Lorenz, 1978	0.05 -0.93	Cantliffe, 1972b
Squash (yellow)	0.09 -0.43	Brown & Smith, 1967	0.003–0.446	Lorenz, 1978	--	
Tomato	0.00 -0.11	Brown & Smith, 1967	0.012–0.024	Lorenz, 1978	--	
Kale	0.002–1.568	Lorenz, 1978	--		--	
Muskmelon	0.018–0.161	Lorenz, 1978	--		--	
Onion	0.020–0.097	Lorenz, 1978	--		--	
Pepper	0.003–0.005	Lorenz, 1978	--		--	
Potato	0.004–0.019	Lorenz, 1978	0.01 -0.05	Carter & Bosma, 1974	0.01 -0.08	Augustin et al., 1977
Turnip	0.016–0.877	Lorenz, 1978	0.068	Natl. Res. Council, 1972	--	
Watermelon	0.018–0.047	Lorenz, 1978	--		--	

(continued on next page)

Table 1—Continued.

Common name	NO_3^--N conc	Ref.	NO_3^--N conc	Ref.	NO_3^--N conc	Ref.
	% of dry wt		% of dry wt		% of dry wt	
			Forages			
Corn fodder	0.06 -0.444	Rhykerd & Noller, 1977	0.223	Schuman & Elliott, 1978	0.04 -0.45	Hicks & Peterson, 1976
Sudangrass	0.7 -1.73	Rhykerd & Noller, 1977	0.05 -0.36	Schneider & Clark, 1970	0.04 -0.68	Pratt et al., 1976
Sorghum-sudangrass	0.001-1.200	George et al., 1971	0.17 -0.50	Natl. Res. Council, 1972	--	
Reed canarygrass	0.10 -1.22	Gomm, 1979a	0.04 -1.25	Gomm, 1979b	0.003-0.183	Ryan et al., 1972
Tall fescue	0.002-0.175	Hojjati et al., 1973	0.003-0.153	Ryan et al., 1972	0.053-0.557	Hojjati et al., 1973
Bermudagrass	0.002-0.077	Hojjati et al., 1973	0.350	Lovelace et al., 1968	--	
Orchardgrass	0.002-0.264	Ryan et al., 1972	0.044	Natl. Res. Council, 1972	--	
Smooth bromegrass	0.001-0.153	Ryan et al., 1972	--		--	
Kentucky bluegrass	0.047-0.333	Hojjati et al., 1973	--		--	
Perennial ryegrass	0.06 -0.99	Bartholomew & Chestnutt, 1977	0.52	Sorenson, 1971	--	
Alfalfa	0.086	Schuman & Elliott, 1978	--		--	
Pearl millet forage	0.110-0.97	Schneider & Clark, 1970	--		--	
Wheat forage	0 -0.380	Baker & Tucker, 1971	<0.5	Strizke et al., 1976	--	
Oats forage	0 -0.400	Baker & Tucker, 1971	0.011	Natl. Res. Council, 1972	0.01 -0.28	Bajpal et al., 1976
Rye forage	0 -0.490	Baker & Tucker, 1971	--		--	
Barley forage	0 -0.580	Baker & Tucker, 1971	0.01 -0.58	Pratt et al., 1976	--	
Slender sedge	0.35 -0.61	Gomm, 1979b	--		--	
Nevada bluegrass	0.66	Gomm, 1979b	--		--	
Beardless wild rye	0.67	Gomm, 1979b	--		--	
Clover	0.0068	Natl. Res. Council, 1972	0.13	Sorenson, 1971	--	
Rape forage	0.05 -0.12	Natl. Res. Council, 1972	--		--	
			Tropical leafy vegetables			
Prince's feather	0.36 -0.98	Schmidt et al., 1971	--		--	
Malabar	0.34 -0.89	Schmidt et al., 1971	--		--	
Collards	1.82 -3.79	Schmidt et al., 1971	--		--	
Swamp morning glory	0.24 -0.53	Schmidt et al., 1971	--		--	

notably low in NO_3^--N, although the vegetative tissues of the plants may contain appreciable NO_3^-. Vegetables reported to contain high levels of NO_3^--N include: cabbage wrapper leaves, lettuce, spinach, celery, parsley, kale, beet roots, radish roots, and the tops of root crops. Potato tubers, onion bulbs, and the fruits of tomato, squash, cantaloupe, melons, peppers, and cucumber are normally low in NO_3^--N (Lorenz, 1978). Forage grasses such as orchardgrass and tall fescue, corn silage, hay made from small grain, forage sorghum, and sugarbeet tops have been found to contain high concentrations of NO_3^--N. Corn, wheat, soybean, and rice grain is usually low in NO_3^--N (Wright & Davidson, 1964). Some weed species also accumulate NO_3^--N.

Cultivars of a given species vary in their ability to accumulate NO_3^-. Early studies on the differential levels of NO_3^--N in 12 oat cultivars grown under identical conditions has been summarized by Wright and Davidson (1964). Recent studies have shown that cultivars of spinach (Barker et al., 1971; Centliffe, 1972c; Cantliffe & Phatak, 1974; Maynard & Barker, 1974; Olday et al., 1976a), lettuce (Cantliffe & Phatak, 1974), radish (Cantliffe & Phatak, 1974), cucumber (Olday et al., 1976b), and pea (Olday et al., 1976b) differ markedly in ability to accumulate NO_3^--N. These cultivar differences should provide plant breeders with opportunities to develop new cultivars that do not contain excessive levels of NO_3^--N.

2. PLANT AGE AND PART OF PLANT

It is well known that NO_3^--N is not uniformly distributed in all parts of plants. In general, NO_3^--N concentrations in plant parts are in the following order: stems > roots > leaves > floral parts (Wright & Davidson, 1964). The lower portion of stems tend to contain higher NO_3^--N concentrations than upper portions of plants (Table 2). The older, outer leaves usually contain higher NO_3^--N levels than younger leaves and leaf petioles are higher in NO_3^--N than leaf blades (Wright & Davidson, 1964). Immature leaves not yet exposed to light have also been reported to accumulate NO_3^--N (Hanway et al., 1963). Patterns of NO_3^--N accumulation in various parts of plants seem to be inversely related to the NR levels and/or photosynthetic activity in the tissues.

Nitrate levels in plants also appear to vary during the plant growth cycle. Generally, NO_3^--N concentrations first increase and then, after reach-

Table 2—Nitrate-N concentrations in varying parts of corn plants (personal communication from C. R. Rhykerd, Purdue Univ.).

Tissue	NO_3^--N Concentration
	mg/kg
Whole plant	978
Leaves	64
Ears	17
Upper one-third of stalk	153
Middle one-third of stalk	803
Lower one-third of stalk	5524

ing a peak value near the prebloom stage, decline as the plant matures (Wright & Davidson, 1964). Recent studies have shown that delaying clipping (allowing greater maturation) of perennial ryegrass (Bartholomew & Chestnutt, 1977), oats (Bajpal, 1976), and sorghum-sudangrass hybrid (George et al., 1971) markedly reduced NO_3^--N levels in tissues. Baker and Tucker (1971) found high NO_3^--N contents at the "joint" and "preboot" stages in the growth of oats, wheat, rye, and barley. Explanations given for the decrease in NO_3^--N content as plants mature are (i) changing proportions of stems, leaves, and fruit with maturation; (ii) heavy demand for N imposed by fruit or seed development; and (iii) levels of NO_3^--N in soil are usually low during the time plants begin maturation (Wright & Davidson, 1964).

5. N FERTILIZATION RATE, TIMING OF APPLICATION, AND FORM OF N

No other factor affects the NO_3^--N concentration in plants more than the rate of applied N (Wright & Davidson, 1964). Most studies have shown a direct relationship between the amount of added N and NO_3^--N content of plants. In fact, the level of NO_3^--N in specified tissues has been used as a test procedure for measuring adequacy of N fertilization in cotton (Sabbe & MacKenzie, 1973), vegetables (Geraldson et al., 1973; Maynard & Barker, 1974), wheat (Brown & Jones, 1979), and mixed pasture grasses (Kemp et al., 1978). Nitrate concentrations in plants are usually excessive when the rate of applied N is higher than that required to give maximum dry matter yield. High NO_3^--N levels are normally associated with high total N concentrations in plants; however, there is no precise relationship between the concentrations of the two N components in plants (Natl. Res. Council, 1978).

Recent studies have confirmed that NO_3^--N concentrations in corn (Hicks & Peterson, 1976; Terman & Allen, 1978b), forage grasses (Bartholomew & Chestnutt, 1977; Friedrich et al., 1977; Gomm, 1979a, b; Hojjati et al., 1972, 1973; Pratt et al., 1976; Ryan et al., 1972; Terman & Allen, 1978b), small grain forage (Bajpal et al., 1976; Baker & Tucker, 1971; Pratt et al., 1976), beets (Cantliffe & Goodwin, 1974; Lorenz, 1978; Peck et al., 1971), spinach (Barker et al., 1971; Barker & Maynard, 1971; Cantliffe et al., 1974; Terman & Allen, 1978a), potato (Augustin et al., 1977; Carter & Bosma, 1974; McDole & McMaster, 1978), and radish (Cantliffe & Phatak, 1974) increase with increasing rate of applied N. Lorenz (1978) and Minotti (1978) have presented similar data for other vegetables and Schmidt et al. (1971) have shown that weed species respond to N fertilization in a similar manner to agronomic crops and vegetables. With excessive rates of applied N, NO_3^--N levels in vegetables become high and the proportion of total N as NO_3^--N also increases (Lorenz, 1978).

Timing of N fertilization in relation to harvest can markedly affect NO_3^--N levels in plants. Peck et al. (1971) have shown that midseason application (side-dressing) of N maintained high NO_3^--N levels in petioles of table beets until harvest. Nitrogen applications relatively soon before harvest can lead to elevated NO_3^--N levels in sugarbeet petioles (Wright & Davidson, 1964) and forage grasses (Bartholomew & Chestnutt, 1977; Hojjati et al., 1972, 1973).

Wright and Davidson (1964) concluded that the form of N added to soil had little effect upon the NO_3^--N content of plants. However, recent studies have shown that use of fertilizers containing NO_3^- give higher NO_3^--N levels in table beets (Cantliffe & Goodwin, 1974; Peck et al., 1971), chard (Terman & Allen, 1978a), and spinach (Barker et al., 1971; Mills et al., 1976a) than did applications of urea or ammonium sulfate. The effects of N form on NO_3^--N levels in plants were most pronounced if fertilizers are applied within 3 weeks of harvest. If fertilizers are applied prior to planting, N form has little effect on NO_3^--N levels in plants, because all of the added NH_4^- has been converted to NO_3^- before plants are actively taking up N from soil. Use of slow-release N fertilizers, such as sulfur-coated urea, has been shown to result in lower NO_3^--N levels in crops than is obtained with soluble fertilizers (Joshi & Prasad, 1977; Terman & Allen, 1978b).

Nitrification inhibitors have been used in an attempt to keep applied N in the NH_4^+ form so that plant uptake of NO_3^- would be suppressed. Application of nitrapyrin with ammoniacal fertilizers has been shown to decrease the NO_3^--N content of lettuce (Moore, 1971); spinach (Mills et al., 1976b), radish (Mills et al., 1976a), and orchardgrass (Chodan & Kucharski, 1977). Turner and MacGregor (1978) showed that Terrazole® (5-ethoxy-3-trichloromethyl-1,2,4-triadiazole) applications reduced the NO_3^--N content of pasture herbage and Joshi and Prasad (1977) showed that sulfathiazole (2-sulfanilamido-thiazole) added to urea resulted in lower NO_3^--N levels in oats.

Although most reports of excessive accumulation of NO_3^--N in crops have been associated with large applications of inorganic N fertilizers, Wright and Davidson (1964) point out the NO_3^- poisoning of livestock has occurred with feed grown on soils that have never been fertilized. Some soils in semiarid regions are naturally high in NO_3^--N and large amounts of NO_3^--N are produced in peat and muck soils each year as a result of organic matter decomposition. Applications of large amounts of animal manure or sewage sludge can also result in high NO_3^--N levels in crops (Jones & Jones, 1977; Pratt et al., 1976; Schuman & Elliott, 1978). The amount of NO_3^--N present in the soil is the major factor influencing NO_3^--N levels in plants rather than the original source of the NO_3^-.

4. WEATHER VARIABLES

Most cases of NO_3^- poisoning of livestock have occurred when plants were grown in semiarid regions or under droughty conditions in more humid regions (Wright & Davidson, 1964). Recent studies have shown that moisture stress markedly increases the NO_3^--N concentration in forage grasses (Gomm, 1979b; Kemp et al., 1978), potato (Augustin et al., 1977; Carter & Bosma, 1974; McDole & McMaster, 1978), and oranges (Shaked et al., 1972). A long, sustained drought is not as likely to bring about NO_3^- accumulation in crops as a brief, intense one (Wright & Davidson, 1964). Although plant uptake of NO_3^- is reduced by moisture stress, the reduction of NO_3^- in tissues is affected by a greater extent by suboptimal moisture and, thus, NO_3^- accumulates.

Light intensity and duration have been found to influence NO_3^--N levels in plants. In general, a reduction in either light intensity or duration results in increased NO_3^--N in plant tissues (e.g., corn plants contain twice as much NO_3^--N when grown under 35% shade as those grown without shade). Recent studies have confirmed that reducing the light intensity increases NO_3^--N concentrations in forage grasses (Gomm, 1979a, b; Stritzke et al., 1976), corn (Hicks & Peterson, 1976), sorghum-sudangrass (George et al., 1971), and spinach (Cantliffe, 1972a). Day-to-day and diurnal variations in NO_3^--N concentrations in plants have been observed. Nitrate concentrations tend to be higher on cloudy days and at the end of the dark period and seem to be inversely related to NR activity in the leaves (Beevers & Hageman, 1969; Wright & Davidson, 1964). Suboptimal light intensity reduces photosynthesis and the requirement for soluble N; however, NO_3^- uptake remains high at most light intensities. A few studies have shown that crops grown under continuous illumination have lower NO_3^--N levels than those receiving normal daylight. However, this effect may arise from insufficient light intensity in greenhouses rather than reflect the length of photoperiod (Wright & Davidson, 1964).

The effects of temperature on NO_3^- accumulation depend upon the N status of the soil and plant species. Forage grasses such as reed canarygrass, Nevada bluegrass, ryegrass, and oats tend to contain highest NO_3^--N concentrations when grown under high temperatures, particularly when coupled with low light intensities (Gomm, 1979a, b; Wright & Davidson, 1964). Conversely, soybeans and sorghum-sudangrass grown under low temperatures often contain high levels of NO_3^--N (George et al., 1971; Wright & Davidson, 1964). Cantliffe (1972b) studied the effects of N rate and temperature on the NO_3^--N content of spinach. With no applied N, the NO_3^--N level increased with temperature from 15 to 25°C, with 50 mg N per pot the NO_3^--N content increased from 10 to 25°C, and with 200 mg per pot the level of NO_3^--N in spinach increased over the range of 5 to 25°C. It appears that the NO_3^--N level in plants depends upon the relative effect of temperature on photosynthesis, NR activity, and NO_3^- uptake rate for a particular species.

5. OTHER NUTRIENTS AND HERBICIDES

Applications of K and P have variable effects upon the NO_3^--N content of plants. Recent studies have shown that K additions increase the NO_3^--N levels in forage grasses (Schneider & Clark, 1970; Sorenson, 1971; Terman & Allen, 1978b), small grains (Schneider & Clark, 1970; Sorenson, 1971), corn (Terman & Allen, 1978b), and spinach and mustard greens (Terman & Allen, 1978a). Schneider and Clark (1970) reported that applications of Ca and Mg also increased the NO_3^--N levels in pearl millet and sudangrass. Additions of P have been reported to increase NO_3^--N levels (Wright & Davidson, 1964) and have no effect or slightly decrease NO_3^--N concentrations in plants (Carter & Bosma, 1974; Sorenson, 1971; Wright & Davidson, 1964). Sorenson (1971) has also reported that addition of Cl^- and SO_4^{2-} decreased the NO_3^--N content of ryegrass grown in pots. Deficiencies of K, P, Ca, and Mg had no effect upon the NO_3^--N levels in spinach (Barker & Maynard, 1971).

Deficiencies of Mo and Mn have been reported to produce accumulations of NO_3^--N in plants because Mo is a component of NO_3^- reductase and Mn is essential for hydroxylamine reductase (an enzyme involved in reducing NO_2^- to NH_3). Levels of NO_3^--N in oats and tomatoes have been reduced by applications of Mn and Mo, respectively (Wright & Davidson, 1964). Recently Cantliffe et al. (1974) reported that addition of Mo to acid (pH 5.0–5.5) soils decreased the levels of NO_3^--N in spinach, but no effect was observed when Mo was added to soil having a pH value > 6.0. They also showed that liming soils to pH 6 decreased NO_3^--N content of spinach by increasing the availability of Mo in the soil.

Wright and Davidson (1964) reviewed the studies on the effects of herbicides on NO_3^--N levels in plants and concluded that no generalization was possible. Some studies showed that 2,4-D (2,4-dichlorophenoxy acetic acid) applications increased NO_3^--N levels in some species (sugarbeets, Russian pigweed), but had no effect on other species. Similar results were found with other herbicides. Little work has been done in recent years on the influence of herbicides and growth regulators on NO_3^--N levels in plants. Klepper (1979) showed that herbicides that are known to inhibit photosynthesis {e.g., atrazine [2-chloro-4-(ethylamino)-6-(isopropylamino)-*s*-triazine], ametryn [2-ethylamino-4-(isopropylamino)-6-(methylthio)-*s*-triazine], and diuron [3-(3,4-dichlorophenyl)-1,1-dimethylurea]} promote accumulation of NO_2^--N in excised wheat leaves by inhibiting nitrite reductase. The combination of photosynthetic inhibitor herbicides with certain other "nonphotosynthetic inhibitor" herbicides caused relatively large amounts of NO_2^--N to accumulate under both dark and light conditions.

6. POSTHARVEST TREATMENTS

Postharvest treatments can have major effects upon the levels of nonprotein N in plant tissues by (i) promoting reduction of NO_3^- to NO_2^-, (ii) reducing the levels of (NO_3^- + NO_2^-)-N through gaseous losses, and (iii) increasing NO_3^--N levels by shattering leaves and increasing the proportion of stems in the forage. Postharvest conversion of NO_3^- to NO_2^- have been reported in oat hay, turnips, corn, and sugarbeets (Phillips, 1968; Wright & Davidson, 1964). More recently Maynard (1978) and Minotti (1978) have reviewed information showing that under certain conditions NO_2^- may be formed in spinach and carrot juice. Formation of NO_2^- occurs only as a result of inadequate or lengthy storage where bacteria have an opportunity for growth. Sanitary, low-temperature storage of short duration eliminates the possibility for NO_2^- formation, even in vegetables such as spinach, which contains high levels of NO_3^- (Maynard, 1978).

Slow drying of forages does not lead to appreciable N losses; however, large losses of NO_3^- occur during ensiling (Rhykerd & Noller, 1977; Wright

& Davidson, 1964). Nitrate is reduced to NO_2^- during ensiling and the acid conditions in the silo promote NO_2^- decomposition as given in Eq. [2]:

$$2HNO_2 \rightarrow NO + NO_2 + H_2O. \quad [2]$$

Loss of NO_3^- during ensiling increases as the initial moisture content of the forage increases (Wright & Davidson, 1964). Not all silage is safe for animal consumption because of insufficient loss of NO_3^- during the ensiling process.

Freezing reduces the NO_3^--N level in vegetables to only a limited extent; however, canning significantly lowers the NO_3^- content (Maynard, 1978). For example, the NO_3^--N level in spinach may be reduced 40% by canning. Leaching of NO_3^- occurs during the canning process, thereby depleting the solids and enriching the liquid portion with NO_3^--N. Because of high NO_3^- levels in the liquid, detinning of cans can occur when NO_3^--rich, acid vegetables such as beans and tomatoes are canned. Maynard (1978) concluded that the solid portion of canned vegetables contain about 50% of the NO_3^--N content of fresh vegetables.

IV. TECHNIQUES TO MINIMIZE NITRATE LEVELS IN PLANTS

A number of management alternatives, which will minimize NO_3^- levels in crops, are available to producers. The primary tool is management of rate and timing of N fertilizer applications. The N rate must be adequate to obtain the anticipated yield, but not above the level required for protein production. Producers should avoid applying N fertilizers shortly (< 3 weeks) before harvest of forage grasses and vegetables, because of the high NO_3^- levels in plants found within a short time after feritlizing. Use of ammoniacal fertilizers, particularly in conjunction with nitrification inhibitors, offers another means to minimize NO_3^- accumulation in plants.

Producers have other means to minimize NO_3^--N levels in crops (for example, selecting cultivars that do not accumulate NO_3^-). Nitrate accumulation in plants is often associated with stress. Stress on plants imposed by nutrient imbalances, diseases, insects, and herbicides can be reduced by appropriately timed applications of agricultural chemicals. When irrigation is available, moisture stress can be minimized by timely applications of water. It may also be possible to select alternative harvest times for some crops such that NO_3^--N levels are at a minimum. Plants harvested in the afternoon and on bright, sunny days will have lower NO_3^- concentrations then those harvested on dark, cloudy days.

Preparation of plants can also affect NO_3^- levels in the food that is consumed. Removal of stalks and petioles from plants such as beets, kale, celery, and spinach will lower NO_3^- ingestion, as will removal of the outer leaves of lettuce and cabbage. Canning vegetables will lower NO_3^--N levels in the solid portion by 50%, whereas freezing may give a 25% reduction in NO_3^--N in plant tissues.

V. USE OF NITRATE-RICH PLANTS IN LIVESTOCK RATIONS

Several management practices can be implemented to reduce the probability of NO_3^- toxicity when NO_3^--rich forages are fed to cattle and sheep. Ensiling forages will normally reduce NO_3^--N concentrations to below toxic levels and the addition of up to 10 kg of ground limestone per tonne of forage going into the silo will reduce NO_3^- contents even further (Rhykerd & Noller, 1977). Producers can also dilute known high NO_3^- feeds with low NO_3^- feeds such as feed grains or legume hay to reduce the percentage of NO_3^--N in the ration. Since NO_3^- tends to accumulate in stalks of drought-stressed corn, high-energy feeds must always be included in rations based on stalkage. High NO_3^--N forages should be introduced into rations slowly over a period of 1 to 2 weeks to allow rumen bacteria to adjust to high NO_3^- intake. Rations should be balanced so that stress on animals is avoided. Additional vitamin A or iodine (I) may be required in the ration to alleviate possible effects of high NO_3^- feeds.

VI. NITRITE, PRUSSIC ACID, AND NITROSAMINES IN PLANT TISSUES

Nitrite is a much greater threat to health than NO_3^-, because it is much more toxic. However, NO_2^- levels in fresh vegetables and agronomic crops appear to be very low (Cantliffe, 1972b; Cantliffe et al., 1974; Maynard, 1978; Minotti, 1978; Wright & Davidson, 1964). Bruising or inadequate, lengthy storage may contribute to NO_2^- formation in vegetables (Heisler et al., 1974; Maynard, 1978; Minotti, 1978; Phillips, 1968). High concentrations of NO_2^- are formed when carrot juice is stored at a warm temperature for 6 hr; however, NO_2^- formation is minimal during storage of spinach and other vegetables (Hall & Hicks, 1977; Lorenz, 1978; Minotti, 1978). Sanitary, short-term, low-temperature storage of vegetables eliminates the possibility of significant NO_2^- accumulation (Maynard, 1978).

Prussic acid content of sorghums is highest in young plants < 60 cm tall, in suckers, in plants that have been frosted or stressed by drought, in plants receiving heavy applications of N fertilizers, and in plants treated with 2,4-D. Prussic acid levels in plant parts normally follow the order: leaf blades > leaf sheaths > stems > heads > seeds. Upper leaves contain higher concentrations of HCN than lower leaves. Prussic acid levels in sorghum types follow the order: sorghum > sorghum-sudangrass hybrid > sudangrass. Johnsongrass may contain toxic concentrations of HCN, but only low levels are found in pearl millet. Sorghum cultivars differ widely in HCN content when grown under uniform conditions (Hankins et al., 1975).

Since sorghum leaves, shoots, and suckers contain high levels of HCN, management procedures have been developed to minimize the proportion of these tissues in forage consumed by livestock. Sorghum should not be grazed unless the plants are 60 cm tall (also included is new growth after a frost). Sorghum that has wilted and dried for 5 to 6 d after being killed by a frost is safe for grazing. Green chop is usually safe because the entire plant is consumed, thereby resulting in a much lower HCN concentration in forage than if only leaves were eaten. Silage is normally safe for consumption, because HCN is released as a gas during ensiling. Curing of hay reduces the HCN concentration in tissues by 75% (Hankins et al., 1975). Thus, simple management procedures eliminate most of the animal health hazards associated with HCN in plant tissues.

Nitrosamines have not been detected in vegetables (Dressel, 1976; Heisler et al., 1974; Gough et al., 1977), fruits (Gough et al., 1977), wheat (Gough et al., 1977), or mixed baby food (Gough et al., 1977). Of particular significance is the finding that nitrosamines were not detected in storage-abused spinach and beets, even though some samples contained 300 mg NO_2^--N/kg (Heisler et al., 1974). Dressel (1976) showed that nitrosamines were rapidly degraded when added to soils and that the only trace amounts were detected in plants for a short time after application to soil. Nitrosamines injected into wheat plants disappeared with 5 d (Dressel, 1976).

VII. ASSESSMENT OF HEALTH IMPLICATIONS OF NITRATE, NITRITE, AND NITROSAMINES IN FOOD AND FEED

White (1975) has used NO_3^- and NO_2^- concentration data and average food and water consumption information to calculate the average daily ingestion of NO_3^- and NO_2^- by persons living in the United States (Table 3). Major sources of NO_3^- in the diet are vegetables and cured meats. Nitrite is produced in the oral cavity through bacterial reduction of NO_3^- and saliva is the major source of NO_2^- entering the stomach. The only other major source of NO_2^- in the diet is cured meats. The amounts of inorganic N compounds ingested are much lower than those known to cause health problems in adult humans or children. Furthermore, it is unlikely that infants under the age of 3 months are at risk from NO_3^- or NO_2^- in baby foods. Almost all cases of infant methemoglobinemia in the United States have originated from consumption of well water containing > 10 mg NO_3^--N/L. Improper-

Table 3—Calculated average daily intake of NO_2^- and NO_3^- by persons living in the United States (calculated from data of White, 1975).

Source	NO_3^--N		NO_2^--N	
	mg/day	% of total	mg/day	% of total
Vegetables	19.1	86.0	0.06	1.8
Fruits, juices	0.2	1.4	0.00	0.0
Milk and products	<0.1	<0.4	0.00	0.0
Bread	0.4	1.8	<0.01	<0.3
Water	0.2	0.9	0.00	0.0
Cured meats	2.1	9.5	0.72	21.1
Saliva	--	--	2.62	76.8
Total	22.2	100	3.41	100.0

ly stored carrot juice containing relatively high levels of NO_2^- has been implicated in one case of infant methemoglobinemia (Keating et al., 1973). There are no other reported cases of methemoglobinemia arising from ingestion of NO_3^-- or NO_2^--rich foods. There are no known chronic effects of long-term, low-level intake of NO_3^- and NO_2^- by adults or children (Natl. Res. Council, 1978).

With livestock, the incidence of methemoglobinemia or other problems induced by NO_3^- ingestion, such as abortions, is thought to be rather high; however, documented reports are rare (Natl. Res. Council, 1978). It appears that publicity concerning NO_3^- poisoning has made livestock producers more aware of the problems and of means to minimize NO_3^- levels in animal rations. Furthermore, veterinarians are better able to diagnose and treat livestock poisoned by NO_3^- or NO_2^-. Although the potential of NO_3^- toxicity exists with ruminant animals fed roughages, the problems seem to be much less in recent years as compared with the 1950's and 1960's. However, producers should endeavor to produce forages that contain minimal levels of NO_3^--N consistent with high yields.

Nitrosamines do not appear to be present in unprocessed foods. Thus, direct ingestion of nitrosamines in fruits, vegetables, or cereals is nil. However, nitrosamines may be formed in the stomach through reaction of HNO_2 and secondary amines. Therefore, it may be prudent to minimize the ingestion of NO_2^- by reducing intake of NO_3^-, because of the reduction of NO_3^- occurring in the mouth. Much additional work is required to assess the significance of in vivo formation of nitrosamines.

APPENDIX

Scientific Names of Plant Species Mentioned in This Chapter

Common name	Scientific name
Alfalfa	*Medicago sativa* L.
Barley	*Hordeum vulgare* L.
Bean, bush	*Phaseolus vulgaris* L.
Beet, common	*Beta vulgaris* L.
Bermudagrass	*Cynodon dactylon* (L.) Pers.
Bluegrass	
Kentucky	*Poa pratensis* L.
Nevada	*Poa nevadensis* Vasey ex. Shribn.
Broccoli	*Brassica oleracea* var. *italica*
Bromegrass, smooth	*Bromus inermis* Leyss.
Brussels sprouts	*Brassica oleracea* var. *gemmifera*
Cabbage	*Brassica oleracea* var. *capitata*
Carrot	*Daucus carota* L.
Cauliflower	*Brassica oleracea* var. *botrytis*
Cantaloupe	*Cucumis melo* L.
Canarygrass, reed	*Phalaris arundinacea* L.
Celery	*Apium graveolens* L.

Chard	*Beta vulgaris* var. *circla*
Clover, red	*Trifolium pratense* L.
Collards	*Brassica oleracea* var. *acephala*
Corn	*Zea mays* L.
Cotton	*Gossypium hirsutum* L.
Cucumber	*Cucumis sativus* L.
Endive	*Cichorium endivium* L.
Fescue, tall	*Festuca arundinacea* Shreb.
Johnsongrass	*Sorghum halepense* (L.) Pers.
Kale	*Brassica oleracea* var. *acephala*
Lemon	*Citrus limon* L.
Lettuce	*Lactuca sativa* L.
Malabar	*Basella alba* L.
Melon	*Cucumis melo* L.
Millet, pearl	*Pennisetum typhoides* (Burm) Staph and Hubbard
Morning glory, swamp	*Ipomoea aquatica* Forsk.
Muskmelon	*Cucumis melo* L.
Mustard greens	*Brassica hirta* Moench.
Oats	*Avena sativa* L.
Onion	*Allium cepa* L.
Orange	*Citrus sinensis* L.
Orchardgrass	*Dactylis glomerata* L.
Parsley	*Petroselinum hortense* Hoffm.
Parsnip	*Pastinaca sativa* L.
Pea	*Pisum sativum* L.
Pepper	*Capsicum frutescens* L.
Pigweed, Russian	*Axyria amaranthoides* L.
Potato	*Solanum tuberosum* L.
Prince's feather	*Amaranthus cruentus* L.
Radish	*Raphanus sativa* L.
Rape	*Brassica napus* L.
Rice	*Oryza sativa* L.
Rye	*Secale cereale* L.
Rye, wild beardless	*Elymus triticoides* Buckl.
Ryegrass, perennial	*Lolium perenne* L.
Sedge, slender	*Carex praegracilis* W. Boott.
Sorghum	*Sorghum bicolor* (L.) Moench.
Sorghum-sudangrass	*Sorghum bicolor* × *Sorghum sudanense*
Soybean	*Glycine max* (L.) Merr.
Spinach	*Spinacia oleracea* L.
Squash	*Circubita maxima* Duchesne
Sudangrass	*Sorghum sudanense* Staph.
Sugarbeet	*Beta saccharifera* L.
Timothy	*Phleum pratense* L.
Tomato	*Lycopersicum esculentum* Mill.
Turnip	*Brassica rapa* L.
Watermelon	*Citrullus vulgaris* Schrod.
Wheat	*Triticum aestivum* L.

REFERENCES

Augustin, J., R. E. McDole, and G. C. Painter. 1977. Influence of fertilizer, irrigation, and storage treatments on nitrate-N content of potato tubers. Am. Potato J. 54:125–136.

Bajpal, L. D., S. P. Arora, L. L. Relwani. 1976. Nitrate accumulation in oats as influenced by nitrogen and stage of growth. Forage Res. 2:175–177.

Baker, J. M., and B. B. Tucker. 1971. Effects of rates of N and P on the accumulation of NO_3^--N in wheat, oats, rye, and barley on different sampling dates. Agron. J. 63:204–207.

Barker, A. V., and D. N. Maynard. 1971. Nutritional factors affecting nitrate accumulation in spinach. Commun. Soil Sci. Plant Anal. 2:471–478.

Barker, A. V., N. H. Peck, and G. E. MacDonald. 1971. Nitrate accumulation in vegetables. I. Spinach grown in upland soils. Agron. J. 63:126–129.

Bartholomew, P. W., and D. M. B. Chestnutt. 1977. The effects of a wide range of fertilizer application rates and defoliation intervals on the dry-matter production, seasonal response to nitrogen, persistence and aspects of chemical composition of perennial ryegrass (*Lolium perenne* cv. S. 24). J. Agric. Sci. 88:711–721.

Beevers, L., and R. H. Hageman. 1969. Nitrate reduction in higher plants. Annu. Rev. Plant Physiol. 20:495–522.

Brown, B. D., and J. P. Jones. 1979. Tissue analysis for nitrogen fertilization on irrigated soft white winter wheat. Current Info. Series no. 461. College of Agric., Univ. of Idaho.

Brown, J. R., and G. E. Smith. 1967. Nitrate accumulation in vegetable crops as influenced by soil fertility practices. Missouri Agric. Exp. Stn. Res. Bull. 920.

Cantliffe, D. J. 1972a. Nitrate accumulation in spinach grown under different light intensities. J. Am. Soc. Hortic. Sci. 97:152–154.

Cantliffe, D. J. 1972b. Nitrate accumulation in spinach grown at different temperatures. J. Am. Soc. Hortic. Sci. 97:674–676.

Cantliffe, D. J. 1972c. Nitrate accumulation in spinach cultivars and plant introductions. Can. J. Plant Sci. 53:365–367.

Cantliffe, D. J., and P. R. Goodwin. 1974. Effects of nitrogen rate, source, and various anions and cations on NO_3 accumulation and nutrient constituents of table beets. Agron. J. 66: 779–783.

Cantliffe, D. J., G. E. McDonald, and N. H. Peck. 1974. Reduction in nitrate accumulation by molybdenum in spinach grown at low pH. Commun. Soil Sci. Plant Anal. 5:273–282.

Cantliffe, D. J., and S. C. Phatak. 1974. Nitrate accumulation in greenhouse vegetable crops. Can. J. Plant Sci. 54:783–788.

Carter, J. N., and S. M. Bosma. 1974. Effect of fertilizer and irrigation on nitrate-nitrogen and total nitrogen in potato tubers. Agron. J. 66:263–266.

Cawley, G. D., D. F. Collins, and D. A. Dyson. 1977. Nitrate poisoning. Vet. Rec. 101:305–306.

Chodan, J., and J. Kucharski. 1977. The influence of 2-chloro-6-(trichloromethyl) pyridine on cocksfoot crop. Pol. J. Soil Sci. 10:71–78.

Dressel, J. 1976. Relationships between nitrate, nitrite, and nitrosamines in plant material and in the soil. Qual. Plant. Plant Foods Hum. Nutr. 25:381–390.

Edwards, W. C., and C. P. McCoy. 1980. Nitrate poisoning in Oklahoma cattle. Vet. Med. Small Anim. Clin. 75:457–458.

Friedrich, J. W., D. Smith, and L. E. Schrader. 1977. Nitrate reductase activity and N fractions in timothy and switch grass as influenced by N and S fertilization. Can. J. Plant Sci. 57:1151–1157.

George, J. R., C. L. Rhykerd, and C. H. Noller. 1971. Effect of light intensity, temperature, nitrogen, and stage of growth on nitrate accumulation and dry matter production of a sorghum-sudangrass hybrid. Agron. J. 63:413–415.

Geraldson, C. M., G. R. Clacan, and O. A. Lorenz. 1973. Plant analysis as an aid in fertilizing vegetable crops. p. 365–379. *In* L. M. Walsh and J. D. Beaton (ed.) Soil testing and plant analysis. Am. Soc. of Agron., Madison, Wis.

Gomm, F. B. 1979a. Accumulation of NO_3 and NH_4 in reed canarygrass. Agron. J. 71:627–630.

Gomm, F. B. 1979b. Herbage yield and nitrate concentration in meadow plants as affected by environmental variables. J. Range Manage. 35:359–364.

Gough, T. A., M. F. McPhail, K. S. Webb, B. J. Wood, and R. F. Coleman. 1977. An examination of some foodstuffs for the presence of volatile nitrosamines. J. Sci. Food Agric. 28:345–351.

Hageman, R. H. 1980. Effect of form of nitrogen on plant growth. p. 47–62. *In* J. J. Meisinger et al. (ed.) Nitrification inhibitors—potentials and limitations. Am. Soc. of Agron. Spec. Pub. no. 38. Am. Soc. of Agron., Madison, Wis.

Hall, C. B., and J. R. Hicks. 1977. Nitrites in inoculated carrot juice as a function of nitrate content and temperature. J. Food Sci. 42:549–550.

Hankins, B. J., D. L. Oswalt, and K. M. Weinland. 1975. Prussic acid in forages. Agron. Guide AY-196. Coop. Ext. Service, Purdue Univ., W. Lafayette, Ind.

Hanway, J. J., J. B. Herricks, T. L. Willrich, P. C. Bennett, and J. T. McCall. 1964. The nitrate problem. Spec. Rep. no. 34. Coop. Ext. Service, Iowa State Univ., Ames, Iowa.

Heisler, E. G., J. Siciliano, S. Krulick, J. Feinberg, and J. H. Schwartz. 1974. Changes in nitrate and nitrite content, and search for nitrosamines in storage-abused spinach and beets. J. Agric. Food Chem. 22:1029–1032.

Hicks, D. R., and R. H. Peterson. 1976. Defoliation and fertilizer nitrogen effects on nitrate-nitrogen profiles in maize. Agron. J. 68:476–478.

Hojjati, S. M., T. H. Taylor, and W. C. Templeton, Jr. 1972. Nitrate accumulation in rye, tall fescue, and bermudagrass. Agron. J. 64:624–627.

Hojjati, S. M., W. C. Templeton, Jr., T. H. Taylor, H. E. McKean, and J. Byars. 1973. Post-fertilization changes in concentration of nitrate nitrogen in Kentucky bluegrass and tall fescue herbage. Agron. J. 65:880–883.

Jones, T. O., and D. R. Jones. 1977. Nitrate/nitrite poisoning of cattle from forage crops. Vet. Rec. 101:266–267.

Joshi, B. S., and R. Prasad. 1977. The effects of rates of application and sources of nitrogen on nitrate concentrations in oat forage. J. Br. Grassl. Soc. 32:213–216.

Keating, J. R., M. E. Lell, A. W. Straus, H. Zachowsky, and G. E. Smith. 1973. Infantile methemoglobinemia caused by carrot juice. N. Engl. J. Med. 288:825–829.

Kemp, A., J. K. Geurink, A. Arolestein, and A. J. Vont Kloaster. 1978. Production of grassland herbage and nitrate poisoning in cattle. Vgeskrift Agronomer Hortonomer Forstandidates Licentiater 128:729–786.

Lorenz, O. A. 1978. Potential nitrate levels in edible plant parts. p. 201–220. *In* D. R. Nielson et al. (ed.) Nitrogen in the environment. Vol. II. Academic Press, New York.

Lovelace, D. A., E. C. Holt, and W. B. Anderson. 1968. Nitrate and nutrient accumulation in two varieties of bermudagrass [*Cynodon doctylon* (L.) Pers.] as influenced by soil applied fertilizer nutrients. Agron. J. 60:551–554.

Maynard, D. N. 1978. Critique of "Potential nitrate levels in edible plant parts." p. 221–233. *In* D. R. Nielson et al. (ed.) Nitrogen in the environment. Vol. II. Academic Press, New York.

Maynard, D. N., and A. V. Barker. 1974. Nitrate accumulation in spinach as influenced by leaf type. J. Am. Hortic. Sci. 99:135–138.

Maynard, D. N., A. V. Barker, P. L. Minotti, and N. H. Peck. 1976. Nitrate accumulation in vegetables. Adv. Agron. 28:71–118.

McDole, R. E., and G. M. McMaster. 1978. Effects of moisture stress and nitrogen fertilization on tuber nitrate-N content. Am. Potato J. 55:611–619.

Mills, H. A., A. V. Barker, and D. N. Maynard. 1976a. Nitrate accumulation in radish as affected by nitrapyrin. Agron. J. 68:13–17.

Mills, H. A., A. V. Barker, and D. N. Maynard. 1976b. Effects of nitrapyrin on nitrate accumulation in spinach. J. Am. Soc. Hortic. Sci. 101:202–204.

Minotti, P. L. 1978. Critique of "Potential nitrate levels in edible plant parts." p. 235–252. *In* D. R. Nielson et al. (ed.) Nitrogen in the environment. Vol. II. Academic Press, New York.

Moore, F. D. 1971. Nitrification suppression and controlled nitrogen release for lettuce. Colorado State Univ. Exp. Stn. Progress Rep. 57–71.

National Research Council. 1972. Accumulation of nitrate. National Academy of Sciences, Washington, D.C.

National Research Council. 1978. Nitrates: an environmental assessment. National Academy of Science, Washington, D.C.

Olday, F. C., A. V. Barker, and D. N. Maynard. 1976a. A physiological basis for different patterns of nitrate accumulation in two spinach cultivars. J. Am. Soc. Hortic. Sci. 101: 217–219.

Olday, F. C., A. V. Barker, and D. N. Maynard. 1976b. A physiological basis for different patterns of nitrate accumulation in cucumber and pea. J. Am. Soc. Hortic. Sci. 101:219–221.

Peck, N. H., A. V. Barker, G. E. MacDonald, and R. S. Shallenberger. 1971. Nitrate accumulation in vegetables: II. Table beets grown in upland soils. Agron. J. 63:130–132.

Phillips, W. E. J. 1968. Changes in nitrate and nitrite contents of fresh and processed spinach during storage. J. Agric. Food Chem. 16:88–91.

Pratt, P. F., S. Davis, R. G. Sharpless, W. J. Pugh, and S. E. Bishop. 1976. Nitrate contents of sudangrass and barley forages grown on plots treated with animal manures. Agron. J. 68:311–314.

Rhykerd, C. R., and C. H. Noller. 1977. Nitrate toxicity—problem and prevention. Agron. Guide ID-128. Coop. Ext. Service, Purdue Univ., W. Lafayette, Ind.

Ryan, M., W. F. Wedin, and W. B. Bryan. 1972. Nitrate-N levels of perennial grasses as affected by time and level of nitrogen application. Agron. J. 64:165–168.

Sabbe, W. E., and A. J. MacKenzie. 1973. Plant analysis as an aid to cotton fertilization. p. 299–313. *In* L. M. Walsh and J. D. Beaton (ed.) Soil testing and plant analysis. Am. Soc. of Agron., Madison, Wis.

Schmidt, D. R., H. A. MacDonald, and F. E. Brockman. 1971. Oxalate and nitrate contents of four tropical leafy vegetables at two soil fertility levels. Agron. J. 63:559–561.

Schneider, B. A., and N. A. Clark. 1970. Effect of potassium on the mineral constituents of pearl millet and sudangrass. Agron. J. 62:474–477.

Schuman, G. E., and L. F. Elliott. 1978. Cropping an abandoned feedlot to prevent deep percolation of nitrate-nitrogen. Soil Sci. 126:237–243.

Shaked, A., A. Bar-Akiva, and K. Mendel. 1972. Effect of water stress and high temperature on nitrate reduction in citrus leaves. J. Hortic. Sci. 47:183–190.

Sorenson, C. 1971. Influence of various factors on nitrate concentration in plants in relation to nitrogen metabolism. p. 229–240. *In* R. M. Samish (ed.) Recent advances in Plant Nutrition. Proc. 6th Int. Coll. Plant Anal. Fert. Problems, Tel Aviv. Vol. 1. Gordon & Breach Sci. Publ., New York.

Stritzke, J. F., L. I. Croy, and W. E. McMurphy. 1976. Effect of shade and fertility on NO_3^--N accumulation, carbohydrate content, and dry matter production of tall fescue. Agron. J. 68:387–389.

Terman, G. L., and S. E. Allen. 1978a. Crop yield—nitrate N, total N, and total K relationships: leafy vegetables. Commun. Soil Sci. Plant Anal. 9:813–825.

Terman, G. L., and S. E. Allen. 1978b. Crop yield—nitrate N and total N and K concentration relationships: corn and fescuegrass. Commun. Soil Sci. Plant Anal. 9:827–841.

Turner, M. A., and A. N. McGregor. 1978. An evaluation of Terrazole as a nitrificide for improving the efficiency of use of nitrogenous fertilizers on pasture. N.Z. J. Agric. Res. 21: 39–45.

White, J. W., Jr. 1975. Relative significance of dietary sources of nitrate and nitrite. J. Agric. Food Chem. 23:886–891.

Wright, M. J., and K. L. Davidson. 1964. Nitrate accumulation in crops and nitrate poisoning in animals. Adv. Agron. 16:197–247.

45

Samuel R. Aldrich
University of Illinois
Urbana, Illinois

Nitrogen Management to Minimize Adverse Effects on the Environment

An excessive or poorly managed supply of N intended for crop production may cause NO_3^- contents in drinking water above the maximum suggested by the U.S. Public Health Service, especially in coarse-textured soils. It may also contribute to excessive plant and algal growth in lakes and reservoirs, which, upon decay, can result in an undesirable odor and an O_2 deficit for fish and other aquatic life. It can increase the amount of N_2O gas lost to the atmosphere, thus contributing to the destruction of O_3 in the stratosphere. Though these effects resulting from N use are not commonly of serious dimensions (except for heavy local runoff of animal manure), they should be avoided or reduced where feasible. These effects constitute additional reasons—beyond efficient N utilization for economic and energy conservation reasons—for a well-managed N program for crop production.

It would be a mistake to begin this chapter by searching for special practices to minimize unwanted environmental effects. That goal is to a considerable extent the by-product of managing N wisely for crop production. The concept is simple. Nitrogen that is taken up and utilized by growing crops that provide human food directly or indirectly as feed for livestock does not cause unwanted environmental effects in the current season. However, the portion that remains as crop residue in the field or is fed to livestock and returned in manure becomes a part of the general pool of N, and some may leach into surface water or groundwater. Several chapters in this book supply information that deals with improved utilization of N by crops. Some of the principles and practices are repeated in this chapter in order to have a complete environmental perspective in one place.

Nitrogen in Crop Production.

By far, the best way to begin a plan to minimize, to the extent feasible, the nitrates in surface and groundwater is to focus on practices that favor high crop yields.

I. MAXIMIZE POSITIVE SUPPORTING PRACTICES

A. Concentrate Crops in Suitable Geographic Regions

Optimum N efficiency can only be achieved by concentrating the production of each crop in geographic regions where the climate and soils are most favorable. No amount of skill in the choice of fertilizers, varieties, tillage methods, planting practices, or pest control practices can offset the effects of trying to grow crops where they are poorly adapted.

In recent years, there has been much discussion in the nonagricultural, technical, and popular press about three possible shifts in agricultural production that relate to the concept of concentrating each crop where it is best suited: organic farming, increased crop diversity on each farm, and producing crops nearer to their markets. These suggestions interfere with a fundamental principle of efficient N utilization that crops be grown where the climate and soils are most favorable.

Opportunity for the economic system to function without outside interference (including governmental policies) provides greatest assurance that crops and livestock will be produced and supplied to consumers with the optimum combination of land use, labor, and energy. It is not an accident, a farmer preference, nor a continuation of historical practices that has resulted in specialized crop-growing regions such as the Corn Belt, the Great Lakes dairy region, the Great Plains wheat-sorghum belt, and the Cotton Belt—formerly in the southeast, but now largely in the southwest.

B. Select the Best Available Hybrids or Crop Varieties

In the case of corn (*Zea mays* L.), for example, the best hybrids possess the optimum combination of length of growing period to reach maturity, standability, resistance to prevailing insects and diseases, and genetic yield potential.

C. Plant at the Optimum Time and Rate (Fig. 1)

Though there are occasional exceptions, early planting of corn generally favors high yield.

There is an optimum plant population for each small grain, for corn, for sorghum [*Sorghum bicolor* (L.) Moench], and other crops that permits the crop to most advantageously utilize the available solar energy. If a small grain like wheat (*Triticum aestivum* L.) or oats (*Avena sativa* L.) is planted too thick, there is increased risk of lodging, which can result in severe loss in yield and low efficiency in the utilization of available N in the soil.

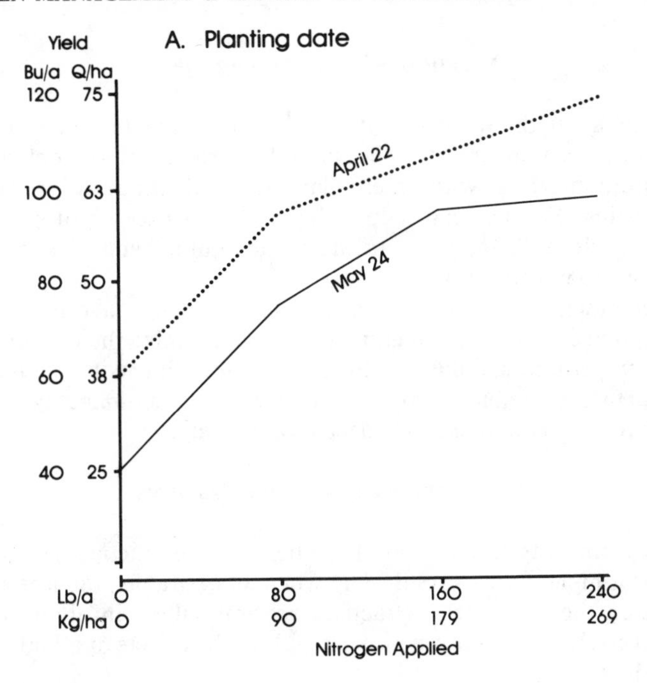

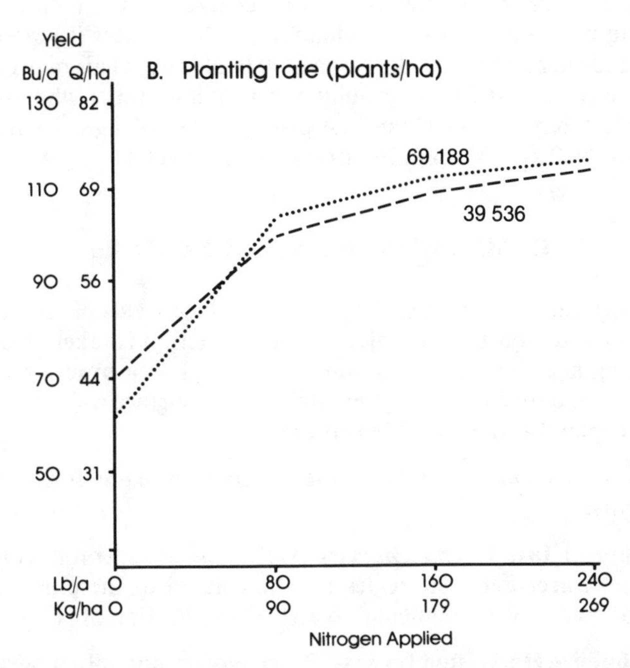

Fig. 1—In this University of Illinois research, the 22 April planting made the most efficient use of applied N (*A*). Likewise, 69 160 plants per hectare (28 000 per acre) utilized N most efficiently (*B*).

D. Improve Water Management

Within a region with a generally suitable climate, there are often periods of excess moisture and other periods when moisture is deficient. Excess moisture interferes with timely planting, cultivating for weed control, and harvesting. Farmers can reduce the effects of excess soil moisture by surface and tile drainage and can often substitute chemical weed control when soil is too wet to cultivate.

Other practices suited to cope with moisture shortage include: early planting; good control of competing weeds; maintaining the soil surface in a condition to permit maximum infiltration of rainfall; leaving plant residues on the surface to reduce evaporation; and using drainage systems that remove excess water and promote deep crop rooting.

E. Combine all Desirable Practices

Most farmers follow several of the best-known practices for crop production. But highest yields result only when all desirable practices are combined within the same field. Practices are mutually reinforcing for high yield; conversely, one poor practice can nullify the effects of all other favorable practices.

The relevance to the mission of this chapter of the concept of high yields resulting from the best combination of all practices is readily apparent when a farmer applies 224 kg N/ha (200 lb N/acre). This may contribute less NO_3^- to surface and underground water than a farmer who applies 168 kg N/ha (150 lb N/acre)—if the first produces 11 300 kg corn/ha and the second only 7900 kg (180 and 126 bu/acre, respectively).

II. MINIMIZE NEGATIVE FACTORS

Negative factors are those that directly reduce yields or the amount of food delivered to consumers, unless corrective action is taken. Controlling or correcting negative factors minimizes the land area of crops needed to supply food, and thus keeps the N contributed to surface water and groundwater by cropland at the lowest feasible level.

a. Control Weeds—Weeds compete with crop plants for light, water, and nutrients.

b. Control Insects and Diseases—There is considerable controversy over the most acceptable strategies for the control of these pests, but the principle of control is fundamental to efficient utilization of N.

c. Minimize Harvesting Losses—This involves not only timely harvesting with well-adjusted machines, but also choice of varieties or hybrids that have good inherent standability and are resistant to insects and diseases that may cause lodging.

d. Dry and Store the Crop to Prevent Spoilage—In order to offset losses that occur in storage, additional land area would be needed and would thus increase the potential for N being introduced into surface water and groundwater.

III. SELECT OPTIMUM N FERTILIZER PRACTICES

It would be fortunate indeed if a relatively simple set of data for a state or a region could be fed into a computer that would then print out a general recommendation for the kind and amount of N and best method of application. But there are great differences among fields in kinds of soils, preceding crops, and in manure and fertilizer treatments. The optimum N fertility program, therefore, needs to be decided individually for each farm and field. The broad general goal is to make the best possible match between the amount of N needed by the crop and the amount that will be available from all sources throughout the growing season.

Nitrogen fertilizer decisions are best reached through a series of sequential steps, some of which are relatively complex. The major steps follow.

1. Estimate the Probable Yield Based upon Previous Experience or Upon Yield Potentials Developed by Colleges of Agriculture.

2. Estimate the Amount of N That Will Become Available From the Soil During the Growing Season.

The latter estimate will include the N inherent in the soil organic matter, the amount from residues of preceding crops, and the carry-over from previous applications. It might be expected that agriculture would have become sufficiently sophisticated to precisely test the soil for N that will become available within a growing season, but research on the subject over the past 50 yr has had only limited success. Irrigated areas are an exception where it is possible to measure the amount of available N in the rooting zone at the beginning of the season and confidently assume that it will remain there during the crop-growing period until removed by the crop.

3. Consider the N That Will be Supplied From Other Outside Sources (Animal Manure, Sludge, Precipitation).

Unfortunately, animal manures vary widely in N content, depending on the class of livestock, amount of bedding materials used, length of time the manure is in storage before being spread on the land, and exposure during storage to rainfall, which leaches out soluble nutrients. Manures vary also in loss of N by volatilization or leaching after having been applied to the land. The usual assumption is that 1 t of large-animal manure supplies 5.0 kg of N (11 lb), 1.1 kg of P (2.2 lb), and 4.0 kg of K (8.8 lb). Atmospheric fallout adds about 10 to 20 kg/ha (9–18 lb/acre) of N annually.

Larger amounts are deposited downwind of large combustion sources and of large animal feedlots.

4. Decide Upon the Amount of Fertilizer N Needed for Optimum Economic Return Modified to the Extent Feasible by Environmental Considerations.

In some cases, the optimum economic rate will exceed the optimum amount from a local environmental viewpoint. If so, whether to regulate the rate of fertilizer application is a complex societal decision that involves a wide variety of considerations such as implications for fairness and equity among farmers, the likelihood of causing problems in other regions, and impacting the amount of crops for export with which to purchase needed oil.

The principal environmental consideration is the influence of N rates used in agriculture on the NO_3^- content of surface water and groundwater. The law of diminishing returns applies to utilization of available N. As additional increments of N are applied to a field, each increment is utilized slightly less than the preceding one. As the rate of yield increase slows, the portion not recovered in the crop (and hence, potentially leachable to surface water and groundwater) increases—slowly at first, but rapidly at the end. The amount not recovered in the crop and potentially leachable would, in fact, become 100% at some extremely high rate, were it not for soil microorganisms that utilize a portion in the process of building soil humus and for an increase in volatilization.

Research in several states indicates that current rates of N application should not cause widespread excessive buildup of nitrates in the subsoil, with the exception of intensive cropping and heavy N applications on coarse-textured soils, especially where irrigation is practiced. These conditions prevail and nitrates above the maximum suggested for drinking water by the U.S. Public Health Service have been found on Long Island, N.Y., California, central Wisconsin, Nebraska, Minnesota, and likely in other localized areas. High nitrates in central Illinois streams (Fig. 2) are likely due to a combination of high organic matter in the soils; intensive tillage, which speeds release of N from organic matter; extensive tile drainage systems, which conduct water quickly to streams; and the relatively large amount of N fertilizer applied. Early reports that 55 to 60% of the nitrates in certain

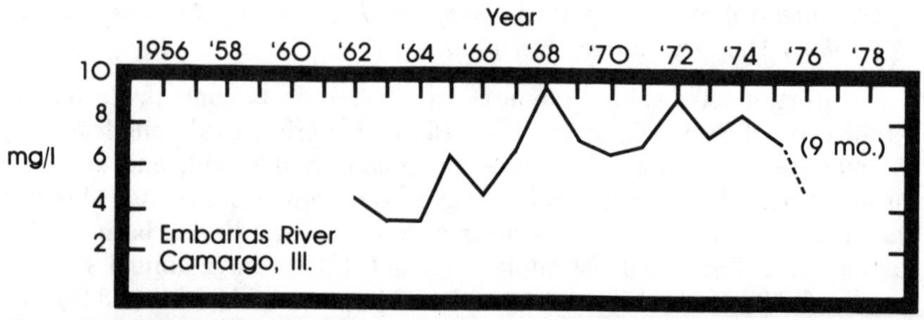

Fig. 2—Trend in NO_3^--N content of a representative central Corn Belt river (Illinois State Water Survey data).

Illinois streams traced directly to N fertilizer (Kohl et al., 1971) have been discredited (Meints et al., 1973).

The future trend in NO_3^- content of surface waters is uncertain. The Embarras River in central Illinois (Fig. 2) is representative of many in the Corn Belt. Nitrate rose dramatically from 1962 to 1968, but shows no clear trend since 1968. It seems reasonable to expect that the increase in NO_3^- content of surface waters will increase much more slowly in the future than during the 1950's and 1960's for several reasons.

It is highly unlikely that the rate of increase of N fertilizer in the next few years will be as great as in the first half of the 1960's. Much of the increase in the future will likely be tied closely to the capability of the crops to efficiently utilize it. There is still a considerable cropland acreage that is fertilized at less than the optimum level of N for current economic conditions. This will gradually be corrected. Yield potential will be increased through genetic improvement, better pest control, and improvements in tillage and water management. In the absence of new breakthroughs in technology, there is relatively little incentive for farmers who have been applying N at the highest rates to increase them.

Surface waters that occasionally exceed the U.S. Public Health Service standard and are used for public drinking water pose a troublesome dilemma. If agriculture is an important contributing cause, changes in cropping systems and N fertilizer practices to remedy the conditions would severely disadvantage local farmers. It might cause erosion and related problems in other feed-dependent regions if they attempt to compensate for lack of grain from a prime-producing region. It would increase the cost of food. It could impact the amount of crops for export, thus affecting balance of trace and money with which to purchase oil.

5. *Select the Best Time and Method of Application.*

The fundamental guiding principle to assure efficient utilization and to minimize loss is to supply N as soon as feasible to the time it is needed by the crop. In order for nitrates to reach groundwater or surface water in substantial quantity, even under excessive rainfall, several conditions must be met: (i) there is adequate time for conversion of NH_4^+ to NO_3^-; (ii) the NO_3^- is not previously immobilized by soil organisms; (iii) NO_3^- is not taken up by the growing crop; and (iv) NO_3^- was not lost to the atmosphere through denitrification.

It has long been known that there is a relatively good match up between the N needs of summer annual crops like corn and the time that N becomes available from applications of animal manures and residues of leguminous crops like alfalfa (*Medicago sativa* L.), clover (*Trifolium* spp.), and lespedeza (*Lespedeza* spp.). But heavy reliance on animal manures, leguminous crop residues, or both together tends to increase the amount of nitrates in the soil in the late fall, the most vulnerable time for loss through leaching.

Practical logistical problems prevent farmers from always applying manure and incorporating crop residues at the ideal time to meet crop needs for N and to minimize losses. For example, delayed plowing of alfalfa and clover aggravates moisture stress in a dry spring. Spreading manure shortly before planting is not always practicable because it requires a long period of storage, does not use labor efficiently, interferes with timely planting, and may cause undesirable soil compaction.

Whereas the amount of N that becomes available from plant residues and humus depends upon soil temperature, moisture, pH, and the nature of the residues, fertilizer N is immediately available to crops. This allows the farmer more control over the time at which he will supply N for his crops. But because of its immediate availability, fertilizer N is not released gradually to the crop throughout the season as was described for crop residues and for the nonliquid portion of animal manure.

A. *Optimum Time of Application*

The optimum time to apply fertilizer N from an environmental standpoint integrates at least four factors.

1. Crop to be Grown—For example, winter wheat needs 10 to 20% of its N in the fall at planting time and the remainder in early spring. Corn requires most of its total supply in midsummer.

2. Climate—Because NO_3^--N is not bound to the soil particles and humus, it can be carried off or through the soil by water. There is a wide range of both surface runoff and percolating water in different climatic regions. On the same type of soil within the central Corn Belt, the range in potential direct runoff per year is 2.5 to 7.5 cm (1–3 inches). The range in mean annual percolation is from 2.5 to 30 cm (1–12 inches).

3. Soil—The likelihood of leaching loss of N depends on soil texture and permeability, which control water-holding capacity and rate of movement. A given amount of rainfall will move nitrates more than four times as far in a coarse sand as in a silt loam and nearly six times as far as in a silty clay soil.

4. Chemical Formulation—Five general levels of chemical stability against leaching and denitrification are recognized.

a) *Nitrate* (NO_3^-)—Nitrate is highly mobile and is immediately susceptible to leaching and denitrification, and so is least suited to application much in advance of the time it will be taken up by the crop.

b) *Urea*—Urea is intermediate in persistence, being leachable until converted to NH_4^+.

c) *Ammonium* (NH_4^+)—Ammonium compounds are bound to the soil until converted to NO_3^-, and hence can be applied in advance of crop need.

d) *Anhydrous Ammonia* (NH_3)—Anhydrous ammonia, though it quickly converts to NH_4^+, is somewhat more stable and persistent in the soil than NH_4^+ compounds, because the high NH_3 concentration within the application zone temporarily kills nitrifying soil bacteria, thus delaying the conversion to NO_3^-.

e) *Ammonia or Ammonium with an Added Nitrification Inhibitor*—This is the most persistent formulation.

B. Seasonal Application

Delayed release of N will likely be advantageous in many years for spring application on coarse-textured soils that are highly susceptible to leaching. Slow-release N fertilizers are highly effective in special situations and moderately effective in fall application for wide areas, but they are no panacea that will dramatically alter the NO_3^- concentration in streams, lakes, and underground waters. They do provide insurance against excessive losses of N in a particular set of conditions (Chapt. 13, this book).

1. Fall—Fall application of N was encouraged for many years by both agricultural extension workers and the fertilizer industry as a means of widening the fertilizer season, thus making more efficient use of storage, transportation, and application facilities.

In general, fall application has been found to be:

a) Relatively unfavorable in southern portions of the United States, because rainfall is moderate to high, and there is little if any winter freezing of soil to prevent winter leaching.
b) Unfavorable in northeastern states where potential for leaching is high during fall and early spring.
c) Very poor on sandy soils.
d) Generally relatively good in the central Corn Belt on medium- to fine-textured soils. Fall-applied N appears to be 75 to 90% as effective as N applied in the spring.

The relative acceptability of fall application of N in the NH_4^+ form can be inferred from Fig. 3. It integrates the effects of soil temperature to convert NH_4^+ to NO_3^-, length of the period when the soil is frozen, and the amount of rainfall.

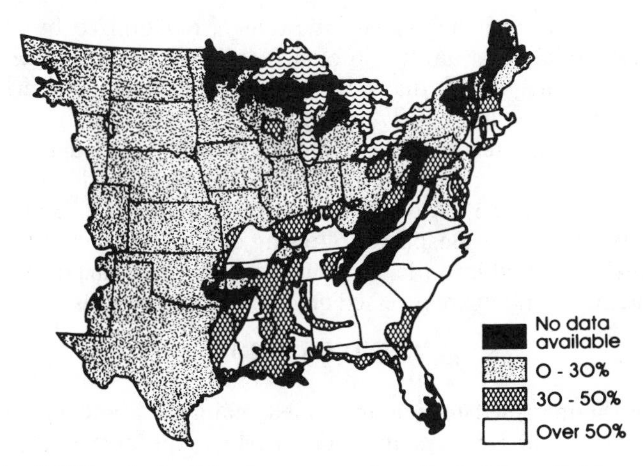

Fig. 3—Potential N loss by leaching from ammonium nitrate fall-applied for corn. (From Stewart et al., 1975.)

With increased awareness in the late 1960's and early 1970's of the potential for leaching of nitrates into surface water and groundwater, more attention was given to the behavior of fall-applied N. Concern for energy conservation became an added factor. As a result, fall application was deemphasized in educational and promotional programs. However, fall application has recently received new impetus with the widespread promotion of a nitrification inhibitor that temporarily blocks the conversion of NH_4^+ to NO_3^-.

2. Winter—Nitrogen applied in winter, especially on frozen sloping land, is least desirable from an environmental viewpoint, because of the risk that N will be washed into surface waters by melting snow or by rainfall.

3. Early to Mid-Spring—Nitrogen applied in early to mid-spring, before planting such annual crops as corn and vegetables, usually results in more efficient use of nitrogen than does fall application. Early spring application has not been encouraged for sandy soils.

4. At Planting Time—Prior to the mid-1950's most fertilizer for row crops and for small grains was packaged in bags and applied through the drill or planter. Because the total amount applied was small, it presented relatively little risk to germination from a "too high" salt concentration near the seed and did not greatly slow the planting operation. A relatively small proportion of the total N for most crops is now applied through the planter.

5. Sidedressing—This term describes the application of fertilizer beside or between row crops after plants have emerged. It is the last opportunity in the season to apply fertilizer with the usual ground equipment, thus most closely meets the goal stated earlier of applying N as shortly as possible before it will be utilized by the crop.

IV. THE FATE OF UNUSED N

It is easy to erroneously conclude that N not utilized by the current crop moves rapidly and directly into surface water, tile drainage water, or groundwater. Fortunately, that does not happen. Nitrogen that is unused during a growing season is likely to be found in these places in the following spring (in approximately decreasing order): (i) left in the leaves, stalks, and the roots of crops that remain in the field; (ii) added to residual NO_3^- in the soil; (iii) tied up in the decay products from residues left from preceding crops; (iv) lost to the atmosphere following denitrification; and (v) leached into groundwater and into tile drains or carried off in surface runoff. The last two items (iv and v) are reversed on coarse-textured soils.

V. COVER CROPS

A well-established technique, aimed mainly at reducing soil erosion and providing some fresh residues in the surface soil for the following crop, is to plant a species capable of fall growth when temperatures are relatively cool. Winter wheat, winter barley (*Hordeum vulgare* L.), rye (*Secale cereale* L.), and ryegrass (*Lolium* spp.) have most commonly been used. Such "cover" crops absorb nitrates from the soil solution and convert them into vegetative parts, thus rendering them nonleachable. Unfortunately, only about 22 to 34 kg of NO_3^--N/ha (20 to 30 lb NO_3^--N/acre) is likely to be tied up in this way; hence, the practice has limited potential for reducing the amount of NO_3^- that is lost to surface water and groundwater.

VI. SUMMARY

The cost of N fertilizer is likely to continue to rise because of the almost certain rise in the cost of energy to manufacture it. Similarly, the cost of home-grown N (legumes and animal manure) will rise because of higher labor and land costs. These increases, combined with the threat of restrictions on N should environmental problems enlarge, provide powerful incentives for agriculture to utilize N efficiently and to minimize losses.

Even in the absence of technological breakthroughs comparable to hybrid corn and the manufacture of low-cost N fertilizer, increases in crop yields can be confidently predicted because of improvements in germplasm, tillage practices, pest control, water management, harvesting methods, and increased use of optimum combinations of these practices. In response to an improvement in each, there is a multiplier effect that ripples through the others, culminating in increasing potential to effectively utilize greater amounts of N from all sources in the future.

The use of slow-release N fertilizers has increased in recent years, but will not likely have a dramatic impact on the NO_3^- content of surface waters or groundwaters.

Over the past 10 yr, crop yields have risen more rapidly than has the NO_3^- content of surface waters in the Corn Belt. This means that, *per unit of food produced,* the relative amount of NO_3^- lost to surface waters is declining.

REFERENCES

Kohl, D. H., G. Shearer, and B. Commoner. 1971. Fertilizer nitrogen: contribution to nitrate in surface water in a Corn Belt watershed. Science 174:1331–1334.

Meints, V. W., G. Shearer, D. H. Kohl, and L. T. Kurtz. 1973. A comparison of unenriched vs. ^{15}N-enriched fertilizer as a tracer for N fertilizer uptake. Soil Sci. 119:421–425.

Stewart, B. A., D. A. Woolhiser, W. H. Wischmeier, J. H. Caro, and M. H. Freere. 1975. Control of water pollution from cropland. USDA Vol. IA, Manual for guideline development. U.S. Government Printing Office, Washington, D.C.

PACIFIC
WASHINGTON
OREGON
CALIF.
MOUNTAIN
MONTANA
IDAHO
WYOMING
NEVADA
UTAH
COLORADO
ARIZONA
NEW MEXICO
WEST NORTH CENTRAL
NORTH DAKOTA
SOUTH DAKOTA
MINNESOTA
NEBRASKA
IOWA
KANSAS
MISSOURI
WEST SOUTH CENTRAL
OKLAHOMA
ARKANSAS
TEXAS
LOUISIANA
EAST NORTH CENTRAL
WISCONSIN
MICHIGAN
ILLINOIS
INDIANA
OHIO
EAST SOUTH CENTRAL
KENTUCKY
TENNESSEE
MISS.
ALABAMA
SOUTH ATLANTIC
WEST VIRGINIA
VIRGINIA
NORTH CAROLINA
SOUTH CAROLINA
GEORGIA
FLORIDA
MD.
DEL.
MIDDLE ATLANTIC
NEW YORK
N. J.
NEW ENGLAND
MAINE
VT.
N. H.
MASS.
R. I.
CONN.

Influence of Climate and Soil on Nitrogen Management

section VII

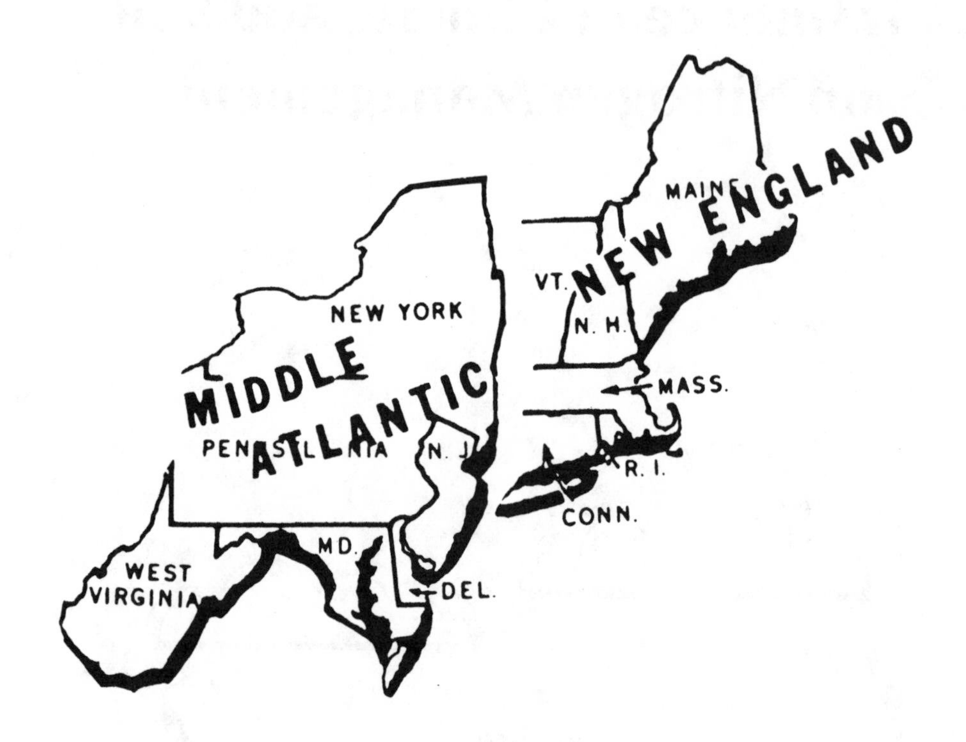
MIDDLE ATLANTIC
NEW ENGLAND
NEW YORK
VT.
N. H.
MASS.
R. I.
CONN.
N. J.
MD.
DEL.
WEST VIRGINIA

46

V. Allan Bandel
University of Maryland
College Park, Maryland

Richard H. Fox
Pennsylvania State University
University Park, Pennsylvania

Management of Nitrogen in New England and Middle Atlantic States

The northeastern United States is a unique and important agricultural region, vibrantly productive in spite of its relatively thin and often rocky soils. Large population centers within its borders provide an excellent market for its agricultural products. Recent census figures indicate there are approximately 178 000 farms in the Northeast consisting of about 14 million ha. Estimates from the 1979 Agricultural Statistics (Table 1) indicate over 5.3 million ha of the major crops { corn (*Zea mays* L.), soybeans [*Glycine max* (L.) Merr.], wheat (*Triticum aestivum* L.), barley (*Hordeum vulgare* L.), oats (*Avena sativa* L.), and hay}. Much of the remaining cropland consists of pasture, fruits, vegetables, and certain specialty crops. Agriculture contributes significantly to the economic welfare of the Northeast.

Northeast agriculture is diversified, consisting primarily of grain, forages, livestock, dairy, poultry, and truck crops. Intensive row-crop agriculture is concentrated largely on the Coastal Plains and to a lesser degree on the Upland Plateaus. A livestock- and dairy-based economy dominates the latter, whereas broilers are an important ingredient in the Middle Atlantic Coastal Plains economy. In the northern part of the region, especially the New England area, crop production is oriented more toward forages and less toward grains.

Nitrogen in Crop Production.

Table 1—Extent of principal crops grown in the New England and Middle Atlantic states (1978). (From *Agricultural Statistics,* 1979.)

State	Crop						
	Corn	Wheat	Barley	Oats	Soybeans	Hay	Total
	ha × 10^3						
Conn.	21.4	--	--	--	--	35.2	56.6
Del.	75.6	12.9	13.3	--	101.0	8.9	211.7
Me.	20.2	--	--	15.8	--	89.3	125.3
Md.	278.8	46.5	42.4	11.3	141.4	100.6	621.0
Mass.	17.4	--	--	--	--	46.9	64.3
N. H.	10.9	--	--	--	--	36.8	47.7
N. J.	54.5	19.4	15.8	3.6	84.8	48.1	226.2
N. Y.	525.3	34.8	4.4	141.4	9.3	1000.1	1715.3
Pa.	652.6	103.0	54.5	145.5	26.3	781.9	1763.8
R. I.	1.6	--	--	--	--	3.2	4.8
Vt.	45.3	--	--	--	--	179.0	224.3
W. Va.	37.6	4.4	4.4	7.3	--	240.4	294.1
Total	1741.1	221.0	135.0	324.9	362.9	2570.3	5355.1

I. SOILS

The soils of the Northeast are complex, ranging from deep soils formed from residuum and alluvial material to the shallow till mantels. Soil drainage varies from the excessively well-drained soils on water-sorted material of glacial deposits, deltas, beaches, and moraines to the very poorly drained clays on glacial lake and marine sediments. Much of this variability is a result of wide differences in the physiographic and topographic nature of the landscape. In the northern part of the region, the younger soils were developed from more recent glacial deposits. The soils are older and generally more strongly acidic in the unglaciated areas. The major physiographic regions include: (i) the mountains, which support only limited agriculture—mostly small farms in the more level valleys of the Allegheny, Catskill, and Taconic mountain ranges; (ii) the Uplands, which include the productive, rolling, nonmountainous sections such as the Piedmont; (iii) the Lowland Plains, which include the intensively cropped Coastal Plain; and (iv) the ridge and valley sections extending from south central Pennsylvania northeasward to the Hudson River valley.

Soils in these areas are weathered primarily from granite, gneiss, phyllite, limestone, sandstone, shale, and siltstone. Soil texture can range from sand to clay, with all intermediate classes represented.

Because of the variable nature of Northeast soils, fertility levels fluctuate considerably. Cation exchange capacities range from < 1 meq/100 g for the sandiest Coastal Plains soil to as high as 35 to 40 meq/100 g for a few rarely occurring high organic matter soils. Certain muck soils may range as high as 150 to 200 meq/100 g cation exchange capacity.

Organic matter levels of mineral soils range from < 1% for some of the highly weathered sandy Coastal Plains soils found in the southernmost part of the region to as high as 15% in the cooler northern areas. The higher

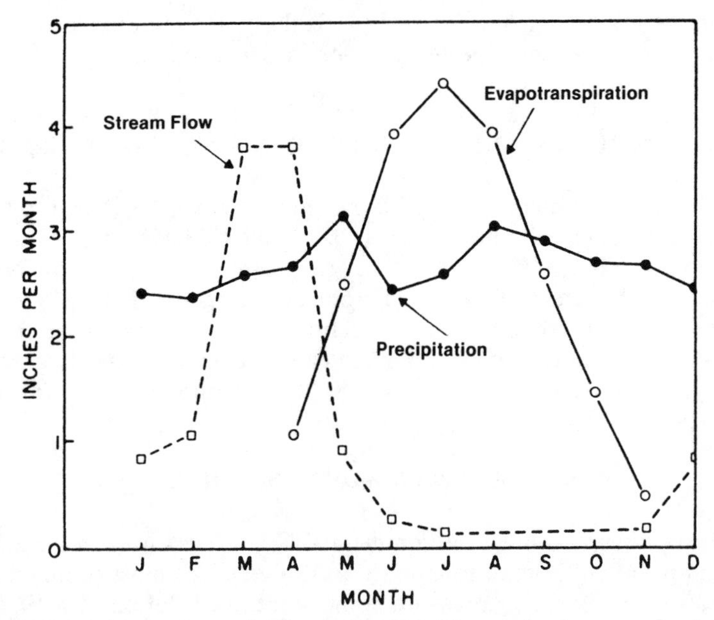

Fig. 1—Estimated evapotranspiration in the Erie-Niagara basin, normal rainfall at Lockport, N.Y., and stream flow of Little Tonawanda Creek at Linden, N.Y. (Harding & Gilbert, 1968).

organic matter levels are more the exception than the rule. Most soils in the Northeast average from 1 to 5% organic matter, increasing from south to north across the region.

Rainfall averages approximately 100 cm/yr across the Northeast region; however, N leaching and/or denitrification can be significant. Subsequently, it is recommended almost universally that timing of fertilizer N applications coincide as closely as possible with the time of the crops' greatest needs in order to minimize such losses. During late winter and early spring, it is likely that precipitation will exceed evapotranspiration in the Northeast (Bouldin et al., 1971). As this excess water moves off the land, it will very probably carry with it soluble and suspended nutrients and other matter from agricultural land. In late fall and winter, when evapotranspiration again falls below precipitation, excess water is not lost to increased stream flow because the soil reservoir is being recharged (Fig. 1) (Harding & Gilbert, 1968). Nevertheless, soluble nutrients could be leached out of the root zone during this period. Obviously, it is essential from an economic and environmental viewpoint that N fertilizers be managed carefully for maximum efficiency and minimum impact on the environment.

Many agronomically important crops grown in the Northeast respond favorably to N fertilization. Because of variable levels of soil fertility and soil texture, soil organic matter and N mineralization rates, relatively abundant precipitation, varying yield goals, and crop management practices, N fertilizers must be managed carefully according to local conditions. As energy costs continue to escalate, N management will require further attention, research, and refinement, particularly as new crop management techniques such as no-tillage become more widely adopted.

II. AMOUNTS AND SOURCES OF N FERTILIZER USED

The total commercial N fertilizer used in the Northeast region from 1 July 1979 through 30 June 1980 was 322 000 t or 3.1% of the U.S. consumption (USDA, 1980; Table 2). Approximately 56 kg of N were purchased per hectare of harvested cropland (Hargett & Berry, 1978). This somewhat low figure is due to the relatively high proportion of cropland that is used for legumes in New York, Pennsylvania, and New England and to the large amount of livestock and poultry manure available for use on field crops.

A. Commercial Fertilizer N

Forty-nine percent of the commercial N fertilizers used in the area are direct application, nonmixed N sources (Table 3). The most common direct application N source is urea-ammonium nitrate solution (UAN) (38%) followed by urea (19%) and anhydrous NH_3 (AA) (15%). Thus, almost 60% of the N used is in a form (UAN or urea) that can lead to NH_3 volatilization losses if not soil incorporated. The preponderance of these sources, combined with the large and expanding use of no-tillage crop management in the more southern states of the region, produces less-than-optimal N fertilizer efficiencies.

Table 2—Total annual quantity of commercial fertilizer N consumed and estimated quantity of manure N available for use on cropland in the New England and Middle Atlantic states.†

State	Commercial fertilizer N	Manure N
	t	
Conn.‡	6 859	3 700
Del.	16 625	1 800
Me.‡	11 444	8 200
Md.	59 489	15 400
Mass.‡	8 248	3 600
N. H.‡	2 696	1 800
N. J.	22 664	3 600
N. Y.	97 294	44 500
Pa.	77 396	43 500
R. I.‡	1 226	0
Vt.‡	7 899	9 100
W. Va.	10 166	4 500
New England Total	38 373	26 400
Middle Atlantic Total	283 633	113 300
Total	322 006	139 700

† From USDA, 1980 and Van Dyne & Gilbertson, 1978.
‡ New England states.

Table 3—Quantity of commercial fertilizer N consumed as direct application material in the New England and Middle Atlantic states from 1 July 1979 to 30 June 1980.†

State	Source									
	NH_3	AN	Ammonium sulfate	Urea	UAN	Sodium nitrate	Ammonium phosphates	Natural organics	Other N sources	Total
	t									
Conn.‡	0	34	0	1 529	0	6	170	202	367	2 308
Del.	543	1 208	735	652	6 936	1	2	18	6	10 102
Maine‡	0	505	93	895	4	0	2	44	0	1 542
Md.	7 759	2 199	593	2 895	21 342	109	0	90	67	35 054
Mass.‡	2	129	5	1 022	70	5	5	181	92	1 510
N. H.‡	0	99	18	214	0	0	5	32	33	400
N. J.	156	1 022	3	1 725	1 546	91	41	222	531	5 335
N. Y.	12 313	3 333	221	2 626	14 171	2 260	1	323	17 971	53 219
Pa.	3 593	1 729	570	15 152	15 210	2	1 182	74	345	37 857
R. I.‡	0	4	0	18	0	5	42	38	0	107
Vt.‡	128	100	11	1 207	1	0	0	0	1 153	4 070
W. Va.	0	1 582	299	1 359	0	18	0	15	3 514	6 788
Total New England	130	870	127	4 885	1 545	16	222	496	1 645	9 937
Total Middle Atlantic	24 364	11 073	2 421	24 409	59 206	2 480	1 226	743	22 434	148 531
Total	24 494	11 943	2 548	29 294	60 751	2 496	1 448	1 239	24 079	158 292

† From USDA, 1980. ‡ New England states.

B. Soil N

Although the residual N content of Northeast soils may be generally low compared with many soils found in the Midwest, Northeast soils contain more N than the more highly weathered soils of the South. Like many soils of the world, most of the residual soil N is organically bound. Thus, soil N buildup largely parallels soil organic matter accumulation.

As commented on previously, the organic matter content of mineral soils in the Northeast may vary from < 1% to > 15%, but most soils average from 1 to 5% organic matter. Assuming that soil organic matter contains approximately 5% N, Northeast soils contain from 1100 to 5500 kg N/ha. However, only a small portion of this total N is mineralized and made available for plant use at any given time. For instance, on the Maryland Coastal Plain, an intensively cropped Mattapex silt loam (fine-loamy, mixed mesic Aquic Hapludult), which received no fertilizer N over a 9-yr period of continuous corn, averaged 3.4 t/ha of grain at 15.5% moisture. The aboveground portions of the corn plants recovered 59.7 kg N/ha per year from the soil. On the Piedmont, a Delanco silt loam (fine-loamy, mixed mesic Aquic Hapludult), which had previously been manured and cropped to several years of alfalfa (*Medicago sativa* L.), after 7 yr of continuous corn averaged 6.1 t/ha of grain at 15.5% moisture. The aboveground portions of the corn plants recovered an average of 101.3 kg N/ha per year from the soil (V. A. Bandel, 1981, unpublished data).

Obviously, the N supplying power of Northeast soils will vary considerably across the region depending upon soil formation factors, soil organic matter content, soil and crop management, etc. These examples are typical for some of the prevalent Coastal Plain and Piedmont soils.

C. Legume N

There are 929 000 ha of alfalfa harvested in the Northeastern and Middle Atlantic states, or approximately 25% of the harvested cropland. From seed sales in the area (Pardee, 1979), it can be calculated that approximately another 405 000 ha of legume hay (red clover, *Trifolium pratense* L.; ladino clover, *T. repens* L.; alsike clover, *T. hybridum* L.; and birdsfoot trefoil, *Lotus corniculatus* L.) are grown. The estimated amount of residual N available from these legume hay crops to subsequent nonlegume crops ranges from zero (several New England states) to a total of 260 kg/ha during a 4-yr period following a sod crop containing 25 to 50% legumes (new York). The most prevalent credit for residual N from alfalfa is approximately 85 kg/ha and for the other legumes 55 kg/ha. Using these figures, it can be calculated that nearly 100 000 t of residual N are available from these legumes for growing nonleguminous cropland harvested in the area. Recent work in Pennsylvania (R. H. Fox, unpublished data) has confirmed the observation in New York that the residual N from alfalfa over a several-year period is more in the 170 to 280 kg/ha range than in the 85 kg/ha range.

Research is needed to more precisely measure the N contribution from these legumes as well as from the nonleguminous sod crops, i.e., the grasses. A procedure for estimating residual N would allow the farmer to apply the minimum quantity of fertilizer N needed for optimum yield of other crops. Much work has been carried out in the past on this need, and an intensive research project is currently being conducted by the USDA Soil Nitrogen Laboratory at Beltsville, Md. and several cooperating State Land Grant Universities.

Researchers in New Jersey and Delaware (Flannery, 1981; Mitchell & Teel, 1977) have shown that planting a hairy vetch (*Vicia villos* Roth.) crop after corn provides approximately 134 kg N/ha to the next season's corn crop. The economics are currently such that the use of hairy vetch as a source of N or buying all the N needed as commercial fertilizer produces about the same profit. However, as the price of N increases and if a large, steady market for hairy vetch seed lowers its price, the economics of using hairy vetch as a N source should improve markedly. Investigations are currently underway in New York (Scott, 1979) to determine the feasibility and N contribution of double cropping legumes with corn in the more northern areas of the region.

D. Manure N

The total annual quantity of manure N voided in the Northeastern and Middle Atlantic states has been estimated at 235 000 t (Van Dyne & Gilbertson, 1978). Of this amount, approximately 140 000 t is believed to be recoverable for application to cropland (Table 2). This is approximately 43% as much N as used in the form of commercial fertilizer. If manure is managed to minimize N losses, manure can be a significant source of N for crop production on the majority of farms in the area. The N fertilizer recommendations from all states in the area make allowances for the N contributed from manure applied during the previous year. For dairy and beef manure, the credit ranges from 1 to 5 kg N/t (2–10 lb N per ton) of manure applied. The most common allowance is approximately 2.5 kg N/t. The low and high figures are used for low level and excellent manure management, respectively. The N available for crop uptake from poultry manure is reported to be from 5 to 25 kg/t (10–50 lb N per ton), depending on the type and moisture content of the manure. The most frequently reported credit is between 15 and 20 kg N/t. In a few states, the credit for residual N from manure in subsequent years is equivalent to a total of 25 to 30% of the first year's credit.

The large variability in the amount of N in manure available for crop uptake due to the wide range of manure collection, storage, and application procedures used makes it difficult for farmers to develop accurate estimates of the contribution of manure N to their total N fertilizer needs. In addition to the above uncertainties, the majority of farmers only have very rough estimates of the rate of manure applied to any given field. Consequently, many farmers tend to underestimate the manure contribution and apply more commercial fertilizer N than is needed for maximum yields. Unfortunately, a dependable soil N availability index does not currently exist that can be used by soil testing laboratories to estimate the N contribution from manure and legume crops. Fox and Piekielek (1978a, b) found some promising N availability indices in Pennsylvania, but further testing has shown that the usefulness of these tests in their present form is limited. Research should continue on the development of a N soil test that can reduce N fertilizer consumption where it is not needed and increase farm profits.

III. MANNER OF N USE

A. Corn (*Zea mays* L.)

The N fertilizer recommendation for corn depends on its yield potential. Yield potential in turn depends on the soil type and area of the state and region. Essentially all corn in the six New England states is grown for silage and has an estimated yield potential of 36 to 49 t/ha (16–22 tons/acre) (70% moisture). The N recommendations in New England states range from 140 to 202 kg N/ha or 3.1 to 4.5 kg N/t of silage with a mean of 3.4 kg N/t. In the six Middle Atlantic states, approximately 70% of the corn is grown for grain and 30% for silage. The corn yield potential in these states ranges from 6.3 to 9.4 t/ha (100–150 bu/acre) for grain and 38 to 56 t/ha for silage. The N fertilizer recommendation for a 7.8 t/ha (125 bu/acre) yield potential in New York and Pennsylvania is approximately 146 kg/ha (130 lb/acre). In the other four states, the recommendation is approximately 168 kg N/ha (150 lb/acre). In general, the N rates recommended for corn grain range from 19 to 23 kg N/t of grain yield (1.0–1.3 lb N/bu). The N recommendations for silage range from 3.5 to 5 kg N/t (7–10 lb N/ton) of silage yield potential. The manure and legume N contributions are subtracted from these rates where applicable.

The recommendation in the New England states and New York is to apply 10 to 20 kg/ha of N in a starter fertilizer at planting with the remainder being applied as a sidedress application when the corn is 30 to 45 cm high. Pennsylvania and West Virginia recommend 10 to 20 kg N/ha in the starter with the remainder applied as a broadcast treatment at planting or as a sidedress. Delaware, Maryland, and New Jersey recommend for their sandy soils, that approximately 50% of the N should be applied as a broadcast treatment at planting and the other half as a sidedress when the corn is 15 to 45 cm tall.

Research in New York (Lathwell et al., 1966, 1970) indicated that fall applications of N to corn were only 50% as effective as sidedress-applied N. This research also showed that N applied in April produced lower yields and N uptake than did sidedress-applied N. Maximum fertilizer N efficiency was obtained by applying a major proportion of the N as a sidedress when the corn was 15 to 45 cm high. Maryland research indicated that fall-applied N was about 50% as efficient as N applied in the spring near planting time (Bandel & Rivard, 1973).

B. Small Grains

The N rates recommended for small grains range from 45 to 111 kg N/ha, depending upon the yield potential and lodging resistance of the variety grown. Most recommendations are in the 55 to 65 kg N/ha range.

For fall-planted grains, the recommended application method is often for approximately 20 kg N/ha with the seed at planting, and depending upon the state, the remainder is broadcasted in late winter or early spring, either before or after the crop has begun to grow. For spring-planted small grains, it is generally recommended that all N be applied at planting.

C. Grass Hay

The N recommended for grass hay throughout the Northeast ranges from 56 to 111 kg N/ha, with half normally applied in the early spring and the other half applied after the first crop is harvested.

D. Soybeans [*Glycine max* (L.) Merr.]

Most states do not recommend N fertilizer for soybeans, although Maryland, New York, and New Jersey recommend up to 20 kg/ha at planting on very sandy soils. These soils normally have lower N levels than medium- to heavy-textured soils, and thus usually require some N in the starter fertilizer, even for legumes.

IV. NO-TILLAGE CORN PRODUCTION

In recent years, reduced tillage and no-tillage crop production systems have gained rapid acceptance by Northeast farmers. This trend has been particularly significant in the southern or Middle Atlantic part of the region. In a 1978 survey conducted across the Northeast, it was estimated that 10, 51, 21, 47, 25, and 23% of the corn cropland in Delaware, Maryland, New Jersey, West Virginia, Pennsylvania, and New York, respectively, was planted no-till or by some form of reduced tillage. Since 1970, most states reported significant increases in no-tillage planting. Before 1970, many of these states indicated < 5% of their corn cropland planted no-tillage. The New England states have not witnessed a rapid increase in no-tillage acceptance. Many New England states continue to report < 10% of their total corn grown by the no-tillage technique.

This rapid increase in no-tillage acceptance was made possible by many factors, but the key elements were the development of effective weed control chemicals and precision no-tillage planters. Prior to the early 1970's, most farmers virtually ignored the no-tillage concept. For instance, in Maryland by 1970, there were < 4000 ha of no-tillage corn, < 2% of the total cropland. But by 1978, no-tillage or reduced tillage accounted for > 50% of Maryland's total corn cropland. And that percentage continues

Table 4—Consumption of selected direct application N fertilizer materials in the New England and Middle Atlantic states during 1955, 1965, and 1975.†

Material	Year		
	1955	1965	1975
	t		
Ammonium nitrate	29 586	33 690	69 816
Anhydrous ammonia	1 820	7 777	22 617
Nitrogen solution	3 192	41 900	131 610
Urea	1 981	21 596	43 668
Ammonium sulfate	--	--	12 407
Sodium nitrate	15 280	9 075	2 809

† From Fertilizer Summary Data, N. L. Hargett, NFDC, TVA, Muscle Shoals, Ala.

to increase. Some counties, particularly in the rolling Piedmont region, report 75 to 80% of their corn planted as no-tillage. Other Northeast states have reported similar trends.

This dramatic change in cultural technique for growing corn and certain other row crops is resulting in some similarly dramatic changes in N fertilizer management. Restrictions in the manufacture, transportation, and storage of certain N fertilizers is also resulting in significant changes in the availability of materials. For instance, although ammonium nitrate (AN) use showed a 136% increase in the Northeast from 1955 to 1975 (Table 4), indications are that AN use will decline in future years as manufacturing facilities become obsolete or wear out and are not replaced. This "vacuum" will be filled primarily by urea, which increased in use 21-fold during that same 20-yr period. Nitrogen solutions have also increased in popularity, demonstrating a 40-fold increase in use from 1955 to 1975.

Much of the increased popularity of N solutions (UAN) has been due to convenience in handling plus uniformity of application. Urea-ammonium nitrate (UAN) solutions have been used extensively for many years as a carrier for corn herbicides. This use, which gained much of its momentum before no-tillage became popular, was quickly adapted by no-till farmers since they could, in one pass over the field, simultaneously apply the required contact and residual herbicides as well as N fertilizer. Subsequent research and farmer experience indicated that this was often a satisfactory means for applying herbicides, but was not always the most efficient method for applying N under no-tillage conditions.

When UAN solutions are sprayed uniformly over no-tillage soil surfaces, particularly in a normal fine spray pattern, it is believed that some material tends to adhere to the cover crop or to other crop residues on the soil surface. Unless sufficient rainfall occurs within a relatively short period of time, a significant portion of the UAN solution apparently does not move down to the soil surface and subsequently to the plant roots. Some of this N may also be lost due to NH_3 volatilization as well as microbial immobilization. Therefore, some changes in N management have become necessary where N solutions were being utilized under no-tillage conditions.

Although no-tillage farmers have recently begun to modify their application methods, research data indicate that UAN solutions are frequently more efficiently utilized by no-tillage corn plants when they are applied in a coarse spray or, preferably, in a band instead of in the traditional uniform spray pattern. It also appears that placement of UAN solutions, as well as urea, beneath the soil surface often significantly improves N uptake efficiency. Such placement apparently reduces NH_3 volatilization losses as well as immobilization by microorganisms active in the organic layer formed at the untilled soil surface.

Another problem related to N management that became apparent under no-tillage conditions was the occurrance of poor residual weed control due to deactivation of the triazine herbicides from acidity created by surface-applied N. Obviously, herbicide deactivation can also be a problem under conventional tillage conditions where the N is surface-applied. Banding, subsurface injection, or other means of N soil incorporation prevents the formation of this acid "umbrella." A frequent (annual) monitoring of soil surface pH (0- to 5-cm depth) is recommended to prevent creation of this acid zone.

Extensive field plot testing of surface-broadcasted urea, AN, and 30% UAN solution at variable N rates on no-tillage corn indicated that of these three materials, AN was superior. Urea resulted in the poorest no-tillage corn yields and UAN was intermediate. A summary of 12 N rate by N source no-tillage corn tests conducted over a 4-yr period in Maryland (Table 5) indicated that, on the average, 180 kg N/ha from surface-broadcasted urea was equivalent to approximately 100 kg N/ha from AN. Similarly, 180 kg N/ha from UAN produced the same corn yields as 135 kg N/ha from AN. Obviously, in cases where urea and UAN solutions are surface-broadcasted, higher rates must be applied compared with AN. Such differences between N sources are not generally a problem when the N is soil incorporated. Some farmers are successfully injecting AA in their no-tillage fields with apparently acceptable results. When available, AA offers a satisfactory solution to the problem. Improved N management and development of more suitable N fertilizers and application equipment is beginning to further minimize these problems.

It is essential that optimum levels of N fertilizer be applied to nonleguminous crops. But the situation is more critical for no-tillage than for conventional-tillage crops. A summary of 40 corn N rate × tillage tests conducted in Maryland from 1973 to 1981 indicated that at suboptimal N

Table 5—Influence of N source and rate of application on no-tillage corn grain yields in Maryland. (Summary of 12 tests, 1976–1979.)†

	N Rate (kg/ha)					
N Source	0	45	90	135	180	Avg
	t/ha					
Ammonium nitrate	5.4	6.7	8.2	8.9	9.3	8.3
Urea	5.4	6.3	7.5	8.0	8.5	7.6
30% UAN	5.4	6.1	7.6	8.4	8.9	7.8
Avg	5.4	6.4	7.7	8.3	8.8	

† Source: V. A. Bandel, Univ. of Maryland, unpublished data.

Table 6—Influence of N rate and tillage on probability of obtaining highest corn grain yields in Maryland. (Summary of 40 tests, 1973–1981.)†

N Rate	Tillage		Tests
	No-tillage	Conventional	
kg/ha	% higher yields		no.
0	32	68	40
45	36	64	25
90	47	53	40
135	62	38	40
180	62	38	40
270	80	20	15
Avg	53	47	

† Source: V. A. Bandel, Univ. of Maryland, unpublished data.

fertilization rates, there was almost twice as great a probability that corn grain yields would be higher under conventional tillage than under no-tillage (Table 6). The probabilities reversed themselves and became more favorable for no-tillage at higher N fertilization levels.

No-tillage corn should be fertilized with somewhat more N than conventional-tillage corn. Although the research data is not yet available, this would also appear to be the case for other nonleguminous no-tillage crops. The additional N can be justified on the basis that no-tillage corn has a higher yield potential than conventional-tillage corn. Currently, it is not clear precisely how much additional N can be economically justified on no-tillage corn. But most data indicate that approximately 35 kg/ha additional N is needed to compensate for the N removed in the higher yield and the greater N retention by the soil.

REFERENCES

Agricultural Statistics. 1979. U.S. Department of Agriculture. U.S. Government Printing Office, Washington, D.C.

Bandel, V. A., and C. E. Rivard. 1973. Influence on corn of fall vs. spring nitrogen fertilization and plowing. Maryland Agric. Exp. Stn. Misc. Pub. MP 843.

Bouldin, D. R., W. S. Reid, and D. J. Lathwell. 1971. Fertilizer practices which minimize nutrient loss. p. 25–35. *In* Agricultural wastes: principles and guidelines for practical solutions. New York State College of Agric. and Life Sci., Ithaca, N.Y.

Flannery, R. L. 1981. Hairy Vetch rye cover crops for corn silage production using no-tillage. Better Crops Plant Food 66:22–25.

Fox, R. H., and W. P. Piekielek. 1978a. Field testing of several nitrogen availability indexes. Soil Sci. Soc. Am. J. 42:747–750.

Fox, R. H., and W. P. Piekielek. 1978b. A rapid method for estimating the nitrogen supplying capability of a soil. Soil Sci. Soc. Am. J. 42:751–753.

Harding, W. E., and B. K. Gilbert. 1968. Surface water in the Erie-Niagara basin, New York. Basin Planning Rep. ENB-2, State of New York. Conservation Dep., Water Resour. Commission, Albany, N.Y.

Hargett, N. L., and J. T. Berry. 1978. 1978 Fertilizer summary data. NFDC, TVA, Muscle Shoals, Ala.

Lathwell, D. J., D. R. Bouldin, and W. S. Reid. 1970. Effects of nitrogen fertilizer applications in agriculture. p. 192–206. *In* Relationship of agriculture to soil and water pollution. Proc. of Cornell Univ. Conf. on Agric. Waste Manage., Ithaca, N.Y. 19–21 Jan. 1970. New York State College of Agric. and Life Sci., Ithaca, N.Y.

Lathwell, D. J., G. R. Free, and D. R. Bouldin. 1966. Efficiency of fall-applied nitrogen in New York for corn and small grains. Agron. Mimeo 66:13. Cornell Univ., Ithaca, N.Y.

Mitchell, W. H., and M. R. Teel. 1977. Winter-annual cover crops for no-tillage corn production. Agron. J. 69:569–573.

Pardee, W. D. 1979. Northeast seed use survey. Small seeded grasses and legumes. P. B. 79-2. Dep. Plant Breeding & Biometry, Cornell Univ., Ithaca, N.Y.

Scott, T. W. 1979. Is there a potential for cover crops with corn. 1979. Forage, Corn and Seed Conf. Pennsylvania State Univ., University Park, Pa.

U.S. Department of Agriculture. 1980. Commercial fertilizers: consumption by class for year ended June 30, 1980. Corp Reporting Board, Economics and Statistics Service, Sp. Cr. 7 (11-80). USDA, Washington, D.C.

Van Dyne, D. L., and C. B. Gilbertson. 1978. Estimating U.S. livestock and poultry manure and nutrient production. USDA Economics, Statistics and Cooperative Service ESCS-12. USDA, Washington, D.C.

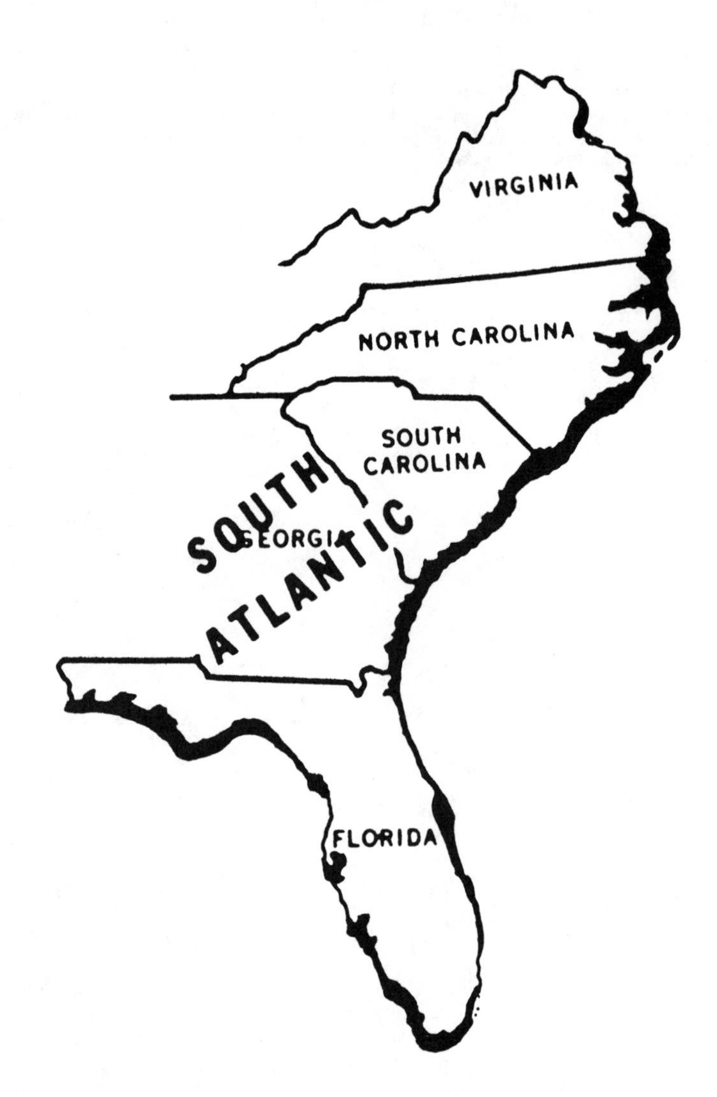
VIRGINIA
NORTH CAROLINA
SOUTH
CAROLINA
GEORGIA
SOUTH
ATLANTIC
FLORIDA

47

J. W. Gilliam
North Carolina State University
Raleigh, North Carolina

Fred Boswell
University of Georgia
Experiment, Georgia

Management of Nitrogen in the South Atlantic States

Proper management of N fertilization in the South Atlantic states is probably more critical than for any region of the United States. The soils of this region generally supply only small amounts of available N to crops through mineralization of organic matter. The large majority of crops, to which fertilizers are applied, is grown on sandy soils, which have a high potential for N leaching because of the annual average rainfall of $\geq$ 120 cm. Nitrogen leaching on most soils is not usually a large problem during the growing season, but most residual inorganic N is removed during the winter and early spring months. This movement of N into the drainage water can be a potential water quality problem in certain areas of the region.

With the high cost of inorganic N and other increased economic input for crop production, along with marginal economic returns, proper management of N fertilizer has become more critical for growers. However, because of a long history of N fertilization in the region, farmers are knowledgeable about proper N management practices and, in general, the average applied N rates are close to the recommended levels.

The region is experiencing a dramatic increase in cropland under irrigation. Total water management, utilizing controlled drainage and subirrigation, also shows considerable promise for the region. As a result of these new water management practices, future N management practices may have to be modified in order to minimize leaching and denitrification and, at the same time, maximize efficiency. Off-site water quality effects must be an inherent consideration in future N fertilization practices.

Nitrogen in Crop Production.

I. SOILS

Three major physiographic divisions—Coastal Plains, Piedmont, and Blue Ridge Mountains—can be used to characterize the important differences in the soils of the South Atlantic states (Daniels et al., 1973). By far the most agriculturally important division is the Coastal Plains, where approximately 70% of the fertilizer N is used. Approximately 25% is used in the Piedmont and the remainder in the Blue Ridge Mountains.

With respect to factors affecting N management, the soils of the region can be divided into three general groups: well-drained Coastal Plains, poorly drained Coastal Plains, and Piedmont and Blue Ridge Mountains. Most of the well-drained soils of the Coastal Plains are fine loamy, siliceous thermic Paleudults with low organic matter content (generally $< 2\%$), and sandy to sandy loam surface soils. These soils are extensively cropped and are generally the most productive of the region. Nitrogen in these soils is highly susceptible to leaching, particularly during the winter and to a lesser extent during the growing season. The depth to the fine loamy textured (18–35% clay) B horizon influences the amount of leaching below the root zone during the growing season, and thus the N recommendations (as will be discussed in later sections). Mineralization of organic soil N provides only a small fraction of the N needed for economic crop growth. Winter rains leach these soils and reduce N carry-over.

Poorly drained soils, Paleaquults–Histosols, occupy a large proportion of the land near the ocean and have surface horizons with relatively high organic matter content (5–95%). Water tables are near the surface much of the year, especially in the winter and early spring unless well-served by drainage systems. Although N leaching occurs on these soils, denitrification during wet periods is generally considered to be an equal problem. These soils furnish more N to commercial crops, via mineralization during the warm growing season, than other Coastal Plains soils and N recommendations reflect this property.

The Piedmont topography is gently rolling, and often much of the original sandy topsoil has been lost in erosion. Soils on the Piedmont (predominately clayey, kaolinitic, thermic Typic Hapludults) have thinner A and B horizons than those in the Coastal Plains, although friable weathered rock (saprolite) below the soil is frequently several meters thick. The cultivated area in the Blue Ridge Mountains physiographic division differs little from those in the Piedmont, except for a somewhat cooler mesic (8–15°C) mean annual soil temperature regime. They are considered similar to Piedmont areas for this discussion. Much of both areas is wooded; however, many rather small fields are utilized for growing forages or row crops. When cultivated or previously eroded, these soils have low soil organic matter contents ($< 2\%$), although those in continuous forage production may contain 4 to 5% organic matter in the surface horizon. Nitrogen leaching losses on these soils are not as great as on Coastal Plains soils, because of their higher clay content and greater water runoff. Piedmont soils are not usually as productive as Coastal Plains soils and N recommendations for crops in these areas are generally lower.

II. CLIMATE

The mean soil temperatures in the South Atlantic states range from approximately 15 to 22°C. Mean winter temperatures remain above freezing; thus, mineralization, nitrification, and denitrification occur throughout the year. Winter rainfall exceeds potential evapotranspiration by approximately 300 to 400 mm/yr; thus, no state recommends fall or winter application of N for spring-planted crops, because of the potential for leaching losses. It has to be recognized, however, that soils and weather patterns differ and Kamprath et al. (1973) have shown winter-applied N to be as effective as spring-applied N on some soils during some years.

The average annual rainfall over much of the region is approximately 122 cm except in Florida, where the average is near 142 cm. Generally, growing-season rainfall is adequate for evapotranspiration requirements, but short droughts reduce crop yields in some locations every year. Growing-season rainfall usually does not leach N below the root zone. When a leaching problem occurs, it most often occurs shortly after planting in the spring.

III. COMMERCIAL FERTILIZER N USED

The total amount of N used and the primary forms of this N are shown in Table 1. The four northern South Atlantic states (Virginia, North and South Carolina, and Georgia), where corn (*Zea mays* L.) receives a high fraction of the fertilizer N, use similar forms of N. In these states, 25 to

Table 1—Total amounts of inorganic N sources used in the South Atlantic states.

Form of N	Fla.		Ga.		N. C.		S. C.		Va.	
	$t \times 10^3$	%	$t \times 10^3$	%	$t \times 10^3$	%	$t \times 10^3$	%	$t \times 10^3$	%
Mixed Fert.†	166.9	73.7	64.0	26.0	77.2	36.0	25.4	30.6	39.6	46.4
NH_3†	4.6	2.0	20.2	8.2	10.4	4.9	2.5	3.1	0.2	0.2
AN†	23.0	10.2	44.5	18.0	16.1	7.5	7.5	9.1	3.5	4.1
Soln†	19.1	8.5	110.3	44.7	100.5	46.9	43.5	52.5	38.0	44.5
Sodium nitrate†	0.3	0.4	0.5	0.2	4.8	2.2	1.0	1.2	0.5	0.5
Urea†	1.4	0.2	1.2	0.5	1.0	0.5	0.3	0.3	1.5	1.8
Other†	--	5.0	--	2.4	--	2.0	--	3.2	--	1.8
Total†	226.4		247.4		214.4		82.9		85.4	2.5
Total—1975	170.0		290.8		199.6		81.2		89.0	
Total—1980	173.3		227.8		177.3		82.2		70.3	
kg N/ha cropland‡	204		128		109		78		78§	
kg N/ha nonlegume‡	253		296		196		216		109§	
kg N/ha nonlegume‡ assuming 34 kg N/ha pasture	161		268		183		193		77§	

† For year ending 30 June 1980. Statistical data from USDA, 1980.

‡ All cropland considered for 1980; only soybeans and peanuts considered as important crop legumes.

§ Cropland statistics for Virginia include a large area of infrequently fertilized hay crops. Fertilization rates for row crops in Virginia are believed to be similar to N. Carolina.

45% of N is used as dry mix fertilizer with most of the application being broadcasted in the spring before planting for annual crops or to perennial crops. The dominant form of N used is solution, which accounts for approximately 50% of the fertilizer N sales in Virginia, North Carolina, South Carolina, and Georgia. Much of this N is used as a side-dressing for corn. Most (74%) of the fertilizer N used in Florida is applied as mixed fertilizer. This is due to a much larger area devoted to pasture, vegetables, and fruit crops where use of complete mixed fertilizer is common. The use of sodium nitrate parallels the area planted to tobacco (*Nicotiana tabacum* L.), because this source is used as a topdressing for tobacco to supply the preferred NO_3^- form of N. A comparison of the sources used in 1975 to those used in 1980 indicated little change during this period.

Total use of N in the South Atlantic states increased only 3% from 1975 to 1980, with essentially all of this increase occurring in Florida (Table 1). There was a substantial decrease (15%) in fertilizer N use in Georgia from 1975 to 1980, because of the decrease of approximately 300 000 ha of corn grown. The rate of N applied per unit area of nonlegumenous crop was substantially greater in Georgia than in other states, which is probably due to a large percentage of the corn being irrigated. If a slight correction is made in rate of N per hectare values (Table 1) due to urban use, it appears that the average fertilizer N rate applied by the farmers is close to that recommended by agronomists in the respective states.

IV. FERTILIZATION OF MAJOR CROPS

A. Corn (*Zea mays* L.)

The South Atlantic states are considered a grain-deficient area. The most limiting factors for grain production are water and N. With the increasing costs associated with energy consumption for production, transportation, and application of N fertilizers, the proper management of this essential element becomes more important.

The South Atlantic states had approximately 1.8 million harvested hectares of corn, with a relatively low average yield in 1980 (3160 kg/ha) due primarily to limited rainfall. In most recent years, the average yield has been 4700 to 5000 kg/ha. If one considers 165 kg/ha as the average N application rate, the total N use for the 1980 corn crop in the South Atlantic states was near 3×10^5 tonnes, an appreciable input cost relative to production, especially in 1980. However, since N is one of the most limiting factors in optimizing production, growers must continue to apply adequate amounts of N along with other recommended production practices.

The South Atlantic states have a relatively wide N recommendation range for corn production, because of the wide range in soil characteristics and a range of management methods often related to environmental factors (Table 2). All five states recommend a split application of N fertilizer on sandy soils, which have a potential for N losses through leaching. On fine-

Table 2—General N fertilizer recommendations for corn in the five South Atlantic states (dryland production).

State†	N (total rate) Recommendation	Comments
	kg/ha	
Fla.	134–202	22 to 45 kg N/ha at planting and remainder as one or two sidedress applications
Ga.	168	168 kg N/ha rate for expected yield of 6272 kg/ha; for each 3135 kg/ha increased yield increment expected, increase the N rate 56 kg/ha
N.C.	112–185	22 to 28 kg N/ha at planting and 90 to 157 kg/ha as sidedress
S.C.	123–168	34 to 45 kg N/ha soil applied and 90 to 123 kg/ha banded beside the row. General recommendation is 134 kg N/ha; reduced to 112 kg/ha if expected yield is $<$ 5000 kg/ha; increase to 168 kg/ha if expected yield is $>$ 8500 kg/ha.
Va.	84–196	Recommendation based on soil productivity groups.

† Source of information: Baird et al. (1979); Committee for Fertilizer Recommendations (1979); Donohue & Hawkins (1979); Plank (1978); and Whitley et al. (1977).

textured soils, it is usually suggested that all of the N may be applied at planting. However, in order to increase the efficiency of N recovery during the season, multiple applications are recommended. A general N recommendation is 60 to 80 kg/ha at planting and the remaining amount when the corn is approximately 45 cm tall.

Numerous N sources are used in the Atlantic Coast states for corn production and are usually effective if optimum environmental conditions exist and proper methods of application are used. The most widely used sources are ammonium nitrate (AN) and N solutions containing AN and urea. Both the NH_4^+ (at pH levels near 7.0) and urea sources of N have a potential to be lost by volatilization, especially when applied to the surface of dry soils. When the N fertilizer is in the NO_3^- form, it has the potential to be lost by leaching and denitrification if excessive water is present.

Because of the short-term droughts and the coarse-textured soils of the Coastal Plains, which may hold $<$ 5 cm of water per 30-cm soil depth, irrigation is becoming common in some areas. Within the five Atlantic states, Georgia has more irrigated cropland than any of the others. The estimate for 1980 in Georgia was 162 000 ha irrigated, or 33% of the harvested cropland.

When N is injected through an irrigation system, application of 45 to 65 kg N/ha should begin when corn plants are about 45 cm tall and continue on a weekly basis until the first tassels appear, resulting in 4 to 5 N applications during the growing season.

In the last decade, great interest has developed in producing corn under conservation tillage (no-till) systems. This practice is rapidly expanding in the South Atlantic states. Recent experiments in Kentucky (Phillips et al., 1980) have shown that when N fertilization is delayed until 30 d after planting, yields of corn under the no-tillage and conventional system are equal.

However, an additional 25 to 35 kg N/ha may be required for certain reduced tillage systems as compared with conventional tillage. Because of organic matter accumulation and possible high soil moisture levels near the soil surface, denitrification can occur, resulting in a significant loss of N. Also, leaching of mobile NO_3^- occurs more readily under reduced tillage than conventional systems, because there is less water evaporation and undisturbed channels allow more rapid and deeper penetration of the NO_3^- associated with the soil water. Additional data are needed for the South Atlantic states under various environmental conditions, soils, and cropping systems to provide sound recommendations for reduced tillage system as compared with conventional tillage.

Crop rotations may influence N rates for corn. If corn follows a legume such as soybeans [*Glycine max* (L.) Merr.], peanuts (*Arachis hypogaea* L.), and clovers (*Trifolium* spp.), the N rate may be reduced 25 to 100 kg/ha (Mitchell & Teel, 1977; Triplett et al., 1979). The rate of reduction will depend upon how the legume crop was managed, i.e., incorporated into the soil or top growth removed. Where corn is included in a double or multiple cropping system, the type and production level of the previous crop should be considered. If high N rates were used with the previous crop and maximum production was not obtained, residual N may need to be considered, especially on soils with fine-textured surface and/or subsurface horizons.

In summary, the state of the art on N use for corn in the South Atlantic states is closely related to adequate rates for maximum production. These application rates are associated with expected yield, the soil type involved, and the management practices that are followed (i.e., split application of N, whether the N is applied by ground equipment or through an irrigation system, conventional or reduced tillage, and rotations and/or multiple cropping systems).

B. Grain Sorghum [*Sorghum bicolor* (L.) Moench]

Although the South Atlantic states produced only about 1.3% of the 1980 estimated U.S. grain sorghum production, it is expected that the area seeded to this crop will increase because of the grain deficit and drought conditions at crucial grain production periods during the summer months. Also, certain cultivars fit well into no-tillage and multiple-cropping systems.

Although the N requirements for grain sorghum are less than for corn, substantial N use may be associated with sorghum, since N is probably the most limiting production factor (Touchton, 1980). Current usage of N for this crop on 72 000 ha in the South Atlantic states is estimated to be approximately 7000 t.

General N fertilizer recommendations for grain sorghum in four of the five South Atlantic states are: Georgia—90 kg/ha, but when high yields are expected, increase rate to 140 kg; South Carolina—90 to 135 kg/ha; North Carolina—56 to 135 kg/ha; and Virginia—84 to 196 kg/ha with recommended rate dependent upon soil productivity group rating. Florida's grain sorghum land area is limited; therefore, a recommendation is not included. However, the recommended rates would be expected to be similar to those of Georgia and South Carolina.

Timing of N applications can be more critical in sorghum than corn. Grain of the sorghum head develops early during the plant growth stages; therefore, adequate N must be present during this development to ensure a high yield potential. On light-textured soils with leaching potential, all states recommended that a portion (up to 50%) be applied at planting and the remainder approximately 30 d after planting. On heavy-textured soils, it would probably be safe to apply all of the N at planting.

As for corn, various N sources are used for grain sorghum, but a major consideration in selecting sources of N is method of application. Anhydrous ammonia is frequently the cheapest source of N, but must be applied with special applicators that many growers do not currently have or are not available for their use. Urea should be incorporated after application. Nitrogen solutions (30–32% N) contain approximately 50% urea N, and like solid urea, the area portion is subject to N loss through volatilization if it is not incorporated soon after application. Solid AN is probably the most frequently used N source for sidedress application for grain sorghum.

Use of winter legumes such as crimson clover (*T. incarnatum* L.) or vetch (*Vicia sativa* L.), or the inclusion of legumes in a rotation can be excellent methods for providing a relatively cheap N source and increase grain sorghum yields. The supply rate may range from 25 to 110 kg N/ha. At the highest level, the legume may supply sufficient N for maximum sorghum production (Baird et al., 1979). One of the major advantages of using winter legumes as the N source is a continuous supply of N throughout the growing season.

Presently, there is little evidence to indicate that N fertilization for grain sorghum under reduced tillage systems is different from conventional tillage. Data on residual N effects for grain sorghum are sparse, but would not be expected to differ greatly from corn.

C. Tobacco (*Nicotiana tabacum* L.)

Flue-cured tobacco in the South Atlantic states is grown on approximately 233 000 ha, resulting in it being fourth or fifth in planted area of major crops. Because of its high gross value (ca. $7 000/ha), it is first in cash value among cultivated crops in the South Atlantic states. Thus, extensive research has been conducted on the fertilization of tobacco.

Nitrogen recommendations are nearly identical among the tobacco-growing states. The suggested range is 56 to 90 kg N/ha, depending upon soil type. The lower rate is suggested for soils having subsoils within 25 to 30 cm of the surface. Recommendations increase in uniform increments with increasing subsoil depths until the highest rate is recommended at a subsoil depth of 50 to 60 cm. Lower rates are suggested when tobacco follows peanuts or soybeans, which results in higher residual levels of N, or when planted on clay soils where leaching is minimal.

The suggested method for application is one-half to two-thirds in one or two bands at transplanting and the remainder as a sidedress within 3 weeks. When a series of leaching rains occurs after the sidedress application, it is recommended that an estimate be made of the N leached and this N be replaced by an additional application.

The control of N availability is very important in tobacco culture. For maximum yields and proper maturity, it is essential that the N supply be adequate during the early stages of growth, or until shortly before flowering, and decrease rapidly during the remainder of the growth period (McCants & Woltz, 1967). Nitrogen deficiency leads to low nicotine and high sugars in cured leaves, while high N results in an undesirable decrease in the reducing sugars/nicotine ratio (Miner, 1980). Simply stated, the best quality tobacco results from a management program that supplies high levels of N during the rapid phase of growth and leads to a N deficiency during the final phase.

Tobacco fertilizers are specially formulated by using sources containing at least 50% of the N as NO_3^- and 2% Mg. These special fertilizers are used almost exclusively for tobacco. By law in some states, the maximum Cl content is 2% when the K content is $\leq$ 10%.

A routine fertilization practice in North Carolina, where approximately two-thirds of the flue-cured tobacco is grown, is the application of 40 to 45 kg N/ha as a mixed fertilizer containing P and K at transplanting, and a side-dress application of 22 to 45 kg N/ha as sodium nitrate or 15-0-14. However, many farmers overestimate the amount of leaching that occurs or simply add too much N as side-dressing in anticipation of leaching. Collins and Peedin (1981) estimated that North Carolina farmers use an average of 123 kg N/ha for tobacco. Overfertilization results in harvesting and curing problems, which are reflected in lower quality of the cured leaf.

Recent work by Williams and Miner (1982) has renewed the interest in using urea as a source of N for tobacco. They have observed that nitrification occurs much faster for urea as compared with ammonium sulfate, which was the reduced source used by McCants and associates (Skogley & McCants, 1963; McCants et al., 1959) when the NO_3^- recommendation was formulated. Also, the rate of fumigants currently used is lower than was common a few years ago, so nitrification is not inhibited as much by fumigants. Williams and Miner (1982) observed reduced tobacco yields when urea was the N source only during a very dry year when the nitrification rate was reduced. The recommendation for 50% of the N to be in the NO_3^- form has not been changed, but must be examined closely in light of continued research in this area. Adding 50 to 100% of the N as NO_3^- is considerably more expensive than if urea were used.

D. Cotton (*Gossypium hirsutum* L.)

Only three of the five South Atlantic states (Georgia, North Carolina, and South Carolina) report cotton hectarage. Production for these three states constitutes only 1.9% of the total U.S. production and 2.6% of the

Table 3—General N fertilizer recommendations for cotton in three (Georgia, N. Carolina, and S. Carolina) of the five South Atlantic states.

State†	N (total rate) Recommended	Comments
	kg/ha	
Ga.	67	0–22 kg N/ha preplant and 34 to 45 kg N/ha sidedress after squaring.
N.C.	95–118	22 kg N/ha at planting and 73 to 95 kg N/ha as sidedress.
S.C.	67–112	39 to 56 kg N/ha at planting and 28 to 34 kg/ha sidedress in the Piedmont. Sidedress with 28 to 56 kg N/ha in the Coastal Plains area. General recommendation is 78 kg/ha and decreased by 20 kg/ha if growth is excessive.

† Source of information: Baird et al. (1979); Committee for Fertilizer Recommendations (1979); Donohue & Hawkins (1979); Plank (1978); and Whitley et al. (1977).

area harvested in 1980. Georgia had the largest hectarage (62 728 ha), while North Carolina had the highest average yield (470 kg/ha).

The general N fertilizer recommendations for Georgia, North Carolina, and South Carolina are shown in Table 3, ranging from 67 to 118 kg N/ha of total N. On sandy soils, it is suggested that 50% of the sidedress N be applied at early square stage and the remainder 3 to 4 weeks later; on fine-textured soils, apply single sidedress N at the early square stage. On fields that have a history of producing rank stalks, specialists in Georgia suggest reducing the total N rate to 45 to 56 kg/ha (Fulford et al., 1980). In addition, they recommend monitoring NO_3^- with leaf petiole test beginning 1 week prior to bloom and adjusting the N supply to the cotton production potential of the particular season by avoiding excess N in early and late season and ensuring adequate N during crucial fruiting periods. Irrigated cotton probably lends itself to this method better than dryland cotton, since N is often applied as a foliar urea application in two or three sprays.

When cotton follows soybeans, peanuts, other legumes, or heavily fertilized corn, total N may be reduced by 22 to 34 kg/ha. This reduction is often accomplished by not applying N at planting.

E. Forages and Pasture

The range for annual N application rates recommended in each state for various forages is wide. Typical ranges are: 'Coastal' bermudagrass [*Cynodon dactylon* (L.) Pers.]—168 to 336 kg/ha; fescue (*Festuca arundinacea* Screb.)—67 to 224 kg/ha; sorghum-sudan [*Sorghum bicolor* (L.) Moench.]—45 to 224 kg/ha; common bermudagrass [*Cynodon dactylon* (L.) Pers.]—84 to 196 kg/ha; and native pasture—0 to 70 kg/ha. The large variance in recommended rates allows for management decisions in grazing intensity, changes in economic situations, and amount of hay needed. The fertilizer practices followed by the farmer vary as much as the recommendations. Forage crop extension specialists agree that forage crops are grossly underfertilized with N, as well as the other major nutrients, with regard to profitable yields in a well-managed forage production system. Coastal bermudagrass, which generally does not do well unless N is applied regularly, is probably the only forage that receives rates near the recommended range.

Much of the fertilizer N applied to forage throughout the region, particularly grazed forage, is applied as a complete mixed fertilizer. An average application would be 30 to 35 kg N/ha as a complete mixed fertilizer. This application is usually made in February–March and/or August–September for cool-season grasses and in April–May for warm-season grasses. In addition, cool-season grasses that are fertilized near the maximum recommended rate may receive additional fertilizer in the early spring and fall with much of the N from solutions. Warm-season grasses that receive maximum N rates are fertilized two to four times per year, usually with AN or N solutions.

One factor that is sometimes overlooked in computing the N applied to forages is the amount of N that is applied as waste from confined poultry and swine. Jim Barker (unpublished data) of North Carolina State University has estimated that in the South Atlantic states, 258 000 t of N is applied annually, mostly to forages, for an average supplement of 50 to 60 kg N/ha per year. Unfortunately, much of the waste is applied at excessive rates and often during the season when the grasses are not actually growing.

Forages are very efficient in the utilization of N fertilizer, and frequently 90% or more of the N is harvested with the crop when it is grazed or removed for hay (Thomas & Gilliam, 1977). However, when grazed by cattle, at least 75% of the N passes through the animal and is excreted back to the land (Azevedo & Stout, 1974), resulting in only a small percentage of N fertilizer removed in the finished product (meat or milk). It has been estimated that 50% of the excreted N is lost through NH_3 volatilization (Stewart et al., 1975). Furthermore, animals are inefficient distributors of the nutrients consumed and excreted (Peterson et al., 1956). A considerable amount of the N may be incorporated into the soil organic matter. However, after a period of time, the soil will tend to establish a new equilibrium level of organic N.

The projected costs of N fertilizer and prices of animals and animal products have convinced many that N fertilization of forages, particularly grazed forages, will decrease in the future. Much of the improved pasture in the region was planted as a clover–grass mixture, but the percentage of clover in older pastures is usually small. Intensive efforts are underway to increase the clover in pastures by sod seeding practices, as well as complete renovation, so that sufficient forage can be obtained with minimum use of N fertilizer.

F. Small Grains

Small grains (for grain) cropland in the South Atlantic states has increased dramatically over the past few years and this increase is projected to continue. The increase is a result of increased wheat (*Triticum aestivum* L.)

prices and the growing trend of planting soybeans or grain sorghum immediately following the harvest of the small grain in a multiple-cropping and/or no-tillage system.

Nitrogen recommendations from each state are similar. All consider 22 to 33 kg N/ha to be the proper amount of fertilizer N to apply at seeding. States that have significant plantings in Piedmont soils indicate that all of the N can be applied at planting in these locations. Most growers apply the N at planting (Sept.–Dec.) as part of a complete mixed fertilizer that is usually bulk spread and disked into the soil. The N applied in later winter or early spring as a topdressing ranges from 33 to 90 kg/ha, depending upon soil type and managerial practices. When the small grain is used only for grazing, the total topdress N rate may be 90 to 112 kg/ha with split applications in January–February and March–April. The most popular sidedress N source is N solution with some AN and urea used. Some of the N added in solutions as urea is lost by volatilization when the urea is hydrolyzed on or very near to the soil surface. However, because the farmer has the equipment to handle solutions and because this form of N may be the least expensive, solution N may be considered economical at the present time, even if a significant amount is lost by volatilization.

G. Soybeans [*Glycine max* (L.) Merr.] and Peanuts (*Arachis hypogaea* L.)

No state agency in the region recommends the use of fertilizer N on these crops, although 15 to 20 kg N/ha is often applied to soybeans at seeding (Thomas, 1976). The reasons for this application are varied but primarily are a result of using the same mixed fertilizer on soybeans as for other crops. Also, a total plant growth response is frequently noted, particularly during early growth, although it commonly does not mean an increased yield. Some farmers feel that this early growth results in the pod set being higher from the soil surface and thus easier to harvest. Also, because of the problem with soil acidity throughout the region, some legumes are grown on soils sufficiently acidic to inhibit N fixation, and a fertilizer N response is sometimes obtained. Because of a difference in fertilization practices, very little N is used on peanuts.

Current research regarding N for legumes is focused toward increasing the efficiency of N fixation and not toward utilization of applied N.

V. PROBLEMS AND OPPORTUNITIES IN N FERTILIZATION

A. Fertilizer N Efficiency

The efficiency of N fertilizer utilization in harvested crops is a problem in the humid South, where only a small fraction of inorganic N remaining in the soil profile is utilized by the succeeding crop. Also, residual N does not become readily associated in the soil organic fraction, as organic matter contents are generally low under all cultivated systems, regardless of management.

An example of the above points can be made with corn, the crop to which approximately 35% of the fertilizer N is applied in the South Atlantic states. The average annual N application rate to nonirrigated corn is, by a conservative estimate, 165 kg/ha. The average grain harvested is approximately 4800 kg/ha, which contains about 77 kg N. Thus, an average of only 47% of the applied N is harvested in the grain. Although 45 to 55 kg N is returned to the soil in the stover, it has little residual fertilizer value and N content of the soil does not increase over time. Gambrell et al. (1974) found on a sandy Coastal Plains soil in North Carolina that only 18% of the applied N not recovered in the corn crop was recovered in succeeding crops. No apparent residual value was observed after 1 yr. It is fortunate that most crops grown in the South Atlantic states are more efficient than corn in their utilization of applied N (e.g., tobacco, small grains, forages).

The problem of efficiency of N utilization is a long recognized problem and many techniques have been employed to improve efficiency. Notable among these are timing of application, slow-release fertilizer, and nitrification inhibitors. In our region, only the proper timing of application is extensively used.

B. Inhibitors and Slow-Release N Materials

Certain compounds are selectively active against bacteria in the soil, which delays the conversion of NH_4^+ to NO_3^-, allowing the NH_4^+ to remain in the soil for a longer period. This can possibly result in decreased losses of NO_3^--N from leaching and/or denitrification, allowing for more efficient plant use of applied N. Some of these compounds are chloropheny isothiocyanates, potassium azide, dinitrophenols, and nitrapyrin [2-chloro-6-(trichloromethyl)pyridine]. The most widely used of these compounds is nitrapyrin. The chemical will retard nitrification and thus conserve soil N by keeping it in the reduced form. At present, none of these states are recommending the use of inhibitors, but neither are they taking a firm negative stand. All agree that relatively warm winter temperatures severely limit their utility for fall and winter applications for spring-planted crops. It is generally surmised that the use of inhibitors may be economically beneficial when high rainfall occurs within 3 to 4 weeks after application. However, available research data to substantiate these points are limited. Regardless of how long inhibitors will prevent NH_4^+-N from nitrifying, crop yield response to inhibitors will not be realized if a combination of soil and climatic conditions do not favor N losses through leaching and/or denitrification.

Much of the slow-release or controlled-release N fertilizer evaluations have been with wax or sulfur-coated urea. The purpose of controlled release N is to increase use efficiency by reducing losses by leaching and/or denitrification, decreasing toxicity, and decreasing frequency of application. With forage crops, early clippings often contain more N than is actually required for optimum crop growth. With normal clipping procedures, this luxury consumption of N results in less available N for crop growth later in the season. Toxicity of soluble N fertilizers to crops, particularly seedlings, may be produced from evolution of NH_3 by hydrolysis of certain salts, particularly urea. When these factors are controlled by a slow-release mechanism, reduction in toxicity may be expected. Split application of soluble N sources often give increased crop yield as well as improvement in the season distribution, especially as related to forage crops. Similar benefits may be expected from controlled-release sources, provided they release plant nutrients in a pattern similar to split applications. At the same time, cost of the application may be reduced by a single application of a controlled-release N source, as compared with split applications of a soluble source. In summary, controlled-release N fertilizers may be beneficial for special situations such as forages, turf, citrus, off-season application, and in areas where leaching during the growing season may be a problem.

C. Soil Acidity

Acid soils are common to essentially all agricultural areas of the South Atlantic states. The high rainfall, warm climate, and generally low effective cation exchange capacity (CEC) all contribute to the widespread problem. It is a well-recognized problem and programs are periodically conducted in each state to make the farmer more aware of the general need for lime. However, the problem is frequently not corrected. More than 80% of the soil samples taken from planned soybean fields and tested by the North Carolina Department of Agriculture in 1978 and 1979 showed a need for lime. Approximately 70% of the samples tested in North Carolina for all crops needed lime applications.

The contribution of N fertilizer to the need for lime is well documented and appears in most undergraduate soils textbooks. Lutz et al. (1977) recently observed that applications of acid-forming N fertilizer may affect the plant root environment markedly in some soils and result in increased Mn and Al concentration and toxicities.

The increased use of high rates of N on irrigated soils will accentuate the problem. It is a problem that agronomists generally know how to solve, but it has been estimated to cost North Carolina farmers alone $50 million/yr in reduced soybean profits.

D. Environmental Concerns of N Fertilization

Because of the high quantities of drainage water from the area and the relatively small percentage of the land mass to which fertilizer N is applied, generally high concentrations of NO_3^- do not occur. However, localized problem areas do exist and fertilizer N contributes to the problem. It is generally accepted that N is the limiting factor in algae growth in estuarine waters of the South Atlantic states. Some economically significant eutrophication problems are occurring in waters in some areas (e.g., Chowan River in North Carolina). There is also concern that high rates of N being applied to irrigated corn on sandy soils in Georgia may lead to problems in some areas. However, because most fertilizer N is used in Coastal Plains soils, which are generally drained by slow-moving swampy streams, much of the N leaving agricultural areas in drainage water may be lost via denitrification before reaching a major stream or estuary.

Environmental problems in the region attributed to excessive nutrients are not large at present or have not been adequately identified, but agronomists must continue to work with environmental agencies and groups for the protection of both agriculture and the environment.

REFERENCES

Azevedo, J., and P. R. Stout. 1974. Farm animal manures: an overview of their role in the agricultural environment. California Agric. Exp. Stn. Ext. Manual 44.

Baird, J. V., F. R. Cox, E. J. Kamprath, S. N. Hawks, E. L. Kimbrough, and G. S. Miner. 1979. Fertilizer suggestions for field pasture and hay crops. p. 353–355. *In* North Carolina Agric. Chem. Manual. North Carolina State Univ., Raleigh, N.C.

Collins, W. K., and G. F. Peedin. 1981. Agronomic production practices. p. 6–32. *In* 1981 Tobacco information. North Carolina Agric. Ext. Circular AG-187.

Committee for Fertilizer Recommendations. 1979. Fertilizer recommendations for South Carolina. Agron. and Hort. Dep. Circular 476. Clemson Univ., Clemson, S.C.

Daniels, R. B., B. L. Allen, H. H. Bailey, and F. H. Beinroth. 1973. Physiography. p. 3–17. *In* S. W. Buol (ed.) Soils of the southern states and Puerto Rico. Southern Coop. Service Bull. no. 174.

Donohue, S. J., and G. W. Hawkins. 1979. Guide to computer programmed soil test recommendations in Virginia. Ext. Div. Pub. 834. Virginia Polytechnic Inst. and State Univ., Blacksburg, Va.

Fulford, A., J. Crawford, M. French, B. Lambert, and G. Sigler. 1980. Cotton production package. Fact Sheet, Coop. Ext. Service. Univ. of Georgia College of Agric., Athens, Ga.

Gambrell, R. P., J. W. Gilliam, and S. B. Weed. 1974. The fate of fertilizer nutrients as related to water quality in the North Carolina Coastal Plain. Water Resour. Res. Inst. of the Univ. of North Carolina Rep. no. 93.

Kamprath, E. J., S. W. Broome, M. E. Raja, S. Tonapa, J. V. Baird, and J. C. Rice. 1973. Nitrogen management, plant population and row width studies with corn. North Carolina Agric. Exp. Stn. Tech. Bull. no. 217.

Lutz, J. A, Jr., W. Kroontje, and H. C. H. Hahne. 1977. Nitrogen fertilization: II. Effect on the soil solution composition, acidity, and nitrate adsorption. Soil Sci. Soc. Am. J. 41: 568–572.

McCants, C. B., E. O. Skogley, and W. G. Woltz. 1959. Influence of certain soil fumigation treatments on the response of tobacco to ammonium and nitrate forms of nitrogen. Soil Sci. Soc. Am. Proc. 23:466–469.

McCants, C. B., and W. G. Woltz. 1967. Growth and mineral nutrition of tobacco. Adv. Agron. 19:211–265.

Miner, G. S. 1980. Effect of harvest method and related management practices on flue-cured tobacco. II. Total N, total alkaloids, reducing sugars, and particulate matter index. Tob. Sci. 182:24–27.

Mitchell, W. H., and M. R. Teel. 1977. Winter-annual cover crops for no-tillage corn production. Agron. J. 69:569–573.

Peterson, R. G., H. L. Lucas, and W. W. Woodhouse, Jr. 1956. The distribution of excreta by freely grazing cattle and its effect on pasture fertility: I. Excreta distribution. Agron. J. 48:440–444.

Phillips, R. E., R. L. Blevens, G. W. Thomas, W. W. Frye, and S. H. Phillips. 1980. No-tillage agriculture. Science 208:1108–1113.

Plank, C. O. 1978. Lime and fertilizer recommendations based on soil tests for all crops grown in Georgia (Code Book). Georgia Coop. Ext. Service Univ. of Georgia College of Agric., Athens, Ga.

Skogley, E. O., and C. B. McCants. 1963. Ammonium and chloride influences on growth characteristics of flue-cured tobacco. Soil Sci. Soc. Am. Proc. 27:391–394.

Stewart, B. A., D. A. Woolhiser, W. H. Wishmeier, J. H. Cart, and M. H. Frere. 1975. Control of water pollution from cropland. Vol. I. A manual for guideline development. USDA-ARS publication. U.S. Government Printing Office, Washington, D.C.

Thomas, G. W. 1976. Development of environmental guidelines for fertilizer use. Southern Coop. Res. Bull. 211.

Thomas, G. W., and J. W. Gilliam. 1977. Agro-ecosystems in the U.S.A. Agro-Ecosystems 4: 182–243.

Touchton, J. T. 1980. Soil fertility management for grain sorghum production. p. 4–12. *In* R. R. Duncan (ed.) Proc. of the Sorghum Shortcourse. Univ. of Georgia College of Agric. Exp. Stn. Spec. Pub. no. 6. Univ. of Georgia, College of Agric., Athens, Ga.

Triplett, G. G., Jr., F. Haghiri, and D. M. VanDoren, Jr. 1979. Plowing effect on corn yield response to N following alfalfa. Agron. J. 71:801–803.

U.S. Department of Agriculture. 1980. Commercial fertilizers: consumption by class for year ended June 30, 1980. Crop Reporting Board, Economics and Statistics Service, Sp. Cr. 7 (11-80). USDA, Washington, D.C.

Whitley, E. B., D. W. Jones, G. Kidder, C. G. Chambliss, D. L. Wright, and J. J. Street. 1977. Fertilization of field and forage crops. Agron. Facts no. 70. Coop. Ext. Service, Univ. of Forida, Gainesivlle.

Williams, L. M., and G. S. Miner. 1982. Effect of urea on yield and quality of flue-cured tobacco. Agron. J. 74:457–462.

WISCONSIN
EAST
NORTH
CENTRAL
MICHIGAN
ILLINOIS
INDIANA
OHIO

48

L. F. Welch
University of Illinois
Urbana, Illinois

Nitrogen Management for the East North Central States

The East North Central states are comprised of Illinois, Indiana, Michigan, Ohio, and Wisconsin. The region extends from 36°58′N Lat in southern Illinois to 48°18′N Lat in northern Michigan and from 84°49′W Long in eastern Ohio to 92°54′W Long in western Wisconsin.

Like many other areas of the world, without proper management N could be the most limiting factor in the growth of nonlegumes in the East North Central states. Intermittent drought during the growing season also looms large as a limiting yield factor for summer annuals. It is considerably more easy for most growers to remove N as a limiting growth factor than to remove shortages or excesses of water. Most growers have a good understanding of the importance of N in plant growth and apply N fertilizers to their nonlegumes.

Nitrogen management is considerably more complex than P and K. This is because of the various transformations that N may undergo once applied to soils. The final state of inorganic N in soils is NO_3^-. This negatively charged, oxidized state is readily mobile in soil water, and is susceptible to reduction to gaseous forms of N. Therefore, N management should include an awareness of the susceptibility of N to loss by leaching and denitrification. The possibility of volatilization of surface-applied N without soil incorporation should also be considered.

Management strategies available to growers include chemical form of N fertilizer, time of application, method of application, the use of chemicals to slow biological activity, and matching rate of N applied to the needs of the crop and the ability of the soil to supply N. The cost of N fertilizer is a large portion of the variable cost for a crop with a high demand for N, like corn (*Zea mays* L.). The high cost of N provides a strong motivational factor for profit-oriented growers to continue to strive for greater N efficiency through management. Furthermore, researchers are obliged to col-

Nitrogen in Crop Production.

Table 1—Precipitation and freeze-free days for the East North Central states (Water Inf. Center, Inc., 1974).

State	Precipitation, cm	Freeze-free days
Illinois	81-122	152-230
Indiana	91-112	147-216
Michigan	71-91	59-190
Ohio	81-112	135-205
Wisconsin	71-91	73-188

lect data that will provide the basis for an even wider array of strategies for grower management. Surely, many of the new concepts and principles to be developed through research will not be limited to only application in the East North Central states, but will find usefulness worldwide.

I. CLIMATE

There is considerable variation in climate within the region and within each of the states, though the amount of precipitation and freeze-free days increase as one moves from north to south. Precipitation in the region varies from 71 to 122 cm (Table 1). Freeze-free days vary from 59 to 230.

Climate is important from the standpoint of cropping practices and management; precipitation is important from the standpoint of N leaching and denitrification; temperature is important as it affects the rate of N transformations such as nitrification, denitrification, volatilization, mineralization, and immobilization.

II. CROPS AND LAND USE

There is intensive agriculture in at least certain areas of each of the East North Central states. Areas in principal crops range from almost 10 million ha in Illinois to about 3 million ha in Michigan (Table 2). (Principal crops include 21 different crops.) About 67% of the total land area in Illinois is planted to principal crops. The similar value for Michigan is 20%. The 26.3 million ha of principal crops in the East North Central states in 1981 accounted for 18.24% of the area devoted to these crops in the United States.

Table 2—Hectares of principal crops harvested and as a percent of total land area, by states (USDA, 1982).

State	Land area, ha × 10^3	Principal crops harvested, ha × 10^3	Cropped as percent of total land
Illinois	14 496	9 761	67.3
Indiana	9 381	5 298	56.5
Michigan	14 775	2 983	20.2
Ohio	10 655	4 412	41.4
Wisconsin	14 177	3 856	27.2

Table 3—Areas by states planted to corn, wheat, and oats and as a percent of area of principal crops, 1981 (USDA, 1982).

State	Corn, wheat, oats ha × 10^3	Area in corn, wheat, and oats as percent of principal crops
Illinois	5464	56.0
Indiana	3076	58.1
Michigan	1757	58.9
Ohio	2411	54.6
Wisconsin	2187	56.7

Table 4—Hectares irrigated in each of the East North Central states, 1981 (Brentwood Publ., Inc., 1981).

State	Irrigated hectares
Illinois	61 100
Indiana	44 500
Michigan	130 800
Ohio	19 800
Wisconsin	102 400

The areas planted to nonleguminous grain crops like corn (*Zea mays* L.), wheat (*Triticum aestivum* L.), and oats (*Avena sativa* L.) influence the amount of N fertilizer used by the states. The proportion of principal crops area devoted to the three nonleguminous grain crops is very similar in the five states (Table 3). The East North Central states had about 37% of the U.S. area devoted to corn grain and about 22% of the U.S. area devoted to corn silage. Comparable figures are 19% for oats and 7% for wheat. These states plant about 28% of the U.S. soybean [*Glycine max* (L.) Merr.] cropland, but this crop receives little or no N fertilizer.

Irrigation ranges from about 20 000 ha in Ohio to about 131 000 ha in Michigan (Table 4). In some areas, irrigation would be largely for vegetable crops rather than agronomic crops. These irrigated areas are indicated because they represent areas where more intense agriculture might be practiced. This suggests that the rate of N fertilizer might be higher, since water stress may potentially be removed as a barrier to high crop yields. Irrigation increased corn yield an average of 65% during 1977–1981 in south central Illinois (M. D. Thorne, Univ. of Illinois, personal communication). Also, these irrigated areas have the option of conveniently applying N fertilizer at the time of water application.

III. FERTILIZER N USE DATA

The East North Central states used about 20 000 t of N in 1940. Usage of N had increased more than 100 times this volume by 1981. The five states have used about 2 million tonnes of N annually since 1978 (Table 5). From 1940 to 1960, N consumption in the East North Central states increased relatively faster than for the United States. The region used about 6 and

Table 5—Nitrogen use in the East North Central states and as a percent of use in the United States for selected years.

Year	N use in East North Central states	
		U.S. usage
	t	%
1940	19 808	5.8
1945	41 511	7.7
1950	85 739	9.9
1955	242 057	13.6
1960	383 883	15.4
1965	863 131	20.5
1970	1 287 339	19.0
1975	1 538 586	19.7
1978	1 881 732	20.8
1979	2 038 772	21.0
1980	2 211 426	21.4
1981	2 169 450	20.3

15% of the total N used in the United States during 1940 and 1960, respectively. Since 1965, the region has consumed about 20% of the N used in the United States, and this value has been fairly constant through 1981.

As in other regions of the United States, there has been a significant increase in the rates of N applied to corn and wheat in the East North Central states (Table 6). Nitrogen use on corn more than doubled in all states from 1964 to 1981. There was more than a fourfold increase in the rate of N applied to corn grown in Wisconsin during the 17-yr period. Increases in the rate of N fertilizer applied for wheat are also striking. If crop yields per hectare continue to increase, future rates of N will likely make the 1981 rates appear low.

Much of the N used in the region is applied to corn (Table 7). The percent of the total N used in each state that is applied to corn was calculated as ranging from 62% for Michigan to 89% for Wisconsin. Similar values for wheat ranged from 6% for Illinois to 13% for Ohio. Corn plus wheat accounted for 72 to 99% of the total N used in each of the East North Central states. Any fluctuation in the land area planted to corn and wheat would be

Table 6—Fertilizer N applied in 1964 and 1981 for corn and wheat, by states (USDA, 1971, 1981).

State	Applied N			
	Corn		Wheat	
	1964	1981	1964	1981
	kg/ha			
Illinois	72	167	32	69
Indiana	76	162	42	94
Michigan	47	122	37	69
Ohio	61	177	28	83
Wisconsin	29	122	--	--

Table 7—Nitrogen applied to corn and wheat as a percent of the total used in each state, 1981.

State	Percent of total N used in state		
	Corn	Wheat	Corn and wheat
Illinois	85	6	91
Indiana	88	11	99
Michigan	62	10	72
Ohio	66	13	79
Wisconsin	89	--	89

expected to have considerable influence on N consumption. This would be especially marked if corn hectarage fluctuated widely. An example of the preceding occurred in 1983 due to the federal government's payment-in-kind program.

IV. SOURCES OF N

There has been a great change in N materials used in the region in the past 40 yr (Table 8). In 1940, ammonium sulfate (AS) and sodium nitrate were the two materials used for direct application of N. During the following 15 yr, other N materials assumed important roles. Ammonium nitrate (AN) accounted for about 46% and anhydrous ammonia (AA) about 30% of the N used for direct application in 1955. The most used source of N for 1955 is in fourth place in 1981. Ammonium nitrate accounted for $< 3\%$ of the N used for direct application in the region in 1981. Anhydrous ammonia accounted for more than twice as much N than the second place material in 1981. Anhydrous ammonia, N solutions, and urea accounted for about 57, 28, and 12% of the total N used for direct application in 1981.

The percent of total N added as mixtures in the East North Central states ranged from about 17 to 28% in 1981 (Table 9). The weighted average of the percent of the total N used as mixtures was similar to the U.S. average, about 21%. The remainder of the N consumed by the states, other than mixtures, would be for direct application. Therefore, direct application in the region accounts for about 80% of the total N used.

Table 8—Selected N materials used in the East North Central states expressed as a percent of the total N used for direct application.

Nitrogen material	Year			
	1940	1955	1970	1981
	percent of total N			
Anhydrous ammonia	--	29.9	62.6	57.0
Aqua ammonia	--	0.1	3.2	0.2
Ammonium nitrate	--	46.4	8.2	2.7
Ammonium sulfate	93.8	13.3	1.0	0.7
Nitrogen solutions	--	5.6	21.2	27.7
Sodium nitrate	6.2	0.3	0.4	--
Urea	--	4.4	3.5	11.7

Table 9—Percent of total fertilizer N added as mixtures, by states, 1981.

State	Percent of total N added as mixtures
Illinois	16.9
Indiana	21.2
Michigan	28.3
Ohio	24.0
Wisconsin	20.0
Weighted average	20.6
United States	21.7

All of the commonly used N materials are readily soluble in water. There is no concern about any of the sources of N being precipitated by other elements in the soil solution to form rather insoluble compounds, such as may occur when P reacts with Fe, Al, and Ca. Some NH_4^+ may be fixed in the interlayers of clay minerals and some N from AA may be fixed by organic matter, but neither of these are important enough to influence the choice of N material. Plants may absorb either NH_4^+ or NO_3^- forms of N. Agronomists and soil scientists indicate that the different sources of N should perform equally well if properly applied to minimize potential losses. Discussion follows on some factors that growers may consider when making a choice of N material.

A. Price of N

Price per kilogram of N applied to the soil is a strong influence when considering a source of N. This consideration is the prime reason that AA is the leading N material. Making AA is the first step in the manufacture of synthetic N fertilizer. Any other N fertilizer requires additional manufacturing, which adds to the cost of the N.

B. Equipment Required for Application

The different N fertilizer materials may be either gas, liquid, or solid at the time of application or shortly thereafter. Since some N material may be broadcasted on the soil surface while others must be incorporated to prevent N losses during application, equipment and power required may differ for application of the various sources of N.

Some growers own the equipment for applying their N fertilizers, while others depend on the fertilizer dealer for custom application. Another alternative is that the fertilizer dealer may supply the applicator equipment, but the grower provides the tractor and driver for operating the applicator.

C. Crop to be Fertilized

There are few limitations in the choice of N for a row crop like corn, provided N is applied when the crop is small enough that it will not be appreciably damaged by certain application equipment. However, postseeding of N applications, for close-seeded crops such as small grains and pastures, are usually limited to those materials that do not require soil incorporation. If a preplant N application is made to these crops, there are few limitations in the choice of N materials.

D. Season of Application

The probability of N loss increases with increased time between N application and its uptake by plants. This means that when there is much time between application and uptake, such as when N is applied in the fall for corn to be seeded the following spring, then a form of N less susceptible to leaching and denitrification may be favored by growers.

Volatilization of N as NH_3 may occur after surface-applied urea has hydrolyzed to ammonium carbonate. The rate of hydrolysis increases with an increase in temperature. The longer the time between urea application and hydrolysis, the greater the likelihood that rain will move the urea into the soil and thus reduce loss from volatilization. Therefore, volatilization losses from urea may be greater for surface-applied urea in July than if top-dressed for wheat in February. This means a grower may choose urea for February application, but not during warm months if there is considerable plant material and no incorporation.

E. Soil Characteristics

Volatilization of surface-applied urea increases with higher pH. Growers with calcareous soils may be sufficiently concerned with potential loss that they may choose some source of N other than urea, or they may incorporate urea.

There are sometimes problems with the knife slit sealing properly on some fine-textured soils. This may result in some choice of N other than AA.

F. Combining Operations

Some growers add N along with another farming operation. This practice eliminates a trip over the field and usually results in improved economy. Irrigators may add N at the time water is being applied. It is a relatively simple matter to inject N into the irrigation system. This procedure should result in improved efficiency of N uptake, since it is added only a short time prior to plant absorption. Certain sources of N are not suited for some irrigation systems. For example, considerable losses of N may occur if AA is injected into a sprinkler system.

Some growers mix herbicides with nonpressure N solutions to apply both materials simultaneously. The ease of doing this is one reason growers may favor N solutions.

The increased cost of fuel and recently depressed grain prices have provided a special incentive for growers to evaluate all cost inputs. Consequently, N application may be combined with a tillage operation. For example, N may be incorporated while using the chisel plow, field cultivator, disc, or other tillage implements. Anhydrous ammonia and N solutions are popular materials for applying into the soil by use of hoses attached to and behind the tillage tool. Low-pressure AA is sometimes used to help prevent potential gaseous losses when the tillage tool may result in shallow placement of the NH_3.

G. Safety

Users of AA should respect it as a hazardous material. Ammonia has a strong affinity for water, and when it removes water from flesh the injury is often called "ammonia burn" because of the desiccation. The eyes are especially susceptible to serious damage from NH_3. People using NH_3 should receive special training in safety precautions. A supply of water, goggles, and gloves should be standard practice for anyone using NH_3. Gas masks would be added to the list by some. The dangerous nature of NH_3 plus the high pressure it produces is sufficient reason for some growers to substitute safety for economy by selecting another source of N.

H. The Fertilizer Dealer

Many dealers sell various sources of N. They will sell whatever material their customers prefer. Nevertheless, dealers may have a preferred source because of profit margin or because of differences required in storage, handling, transport, and application of the various materials.

Some dealers may limit their stock of sources of N materials. Personality of the dealer, dependability, promptness of fulfilling orders, services provided, and credit terms may attract customers to a dealer and thus to the particular sources of N stocked.

V. N MANAGEMENT

The sources, rates, time, and methods of N application used in the region have evolved through the years with consideration for increased N efficiency, environmental aspects, and profit for the users. The management practices adopted are based on principles and concepts developed from theoretical considerations and research results for various soils, crops, and N materials. Some of the characteristics of N fertilizer that have influenced management practices are: (i) the numerous transformations that N may undergo once applied to soils; (ii) the relative immobility of NH_4^+ in most soils and the high mobility of NO_3^-; (iii) the difficulty of "storing" N in the soil for long periods without it being lost, especially in regions of high precipitation; (iv) the volatility of certain NH_4^+ sources of N when surface broadcasted without incorporation, especially on soils with high pH; and (v) economic considerations that include the cost per kilogram of N applied to the soil.

None of the East North Central states measure inorganic N as a basis for making N fertilizer suggestions. However, certain laboratory measurements such as organic matter content and cation exchange capacity of soils are used by some states to adjust the rate and other N management practices. Results of field research with various crops are relied on heavily by all states in the region for making N fertilizer suggestions. The following sections include discussions of factors that are given consideration when making N suggestions in the various states.

A. Crop to be Grown

Nitrogen is generally suggested for nonlegumes, but not for legumes (Welch, 1979). In some cases, starter fertilizer for legumes may include N. Nitrogen is recommended for soybeans in Michigan when grown north of Lansing, but not when they are grown south of Lansing. A higher rate of N is suggested for corn than for wheat or oats. Also, more N is suggested for a stiff-strawed variety of wheat than for one more susceptible to lodging (Polizotto & Mengel, 1981; Vitosh & Warncke, 1977; Illinois Coop. Ext. Service, 1980; Ohio Coop. Ext. Service, 1980).

As indicated earlier, most of the N used in the region is applied for corn. This is because of both the large land area and the high rate per hectare. The rates of N suggested by the five states range from 190 to 224 kg/ha for a corn grain yield goal of 10 t/ha when corn was also the previous crop (Table 10). This rather narrow range is equivalent to 1 kg of N fertilizer for 44.6 to 52.6 kg of corn grain.

B. Yield Goal

Numerous studies have shown that there is a strong positive relation between corn yield level and optimum fertilizer N rate. High yield levels may be the result of such factors as early planted corn, high-yielding hybrids, and better water relations because of irrigation, more favorable natural precipitation, or improved drainage. For a mobile nutrient like N (when present as NO_3^-), it is not surprising that optimum N rate increases as the yield level increases.

Table 10—Fertilizer N suggested for corn by the East North Central states when the grain yield goal is 10 t/ha and the organic matter content of soil is 2 to 4%.

Previous crop or treatment	N suggested				
	Ill.	Ind.	Ohio	Mich.	Wis.
	kg/ha				
Corn	215	224	202	224	190
Soybeans	170	213	179	190	146
Forage legume†	103	134	101	112	78
Manure‡	159	179	146	185	146

† Assumes a good stand. ‡ Cattle manure applied at 22.4 t/ha.

C. Soil Properties

Although none of the East North Central states have laboratory soil N tests, consideration is given to the general ability of the soil to supply available N. Soil texture and organic matter are often considered when suggesting a rate of N fertilizer. Rate of N is decreased as the organic matter content of soil increases. It is suggested that N rates be reduced 45 kg/ha when wheat is grown on high organic matter soils in Illinois, as compared with low organic matter soils.

In coarse-textured soils, leaching is likely to be greater than with fine-textured soils. Conversely, because of poor drainage and anaerobic conditions, denitrification is likely to be more of a problem on fine-textured soils. Soil texture and organic matter content also influence the water-holding capacity of the soils, and hence, the yield potential. Soil texture can be a big factor in N management, especially with respect to time of N fertilizer application.

D. Previous Crop

The previous crop influences the amount of available N that may be supplied by the soil system (Kelling et al., 1981; Spies & Mengel, 1981; Vitosh et al., 1979). The highest rate of N fertilizer is suggested when a nonlegume follows a nonlegume. The N rate is reduced to account for N that may have been fixed by previous legume crops (Table 10). The N rate is lower when corn follows a good stand of forage legume than when it follows a mediocre stand of forage legumes. Credit is also given when corn follows soybeans as contrasted to when corn follows corn.

E. Time of N Application

Personnel in the five states recognize that the potential for loss of N fertilizer is greater when there is considerable time between N application and crop uptake (Welch et al., 1971). Sidedress application of N for corn often results in the highest percent of the applied N being absorbed by the crop. Spring preplant applications are the next most efficient, and fall application the least efficient from the standpoint of N uptake. In certain years, all three times of application may perform equally well. This would be when the potential for N loss is low during the fall, winter, and early spring months. When climatic conditions are favorable for high loss of N, there may be considerable difference due to time of application.

Even though spring preplant and fall application may be less efficient than sidedress when evaluated as individual operations, the earlier applications may prove to be best for some growers when viewed with respect to their total farm operation. Fall application of N may result in less efficient use of N than spring preplant, in that a lower percent of the fall-applied N is absorbed by the corn plants. However, if spring preplant application presents problems, the grower may opt for fall application. Planting corn early is an important factor in high yields. If corn planting is delayed because of spring preplant application of N, this is sufficient justification for considering another time of N application. Similarly, sidedress N may not be chosen because of damage to growing corn during application, or because of the risk of undue delay in N application because of wet soils.

F. Nitrification Inhibitors

The commercial use of nitrification inhibitors in recent years has added new flexibility to the time of N application (Johnson, 1981; Walsh, 1977). Theoretically, maintaining the N in the NH_4^+ form until shortly before N is absorbed by plants should result in increased efficiency. All of the states in the East North Central region suggest the use of nitrification inhibitors under certain conditions. Ohio suggests the use of a nitrification inhibitor in all cases when N is fall-applied for corn. Ohio also suggests the use of nitrification inhibitors when spring-applied N is added to wet soils where the potential for denitrification is high. The other four states suggest that nitrification inhibitors are more likely to be a good management practice on soils where the potential for N loss is high. Research results in Indiana have often shown that use of inhibitors significantly increased crop yields (Huber et al., 1977). A more detailed discussion of the roll of nitrification inhibitors and N management is presented elsewhere in this book.

G. Use of Manure

The suggested rate of N fertilizer is adjusted downward to compensate for N added in manure. Wisconsin traditionally has applied larger amounts of manure than have some of the other states. This may partially account for the lower rate of N added for corn by Wisconsin growers (Table 6). The N content of manure may vary widely, depending on the animals involved, amount of bedding used, and storage and handling methods. The actual N content of manure would be the best method for making adjustments in the rate of N fertilizer. In the absence of such data, the states provide general guidelines for N from various manure-handling systems (Table 11). In some

Table 11—Average composition of some farm manures.

Kind of animal	Nutrients in manure N	P	K
		kg/t	
Dairy cattle	5.5	1.10	4.56
Beef cattle	7.0	1.98	4.56
Hogs	5.0	1.54	3.32
Chicken	10.0	3.52	3.32
Dairy cattle (liquid)	2.5	0.44	1.66
Beef cattle (liquid)	2.0	0.22	1.24
Hogs (liquid)	5.0	1.10	1.66
Chicken (liquid)	6.5	2.64	2.08

situations, considerable amounts of N may be lost once the manure is applied to the soil, unless incorporated.

H. Reduced Tillage

Tillage practices are often aimed at reducing the potential for soil erosion and the use of petroleum products. These practices may leave increased amounts of crop residues on the soil surface. Personnel in the East North Central states recognize the potential for N loss when solid urea and liquids that contain urea are surface-applied without incorporation, especially with large amounts of residue on the soil surface (Spies, 1981; Vitosh & Warncke, 1976). Knifing N into the soil avoids the potential for volatilization loss. However, knifing into the soil with large amounts of surface residues requires some special mechanical considerations. But such equipment is now readily available.

Surface-applied N, irrespective of chemical form, may be at least temporarily immobilized by crop residues on the soil surface. Also, the surface mulch may increase the water content of the soil and increase the potential for N loss by denitrification and leaching.

I. Price of N and Crop

Economics are always important when making N rate suggestions. The lower the price of N in relation to the unit value of output, the higher the N fertilizer suggestion. Because of the nature of the yield response curve, there may be considerable fluctuation in the prices of N and unit output with only a minimal effect on the most profitable rate of N. For example, if the prices of corn and N are such that 5.6 kg of corn are required to purchase 1 kg of N, then the N fertilizer suggestion for corn following corn in Illinois is 2.14 kg of N for each 100 kg of expected corn grain yield. However, if only 2.8 kg of corn are required to buy 1 kg of N, then the N fertilizer is increased to 2.32 kg of N per 100 kg of expected grain yield (Illinois Coop. Ext. Service, 1980).

ACKNOWLEDGMENT

The author especially thanks Clifford D. Spies, Purdue University; Maurice L. Vitosh, Michigan State University; Jay W. Johhnson, Ohio State University; and Keith A. Kelling, University of Wisconsin for their personal communications with me while this chapter was in preparation. Helpful suggestions by Robert G. Hoeft, my colleague here at Illinois, is acknowledged.

REFERENCES

Brentwood Publications, Inc. 1981. 1981 Survey. Irrig. J. 31(6) (insertion). Brentwood Publ., Inc., Elm Grove, Wis.

Huber, D. M., H. L. Warren, D. W. Nelson, and C. V. Tsai. 1977. Nitrification inhibitors—new tools for food production. BioScience 27:523–529.

Illinois Cooperative Extension Service. 1980. Illinois agronomy handbook 1981–1982. College of Agric. Circ. 1186.

Johnson, J. W. 1981. Nitrification inhibitors potential use in Ohio. Agronomic tips, SFT-24. Ohio State Univ., Columbus, Ohio.

Kelling, K. A., P. E. Fixen, E. E. Schulte, E. A. Ligel, and C. R. Simson. 1981. Soil test recommendations for field, vegetable, and fruit crops. Coop. Ext. Programs, A2809. Univ. of Wisconsin, Madison, Wis.

Ohio Cooperative Extension Service. 1980. 1981–82 Agronomy guide. Ohio State Univ. College of Agric. Bull. 472.

Polizotto, K. R., and D. B. Mengel. 1981. Wheat production and fertilization in Indiana. Coop. Ext. Service (Small Grains) AY-244, Purdue Univ., W. Lafayette, Ind.

Spies, C. D. 1981. Types and uses of nitrogen fertilizers for crop production. Coop. Ext. Service (Fertilization) AY-204. Purdue Univ., W. Lafayette, Ind.

Spies, C. D., and D. B. Mengel. 1981. Corn fertilizer in Indiana. Coop. Ext. Service (Corn) AY-171. Purdue Univ., W. Lafayette, Ind.

U.S. Department of Agriculture. 1971. Cropping practices. Statistical Reporting Service, SRS-17. USDA, Washington, D.C.

U.S. Department of Agriculture. 1981. Fertilizer outlook and situation. Economics Research Service, FS-12. USDA, Washington, D.C.

U.S. Department of Agriculture. 1981. Crop production 1981 annual summary, acreage, yield, production. Crop Reporting Board, Statistical Reporting Service, Cr. Pr. 2-1 (82). USDA, Washington, D.C.

Vitosh, M. L., R. E. Lucas, and R. J. Black. 1979. Effect of nitrogen fertilizer on corn yield. Coop. Ext. Service, Ext. Bull. E-802. Michigan State Univ., East Lansing, Mich.

Vitosh, M. L., and D. D. Warncke. 1976. No till corn: 2, Fertilizer and liming practices. Coop. Ext. Service, Ext. Bull. E-905. Michigan State Univ., East Lansing, Mich.

Vitosh, M. L., and D. D. Warncke. 1977. Fertilization of wheat. Coop. Ext. Service, Ext. Bull. E-1067. Michigan State Univ., East Lansing, Mich.

Walsh, L. M. 1977. Should you be using nitrapyrin? Crops Soils 30(1):8–10.

Water Information Center, Inc. 1974. Climates of the states. Water Inf. Center, Inc., Port Washington, N.Y.

Welch, L. F. 1979. Nitrogen use and behavior in crop production. Illinois Agric. Exp. Stn. Bull. 761.

Welch, L. F., D. L. Mulvaney, M. G. Oldham, L. V. Boone, and J. W. Pendleton. 1971. Corn yields with fall, spring, and sidedress nitrogen. Agron. J. 63:119–123.

NORTH DAKOTA
MINNESOTA
WEST NORTH CENTRAL
SOUTH DAKOTA
IOWA
NEBRASKA
KANSAS
MISSOURI

49

G. A. Peterson
*University of Nebraska
Lincoln, Nebraska*

Regis D. Voss
*Iowa State University
Ames, Iowa*

Management of Nitrogen in the West North Central States

Nitrogen fertilizers are key inputs in the cropping systems of the West North Central states. Proper use of N is not only a factor in plant nutrition, but it is highly linked to efficient water use by plants and to the profitability of crop production throughout the entire region. The western half of this region lies in the Great Plains area, where precipitation levels are as low as 200 mm/yr. In contrast with these N-moisture critical areas, a part of the region receives > 1000 mm of precipitation per year. These areas have a different set of N management problems. Denitrification losses and leaching losses of N are two examples. Irrigation is widely practiced in the western half of the region, which adds a third set of management problems. Because irrigation is practiced on soils with diverse physical characteristics ranging from highly permeable sandy soils, to poorly drained clay-textured soils, to soils with slowly permeable clay pans, proper N and water management can differ drastically within any given state.

Climatic differences in rainfall and temperature have greatly affected native soil organic matter levels in the region. This in turn affects soil N mineralization potentials. As we proceed to discuss specific management practices, one can observe how these soil and climatic differences are related to efficient fertilizer N use. Corn (*Zea mays* L.) and wheat (*Triticum aestivum* L.) are the crops occupying the greatest area in the region. Soybeans [*Glycine max* (L.) Merr.] and sorghum [*Sorghum bicolor* (L.) Moench] are third and fourth in area. Other crops grown in the region include sugarbeets (*Beta vulgaris* L.), potatoes (*Solanum tuberosum* L.), sun-

Nitrogen in Crop Production.

flowers (*Helianthus annuus* L.), and forages. Our discussion will proceed by crop from this point.

I. FERTILIZATION OF MAJOR CROPS

A. Corn (*Zea mays* L.)

Corn is grown under rain-fed conditions in much of the region, but in Kansas, Nebraska, and South Dakota there are sizeable areas under irrigation. Direct N application accounts for 75 to 80% of the total fertilizer N used annually. Of the direct application materials, NH_3 accounts for the largest percentage of the market in all states (Table 1) (USDA, 1980). The remaining 20 to 25% of the total N used is applied indirectly with other nutrients in mixed fertilizers. These mixtures vary considerably in composition, but the primary compounds are ammonium phosphates.

Agronomists in each state indicated that the bulk of the N fertilizer used for corn in the 1970's was NH_3. The primary reason given for the popularity of NH_3 was its lower price per unit of N. It is being applied preplant either in the fall or spring. Fall application is recommended by university agronomists only after soil temperatures are below 10°C. This is especially emphasized in areas of higher rainfall where leaching or denitrification of NO_3^--N is a potential problem. Some use of nitrification inhibitors is occurring, but is not a widespread practice, although their recommended use with fall-applied NH_3 is becoming more frequent.

Agronomists continue to recognize the N use efficiency advantages of side-dressing or applying fertilizer N close to the period of maximum N uptake by the corn plant. Neither the growers nor the fertilizer industry are prepared, however, to wait and apply all of the N on corn at side-dressing time. The risks of not being able to get all hectares fertilized in a relatively short time and the subsequent yield loss are too great for many large operations.

For nonirrigated corn, an extended rainy period may prevent N side-dressing at the proper time, and the corn may attain a growth stage that pro-

Table 1—Relative amounts of directly applied N sources for each state in the West North Central region.

State	N Source						
	NH_3	AN	Solu-Solu-tions	Urea	Am-monium sulfate	Sodium nitrate	AN Lime
	%						
N. D.	58.9	5.2	10.9	24.8	0.2	--	--
S. D.	43.0	20.0	13.2	23.7	0.1	--	--
Minn.	68.3	2.7	9.8	18.8	0.3	--	--
Iowa	68.0	2.9	17.7	11.7	0.2	--	--
Neb.	73.3	4.8	18.3	3.1	0.4	--	--
Mo.	51.3	17.8	16.8	14.0	<0.1	<0.1	<0.1
Kan.	68.2	11.0	15.1	5.6	0.1	--	--

hibits use of standard side-dressing equipment without breaking over a high percentage of corn plants. In this situation, alternative methods are to use high-clearance equipment to apply nonpressure N solutions to the soil surface (not on the crop foliage) and to aerially apply urea. Ammonium nitrate (AN) or materials containing it are not suggested to be applied to corn that has a canopy cover of the soil. Applications of N at this growth stage may not produce the yield increase obtained with a normal sidedress application. Profitable yield increases have been obtained, however, from fertilizer N applied at the silking stage, but the economical optimum rate is lower than that at normal side-dressing time.

Of course, growers using irrigation have an opportunity to apply N throughout the growing season in the irrigation water. If rain occurs during the planned N application period, that opportunity is lost. Growers with gravity irrigation systems find, however, that N application in irrigation water is not a very convenient method.

Corn growers using center pivot irrigation systems have made excellent use of in-season applications. Soils under these systems frequently are coarse-textured, and preplant applications can be quite inefficient. The primary disadvantage of N application in irrigation water has been the need to use a more expensive source of N. Nonpressure solutions, which are mixtures of AN and urea, are essentially the only option open to a grower for this purpose. Research is being conducted to develop technology that will allow injection of NH_3 in water. The problem to be solved is the initial rise in pH at the NH_3 injection point. The pH rise causes precipitation of $CaCO_3$, which in turn plugs the sprinkler system. Techniques to simultaneously inject sulfuric or phosphoric acid to prevent the pH rise are being researched (Bock, 1978). At this writing, the option of application of NH_3 in irrigation water is not available to growers.

To minimize the cost and still efficiently supply adequate N for irrigated corn grown on coarse-textured soils, most growers use a combination of spring preplant NH_3 and nonpressure N solutions in the irrigation water during the growing season. A common division is 50% for each method. A modification of this may occur with the use of nitrification inhibitors.

The amount of N being recommended on corn in the West North Central states is generally related to yield goal or potential yield (Brown et al., 1981; Buchholz et al., 1981; Gelderman et al., 1977; Jokela et al., 1981; N. D. Agric. Assn. : NDSU, 1981; Voss et al., 1974; Whiteney, 1976; Wiese & Penas, 1974). Yield goals in turn differ across the region as a function of rainfall and irrigation capability. Within state and even within a county, yield goals differ due to individual growers management. Examples of management differences include plant population, weed control, irrigation system capacity, and timeliness.

Agronomists agree that corn requires between 18 to 27 g of N/kg of corn grain. Some states increase N recommendations per kilogram of corn grain as yield goal increases, and some use a fixed amount per kilogram at all yield levels. How the actual recommendation is made once the yield goal N requirement is established differs widely from state to state (Table 2).

The specific amounts of N being recommended varies from none to 280

Table 2—Basis for fertilizer N recommendations with emphasis on corn for each state in the West North Central region.

State	Soil test and equation	Adjustments
N. D.	NO_3^--N: one unit of NO_3^--N in surface 60 cm of soil equals one unit of fertilizer N. Recommendation = (Yield goal N requirement) − (NO_3^--N in surface 60 cm of soil)	Equation holds for samples taken from 1 Sept. to 1 Apr. After 1 Apr. add 11 to 22 kg N/ha to recommendation. Adjustments for previous crops are minus 40 kg N/ha for alfalfa and minus 22 kg N/ha for soybeans.
S. D.	NO_3^--N: one unit of NO_3^--N in surface 60 cm of soil equals one unit of fertilizer N. Recommendation = (Yield goal N requirement) − (NO_3^--N in surface 60 cm of soil) Alternative recommendations based on organic matter with base recommendation starting at ≤ 2% organic matter.	Alternative recommendations are based on organic matter. Adjustments for corn average minus 34 kg N/ha for each 1% organic matter > 2%.
Minn.	Two tests: 1. General; by yield goal, previous crop history, and organic matter; 2. Western Minnesota; by NO_3^--N in surface 30 to 60 cm. One unit of NO_3^--N in surface 30 cm = 0.85 units of fertilizer N; and one unit of NO_3^--N in 30 to 60 cm layer = 0.25 units of fertilizer N.	Adjustments for previous crops are minus 11 to 45 kg N/ha for soybeans and 45 to 110 kg N/ha for alfalfa (*Medicago sativa* L.), depending on stand.
Iowa	No soil test used. Recommendations vary by crop for major soil areas and are based on N-rate studies conducted in major soil areas.	Adjustments for previous crops are minus 45 kg N/ha for soybeans or 17 g N/kg soybean/ha, 155 kg N/ha for 50 to 100% legume meadow, and 110 kg N/ha for 20 to 50% legume meadow.
Neb.	NO_3^--N: recommendation based on calibrated soil NO_3^--N index. Differs for each crop and for each area of state. Usually surface 30 cm to 45 cm samples unless NO_3^- accumulation is suspected due to past management.	On highly eroded soils, 22 kg N/ha are added.
Mo.	Soil organic matter and cation exchange capacity are used in association with yield goal and crop.	N rate and plant population are increased for irrigated and high yield situations.
Kan.	NO_3^--N + NH_4^+-N: based on average mg/kg (ppm) of NO_3^--N + NH_4^+-N in surface 60 cm of soil. Recommendation = $[(X_1)\text{ (yield goal)} - (7.5)(X_2)](X_3) + (X_4) + (X_5)$, where X_1 = units of N per unit of crop produced; X_2 = soil test value; X_3 = soil texture adjustment factor; X_4 = legume N adjustment; X_5 = manure N adjustment.	Texture adjustment = 1.1, sandy; 1.0, others. Adjustments for previous crops are minus 140 kg N/ha for 100% legume meadow, 34 kg N/ha for weedy legume meadow, 34 kg N/A for second year after 100% legume meadow, and 34 kg N/ha for soybeans. Adjustments for manure are minus 55 kg N/ha for first year of application and 28 kg N/ha for second and third year after application.

kg/ha. The higher rates are associated with yield goals greater than 12 500 kg/ha. This can occur under irrigation or dryland, depending on the state, but are usually associated with irrigation. In other states, yield goal is not specifically used, but N recommendations are on the basis of potential yield for a given area of the state. Plant population is also used in conjunction with yield goal, and adjustments in the N recommendations are made for irrigated and nonirrigated situations (Buchholz et al., 1981). Most states make adjustments for N additions from previous legume crops and/or manure applications. Forage legumes and soybean grown in rotation with corn affects both the corn yield and yield response to fertilizer N (Voss & Schrader, 1979). It is a general concensus that a soybean crop proceeding corn has the effect of supplying 22 to 45 kg N/ha, and fertilizer rates are adjusted accordingly. Specific adjustments by state are shown in Table 2.

B. Wheat (*Triticum aestivum* L.)

Wheat is grown in all of the states of this region and is a major crop in all of the states except Iowa. Much of the wheat in the Dakotas, Nebraska, and Kansas is grown in summer fallow systems. Nitrogen fertilizer management may be altered by the type of system used. In our discussion, a distinction for cropping system will be made where appropriate.

As with corn, anhydrous NH_3 (AA) is the primary N source used on wheat. Price per kilogram of N is the primary factor in this choice. Of course in wheat systems (nonrow crop) this usually means preplant application. The state universities of the region recommend spring topdressing as the most efficient time of application, but most commonly, preplant NH_3 is used. One state, Kansas (Whitney, 1976), indicates that a grower could use a lesser amount (11 kg/ha) if topdressing is chosen over preplant on medium- and fine-textured soils. Spring-applied N in general has increased grain protein content more than fall-applied N (Whitney & Murphy, 1969).

Two states, Iowa and Missouri, are exceptions to the dominance of AA in wheat fertilization. Iowa, a minor wheat-producing state, estimates that 50% of the N is applied as AN or urea and 30% as nonpressure N solutions. Missouri indicates that 85% as dry sources (AN and urea). The remaining 15% is 10% solutions and 5% AA. Current emphasis on reducing N in runoff waters in Missouri probably will change this distribution in the future.

Growers practicing fallow find that AA application fits well with tillage events that are necessary for weed control during the fallow period. Almost every type of equipment, from moldboard plows to large sweep machines to rod weeders, are used for NH_3 applications during fallow. In Nebraska and Kansas, much of this NH_3 is applied in August, 4 to 6 weeks before seeding time. Dual application of N and P fertilizers is becoming popular in western Kansas. Kansas State University has shown that the dual preplant application of NH_4^+-N and P, e.g., AA and 10–34–0 solution replaces the need to use phosphate fertilizers at planting time (Kissel & Whitney, 1979).

Yield goal directly or "indirectly" controls the amount of N recommended by university agronomists in this region (Buchholz et al., 1981; Gelderman et al., 1977; Jokela et al., 1981; N.D. Agric. Assn. & NDSU, 1981; Sander et al., 1973; Voss, 1976; Whitney, 1976). In states like North and South Dakota, Missouri, and Kansas, yield goal is used directly in making the recommendation. A value of approximately 42 g N/kg of wheat is used in the calculation. In other states, crop regions delineated on the basis of precipitation are used to adjust recommendations. It is assumed that the precipitation level determines the yield goal. Nebraska uses the regional approach and the recommendations in each region are expected to produce 11% grain protein at the anticipated yield level for that region.

Actual rates of N being used vary widely, even on a local basis. Generally, the dryest regions tend to receive the lowest rates. Because fallow is practiced in the lowest rainfall areas, however, the yield potential for the cropped year is quite high and so N recommendations are not greatly different than for the more humid portions of the region. Table 3 gives a summary of the maximum recommendations by state and how they are determined.

Adjustments for irrigated wheat are made according to yield goal changes associated with irrigation. Timing of the N application under irri-

Table 3—Maximum N recommendations for wheat by state in the West North Central region.

State	Maximum rates	Remarks
N. D.	225 kg N/ha total of soil NO_3^--N and fertilizer N for 5400 kg/ha yield goal.	Rates adjusted for soil NO_3^--N as shown in Table 2.
S. D.	215 kg N/ha total of soil NO_3^--N and fertilizer N for 5400 kg/ha yield goal. Alternative recommendations based on organic matter with base recommendation starting at $\leq$ 2% organic matter	Rates adjusted for soil NO_3^--N as shown in Table 2. Alternative recommendations are based on organic matter. Adjustments for wheat average minus 22 kg N/ha for each 1% organic matter > 2%.
Minn.	170 kg N/ha total of soil NO_3^--N and fertilizer N for yield goal > 4700 kg/ha in western Minn. Alternative recommendations based on previous crop and soil texture in other areas of state.	Rates adjusted for NO_3^--N in surface 60 cm in western Minn. Adjustments made in alternative recommendations for previous crop and soil texture.
Iowa	55 to 67 kg fertilizer N/ha.	Over all areas of state.
Neb.	80 to 100 kg fertilizer N/ha.	Dependent on area of state. Adjusted for NO_3^--N soil test as stated in Table 2. Additional 22 kg N/ha added per 1% protein increase desired. (Base level = 11% protein.)
Mo.	130 kg N/ha at low organic matter, cation exchange capacity > 18 meq/100 g, and for 5400 kg/ha yield goal.	Rates adjusted for factors stated in Table 2.
Kans.	45 kg N/ha for fallow and 80 kg N/ha for continuous wheat.	Yield goal, soil NO_3^--N, and previous crop affected recommendations as shown in Table 2.

gation can be critical. Lodging difficulties are minimized if both N and water applications can be delayed until the wheat plant is in the boot stage.

C. Sorghum [*Sorghum bicolor* (L.) Moench]

Grain sorghum is considered a major crop in Kansas and Nebraska and it is usually produced under rain-fed conditions in both states. Whenever adequate irrigation is possible, corn is usually substituted for grain sorghum. In South Dakota, Minnesota, Iowa, and Missouri, it is a minor crop compared with corn and/or wheat. In Missouri, for example, it is generally grown on problem soils with clay pans and/or steep slopes. It is also used in rotations with soybeans in southeastern Missouri to minimize soybean cyst nematode problems. There have been recent increases in the amount of sorghum grown in Missouri among livestock producers, because of its greater yield stability than for corn in drought years.

About 50% of the N used on sorghum in Kansas is applied as AA. In Nebraska, AA probably accounts for 75% of the N fertilizer used on sorghum. Amounts of N recommended for grain sorghum by agronomists in this region are shown in Table 4 (Buchholz et al., 1981; Gelderman et al., 1977; Jokela et al., 1981; Sander & Frank, 1974; Voss, 1976; Whitney, 1976). One concludes from this table that N recommendations tend to be lower for sorghum than corn. However, two states (Kansas and Iowa) do recommend equal amounts of N for sorghum and corn.

Table 4—Maximum N recommendations for grain sorghum by state in the West North Central region.

State	Maximum rates	Remarks
S. D.	180 kg N/ha total of soil NO_3^--N and fertilizer N for 10 000 kg/ha yield goal. Alternative recommendations based on organic matter with base recommendation starting at ≤ 2% organic matter.	Rates adjusted for soil NO_3^--N as shown in Table 2.
Minn.	155 kg N/ha total of soil NO_3^--N and fertilizer N for yield goal > 6300 kg/ha in western Minn. Alternative recommendations based on previous crop and organic matter level in other areas of the state.	Rates adjusted for soil NO_3^--N as shown in Table 2. Adjustments made in alternative recommendations for previous crop and organic matter.
Iowa	200 kg fertilizer N/ha.	Same as corn. Adjustments downward made for soil area and previous legume crop.
Neb.	185 kg fertilizer N/ha for 9400 kg/ha yield goal.	Recommendation changes with organic matter and soil NO_3^--N research.
Mo.	195 kg fertilizer N/ha for 10 000 kg/ha yield goal.	Recommendation changes with cation exchange capacity and organic matter.
Kan.	Varies with yield goal. Use 23 g N/kg grain sorghum.	Rate adjustments for soil NO_3^--N are similar to corn.

D. Soybeans [*Glycine max* (L.) Merr.]

Soybeans are grown on a high number of hectares in this region. Because they are a legume, they are not frequently thought to need N fertilization. Some states do recommend N application on soybeans, however. In Kansas, N is recommended in two cases: (i) newly leveled land with low available N and no history of previous soybeans, and (ii) incorporation of heavy straw occurring ahead of double-crop soybeans (Whitney, 1976). Amounts of N recommended vary depending on the situation. South Dakota recommends a 11 kg/ha N rate in any starter fertilizer program. In Nebraska (Penas & Wiese, 1973), agronomists recommend 67 to 135 kg N/ha when one or more of the following conditions exist: (i) plants do not have a dark green color and the chlorosis is not attributable to wetness, salinity, or Fe deficiency; (ii) acid soils with lime requirements of 4.5 t/ha; (iii) light colored, eroded soils; and (iv) fields that have not grown legumes or have not had adequate N applied on grain crops. No N recommendation is made in other states of the region, and the only N received by soybeans in these states is incidental in fertilizer materials such as ammonium phosphates.

E. Forages

Pastures are an important part of agriculture in all states of the region. However, the amount of N fertilizer used on them is small compared with that used on grain crops. A large portion of the pastures in the region lie in the dryest areas of the Dakotas, Nebraska, and Kansas. These native ranges are essentially unfertilized. Seeded pastures do receive N fertilizer, and the rates vary directly with amount of rainfall and/or availability of irrigation. Table 5 shows the recommended rates by state (Buchholz et al., 1981; Gelderman et al., 1977; Jokela et al., 1981; N. D. Agric. Assn. & NDSU, 1981; Rehm & Knudsen, 1978; Schaller et al., 1979; Whitney, 1976). The potential for water is the dominant feature of the table. Other considerations involve timing of the N application. At the highest rates, which are

Table 5—Nitrogen recommendations for established forages by state in the West North Central region.

State	Recommendation	
N. D.	Dryland†	Irrigated
	Red River Valley 105 to 130 kg/ha, East Central 80 to 100 kg/ha, West Central 65 to 80 kg/ha, West 0 to 55 kg/ha.	170 kg/ha all areas with 50% applied early spring and 50% divided among grazing rotations.
S. D.	Native grass†	No N up to yield goal of 4.5 t/ha. 15 kg/t per ha of yield goal > 4.5 t/ha with maximum of 80 kg/ha.

(continued on next page)

Table 5—Continued.

State	Recommendation	
	Cool-season grass†	55 kg/ha up to yield goal of 4.5 t/ha. 15 kg/t per ha of yield goal > 4.5 t/ha.
	Warm-season grass†	0 to 45 kg/ha up to yield goal of 4.5 t/ha. 10 kg/t per ha of yield goal > 4.5 t/ha with maximum of 270 kg/ha.
Minn.	Rotational grazing and adequate rain	170 kg/ha split applied equally
	Continuous grazng and adequate rain	110 kg/ha split applied equally
	Moderate rain	55 kg/ha
	Sandy soils	35 kg/ha
	Organic soils	55 kg/ha

Iowa	Cool-season grasses‡		Warm-season grasses
Area	Bluegrass	Tallgrass	
	kg N/ha		
Northwestern Iowa	65	90	90
Western Iowa	80	110	125
Other areas	90	135	170

Neb. Area	Cool-season grasses§	Warm-season grasses¶	Irrigated nonnative grasses
	kg N/ha		
Eastern	90–110	65–90	280 kg all areas with 100 kg applied early spring and remainder divided during growing season.
Central	65–90	45–65	
Western	45–65	0–34	

Mo.	Use	
	Pasture	Hay
Bluegrass	0.7 kg/cow day	--
Bermudagrass	0.7 kg/cow day	25 kg/t
Cool season grass‡	0.7 kg/cow day	20 kg/t
Sudangrass	0.7 kg/cow day	20 kg/t
Warm-season grass	65 kg/ha	65 kg/ha

Kan. Area	Dryland††	Irrigated
Eastern	90–135 kg/ha applied ⅔ in early spring and ⅓ mid-August.	140–225 kg/ha; ⅓ early spring, ⅓ June, ⅓ August, or
Central	45–90 kg/ha applied ⅔ in early spring and ⅓ mid-August.	140–225 kg/ha; ⅔ early spring and ⅓ August.
Western	--	140–225 kg/ha for all areas.

† Applied fall–early spring at rates > 170 kg/ha. Use 50% early and 50% divided among grazing rotations.

‡ Additional N up to double these rates can be used, but as a split application.

§ Early spring application with possible small dose in August.

¶ Definite late spring application.

Seed plus residue hay or pasture—110–145 kg/ha; stockpile fall growth—180 kg/ha. Bluegrass, *Poa annua* L.; bermudagrass, *Cynodon dactylon* (L.) Pers.; sudangrass, *Sorghum sudanense* (Piper) Stapf.

†† Cool season grasses only.

used in the high rainfall or irrigation areas, the fertilizer should be applied in two or more portions to ensure most efficient usage and to reduce the potential for NO_3^- toxicity. With warm-season grasses, it is important to delay application until late spring to avoid stimulating cool-season grasses and weeds. Repeated early applications will often completely change the species distribution in a pasture.

Materials used are AN, urea, and nonpressure N solutions. The latter are commonly used in irrigation systems for growing-season applications.

F. Sugarbeets (*Beta vulgaris* L.)

Four states (North Dakota, Minnesota, Kansas, and Nebraska) produce sugarbeets. In all four states, the N recommendations are based on soil NO_3^- tests (Daigger et al., 1973; Jokela et al., 1981; N. D. Agric. Assn. & NDSU, 1981). Extensive research programs in Nebraska have shown that sugarbeet quality, percent sucrose, and clear juice purity can be improved by using soil NO_3^- testing. Maximum recommended N rates on soil with low organic matter and low soil NO_3^- range from 250 kg/ha in Nebraska to 167 kg/ha in Minnesota and North Dakota. Often times the actual recommendations are less than 110 kg/ha, because very few extremely deficient soils are found in the sugarbeet-growing areas.

Less AA is used on sugarbeets than other crops. Nonpressure N solutions, dry urea, and AN make up the bulk of the N used on sugarbeets. These sources are used because soils in the northern areas are often too wet in the spring for NH_3 injection. In Nebraska, the primary reason for fertilizer source choice is custom or habit. Ammonia could certainly be substituted in the majority of cases in Nebraska. Higher fertilizer N prices may encourage this switch.

G. Sunflower (*Helianthus annuus* L.)

Sunflowers are an important crop in North and South Dakota and Minnesota. Nitrogen fertilizer recommendations are based on soil NO_3^- levels and organic matter in these states (Gelderman et al., 1977; Jokela et al., 1981; N. D. Agric. Assn. & NDSU, 1981). To compare recommendations, a yield goal of 3350 kg/ha was chosen. At this level, North Dakota's recommendation is that soil NO_3^--N in the top 60 cm of soil plus fertilizer N should be 180 kg/ha. South Dakota and Minnesota indicate a total of 155 kg/ha at the same yield level. Nonpressure solutions and dry materials predominate as N sources. This may be due to wet soil problems in the spring in these northern areas, which makes NH_3 injection more difficult.

H. Potatoes (*Solanum tuberosum* L.)

Potatoes are only of significant importance in three states of the region —North and South Dakota and Minnesota. At a yield goal of 50.5 t/ha, soil NO_3^--N plus fertilizer N is recommended to be 180, 180, and 225 kg/ha for

North and South Dakota and Minnesota, respectively (Gelderman et al., 1977; Jokela et al., 1981; N. D. AGric. Assn. & NDSU, 1981). As with sunflowers, nonpressure N solutions and dry materials are the sources most commonly applied.

II. GENERAL DISCUSSION AND RESEARCH NEEDS

Agronomists of the region indicate a fair to excellent acceptance of the N recommendations made by the university soil testing laboratories. When recommendations are not followed, it is usually on the overapplication side. Several reported that the most difficult recommendation to get growers to accept is zero application in cases where soil tests for N are high. These same agronomists report, however, that once a grower tests this recommendation for himself, he is most often convinced that the N soil test really does accurately predict N need. Sugarbeet growers have observed improvements in their sugarbeet quality when using N soil testing and have accepted N soil testing more widely than grain farmers. In wheat-fallow areas, little soil testing is being done compared with irrigated areas and/or areas where high-value crops like potatoes and sugarbeets are grown.

In the cotton (*Gossypium hirsutum* L.)-producing areas, soil and petiole NO_3^- concentrations are used to determine fertilizer N requirements. Although plant analysis is not used widely for the major grain crops to adjust fertilizer N rates, research is needed to improve the interpretation of this index of N sufficiency. In Iowa, research is being done on using the N concentration in mature corn grain as an index for percent of maximum yield. Factors affecting N concentration in corn grain have been identified, and a major factor is the hybrid. The N concentration for a particular hybrid seems to be predictable if the N concentration of the inbreds at maximum yield is known. The methodology for using plant analysis indexes to adjust fertilizer N rates has been derived for the percent of maximum yield concept (Capurro & Voss, 1981; Pierre et al., 1977).

None of the states are adjusting N recommendations for reduced-tillage and/or no-till cropping systems. Apparently, the data base on which these decisions must be made is not adequate or existent at the present time. A survey of state specialists shows an interest in improving the management of N use in no-till systems. Specific mention was made of needed research in N cycling. Reduced tillage systems are suspected to cycle N differently than conventional systems. This may mean differences in the N fertilizer use. An understanding of the N cycle also includes further study of the times and amounts of N losses occurring by denitrification, N volatilization from surface applications of urea-containing fertilizers, and leaching potentials.

Another research need specifically mentioned was an improved evaluation of N release from organic matter. North Dakota scientists have been studying a "quick test" involving organic matter extraction with a weak sodium bicarbonate solution. They report that it has worked well in N experiments with barley (*Hordeum vulgare* L.) and flax (*Linum usitatissimum*

L.). If further testing confirms these preliminary results, they will be making the test routinely in the near future. Several other agronomists expressed the need for this kind of evaluation, but no other research is underway.

It is recognized that much progress has been made in predicting N fertilizer needs in this region. Nitrate testing of soil profiles has worked very well in five of the seven states. Missouri has used it in cotton-producing areas. Missouri and Iowa have certain regions where it may not be feasible to use NO_3^- testing. For example, soils with tile-drain systems in high-rainfall areas probably do not accumulate NO_3^-. These states may also have some regions where NO_3^- testing could be beneficial, such as in areas of lower rainfall and on soils containing no artificial drainage systems.

Management of N in irrigated regions cannot be separated from water management if we are to achieve efficient N use. Irrigation scheduling can reduce wasteful leaching of NO_3^--N and help prevent NO_3^- pollution of aquifers. This is a real challenge in areas of high water tables that are overlain by coarse-textured soils. A cooperative project in central Nebraska between corn growers and the Extension Service is promoting the joint use of irrigation scheduling, amount and timing of N fertilization, and N analysis of groundwater. They are reporting that some corn growers have irrigation water that contains 22 to 55 kg of N per 30 cm of irrigation water per hectare. By correct timing, soil testing, and irrigation scheduling, these growers are improving fertilizer N use efficiency and preventing further pollution of the groundwater.

The future for continued improvement in N management seems bright. More reliance on NO_3^- soil testing appears to be the way of the future. Extension programs to educate fertilizer dealers and growers regarding proper use of NO_3^- testing are the key. Research has shown the usefulness of this soil test. As the educational background of our dealers and growers improves, it should be possible to obtain wide usage of this powerful tool.

REFERENCES

Bock, B. R. 1978. Presoil-contact NH_3 losses and precipitation from ammoniacal sources in sprinkler water. Ph.D. Diss. Univ. of Nebraska. Univ. Microfilms, Ann Arbor, Mich. (Diss. Abst. 39:4133B).

Brown, J. R., R. G. Hanson, and D. D. Buchholz. 1981. Interpretation of Missouri soil test results. Univ. of Missouri, Agron. Misc. Publ. 80-04.

Buchholz, D. D., J. R. Brown, R. G. Hanson, H. N. Wheaton, and J. N. Garrett. 1981. Soil test interprtations and recommendations handbook. Univ. of Missouri, Columbia.

Capurro, E., and R. Voss. 1981. An index of nutrient efficiency and its application to corn yield response to fertilizer N. I. Derivation, estimation, and application. Agron. J. 73: 128–135.

Daigger, L. A., F. N. Anderson, D. Knudson. 1973. Fertilizing sugar beets. Coop. Ext. Service, NebGuide G73-8. Univ. of Nebraska, Lincoln.

Gelderman, R., E. Adams, E. Williamsen, and P. Carson. 1977. Fertilizer guide for soil test computerized recommendations. Agric. Exp. Stn., Plant Sci. Pm34. South Dakota State Univ., Brookings.

Jokela, W. E., W. E. Fenster, C. J. Overdahl, C. A. Simkins, and J. Grava. 1981. Guide to computer programmed soil test recommendations for field crops in Minnesota. Agric. Ext. Service, Univ. of Minnesota Ext. Bull. 416.

Kissel, D. E., and D. A. Whitney. 1979. Phosphorus placement for wheat. Coop. Ext. Service, AF-40. Kansas State Univ., Manhattan.

North Dakota Agricultural Association and North Dakota State University. 1981. Crop production guide. Fargo, N.D.

Penas, E. S., and R. A. Wiese. 1973. Fertilizing soybean fields. Coop. Ext. Service, NebGuide G73-1. Univ. of Nebraska, Lincoln.

Pierre, W. H., L. Dumenil, and J. Henao. 1977. Relationship between corn yield, expressed as a percentage of maximum, and the N percentage in the grain. II. Diagnostic use. Agron. J. 69:221–226.

Rehm, G., and D. Knudsen. 1978. Fertilizing irrigated pastures. Coop. Ext. Service, NebGuide G73-3. Univ. of Nebraska, Lincoln.

Sander, D. H., L. A. Daigger, and G. A. Peterson. 1973. How much fertilizer on wheat? Coop. Ext. Service, NebGuide G73-37. Univ. of Nebraska, Lincoln.

Sander, D. H., and K. D. Frank. 1974. Fertilizing grain sorghum. Coop. Ext. Service, NebGuide G74-112 (rev.). Univ. of Nebraska, Lincoln.

Schaller, F. W., R. D. Voss, and J. R. George. 1979. Fertilizing pasture. Coop. Ext. Service, Pm-869. Iowa State Univ., Ames.

U.S. Department of Agriculture. 1980. Commercial fertilizers: consumption by class for year ended June 30, 1980. Crop Reporting Board, Economics and Statistics Service, Sp. Cr. 7 (11-80). USDA, Washington, D.C.

Voss, R. D. 1976. General guide for fertilizer recommendations in Iowa. Coop. Ext. Service, Mimeo. AG-65 (rev.). Iowa State Univ., Ames.

Voss, R. D., and W. D. Shrader. 1979. Crop rotations: Effect on yields and response to nitrogen. Coop. Ext. Service, Pm-905. Iowa State Univ., Ames.

Voss, R. D., W. D. Shrader, J. R. Webb, L. C. Dumenil, J. T. Pesek, G. O. Benson, and H. E. Thompson. 1974. Getting the most out of N for corn. Coop. Ext. Service, Pm-585. Iowa State Univ., Ames.

Whitney, D. A. 1976. Soil test interpretations and fertilizer recommendations. Coop. Ext. Service, Pm C-509. Kansas State Univ., Manhattan.

Whitney, D. A., and L. S. Murphy. 1969. Lime and fertilizer recommendations. Coop. Ext. Service, Pm C-352. Kansas State Univ., Manhattan.

Wiese, R. A., and E. J. Penas. 1974. Fertilizer recommendations for corn. Coop. Ext. Service, NebGuide G74-174. Univ. of Nebraska, Lincoln.

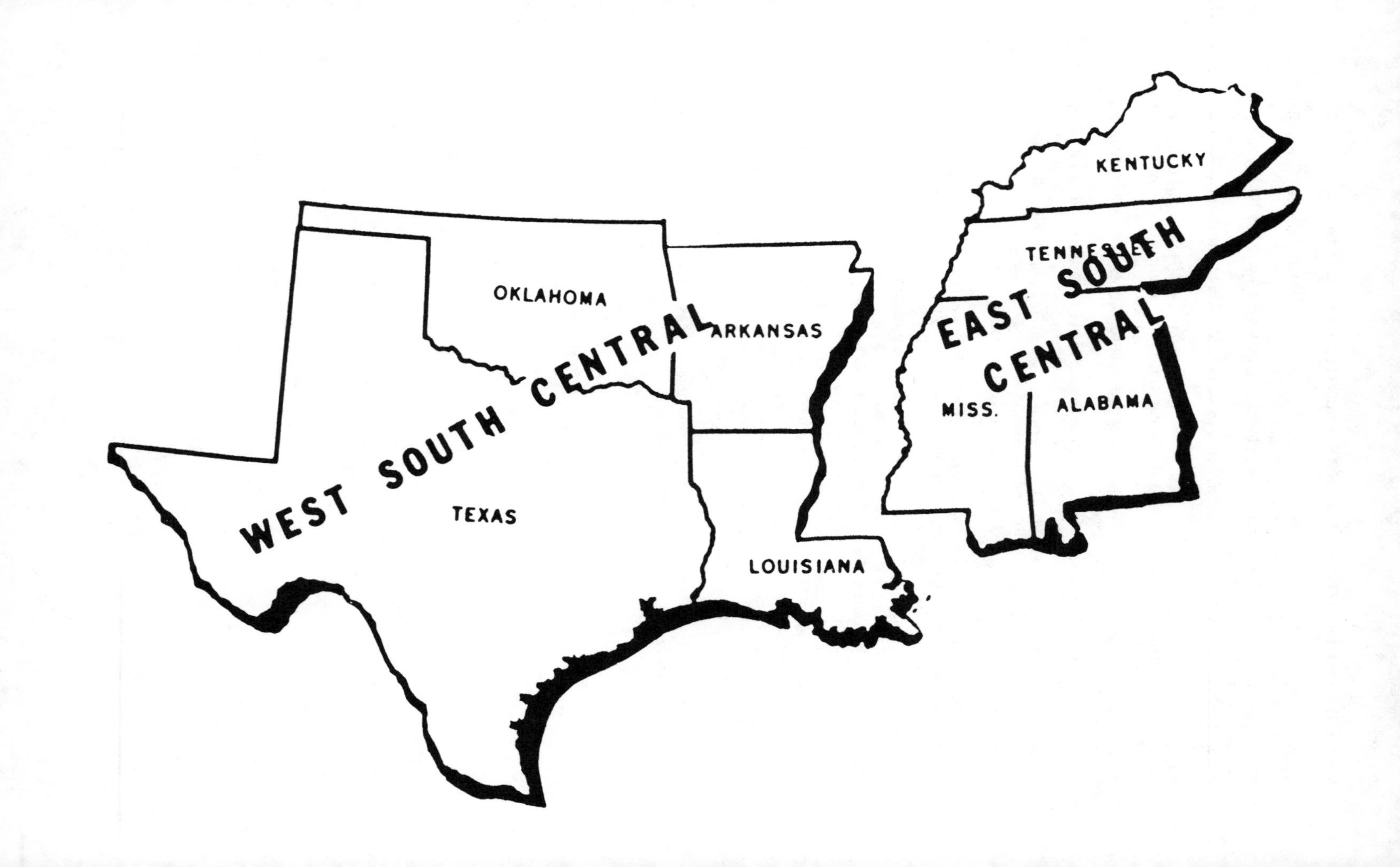
WEST SOUTH CENTRAL
OKLAHOMA
ARKANSAS
TEXAS
LOUISIANA
EAST SOUTH CENTRAL
KENTUCKY
TENNESSEE
MISS.
ALABAMA

50

Billy B. Tucker
Oklahoma State University
Stillwater, Oklahoma

L. W. Murdock
University of Kentucky
Lexington, Kentucky

Nitrogen Use in the South Central States

Nitrogen fertilization has become extremely important in crop production throughout the South Central states. In general, soils of the South Central region were lower in innate soil organic matter, compared with those regions with shorter summers. Upon cultivation, soil organic matter in the region rapidly oxidized because of the long warm season. This rapid decrease in soil organic matter resulted in adequate N for cultivated crop production during the first years of farming. However, as innate organic matter reserves were depleted to low levels, N shortages became severe. Legumes in crop rotations were used almost exclusively to supply needed N until after the World War II period, at which time N fertilization became widespread.

Even though many soils of the South Central states were initially more deficient in available soil P than N, upon cultivation, N soon became the most limiting plant food element for production of nonlegumes. Because N fertilization is such an important crop production practice in the region, a judicious choice of the kind of N fertilizer and dosage is essential in order to obtain high yields. Proper rates not only maximize profits, but they also prevent an accumulation of excess NO_3^--N in the soil profile and leakage into water supplies.

It is the purpose of this chapter to take a look at the current state-of-the-art of N use in the South Central states and to present a summary of the knowledge on the practical aspects of using N efficiently in agricultural soils of the region.

I. DESCRIPTION OF THE REGION

The South Central area comprises the states of Oklahoma, Texas, Louisiana, Arkansas, Kentucky, Tennessee, Mississippi, and Alabama. Fertilizer use ranges from a long history in the southern and eastern portions of the area to a relatively recent history in the western portion. Nearly 1.8 million tonnes of N are consumed in these states annually, which is approximately one-fifth of the total U.S. consumption. On a plant-food basis, anhydrous ammonia is the leading material followed by ammonium nitrate (AN), N solutions, and urea. Small quantities of ammonium sulfate, sodium nitrate, calcium nitrate, and aqua ammonia are used. About one-third of the N applied is in fertilizer mixtures.

The total cropland in the area is about 25.7 million ha, with an average application of approximately 45 kg of total plant food per hectare. Cash receipts from livestock and crops are about equal. Almost $2 billion are spent annually for fertilizer and lime.

The South Central area is diverse in climate, soils, and crops; therefore, for our purposes it is necessary to discuss the South Central states by the following broad subregions. The Southern Great Plains, the Coastal Plains, the Mississippi Delta, and the East Central Uplands.

A. Southern Great Plains

For this discussion, the area lying west of the humid, forested part of eastern Texas and Oklahoma is designated the Southern Great Plains. It comprises 75% of the states of Oklahoma and Texas; it excludes the Coastal Plains and the Ozark and Quachita Highlands. The climate varies from semiarid to humid, but the summers are generally hot and dry. The winter temperatures vary greatly across the region with a growing season of 180 to 200 d. Soils are quite variable with large extremes in pH, texture, and depth. Although 90% of the area is dryland, agriculture responses to irrigation are large.

The primary difference in N fertilizer use between irrigated and dry land in the Southern Great Plains is the marked differences in yields, and consequently, higher rates of use under irrigation. Rates used for fertigation varies from 35 to 70 kg N/ha per application.

In general, the area is dry enough that N leaching losses are not great. However, under irrigation on permeable soils, split N applications are necessary. Therefore, some fertigation is practiced. Most fertigation in the Southern Great Plains is in connection with sprinkler irrigation. Surface and furrow irrigation water distribution over the field is often uneven and may result in uneven fertilizer distribution in such systems: the first one-third of the water application is without fertilizer; the second one-third of the water is applied with fertilizer injected; and the last one-third is applied without fertilizer. More uniform water and N distribution can be achieved with fertigation using sprinkler and drip irrigation, but it varies with water management.

Water quality plays an important role in choice of N sources for fertigation. Nitrogen solutions and AN are the primary N sources, since they are so readily soluble in water. Materials containing free ammonia, such as NH_3, may cause precipitation of salts, thereby clogging pipes and nozzles.

B. Gulf Coastal Plains

For the purposes of this chapter, the Coastal Plains includes those parts of the states of Alabama, Mississippi, Louisiana, and Texas adjacent to but not including the coastal prairies. In addition, the area includes the hilly sections (but not mountains) of southeastern Oklahoma, eastern Texas, southwestern Arkansas, western Louisiana, eastern Mississippi, and southern Alabama. The soils are of the red-yellow Podzolic group and vary from sandy to fine texture. The soils are productive, but were inherently low in major fertilizer elements. The climate is mild, with a growing season of 200 to 280 d. Precipitation varies from 100 to 150 cm annually, but summer droughts are common.

C. Mississippi Delta and Gulf Coastal Prairies

The region consists of the flood plains of the Mississippi River and associated rivers alluvium. In this chapter, it also includes the Mississippi Valley silty uplands. Many soils in the flood plains and low terraces along the Mississippi are fine-textured, but other rivers' alluvium, like the Arkansas and Red rivers, tend to be more coarse-textured and permeable.

Average annual precipitation varies from 165 cm in the south to 115 cm in the north. The average frost-free period is from 200 to 280 d.

D. East Central Uplands

This broad region is the border line between the northern and southern United States and is sometimes referred to as the Upper South. The specific area referred to in this chapter includes the limestone valley and dissected plateaus of Tennessee and Kentucky, along with the Cumberland Plateau and mountains in Kentucky and Tennessee and the Ozark regions of Arkansas and Oklahoma. In addition, the sandstone and shale hills and valleys of northwestern Kentucky are included.

Because of its great variability in topography and geologic formation, soils vary widely in depth, texture, and innate fertility. In much of the area, soils are low in major fertilizer elements, with P being very deficient; but in the Kentucky Bluegrass region, native soil P is high.

II. N FERTILIZATION PRACTICES OF IMPORTANT CROPS

A. Forages

Because of the large number of livestock, forages are perhaps the most important commodity produced in the region. Both ranges and improved pastures comprise vast land areas of the South Central states.

1. RANGES

In the Southern Great Plains, > 50% of the agricultural land area is rangeland and much range is also present in other parts of the South Central states, which is mostly mixed with brush and forest. However, very little of the rangeland is fertilized. Yields are low and N fertilizer can increase weed competition. Some ranges and native pasture are fertilized with N to improve pasture grass utilization. Native grass meadows cut for hay do receive fertilizer applications. Nitrogen rates vary from 40 to 60 kg/ha.

2. IMPROVED PASTURES

Improved forage species are more efficient in utilizing N than native species and have higher forage potentials. As pressures for more food production increase, ranges will continue to be replaced by more productive species, and these improved grasses will have higher N requirements.

a. Warm-Season Pastures—Bermudagrass [*Cynodon dactylon* (L.) Pers.] is the predominate improved warm-season pasture grass in the region. Varieties vary depending primarily upon winter temperatures. However, weeping lovegrass [*Eragrostis curvula* (Schrad.) Nees], kleingrass (*Panicum coloratum* L.), buffelgrass (*Cenchrus ciliaris* L.), Asiatic bluestem (*Bothriochloa* spp.), bahiagrass (*Paspalum notatum* Fluegge), dallisgrass (*Paspalum dilatatum* Poir.), and johnsongrass [*Sorghum halepense* (L.) Pers.] are among other species used extensively for forages in some parts of the region.

Each of the improved warm-season grasses respond to N fertilization commensurate with yield potentials. One kilogram of N will produce from 15 to 25 kg of dry forage for most of these species. This quantity of forage can add 1 kg of weight to a beef animal with average management. Therefore, 1 kg of added N is expected to produce 1 kg of beef within the responding range of the yield curve.

Much research during the past half century has shown bermudagrass to respond very dramatically to N fertilization. The amount of N fertilizer needed depends upon yield potentials. Bermudagrass requires approximately 25 kg N/t of forage.

Bermudagrass pastures can utilize 20 to 30 kg of N monthly during hot weather if moisture is adequate. Therefore, for grazing, these rates are often applied at the beginning of spring growth and continued as long as soil moisture is available. In much of the area, 3 to 4 dressings will be made each year. Where harvest by haying is intended, single doses from 110 to 225 kg/ha are recommended.

Kleingrass and buffelgrass are grown in the southern portions of the region where, with the longer growing season, higher yields are obtained. Therefore, N rates up to 170 kg N/ha are not uncommon. For weeping lovegrass and Asiatic bluestem, lesser quantities of N are usually applied, ranging from 35 to 100 kg/ha in the Southern Great Plains. In the Coastal Plains where johnsongrass, dallisgrass, and bahiagrass is grown extensively, it requires from 100 to 225 kg N/ha to produce yields ranging from 5000 to 9000 kg/ha of dry forage. The N requirement per unit of dry forage does not vary markedly between perennial warm-season species. The higher the potential yield for a species, the higher the N requirement.

Major problems with N use on improved warm-season pastures in the South Central states are (i) the capability of producing large quantities of forage whenever soil moisture is adequate and (ii) the slowing down of production under moisture stress. It is generally impossible to stock pastures to adequately utilize the forage during rapid growth. Warm-season grasses, like bermudagrass and weeping lovegrass, deteriorate in quality unless grazed closely. Therefore, in the management scheme, rotational grazing and haying become necessary with a fertilization program.

b. Cool-Season Pastures—Cool-season forages consist mainly of fescue (*Festuca arundinacea* Schreb.) and small grains; however, in the East Central Uplands, bluegrass (*Poa pratensis* L.) occupies a significant land area in some parts of the area.

Each kilogram of N will increase cool-season dry matter production from 20 to 25 kg. Cool-season grasses are high-quality forages tending to be high in protein.

Since production and quality of cool-season grasses (fescue and bluegrass) is greatest in the spring and fall, this is when most of the N is added. A topdressing in late winter or early spring—just before growth begins—will increase growth so that grazing can begin about 2 weeks earlier than pastures receiving no N. This effect can be used as a means of lowering over-wintering feed costs. Another topdressing in late spring following a graze down or hay clipping will stimulate an increased amount of growth and will carry over into the low-production period of summer. Another topdressing following a graze down or clipping in late summer will stimulate fall growth, which, if properly utilized, will extend grazing several weeks later into the fall and winter. This is a practical way of lowering winter feed costs.

The amount and time of N additions are determined by how much and when the production is needed in the overall feed production system. As much as 225 kg N/ha are added throughout the growing season with only 55 to 100 kg added at each topdressing.

The principal sources of N for pastures are ammonium nitrate (AN), and N solutions. However, the use of urea is increasing.

There is some concern about N losses from topdressing forages with urea during high temperatures, especially on dense sods. Nevertheless, urea is used for pasture fertilization with obvious success. Most research shows urea to be from 80 to 95% as effective as AN on pasture sods. Many research experiments have been conducted comparing sulfur-coated urea (SCU) with other straight N sources. Even though SCU does help in more even production during the growing season, more total forage is not produced. Consequently, the additional cost is not warranted.

Anhydrous ammonia is used to some extent on sod crops. There is a lag in production when NH_3 is applied during the growing season. This lag is apparently due to NH_3 "burn" of the roots, which must recover for growth to resume. Cold-Flo® NH_3 has been tried on sod crops in the region with varying results. It is theoretically easier to apply, because it can be applied shallower for retention. The use of NH_3 on sods is not expected to increase significantly, due to the power requirement and difficulty in accomplishing proper injection for full retention of the NH_3.

c. Grass-Legume Mixtures—There is renewed interest in pasture improvement in the South Central states by the establishment of legumes, mainly clovers (*Trifolium* spp.), in grass pastures. This greatly improves yield and pasture quality and reduces the need for N fertilization.

Legume forages most frequently grown with bermudagrass in the South are white (*T. repens* L.), crimson (*T. incarnatum* L.), and arrowleaf (*T. vesiculosum* Savi) clovers, and hairy vetch (*Vicia villosa* Roth). Good stands of these legumes growing in bermudagrass will provide from 55 to 110 kg N/ha to the bermudagrass. Additional N is generally required for full production.

In the northern portions of the South Central states, particularly in the East Central Uplands where fescue and bluegrass are the predominate cool-season grasses, clovers such as red (*T. pratense* L.), ladino (*T. repens* L.), and alsike (*T. hybridum* L.) are grown in the sward. These clovers provide much of the N needed and add to the quality of forage and extend the grazing season.

One of the major problems with grass-legume mixtures in the South Central states is keeping adequate stands of legumes in the sward. The grasses persist after the associated legume has been killed by competition, drought, or disease. When this happens, reseeding the legume may be necessary.

Nitrogen use on improved pastures in the South Central states has not nearly reached full potential. In spite of large acreages of properly managed pastures, much of the area pasture land is underfertilized; thus, it has rather low production.

3. HAY CROPS

Most improved pasture grasses grown in the region, along with native species, are used for hay.

a. Bluestems—Most native grass hays consist of bluestems. These bluestem meadows are now generally fertilized with N. Because only one cutting is taken, the production is low and N rate varies from 30 to 60 kg N/ha.

b. Bermudagrass—For bermudagrass, whenever harvest by haying is intended as contrasted with grazing, higher rates of N are added at the beginning of the growing season. Common rates of N for bermudagrass for hay production range from 125 to 200 kg/ha.

c. Cool-Season Forages—For cool-season forages such as fescue, rates per cutting for haying are less than for bermudagrass, ranging from 80 to 100 kg N/ha.

d. Forage Sorghums and Sudans—In the Southern Great Plains, forage sorghums and sudans are frequently grown as *opportunity forages,* i.e., needed forage is produced whenever sufficient soil moisture is present and forage is vitally needed. These species produce large yields in short periods of time and make high-quality hay and/or ensilage.

There are two major problems associated with N fertilizer use on these crops.

1) N fertilization tends to increase concentrations of prussic acid (HCN). Prussic acid is toxic to grazing animals. Concentrations in the forage $>$ 1000 mg/kg are nearly always lethal to grazing cattle.
2) N fertilization tends to increase concentrations of nitrates. Excessive nitrates can accumulate in the forage and be toxic for grazing animals, and it persists in the hay. Nitrate concentrations $>$ 6000 mg/kg can be lethal to any type of cattle.

Both nitrates and prussic acid are generally associated with droughty conditions or other conditions that induce plant stress.

In spite of the above problems, the production of forage sorghums and sudans requires from 20 to 30 kg of N to produce a tonne of oven-dry forage. Common N fertilization rates in the area are from 67 to 100 kg N/ha, with NH_3 being the most common source.

B. Wheat (*Triticum aestivum* L.)

Wheat yields have increased substantially during the last 20 yr and it is no coincidence that yields have increased simultaneously with increased fertilization. Wheat requires approximately 30 kg of available soil N for each 1000 kg of grain produced. In the Southern Great Plains, much of the wheat is grazed during the fall and winter and additional N (about 36 kg/1800 kg of forage on a dry-wt basis) is required to satisfy the grazing requirement. Therefore, N rates depend upon expected yield goals of both grain and forage. Since moisture is usually the limiting factor in grain production, N rates vary from 110 to 135 kg/ha in the more humid eastern areas to essentially none on the western fringes where yields are low. The major problem in selecting N rates is the widely fluctuating yields from year to year, primarily a function of available soil moisture.

Anhydrous ammonia is by far the major N source for wheat production in the Southern Great Plains, being applied preplant in July, August, and September. Ammonium nitrate, urea, and N solutions are used for topdressing whenever NH_3 is not applied or whenever improved crop prospects during the season warrant additional N. Anhydrous ammonia is used because of the lower cost per unit of N. A large portion of the NH_3 is applied while sweeping or chiseling for tillage purposes—saving a field operation.

Irrigated wheat in the Southern Great Plains has increased substantially as a result of the introduction of semidwarf varieties of hard red winter wheats. These cultivars possess high yield potential. They respond well to water and N without being so prone to lodging as previously grown varieties. Yields up to 6750 kg/ha have been produced in addition to considerable grazing production. Nitrogen rates from 135 to 180 kg/ha are not uncommon. The measure of preplant NO_3^--N in the top 60 cm of the soil profile is a great aid in determining the fertilizer requirement.

In the Reddish Prairie, continued relatively high N fertilization on some soils has resulted in excess soil acidity. Low soil pH in the surface (down to 4.0) has been found in spite of an accumulation of $CaCO_3$ in the lower horizons of the soil profile (varying from 50 to 150 cm). Agricultural lime applications are required to prevent yield reductions.

C. Cotton (*Gossypium hirsutum* L.)

The N requirement of cotton during heavy fruiting is large, but excess N during maturity results in poor quality and even reduced yields. In addition, excess N prior to fruiting can cause excessive vegetative growth, which interferes with insect and harvest management. This makes N management for cotton production difficult at best.

In the Mississippi Delta, efforts to develop techniques to properly manage cotton include petiole NO_3^- analyses during fruiting. By testing nitrates of cotton petioles beginning near first bloom and continuing through about the ninth week of blooming, N can be adjusted to needs during fruiting as affected by varying seasonal conditions. The purpose is to avoid excessive N and prevent N deficiency during crucial heavy fruiting.

In some programs, if petiole N is excessive early in the season, foliar boron (B) is recommended to help translocate N and sugars and enhance fruiting. When the petiole tests show N deficiency, soil N is recommended during early bloom, but after the third or fourth week of blooming, foliar N applications are recommended. Recommendations for small amounts of sugar (sucrose) have been suggested in late season if, in spite of B application, excessive N in the petiole persists.

The petiole NO_3^- program seems to work better in the Mississippi Delta and Coastal Prairie region and far west than in the Southern Great Plains, probably due to the longer growing season.

General N rates suggested for cotton in the Mississippi Delta varies from 55 to 85 kg N/ha. Anhydrous ammonia is the most common preplant N source.

The N requirement for dryland cotton is relatively high and soils on which cotton is grown in the Southern Great Plains are more often deficient in N than any other plant nutrient. Even though high yields of cotton require large amounts of available N, cotton plants do have the capability of using small amounts of N very efficiently A bale (225 kg) per 0.5 ha is a good dryland yield in the Southern Great Plains. This yield requires approximately 30 kg of N.

Irrigated cotton in the Southern Great Plains receives from 60 to 125 kg N/ha. This is less than amounts used previously. Cotton requires adequate but not excessive amounts of N early in the seedling to squaring stage, but large amounts of soil N are desirable during rapid fruiting; for best results, it should be depleted by maturity. Excess N tends to keep cotton in the vegetative stage and interferes with proper maturity. This presents special problems in N management on irrigated cotton. It would appear that split applications or fertigation would aid production, but attempts to improve N nutrition by these means have been disappointing. Measuring accumulated residual NO_3^- in the upper 60 cm of the soil profile aids growers in avoiding excessive N applications. Anhydrous ammonia is the most common N source used on cotton in the area.

Cotton N requirements in the Gulf Coastal Plains are similar to those described for the Southern Great Plains under dryland conditions, except yield potentials are higher. Rates of N commonly used vary from 35 to 70 kg/ha. In this area, a much greater portion of the N is applied in combination with P and K. In addition, split N applications are more common due to risks from leaching below the root zone. Here again, AN and N solutions are the most used sources.

D. Corn (*Zea mays* L.)

Corn is grown throughout this region. In the East Central Uplands, yields are good and the crop is usually fertilized to exceed a yield of 6300 kg/ha. Most N is applied at planting and the most common rates are 140 to 200 kg/ha, with rates as high as 225 kg/ha on high-producing, poorly drained soils and irrigated systems.

Excessive soil moisture is common prior to June. This causes N losses on moderately to poorly drained soils in the range of 30 to 55 kg/ha. Some farmers delay the application of most or all of the N until 4 to 6 weeks after planting. This prevents the early loss and reduces the total amount of N needed by an average of 40 kg/ha.

Nitrification inhibitors have been found to be quite effective in preventing much of the early season N loss in no-tillage corn. This is helpful to farmers who are unable to delay N application when no-tilling the imperfectly drained soils where N losses are highest.

Anhydrous ammonia, N solutions, urea, and AN are all common sources of N used in both conventional and no-tillage corn. Anhydrous ammonia is less popular with no-tillage because of sealing difficulties. Ammonium nitrate is preferred over urea due to the risks of N loss by volatilization from surface-applied urea. Some farmers prefer N solutions, because herbicides can be mixed with the fertilizer and applied in one application.

Yields are variable, ranging from very good to rather low on corn grown in the Mississippi Delta and Coastal Plains area. Relatively low water-holding capacities of many Coastal Plains soils causes severe soil moisture stress in some years. Nitrogen requirements for corn for various yield levels are the same as for other corn-growing regions, but little innate soil N is available.

Considerable interest is developing in no-till corn. With these systems, N requirements are increased. This is especially true for corn planted in unkilled sod. When corn is seeded in unkilled bermudagrass sod, competition for N from the bermudagrass is great, even though the corn may be depressing bermudagrass growth by shading. Both the growing grass and decomposition of organic residue compete with the corn for N. Therefore, adequate N fertilization is critical for optimum production.

No-tillage corn differs from conventional corn in two ways.

1) Due to greater amounts of leaching, denitrification, and immobilization, more N (30–55 kg/ha) is lost or rendered unavailable for plant growth. Much of this loss is avoided by delaying the application of part or all of the N until 4 to 6 weeks after planting.
2) Rapid acidification of the surface 5.1 cm of soil. Surface application of N in no-till systems causes development of very acid surface soils. These very acid surfaces not only interfere with plant nutrition, but certain herbicides are inactivated. This problem can develop rather rapidly so that more frequent and systematic soil testing is necessary. The amount and frequency of liming will not necessarily be increased, but one must be more cognizant of the soil pH. Soil testing should be frequent and the sampling depth should be more than 10 cm.

Corn is grown under irrigation in the Southern Great Plains for both ensilage and grain. The N requirement closely parallels N use in other corn-growing regions. Estimated N requirements range from 18 to 26 g N/kg grain or 33 g N/kg forage. After a yield goal has been established, the preplant soil NO_3^--N in the 60-cm soil profile is subtracted from the total N requirement to determine fertilizer applications. For full production, i.e., favorable conditions, N application rates of 180 to 270 kg N/ha as NH_3 are common.

E. Grain Sorghum [*Sorghum bicolor* (L.) Moench]

Yield levels of 2250 to 4500 kg/ha are normal dryland grain sorghum yields in the Southern Great Plains. Estimated N requirements range from 18 to 26 g N/kg grain. Nitrogen fertilizer rates commonly used under dryland vary from 18 to 60 kg/ha. Under irrigation, yields up to 9000 kg/ha are achieved, which requires 240 kg N/ha. The N requirement for grain sorghum is somewhat less than corn, due to its more extensive fibrous root system and the lower grain/forage ratio.

There are no special N fertilizer problems on dryland grain sorghum in the area. Anhydrous ammonia applied preplant is the most commonly used N source.

F. Soybeans [*Glycine max* (L.) Merr.]

The acreage in soybeans has increased rapidly and high yields are common. Little or no N is used on soybeans and, if used, is generally restricted to a few kilograms per hectare in the starter fertilizer. This starter fertilizer often consists of 1-4-4 and 1-5-5 grades. Only 10 to 35 kg/ha of N are usually added. The application of small amounts of N is probably most common on soybeans planted in a double-cropped system after small grain. Although yields are not usually increased, the quick early growth and canopy closure is desired for weed control.

In areas where lime is quite expensive, Mo is added when the pH drops below 6.2. This stabilizes yield by maintaining the level of N fixed from the atmosphere by the rhizobia bacteria. As the pH drops below 5.5, the yield may begin to decrease regardless of Mo additions.

Efforts to increase soybean yields by foliar fertilization late in the season have been disappointing.

G. Legumes

Principal legumes grown in the region are alfalfa (*Medicago sativa* L.), peanuts (*Arachis hypogaea* L.), and pasture legumes such as vetch and clovers. The only N used on alfalfa and peanuts is a small amount in the starter fertilizer (10-20 kg/ha). Principal starter fertilizers are ammonium phosphates and complete mixes or blends. Sporatically, there are reports of large yield increases from applications of larger quantities of N. The conditions are nearly always traced to a lack of nodulation. Much research has been conducted on efforts to improve production with N fertilization. These efforts have been centered upon late-season applications, deep placement, slow-release materials, and even foliar application. The probability of response has been too low to cause adoption of the practice.

With legume-grass pasture combinations, N applications favor grass competition, but additional N is generally needed for optimum forage production. The problem of adjusting N rates to ensure adequate grass production while keeping a desired proportion of legumes has not been solved.

H. Tobacco (*Nicotiana tabacum* L.)

Tobacco is the most important cash crop in the East Central Uplands, and no expense is spared on fertilization. Common rates of N use are 280 to 335 kg/ha on burley and 170 to 225 kg/ha on dark tobacco. The N is usually applied just prior to or at transplanting time. On soils where high amounts of N loss is a good possibility due to excessive spring moisture, some farmers will split the N application. About 50% will be applied at transplanting and 50% 2 to 3 weeks later. The most common source of N is AN. However, all of the commonly available N sources are used with good results.

The greatest problem of N use with tobacco is the precipitous pH drop, which takes place with the use of such large amounts of NH_4^+-N. It is not uncommon to experience a drop of one complete pH unit during the growing season. This usually increases Mn uptake and reduces Mo availability and yield if the fields do not have a pH of $\geq$ 6.4 prior to transplanting. In addition to proper liming, some growers use NO_3^- forms of N, such as calcium nitrate to help reduce the pH change.

I. Rice (*Oryza sativa* L.)

Rice is grown in the Mississippi Delta and Gulf Coastal Prairies. Both proper rate and timing of N fertilizer are essential for high rice yields. Excessive quantities of N induces lodging and results in a more favorable environment for diseases. Rice uses NH_4^+ forms of N, which are stable under flooded soil conditions. Ammonium N can be oxidized to NO_3^--N in a nonflooded soil and subsequently be denitrified under reflooding. Therefore, the NH_4^+-N should be applied while the soil is flooded or just prior to flooding. Approximately 50% of the N needed for rice is for vegetative growth. To meet this requirement, N can be applied preplant and incorporated just ahead of seeding or broadcast on dry soil prior to flooding (usually about 3 weeks after seedling emergence). The additional N needed for grain development and production is applied during midseason. Either one or two applications is made soon after panicle initiation and internode elongation. Usually one application suffices if the total N requirement is < 45 kg/ha. Whenever two applications are necessary the second application is applied about 2 weeks after the first.

Total N requirements for rice vary from 110 to 155 kg N/ha, with 50% being applied near seeding and the other 50% split in two midseason applications.

There is a growing trend to base N applications on growing degree days. For rice, growing degree day accumulations are the number of degrees that the daily mean temperature exceeds 10°C. Midseason N should be applied commensurate with internode lengths, which vary for each variety. The growing degree days predicts fairly accurately the internode length for each variety.

Because of the need for NH_4^+-N on rice, urea and NH_3^- are the common sources, with ammonium sulfate being popular in some areas.

Nitrification inhibitors are useful in rice production if the NH_4^+ fertilizer is applied in advance of flooding; otherwise, no benefits from these inhibitors have occurred.

J. Sugarcane (*Saccharum officinarum* L.)

Sugarcane is an important crop in the lower Delta and Coastal Prairies. Sugarcane is a tropical plant that produces a large amount of dry matter. It demands high amounts of N and water. During rapid growth it can utilize

4.5 to 7 kg N/ha per day. However, as harvest approaches, it is desirable to have much of the soil N depleted. Excessive N limits the conversion of reducing sugars to recoverable sucrose. Therefore, sufficient N must be present to ensure rapid growth during the early growth stage and to ensure continued growth for a period of several months. However, N rates must be commensurate with expected uptake during the growing period so as not to leave substantial quantities of available residual N during maturity. In an effort to ensure low soil levels of N near maturity, applications are often split. The last application may be omitted if conditions dictate. Many growers apply 65 to 110 kg N/ha after harvest or at establishment to ensure rapid growth of the ratoon crop and side-dress if additional quantities are needed. Usually two-thirds to three-fourths of the N is applied at planting in 2-yr cropping with the balance being applied at "closing in" by airplane if dry land, and by fertigation if irrigated.

The total N requirement for sugarcane is from 3 to 4 kg/t of dry matter. Dry-matter weight varies widely, but yields of 7 t/ha are common. Total N fertilizer rates average from 200 to 270 kg N/ha, but very high yields require much higher N rates.

Anhydrous ammonia is a common source of N for preplant applications with AN being the most widely used side-dressing source.

K. Forest

Tree growth in the Gulf Coastal Plains is quite rapid, because of favorable soils and climate. Commercial timber production is vital to the economy of the region. Large tracts of company-owned land, as well as individually owned small land areas, are devoted to commercial woodlands. Pine tree (*Pinus* spp.) production accounts for the majority of forested products.

As a general rule, pine seedlings do not require N fertilization, but older stands respond well to N additions. The general recommendation is to apply the first N application as soon as the crown closes (solid canopy), which is usually about the eighth year after transplanting.

The traditional way is to apply the N by helicopter. Because of the necessity of aerial application, urea is the source used. The loss from volatilization apparently is not large under the tree canopy, because it holds soil temperatures down.

Rates of N vary from 70 to 180 kg/ha per year. Some commercial lumber companies apply N every 3 to 4 yr. By applying a 3- to 4-yr supply in one dose, tree growth may be adequately stimulated to last for a while, and this procedure eliminates the very expensive application costs each year.

Phosphorus is the most critically deficient element for pines in the Coastal Plains. Best results from P applications have been obtained with ammonium phosphates, with diammonium phosphate being the commonly applied ammonium phosphate. The use of ammonium phosphates as the P source results in the simultaneous application of considerable quantities of N.

III. MAJOR PROBLEMS AND CONCERNS FOR N FERTILIZER USE

Because forages comprise so much of the land area of the South Central states, fertilizer use is tied to that crop. One of the major problems facing cattlemen of the region is oscillating cattle prices. During periods of relatively high cattle prices, it becomes profitable to use pasture management schemes that maximize production. This usually includes applications of N fertilizer at rather high rates, but during relatively low prices, production costs must be reduced. Frequent changes in optimum N rates makes planning fertility programs difficult.

Another problem facing cattlemen in the region is the necessity to alter forage production management resulting from short periods of very high production caused by fertilizer application. This is particularly true with bermudagrass pastures, where very high yields are obtained with high N rates whenever temperatures are high and soil moisture is adequate. Later in the season, forage production usually decreases due to moisture stress. Stocking rates for the peak production period will be excessive as production declines, and stocking for the lower production periods leaves too much growth and poor quality forage during peak growth.

A major problem with N fertilizer use in the region is soil acidification resulting from the continued use of the ammoniacal form of N. This is a special problem on pastures and no-till crop production. Once acidity develops in a no-till or pasture system, it cannot be corrected with only surface applications of lime. Increasing soil acidity is certainly a high-priority concern for the future, because aglime applications are not keeping pace with the needs resulting from cropping intensification.

Anhydrous ammonia is a major fertilizer source for much of the area. Concerns about the hazards in transportation and handling are threatening its continued use. Its lower price per unit of N is an economic stimulus to crop production, which, if eliminated, would further enhance the cost-price squeeze facing farmers of the region. Research and development is vitally needed on the application of NH_3 during other essential tillage or field operations in order to reduce energy requirements.

Research and development is needed in several other subject matter areas in order to help farmers and ranchers in the South Central states. These include:

1) Improved management techniques for N application on crops like cotton and sugarcane. On these crops, adequate N is essential for early growth, followed by a high demand during rapid growth and/or fruiting; but best final results are obtained if the N is depleted for maturing the crop. Presently, satisfactory economical management schemes are not available to adequately cope with the problem.
2) Much of the fertigation being done in the region is based upon trial and error as to compatibility of sources and amounts with the quality of the irrigation water. In addition, N losses under different fertigation schemes have not been defined.

3) Excessive NO_3^- in forages and associated problems induced by N fertilization continues to reduce farmer and rancher profits. Very little basic information concerning plant physiological processes responsible for accumulations of NO_3^- is known.
4) In spite of much research effort, it is difficult to adjust N and other fertilizer rates adequately to keep the desired proportion of grasses and legumes in mixed pasture swards. The whole picture of N fertilization of legumes needs clarification. Presently there are many gaps in our knowledge on the effects of N fertilization of legumes. This is especially true of crops such as soybeans, where N may be the limiting production factor.
5) As natural gas prices increase, the price of N may rise faster than the price of grain and hay. This change will demand more efficient use of N. Research methods for increasing N efficiency by all types of methods will be needed. Simple things such as the refinement of the most economic rate of N application by crop, soil type, etc. will be very helpful.

In conclusion, it should be pointed out that for parts of the South Central states many farmers and ranchers are using rates of N and other fertilizers that are too low for maximizing net returns. Innovative educational programs are needed to create an awareness and provide information that results in improved changes in soil fertility programs used on the land.

MONTANA
IDAHO
WYOMING
MOUNTAIN
NEVADA
UTAH
COLORADO
ARIZONA
NEW MEXICO

51

D. G. Westfall
Colorado State University
Fort Collins, Colorado

Management of Nitrogen in the Mountain States

The management and use of N in the Rocky Mountain states varies considerably, because the geographic area is very diverse. The agriculture production area varies in elevation from about 42 m (140 feet) along the Colorado River in Arizona to over 3000 m (10 000 feet) in the mountain meadow areas of the Rocky Mountains; annual rainfall varies from 10 cm to > 76 cm. Great land areas of native grass, sagebrush (*Artemisia tridentata* Nutt.), timber, and improved grassland occur throughout the Mountain states area and the area of agricultural land is relatively small. Alfalfa (*Medicago sativa* L.), corn (*Zea mays* L.), sugarbeet (*Beta vulgaris* L.), small grain, cotton (*Gossypium hirsutum* L.), potato (*Solanum tuberosum* L.), vegetable, and citrus crops are grown under irrigated conditions—both furrow and sprinkler. Alfalfa, small grains, and corn are grown in all these states, while other crops are more area specific. For example, citrus and cotton production is isolated in areas where the climate is warmer, and sugarbeet production is isolated in areas close to processing plants. Fruits such as apples (*Malus sylvestris* Mill.), peaches [*Prunus persica* (L.) Batsch], pears (*Pyrus communis* L.), apricots (*Prunus armeniaca* L.), and cherries [*Prunus avium* (L.) L.] are grown in isolated areas of Colorado, New Mexico, Idaho, and Utah and contribute significantly to the local economy.

Wheat (*Triticum aestivum* L.) is the only major crop grown under non-irrigated (dryland) conditions in this region, although barley (*Hordeum vulgare* L.) is an important dryland crop in Idaho and Montana. Dryland small grain production is generally restricted to rainfall areas of 33 to 68 cm annual precipitation. It is generally grown under summer-fallow cropping, where the land is left uncropped and maintained weed-free for one growing season. This facilitates the accumulation and storage of moisture in these low-rainfall areas and also provides a time period for mineralization of plant nutrients. The major summer-fallow area is the western edge of the

Nitrogen in Crop Production.

Great Plains that includes the eastern part of Montana, Wyoming, Colorado, and New Mexico, although other small isolated areas exist in the Mountain states. A small land area of safflower (*Carthamus tinctorius* L.), grain sorghum [*Sorghum bicolor* (L.) Moench], sunflower (*Helianthus annuus* L.), and corn is also grown under summer-fallow conditions.

On the east edge of Palouse area in northern Idaho, where the annual rainfall is > 51 cm, winter wheat, barley, and peas (*Pisum sativum* L.) are produced under nonirrigated annual cropping. The wheat yields are considerably greater in the Palouse area, averaging about 4000 kg/ha (60 bu/acre), with yields in excess of 6700 kg/ha (100 bu/acre) not unusual. In the western Great Plains area, yields average about 2000 kg/ha (30 bu/acre).

I. N SOURCES

The diversity in crop production leads to a range in N-source usage. This varied usage is related to source advantages from the agronomic and economic, as well as the application and availability standpoint. Tradition also plays a major part in farmer preference for various N sources.

The use of various N fertilizer materials as direct application in the Rocky Mountain states from 1955 to 1980 is shown in Table 1. These figures represent about 75 to 80% of the total N fertilizer use and do not include the use of mixed fertilizer materials, which would be the remaining 20 to 25%. In 1955, ammonium nitrate (AN) and ammonium sulfate (AS) were the dominant N sources in every state except New Mexico, where 52% of the N applied was anhydrous ammonia (AA). By 1980, AA and N solution usage had increased substantially, while there was a general trend of decreased percent use of AN and AS. It is interesting to note that in the northern states of Montana, Idaho, and Wyoming, AN is still one of the most widely used N sources in 1980, while in the southern states of New Mexico and Arizona, AA and N solutions represent the most widely used N sources. The exact reason for this trend is not known, but it is believed to be related to source availability, transportation costs, marketing pressures by fertilizer companies, and farmer tradition.

Table 1—Trend in the use of selected direct application N fertilizer sources in the Mountain states, 1955–1980.†

N Fertilizer	1955	1960	1965	1970	1975	1980
	Total direct application (%)					
Arizona						
Ammonium nitrate	15	10	7	7	6	2
Anhydrous ammonia	14	33	30	40	38	26
Aqua ammonia	10	12	9	8	8	8
Nitrogen solutions	9	7	18	22	27	44
Urea	13	18	23	16	17	16
Ammonium sulfate	39	20	13	7	4	4

(continued on next page)

Table 1—Continued.

N Fertilizer	1955	1960	1965	1970	1975	1980
	Total direct application (%)					
Colorado						
Ammonium nitrate	54	54	57	35	28	20
Anhydrous ammonia	20	21	15	39	37	31
Aqua ammonia	0	0	0	<1	<1	<1
Nitrogen solutions	0	9	8	15	18	36
Urea	4	5	7	1	3	4
Ammonium sulfate	22	11	13	10	14	9
Idaho						
Ammonium nitrate	35	39	24	23	23	20
Anhydrous ammonia	7	10	8	11	16	17
Aqua ammonia	3	23	11	3	5	4
Nitrogen solutions	<1	3	14	14	17	18
Urea	<1	1	6	7	6	10
Ammonium sulfate	55	24	37	42	33	31
Montana						
Ammonium nitrate	68	74	65	73	73	49
Anhydrous ammonia	5	18	10	7	9	11
Aqua ammonia	0	0	0	1	<1	1
Nitrogen solutions	0	<1	16	12	3	8
Urea	1	<1	3	<1	5	18
Ammonium sulfate	26	6	6	6	9	13
Nevada						
Ammonium nitrate	2	16	19	17	6	12
Anhydrous ammonia	4	47	39	32	17	8
Aqua ammonia	0	0	0	0	0	0
Nitrogen solutions	0	5	1	<1	8	45
Urea	3	4	10	22	3	7
Ammonium sulfate	91	28	31	28	65	28
New Mexico						
Ammonium nitrate	21	14	10	6	7	5
Anhydrous ammonia	52	33	33	43	47	37
Aqua ammonia	0	0	0	1	<1	0
Nitrogen solutions	0	16	13	24	26	38
Urea	17	15	20	18	4	10
Ammonium sulfate	10	22	24	8	5	9
Utah						
Ammonium nitrate	47	67	37	42	58	45
Anhydrous ammonia	14	2	15	14	13	7
Aqua ammonia	0	0	0	<1	0	0
Nitrogen solutions	0	<1	<1	5	5	3
Urea	1	2	7	4	5	6
Ammonium sulfate	38	29	40	35	19	39
Wyoming						
Ammonium nitrate	53	53	73	66	72	43
Anhydrous ammonia	13	14	12	3	8	21
Aqua ammonia	0	0	0	0	<1	<1
Nitrogen solutions	0	17	6	26	16	29
Urea	15	7	<1	<1	4	7
Ammonium sulfate	19	9	8	4	1	<1

† Hargett (1976) and USDA (1980).

In general, aqua ammonia is used in the smallest quantity with many states not reporting any sales. The use of other N solutions has increased in recent years. This is partly due to recent availability of liquid handling equipment that reduces labor requirements as well as farmer acceptance of N solutions and their adaptation to production programs. Nitrogen solution usage is becoming more popular under sprinkler irrigation, where it can be injected directly into the sprinkler system during irrigation, thereby saving the cost of ground application. Solutions are also used to "spoon feed" crops where small applications of N during the growing season are thought to be beneficial or where leaching losses can occur on light-textured soils.

As the cost of fertilizer increases, it is felt that the use of AA will increase due to its lower price. Some major fertilizer manufacturers are trying to direct consumers away from AA, because of safety problems and liability in case of accidents. Some fertilizer dealers are persuading farmers to use more solutions because of its ease of handling, use in irrigation systems, and potential for use in weed-control programs. Farmers are showing some resistance because N solutions are generally more expensive per unit of N.

II. N FERTILIZER RECOMMENDATIONS

The major objective of this chapter is to identify the N management practices used in this region. An efficient N management program requires that the proper rate and method of application be used for the crop and conditions under which it will be grown. This can only result from soil testing and proper N fertilizer recommendations that are actually used by farmers. Residual soil NO_3^--N testing is the basis by which most states in the Mountain region make N fertilizer recommendations. Previous cropping history and soil organic matter input are also generally considered to be important. A summary of the basis upon which each state university makes N fertilizer recommendations is given in Table 2. Many private soil testing laboratories operating in the region follow the N recommendation guidelines given by universities, but higher recommendations are not uncommon by some private laboratories. Colorado, Idaho, Montana, New Mexico, and Wyoming use what is commonly referred to as the "N budget" in determining N fertilizer recommendations. This involves the process of conducting an inventory of all N inputs and requirements of production. The inputs include residual soil N in the profile to the effective depth of rooting, N mineralization of soil organic matter, effective N carry-over from previous legume crops, and N contributions from manures. The production needs include N requirement of the crop and, in some states, an additional input of N needed if straw or nonlegume residue is plowed down.

Idaho is the only state that includes NH_4^+-N in the N budget. Research in other states, particularly Colorado, has shown that the inclusion of NH_4^+-N in the N budget equation does not increase the accuracy of the N fertilizer recommendation. Idaho researchers have found the NH_4^+-N input is usually very small, generally < 1 mg/kg (R. E. McDole, Univ. of Idaho, Moscow, personal communication). Significant levels of NH_4^+-N can occur

Table 2—Basis for N fertilizer recommendation by state in the Mountain states.

State	N Recommendation basis†	Adjustments
Ariz.	Soil test NO_3^--N for preplant N plus NO_3^- tissue testing in-season for additional application of N. Critical tissue levels established for all major crops.	Reduce N recommendation by amount of NO_3^--N applied in irrigation water.
Colo.	Soil test NO_3^--N (effective rooting depth) plus O.M. and yield goal. N Budget Equation: Total N requirement (based on yield goal) – Residual soil NO_3^--N (effective rooting depth) – N from O.M. (33 kg/ha × % O.M.) – N from manure – N from previous crop = N fertilizer recommendation (kg/ha)	Previous history — kg N/ha Alfalfa −56 Beans −34 Grass Legume −34 Manure −5.6/t
Idaho	Soil test NO_3^--N + NH_4^+-N (effective rooting depth) plus mineralizable N and yield goal or based on previous crop and yield goal. N Budget Equation: Total N requirement (based on yield goal) – Residual soil NO_3^--N + NH_4^+-N (effective rooting depth) – Mineralizable N + N required for residue incorporation ÷ 0.65 (efficiency of fertilizer N recovery) = N fertilizer recommendation (kg/ha)	Add 17–22 kg N/t nonlegume residue plowed down to max. of 56–90 kg N/ha, depending on area of state. Adjustment made on dryland wheat for stored soil moisture, add 22 kg N/ha if above average moist soil depth of 46 cm exists in spring, total fertilizer N not to exceed 62 kg/ha.
Mont.	Soil test NO_3^--N (in effective rooting zone) plus O.M. and yield goal. N Budget Equation: Total N requirement (based on yield goal) – Residual soil NO_3^--N (effective rooting zone) – N from O.M. (33 kg/ha × % O.M.) – N from manure ± N from previous crop = N fertilizer recommendation (kg/ha)	Previous history — kg N/ha Peas or beans −45 Alfalfa or clover −56 Legume plowed down −112 Straw or stover plow down +22/t Manure −9/t first year Manure −4.5/t second year Complex adjustment made on dryland wheat for stored soil moisture.
Nev.	Soil texture, management level, crop to be grown, climatic zone, and frost-free growing season.	Reduce N recommendation by 90 kg/ha first year and 45 kg/ha second year after plow down of alfalfa.
N. M.	Soil test NO_3^--N (effective rooting depth) plus O.M. and yield goal. N Budget Equation: Total N requirement (based on yield goal) – Residual soil NO_3^--N (effective rooting depth) – N from O.M. (33 kg/ha × % O.M.) – N from manure – N from previous crop = N fertilizer recommendation (kg/ha)	Previous history — kg N/ha Alfalfa −56 Beans −34 Grass Legume −34 Manure −5.6/t

(continued on next page)

Table 2—Continued.

State	N Recommendation basis†	Adjustments	
Utah	Crop to be grown and location in state	Complex adjustment made based on previous crop, previous N applied and soil texture. Subtract 6 kg N/t manure applied.	
Wyo.	Soil test NO_3^--N (effective rooting depth) plus O.M. and yield goal.	Previous history	kg N/ha
	N Budget Equation: Total N requirement (based on yield goal)	Alfalfa	−56
	− Residual soil NO_3^--N (effective rooting depth)	Beans	−34
	− N from O.M. (22 kg/ha × % O.M. for < 1828 m elev.)	Grass Legume	−34
	− N from manure	Manure	−5.6/t
	− N from previous crop		
	= N fertilizer recommendation (kg/ha)		

† O.M. = organic matter.

following application of ammoniacal fertilizers. Idaho also evaluates the contribution from organic matter as mineralizable N. This value is much larger than the credit given to soil organic matter (22 to 33 kg N/ha × % organic matter) by the other states. The mineralizable N input into the N budget ranges from as high as 168 kg N/ha (Painter et al., 1977) under sugarbeet production in southern Idaho to as low as 45 kg N/ha under wheat production (McDole et al., 1978) in northern Idaho. Idaho uses a N efficiency factor of 65% in their N budget equation. No other state in this region uses a N efficiency factor. It appears that the larger N contribution from mineralization used in Idaho is offset by the N efficiency factor; thus, all states using the N budget system would give similar N fertilizer recommendations.

Arizona considers the residual NO_3^--N soil test for preplant N recommendations, but relies heavily upon tissue testing to determine the need of additional in-season N (Table 2). On some shallow-rooted vegetables, such as lettuce (*Lactuca sativa* L.), they do not recommend preplant N because of leaching; all N is recommended in-season. Arizona subtracts any N applied in irrigation water from the fertilizer recommendation. No other state considers this potential source of N in making recommendations presently, but interest is being generated on this subject. Colorado is presently evaluating the effect of irrigation water N input on the N fertilizer requirement of onions (*Allium cepa* L.). In some areas of Colorado it can be significant, but generally this source accounts for < 24 kg N/ha-m (about 6.5 lb N/acre-ft) of water under deep well irrigation. The Yellowstone River at Huntley, Mont. contains about 4.8 kg N/ha-m (1.3 lb N/acre-ft) (V. A. Haby, Montana State Univ., personal communication).

The N fertilizer recommendations in Utah (Table 2) are made based on the crop to be grown and location in the state, with detailed adjustment being made depending on the previous crop, the previous year's N rate, and soil texture. Nevada uses soil texture, management level, climatic zone, and frost-free season as their basis for N recommendations, with an adjustment made for plow down of alfalfa.

The N budget is the most widely used method of making N recommendations in this region, with five of the eight states using this procedure. Under the short growing-season conditions that exist in most of this area, it is an effective method of making accurate N recommendations. Arizona has a longer growing season and therefore can effectively use tissue testing to evaluate the N status of various crops. Tissue testing has been tried on sugarbeets (Gilbert et al., 1981) in conjunction with the preplant residual soil NO_3^--N with some success, but it is not believed that it will gain wide acceptance in the area due to the short time available for sample collection, analysis, and application of in-season N in the short growing season. Tissue testing is used by some private consulting firms in states other than Arizona. The criteria used to evaluate N critical or sufficiency levels is generally not based on local data, and its applicability to local conditions is often questioned by university researchers.

III. N MANAGEMENT ON VARIOUS CROPS

The N management practices used on various crops in this region are diverse. An attempt has been made to summarize the management practices used, but large deviations from those mentioned do occur.

A. Small Grains

Wheat production occupies the largest land area in the Mountain states region. Spring and winter wheat are widely grown under summer-fallow cropping conditions, but some states have large areas of irrigated wheat. Dryland spring wheat is grown in northern areas where winters are too severe to allow the production of winter wheat. Ammonium nitrate is the major source of N fertilizer for wheat in Montana. Urea is also used with caution given as to its potential loss if it is left unincorporated on the surface of soils containing free calcium carbonate. Anhydrous ammonia is the most widely used source of N in Idaho, Colorado, and New Mexico. Nitrogen is applied during the fallow period, usually shortly before seeding.

In summer-fallow areas where stand loss is common due to inadequate fall and winter moisture or wind erosion, growers usually delay fertilizer application until spring. At this time, they evaluate their stand and stored soil moisture situation to determine if an investment in N is warranted. If it is, they broadcast dry fertilizer in the spring before jointing. Some liquid N is also used. Equal response to fall or spring applications will generally occur if the spring application is made before jointing.

Ammonium nitrate, anhydrous ammonia, urea, and urea-ammonium nitrate solutions have all been found to be equally effective N sources (Goos et al., 1981) if some simple precautions that have been known for some time are followed. Anhydrous ammonia should not be injected into very dry soil and should be applied 4 to 5 d before planting to prevent reduction in seed germination. However, the author has not observed a reduction in germination when AA was injected about 13 cm deep followed by planting within a few hours. Urea and urea-containing solutions should not be topdressed during hot weather, because volatilization losses can occur. Usually, spring surface applications made before mid-March will not result in volatilization losses. Surface applications of urea should not be made on fallow with heavy amounts of prostrate stubble, unless it is immediately incorporated.

In the intramountain areas of Idaho, Montana, and Utah, dryland wheat production is usually confined to the foothill areas that are too high or too steep for irrigation. In Idaho, AA is the most widely used source of N, while in Montana and Utah, AN dominates.

The N recommendations are primarily based on residual NO_3^--N levels, as determined by soil testing and yield goal, but other past production factors are considered by each state (Table 2). The amount of N commonly recommended for rainfall areas < 50 cm vary from 0 to 80 kg/ha for yield goals in the 2000 to 4000 kg/ha (30–60 bu/acre) range. The protein content has recently been found to be an excellent postharvest indicator of N nutrition of dryland winter wheat (Goos et al., 1982). If protein contents for a particular field have been < 11% for several years, yields may be limited by N deficiencies and N applications would undoubtedly increase yield and protein. When protein content is in the 11 to 12% range, yields may or may not have been limited by N deficiency. Application of N may increase yield, but probably will increase protein content. When protein content is greater than 12%, N deficiency probably was not a problem, but the addition of more N should increase protein without affecting yield.

The percentage of dryland growers that have their fields' soil sampled for N recommendations is relatively small and is estimated to be about 15%. It is further estimated that from 15 to 60% use the general fertilization guidelines suggested by universities. Obviously, there is a need for educational programs for dryland grain growers to inform them of opportunities for greater economic returns with proper N management.

Under irrigation, N rates range as high as 180 kg/ha. Split applications of N are common. Split applications can include preplant N, band-applied N at seeding, or broadcasted N before jointing. Under sprinkler irrigation, one or more broadcast applications are made, usually through the irrigation system using N solutions. Some growers inject N solutions into the irrigation ditch with furrow irrigation, but this practice is not commonly used or recommended, because of inefficiency due to N loss in tail (drainage) water. Injection of AA into irrigation ditches is also practiced by a very small percentage of the growers. Use of AA in this method can result in large losses if alkaline water containing high salt is used for irrigation. The acceptance of soil testing by irrigated wheat growers is much higher than dry land, probably approaching 60%.

Irrigated barley and oats (*Avena sativa* L.) are grown throughout the Mountain states. Their land area is considerably less than irrigated wheat, but nonetheless are important crops. The N management practices are similar to those used on irrigated wheat. Recommended N rates are usually lower for barley, due to lodging problems and a desire for lower protein content in malting varieties.

Malting barley is grown in isolated areas. Its production area is strictly dependent upon the brewing industry's desire to contract land for malting barley in "desirable" production areas. Brewing quality is somewhat environmentally controlled and the brewing industry has identified these "desirable" areas in Montana, Wyoming, Colorado, and Idaho. Consequently, they contract land only in these areas. Nitrogen management is critical in malting barley production. Barley with high protein content is unacceptable for brewing. Since high N rates result in high protein content, the amount of N applied is controlled by contractual agreement between the brewer and the grower. If the protein content exceeds 12.5 to 13.0%, the brewer will not accept the barley. The grower must then market it as feed barley, which sells for considerably less than brewing barley. This price premium for quality brewing barley is a motivating factor for growers to accept less than maximum yields in order to keep protein content sufficiently low to pass brewing standards. Price premiums are given to growers who produce malting barley with protein lower than the maximum allowed.

The N rates are determined by NO_3^--N soil testing, organic matter, etc. as outlined in Table 2. The original "N Budget" concept was developed for malting barley (Reuss & Geist, 1970; Geist et al., 1970) and has subsequently been adapted to several crops in the west. Nitrogen application rates for malting barley range from 0 to 135 kg/ha. Ammonium nitrate and urea are the major sources of N.

B. Corn (*Zea mays* L.)

Corn, for both grain and silage, is almost exclusively grown under irrigation in the Mountain states. Furrow and center-pivot sprinkler irrigation systems are primary irrigation methods. In the northern area, dry fertilizers such as AN and urea are the major sources of N. Further south, AA is usually used as the preplant source and AA or liquid N as sidedress sources. Some N is always applied preplant. Sidedress applications are usually made well before tasseling in order to supply the N during the period of maximum need by the plant.

Under center-pivot sprinkler irrigation, growers apply N through the center-pivot during the growing season, after an initial preplant application of AA. Liquid N sources are very well-adapted to this method of application and are exclusively used. Much of the center-pivot–irrigated corn is grown on light-textured soils and small quantities of N (20–40 kg/ha) are applied through the center-pivot several times during the growing season in an attempt to increase N efficiency.

Various minimum tillage systems are used to prevent wind erosion. The influence of minimum tillage on N fertilizer need has not been widely investigated. It is generally assumed the N requirements are not substantially different under this system, but this is not quantitatively known. Several studies are under way to investigate N fertilization needs under various tillage methods.

The amount of N recommended varies from 0 to 330 kg/ha, depending on yield goal and residual soil NO_3^--N level. These two factors are the most important input in determining the N recommendation. Other factors shown in Table 2 are also important.

Most growers use a soil test and follow the fertilizer recommendations closely. Some apply more than is recommended by the universities. This is primarily due to private soil testing laboratories and consultants encouraging growers to shoot for excessively high yields that are only rarely obtained. This is resulting in the recommendation of excessive rates of fertilizer that do not increase yields and cost the farmers many dollars in unnecessary fertilizer costs (Soil Fertility Staff, 1980; Whitney, 1980; Westfall et al., 1981).

C. Sugarbeets (*Beta vulgaris* L.)

Sugarbeet production, like malting barley, is controlled by the industry through contractual agreements with growers. The areas of production are controlled by proximity to sugar-refining facilities. The land area of sugarbeets has declined substantially over the last 30 yr, with major area cuts coming in the mid-1970's. This resulted in one of the oldest companies, U and I Sugar Co., going out of business and the closing of several Great Western Sugar Co. refineries in Colorado. Low sugar prices, expiration of the Sugar Act, inefficient sugar recovery in the factories, increased transportation costs, and increased farm production costs all contributed to these closings.

Urea and ammonium nitrate are the primary sources used for sugarbeet fertilization. Proper N management is of critical importance, because growers are paid based on beet weight and beet sucrose content, with higher prices being paid for high sucrose-content sugarbeets. As the available N increases, the weight increases and the sucrose content decreases. Therefore, a compromise between weight and sucrose content must be made. Considerable research has been conducted over the years to establish this "most desirable" N rate. The earliest research (James et al., 1971; Reuss & Rao, 1971) provided the basic framework for development of the "N budget" for sugarbeets. It is used very successfully by growers to arrive at the optimum compromise in their N management program between weight and sucrose content.

The N rates recommended range from 0 to 260 kg/ha. The majority of sugarbeet growers soil test, usually to a depth of 120 cm, and follow the recommendations in Montana, Wyoming, and Colorado. The sugarbeet industry encourages their growers to soil test and commonly do the deep soil sampling at a minimal cost. This is the main reason for the high acceptance rate of soil testing by growers in Montana, Wyoming, and Colorado. Idaho agronomists report that growers often exceed recommended rates by 100 to 150 kg/ha. This is a result of the sugarbeet companies' practice of paying growers based on weight rather than on sucrose content in Idaho.

Generally, all the N is applied preplant, but some in-season applications are made. Growers in the northern area are contractually prohibited from applying N after 15 July. Application of N after this date decreases sucrose content substantially. In Arizona, sugarbeet harvest is not dictated by freezing weather and a growing cycle may be 10 to 12 months. Under these conditions, petiole analysis is used extensively to make N recommendations in season.

D. Mountain Meadows

Fertilization of mountain meadows in this region is not common. Ranchers usually rely on organic matter decomposition to supply N to the plant. Educational programs are being conducted to convince the rancher of the potential benefit of N fertilization of mountain meadows and the practice is becoming more widespread every year.

Nitrogen is generally broadcast-applied in the spring. Fall applications can be made after the maximum daily temperature remains below 10°C (50°F). Grass meadow recommendations vary depending on soil test ranging from 0 to 180 kg/ha. Dry fertilizer materials such as AN, AS, and urea are used; the selection is made based on availability and price. All three sources normally perform equally. Proper N management is difficult because of the wide range of species present in meadow stands, unaccessibility of all fields with fertilizer equipment, and poor water management.

E. Native and Improved Rangeland

As with mountain meadows, N fertilization of native or improved rangeland is not common. Ranchers fail to see the economics of fertilization. More ranchers are accepting the practice, but the acceptance rate is very slow. Soil testing is the basis for N recommendations. The rate varies from 0 to 40 kg/ha. A profitable response to N can be expected only when soil moisture is favorable and medium and tall grass species dominate the pasture. Dry N sources are broadcast-applied, usually in the spring.

F. Alfalfa (*Medicago sativa* L.) and Other Legumes

Alfalfa is grown extensively in the Mountain states and probably occupies the largest land area of any irrigated crop. Nitrogen is not recommended on established stands, but it is recommended from 0 to 45 kg/ha for new seeds when grown with a companion crop such as small grains. Some states (Utah and New Mexico) recommend N when grown without a companion crop. The use of a companion crop during stand establishment is almost exclusively used in order to protect the new seeding from high temperatures and excessive soil moisture evaporation.

Peas, beans (*Phaseolus vulgaris* L.), and lentiles (*Lens culinaris* Medik) are grown in some areas. Because of their symbiotic N_2-fixing capacity no N is recommended, except small amounts to establish the crop.

IV. RESEARCH NEEDS

A survey of state specialists revealed several areas of research needed. Research is needed on the N-fertility requirements of some relatively new crops to the region, such as safflower and sunflower. Very little work has been done on these crops in this region, since they have only been grown for a few years. Nitrogen–variety–water efficiency research is needed, particularly on corn. With the rising cost of energy to pump irrigation water and the rising cost of fertilizer, this is of vital importance. Research is needed to determine the N fertility requirements under modified, minimum, or no-till cropping systems under both dryland and irrigated production. The effect of these tillage systems on N requirements is not extensively known. It is assumed that they are the same as under conventional-tillage systems, which may or may not be correct. The selection and evaluation of new varieties of all major crops to increase N efficiency and to obtain optimum production is an area that needs investigation. Plant breeders and soil fertility researchers need to work together to ensure that new varieties are evaluated for relative N efficiency before release. Low N efficiency is a problem where adequate progress is not being made. This problem is particularly acute under irrigation where leaching is a potential. Continuous cropping of dryland wheat and barley is gaining more interest. The N needs under this continuous-cropping situation under the low rainfall conditions that exist in this region need investigation. Presently, very little is known about how this cropping system effects N needs.

A general consensus existed that low N efficiency was a major problem and more research effort should be expended in this area. Some specialists thought it should be approached from the irrigation standpoint, while others thought other approaches were best. Regardless of the approach, this appears to be of major concern.

REFERENCES

Geist, J. M., J. O. Reuss, and D. D. Johnson. 1970. Prediction of the nitrogen requirements of field crops: Part II. Application of theoretical model to malting barley. Agron. J. 62: 385–389.

Gilbert, W. A., A. E. Ludwick, and D. G. Westfall. 1981. Predicting in-season N requirements of sugarbeets based on soil and petiole nitrate. Agron. J. 73:1081–1023.

Goos, R. J., D. G.Westfall, and A. E. Ludwick. 1981. Nitrogen fertilization of dryland winter wheat. Colorado State Univ. Ext. Service "Service In Action" sheet no. 544.

Goos, R. J., D. G. Westfall, A. E. Ludwick, and J. E. Goris. 1982. Grain protein content as an indicator of N sufficiency for winter wheat. Agron. J. 74:130–133.

Hargett, N. L. 1976. Fertilizer summary data. NFDC, TVA, Muscle Shoals, Ala.

James, D. W., A. W. Richards, W. H. Weaver, and R. L. Reeder. 1971. Residual soil nitrate measurements as a basis for managing nitrogen fertilizer practices for sugarbeets. J. Am. Soc. Sugar Beet Technol. 16:313–322.

McDole, R. E., J. P. Jones, and R. W. Harder. 1978. Northern Idaho fertilizer guide—wheat. Idaho Coop. Ext. Service, Agric. Exp. Stn. Current Inf. Services no. 453.

Painter, C. G., J. P. Jones, R. D. Johnson, D. T. Westerman, and J. N. Carter. 1977. Idaho fertilizer guide—sugarbeets. Idaho Coop. Ext. Service, Agric. Exp. Stn. Current Inf. Series no. 217.

Reuss, J. O., and J. M. Geist. 1970. Prediction of the nitrogen requirements of field crops: I. Theoretical models of nitrogen release. Agron. J. 62:381–384.

Reuss, J. O., and P. S. C. Rao. 1971. Soil nitrate nitrogen levels as an index of nitrogen fertilizer needs of sugarbeets. J. Am. Soc. Sugar Beet Technol. 16:461–470.

Soil Fertility Staff. 1980. A comparison of suggested fertilizer programs obtained from several soil testing laboratory services. Univ. of Nebraska, Lincoln, Dep. no. 33.

U.S. Department of Agriculture. 1980. Commercial fertilizers: consumption by class for year ended June 30, 1980. Crop Reporting Board, Economics and Statistics Service, Sp. Cr. 7 (11-80). USDA, Washington, D.C.

Westfall, D. G., J. A. Cruz, E. E. Rothman, T. J. Doherty, C. M. Richardson, and H. M. Golus. 1981. Soil testing laboratory fertilizer recommendation comparisons. Colorado State Univ. Exp. Stn. PR-12.

Whitney, D. A. 1980. Soil test recommendation comparison studies—1980. Kansas State Univ. Coop. Ext. Service Mimeo Sheet.

WASHINGTON
OREGON
PACIFIC
CALIF.

52

Roy S. Rauschkolb
University of Arizona
Tuscon, Arizona

T. L. Jackson
Oregon State University
Corvallis, Oregon

A. I. Dow
Washington State University
Prosser, Washington

Management of Nitrogen in the Pacific States

It is well recognized that diversity of cropping patterns and management practices in an agricultural area are dictated by variations that exist in climate and soils. The Pacific states are especially varied in both climate and soils, the latter a function of the former. This variability leads to the great differences that are encountered in the crops adapted to the region and affects the practices one uses to increase the efficiency of N use in agricultural production.

I. THE REGION

A. Climate

The climate for a major portion of the Pacific states is predominantly "Mediterranean." There is usually little or no summer rainfall and the temperatures seldom reach the extremes of hot or cold. For other areas, there are wide climatic differences ranging from typically alpine to hot desert conditions. For a major portion of the crop-production areas of the region the soils are seldom frozen, allowing nitrification to occur year-round, thus increasing the potential for both denitrification and leaching

Nitrogen in Crop Production.

losses of N. This dictates management practices that should be followed to conserve N in the region.

The Cascade Mountains divide Oregon and Washington into two strikingly different climatic areas. The major cultivated area west of the Cascades, consisting of the Willamette Valley of Oregon and the Columbia Basin of Oregon and Washington, has in excess of 90 cm (35 inches) of rainfall each year as shown in Fig. 1. About 75% comes between 15 October and 15 March, when evaporation rates and plant growth are limited. The temperatures are moderated by the influence of ocean currents, which prevent extreme heat or cold. The soil tempratures rarely are below freezing for any extended period of time. The rainfall in central Oregon and Washington, the areas immediately east of the Cascades, varies from 20 to 40 cm (8–16 inches) each year and increases to 40 to 80 cm (16–32 inches) in some eastern areas of these states. The winter temperatures are lower and the soils are frozen for extended periods of time. Temperatures drop below freezing on most nights from early December through February.

In California, there is a relatively low mountain range near the coast that extends the entire length of the state and a relatively high mountain range inland about 200 km that extends nearly two-thirds the length of the state. The central valley between these mountain ranges is comprised of the Sacramento Valley in the north and the San Joaquin Valley in the south. These two valleys account for nearly 75% of the cropland in the state. The southern coastal valleys and foothills, which includes the Salinas Valley, a major vegetable-growing area, along with the Imperial Valley account for an additional 20% of the cropland in California. Annual rainfall for these cropping regions are also shown in Fig. 1. Over 80% of the precipitation occurs during the months October through March and most of that in December, January, and February.

Temperature variations are considerable. As in Washington and Oregon, the coastal areas of California are influenced by ocean currents, which prevent extremes in temperature fluxation. In most years there will be a 365-d frost-free period in these areas. Temperatures in the inland valleys are more extreme. Winter temperatures seldom reach freezing in the Imperial Valley, occasionally fall below freezing in the San Joaquin Valley and southern Sacramento Valley, and will frequently fall below freezing in the northern-most part of the Sacramento Valley. However, these low temperatures are rarely sustained for extended periods of time. As a consequence, soil temperatures seldom fall below 7°C (45°F), except at the surface. The high summer temperatures are somewhat similar for all three of these areas, with the Imperial Valley having a higher daily average. The delta region of the San Joaquin and Sacramento rivers is an exception, because the temperature is moderated by sea breezes reaching inland through the straits, providing for the outflow of the confluence of the two rivers.

B. Soils

Soils for the Pacific states have developed under somewhat similar conditions. Major cropping areas consist of soils developed from lacustrine de-

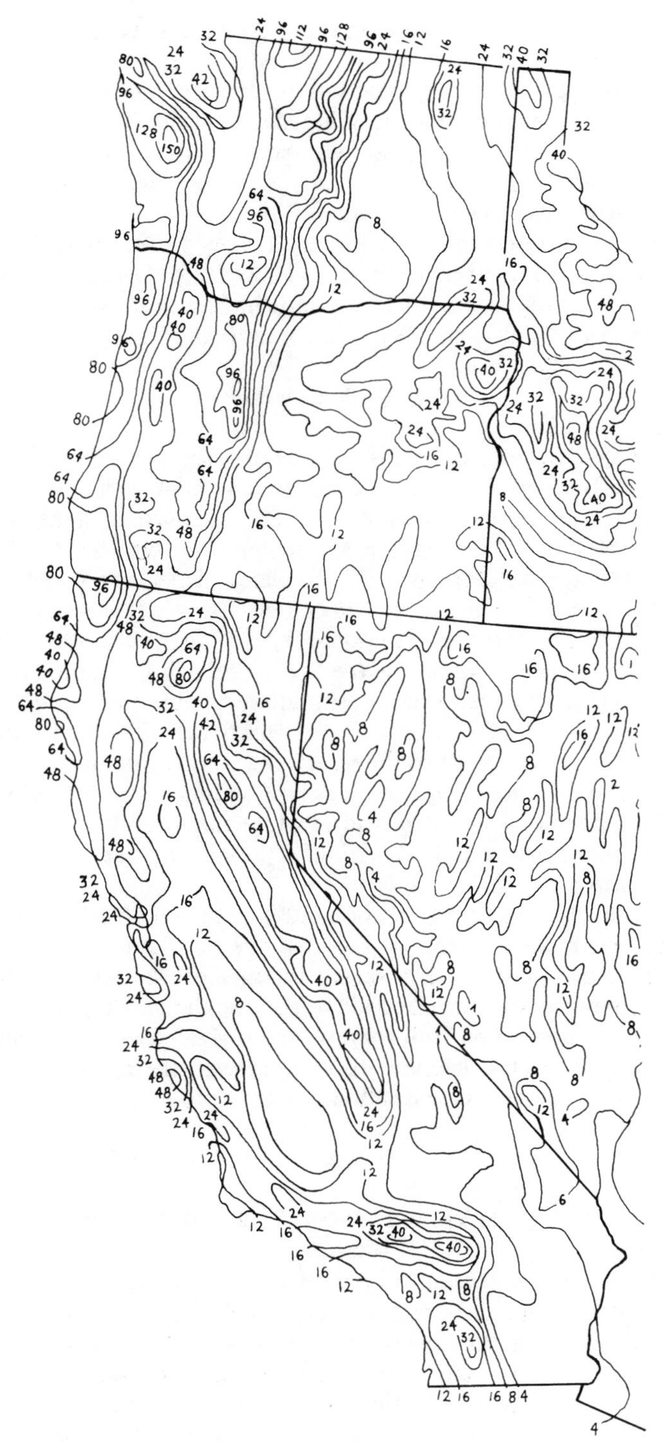

Fig. 1—Normal annual total precipitation in inches (Natl. Oceanic & Atmos. Admin., 1965).

posits, as well as alluvial deposits, on valley bottoms and low terraces. Some of the older Alluvial soils have well-developed profiles.

There are some unique regional soil differences. In eastern Oregon and Washington, there are major areas of loessian deposits. There is also a large area of high organic matter and peat soils in the delta of the Sacramento–San Joaquin rivers in California and in the Klamath Basin and low elevation valleys of southeast Oregon.

Except for the peat soils, most of the formerly arid soils that are now irrigated have < 1% organic matter. In western Oregon well-drained mineral valley soils average 2 to 3% organic matter, Red Hill soils 5 to 7%, and Palouse soils, with 45 cm (18 inches) or more of rainfall, 2 to 3%.

C. Irrigation Methods

About 50% of the cropland in the Pacific region is subject to either a seasonal insufficiency of rainfall or a low annual amount, making irrigation an essential feature of crop production. There are approximately 1.2 million ha under irrigation in east central Washington and Oregon. Most of this area is under intensive cropping and completely dependent upon irrigation for production; the remainder will utilize supplemental irrigation to compensate for the vagaries of seasonal rainfall distribution and amounts. Most of this area grows wheat. Of the approximately 1.2 million ha of irrigated land in eastern Oregon and Washington, roughly 50%, or 600 000 ha, are sprinkler irrigated. Out of this, slightly more than 200 000 ha are under center-pivot irrigation. Most of the remaining 50% is irrigated by the rill or furrow method. The area irrigated by basin flooding is small. California has about 3.8 million ha of irrigated land of which nearly 80% is surface–flood basin or furrow irrigated, 18% sprinkler, 1.5% drip, and 0.5% subirrigated.

In Washington and Oregon, about 20% of the cropland is irrigated; in California, the reverse is true with almost 80% of the total cropland irrigated. Because of the much greater amounts of rainfall that occur in portions of western Washington and Oregon, nonirrigated crop production is greater per unit area of crop than the nonirrigated areas in California. Since water is not such a yield limitation in these northern Pacific states, the N fertilizer requirement is greater than for the dryland production areas of California.

II. MANAGEMENT VARIABLES

Within a crop production system, there exists a given set of conditions of soil, climate, crop, and irrigation method that predetermines the management practices that may be used to achieve peak N efficiency. The management variables one has to choose from are *placement,* the type of *equipment* available for application of a material, the *rate* of application, the *source* of N, *irrigation* management, *timing* of N application, and *energy* for soil microorganisms (organic matter). The acronym *PER SITE* for these management variables also emphasizes the site specificity in the choice of management practices. None of these variables are independent of the others. Selecting the best combination of practices under a specific set of conditions is where the skill lies in achieving N use efficiency in crop production. In the Pacific states, certain practices have evolved over several years of research and investigation that provides improved N efficiency. The cultural practices used with respect to each of the management variables are integrated to provide an overall management of N adapted to each set of conditions. For this reason, each of the variables will be discussed somewhat separately, thus providing all the elements for recommending a set of practices to meet the requirements for individual fields.

A. Placement

Most, if not all, of the major factors that favor volatilization losses of NH_3-N from soil surfaces are found in combination in the Pacific states. These factors are calcareous soils, optimum soil moisture, medium and coarse texture of soils, and high summer temperatures. Proper placement of the various N sources will minimize or eliminate the loss. A general guide for evaluating the effect of placement on the magnitude of NH_3-N loss for different sources of N is provided in Table 1. Volatilization losses can essentially be eliminated from a susceptible N source by immediate incorporation after surface application (mechanically or with irrigation application, whichever is available and appropriate) or by a band application be-

Table 1—Relative N losses by NH_3 volatilization for different application methods and fertilizer materials.†

N Source	N Content	Surface broadcasted with method of incorporation					Apply in irrigation		Below surface
		Without water		With water		Mechanical			
	%	<7‡	>7§	<7	>7		<7	>7	
Anhydrous ammonia	82	--¶	--	--	--	--	L#	H	V
Aqua ammonia	20	--	--	--	--	--	L	H	V
Ammonium sulfate	21	L	H	L	H	L	L	H	V
Ammonium phosphate	11	L	H	L	M	L	L	M	V
Ammonium nitrate	33.5	V	L	V	L	V	V	V	V
Urea-ammonium nitrate	32	V	L	M	V	V	V	V	V
Urea	45	M	H	V	V	L	V	V	V
Potassium nitrate	13	V	V	V	V	V	V	V	V

† Rauschkolb et al., 1979.
‡ Soil pH <7.
§ Soil pH >7.
¶ Dash line indicates that surface application of these materials is not a normal cultural practice used, because the physical state of the material leads to extreme volatilization losses.
# H = losses >40%; M = losses between 20 and 40%; L = losses between 5 and 20%; V = losses <5%.

low the soil surface. The general practice is to provide mechanical incorporation as rapidly as possible where materials susceptible to volatilization are used.

B. Equipment

The equipment that is available or will be used for applying N fertilizer determines the source that can be used. Because of the prevalence of large fields in the west, one method frequently used is aerial application. In this method, it is preferred to use a high analysis material to keep unit costs of application down. For nonaerial methods, the type of equipment a grower owns or has access to on a custom basis will be a major factor in determining the source of N.

C. Rate

A range of rates of N application for several major crops in the Pacific states are presented in Table 2. Because of the differences in yield potential as a function of climate and soils, the ranges for most of the crops are quite large. When distinction can be made between dryland (where water is limiting yield) and irrigated or adequate rainfall conditions, the N ranges are given. Fertilizer recommendations provided by the universities are based on results from field experiments designed with a range of N rates to evaluate the response for soil conditions and previous cropping practices. This provides a basis for predicting N application rates. Only in the case of a few crops are optimum rates of application predicted from soil or tissue analyses.

D. Source

There are considerable differences that exist in the magnitude of the various N sources used by each state in the Pacific region (Table 3). California and Washington both use proportionately greater amounts of anhydrous ammonia and N solutions than is used in Oregon. California has the largest proportion of liquid N sources and, while using the greatest absolute amount of ammonium sulfate, this is proportionately less than is used in Oregon.

As indicated, the selection of a N source is a critical part of the N management choices available to a grower. Nitrogen sources with potentially high NH_3 volatilization losses necessitates incorporation after surface broadcasting or by banding below the soil surface. An increasing amount of N is applied in irrigation water. The relatively low NH_3 volatilization potential of urea–ammonium nitrate solution and its adaptability to application either by irrigation, aerial application, or banding make this a very versatile material for irrigated crop production. Urea is also well adapted for application in irrigation water; its high solubility and mobility in soils before it becomes hydrolyzed permits movement below the soil surface before hydrolysis to ammonium carbonate can occur, thus reducing the po-

Table 2—Range of N application rates for selected crops by crop category for the Pacific states.

Crops	Latin name	Calif.†	Ore.‡	Wash.‡
		kg of N/ha		
	Field crops, pasture, & range			
Annual and perennial grasses	*Festuca* sp., *Lolium* sp., and *Poa* sp.	35–120	80–150	30–80
Barley (nonirrigated)	*Hordeum vulgare* L.	25–70	0–50	20–80
Barley (irrigated)	*Hordeum vulgare* L.	35–145	30–120	60–160
Beans (dry)	*Phaseolus vulgaris* L.	20–65	--§	0–100
Corn	*Zea mays* L.	125–255	100–200	80–200
Cotton	*Gossypium hirsutum* L.	64–185	--	--
Rice	*Oryza sativa* L.	55–130	--	--
Sugarbeets	*Beta vulgaris* L.	85–205	85–205	80–160
Wheat (nonirrigated)	*Triticum aestivum* L.	25–65	0–80	20–100
Wheat (irrigated or adequate seasonal rainfall)	*Triticum aestivum* L.	50–165	80–180	80–180
	Fruits & nuts			
			(All 50–100)	(All 50–100)
Almonds	*Prunus amygdalus*	80–120	--	--
Apples	*Malus sylvestris* Mill.	--	--	37–120
Caneberries	*Rubus loganobaccus*	90–185	60–90	--
Grapes	*Vitis vinifera* L.	35–110	--	--
Peaches	*Prunus persica*	85–200	--	--
Prunes	*Prunus domestica*	50–155	--	--
Walnuts	*Juglans regia* L.	75–210	--	--
	Citrus & subtropicals			
Avocado	*Persea americana* Mill.	90–235	--§	--
Grapefruit	*Citrus paradisi* Macfad.	125–280	--	--
Lemon	*Citrus limon* (L.) Burm. f.	130–270	--	--
Olive	*Olea europaea* L.	45–150	--	--
Orange	*Citrus sinensis* (L.) Osb.	95–220	--	--
	Vegetables			
Asparagus	*Asparagus officinalis* var. *atilis*	110–210	--	--
Beans (snap)	*Phaseolus lunatus* L.	60–125	50–75	--
Broccoli	*Brassica oleracea*	140–280	150–250	--
Cantaloupe	*Cucumis melo* var. *cantalupensis*	75–170	--	--
Lettuce	*Lactuca sativa* L.	130–230	--	--
Onions	*Allium cepa* L.	125–255	130–225	--
Peas	*Pisum sativum* L.	45–100	0–30	--
Peppermint	*Mentha* X *piperita* L.	--	150–200	120–190
Potatoes	*Solanum tuberosum* L.	160–265	50–100a¶ 120–400b	200–300
Sweet corn	*Zea mays* var. *rugosa*	115–220	120–160	--
Tomatoes	*Lycopersicon esculentum*	100–210	--	--

† Values adapted from Univ. of Calif. Div. of Agric. Sci., 1978.
‡ Based on estimated N use by the authors for the states indicated.
§ Denotes a relatively small area of production, or the crop is not grown in the state.
¶ N rates are for (a) west of the Cascade Mountains and (b) east of the Cascade Mountains.

Table 3—Amounts of N from reported sales of various fertilizer materials for the Pacific states.†

N Sources	Calif.	Ore.	Wash.
	tonnes of N		
Anhydrous ammonia	141 470	18 288	110 020
Ammonium nitrate	27 259	8 285	13 720
Ammonium sulfate	51 221	15 423	3 154
Urea	39 257	37 004	26 725
Nitrogen solutions:			
Ammonium nitrate	21 796	--‡	--
Aqua ammonium	64 938	5 965	13 801
Urea	2 570	--	--
Urea-ammonium nitrate	72 407	17 838	28 795
Mixed nutrient sources:			
30-0-0-6	--	4 661	--
40-0-0-6	--	3 169	--
Other	140 600	22 597	16 285
Total	566 940	133 230	212 500

† Values are yearly totals for each material adopted from state fertilizer tonnage reports on sales for 1979.
‡ Data not reported for these materials in the state.

tential for NH_3 volatilization. Preplant applications of fertilizer are frequently combinations of N and P in various ratios, depending upon the crop. Most of the combinations are liquid and are applied primarily for their P content for a starter material.

E. Irrigation Method

An important factor in N use efficiency is water management. Water applied in excess of crop needs is subject to either surface runoff or deep percolation. Nitrate present in the soil is subject to leaching when excess water percolates through the soil profile. In very sandy soils it is nearly impossible to manage water applications under furrow/flood irrigation or high-volume sprinklers to avoid the loss of water and some N below the root zone. Under such conditions, it is common for small quantities of N (usually < 25 kg/ha) to be applied in each irrigation to supply the crop throughout the season. Banding of anhydrous or aqua ammonia is also recommended to minimize leaching by taking advantage of the lack of mobility of NH_4^+ in soils and the delayed nitrification.

On other soils not subject to severe leaching problems, N applications can be made prior to either sprinkler or surface irrigation. Also, larger, less frequent applications of N may be made in the irrigation on these soils, providing care is taken to minimize volatilization losses from the water. Where furrow irrigation is practiced, it is recommended that N be incorporated before planting, preferably by banding, to prevent positional unavailability of N from occurring where NO_3^- accumulates in the top of the beds.

F. Timing

The proper timing of N applications is an important means of improving the efficiency of N use. Losses from leaching, volatilization of NH_3, or denitrification can be minimized by applying and incorporating N as close to the time plants need it as possible. How close is determined by climate and crop growth characteristics, as well as the choices one has with regard to the other N management variables.

In eastern Washington and Oregon, as well as higher elevations of California, cold temperatures and frozen soils permit fall incorporation and/or banding of NH_4^+ fertilizers. Most of the NH_4^+ is expected to remain in that form throughout the winter. In addition, the low amount of precipitation in Washington and Oregon east of the Cascades during the winter months limits NO_3^- leaching below the root zone of most crops grown in the region.

Where climatic conditions and other factors are contributing to N losses, it becomes more critical to make N applications more closely to the time when the plant needs it to enhance efficiency. In order to accomplish this, split applications are frequently recommended. The applications subsequent to preplant are applied in a variety of ways: aerial, water run (drip, sprinkler, or surface flood), broadcast, or banded. The timing of availability of N to the plant then becomes a function of the source chosen and placement achieved. Allowances for conversion and/or movement to achieve positional availability to the plant must be made depending upon how the N is applied.

In split applications, one-third to one-half of the total expected fertilizer N requirement is applied preplant, with the remainder applied prior to the rapid growth period of the plant. Exceptions occur where a grower has substantial control over the timing and amount of water that can be applied, in which case several small applications of N may be made or all the N applied at preplant. Also, where the soils are very sandy, the grower may make several small applications of N in the irrigation water as previously discussed.

G. Energy (Organic Matter)

The mild climate permits soil microorganisms to utilize organic matter as an energy source most, if not all, of the year. When large quantities of crop residue with a wide C/N ratio are incorporated before planting, there is a competition between soil microorganisms carrying out decomposition and the plant for the soil N. As a result, the plants are unable to compete satisfactorily with the soil microorganisms for N, frequently causing the plants to become N-deficient. Incorporation of crop residues 2 to 4 months prior to planting a crop will usually provide adequate time for decomposition to proceed to the point where net mineralization of N is occurring. In cases when it is not possible to permit such a waiting period between incorporation of the crop residue with a high C/N ratio and planting, it is advised to band-apply the fertilizer N in the amounts required by the subsequent crop. This reduces the amount of N consumed by the soil microorganisms and provides the growing crop a greater opportunity to compete more favorably for the fertilizer. If banding is not possible, then an amount of N slightly in excess of the subsequent crops requirement can be applied. This is intended to provide for both decomposition and plant uptake, recognizing that some immobilization will occur, but that it will be compensated by the net mineralization of N during the growing season. In some cases, leguminous crops are purposely utilized in a cropping sequence to provide some N for the subsequent crop. Because of the relatively high N content (low C/N ratio), net mineralization occurs almost immediately as decomposition takes place. The same is also true of other crop residues with low C/N ratios.

Availability of an energy source during periods of possible waterlogging can contribute to increased potential for denitrification losses. This is a situation that frequently occurs in the coastal and some inland areas of northern California, Oregon, and Washington. The combination of heavy soils, mild temperatures, and heavy winter rainfall creates a condition conducive to high denitrification losses. This is enhanced by the presence of crop residues.

III. MANAGEMENT TOOLS

Through a combination of factors such as crop yield, temperature, winter rainfall, soil texture, treatment of crop residues and application of manures, and use of leguminous crops in rotation, it is impossible to estimate N carry-over from one crop season to the next. Some measurement of soil N supply or plant N status is required. The methods most frequently used to estimate fertilizer N requirements for the plant are the soil test and a tissue test for NO_3^-. These may be used alone or in combination. The extent that these are relied upon as a means for predicting N fertilizer needs differs among the states in the Pacific region.

A. Soil Testing

In Washington and Oregon, The NO_3^- found in soil samples is utilized to calculate the amount of available N in the soil profile. This amount is then deducted from the projected N requirement of the plant to give the N fertilizer recommendation. Soil samples are taken prior to planting in 30-cm increments to a depth of 1.5 m or to the bottom of the root zone. Research in these states has shown this to be a good basis for assessment of soil NO_3^- and prediction of the amount of fertilizer N needed. Fertilizer recommendations are made by both the state university and commercial laboratories in Washington and Oregon.

In California, experience with predicting N fertilizer requirements from soil tests have been much less satisfactory. Nevertheless, the soil test value from the top 30 cm of soil profile is frequently used to evaluate the general N status of the soil. On the basis of a low, medium, or high value, a determination is made regarding the need for a preplant N application. At low or medium ranges, some N is applied. The purpose is to ensure adequate N to carry the plant to the point where tissue tests can be performed. One exception to this is for sugarbeet production. A procedure has been researched for several years, which is similar to the procedure used in Washington and Oregon.

When sampling furrow-irrigated fields, an adjustment in the sampling pattern must be made to account for the movement of NO_3^- to the top of the beds with the water. Usually the first 30-cm increment samples are taken on the shoulder of the bed toward the center at a 30 to 40° angle from the perpendicular. Fertilizer recommendations in California are usually made by commercial laboratories. The University of California provides calibration data and standardized testing procedures for the laboratories and makes fertilizer recommendations based upon empirical studies in the field conducted by farm advisors in the locale where the crops are grown.

B. Tissue Testing

Experience in the Pacific states has indicated that tissue tests for NO_3^- to evaluate N supply to the plant are much more reliable than soil testing. Presumably, this is because the root system of the plant has a tendency to integrate the spatial variability that exists in N supply over a larger volume of soil. Tissue tests are most commonly used to indicate a sufficiency/deficiency. They are seldom used to predict the amount of fertilizer N needed by the plant, except on a very few crops, and even more seldom to signal an excess of available N. Emperical field studies have been used as a basis for determining the amount of fertilizer N to be applied once a deficiency is identified. Tissue testing is used routinely for evaluating the N nutritional status of potatoes and citrus. It is used less routinely for evaluating the N nutritional status of vineyard and orchard crops and even to a lesser extent for wheat, sugarbeets, cotton, corn, and tomatoes. For most other crops, tissue analyses are used as a diagnostic tool rather than a basis for adjusting fertilizer practices.

IV. N FERTILIZER AND ENERGY USE

Because of the important role that energy plays in the agricultural production system and the current concern about the future availability of energy, this separate discussion will deal with agricultural energy use in the Pacific states with particular attention to N fertilizer use. According to a 1974 study, approximately 5% of the total energy consumed in California was attributed to agriculture. For the other Pacific states, this value may be higher because of differences in metropolitan and industrial demand. Of the 5%, only 14% was attributed to fertilizer manufacture, distribution, transportation, and application (i.e., about 0.7% of the total energy consumed in the state). This is for all forms of fertilizer and not just N materials.

However, substantial amounts of N are imported, so an upward adjustment in the proportion of actual fertilizer energy use could be made. The point is that a small energy investment pays very large dividends. Nevertheless, it still is essential that energy efficiencies in crop production be accomplished when possible. Practices that achieve N use efficiency for little or no additional energy use should be encouraged. For example, if a single splitting of a N fertilizer application would result in a savings of as little as 5 kg of N, then the energy cost in terms of the diesel for the tractor to make the second application is about equivalent. Therefore, such a practice would have to save > 5 kg/ha of N to become energy efficient. Savings considerably in excess of 5 kg/ha are frequently attainable by split application of N fertilizers.

Figure 2 shows the relationships between total energy input, fertilizer energy, crop yield, and net energy return by the crop. It is particularly important to note that peak energy return occurs at the rate of N that provides maximum yield. The same principal applies for other crops. The principal difference among crops is the amount of caloric energy contained in the crop. There may not be a net energy return. This brings out a critical issue concerning energy return for energy invested.

Frequently, it is suggested that only those crops for which there is a net energy return should be grown using our precious energy supplies. People

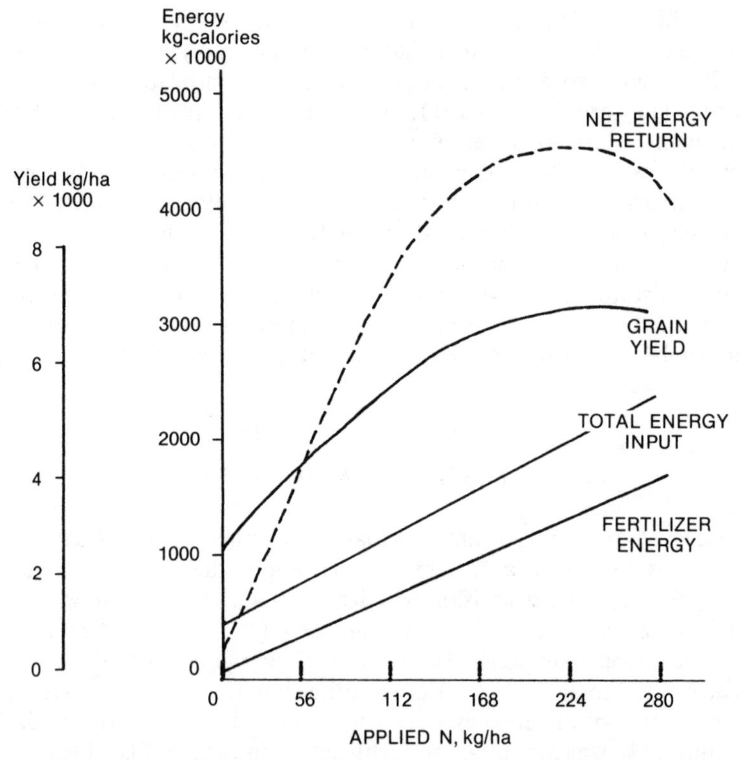

Fig. 2—The relationship between use of N fertilizer and yield of grain sorghum, energy inputs, and energy returns.

who make that contention, either fail to consider or disregard the need for a variety of foods to meet human nutritional requirements for vitamins, minerals, and protein. It would be foolhardy to only grow those plants that provide a net caloric energy return.

Another aspect of the relationship between N use and energy is the great contribution N fertilizer makes to the energy efficiency for other practices. By increasing the crop yields per unit area through the use of N fertilizer, the efficiency of other inputs is improved. A corollary of this is a minimization of an adverse impact on the environment, because of the need to use more land and water resources to grow the same amount of food if N fertilizers were not used.

V. RESEARCH NEEDS

The problems associated with improving N fertilizer efficiency include a combination of factors—such as predicting N use by the plant, spatial variability of N in soils, and application of N in irrigation water. Research must address these factors. Reliable methods for predicting N requirements and rates of application at different stages of plant growth and levels of N in the tissue is a prerequisite for efficient N use. Current testing procedures have not been calibrated to provide this information for most crops in the Pacific states.

The spatial variability of N in soils must be accounted for in order to predict rates of preplant N applications. Currently, soil tests are not a satisfactory means of estimating preplant N applications. Even if these can be improved, there is the additional uncertainty of seasonal influences on potential yield. A corollary of the soil test research should be a prediction of yield possibility involving as many discernible factors as possible. With the computer to assist, this may be more likely to succeed than in the past.

Finally, a third broad area is to examine the techniques for water application of N in order to achieve placement, proper timing, uniformity, and energy efficiency. Because of the unique aspect of water application in irrigated agriculture, this avenue of N application offers particular promise to provide N and energy efficiency. The selection and/or development of an N source to fully utilize this practice should be investigated concomitantly.

REFERENCES

Cervinka, W., W. J. Chancellor, R. J. Coffelt, R. G. Curley, and J. B. Dobie. 1974. Energy requirements for agriculture in California. Joint study, California Dep. of Food & Agric. and Univ. of California-Davis. Univ. of California, Davis.

National Oceanic and Atmospheric Administration. 1965. Climate atlas of the United States. Natl. Oceanic & Atmos. Admin., Washington, D.C.

Rauschkolb, R. S., F. J. Hills, A. B. Carlton, and R. J. Miller. 1979. Nitrogen management relative to crop production factors. p. 647–689. *In* P. F. Pratt (ed.) Nitrate in effluents from irrigated lands. Univ. of California final report to the National Science Foundation. Univ. of California, Riverside.

University of California Division of Agricultural Sciences. 1978. Survey of fertilizer use in California—1973. Univ. of California Bull. 1887.

A. G. Norman
Emeritus Professor
University of Michigan
Ann Arbor, Michigan

PERSPECTIVE

Nitrogen is the quintessential element in all living organisms. The units of deoxyribonucleic acid (DNA) are linked through the nitrogen atom, as also are the amino acid components of all proteins. The green plant uniquely combines nitrogen and carbon in a multiplicity of products, but few of these are outside the dynamic turnover, replacement, or translocation phenomena of living cells, tissues, and organs. Similarly, in the soil-root environment, the forms of nitrogen are mutable and complex, and only incidentally may meet the nutritional needs of the plant. The conventional diagrams representing the so-called nitrogen cycle are accurate in that they portray the major pathways involving the oxidation or reduction of the nitrogen atom, valency changes, and formation of N–C linkages in organic compounds. They do not show the physiological, biochemical, or microbial steps, the subpathways, or alternatives inherent in assimilation and dissociation. They do not portray the complexity of the two-way traffic or reveal the lability of the nitrogen atoms in the plant, entering maybe as nitrate and ending up in an amino acid component of a characteristic protein in the grain. In between, in the physiological processes of tissue and organ growth, assimilate partitioning, translocation, and seed development, the same nitrogen atom may have been incorporated in a diversity of compounds in the recycling process. Similarly, the events in or on the soil—involving the decomposition of crop residues, assimilations into microbial cells, and subsequent mineralization—are complex with multiple alternative pathways.

If one recognizes and understands the nitrogen switchyard processes in plant and soil, the next question is to assess the potential for management of nitrogen utilization. The overriding criterion is that its use be efficient and economic in relation to the yield and quality of the harvested crop. But the conditions under which that crop is produced in most locations in the world are neither stable nor assuredly predictable, except within rather broad limits. Seasonal weather differences in the temperate zone, where most agricultural production occurs, can be substantial. Because yield depends directly on the fixation of carbon in the photosynthetic process, one would

Nitrogen in Crop Production.

strive for the maximum utilization of light energy permitted by the ambient temperature and available water. This would include the assurance of a supply of nitrogen so that at no time would there be constraint on the synthesis of N–C compounds. Some steps in the biochemical events involving nitrogen may be directly photosynthetically dependent on light energy in the chloroplasts of the leaves; others may be dependent on the energy supplied from previously synthesized carbon compounds, as in the root where these compounds must also provide the acceptor carbon structures leading to N–C products, such as glutamine, for export to the tops. Nonlegume crop yields under normal conditions can usually be shown to be dependent on the level of available soil and fertilizer nitrogen. The amount of protein present in the whole crop, per acre or per hectare, similarly is relatable to the nitrogen supply. However, the protein content of a harvested crop, such as cereal grain, is not so relatable, because the redistribution of nitrogen to the grain is a genetically controlled process.

The management of nitrogen, aspects of which are reviewed in many of the chapters of this book, although scientifically based in that the underlying processes are understood, is still an inexact science. Quantifying the many steps in the nitrogen transformations in soil and plant is dismayingly difficult, even with the use of nitrogen-15. Many attempts at constructing a balance sheet for fertilizer nitrogen applications have ended in significant under-recovery. Losses then have to be ascribed to be through leaching out of the root zone, or to the atmosphere as ammonia, or through microbial or chemical denitrification to dinitrogen and nitrogen oxides. In the management of fertilizer nitrogen for crop production, there are, therefore, some limitations. Similarly, in the management of the crop for nitrogen utilization, there are limitations inherent in the expression of the genetic control of plant growth and characteristics. Both of these management scenarios have to include the effect of climatic variables on their outcome. The technology to be adopted, therefore, has to be a compromise, but, as our information becomes more precise, so can the goal of efficient nitrogen management be more assuredly attained.

Roland D. Hauck
National Fertilizer Development Center
Tennessee Valley Authority
Muscle Shoals, Alabama

EPILOGUE

World human population will approach 7 billion within the next 20 yr. The N needed to grow the food and fiber for this rapidly expanding world population will be obtained mostly from increased production of grain legumes, which obtain part of their N directly from the atmosphere; and from increased production of N fertilizers, for use mainly on cereal grains and fiber crops. The total amount of N that will be needed for crop production by the year 2000 may exceed 250 Mt (million metric tonnes) annually, of which 110 to 160 Mt would be obtained from fertilizer and 80 to 125 Mt from the production of grain legumes. An additional 50 to 75 Mt would probably be added annually through biological N_2 fixation in permanent meadows and pastures. This large amount of N would be concentrated in 11% of the earth's total land surface, resulting inevitably in an increase in the total N contents of soils, waters, crop residues, and municipal and industrial wastes. A primary challenge, therefore, is to meet the urgent need for food and fiber, while at the same time keeping agricultural N in harmonious balance with the environment.

Clearly, a major objective is to maximize the efficiency of agricultural N in crop production. Increasing efficiency will (i) increase the value of N as a crop production factor, and therefore, its profitability to the farmer, (ii) will conserve the energy and raw materials needed to produce N, and (iii) will minimize the risk of producing undesirable effects in the environment that can result when N is used inefficiently.

What happens to agricultural N added to soils? Generally, about 50% of the N applied is taken up by crop plants during the season of application. This value may be as low as 25% in poorly managed flooded rice systems or as high as 80% for pasture grasses or crops growing on soils of very low N status, but a recovery of 50 to 60% of the N applied is representative of the average cropping situation. About 25% of the applied N is lost from the soil-plant system. This value may be $<$ 10% or as high as 50%, but perusal of the world's literature suggests that a range of 25 to 30% can be adopted for use in working models. After cropping, about 25% of the applied N remains in soil in inorganic and organic forms. The amount of residual inor-

Nitrogen in Crop Production.

ganic N and N immobilized in the biomass may be < 15% or > 45% of the amount applied, but 25% can be used as an average value. These numbers represent our current conception of the partitioning of applied N among plants, soil, waters, and atmosphere during and immediately after the growing season. How can the amount of N taken up and used to produce a harvestable crop be increased, the amount lost be decreased, and the amount of N immobilized in the soil biomass be maintained in active forms of immediate or potential benefit to the plant? The topics discussed in the preceding chapters of this book suggest answers to this complex question.

Research within these topic areas has generated much information, which is being applied to local farming situations and is continually being refined. What new information is needed and what new ideas should be developed to dramatically increase crop productivity? The steady increases in grain yields that have occurred during the past 25 to 50 yr have largely been the result of improved genotypes responsive to N applications and other management factors. During the last 30 to 50 yr, rice yields have increased by 60 to 100% (depending on world region), maize by 60%, and wheat by 35 to 60%. The increase in soybean yields (25%) has been less dramatic. No one knows what the upper limits of production are, but statistics show that yields for maize and soybeans continue to increase slowly while rice yields appear to be reaching a plateau.

Regarding the efficient use of N, the ideal crop plant produces photosynthate at a high rate and readily takes up N and converts it to grain components. Its characteristics are (i) the ability to produce high levels of dry matter and to accumulate reduced forms of N, (ii) a high rate of N uptake and high photosynthetic rate during grain filling, and (iii) a long grain-filling time without delayed maturity.

Improvements in the efficiency of biological N_2 fixation can be achieved through better matching of the legume host and its microbial symbiont, by improving the transfer of photosynthate from the leaf to the nodule without the accumulation of undesirable metabolites, such as β-hydroxybutyrate, and by genetic selection for high nitrogenase and hydrogenase activity. Of high priority, in my opinion, is research on elucidating the mechanism that shuts off the production of nitrogenase as the level of fixed N that is available to the legume-*Rhizobium* symbiosis increases. At present, N_2 fixation by soybeans in a bean-maize rotation is restricted by the ability of the soybean to scavenge for and use the N that was applied to, but not used by, the preceding maize crop.

Improved genotypes that can respond to improvements in management practices and make use of technological advances most probably can be developed; these new plants will take up a larger proportion of applied N than plants currently available do and will translate this increased uptake into increased production of harvestable crop. Some of the new plants will be produced in conventional breeding programs using germ plasm on hand. Dramatic progress in the development of improved genotypes and adaptive hybrids probably will not occur until the physiological and biochemical characteristics leading to high levels of production are more clearly defined.

Increasingly, agricultural scientists and those interested in the nutritional and environmental effects of increased agricultural production engage in multidisciplinary efforts, and by so doing, work on several levels of integration. For example, consider the efforts directed toward understanding the workings of a bacterial cell and its ecological relationships. This cell is a self-contained unit, functioning in the surrounding medium. It is visible to the bacteriologist, who can observe its gross physical structure, mobility in the medium, and its reaction to external stimuli. Within the bacterial cell are subcellular constituents of interest to the molecular biologist and others who practice the arts of recombination and chromosome husbandry, which are associated with genetic engineering. Biochemical activities cannot readily be monitored for a single cell, because their substrates and products lie below our limits of detection—but they can be measured in a colony of cells. Such measurements are of interest, for example, to the soil biochemist and microbiologist. The single cell, which is the environment of its subcellular components, is at the same time part of the environment of the colony of cells. The colony is part of an environment filled with a variety of living and nonliving entities, and this environment is the area of interest to the microbial ecologist. In turn, this localized assortment of entities in its own microenvironment is part of a small, more complex macroenvironment, and so on up the various levels of integration to all entities in the global environment. The deoxyribonucleic acid (DNA) is linked to the cell, and the cell is linked to the tissue; the leaf, the plant, and the field of plants are ascending levels of integration, each level being studied by different techniques and for different reasons by human beings with widely different aptitudes, interests, and thought patterns.

What are necessary steps for these humans to help each other produce information that will prove useful in the quest for ever higher crop productivity? A first step is for interacting scientists to learn each other's specialized technical language and to understand its meanings, have a better appreciation of each other's problems, and truly understand the objectives of each other's research. Increased communication among scientists of different disciplines but with the same major objective will increase cooperation. This, in turn, increases the likelihood that results produced at one level of integration will be useful on other levels. A consistent body of knowledge is based on such a hierarchy of information. The management of N for crop production is clearly based on science, but on a science that is still very inexact. Because the power of a science-based technology is weakened when the science is inexact, increasing the accuracy of plant-related information on all levels of integration is an important approach to increased crop productivity.

Some characteristics of the ideal crop plant were given earlier. The heritable, physiological, and biochemical traits of new genotypes are not easy to quantify. How much N should be available in different plant parts at different stages of growth? How can desirable traits that have been identified be measured? Only because a method for assaying nitrate reductase is available, is it possible to survey plants for the presence of this trait and correlate its activity with yield potential. The plant physiologist working at his or her level of integration investigates the role of NO_3^--N as a crop production factor. The fertilizer technologist, in order to reduce loss of N via leaching or denitrification, attempts to retard the production of NO_3^- in soil. Close communication between the practitioners of these two disciplines is desirable so that the absence of NO_3^-, when and where needed, does not become a factor limiting crop production.

Nitrogen transformations and their interrelationships in soil are no less complex than those occurring within the plant. Additional detailed information about these transformations within both plant and soil will be needed before new ideas for managing soils and plants can be generated and placed into practice. Our current concept of soil organic matter portrays immobilized N as cycling through the biomass, and, during each cycle of biomass generation and decay, some part of the labile organic N participates in oxidative polymerization reactions that convert it to stable organic forms of little value as a source of plant nutrient. Can the sequence of immobilization-mineralization reactions be changed qualitatively and quantitatively in ways beneficial to plant growth? Are there differences in the ways in which different crop residues affect the quality of soil organic matter? How does change in the microbial ecology of the rhizosphere affect N supply and use by the crop plant? Can efforts to control nitrification (e.g., through use of nitrification inhibitors) sometimes result in an undesirable sequence of soil N transformations? For example, does blocking nitrification promote uptake of NH_4^+ by heterotrophic microorganisms, thereby decreasing the supply of crop-available N? Does N that is immobilized remineralize, nitrify, and leach or denitrify after the growing season? How does N immobilization under reduced tillage differ from that occurring in mechanically cultivated soil? These are but some of numerous questions that will be answered as we integrate our studies on, and more precisely measure the complex course of transformations within, soil organic matter.

Soil is the focal point of N cycle processes. The information obtained from studies of N in soils and waters now is of interest to ecologists, environmentalists, modelers of N transformations and transfers, and all those interested in the global N cycle. What processes are common to the internal N cycles of different agricultural systems? How do these processes and their interrelationships differ quantitatively? Quantifying and comparing the processes occurring in diverse ecosystems requires measurements made in ever-increasing detail; this kind of information will be generated only through multidisciplinary effort.

What is the prognosis for changing the average proportion of agricultural N that is partitioned among the plant (50–60%), soil (25%), and avenues of N loss (25–30%)? Crop use of N can reasonably be expected to increase to 60 to 65% of the N applied. Perhaps new factors previously overlooked may be limiting yields. For example, there is mounting evidence that significant amounts of NH_3, amines, and other N-containing gases are volatilized from the plant leaf surface and that this avenue of N loss may be species-dependent, as well as being affected by the external environment.

Should N loss from the phyllosphere be appreciable, intensive study of this phenomenon by plant physiologists will be needed and soil chemists and microbiologists will need to reassess the extent of N loss caused by processes occurring in soil.

Theoretically, use of urease and nitrification inhibitors and judicious use of applied N and irrigation water should virtually eliminate loss of N via NH_3 volatilization from the surfaces of soils and waters, and via leaching and runoff. Although some denitrification loss can be eliminated through use of technological advances and improved agronomic practices, some may be beyond our practical control. Little is known about the extent of N loss via chemodenitrification or as nitrous oxide evolution during nitrification. Should N loss by these avenues prove to be significant, methods for reducing this loss are available. Reducing N loss does not necessarily result in more efficient N uptake and use by the plant. But when N loss is decreased, less fertilizer N is needed to produce the same yield, resulting in a corresponding increase in the percentage of applied N used by the crop.

Whether soil organic matter can be manipulated to keep more N in readily mineralizable forms is not known. Since about 25% of applied N typically is stored temporarily in the soil organic complex, a relatively small change in the sequence of immobilization-mineralization may benefit plant growth. Research on the chemistry and microecology of soil organic matter requires attention at several levels of integration. Useful questions to ask oneself are: What research is being conducted in disciplines other than one's own, knowledge of which could be used to increase the relevance of one's own work? What is one doing that might be of interest to another scientist working toward the same overall objective, but from a different point of view?

No more appropriate area for joint study comes to mind than soil chemistry-fertility and biological N_2 fixation. All plants use N derived from the atmosphere, fixed either biologically or industrially. Now is the time for increased dialogue and joint effort between those who study the efficient use of fertilizer N in crop production and those who seek to increase the contribution of biological N_2 fixation in agricultural systems. Such interdisciplinary research addresses the N needs of agriculture in developing countries and the challenges provided by changing crop and soil management practices of developed agriculture.

Six people were walking through a farmer's corn field. One was a neighboring farmer, the others a crop production specialist, an entomologist, a plant pathologist, the county agent, and a local businessman. The latter noted that the maize grew poorly in crooked rows and wondered why. The crop production specialist noted the firing of the lower leaves and diagnosed a N deficiency. The entomologist noted the insect-ridden leaves, and the plant pathologist noted the smut on the emerging ears. The county agent remembered that the local farm cooperative had been unable to obtain the seed recommended for that region and also had been short on fertilizers and pesticides. The farmer saw the crooked rows and wondered whether his neighbor had problems with his balance and orientation. He resolved to persuade his neighbor to consult a physician. That autumn, the local businessmen called the group together for discussion, then arranged for the farm cooperative to be well-supplied with fertilizers, pesticides, and appropriate seed. The farmer took his neighbor to the clinic for treatment during the winter months. During the following season, the maize grew tall and green and was most productive. An interdisciplinary effort on several levels of integration succeeded.

An underlying objective of research on the agricultural use of N is to make crop plants as self-sufficient as possible in meeting their nutrient needs, especially N needs, while achieving maximum economic yields. Concurrently, our challenge is to meet food and fiber needs with minimum undesirable disruption of the environment. With focus on this objective and challenge, food will be available to sustain a growing population in a safe environment.

Glossary of Fertilizer Nomenclature and Abbreviations

Ammonium chloride	AC
Ammonium nitrate	AN
Ammonium nitrate sulfate	ANS
Ammonium polyphosphate	APP
Ammonium sulfate	AS
Anhydrous ammonia	AA
Aqua ammonia	AQA
Calcium ammonium nitrate	CAN
Calcium nitrate	CN
Diammonium phosphate	DAP
Dicyandiamide	DCD
Isobutylidene diurea	IBDU
Monoammonium phosphate	MAP
Nitric phosphate (fertilizers containing N and P_2O_5)	NP
Potassium nitrate	KN
Potassium polyphosphate	KPP
Slow-release nitrogen fertilizers	SRN
Slow-release sulfur fertilizers	SRS
Sodium nitrate	SN
Sulfur-coated urea	SCU
Urea–ammonium nitrate solution	UAN
Urea–ammonium phosphate	UAP
Urea–ammonium polyphosphate	UAPP
Urea-formaldehyde	UF
Urea nitrate	UN
Urea phosphate	UP

SUBJECT INDEX